GW00602801

Shelf Mark: VRH.F

Biotechnology of
Industrial Antibiotics

DRUGS AND THE PHARMACEUTICAL SCIENCES

A Series of Textbooks and Monographs

Edited by
James Swarbrick
School of Pharmacy
University of North Carolina
Chapel Hill, North Carolina

Volume 1. PHARMACOKINETICS, *Milo Gibaldi and Donald Perrier* (out of print)

Volume 2. GOOD MANUFACTURING PRACTICES FOR PHARMACEUTICALS: A PLAN FOR TOTAL QUALITY CONTROL, *Sidney H. Willig, Murray M. Tuckerman, and William S. Hitchings IV* (out of print)

Volume 3. MICROENCAPSULATION, *edited by J. R. Nixon*

Volume 4. DRUG METABOLISM: CHEMICAL AND BIOCHEMICAL ASPECTS, *Bernard Testa and Peter Jenner*

Volume 5. NEW DRUGS: DISCOVERY AND DEVELOPMENT, *edited by Alan A. Rubin*

Volume 6. SUSTAINED AND CONTROLLED RELEASE DRUG DELIVERY SYSTEMS, *edited by Joseph R. Robinson*

Volume 7. MODERN PHARMACEUTICS, *edited by Gilbert S. Banker and Christopher T. Rhodes*

Volume 8. PRESCRIPTION DRUGS IN SHORT SUPPLY: CASE HISTORIES, *Michael A. Schwartz*

Volume 9. ACTIVATED CHARCOAL: ANTIDOTAL AND OTHER MEDICAL USES, *David O. Cooney*

Volume 10. CONCEPTS IN DRUG METABOLISM (in two parts), *edited by Peter Jenner and Bernard Testa*

Volume 11. PHARMACEUTICAL ANALYSIS: MODERN METHODS (in two parts), *edited by James W. Munson*

Other Volumes in Preparation

Biotechnology of Industrial Antibiotics

Edited by

ERICK J. VANDAMME

Laboratory of General and
Industrial Microbiology
State University of Ghent
Ghent, Belgium

MARCEL DEKKER, INC.   New York and Basel

LIBRARY OF CONGRESS CATALOGING IN PUBLICATION DATA

Vandamme, Erick J.
   Biotechnology of industrial antibiotics.

   (Drugs and the pharmaceutical sciences ; v. 22)
   Includes index.
   1. Antibiotics--Industrial applications. 2. Bio-
technology. I. Title. II. Series. [DNLM: 1. Antibiotics.
2. Technology, Pharmaceutical. W1 DR893B v.22 / QV 350
B6165]
TP248.A62V36 1984   660'.63   84-1813
ISBN 0-8247-7056-0

COPYRIGHT © 1984 BY MARCEL DEKKER, INC. ALL RIGHTS RESERVED

Neither this book nor any part may be reproduced or transmitted in any form
or by any means, electronic or mechanical, including photocopying, micro-
filming, and recording, or by any information storage and retrieval system,
without permission in writing from the publisher.

MARCEL DEKKER, INC.
270 Madison Avenue, New York, New York 10016

Current printing (last digit):
10 9 8 7 6 5 4 3 2 1

PRINTED IN THE UNITED STATES OF AMERICA

To Mireille

Microorganisms are among man's best friends and his worst enemies, but it took him a million years to find out.

E. C. Stakman, 1964

PREFACE

Synthesis of bioactive and chemotherapeutic compounds with the aid of micro-organisms is an ever-increasing and most important branch of microbial biotechnology and the fermentation business. This is apparent from the industrial fermentation activities dealing with antibiotics, antitumor agents, vitamins, biopolymers, vaccines, amino acids and from industrial bioconversion of steroids, antibiotics, vitamins, amino acids, and carbohydrates.

Antibiotics, by far, are the most important compounds made by microbiological synthesis. Since the discovery of early antibiotics and of penicillin, more than 5500 natural microbial compounds which display antibiotic activity have been described.

Only a relatively small number of these products has found practical use; indeed about 150 antibiotic compounds are currently being produced—exclusively by fermentation processes—and find application in medical, veterinary, and agricultural practice as products with antibacterial, antifungal, anticoccidial, antiprotozoal, anthelmintic, pesticidal, antiviral, antitumor, piscicidal, or insecticidal activity.

The last decade displayed a revival of antibiotic research and important discoveries, after a period of satisfaction with—and retrospection upon—the "Antibiotic Golden Era" of the 1940s and 1950s.

Antibiotic compounds are now a *conditio sine qua non* in our society. Not only in the medical field, but also in agriculture, in the food industry, and in basic scientific research are these compounds essential and upon proper use, they have proven their uniqueness. They have, directly or indirectly, contributed much to the improvement of our standard of living, and toward an increase in the average human lifespan.

The idea for publishing and editing a book on the production of antibiotics by fermentation biotechnology was born partly out of frustration in attempting to keep abreast of current literature and of advancements in this fascinating area. Also, much secrecy embodies this vital and extremely important research—and especially—production field. It was also a chief aim to present a balanced picture of microbiology, biochemistry, genetics, and engineering, needed for a better understanding of the underlying principles of antibiotic overproduction.

This book hopes to "lift the curtain" (or at least the bottom of it) and to fill both a gap and a need in industrial fermentation literature. Thus far, antibiotics have been mainly treated from their biochemistry or chemistry side, or, only a few examples of fermentation processes are described as representative of the whole field. Furthermore, for other antibiotic compounds, industrial production has been achieved, yet basic understanding of their chemistry, biosynthesis, and mode of action is scarce and difficult to gain access to.

In this respect, a serious attempt has been made to gather the latest information on the microbiology, genetics, and biochemistry, as well as the technology of industrial antibiotic fermentations.

Now that biotechnology is on the march, I sincerely hope that this book will present an overview of economic antibiotic fermentations and serve as an example for the power of microbial fermentation and biotechnology for the benefit of humankind in the coming decades.

In all areas of scientific enterprise, people with insight conceive novel ideas and lead a scientific field into new directions; their faith and vision force them to critically test their theoretical concept and hypothesis at the laboratory bench, and they eventually see their efforts realized in practice. Invariably, these scientists inspire a new generation of students to enthusiastically enter into their research field and establish an atmosphere of enthusiasm, devotion, and interdisciplinary cooperation in favor of science and its applications. They deserve recognition and respect and as only a small part of my esteem for them, I am pleased to dedicate this book to my inspiring teachers, Sir Edward P. Abraham, Sir William Dunn School of Pathology, University of Oxford; Professor Arnold L. Demain, Fermentation Microbiology Laboratory, Department of Nutrition and Food Science, Massachusetts Institute of Technology, Cambridge, and Professor S. John Pirt, Department of Microbiology, Queen Elizabeth College, University of London, in appreciation of their leadership and invaluable contributions to many facets of antibiotic fermentation research and application in particular and to microbial productivity in general.

This book would not exist if it were not for the cooperative effort and continuous advice of Dr. A. L. Demain in suggesting potential authors and chapter subjects. My gratitude goes to all contributing authors, who were cooperative and understanding during the handling of their manuscripts. I would like to thank Dr. Maurits Dekker of Marcel Dekker Inc., New York, for his inspiration, enthusiasm, and invaluable hints, which facilitated the editorial work. It was a real pleasure working and meeting with you, Mau, while you were practicing your dutch! Thanks also to the staff of Marcel Dekker, Inc. whose help was invaluable to me. Not least, I am indebted to my wife, Mireille, for her continued help and moral support throughout the duration of my editorial task.

Erick J. Vandamme

CONTRIBUTORS

JOHN A. L. AUDEN Fermentation Pilot Plant, Central Research Laboratories and Pharmaceutical Research Department, Ciba-Geigy Ltd., Basel, Switzerland

S. BHUWAPATHANAPUN* School of Biotechnology, University of New South Wales, Sydney, Australia

ANDRÉ M. BIOT Process Development Department, Antibiotic Plant, Smith-Kline-RIT, Rixensart, Belgium

MARGITA BLUMAUEROVÁ Department of Biogenesis of Natural Substances, Institute of Microbiology, Czechoslovak Academy of Sciences, Prague, Czechoslovakia

DENIS BUTTERWORTH Microbiological Pilot Plant, Beecham Pharmaceuticals Research Division, Chemotherapeutic Research Centre, Betchworth, Surrey, England

JERRY L. CHAPMAN Antibiotic Culture Development Division, Eli Lilly and Company, Indianapolis, Indiana

CHARLES A. CLARIDGE Antitumor Biology Department, Pharmaceutical Research and Development Division, Bristol-Myers Company, Syracuse, New York

KAREL ČULÍK Research Institute of Antibiotics and Biotransformations, Roztoky-near-Prague, Czechoslovakia

ARNOLD L. DEMAIN Department of Nutrition and Food Science, Fermentation Microbiology Laboratory, Massachusetts Institute of Technology, Cambridge, Massachusetts

RICHARD P. ELANDER Fermentation Research and Development, Industrial Division, Bristol-Myers Company, Syracuse, New York

ØYSTEIN FRØYSHOV Department of Research and Development, A/S Apothekernes Laboratorium for Specialpraeparater, Oslo, Norway

---

Present affiliation: Department of Biotechnology, Kasetsart University, Bangkok, Thailand

YASUO FUKAGAWA Central Research Laboratories, Sanraku-Ocean Company, Ltd., Fujisawa, Japan

ORESTE GHISALBA Central Research Laboratories and Pharmaceutical Research Department, Ciba-Geigy, Ltd., Basel, Switzerland

JOSÉ A. GIL Department of Microbiology, University of León, León, Spain

PETER P. GRAY School of Biotechnology, University of New South Wales, Sydney, Australia

GEORGE J. M. HERSBACH Department of Fermentation Process Development, Research and Development Division, Gist-Brocades N.V., Delft, The Netherlands

EIJI HIGASHIDE Applied Microbiology Laboratories, Central Research Division, Takeda Chemical Industries, Ltd., Osaka, Japan

FLOYD M. HUBER Fermentation Technology Department, Eli Lilly and Company, Indianapolis, Indiana

A. HURST Microbiology Research Division, Bureau of Microbial Hazards, Health and Welfare Canada, Ottawa, Ontario, Canada

GREGORY S. HYATT Department of Chemical Engineering, University of Michigan, Ann Arbor, Michigan

TOMOYUKI ISHIKURA Central Research Laboratories, Sanraku-Ocean Company, Ltd., Fujisawa, Japan

SVERRE JAHNSEN Fermentation Department, Leo Pharmaceutical Products, Ballerup, Denmark

JOHN L. JOST* Fermentation Research and Development, The Upjohn Company, Kalamazoo, Michigan

ISAO KARUBE Research Laboratory of Resources Utilization, Tokyo Institute of Technology, Yokohama, Japan

INGER KIRK Microbiological Research Laboratory, Leo Pharmaceutical Products, Ballerup, Denmark

LEO A. KOMINEK Fermentation Research and Development, The Upjohn Company, Kalamazoo, Michigan

ROLAND KURTH† Department of Nutrition and Food Science, Fermentation Microbiology Laboratory, Massachusetts Institute of Technology, Cambridge, Massachusetts.

Present affiliations:
*Fermentation Research and Development, Genentech, Inc., San Francisco, California
†Biotechnologie, BASF, Ludwigshafen, Federal Republic of Germany

ROBERT LARSEN, Fermentation Department, Leo Pharmaceutical Products, Ballerup, Denmark

PALOMA LIRAS Department of Microbiology, University of León, León, Spain

HENNING LÖRCK Microbiological Research Laboratory, Leo Pharmaceutical Products, Ballerup, Denmark

JUAN F. MARTIN Department of Microbiology, University of León, León, Spain

JAMES P. MCDERMOTT Biochemical Development Division, Eli Lilly and Company, Indianapolis, Indiana

TAKASHI NARA Professional Relations Department, Pharmaceuticals Division, Kyowa Hakko Kogyo Company, Ltd., Tokyo, Japan

CLAUDE NASH Antibiotic Fermentation Technology Department, Eli Lilly and Company, Indianapolis, Indiana

JAKOB NÜESCH Central Research Laboratories and Pharmaceutical Research Department, Ciba-Geigy, Ltd., Basel, Switzerland

RYO OKACHI Pharmaceuticals Research Laboratory, Kyowa Hakko Kogyo Company, Ltd., Shizuokaken, Japan

NOBORU ŌTAKE Institute of Applied Microbiology, University of Tokyo, Tokyo, Japan

CHESTER PLATT Fermentation Technical Services, Eli Lilly and Company, Indianapolis, Indiana

MILOSLAV PODOJIL Department of Biogenesis of Natural Substances, Institute of Microbiology, Czechoslovak Academy of Sciences, Prague, Czechoslovakia

KENNETH E. PRICE Pharmaceutical Research and Development Division, Bristol-Myers Company, Syracuse, New York

STEPHEN W. QUEENER Biochemical Development Division, Eli Lilly and Company, Indianapolis, Indiana

KHALIL RAYMAN Microbiology Research Division, Bureau of Microbial Hazards, Health and Welfare Canada, Ottawa, Ontario, Canada

T. A. SAVIDGE Fermentation Development Department, Beecham Pharmaceuticals, Worthing, Sussex, England

THOMAS SCHUPP Central Research Laboratories and Pharmaceutical Research Department, Ciba-Geigy, Ltd., Basel, Switzerland

S. N. SEHGAL Department of Microbiology, Ayerst Research Laboratories, Princeton, New Jersey

RONALD M. STROSHANE* FCRC Fermentation Program, Frederick Cancer Research Center, Frederick, Maryland

SHUICHI SUZUKI Research Laboratory of Resources Utilization, Tokyo Institute of Technology, Yokohama, Japan

TOMOHISA TAKITA Department of Chemistry, Institute of Microbial Chemistry, Tokyo, Japan

HEINZ THRUM Department of Antibiotic Chemistry, Central Institute of Microbiology and Experimental Therapy, Academy of Sciences of the German Democratic Republic, Jena, German Democratic Republic

ANTHONY J. TIETZ Fermentation Technology Department, Eli Lilly and Company, Indianapolis, Indiana

DONALD R. TUNIN Antibiotic Culture Development Division, Eli Lilly and Company, Indianapolis, Indiana

ERICK J. VANDAMME Laboratory of General and Industrial Microbiology, Faculty of Agricultural Sciences, State University of Ghent, Ghent, Belgium

CEES P. VAN DER BEEK Department of Yeast and Fungal Genetics, Research and Development Division, Gist-Brocades N.V., Delft, The Netherlands

ZDENKO VANĚK Department of Biogenesis of Natural Substances, Institute of Microbiology, Czechoslovak Academy of Sciences, Prague, Czechoslovakia

PIET W. M. VAN DIJCK Department of Cell Biology, Research and Development Division, Gist-Brocades N.V., Delft, The Netherlands

CLAUDE VÉZINA† Department of Microbiology, Ayerst Research Laboratories, Montreal, Quebec, Canada

LEO C. VINING Department of Biology, Dalhousie University, Halifax, Nova Scotia, Canada

WELF VON DAEHNE Chemical Research Laboratories, Leo Pharmaceutical Products, Ballerup, Denmark

HENRY Y. WANG Department of Chemical Engineering, University of Michigan, Ann Arbor, Michigan

DONALD W. S. WESTLAKE Department of Microbiology, University of Alberta, Edmonton, Alberta, Canada

---

Present affiliations:
*Department of Drug Metabolism and Distribution, Sterling-Winthrop Research Institute, Renssalaer, New York
†Institut Armand-Frappier, Ville de Laval, Quebec, Canada

JANET WESTPHELING* Biochemical Development Division, Eli Lilly and Company, Indianapolis, Indiana

RICHARD J. WHITE† FCRC Fermentation Program, Frederick Cancer Research Center, Frederick, Maryland

SARAH WILKERSON Biochemical Development Division, Eli Lilly and Company, Indianapolis, Indiana

HIROSHI YONEHARA Kaken Chemical Company, Ltd., Tokyo, Japan

Present affiliations:
*Biological Laboratories, Harvard University, Cambridge, Massachusetts
†Infectious and Neoplastic Diseases Research Section, Medical Research Division, Lederle Laboratories, Pearl River, New York

CONTENTS

ANTIBIOTICS USED IN MEDICAL OR AGRICULTURAL PRACTICE: ANTI-FUNGAL ANTIBIOTICS

ANTIBIOTICS USED IN MEDICAL PRACTICE: ANTITUMOR AND ANTIVIRAL ANTIBIOTICS

ANTIBIOTICS USED IN AGRICULTURAL PRACTICE

NOVEL TRENDS IN MICROBIAL ANTIBIOTIC PRODUCTION

Biotechnology of
Industrial Antibiotics

Introduction

1

ANTIBIOTIC SEARCH AND PRODUCTION: AN OVERVIEW

ERICK J. VANDAMME *Laboratory of General and Industrial Microbiology, State University of Ghent, Ghent, Belgium*

I. INTRODUCTION

The term *antibiotic* appeared as early as 1928 in the French microbiological literature on antibiosis (Burkholder, 1952). This phenomenon of antagonism between living organisms was frequently observed—ever since 1877, when Pasteur and Joubert noticed that aerobic bacteria antagonized the growth of *Bacillus anthracis*.

However, the word in its present restrictive meaning—"a chemical substance derived from microorganisms which has the capacity of inhibiting growth, and even destroying, other microorganisms in dilute solutions"—was introduced by Selman Waksman in 1942.

In 1940, Waksman had forecasted: "We are finally approaching a new field of domestication of microorganisms for combating the microbial enemies of man and of his domesticated plants and animals. Surely, microbiology is entering a new phase of development."

Antibiotics have indeed found extensive application in the treatment of infectious diseases of man, animals, and, to a smaller extent, plants. Several compounds are used in animal nutrition, and a few in food conservation. The successful introduction of antibiotics for the treatment of infectious diseases caused by bacteria, fungi, and other pathogenic microorganisms has now aroused great hope that some of the more important nonmicrobial diseases will soon become subject to microbial therapy.

Today Waksman's definition might be broadened to include all microbial compounds which are able to selectively affect various biochemical growth processes at minimal concentration in humans, animals, plants, or microorganisms.

II. HISTORY

When Pasteur noticed that experimental anthrax in susceptible animals could be repressed by simultaneous inoculation of various nonpathogenic bacteria, he stated that this phenomenon of interaction between microbial species might ultimately find therapeutic application.

In 1885 Babes was the first to interpret microbial antibiosis as being due to production of an inhibitory chemical substance by the antagonistic organism. Some years after Babes' work, Bouchard, and Emmerich and Low, prepared an extract of *Pseudomonas aeruginosa* which they called "pycocyanase". In very high dilution, this compound was inhibitory to *Corynebacterium diphteriae, Salmonella typhi,* and *Pasteurella pestis,* as well as to pathogenic cocci. For approximately 20 years, pyocyanase was used in therapy in treating a variety of infectious diseases, but its toxicity prevented it from further extensive use. Many reports dealing with inhibitory effects of bacteria upon other microorganisms were written during this period. In 1907, Nicolle gave one of the earliest reports of antibacterial action of *Bacillus subtilis,* and since then the antibiotic activity of sporeforming bacilli has become well known (Katz and Demain, 1977). Around 1925, Gratia and his collaborators in Belgium extracted a lytic agent from a mold and used it successfully for treating staphylococcal skin infections. They were among the first to start a systematic search (a screening!) for antagonistic organisms from natural sources. This type of work rapidly spread to soil and plant microbiologists and several reports emerged confirming antagonism by fungi. Fleming's important observation in 1928 fits into this research atmosphere, and had he not written such a detailed analytical report of the inhibitory action of his *Penicillium notatum* strain, the "penicillin story" would perhaps never have followed!

Around 1938–1939, Florey and collaborators undertook the reevaluation of the therapeutic possibilities of Fleming's penicillin at the Sir William Dunn School of Pathology at Oxford University; the outcome of this fortunate choice is now a milestone in the history of humankind (Abraham, 1974). At about the same time, Dubos, at the Rockefeller Institute for Medical Research in New York, observed that certain pathogenic microorganisms disappeared from infected soils, and this led to the discovery of the antibiotic complex (called tyrothricin) produced by the soil bacterium *Bacillus brevis.* This work contributed much to the beginning of the "Golden Era of Antibiotics" which led to antibiotic application.

Following the work of Dubos, Waksman and Woodruff described the first antibiotic—actinomycin—obtained from a culture of an actinomycete. Further actinomycete research culminated in the discovery of streptomycin, the miracle drug in the treatment of tuberculosis. The impact of the discovery of this clinically important compound on the search for antibiotics from actinomycetes was enormous, and ever since then actinomycetes have been a primary reservoir for new antibiotics. A frenzied search for antibiotics from all types of microorganisms was initiated throughout the world and proved highly successful. In 1945, the discovery of bacitracin occurred, followed by chloramphenical and polymyxin in 1947, chlortetracycline and neomycin in 1948, oxytetracycline in 1950, and erythromycin in 1952, all of these being very important antibiotic compounds. A detailed chronological listing of economically important antibiotic discoveries has been compiled by Berdy (1974).

During the 1960s, the pace of discovery of antibacterials slackened, but efforts were then made to search also for antifungal, antimycoplasmal, antispirochetal, antiprotozoal, antitumor, antiviral, and antiphage compounds, as well as for antibiotics for nonmedical use (Waksman, 1969; Perlman, 1977). Also, the problem of bacterial resistance to antibiotics had evolved, and new compounds or derivatives of known antibiotics had to be found to replace existing ones.

The isolation of 6-aminopenicillanic acid (6-APA) in the late 1950s opened the way to the semisynthetic penicillins and provided an example for development of semisynthetic antibiotics in general (Rolinson, 1979; Vandamme, 1980).

In the late 1960s, a revival of antibiotic discovery occurred, owing to the application of novel screening programs, supersensitive test organisms, new antibiotic sources, and the broadening of the search for novel microbial products (classically encompassing antibacterial, antifungal, and antiviral agents) to include agents with pesticidal, antitumor, insecticidal, herbicidal, anticoccidal, cytotoxic, anthelminthic, hormonal, immunoregulatory, food-preserving, growth-promoting, and enzyme-inhibiting activities, as well as products with pharmacological activity (Berdy, 1980; Nisbet, 1982; Umezawa, 1977, 1983; Iwai and Omura, 1982; Miller et al., 1983; Demain, 1983).

Apart from the use of sophisticated pilot plant and production facilities, including computer-coupled fermentations, the impressive increase in antibiotic productivity in present industrial fermentations is largely a result of forcing the microorganism to overproduce the useful metabolite. This has been accomplished by mutation, protoplast fusion, and other types of genetic recombination, or by directly influencing cell metabolism, for example, through nutritional control or precursor addition; recently plasmid transfer and genetic engineering have been considered (Pirt, 1975; Perlman, 1977, 1978; Claridge, 1983; Daum and Lemke, 1979; Demain, 1977, 1980; Ninet et al., 1981; Normansell, 1982; Chater et al., 1982; Hopwood and Chater, 1982; Vournakis and Elander, 1983). Recent trends in antibiotic production relate also to total enzymatic synthesis of antibiotics (Demain and Wang, 1976), chemical derivatization and bioconversion of natural compounds yielding semisynthetic antibiotics (Sebek, 1980; Vandamme, 1980), and use of immobilized cell technology (Vandamme, 1981, 1983).

III. PRESENT STATUS OF COMMERCIAL ANTIBIOTIC PRODUCTION—1983

An overview of presently (1983) commercialized antibiotic compounds and their producer/fermentation companies is given in Table 1. Although Table 1 was designed to be as comprehensive and accurate as possible, the author realizes that assuring completeness is a mammoth task despite the generous cooperation of numerous colleagues and friends from all over the world in supplying pertinent information.

Table 1 Survey of Economically Important Antibiotic Compounds and Their Producer Companies

Antibiotic compound	Producer companies[a]	Producer microorganisms[b]	Activity spectrum	Chemical type
Aclacinomycin	129	*Streptomyces galilaeus*	Antitumor	Anthracycline
Actinomycin C,D,I,V	21,96,98,122	*S. antibioticus, S. chrysomallus*	Antitumor, G+	Peptide
Adriamycin	54,86,124	*S. peuceticus* var. *caesius*	Antitumor, G+,G-	Anthracycline
6-Aminopenicillanic acid (6-APA)[c]	2,10,21,22,62,63,86 143,155,158	*Escherichia coli, Bacillus megaterium, Pleurotus ostreatus* (from penicillin G or V)		β-Lactam
Amphomycin	95	*S. canus, S. violaceus*	G+	Polypeptide
Amphotericin B	143	*S. nodosus*	Yeast, fungi	Polyene
Anisomycin (Flagecidin)	110	*S. griseolus, S. roseo-chromogenes*	Herbicidal, fungi	
Antimycin A	19,86	*S. griseus, S. antibioticus, S. kitasawaensis*	Fungi, piscicidal	
Apramycin	52	*S. tenebrarius*	G+,G-, feed additive	Aminoglycoside
Ara-A (9-β-D-arabinofuranosyladenine)	108	*S. antibioticus*	Antiviral	Nucleoside
Asparaginase	21,86,98	*Erwinia* sp., *E. coli*	Antileukemia	Protein (enzyme)
Avermectins (+Ivermectin)[d]	98	*S. avermitilus*	Insecticidal, antihelmintic	
Avoparcin	7	*S. candidus*	Feed additive	Glycopeptide
Azalomycin F	128	*S. hygroscopicus* var. *azalomyceticus*	Fungi, yeast, G+, protozoa	Polypeptide
6-Azauridine	86	*E. coli* (from 6-azauracil)	Antiviral	Nucleoside
Bacitracins	12,15,51,61,74,77,83, 95,106,109,110,113, 115,137,142	*Bacillus subtilis, B. licheniformis*	G+, feed additive	Polypeptide
Bambermycins (Moenomycin, Flavomycin)	67	*S. bambergiensis, S. ghanensis*	G+, G-, feed additive	Phosphoglycolipid
Bicyclomycin	60	*S. sapporoensis*	G-, feed additive	Peptide

			Protein
Bioinsecticides	1,62,86,124,126,127, 140,160,161	*Bacillus thuringiensis var. kurstaki, var. israelensis, var. aizawae*	Insecticidal	Protein
		Baculoviruses (Nuclear polyhedrosis virus, NPV; Granulosis virus, GV) (Elcar)		
		Beauveria bassiana (Boverin), Metarhizium anisopliae (metaquino), Hirsutella thompsonii (Mygar), Aschersonia, Verticillium lecanii, Nomuraea rileyi		
Blasticidin S	76	*S. griseochromogenes, S. globifer*	Agricultural, fungi G+, G- (*Piricularia oryzae*)	
Bleomycin (+pepleomycin)	104	*S. verticillus*	Anticancer, G+, G-	Peptide
Butirosin	86	*Bacillus circulans*	G+, G-	Aminoglycoside
Candicidin B	51,95,109,114	*S. griseus, S. globisporus*	Yeast, fungi	Polyene
Candidin	95	*S. viridoflavus*	Fungi	Polyene
Capreomycin	46	*S. capreolus*	Tuberculosis	Peptide
Carzinophilin	86	*S. sahachiroi*	G+, mycobacteria	
Cellocidin (Aquamycin, Lenamycin)	105	*S. chibaensis, S. reticuli*	G+, G-, phytopathogenic bacteria	Polyene
Cephalosporin C (+semisynthetic cephalosporins: cephalotin, cephaloridine, ceph alexin, cefazolin, cephacetrile, cephapirin, cephradine,	8,10,15,27,29,33,34, 37,39,47,52,54,55, 57,58,59,60,63,67, 90,98,106,116,118, 123,128,146,151,160	*Acremonium chrysogenum (Cephalosporium acremonium)*	G+, G-	β-Lactam

Table 1 (continued)

Antibiotic compound	Producer companies[a]	Producer microorganisms[b]	Activity spectrum	Chemical type
cefamandole, cefaclor, cefuroxime, cefadroxil, cefotaxime, cefonicid, ceftizoxime...)				
Cephamycin C (+cefoxitin)	98	*S. lactamdurans*	G+,G-	β-Lactam
Cerulenin	96	*Cephalosporium caerulens*	G+,G-	
Chloramphenicol[c]	18,44,49,50,53,54,55, 84,99,108,128,130, 134,144,151,159,160, 161	*S. venezuelae*	G+,G-, rickettsia, antiviral	
Chromomycin A$_3$ (Toyomycin)	146	*S. griseus, S. olivochromo-genes*	Antitumor, G+	
Clavulanic acid	22	*S. clavuligerus*	G-	β-Lactam
Colistin	15,18,20,51,68,77,80, 86,88,124	*Bacillus polymyxa var. colistinus*	G-	Peptide
Coumermycins	29	*S. rishiriensis*	G+,G-, mycobacteria	
Cycloheximide (Actidione)	76,153	*S. griseus*	Fungi, agricultural (*Piricularia oryzae*)	
Cycloserine (Oxamycin)	3,4,74,97	*S. orchidaceus, S. lavendulae, S. garyphalus, Pseudomonas fluorescens*	G+,G-, mycobacteria	
Cyclosporine A	26,126	*Tolypocladium inflatum*	Immunosuppressant	
Daunorubicin	54,97,124	*S. peuceticus, S. coeruleo-rubidus*	Antitumor, protozoa	Anthracycline
Destomycin A	97	*S. rimofaciens*	G+,G-, mycobacteria, antihelminthic, feed additive	Aminoglycoside
Distamycin	54,124	*S. distallicus*	Antitumor, antiviral	Polyene
Diumycin	143	*S. umbrinus*	Feed additive	Phosphoglycolipid

Antibiotic	References	Producing organism	Spectrum / Use	Class
Enduracidin (enramycin)	146	S. fungicidus	G+, mycobacteria, feed additive	Peptide
Erythromycin	1,5,6,8,13,17,18,31,33, 39,40,42,43,46,52,57, 73,76,100,102,104, 110,111,116,117,118, 125,134,139,146,148, 153,160	S. erythreus	G+, mycobacteria, antiviral	Macrolide
Fortimicins	1,86	Micromonospora olivoasterospora	G+, G-	Aminoglycoside
Fosfomycin	40,97,98	S. fradiae, S. wedmorensis	G+, G-	
Fumagillin (Aminomycin, Perimycin)	1,34	Aspergillus fumigatus	Amoebicidal, antiprotozoal	Polyene
Fungimycin	95	S. coelicolor var. aminophilus	Fungi	Polyene
Fusidic acid	91	Fusidium coccineum	G+	Steroid
Gentamicins	18,23,32,33,34,49,57, 113,131,132,133,134, 158,160	Micromonospora purpurea	G+, G-, mycobacteria	Aminoglycoside
Gramicidin A,C,D	26,62,95,109,138,154	Bacillus brevis	G+	Peptide
Gramicidin S	97,103	Bacillus brevis	G+	Peptide
Griseofulvin	16,18,33,36,43,60,63, 71,71,91,104,128, 133,144,146,156,160, 161	Penicillium griseofulvum	Fungi, dermatophytes, mycoses	Polyene
Grisin (Grisemin, Kormogrisin)	161	Actinomyces griseus	Feed additive, agricultural	Streptothricin type
Hamycin (Trichomycin)	66	S. primprina	Fungi	Polyene
Hygromycin B (Marcomycin)	27,52,146	S. rimofaciens, S. hygro-scopicus, S. noboritoensis	G+, G-, anthelmintical, feed additive	Aminoglycoside
Josamycin (leucomycin A3)	129,156	S. narbonensis	G+, mycobacteria	Macrolide

Table 1 (continued)

Antibiotic compound	Producer companies[a]	Producer microorganisms[b]	Activity spectrum	Chemical type
Kanamycins (+Amikacin, Kanendomycin B, Dibekacin)	5,18,20,29,30,32,33,47, 97,104,116,128,134, 146,147,149,156,158, 159,160	S. kanamyceticus	G$^+$, G$^-$	Aminocyclitol
Kasugamycin	20,69,76,97,104,129	S. kasugaensis	G$^-$, fungi, phytopathogenic bacteria, rice blast disease (Piricularia oryzae)	Aminoglycoside
Kitasamycin (Leucomycin)	151	S. kitasatoensis	G$^+$, G$^-$	Macrolide
Krestin	85,145	Basidiomycete	Antitumor, immune modulator	Polysaccharide
Lasalocid	68	S. lasaliensis	Coccidiostat	Polyether
Lincomycin (+Clindamycin)	144,153	S. lincolnensis	G$^+$, G$^-$, rickettsia	
Lividomycin (Quintamycin)	81,124	S. lividus	G$^+$, G$^-$, tuberculosis	Aminoglycoside
Lucensomycin	54	S. lucensis	Fungi	Polyene
Macarbomycin	97	S. phaeochromogenes	G$^+$, G$^-$, mycobacteria, feed additive	Phosphoglycolipid
Maridomycin	146	S. hygroscopicus	G$^+$	Macrolide
Midekamycin	97	S. mycarofaciens	G$^+$	Macrolide
Mikamycins (Pristinamycins)	20,79,124	S. mitakaensis, S. pristinae spiralis	G$^+$, feed additive	Peptide
Mithramycin	96,110	S. tanashiensis	G$^+$, mycobacteria, antitumor	
Mitomycin C	25,29,86	S. caespitosus, S. fervens, S. verticillatus	Antitumor, G$^+$, G$^-$, mycobacteria, antiviral	
Mocimycin (Kirromycin)	62	S. ramocissimus	Feed additive	
Monensin	52	S. cinnamonensis	G$^+$, coccidiostat (Eimeria)	Polyether
Mutamycin	29	Streptomyces sp.	Anticancer	

Name	References	Organism	Activity	Class
Mycoheptin	66	S. netropsis, Streptovertcillium mycoheptinicum	Fungi	Polyene
Myxin	68	Sorangium sp.	Antihelmintic, G+,G-	Phenarine
Natamycin (pimaricin)	62	S. natalensis	Fungi, food preservative	Macrolide
Neocarzinostatin	77	S. carcinostaticus	Fungi, food preservative	Macrolide
Neomycins (Neamine = A, Framycetin = B)	10,12,27,33,54,95,104,109,110,124,125,134,143,148,153	S. fradiae	G+,G-, mycobacteria	Aminoglycoside
Nigericin (Polyetherin A)	27	S. violaceoniger	G+, mycobacteria, fungi, coccidiostat	Polyether
Nisin	11,82,152,161	Streptococcus lactis, Streptococcus cremoris	G+, food preservative	Polypeptide
Nosiheptide (Multhiomycin)	124	S. actuosus	Feed additive	
Novobiocin	97,124,153	S. spheroides, S. niveus	G+	
Nystatin	7,33,34,97,116,117,124,130,142,143	S. noursei	Fungi, yeast	Polyene
Oleandomycin	87,110,111,113,128,156	S. antibioticus	G+, antiviral, antiprotozoal	Macrolide
Oligomycin	122	S. diastatochromogenes	Fungi	
Paromomycins (Neomycin E,F, Zygomycin A, Hydroxymycin, Aminosidine, Catenulin, Crestomycin, Farmiglucin)	54,86,108	S. fradiae, S. chrestomyceticus	G+,G-, protozoa, antihelmintic	Aminoglycoside
Penicillin G	2,3,4,5,10,18,20,21,22,26,28,29,33,38,43,52,58,62,63,65,66,67,75,91,97,98,104,106,110,111,113,116,124,130,135,143,151,153,155,160	Penicillium chrysogenum	G+	β-Lactam

Table 1 (continued)

Antibiotic compound	Producer companies[a]	Producer microorganisms[b]	Activity spectrum	Chemical type
Penicillin V	1,2,3,4,18,20,26,27,29, 38,46,52,58,62,63,67, 91,97,106,110,124, 126,134,143,151,155	*Penicillium chrysogenum*	G+	β-Lactam
Semisynthetic penicillins (pheneticillin, propicillin, azidocillin, carbenicillin, carindacillin, carfecillin, ticarcillin, sulbenicillin, azlocillin, apalcillin, piperacillin, mezlocillin, amoxicillin, augmentin (= amoxicillin + clavulanic acid), cyclacillin, epicillin, ampicillin, pivampicillin, bacampicillin, talampicillin, methicillin, nafcillin, oxacillin, cloxacillin, dicloxacillin, flucoxacillin, mecillinam, pivmecillinam, etc....)	2,5,10,20,21,22,24,26, 29,33,39,43,48,55, 57,58,60,62,73,75, 91,97,98,107,110,111, 116,119,120,126,128, 132,133,143,144,146, 151,155,158,160		G+, G-	β-Lactam
Pentamycin	103	*S. pentaticus*	Fungi	Polyene
Picibanil	36	*Streptococcus* sp.	Antitumor, immune modulator	
Polymyxins	51,106,110,111	*Bacillus polymyxa*	G-, anti-*Pseudomonas*	Peptide

Antibiotic	References	Source	Activity	Class
Polyoxins	76,129	S. cacaoi var. asoensis	Fungi	Nucleoside
Puromycins	96	S. alboniger	G⁺, antitumor	
Pyrrolnitrin	60	Pseudomonas pyrrocinia, Pseudomonas schuylkil-liensis	Fungi	
Quebemycin	86	S. canadiensis	G⁺, feed additive	
Ribostamycin	97	S. ribosidificus, S. thermoflavus	G⁺,G⁻	Aminoglycoside
Rifamycins	13,18,25,37,41,49,50, 59,78,84,92,157,160	Nocardia mediterranei	G⁺, tuberculosis, antiviral	Ansamycin
Ristocetin	95	Nocardia lurida	G⁺, mycobacteria	
Rosaramicin	131	Micromonospora rosaria	G⁺,G⁻	Macrolide
Sagamicin	86	Micromonospora sagamiensis var. nonreductans, Micromonospora echinospora, Micromonospora purpurea	G⁺,G⁻	Aminoglycoside
Salinomycin	76,111	S. albus	G⁺, protozoa, coccidiostat	Polyether
Saramycetin	68	S. saraceticus	Fungi	Peptide
Sarkomycin	20,97	S. erythrochromogenes	Antitumor	
Siccanin	128	Helmintosporium siccans	Fungi	
Sideromycin (grisein)	37	S. griseus	G⁺	
Siomycin	133	S. sioyaensis	G⁺, mycobacteria, feed additive	Peptide
Sisomicin (+Netilmicin)	21,34,131	Micromonospora inyoensis	G⁻,G⁺	Aminoglycoside
Spectinomycin (Actino-spectatin, Trobicin)	1,76,153	S. spectabilis, S. flavo-pericus	G⁺, Neisseria, spirochetes	Aminoglycoside
Spiramycin	86,124	S. ambofaciens	G⁺,G⁻,rickettsia,protozoa	Macrolide
Stendomycin	52	S. endus, S. antimycoticus	G⁺,G⁻, mycobacteria, fungi, agricultural	Peptide
Streptomycin and dihy-drostreptomycin	5,10,20,33,34,42,43,51, 62,63,86,89,97,98,99, 106,110,111,128,130, 149,151,155,160	S. griseus, S. bikiniensis	G⁺,G⁻, tuberculosis, agricultural	Aminoglycoside

Table 1 (continued)

Antibiotic compound	Producer companies[a]	Producer microorganisms[b]	Activity spectrum	Chemical type
Tetracyclines (Methacycline, Doxycycline, Minocycline, Rolitetracycline)	2,7,9,10,13,14,16,17, 18,20,29,31,33,35,39, 40,44,45,54,56,57,64, 66,67,69,70,73,75,83, 92,93,97,101,106,110, 113,117,118,119,121, 125,130,134,137,141, 143,146,148,149,153, 160,161	S. aureofaciens	G^+, G^-, rickettsia, antitumor, mycobacteria, feed additive, plant-Mycoplasma	Polyketide
Chlortetracycline (Aureomycin)	7,9,28,33,45,54,66,75, 83,94,118,121,124,133, 136,146	S. aureofaciens		Polyketide
Demeclocycline	7,39,116,118,124,146	S. aureofaciens		Polyketide
Oxytetracycline (Terramycin)	9,18,27,33,34,35,36, 39,51,57,62,71,73,82, 83,84,86,87,90,110, 111,112,113,115,117, 118,121,137,138,148	S. rimosus		Polyketide
Tetranactin	36	Streptomyces aureus, S. flaveolus	G^+, G^-, fungi, miticidal, insecticidal, pesticidal	Macrolide
Thienamycins	98	S. cattleya	G^+, G^-	β-Lactam
Thiopeptin	60,76	S. tateyamaensis	G^+, mycobacteria, feed additive	Peptide
Thiostrepton (Bryamycin)	143	S. azareus, S. hawaiiensis	G^+, mycobacteria	Polypeptide
Tobramycin	27,52,133	S. tenebrarius	G^-	Aminoglycoside
Trichomycin (Hamycin, Candicidin)	60	S. hachijoensis	Fungi, yeast, protozoa	Polyene

Tylosin	46,52	*S. fradiae, S. hygroscopicus*	G+, mycobacteria, antiviral	Macrolide
Tyrocidine	95,138	*Bacillus brevis*	G+	Oligopeptide
Tyrothricin	26,62,95,109,138,154	*Bacillus brevis*	G+	Oligopeptide
Uromycin	122	*Streptomyces sp.*		
Validamycin	146	*S. hygroscopicus var. limoneus*	Fungi, sheath blight disease, pesticidial (*Pellicularia sasakii*)	Aminoglycoside
Vancomycin	52	*S. orientalis*	G+, mycobacteria	
Variotin	91,104	*Paecilomyces varioti*	Fungi, yeast	
Viomycin	111,128,151	*S. puniceus, S. floridae*	G+,G-, mycobacteria	Peptide
Virginiamycin	123	*S. virginiae*	G+, mycobacteria	Depsipeptide
Zearalenone (+Zeranol)	74	*Fusarium roseum, Gibberella zeae, F. graminearum*	Estrogenic	Macrolide

[a]See list below.

[b]*S.* = *Streptomyces.*

[c]Compound also produced by chemical synthesis.

[d]+ Semisynthetic derivative.

Fermentation Companies Producing Antibiotic Compounds

1. Abbott Laboratories, North Chicago, Illinois
2. Aktiebolaget Astra, Södertalje, Sweden
3. Aktiebolaget Fermenta, Strängnäs, Sweden
4. Aktiebolaget KABI, Stockholm, Sweden
5. Alembic Chemical Works Company, Ltd., Baroda, India
6. Alkem Laboratories Pvt. Ltd., Bombay, India
7. American Cyanamid (Lederle), Wayne, New Jersey
8. Anheuser-Busch, Inc., St. Louis, Missouri
9. Ankerfarm S.p.A., Milano, Italy
10. Antibioticos S.A., Leon, Spain
11. Aplin & Barrett Ltd., Trowbridge, England
12. Apothekernes Laboratorium for Specialpraeparater A/S, Harbitzalleen, 3, Oslo, 2, Norway
13. Archifar S.p.A., Milano, Italy
14. Armour Pharmaceuticals, Phoenix, Arizona
15. Asahi Chemical Industry Co., Shin-Osaka Bldg; 1-15-1, Dojumahamadori, Osaka, Japan

Table 1 (continued)

Fermentation Companies Producing Antibiotic Compounds (continued)

16. Ashford NDC Corporation, Hong Kong
17. Asia Pharmaceuticals Ltd., Hong Kong
18. Atlantic Laboratories Ltd., Hong Kong
19. Ayerst-Laboratories, Montreal, Quebec, Canada
20. Banyu Pharmaceutical Company, 2-7, Nihonbashi Honcho, Chuo-ku, Tokyo, 103, Japan
21. Bayer AG, Wuppertal, West Germany
22. Beecham Pharmaceutical Company Ltd., Surrey, England
23. Beneficiadora e Industrializadora, S.A. de C.V., Carretera Mexico-Laredo, Mexico
24. Bengal Chemical & Pharmaceutical Works Ltd., Calcutta, India
25. Biochemical & Pharmaceutical Industries, Bombay, India
26. Biochemie GmbH, Kundl, Austria
27. Biogal, Debrecen, Hungary
28. Biotika, Slovenska Lupca (Spofa), Czechoslovakia
29. Bristol-Myers Company, Syracuse, New York
30. Charoen Bhaesaj Laboratories, Bangkok, Thailand
31. Chembiotic Ltd., Inishannon, Ireland
32. Chi Cheng Chemicals, Bangkok, Thailand
33. China National Chemicals Import and Export Corporation, Peking, People's Republic of China
34. Chinoin Pharmaceutical and Chemical Works, Ltd., Budapest, Hungary
35. Chong-Kun-Dang Corporation, Seoul, South Korea
36. Chugai Pharmaceutical Company, 1-10-6, Iwamoto-Cho, Chiyoda-ku, Tokyo, 101, Japan

60. Fujisawa Pharmaceutical Company, 4-3, Doshomachi, Higashi-ku, 541, Osaka, Japan
61. Gedeon Richter Chemical Works, Budapest, Hungary
62. Gist-Brocades, Delft, The Netherlands
63. Glaxo Laboratories Ltd., Greenford, England
64. Griffon Laboratories PVT., Ltd., Bombay, India
65. Grünenthal AG, Stolberg, West Germany
66. Hindustan Antibiotics Ltd., Pimpri, India
67. Hoechst AG, Frankfurt, West Germany
68. Hoffman-La Roche Inc., Nutley, New Jersey
69. Hokko Kagaku Kogyo Company, 4-2, Nihonbashi Hongoku-Cho, Chuo-ku, Tokyo, Japan
70. ICN-Chimica S.p.A., Milano, Italy
71. Imperial Chemical Industries Ltd. (ICI), Manchester, England
72. Indian Drugs & Pharmaceuticals Ltd., Guragaon, India
73. Instituto Biochemico Italiano, Milano, Italy
74. International Minerals and Chemical Corporation, Terre Haute, Indiana
75. I.S.F. S.p.A., Rome, Italy
76. Kaken Chemical Company, 2-28-8, Honkomagone, Bunkyo-ku, 113, Tokyo, Japan
77. Kakenyaku Kako Company, Ltd., Nihonbashi Honcho, Tokyo, 103, Japan
78. Kanebo Ltd., Osaka, Japan
79. Kanegafuchi Chemical Industries, 3-3, Nakanoshima, Ita-ku, Osaka, Japan

80. Kayaku Antibiotics Research Company Ltd., 2-16-23, Meijiro, Toshima-ku, Tokyo, Japan

81. Kowa Company, Nagoya, Japan

82. Krakow Pharmaceutical & Chemical Works (Polfa), Krakow, Poland

83. KRKA Pharmaceutical & Chemical Works, 68000-Novo Mesto, Yugoslavia

84. Kuk Je Yak Pum Industry, Ltd., Seoul, South Korea

85. Kureha Chemical Industry, Ltd., Nihonbashi Hozitomecho, Tokyo, 103, Japan

86. Kyowa Hakko Kogyo Company, Otemachi Bldg., 1-6-1, Otemachi, Chujoda-ku, Tokyo, 100, Japan

87. Laboratoires Pfizer, S.A.R.L., Orsay, France

88. Laboratoires Roger Bellon, Monts, France

89. Laboratoires Mac Pvt., Ltd., Bombay, India

90. Lark S.p.A., Milano, Italy

91. Leo Pharmaceutical Products, Ltd. A/S, Industriparken, 55, DK-2750, Ballerup, Denmark

92. Lepetit S.p.A., Milano, Italy

93. Linson Ltd., Dublin, Ireland

94. Lohmann and Company AG, Cuxhaven, West Germany

95. Lundbeck H. and Company A/S, Ottiliavej 7-9, DK-1500, Valby, Denmark

96. Makor Chemicals Ltd., Jerusalem, Israel

97. Meiji Seika Kaisha Ltd., 2-8, Kyobashi, Chuo-ku, Tokyo, 104, Japan

98. Merck and Company Inc., Rahway, New Jersey

99. Mercury Pharmaceuticals Industry, Vadodara, India

100. Mochida Pharmaceutical Company, Ltd., 1-7, Yotsuya, Shinjiukuku, Tokyo, 160, Japan

101. National Fermentation Products, Industria, South Africa

102. National Pharmaceutical Company, Ltd., Hong Kong

37. Ciba-Geigy, Basel, Switzerland

38. Commonwealth Serum Laboratories, Parkville, Victoria, Australia

39. Companhia Industrial Produtora de Antibioticos S.A.R.L. (CIPAN), Lisboa, Portugal

40. Compania Espanola de Penicillina y Antibioticos S.A., Madrid-Arenjulz, Spain

41. Daiichi Seiyaku Co., Ltd., 3-14-10, Nihonbashi, Chuoku, Tokyo, 103, Japan

42. Dainippon Pharmaceutical Co. Ltd., 3-25, Doshomachi, Higashiku, Tokyo, 103, Japan

43. Dalian Zhiyao Chang, Dalian City, China

44. Dey's Medical Stores (Mfg) Ltd., Calcutta, India

45. Diaspa S.p.A., Coranna, Italy

46. Dista Products Ltd., Liverpool, England

47. Dong A Pharmaceutical Company, Ltd., Seoul, South Korea

48. Dong Shin Pharmaceutical Company, Ltd., Seoul, South Korea

49. Dongbei Shiyo Zongchang, Sheyang City, China

50. Dow Lepetit, Milano, Italy

51. Dumex Ltd., Prags Boulevard, 37, DK-2300 Copenhagen, Denmark

52. Eli Lilly and Company, Indianapolis, Indiana

53. Ethica, P.T., Djakarta, Indonesia

54. Farmitalia Carlo Erba, Milano, Italy

55. Fermentaciones y Syntesis, S.A., Mexico

56. Fermentfarma S.p.A., Milano, Italy

57. Fermic S.A. de C.V., Ixapalapa, Mexico

58. Fermion Oy, PI 28, Koivumankhaanhiya, SF-0-2101, Espoo, 10, Tapiola, Finland

59. Fervet S.p.A. (Ciba-Geigy Ltd.), Torre Annunziata, Italy

Table 1 (continued)

Fermentation Companies Producing Antibiotic Compounds (continued)

103. Nikken Chemicals Company Ltd., 5-4-14, Tsukiji, Chuo-ku, Tokyo, 104, Japan
104. Nippon Kayaku Company, 1-2-1-Marunouchi, Chiyoda-kun, Tokyo, 100, Japan
105. Nippon Nohuyaku Company, Tokyo, Japan
106. Novo Industri A/S, Novo Allé, DK-2880, Bagsvaerd, Denmark
107. Orsabe, S.A., Cuernavaca, Mexico
108. Parke Davis/Warner-Lambert, Detroit, Michigan
109. Penick and Company, Lyndhurst, New Jersey
110. Pfizer Inc., New York
111. Pfizer Korea Company Ltd., Seoul, South Korea
112. Pfizer Taito Company, 2-1-10, Nihonbashi, Chuo-ku, Tokyo, Japan
113. Pharmachim Antibiotic Works, Razgrad, Bulgaria
114. Pharmax International Medical Co., Hong Kong
115. Phylaxia Veterinary Biologicals and Feedstuffs Ltd., Budapest, Hungary
116. Pierrel, S.p.A., Milan, Italy
117. Pliva Pharmaceutical and Chemical Works, 41000-Zagreb, Yugoslavia

136. Societa Prodotti Antibiotici (SPA), Milan, Italy
137. Société Chimique Pointet Girard, Villeneuve la Garenne, France
138. Société Rapidase, Séclin, France
139. Soho Industri Pharmasi, Djakarta, Indonesia
140. Solvay, Brussels, Belgium
141. South Africa Cyanamid, Witbank, South Africa
142. Spofa (United Pharmaceutical Works), Prague, Czechoslovakia
143. Squibb and Sons Inc., Princeton, New Jersey
144. Sumitomo Chemical Industries Company Ltd., Kita-hama, Osaka 541, Japan
145. Takara Shuzo Company Ltd., Shijodori, Kyoto, 600, Japan
146. Takeda Chemical Industries, 2-27, Doschomachi, Higashi-ku, Osaka, 541, Japan
147. Tanabe Seiyaku, Ltd., 3-21, Doshomachi, Higashi-ku, Osaka, 541, Japan
148. Tarchomin Pharmaceutical Works (Polfa), Warsaw, Poland

118. Proter S.p.A., Milan, Italy
119. Quimasa S.A., Sao Paulo, Brazil
120. Quinonas de Mexico, Mexico City, Mexico
121. Rachelle Laboratories Inc., Long Beach, California
122. Reanal Fine Chemicals, Budapest, Hungary
123. Recherches et Industries Thérapeutiques (RIT) – Smith-Kline, Rixensart, Belgium
124. Rhône-Poulenc S.A., Paris, France
125. Roussel-Uclaf, Romainville, France
126. Sandoz Inc., Basel, Switzerland
127. Sandoz Inc., Hanover, New Jersey
128. Sankyo Company Ltd., 2-7-12, Ginza, Chuo-ku, Tokyo, 104, Japan
129. Sanraku Ocean Company Ltd., 1-7, Takara-cho, Chuo-ku, Tokyo, 104, Japan
130. Sarabhai Chemicals, Baroda, Indiana
131. Schering Corporation, Bloomfield, New Jersey
132. Seoul Pharmaceutical Company Ltd., Seoul, South Korea
133. Shionogi and Company Ltd., 3-12, Dosho-Machi, Higashi-ku, Osaka, 104, Japan
134. Silom Medical Company, Bangkok, Thailand
135. Sociedade Produtora de Leveduras Seleccionadas, Mastosinhus, Portugal

149. Thai Meiji Pharmaceutical Company, Ltd., Bangkok, Thailand
150. Toyama Chemical Company Ltd., 1-18, Nihonbashi-Kayatacho, Chusku, Tokyo, 103, Japan
151. Toyo Jozo Company Ltd., Mifuku, Ohitocho, Tagata-Gun, Shizuoka, 610-23, Japan
152. Unigate Ltd., London, England
153. Upjohn Company, Kalamazoo, Michigan
154. Wallerstein Laboratories, Inc., Morton Grove, Illinois
155. Wyeth Laboratories, Philadelphia, Pennsylvania
156. Yamanouchi Pharmaceutical Company, 2-5, Honcho, Nihonbashi, Chuo-ku, Tokyo, Japan
157. Yuhan Company, Seoul, South Korea
158. Yung Jin Pharmaceutical Industry Ltd., Seoul, South Korea
159. Zambon (United Italian Corp. Ltd.), Milan, Italy
160. Antibiotics produced in China (location not exactly known): gentamicin in factories in Hopei-Sheng and Sheyang Yaochang; ampicillin and carbenicillin in factories in Shanshi-Sheng and Shangai City
161. Antibiotics produced in the Soviet Union (location not exactly known)

IV. TRENDS IN ANTIBIOTIC SEARCH AND APPLICATION

A. General Remarks

In the search for new antibiotics, the leading position of Japan, the United States, and England remains unchallenged. Furthermore, it cannot be denied that the majority of recently introduced new antibacterial agents are so called semisynthetic derivatives of a few classic fermentation compounds.

The semisynthetic β-lactams (penicillin, cephalosporins, cephamycins) continue to dominate the medical market; improved derivatives of tetracyclines (minocycline, doxycycline) and of the aminoglycoside kanamycins (amikacin, dibekacin, kanendomycin) increase in importance; the semisynthetic rifamycin, rifampicin and lincomycin and clindamycin still remain important. These successes have stimulated attempts to modify antibiotic groups such as macrolides, antiviral agents, the antifungal polyenes, and the antitumor anthracyclines. As an example, the semisynthetic derivative of adriamycin, 4'-O-tetrahydro-pyranyladriamycin (THP-ADM), promises to be an important contribution to cancer chemotherapy.

There has recently been a dramatic increase in the number of analogs, minor components, or biosynthetic modifications of earlier known antibiotics, mainly as a result of novel strain isolation and selection methods, refined compound isolation and characterization procedures, and in vivo assay systems. Completely new compounds still emerge at a slower pace though (Umezawa, 1982a,b; Demain, 1983). The discovery of the new streptomycete β-lactams (Aoki and Okuhara, 1980) and more recently of the bacterial monobactams from *Gluconobacter, Pseudomonas, Agrobacterium,* and *Chromobacterium* are important examples in this respect (Cassidy, 1981; Sykes et al., 1981a,b).

Several new aminoglycosides have reached clinical practice rather quickly (sisomicin, fortimicin A, tobramycin, ribostamycin) or found use in agricultural practice (apramycin).

Precursor fed fermentation of *Streptomyces verticillus* recently yielded a new bleomycin, pepleomycin, which displays improved antitumor activity (Umezawa, 1982a).

Rare, neglected, or fastidious microorganisms, algae, marine organisms, invertebrates, and higher plants are now routinely screened for antibiotic compounds. Non-*Streptomyces Actinomycetales* (*Actinoplanes, Micromonospora, Saccharopolyspora, Actinomadura, Dactylosporangium, Ampulariella*) recently have yielded a considerable number of interesting compounds, including antibacterial and antitumor products (Wagman and Weinstein, 1980; Parenti and Coronelli, 1979; Iwai and Omura, 1982).

There exists a continuous and urgent need for new drugs, especially those effective against resistant bacteria, anaerobes, opportunistic pathogens, viruses, protozoa, fungi, and tumors, The synergistic antibacterial activity of the antibiotic mixture clavulanic acid plus amoxicillin, called augmentin, has recently been exploited to combat β-lactamase producers in clinical practice.

New animal feed additives, pesticides, and food preservatives might soon result from an intensified search for selective agricultural antibiotics (Woodbine, 1977; Burg, 1982; Miller et al., 1983; Demain, 1983).

Environmental (with precursors) and genetic manipulation (recombinant DNA, protoplast fusion, mutational biosynthesis, mutation, plasmid transfer) of known strains are bound to yield new compounds (Demain, 1980; Normansell, 1982; Vournakis and Elander, 1983).

Although chemical derivatization or bioconversion of natural antibiotic compounds offers more potential to yield useful drugs than testing synthetic compounds (Rosazza, 1978; Sebek, 1980), finding new natural compounds remains the most desirable objective.

B. Medically Useful Antibiotic Compounds

Apart from continued use of the now classic medical antibacterials (see Table 1), current interest is largely concentrated on the novel streptomycete β-lactams (nocardicins, thienamycins, olivanic acids, clavulanic acid, PS-5 series). (Aoki and Okuhara, 1980), the bacterial monocyclic β-lactams, called monobactams (Imada et al., 1981; Sykes et al., 1981a,b) and the semisynthetic penicillins, cephalosporins, cephamycins, and nocardicins (Raff and Summers-Gill, 1981).

In analogy to the penicillin nucleus, 6-APA, and the cephalosporin nucleus, 7-ACA, the nocardicin and monobactam nuclei have been prepared and substituted to yield interesting semisynthetic derivatives. Among the aminoglycosides, new compounds such as sagamicins, lysinomicin, butirosins, and fortimicins (sporaricins, sannamycins) are the subject of intense research and clinical evaluation; semisynthetic derivatives of several aminoglycosides have gained clinical application.

Table 2 Important Antifungal Antibiotics

Antibiotic	Producer strain (S = Streptomyces) (P = Penicillium)
Amphotericin B	*S. nodosus*
Ascosin	*S. canescus*
Candicidin	*S. griseus*
Candidin	*S. viridoflavus*
Cyanein	*P. cyaneum*
Cycloheximide[a]	*S. griseus*
Durhamycin	*S. durhamensis*
Eulicin	*S. parvus*
Ezomycin A	*S. kitazawaensis*
Filipin	*S. filipinensis*
Griseofulvin[a]	*P. griseofulvum*
Hamycin	*S. pimprina*
Levorin	*S. levoris*
Natamycin[b] (Pimaricin)	*S. natalensis*
Nystatin	*S. noursei*
Pyrrolnitrin	*Pseudomonas pyrrocinia*
Saramycetin	*S. Streptomyces*
Scopafungin	*S. hygroscopicus*
Trichomycin	*S. hachijaensis*
Variotin	*Paecilomyces varioti*

[a]Used in agriculture.
[b]Used as food preservative.

Promising macrolide antibiotics, include rosaramicin, juvenimicin, miokamycin, and mycinamicin. A series of phosphonic acid antibiotics has recently also aroused much interest (Okachi and Nara, 1980).

Medically important antifungal compounds are listed in Table 2. Progress in the field of antifungal antibiotics is slow compared to the dramatic evolution of antibacterial compounds.

So far, progress in the field of antiviral agents has been minimal, despite the fact that numerous products have been detected with distinct antiviral action. Several of these compounds are listed in Table 3, while the few products of practical interest can also be found in Table 1. The discovery of even one new and effective agent for the treatment of one of the many viral infections would have an enormous impact on public health and also provide tremendous stimulus to the search for such compounds (Grunert, 1979).

Apart from established antiprotozoal compounds (azalomycin F, paromomycin, pentamycin), antirickettsial compounds (spiramycin) and antispirochetal antibiotics (spectinomycin), new antiprotozoal products have emerged only very slowly from fermentation broths. Antileishmanial, antitrichomonal, antiamoebal, and antinematodal activities have already been demonstrated as secondary properties of known antibiotics, but a systematic and selective search should be undertaken to obtain superior compounds.

Antitumor antibiotics can be defined as microbial products that are able to suppress or retard the growth of tumor cells and inhibit experimental animal tumors (Oki, 1980).

Only recently, progress has been made in the exciting search for anticancer drugs, though the screening for microbial metabolites for anticancer activity began around 1953 (Oki, 1980; Umezawa, 1982a,b). New screening

Table 3 Some Antiviral Agents

Antibiotic	Producer strain[a]
Aabomycin A	S. hygroscopicus var. aabomyceticus
Abikoviromycin	S. abikoensis
Aklavin	Streptomyces sp.
Alanosine	S. alanosinicus
Amidinomycin	Streptomyces sp.
9-β-D-ARA-A	S. antibioticus
(arabinofuranosyladenine)	
8-Azaguanine	S. albus
Chloramphenicol	S. venezuelae
Distamycin A	S. distallicus
Formycin B	Nocardia interforma
Galirubins	S. galilaeus
Gliotoxin	Asperfillus fumigatus
Kikumycins	S. phaeochromogenes
Mitomycins	S. caespitosus
Mycophenolic acid	Penicillium brevicompactum
Pyrazomycin	S. candidus
Rifamycin B	S. albovinaceus
Tetracyclines	S. aureofaciens
Threomycin	S. threomyceticus
Virothricin	S. lavendulae var. virothricinus

[a]S. = Streptomyces.

systems, such as described by Hanka et al. (1978) and by Coetzee and Ove (1979), are being introduced to assist in selecting clinically important candidates. A list of interesting compounds is given in Table 4, while products currently in practical use are also listed in Table 1.

Several well-known antibiotics display pharmacological activity without ever being used as such, however. A few examples are presented in Table 5.

Well documented microbial metabolites display hypoglycemia-inducing, estrogenic, ACTH-like, anticoagulant, anti-inflammatory, cardiotonic, emetic, sedative, antispasmodic, myolytic, salivation-inducing, epinepherine-like, hallucinogenic, toxinlike, neuromuscular blocker, diuretic, hypolipidemic, and enzyme-inhibiting properties (Perlman and Peruzzoti, 1970, Brannon and Fuller, 1974; Umezawa, 1972; Matthews and Wade, 1977; Woodruff, 1980; Demain, 1983). Many of these products were isolated on the basis of antibiotic

Table 4 Cytotoxic and Antitumor Antibiotics

Antibiotic	Producer strain[a]
Aclacinomycin A,B[b]	S. galileus
Actinomycin D[b]	S. antibioticus
Adriamycin[b],[c]	S. peucetius
Ansamitocin	Nocardia sp. C-15003(N-1)
L-Asparaginase[b]	E. coli
Azaserine	S. fragilis
Bactobolin	Pseudomonas sp.
Bestatin	S. olivoreticuli
Bleomycin[b]	S. verticillus
Carcinomycin	S. carcinomyceticus
Carzinophilin[b]	S. sahachiroi
Chromomycin A_3[b]	S. griseus
Coriolin B	Coriolus consor
Cyclosporin A[b]	Tolypocladium inflatum
Daunomycin[b]	S. coeruleorubidus
Macromomycin	S. macromomyceticus
Mitomycin C[b]	S. caespitosus
Krestin (PS-K)[b]	Basidiomycete
Neocarzinostatin[b]	S. carcinostaticus
Neothramycin A,B	Streptomyces sp. MC 916-C4
Oxanosine	Actinomycetes
Picibanil[b] (OK-432, PCB)	Streptococcus sp.
Pepleomycin	S. verticillus
Sarkomycin	S. erythrochromogenes
Soedomycin	S. hachijoensis
Spergualin	Bacillus laterosporus
Sporamycin	Streptosporangium pseudovulgare
Streptozotocin	S. achromogenes
Tallysomycin	Streptoalloteichus hindustanus

[a]S. = Streptomyces.
[b]Commercial products in 1983.
[c]4'-O-Tetrahydropyranyladriamycin (THP-ADH or theprubicin) (semisynthetic adriamycin).

Table 5 Antibiotics Displaying Physiological and/or Pharmacological Activity

Physiological or Pharmacological activity	Antibiotic	Producer microorganisms[a]
Antispasmodic	Colisan	*B. brevis*
	Colistin	*B. colistinus*
	Patulin	*P. patulum*
Tranquilizer	Monorden	*Monosporium bonorden*
Immunosuppressive	Mycophenolic acid	*P. brevicompactum*
	Alanosine	*Streptomyces alanosinicus*
	Cyclosporin A	*T. inflatum* (see Table 4)
	Pluramycin	*S. pluricolorescens*
Anti-inflammatory	Griseofulvin	*P. griseofulvum*
	Amicomacin A$_1$	*B. pumilus*
Hypotensive	Aquayamycin	*S. reticuli*
	Dopastin	*Pseudomonas* sp.
	Erythromycin	*S. erythreus*
	Oleandomycin	*S. antibioticus*
	Oudenone	*Oudemansiella radicata*
	Spiramycin	*S. ambofaciens*
	Leucomycin	*S. kitasatoensis*
Hypertensive	Aeruginic acid	*Pseudomonas aeruginosa*
Hypocholesterimic	Compactin	*P. brevicompactum*
	Citrinin	*Pythium ultimum*
	Monacolin K (Mevinolin)	*Monascus ruber, Aspergillus terreus*
Hyperlipidemic	Ascofuranone	*Ascochyta viciae*
Diabetogenic	Streptozotocin	*S. achromogenes*
Cardiotonic	Adriamycin	*S. peucetius* var. *caesius*
	Fungichromin	*S. cellulosae*
	Lasalocid	*S. lasaliensis*
	Endomycin	*S. endus*
	Eurocidine	*S. eurocidicus*
	Trichomycin	*S. hachijoensis*
	Hamycin	*S. pimprina*
	Natamycin	*S. natalensis*
	Amphotericin B	*S. nodosus*
	Nystatin	*S. noursei*
	Pentamycin	*S. penticus*
	Candidin	*S. viridoflavus*
	Lagosin	*Streptomyces* sp.
Anticoagulant	Actinomycin D	*S. antibioticus*
	Nogalamycin	*S. nogalater*
	Puromycin	*S. alboniger*
	Cycloheximide	*S. griseus*
	Filipin	*S. filipinensis*
Neuromuscular blocker	Neomycin	*S. fradiae*
	Streptomycin	*S. griseus*
	Kanamycin	*S. kanamyceticus*
	Gentamicin	*Micromonospora purpurea*
	Polymyxin	*B. polymyxa*
	Colistin	*B. polymyxa* var. *colistinus*

Table 5 (continued)

Physiological or Pharmacological activity	Antibiotic	Producer microorganisms[a]
Smooth muscle relaxation (myolytic)	Bacitracin	B. licheniformis
	Viomycin	S. puniceus
	Chloramphenicol	S. venezuelae
	Gentamicin	Micromonospora purpurea
	Paromomycin (Aminosidine)	S. fradiae
	Ampicillin (semisynthetic from 6-APA)	
Enzyme-inhibitory		
Protease	Leupeptin	Streptomyces sp.
	Pepstatin	
β-Lactamase	Clavulanic acid	Streptomyces clavuligerus
Fatty acid synthetase	Cerulenin	Cephalosporum caerulens
Aminoglycoside-antibiotic inactivating enzymes	7-Hydroxytropolon	Streptomyces sp.

[a]S. = Streptomyces, B. = Bacillus, P. = Penicillium.

action, but their pharmacological action was indirectly demonstrated during clinical or toxicity testing. Direct screening for pharmacologic/biologic activities, rather than antibiotic activities, would certainly yield a greater variety of compounds with practical application. The main reason for hesitating to use this approach is that microbial agents against nonmicrobial disease are not as easily detected as is antibiotic activity, and would thus require screening of fermentation broths in in vivo assays (Brannon and Fuller, 1974), unless indirect in vitro assays for such compounds are designed, as pioneered by Umezawa (1972, 1977, 1983).

C. Agricultural Antibiotic Compounds

Until the 1960s, antibiotics established in human chemotherapy, such as penicillins, tetracyclines, streptomycin, and erythromycin, were also utilized in agricultural applications such as food preservation, animal nutrition, and plant protection. Serious problems claimed to be due to this indiscriminate use might be related to the development and spread of antibiotic-resistant microorganisms. Presently, many countries limit legally the agricultural utilization of therapeutic antibiotics (Swann, 1980; Anonymous, 1981; Burg, 1982). This also forced the current development of three main groups of antibiotics: the therapeutic, feed-grade, and pesticidal antibiotics. As to food preservation, only nisin and natamycin (pimaricin) are used (Hurst, 1981).

The nonmedical uses of antibiotics are mainly agricultural (Misato et al. 1977; Woodbine, 1977), and such examples are summarized in Tables 6, 7, and 8. Apart from applications as antihelmintics, herbicidals, insecticidals, miticidals, teleocidals, anticoccidials, antiprotozoals, hormonal compounds, and food preservatives, a considerable number of antibiotics find a selective use as feed additives to promote animal growth or as plant disease control

Table 6 Examples of Antibiotic Compounds of Agricultural Interest

Activity	Compound	Producer strain[a]
Anthelmintic	Avermectin	*S. avermitilis*
	Hygromycin	*S. rimofaciens*
Herbicide	Herbicidin	*S. saganonenses*
	Anisomycin (NK-049)	*Streptomyces* sp.
Insecticide	Piericidin	*S. mobaraensis*
	Tetranactin	*S. aureus*
Miticide	Tetranactin	*S. aureus*
	Milbemycins	*Streptomyces* B-41-146
Plant hormone	Gibberellins	*Gibberella fujikuroi*
Food pigment	Monascin	*Monascus* sp.
Detoxicant	Detoxin	*S. caespitosus* var. *detoxicus, S. mobaraensis*
Coccidiostat	Monensin	*S. cinnamonensis*
	Lasalocid	*S. lasaliensis*
Animal growth promotant (see also Table 8)	Virginiamycin	*S. virginiae*
	Avoparcin	*S. candidus*
Antiprotozoal	Azalomycin F	*S. hygroscopicus*
Plant disease controller (see also Table 7)	Blasticidin S	*S. griseochromogenes*
	Validamycin	*S. hygroscopicus* var. *limoneus*
Food preservative	Nisin	*Streptococcus lactis*
	Natamycin	*S. natalensis*
Piscicidal	Antimycin A	*S. griseus*
Growth factors	Zearalanol	*Gibberella zeae*
Antiviral (plants)	Aabomycin A	*S. hygroscopicus* var. *aabomyceticus*
Abscission agent	Cycloheximide	*S. griseus*
Fungicide	Cycloheximide	*S. griseus*

[a]*S.* = *Streptomyces.*

agents. These applications, however, are not universal and depend upon national governmental regulations and still contribute to debates on public health concern, spread of antibiotic resistance, and residue problems (Anonymous, 1981). Recently, these concerns have been mainly overcome by the selective screening, development, and use of typical agricultural (or pesticidal or feed additive) antibiotics.

The use of medical antibiotics as pesticidal (agricultural) chemicals is now limited to streptomycin, oxytetracycline, and chloramphenicol to control several bacterial diseases, and to cycloheximide and griseofulvin to combat fungal plant diseases. In several countries, medical antibiotics are still in use as feed additives, but this practice is declining rapidly.

This development of nonmedical antibiotic application has not been limited to controlling plant diseases, but has been extended to include pesticides, microbial insecticides, herbicides, and plant growth regulators, especially in Japan (Misato et al., 1977). Important examples of "pesticidal antibiotics" are presented in Table 7, while such compounds of economic importance are listed in Table 1.

Table 7 Important Pesticidal Antibiotic Compounds

Compound	Target
Antifungal	
Blasticidin S	Rice blast (*Piricularia oryzae*)
Kasugamycin	Rice blast (*Piricularia oryzae*)
Ezomycin A	Stem rot
Polyoxins	Rice sheath blight
Validamycin	Rice sheath blight
Cycloheximide (Actidione)	Fungal plant diseases
Griseofulvin	Fungal plant diseases
Antibacterial	
Cellocidin	
Chloramphenicol	
Streptomycin	Pear fire blight (*Erwinia, Xanthomonas*)
Tetracycline	
Novobiocin	Bacterial plant diseases
Insecticidal	
Tetranactin	Carmine mite of fruits and tea
Bacillus thuringiensis toxins	
var. *israelensis*	
var. *kurstaki*	
Baculoviruses	
Antiviral (plants)	
Aabomycin A	Tobacco mosaic virus, tomato mosaic virus
Herbicidal	
Anisomysin (NK-049)	
Plant growth regulators	
Gibberellins	
Abscission agents	
Cycloheximide	

The application of antibiotics in animal nutrition as feed additives has occurred along the same pattern: today typical feed additive antibiotics are available which display growth-promoting activity in addition to anticoccidial, anthelmintic, antiprotozoal, or hormonal properties (Burg, 1982). Important examples of antibiotic feed additives are given in Table 8; commercial feed additive antibiotics are listed in Table 1.

D. Antibiotics as Research Tools

Antibiotics continue to play a crucial role in the development of tissue culture techniques and basic sciences, primarily in biochemistry, molecular biology, microbiology, and genetics, including genetic engineering and, to a lesser degree, pharmacology and organic chemistry (Gale et al., 1981).

Table 8 Important Feed Additive Antibiotics

Compound	Producer strain[a]
Apramycin	*S. tenebrarius*
Aureofungin	*S. cinnamonensis var. terricola*
Avermectins[b]	*S. avermitilus*
Avoparcin	*S. candidus*
Bacitracin	*Bacillus licheniformis*
Bambermycin (Flavophospholipol, Flavomycin, Moenomycin)	*S. bambergiensis*
Bicyclomycin	*S. sapporoensis*
Delvomycin	*Streptomyces* sp.
Destomycin A[b]	*S. rimofaciens*
Diumycin	*S. umbrinus*
Enduracidin	*S. fungicidus*
Grisin	*Streptomyces griseus*
Hygromycin A,B[b]	*S. hygroscopicus*
Lasalocid[c]	*S. lasaliensis*
Lincomycin	*S. lincolnensis*
Macarbomycin	*S. phaeochromogenes*
Mikamycin	*S. mitakaensis*
Mocimycin	*Streptomyces* sp.
Monensin[c]	*S. cinnamonensis*
Nigericin[c]	*S. violaceoniger*
Nosiheptide	*S. actuosus*
Parvullin	*S. parvullus*
Quebemycin	*S. candadiensis*
Salinomycin[c]	*S. albus*
Siomycin	*S. sioyaensis*
Spiramycin	*S. ambofaciens*
Stendomycin	*S. endus*
Thiopeptin	*S. tateyamaensis*
Thiostrepton	*S. azareus*
Tylosin	*S. fradiae*
Virginiamycin	*S. virginiae*
Zearalanol[d]	*Gibberella zeae*

[a]*S.* = *Streptomyces*.
[b]Antihelmintic.
[c]Anticoccidial.
[d]Hormonal.

V. CONCLUSION

The continuing success of microbiologists in the search among microbial metabolites for antibiotic compounds useful in combating human, animal, and plant diseases has stimulated the belief that microorganisms constitute an inexhaustable reservoir of new types of interesting compounds, with pharmacological, physiological, medical, or agricultural application, other than combat pathogenic microorganisms. This has already been proven extensively.

As judged from the current academic and industrial interest in the search for and production of such compounds, more than ever before, microorganisms provide a unique source of unexpected useful products, and further study is bound to lead to further extension and broadening of the application of the microbe.

ACKNOWLEDGMENTS

The author gratefully acknowledges the help of numerous colleagues and friends in collecting pertinent data, summarized in Table 1. The ground work for Table 1 was provided by D. Perlman (*ASM News*, 1977, 43(2):82–89) and the information presented here is an extension and updating of his work. In this respect, I am particularly indebted to J. A. L. Auden, K. Aunstrup, D. Behrens, S. Bhuwapathanapun, A Biot, G. Bisiaux, M. Blumauerova, Z. Bohak, S. K. Bose, M. Bosnjak, P. E. Bost, G. B. Carter, T. S. Chen, M. Cole, A. J. Collados, J. Debacq, L. Delcambe, A. L. Demain, J. G. Dubois, Z. Er-El, G. G. Fowler, G. Garian, A. K. Goel, P. Gray, I. H. Hamdam, E. Higashide, F. Huber, M. Jarai, B. Johannesen, K. Kieslich, G. Kleiner, Y. K. Lee, T. Nara, R. Okachi, T. Oki, N. Otaki, K. L. Perlman, S. J. Pirt, P. Prave, P. E. Pyndt, R. Quintero, G. Rand, C. Rolz, C. Spalla, M. Spivey, R. M. Stroshane, A. Tietz, H. Umezawa, A. Vatanen, C. Vezina, W. Von Daehne, K. Yamada, and many others, who I hope will forgive me for having forgotten to mention them explicitly.

REFERENCES

Abraham, E. P. 1974. Some aspects of the developments of the penicillins and cephalosporins. *Dev. Ind. Microbiol. 15*:3–15.

Anonymous. 1981. Saving antibiotics from themselves. *Nature (London) 292*: 661.

Aoki, H., and Okuhara, M. 1980. Natural β-lactam antibiotics. *Annu. Rev. Microbiol. 34*:159-181.

Berdy, J. 1974. Recent developments of antibiotic research and classification of antibiotics according to chemical structure. *Adv. Appl. Microbiol. 14*:309–406.

Berdy, J. 1980. Recent advances in and prospects of antibiotic research. *Proc. Biochem. 15*:28–35.

Brannon, D. R., and Fuller, R. W. 1974. Microbial production of pharmacologically active compounds other than antibiotics. *Lloydia 37*:134–146.

Burg, R. P. 1982. Fermentation products in animal health. *ASM News 48*: 460–463.

Burkholder, P. R. 1952. Cooperation and conflict among primitive organisms. *Am. Sci. 40*:601–631.

Cassidy, P. J. 1981. Novel naturally occurring β-lactam antibiotics: A review. *Dev. Ind. Microbiol. 22*:181–209.

Chater, K. F., Hopwood, D. A., Kieser, T., and Thompson, C. J. 1982. Gene cloning in streptomyces. In *Current Topics in Microbiology and Immunology*. Springer Verlag, Berlin, Vol. 96, pp. 69–95.

Coetzee, M. L., and Ove, P. 1979. In vitro assays as a screening test for anti-tumor agents. *Proc. Biochem. 14*:26–37.

Claridge, C. A. 1983. Mutasynthesis and directed biosynthesis for the production of new antibiotics. In *Basic Biology of New Developments in Biotechnology*. A. Hollaender, A. I. Laskin, and P. Rogers (Eds.). Plenum, New York, pp. 231–269.

Daum, S. J., and Lemke, J. R. 1979. Mutational biosynthesis of new antibiotics. *Annu. Rev. Microbiol. 33*:241–265.

Demain, A. L. 1977. The health of the fermentation industry: A prescription for the future. *Dev. Ind. Microbiol. 18*:72–77.

Demain, A. L. 1980. The new biology: Opportunities for the fermentation industry. *Annu. Rep. Ferment. Proc. 4*:193–208.

Demain, A. L. 1983. New applications of microbial products. *Science 219*: 709–714.

Demain, A. L., and Wang, D. I. C. 1976. Enzymatic synthesis of gramicidin S. In *Second International Symposium on the Genetics of Industrial Microorganisms*. K. D. MacDonald (Ed.). Academic, New York, pp. 115–128.

Gale, E. F., Cundliffe, E., Reynolds, P. E. Richmond, M. H., and Waring, M. J. 1981. *The Molecular Basis of Antibiotic Action*. Wiley, London.

Grunert, R. R. 1979. Search for antiviral agents. *Annu. Rev. Microbiol. 33*:335–353.

Hanka, L. J., Martin, D. G., and Neil, G. L. 1978. In vitro methods used in detection and quantitation of antitumor drugs produced by microbial fermentations. *Lloydia 41*:85–97.

Hopwood, D. A. and Chater, K. F. 1982. Cloning in streptomyces-systems and strategies. In *Genetic Engineering*, Vol. 4. J. K. Setlow and A. Hollaender (Eds.). Plenum Publishing, New York, p. 119.

Hurst, A. 1981. Nisin. *Adv. Appl. Microbiol. 27*:85–123.

Imada, A., Kitano, K., Kintaka, K., Muroi, M., and Asai, M. 1981. Sulfazecin and isosulfazecin, novel β-lactam antibiotics of bacterial origin. *Nature (London) 289*:590–591.

Iwai, Y., and Omura, S. 1982. Culture conditions for screening of new antibiotics. *J. Antibiot. 35*:123–141.

Katz, E., and Demain, A. L. 1977. The peptide antibiotics of *Bacillus*: Chemistry, biogenesis and possible functions. *Bacteriol. Rev. 41*:449–474.

Matthews, H. W., and Wade, B. F. 1977. Pharmacologically active compounds from microbial origin. *Adv. Appl. Microbiol. 21*:269–288.

Miller, L. K., Lingg, A. I., and Bulla, L. A. 1983. Bacterial, viral and fungal insecticides. *Science 219*:715–721.

Misato, T., Ko, K., and Yamaguchi, I. 1977. Use of antibiotics in agriculture. *Adv. Appl. Microbiol. 21*:53–88.

Ninet, L., Bost, P. E., Bouanchaud, D. H., and Florent, J. 1981. *The Future of Antibiotherapy and Antibiotic Research*. Academic, New York.

Nisbet, L. J. 1982. Current strategies in the search for bioactive microbial metabolites. *J. Chem. Technol. Biotechnol. 32*:251–270.

Normansell, I. D. 1982. Strain improvement in antibiotic producing microorganisms. *J. Chem. Technol. Biotechnol. 32*:296–303.

Okachi, R., and Nara, T. 1980. Current trends in antibiotic fermentation research in Japan. *Biotechnol. Bioeng. Suppl. 1*:65–81.

Oki, T. 1980. Cytotoxic and antitumor antibiotics produced by microorganisms. *Biotechnol. Bioeng. Suppl. 1*:83–97.

Parenti, F., and Coronelli, C. 1979. Members of the genus *Actinoplanes* and their antibiotics. *Annu. Rev. Microbiol. 33*:389–411.

Perlman, D. 1977. The fermentation industries—1977. *ASM News 43*:82–89.

Perlman, D. 1978. Stimulation of innovation in the fermentation industries 1910—1980. *Proc. Biochem. 13*:3–5.

Perlman, D., and Peruzotti, G. P. 1970. Microbial metabolites as potentially useful pharmacologically active agents. *Adv. Appl. Microbiol. 12*:277–294.

Pirt, S. J. 1975. *Principles of Microbe and Cell Cultivation*. Blackwell, Oxford.

Raff, P. J., and Summersgill, J. T. 1981. Functional chemical alterations in non-penicillin β-lactam compounds. *Proc. Biochem.* 16:15–23.

Rolinson, G. N. 1979. 6-APA and the development of the β-lactam antibiotics. *J. Antimicrob. Chemother.* 5:7–14.

Rosazza, J. P. 1978. Microbial transformations of natural antitumor agents. *Lloydia* 41:297–311.

Sebek, O. K. 1980. Microbial transformations of antibiotics. In *Economic Microbiology*, Vol. 5. A. H. Rose (Ed.). Academic, New York, pp. 575–612.

Swann, M. 1980. The biological credit–debit balance. *Interdiscip. Sci. Rev.* 5:2–5.

Sykes, R. B., Cimarusti, C. M., Bonner, D. P., Bush, K., Floyd, D. M., Georgopapadakou, N. H., Koster, W. H., Liu, W. C., Parker, W. L., Principe, P. A., Rathnum, M. C., Slusarchyk, W. A., Trejo, W. H., and Wells, S. S. 1981a. Monocyclic β-lactam antibiotics produced by bacteria. *Nature (London)* 291:489–491.

Sykes, R. B., Bonner, D. P., Bush, K., Georgopapadakou, N. H., and Wells, S. S. 1981b. Monobactam—Monocyclic β-lactam antibiotics produced by bacteria. *J. Antimicrob. Chemother. Suppl. E* 8:1–16.

Umezawa, H. 1972. *Enzyme Inhibitors of Microbial Origin*. University Park Press, Baltimore.

Umezawa, H., 1977. Recent advances in bioactive microbial secondary metabolites. *Jap. J. Antibiot. Suppl.* 30:138–163.

Umezawa, H. 1982a. Recent studies on antibiotics and small molecular immunomodulators with potential usefulness in treating lung cancer: Part I— Antitumor antibiotics and their derivatives. *Int. J. Clin, Pharmacol. Ther. Toxicol.* 20:12–18.

Umezawa, H. 1982b. Recent studies on antibiotics and small molecular immunomodulators with potential usefulness in treating lung cancer: Part II— Small molecular weight immunomodulators produced by microorganisms. *Int. J. Clin. Pharmacol. Ther. Toxicol.* 20:19–23.

Umezawa, H. 1983. Studies of microbial products in rising to the challenge of curing cancer. *Proc. R. Soc. London, B* 217:357–376.

Vandamme, E. J. 1980. Penicillin Acylases and β-lactamases. In *Economic Microbiology*, Vol 5. H. Rose (Ed.). Academic, New York, pp. 467–522.

Vandamme, E. J. 1981. Use of microbial enzyme and cell preparations to synthesize oligopeptide antibiotics. *J. Chem. Technol. Biotechnol.* 31:637–659.

Vandamme, E. J. 1983. Peptide antibiotic production through immobilized biocatalyst technology. *Enz. Microbiol. Technol.* 5:403–416

Vournakis, J. N., and Elander, R. P. 1983. Genetic manipulation of antibiotic producing microorganisms. *Science* 219:703–709.

Wagman, G. H., and Weinstein, M. J. 1980. Antibiotics from *Micromonospora*. *Annu. Rev. Microbiol.* 34:537–557.

Waksman, S. A. 1969. Successes and failures in the search for antibiotics. *Adv. Appl. Microbiol.* 11:1–16.

Woodbine, M. 1977. *Antibiotics and Antibiosis in Agriculture*. Butterworths, London.

Woodruff, H. B. 1980. Natural products from microorganisms. *Science* 208:1225–1229.

2

BIOLOGY OF ANTIBIOTIC FORMATION

ARNOLD L. DEMAIN *Fermentation Microbiology Laboratory, Massachusetts Institute of Technology, Cambridge, Massachusetts*

I. INTRODUCTION

> Antibacterial substances are, so to speak, charmed bullets which strike only those objects for whose destruction they have been produced.
>
> Paul Ehrlich

The remarkable group of compounds known as antibiotics form a heterogeneous assemblage of biologically active molecules with different structures (Bérdy, 1974) and modes of action (Hash, 1972; Demain, 1975) (Table 1). Since 1940, we have witnessed a virtual explosion of new and potent molecules which have been of great use in medicine, agriculture, and basic research. Over 20,000 tons of these metabolites are produced annually around the world. However, the search for new antibiotics continues in order to combat naturally resistant bacteria and fungi, as well as those previously susceptible microbes that have developed resistance (Davies and Smith, 1978); improve the pharmacological properties of antibiotics; combat tumors, viruses, and parasites; and discover safer, more potent, and broader spectrum antibiotics. All commercial antibi-

Table 1 Mode of Action and Structure of Selected Antibiotics

Target	Antibiotic	Structure
DNA replication	Bleomycin	Peptide
	Griseofulvin	Condensed aromatic
Transcription	Actinomycin	Peptide
	Rifamycin	Ansamycin
Translation		
70-S ribosomes	Chloramplenicol	Aromatic
	Tetracycline	Polyketide
	Lincomycin	Sugar ester
	Erythromycin	Macrolide
	Streptomycin	Aminocyclitol
70-S and 80-S ribosomes	Puromycin	Nucleoside
	Fusidic acid	Steroid
80-S ribosomes	Cycloheximide	Glutarimide
Cell wall synthesis	Cycloserine	Amino acid
	Bacitracin	Peptide
	Penicillin	β-Lactam
	Vancomycin	Oligosaccharide aromatic
Cell membrane		
Surfactants	Polymyxin	Peptide
	Amphotericin	Polyene
Ionophores		
Channel formers	Linear gramicidin	Peptide
Mobile carriers	Monensin	Polyether

otics in the 1940s were natural, but today most are semisynthetic, that is, chemical modifications of natural antibiotics. Indeed, over 30,000 semisynthetic β-lactams (penicillins and cephalosporins) have been synthesized.

Antibiotics are extremely important to the health and nutrition of our society and have tremendous economic importance. They are produced in batch culture during the idiophase of development (a phase which may overlap but often follows the growth phase or trophophase); have no function in growth processes, although they probably contribute to survival of a particular producing organism in nature; are produced by certain restricted taxonomic groups of organisms; and are usually formed as a mixture of closely related members of a chemical family (Weinberg, 1970). Production ability is easily lost by mutation or plasmid loss ("strain degeneration").

About 5500 antibiotics have been described, 4000 from actinomycetes alone, and they still are being discovered at a rate of about 300–400 per year.

II. FUNCTION IN NATURE

Antibiotic function in the metabolism of the producing organism has been the subject of considerable speculation and discussion (Demain, 1980a). However, we still do not completely understand the role they actually play. Several proposed functions for antibiotics are no longer accepted, whereas other ideas are still under consideration in several laboratories. Those possible functions that have been discarded involved antibiotics as evolutionary relics, waste products of cellular metabolism, reserve food materials, spore coat components, or breakdown products derived from cellular macromolecules. Other classical hypotheses that lack support are: (1) secondary metabolism serves to maintain the enzymatic machinery of the cell in working order until conditions favorable for growth are found and (2) antibiotics are detoxification products.

One possibility still being considered is that antibiotics function to kill or inhibit the growth of other organisms in nature, thereby providing a competitive advantage to the producing species (Gottlieb, 1976; Demain, 1980a). The capacity to produce antibiotics is quite common among soil saprophytes and, in fact, the most capable saprophytic antagonists of root parasites are often antibiotic producers. Most antibiotic-producing microorganisms can form their antibiotics in sterilized soil supplemented with organic material. A few antibiotically active cultures can synthesize in unsupplemented sterilized soil or in unsterilized supplemented soil. There are also examples of antibiotic production in unautoclaved, unsupplemented soil. Clearly, antibiotics can be produced in soil; the main limiting factor is the nutrient concentration. Although the average organic content of soil is low, certain microenvironments (e.g., dead plant debris, seed coats, and the rhizosphere of plants) provide enough carbon for antibiotic production. Furthermore, antibiotics are antagonistic when present in soil. Although competition is usually thought to be between bacterial species A and B, or between bacterium and fungus, it could involve bacteria versus protozoa, since protozoa use bacteria as a major source of food (Singh, 1942; Habte and Alexander, 1977).

One function postulated for antibiotics that has received the most attention recently is that antibiotics are important compounds in cellular differentiation, that is, in the transition from vegetative cells to spores or from spores to vegetative cells. Numerous antibiotics are elaborated by microorganisms that undergo sporulation (Katz and Demain, 1977). Since the majority of antibiotics produced by bacilli are peptidic in nature, these peptides may play a key role in the termination of vegetative growth, allowing sporulation to take place (Hodgson, 1970; Sadoff, 1972). It has been suggested (Sarkar and Paulus, 1972) that antibiotic production is required for sporulation; peptide antibiotics might be used in several ways by an organism during sporulation to modify the cell membrane, for example, as detergents disrupting structural components or as ion carriers modifying permeability properties. It is argued that by the antibiotic selectively functioning at certain stages, the sporulation process is able to proceed normally.

Although many observations point to an intimate relation between antibiotic formation and sporulation, by no means do they prove that the antibiotic is necessary for sporulation. The most damaging evidence to the hypothesis involving antibiotics in sporulation is the existence of mutants that are antibiotic negative but which can still sporulate (Demain and Piret, 1979). Such mutants have been obtained from bacteria, actinomycetes, and fungi producing gramicidin S, linear gramicidin, tyrocidine, mycobacillin, bacitracin, methylenomycin A, and patulin.

All the experimental data described to date could fit well with the hypothesis that sporulation and antibiotic formation, although independent phenomena, are regulated by a common or similar regulatory mechanism. There would be several advantages to such a coordinated control. For example, the antibiotic might be packaged in the spore to inhibit germination until conditions became favorable for growth. Certain antibiotics in bacilli and actinomycetes are indeed found in spores, indicating a possible function in the survival of dormant or germinating spores. Both the antibiotic in *Streptomyces viridochromogenes* spores and gramicidin S in *Bacillus brevis* inhibit germination or outgrowth of the producer's spores (Hirsch and Ensign, 1978; Piret and Demain, 1981). Such roles emphasize the selective advantage a spore-former would have in producing an antibiotic in nature, without invoking an obligatory function for the antibiotic in sporulation.

Not all antibiotics are involved in sporulation; however, it is highly probable that all antibiotics have some functional role in the survival of the producing organism. It is inconceivable that the multienzyme reaction sequences of antibiotic biosynthesis have been retained in nature without some beneficial effect on survival. Other activities of antibiotics include phytotoxicity, symbiosis of microbes with plants, and transport of metals into microbial cells (Demain and Piret, 1981).

III. SUICIDE AVOIDANCE BY ANTIBIOTIC-PRODUCING MICROORGANISMS

Because antibiotics are among the most potent compounds made by living organisms, it is remarkable that producing strains can remain metabolically active and viable in their own environment (Demain, 1974; Vining, 1979). Despite this apparent insensitivity, when a producing strain is inoculated into a fresh medium that contains its antibiotic, adverse effects on growth are observed.

Microorganisms do grow and synthesize secondary metabolites, such as antibiotics, to which they are sensitive, because these organisms are programmed to elaborate their secondary products only after having passed through part or all of their growth phase. The usual onset of the production phase varies in different organisms, from the latter part of the growth phase to many hours after rapid growth has ceased. In these cases, antibiotic formation is delayed because enzymes specifically involved in antibiotic biosynthesis are repressed or inhibited during growth.

Although microorganisms perform most of their growth processes before they produce antibiotic, they might be killed by their own antibiotic during production. We know, however, that industrial fermentations are usually conducted for many days after the onset of antibiotic production. Thus the synthesizing organism develop resistance during production.

The resistance mechanisms developed by antibiotic-producing microorganisms against their own antibiotic are no different from those in clinically resistant bacteria. Permeability modifications are involved in many instances. Antibiotics are pumped out of cells against a concentration gradient. A decrease in inward permeability during the idiophase protects the organism from high extracellular concentration of its own antibiotic. Additional mechanisms exist to protect cells from internal antibiotic that is not excreted or from antibiotic that escapes from an antibiotic production compartment (if such compartmentation exists). One such mechanism is the synthesis of enzymes that modify the antibiotic; many antibiotic producers possess enzymes capable of converting their antibiotic into inactive or less active derivatives. Another mechanism involves a modification in the machinery of the producer, such as in the ribosomes, which serve as targets of the particular antibiotic. Another means by which antibiotic producers protect themselves is by feedback inhibition or repression of antibiotic production.

IV. ANTIBIOTIC PRODUCTION AND REGULATION

Environmental manipulation of culture media in any development program (Woodruff, 1961) often involves the testing of hundreds of additives as possible precursors of the desired product. Occasionally, a precursor that increases production of the secondary metabolite is found. The precursor may also direct the fermentation toward the formation of one specific desirable product; this is known as directed biosynthesis. In many fermentations, however, precursors show no activity because their syntheses are not rate limiting. In such cases, screening of additives has often revealed dramatic effects—both stimulatory and inhibitory—of nonprecursor molecules on the production of secondary metabolites. These effects are usually due to interaction of these compounds with the regulatory mechanisms existing in the fermentation organism (Bu'Lock, 1965; Martin and Demain, 1980).

Antibiotics exert feed-back regulation on their own formation. Some act by repressing one or more antibiotic synthetases. Other antibiotics interfere with their own formation by inhibiting an enzyme involved in their biosynthesis.

A second type of feed-back regulation is involved in a branched pathway leading to a primary and secondary metabolite. In such cases, negative feedback regulation of an early common enzyme by the primary end product might be expected to diminish antibiotic production. For example, lysine interferes with penicillin and cephalosporin biosynthesis by feed-back inhibition and repression of homocitrate synthase, the first enzyme of lysine biosynthesis. The resulting decrease in the intracellular concentration of L-α-aminoadipate, an antibiotic precursor, reduces antibiotic formation.

Catabolite inhibition or repression of enzymes by a rapidly used carbon source, usually glucose, affects secondary metabolism. After years of empirical development, most antibiotic fermentations are now conducted with sources of carbon and energy other than glucose. If glucose is used, it is usually fed at a slow, continuous rate so that catabolites do not accumulate.

The use of different carbon sources for production of a given antibiotic results in different rates and extents of production. Thus, it is important in any development program for the industrial production of a new antibiotic to assess the best carbon source, or mixture of carbon sources, from an economic point of view. However, the mechanisms by which a given carbon source affects positively or negatively the production of a secondary metabolite are not clearly understood.

At one time antibiotic formation was not considered to be regulated by carbon sources, or by any other type of effector. Since antibiotics can inhibit growth of their producers, it makes biological sense for cells to suppress the formation of these toxic compounds until rapid growth nears completion. Viewed from this angle, carbon catabolite regulation of secondary metabolism, like that of primary metabolism, offers the cell a survival advantage (Demain et al., 1979). Thus it is not surprising that glucose suppresses production of a large number of antibiotics. Although the exact mechanism by which the carbon source controls secondary metabolism is unknown, repression of enzyme synthesis has been shown to be involved in a number of cases.

The earliest recognition of a negative effect of glucose on a secondary process involved benzylpenicillin production (Soltero and Johnson, 1953). Glucose was found to be excellent for the growth of *Penicillium chrysogenum*, but poor for penicillin production; lactose showed the opposite pattern. A medium containing both glucose and lactose was devised in which growth occurred only on glucose until the hexose was exhausted. At that point, the extensive mycelial mass that developed on glucose began to produce antibiotic on lactose. Penicillin production could be increased simply by feeding glucose intermittently so that its level in the medium never became high enough to interfere with antibiotic production. Continuous feeding of glucose or other sugars was soon established as a routine practice in industry.

The discovery that nitrogen metabolites regulate primary metabolism is relatively recent; it involves repression of enzymes that act on nitrogen-containing substrates by rapidly used nitrogen sources, especially ammonia. The use of soybean meal in streptomycin and other actinomycete fermentations is probably due to its ability to avoid nitrogen metabolite repression of antibiotic biosynthesis (Aharonowitz, 1980). Of the chemically defined sources of nitrogen for streptomycin production, proline is best, because it is very slowly utilized as a sole nitrogen source. Although ammonium salts can be used, they are not as good for streptomycin production as proline or soybean meal. Long before the discovery of nitrogen metabolite regulation of primary metabolism, it was realized that soybean meal and proline were effective because they yield ammonia at a slow rate over the course of a long fermentation.

In certain antibiotic fermentations, the response to stimulatory additives is due to enzyme induction; one example is methionine stimulation of cephalosporin C biosynthesis in *Cephalosporium acremonium*. Although this example involves induction by an amino acid precursor, streptomycin formation is induced by a nonprecursor molecule called A factor (2-isocapryloyl-3-hydroxymethyl-4-hydroxybutanoic acid lactone) (Khokhlov and Tovarova, 1979), whose structure is

Another type of regulation involves inorganic phosphate (Martin, 1977). Many fermentations must be conducted in the presence of inorganic phosphate levels that are suboptimal for growth. In some cases, the diminution in product formation caused by phosphate involves feed-back regulation of phosphatases by inorganic phosphate. Phosphatases participate in biosynthesis because biosynthetic intermediates of many secondary pathways are phosphorylated, although the ultimate products are not. For example, the final enzymatic step in streptomycin production is the cleavage of phosphate from streptomycin phosphate. The enzyme catalyzing this reaction is inhibited by inorganic phosphate. Thus when a streptomycin fermentation is conducted in a complex medium containing a 10 mM phosphate supplement, streptomycin production is poor, while the biologically inactive streptomycin phosphate accumulates extracellularly.

Formation of other antibiotics is also markedly reduced by inorganic phosphate; in batch fermentations, production begins when phosphate in the medium is exhausted. Since these biosyntheses involve no known phosphorylated intermediates, the detrimental effect of phosphate probably does not involve feed-back regulation of phosphatases. Phosphate inhibition in these cases may involve regulation by ATP. In these fermentations, ATP concentration increases during growth, rapidly decreases, and then remains at a low level during antibiotic production. In one case, it has been shown that addition of inorganic phosphate leads to an immediate threefold increase in the intracellular level of ATP and concomitantly stops antibiotic formation.

In certain fermentations, growth rate regulates the formation of antibiotic synthetases.

V. GENETIC ASPECTS OF ANTIBIOTIC FORMATION

The intense development of basic microbial genetics began in the 1940s at the same time that the fermentative production of penicillin became an international necessity because of World War II. These early basic studies concentrated heavily on the production of mutants and the study of their properties. The simplicity of the mutation technique had tremendous appeal to microbiologists. Thus began the cooperative "strain selection" program involving the U.S. Department of Agriculture, the Carnegie Institute, Stanford University, and the University of Wisconsin, followed by extensive individual programs that still exist today in industrial laboratories throughout the world (Elander, 1966; Alikhanian, 1970).

Since antibiotic production is affected by the same regulatory mechanisms that control primary metabolism (i.e., induction, feed-back regulation, and catabolite regulation) and because each of these mechanisms is genetically determined, it is easy to understand why mutation procedures have had such a major effect on secondary metabolite production. Indeed, mutation is the chief factor responsible for the 100- to 1000-fold increases obtained in antibiotic production from the time of their initial discovery to the present (Demain, 1973). These dramatic improvements mainly resulted from procedures involving mutagenesis, followed by testing random survivors and morphological and color mutants. Although the basic rationales are still unclear, the following additional mutant types have been shown to include a significant proportion of improved antibiotic producers: auxotrophs, "revertants" of auxotrophs, "revertants" of nonproducing mutants, amino acid analog-resistant mutants, and mutants resistant to the antibiotic which the organism produces (Demain, 1973).

Mutation has also been used to eliminate undesirable antibiotics from a production culture and to produce new antibiotics such as 6-demethyltetracyclines and adriamycin. A recent development in the use of mutation for the production of new antibiotics is "mutational biosynthesis" (Shier et al., 1969). In this technique, a mutant ("idiotroph") is isolated which cannot make the antibiotic unless supplemented with a moiety of the antibiotic. Certain analogs of the missing moiety can also be used by idiotrophs to form new antibiotics. Use of mutational biosynthesis has led to the discovery of many new antibiotic derivatives (Daum and Lemke, 1979).

Mutant methodology is valuable in determining how antibiotics are made (Nüesch, 1979; Queener et al., 1978). In several antibiotic pathways, intermediates that accumulate in blocked mutants diffuse out of the cell and can be used by other mutants blocked earlier in the pathway. Identification of these intermediates allows elucidation of the biosynthetic path (Hopwood and Merrick, 1977).

In contrast to the extensive use of mutation in industry, genetic recombination has been used very little (Hopwood, 1977). What is often unappreciated is that recombination should not be looked upon as an alternative to mutation, but as a method to complement mutation programs. By crossing mutants, genotypes are generated that never occur as strictly mutational descendants of parent. Furthermore, the recombination techniques are useful in providing a means of incorporating two beneficial mutations into a single strain.

The availability of a genetic recombination system is also useful in constructing a genetic map of the culture. Genetic maps of various antibiotic-producing actinomycetes are known to a limited extent. The model for such investigations is the map of *Streptomyces coelicolor*. It is interesting that the maps of *Streptomyces bikiniensis, Streptomyces olivaceus,* and *Streptomyces glaucescens* appear similar to that of *S. coelicolor*. More important, the maps of important commercial species such as *Streptomyces rimosus* (oxytetracycline producer) and *Nocardia mediterranei* (rifamycin producer) are also similar. If these maps only contained the location of genes of primary metabolism, they would have only limited use for an antibiotic-producing company. However, when they also contain genes of antibiotic synthesis, both structural and regulatory, they have much greater use. For example, if the genes of antibiotic formation were clustered (as they seem to be), mutagenic activity might be directed to that part of the chromosome where the production genes are clustered using the technique of directed mutation in synchronized cultures.

In industrial genetics, there is excitement over the recent finding that actinomycetes contain plasmids, or extrachromosomal DNA. In *S. coelicolor,* the plasmid is a sex factor (SCP 1). The plasmid can be transferred to other species of *Streptomyces* and, in doing so, can carry along with it some chromosomal genes. This will lead the way to the amplification of antibiotic production genes in industrial actinomycetes. Plasmids have been found in many antibiotic producers and in one case, *S. coelicolor,* the antibiotic (methylenomycin A) synthetases are coded by plasmid genes (Hopwood, 1978). In most cases, antibiotic synthetase genes are chromosomal, but the plasmid is involved in gene expression. The clustering of genes of antibiotic synthesis allows us to be optimistic about the application of recombinant DNA and protoplast fusion techniques (Hopwood et al., 1977; Baltz, 1980) to improve strains and produce new antibiotic molecules (Hopwood, 1978, 1981; Demain, 1980b).

REFERENCES

Aharonowitz, Y. (1980). Nitrogen metabolite regulation of antibiotic biosynthesis. *Annu. Rev. Microbiol. 34*:209–233.

Alikhanian, S. (1970). Applied aspects of microbial genetics. *Curr. Top. Microbiol. Immunol. 53*:91–148.

Baltz, R. H. (1980). Genetic recombination by protoplast fusion in *Streptomyces. Dev. Ind. Microbiol. 4*:43–54.

Bérdy, J. (1974). Recent developments of antibiotic research and classification of antibiotics according to chemical structure. *Adv. Appl. Microbiol. 18*:309–406.

Bu'Lock, J. D. (1965). Aspects of secondary metabolism in fungi. In *Biogenesis of Antibiotic Substances.* Z. Vanek and Z. Hostalek (Eds.). Czechoslovak Academy of Sciences, Prague, pp. 61–72.

Daum, S. J., and Lemke, J. R. (1979). Mutational biosynthesis of new antibiotics. *Annu. Rev. Microbiol. 33*:241–265.

Davies, J., and Smith, D. I. (1978). Plasmid-determined resistance to antimicrobial agents. *Annu. Rev. Microbiol. 32*:469–518.

Demain, A. L. (1973). Mutation and the production of secondary metabolites. *Adv. Appl. Microbiol. 16*:177–202.

Demain, A. L. (1974). How do antibiotic-producing microorganisms avoid suicide? *Ann. N.Y. Acad. Sci. 235*:601–612.

Demain, A. L. (1975). Why mode of action studies? *Chem. Technol. 5*:287–289.

Demain, A. L. (1980a) Do antibiotics function in nature? *Search 11*:148–151.

Demain, A. L. (1980b). The new biology: Opportunities for the fermentation industry. *Annu. Rep. Ferment. Proc. 4*:193–208.

Demain, A. L., and Piret, J. M. (1979). Relationship between antibiotic biosynthesis and sporulation. In *Regulation of Secondary Product and Plant Hormone Metabolism*. M. Luckner and K. Schreiber (Eds.). Pergamon, Oxford, pp. 183–188.

Demain, A. L., and Piret, J. M. (1981). Why secondary metabolism? In *Microbiology 1981*. D. Schlessinger (Ed.). American Society for Microbiology, Washington, D.C.

Demain, A. L., Kennel, Y. M., and Aharonowitz, Y. (1979). Carbon catabolite regulation of secondary metabolism. *Symp. Soc. Gen. Microbiol. 29*:163–185.

Elander, R. P. (1966). Two decades of strain improvement in antibiotic-producing microorganisms. *Dev. Ind. Microbiol. 7*:61–73.

Gottlieb, D. (1976). The production and role of antibiotics in soil. *J. Antibiot. 29*:987–1000.

Habte, M., and Alexander, M. (1977). Further evidence for the regulation of bacterial populations in soil by protozoa. *Arch. Microbiol. 113*:181–183.

Hash, J. H. (1972). Antibiotic mechanisms. *Annu. Rev. Pharmacol. 12*:35–56.

Hirsch, C. F., and Ensign, G. C. (1978). Some properties of *Streptomyces viridochromogenes* spores. *J. Bacteriol. 134*:1056–1063.

Hodgson, B. (1970). Possible roles for antibiotics and other biologically active peptides at specific stages during sporulation of *Bacillaceae*. *J. Theor. Biol. 30*:111–119.

Hopwood, D. A. (1977). Genetic recombination and strain improvement. *Dev. Ind. Microbiol. 18*:9–12.

Hopwood, D. A. (1978). Extrachromosomally determined antibiotic production. *Annu. Rev. Microbiol. 32*:373–392.

Hopwood, D. A. (1981). Genetic studies of antibiotics and other secondary metabolites. *Symp. Soc. Gen. Microbiol. 31*:187–218.

Hopwood, D. A., and Merrick, M. J. (1977). Genetics of antibiotic production. *Bacteriol. Rev. 41*:595–635.

Hopwood, D. A., Wright, H. M., Bibb, M. J., and Cohen, S. N. (1977). Genetic recombination through protoplast fusion in *Streptomyces*. *Nature 268*:171–174.

Katz, E., and Demain, A. L. (1977). The peptide antibiotics of *Bacillus*: Chemistry, biogenesis and possible functions. *Bacteriol. Rev. 41*449–474.

Khokhlov, A. S., and Tovarova, I. I. (1979). Autoregulator from *Streptomyces griseus*. In *Regulation of Secondary Product and Plant Hormone Metabolism*. M. Luckner and K. Schreiber (Eds.). Pergamon, Oxford, pp. 133–145.

Martin, J. F. (1977). Control of antibiotic synthesis by phosphate. *Adv. Biochem. Eng. 6*:105–127.

Martin, J. F., and Demain, A. L. (1980). Control of antibiotic biosynthesis. *Microbiol. Rev. 44*:230–251.

Nüesch, J. (1979). Contribution of genetics to the biosynthesis of antibiotics. In *Genetics of Industrial Microorganisms*. O. K. Sebek and A. I. Laskin (Eds.). American Society for Microbiology, Washington, D.C., pp. 77–82.

Piret, J. M., and Demain, A. L. (1981). Role of gramicidin S in the producer organism, *Bacillus brevis*. In *Sporulation and Germination*. H. S. Levinson, A. L. Sonenshein, and D. J. Tipper (Eds.). American Society for Microbiology, Washington, D.C., pp. 243–245.

Queener, S. W., Sebek, O. K., and Vézina, C. (1978). Mutants blocked in antibiotic synthesis. *Annu. Rev. Microbiol. 32*:593–636.

Sadoff, H. L. (1972). The antibiotics of *Bacillus* species: Their possible roles in sporulation. *Prog. Ind. Microbiol. 11*:2–27.

Sarkar, N., and Paulus, H. (1972). Function of peptide antibiotics in sporulation. *Nature New Biol. 239*:228–230.

Shier, W. T., Rinehart, K. L., Jr., and Gottlieb, D. (1969). Preparation of four new antibiotics from a mutant of *Streptomyces fradiae*. *Proc. Nat. Acad. Sci. U.S.A. 63*:198–204.

Singh, B. N. (1942). Toxic effects of certain metabolic products on soil protozoa. *Nature 149*:168.

Soltero, F. V., and Johnson, M. J. (1953). Effect of the carbohydrate nutrition on penicillin production by *Penicillium chrysogenum* Q-176. *Appl. Microbiol. 1*:52–57.

Vining, L. C. (1979). Antibiotic tolerance in producer organisms. *Adv. Appl. Microbiol. 25*:147–168.

Weinberg, E. D. (1970). Biosynthesis of secondary metabolites: Roles of trace metals. *Adv. Microbiol. Physiol. 4*:1–43.

Woodruff, H. B. (1961). Antibiotic production as an expression of environment. *Symp. Soc. Gen. Microbiol. 11*:317–342.

Antibiotics Used in Medical or
Agricultural Practice:
Antifungal Antibiotics

3

THE PENICILLINS: PROPERTIES, BIOSYNTHESIS, AND FERMENTATION

GEORGE J. M. HERSBACH, CEES P. VAN DER BEEK, AND PIET W. M. VAN DIJCK *Research and Development Division, Gist-Brocades N. V., Delft, The Netherlands*

I. INTRODUCTION

Major scientific discoveries are often based on a mixture of vision, luck, and mistakes. The discovery of penicillin is a good example.

In 1928, in the process of cleaning his laboratory, Alexander Fleming's interest was aroused by a Petri dish with staphylococci that had been contaminated by a mold. This mold, *Penicillium notatum*, had caused a clear zone of lysis in the vicinity of the colony. Fleming concluded that the mold had secreted a compound, lysogenic toward staphylococci, and his other tests also showed activity toward several other pathogenic bacteria. He termed the lytic mold broth filtrate penicillin (Fleming, 1929). Fleming's results almost passed into oblivion, especially when Clutterbuck et al. (1932) failed to isolate the substance. This group of well-known natural products chemists tried to isolate penicillin, but the material already lost the bacteriostatic properties in the first acid ether extraction step.

In 1938 Chain made a literature search on old data on lysogenic compounds and was struck by Fleming's report. He persuaded Florey, the head of his department, to let him work on penicillin. Having *P. notatum* at hand, he could start immediately. In the first year of World War II, Chain and his co-workers had isolated and purified enough material to successfully perform animal tests (Chain et al., 1940). The next year the first clinical trials were carried out on patients who had been given up as hopeless by the surgeons. They were all successful in that only one patient died, but this was due to the limited amounts of penicillin available (Abraham et al., 1941). Owing to the situation at that time, no funds became available to scale up the production process in spite of these results. The government's main interest was to let Britain survive this critical war period. Therefore Florey was sent to the Northern Regional Research Laboratory of the Department of Agriculture in Peoria, where he found a penicillin-minded atmosphere.

Several U.S. industrial laboratories had already started to investigate the production of penicillin in surface cultures, research which had been initiated by the paper of Clutterbuck et al. (1932). In Peoria the governmental activities of the penicillin war project became concentrated. They supplied the industry with strains and fermentation advice. A medium was developed consisting of hydrolyzed starch, lactose, and corn-steep liquor. Kluyver and Perquin (Perquin, 1938) had shown that it was possible to cultivate molds in submerged culture. The Peoria laboratory adapted this technique in 1942 for growing *Penicillium*. From a moldy cantaloupe the strain *Penicillium chrysogenum* NRRL 1951 was isolated. Irradiation and selection yielded Wisconsin Q-176, which became the ancestor of all major industrial strains. This high-production strain, however, produced penicillin K in submerged culture instead of penicillin G. This could be corrected by adding extra phenylacetic acid to the lactose corn-steep liquor medium (Coghill, 1944; Helfand et al., 1980; also see Sec. VII).

The development of the industrial production of penicillin was an Anglo-American war project with top priority. In close cooperation, governmental, university, and industrial laboratories solved within a few years major problems in fermentation, fermenter design, recovery, and so on. In 1943 Abraham et al. and, independently, a team at Merck (Chain, 1948) proposed the correct chemical structure for penicillin to be a fused thiazolidine β-lactam ring system. Detailed accounts of the progress in these first years have been given by many of the pioneers in this Anglo-American endeavor, particularly on the occasion of the fiftieth anniversary of Fleming's publication in 1979 (Chain, 1980; Helfand et al., 1980; Selwyn, 1980).

Industrial companies like Pfizer, Merck, Squibb, Eli Lilly, Abbott, Upjohn, Lederle, and Hoffmann-La Roche in the United States and Beecham, Glaxo, and ICI in Great Britain took advantage of these combined efforts during the war period (Perlman, 1974). The Koninklijke Nederlandsche Gisten Spiritusfabriek (now known as Gist-Brocades) in Delft, The Netherlands, was an exception, in that, because of the war situation, it could not profit from this free flow of information (Hoogerheide, 1980). Before World War II Gist-Brocades had produced yeast, alcohol, vitamins, and chemicals by a combination of fermentative and chemical processes (Elema, 1970). During the war its research activities became severely restricted. The researchers spent most of their time in the library planning postwar activities, and they also listened, clandestinely, to radio broadcasts from England. In 1943 they heard of the discovery of a wonder drug called "penicillin," understood that it must have something to do with the mold *Penicillium*, and soon found Fleming's publication of 1929. Their suppositions were confirmed when they managed to get hold of an article by Kiese (1943) containing some information about the discovery and properties of penicillin. Having access to a number of *Penicillium* strains which had been deposited by Fleming at the Centraal Bureau Voor Schimmelcultures in Baarn, The Netherlands, they started their secret research on penicillin. Totally isolated from all sources of information, they performed their research. Under the code name Bacinol (from *Penicillium baculatum*), they succeeded in isolating limited amounts of penicillin in a fairly pure form. It was sufficient for the treatment of a number of patients, with excellent results. After the winter of 1944, an ampoule with American penicillin was acquired from the Allied food and medicine droppings which proved that indeed the researchers had isolated penicillin and not a related antibiotic. Scaling up the experiments, they reached a stage in 1948 where the penicillin production already met the total Dutch requirements. Figure 1 shows the increase in the penicillin production at Gist-Brocades; the increase in the world production will show a similar pattern. As a consequence of the rapidly increasing penicillin yields, the production costs simultaneously decreased to a few percent of the original level. According to Swiss investigations, which also have been published by Queener and Swartz (1979), in 1975 Gist-Brocades was the largest producer of penicillin G, and responsible for 15–20% of the total world production of 10,000 tons. Demain (1981) reported a value of 17,000 tons of penicillin produced in 1980, but his estimate is probably based on production capacities and not on the actual production; when corrected for this, a value of around 12,000 tons is found, which is in good agreement with our estimates of the world production of 11,500 tons in 1981 (Table 1) and 12,000 tons in 1982 (Hersbach, 1982). The major producers account for over 80% of this production (Table 1).

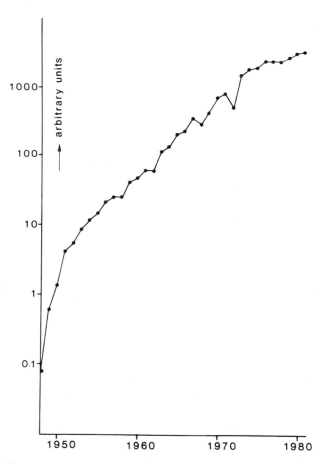

Figure 1 Annual increase in the penicillin production of Gist-Brocades.

Table 1 Major World Producers of Penicillins G and V[a]

Company	Rank—1975[b]	Rank—1981	Estimated share of market in 1975[b] (%)
Gist-Brocades	1	1	15−20
—	—	—	10−15
Glaxo	2	2	5−10
Beecham	5	3	5−10
Bristol	7	4	5−10
Squibb	3	—	5−10
Pfizer	4	5	5−10
Eli Lilly	6	—	5−10
Antibioticos	—	6	—
Rhône-Poulenc	8	7	1−5
Meiji Seika	11	8	1−5
Hoechst	10	9	1−5
Wyeth	9	10	1−5
Total production	10,000 tons	11,500 tons	

[a]East European countries not included.
[b]SoGen-Swiss estimates.

II. STRAIN SELECTION

Today's commercial penicillin-producing *P. chrysogenum* strains have been developed in large-scale strain development programs. The initial work was carried out at the Carnegie Institute in Washington, the University of Minnesota, and the University of Wisconsin (Backus and Stauffer, 1955). Later the penicillin-producing companies took over and started their own strain selection programs (Elander, 1967a). Since 1973, a non-penicillin-producing company (Panlabs Inc.) has developed a series of penicillin-producing strains which are supplied to subscribers at regular intervals (Queener and Schwartz, 1979).

Strain development programs generally pass through five stages:

Induction of variation
Preselection
Selection in shake flasks, possibly followed by tests in laboratory fermenters
Evaluation of selected strains on a pilot-plant scale
Introduction on a production scale in the main plant

Variation can be induced either by mutation or by recombination. As the construction of a *Penicillium* genetic library has already started (Ratzkin, 1981), it is expected that recombinant DNA techniques will also be used in the near future for the generation of altered *Penicillium* strains. Several mutagens have been used for the induction of variation in *P. chrysogenum* (Backus and Stauffer, 1955; Elander, 1967a; Ball, 1973; Tien, 1981), but no special preference has emerged as to which mutagens are most suitable for strain improvement. Since the first description of the parasexual process by Pontecorvo and Sermonti (1953), many attempts have been made to use this technique for the improvement of penicillin production (Ball, 1978), In two cases an improvement over the prototrophic parental titer was reported (Elander, 1967b; Ball, 1982). The availability of externally obtained penicillin-producing strains (Panlabs Inc.), containing independently obtained improvements in penicillin production properties, will undoubtedly increase interest in this technique. However, for selective isolation of all possible recombinants, it will be necessary to construct "commercial master strains" in which each chromosome contains at least one marker that does not affect penicillin production (Ball, 1973).

Preselection methods are used to reduce the number of strains to be tested in shake flasks. When the preselection is based on penicillin production on agar plates, the mutated colonies are assayed for penicillin production either by spraying the Petri dishes with a penicillin-sensitive bacterium, which results in a growth inhibition zone (Ball and McGonagle, 1978), or by incorporating a penicillin-specific dye in the agar (Ray Chowdhurry et al., 1980). Some practical problems are associated with this technique:

1. Depending on the fermination rate, growth rate, and available nutrients, colonies of different sizes are produced on the plates, resulting in differences in "clearing zones," which are dependent on differences in the amount of biomass produced. Some type of correction can be made by using a "potency index," which is the diameter of the clearing zone divided by the diameter of the colony (Ball and McGonagle, 1978). Another possibility is the use of the so-called agar-piece method, in which each mutant receives the same amount of substrate to develop a colony (Chang and Elander, 1979; Ditchburn et al., 1974; Ichikawa et al., 1971; Trilli et al., 1978).

2. The diameter of the clearing zone is proportional to the logarithm of the amount of penicillin plus a constant factor (Ingram et al., 1953). For this reason small improvements are difficult if not impossible to detect.

As a consequence, the agar plate preselection technique is useful in eliminating low-production or nonproducing strains, rather than for selecting improved strains. Preselection methods can also be based on knowledge of (or guesses at) processes which influence penicillin production or yield. Specific changes which were investigated for their effect on penicillin production or yield are not published in great abundance, for the obvious reason that production improvements are kept as company secrets. However, some rational attempts at strain improvement have been described for *Penicillium*:

1. Elimination of feed-back inhibition of lysine on penicillin biosynthesis was accomplished by isolating regulatory mutants secreting lysine (Masurekar and Demain, 1974) and by isolating lysine-auxotrophic mutants blocked after α-aminoadipic acid which were fed limiting amounts of lysine (O'Sullivan and Pirt, 1973). In both cases no improvement of penicillin production was observed, indicating that lysine feed-back regulation was not the rate-limiting factor in the strains that were used in these investigations. Alteration of lysine regulation may, however, have been of some importance in the development of high-production strains (Luengo et al., 1979a).

2. Comparable studies on the other two amino acid building stones of penicillin, namely, valine and cysteine, have not been published, although the loss of a valine binding site on acetohydroxy acid synthetase (Goulden and Chattaway, 1969) and the increased ability to accumulate intracellular sulfate (Tardew and Johnson, 1959; Segel and Johnson, 1961) in high-production strains suggest that changes in the regulation of the synthesis of these two amino acids might be worth investigating.

3. Changes in the permeability of the cell membrane were induced by selecting mutants which were resistant against polyene macrolide antibiotics (Luengo

et al., 1979b). Both positive and negative effects were seen. Improvements of penicillin production were suggested to be caused by enhanced secretion of penicillin, resulting in decreased feed-back inhibitory levels of penicillin. Our present production strain probably has no restrictions with regard to this property, as no mutants with increased penicillin production or yield in shake flasks were found among 100 fungimycin-resistant isolates and 100 nystatin-resistant isolates.

4. *Penicillium* strains (and corresponding culture conditions) have been selected which gave a lower viscosity broth (van der Waard, 1976; Bartholomew et al., 1977, cited by Queener and Swartz, 1979). As a result, the amount of biomass, which is normally limited by the oxygen transfer capacity, could be increased, resulting in a higher yield per fermenter.

5. Panlabs Inc. has recently selected and issued a *Penicillium* mutant with reduced levels of phenylacetic acid oxidase, thereby reducing the amount of phenylacetic acid that had to be supplied to the culture as compared to the parent strain.

6. Several preselection methods were tested by Elander (1980), using the percentage of mutants retained in secondary screening tests as a criterion. Biochemically deficient mutants and strains resistant to a variety of mitotic inhibitors, heavy metals, and sulfur and amino acid analogs yielded a higher percentage of superior strains compared to random survivors.

7. Several other properties that have not been mentioned yet, have been correlated with strain improvement: tolerance to phenylacetic acid (Fuska and Welwardova, 1969), ability to assimilate carbohydrate and precursor (Pan et al., 1972), sensitivity to iron (Pan et al., 1975) penicillin acylase activity (Erickson and Dean, 1966), acyltransferase activity (Preuss and Johnson, 1967), and feed-back inhibition of penicillin biosynthesis by penicillin (Gordee and Day, 1972; Martin et al., 1979). As these properties were detected in strains that were already high producers, it is impossible to say whether changes in these properties were the direct cause of the improvement. If, however, mutants specifically selected for such properties should demonstrate increased penicillin production, further accentuation of the property in question might improve the penicillin production even more.

The actual selection of strains is usually done in shake flasks. This yields candidates for further testing in laboratory fermenters, on a pilot-plant scale and eventually on a production scale.

III. BIOSYNTHESIS AND THEORETICAL CONVERSION YIELD

As the biosynthesis of penicillin has been reviewed several times recently (Aberhart, 1977; O'Sullivan and Abraham, 1981; Queener and Neuss, 1982), we will focus our attention on the consequences that recent biochemical knowledge concerning penicillin biosynthesis might have on the theoretical maximum conversion yield. In addition, we will discuss the analytical aspects of the individual reactions in the penicillin biosynthetic pathway in Sec. V. Briefly, penicillin is synthesized as follows: α-Aminoadipic acid, an intermediate in the fungal biosynthesis of lysine, and cysteine and valine are condensed to the tripeptide δ-(α-aminoadipyl) cysteinylvaline. Subsequently the lactam and thiazolidine rings are closed, yielding isopenicillin N. Exchange of the α-aminoadipyl moiety for phenylacetic or phenoxyacetic acid yields penicillin G or V (Fig. 2).

Based on the knowledge of that time Cooney and Acevedo (1977) made a calculation of the theoretical maximum conversion yield of penicillin from

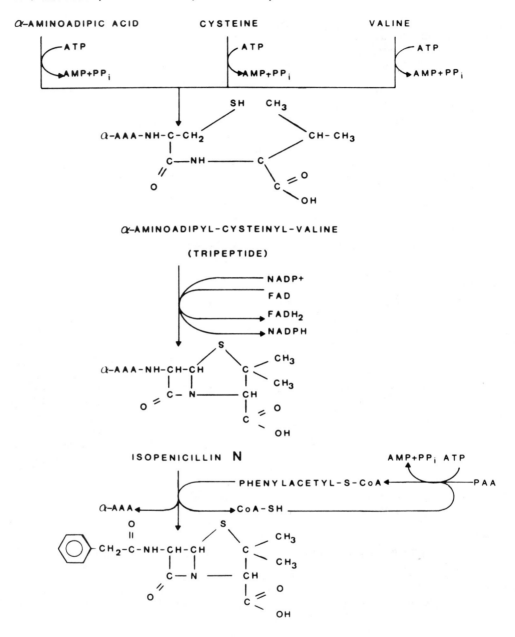

penicillin G

Figure 2 Biosynthesis of penicillin G.

glucose. Their maximum yield was 1980 or 1100 units of penicillin per milligram of glucose, depending on whether α-aminoadipic acid was recycled or not. We have extended these calculations using recent biochemical knowledge on cysteine biosynthesis in *Cephalosporium* and *Penicillium* (Treichler et al., 1979) which suggests that cysteine is produced by cleavage of cystathionine (Figs. 3, 4) rather than by condensation of sulfide on O-acetylserine (Fig. 5). When cystathionine is cleaved by the enzyme cystathionine-γ-lyase, it yields cysteine, α-ketobutyrate, and NH_3.

α-Ketobutyrate either may be transformed into isoleucine by the same enzymes that take care of the valine biosynthesis (Fig. 4) or it may yield propionyl-S-coenzyme A (CoA) when it is catabolized (Fig. 3). When propionyl-S-CoA is carboxylated to methylmalonyl-S-CoA, racemized into succinyl-S-CoA, and further metabolized in the tricarboxylic acid (TCA) cycle to oxaloacetate, a cycle is generated in which α-ketobutyrate yields homoserine, one of the precursors for cysteine biosynthesis via cystathionine. If homoserine is indeed resynthesized from α-ketobutyrate, there is no need for anaplerotic biosynthesis of this compound for cysteine biosynthesis. The tripeptide precursor of penicillin may be synthesized on a multienzyme complex by means of the thiotemplate mechanism, as suggested by Martin et al. (1979). Its synthesis may, on the other hand, resemble the glutathione biosynthetic pathway (Kaszab and Enfors, 1981, Lara et al., 1982). We have assumed that each amino acid has to be activated by adenosine 5'-triphosphate (ATP) at the cost of two high-energy phosphate bonds:

Cysteine + valine + α-aminoadipic acid (α-AAA) + 3ATP → tripeptide

+ 3AMP + $3PP_i$

Isopenicillin N formation is thought to be mediated by NADP and FAD (Demain, 1966). Phenylacetic acid (PAA) enters the cell by diffusion and is activated by CoA-SH at the expense of two high-energy phosphate bonds. With these assumptions the overall stoichiometry for the synthesis of penicillin becomes

Cysteine + valine + α-AAA + PAA + 4ATP + $NADP^+$ + FAD → penicillin G

+ α-AAA + 4AMP + $4PP_i$ + NADPH + $FADH_2$ (I)

Owing to a lack of detailed knowledge, we have not included energy consumption due to intracellular compartmentation and transport of penicillin from the cell to the fermentation broth. Pathways and energy balance schemes for the biosynthesis of α-aminoadipic acid and valine are given in Figures 6 and 7. Three different routes of cysteine biosynthesis and α-ketobutyrate transformation are given in the Figures 3, 4, and 5. In these schemes we have taken into account that NADPH, which is needed in many biosynthetic reactions, is synthesized from NADH in the mannitol cycle at the cost of one molecule of ATP for each molecule of NADPH (Hult et al., 1980). Equation (I) then becomes

Cysteine + valine + α-AAA + PAA + ADP + P_i + 3ATP + NAD^+ + FAD

→ penicillin G + α-AAA + 4AMP + $3PP_i$ + NADH + $FADH_2$

In addition, we have assumed that active transport of sulfate and urea is carried out at the expense of one high-energy phosphate bond per molecule. Active transport of glucose yields glucose-6-phosphate, the second intermediate in the metabolism of glucose. For this reason no energy requirements are included for active transport of glucose.

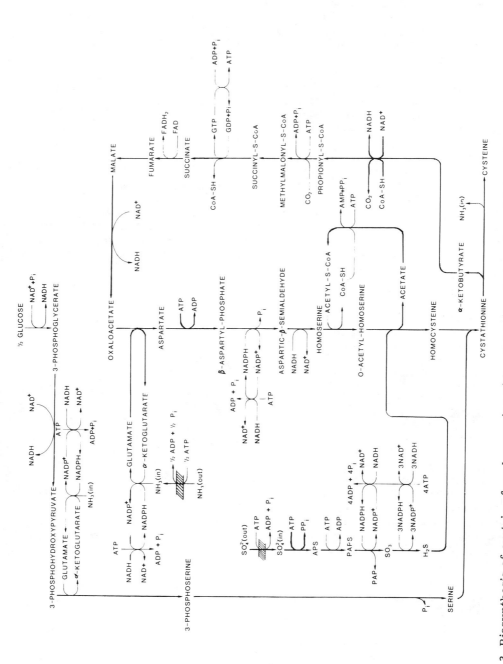

Figure 3 Biosynthesis of cysteine from homoserine with α-ketobutyrate transformed into oxaloacetate:
$\frac{1}{2}$glucose + NH_3 + SO_4^{2-} + $12\frac{1}{2}$ATP + 4NADH + FAD → cysteine + AMP + PAP + $10\frac{1}{2}$ADP + $10\frac{1}{2}P_i$ + $2PP_i$ + 4NAD$^+$ + FADH$_2$. (V)

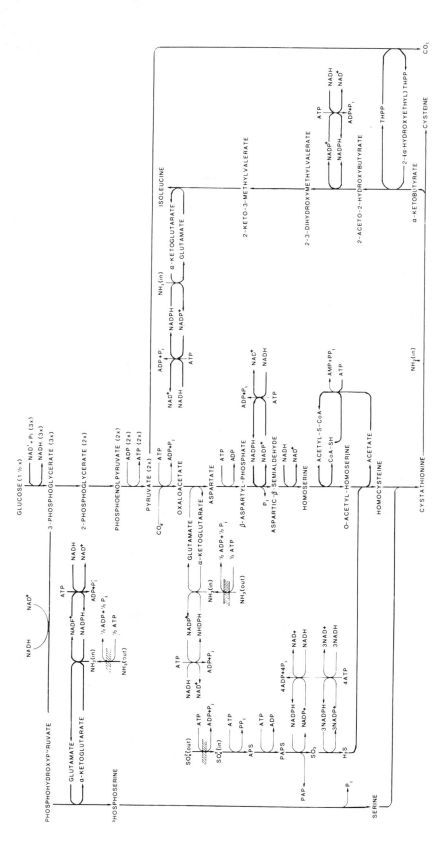

Figure 4 Biosynthesis of cysteine from homoserine with α-ketobutyrate transformed into isoleucine:

$1\frac{1}{2}$glucose + $2NH_3$ + SO_4^{2-} + 14ATP + 6NADH $\rightarrow$ cysteine + isoleucine + AMP + PAP + $2PP_i$ + 12ADP + $12P_i$ + $6NAD^+$.   (VI)

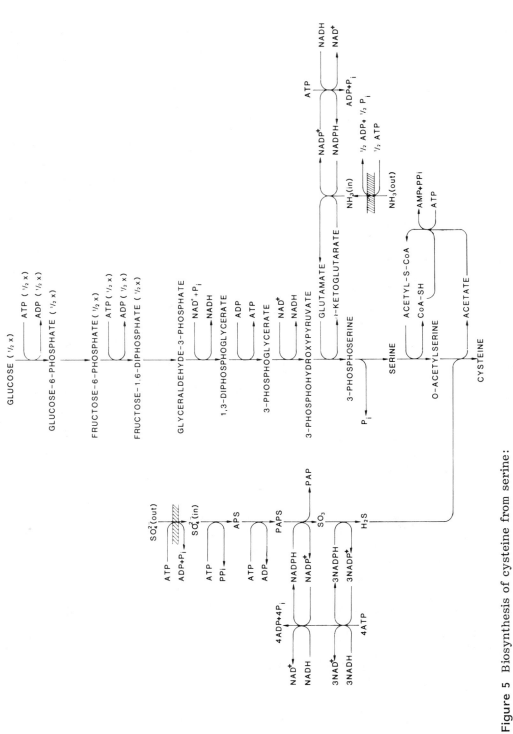

Figure 5 Biosynthesis of cysteine from serine:

$$\tfrac{1}{2}\text{glucose} + NH_3 + SO_4^{2-} + 9\tfrac{1}{4}ATP + 3NADH \rightarrow \text{cysteine} + AMP + PAP + 2PP_i + 7\tfrac{1}{2}ADP + 7\tfrac{1}{2}P_i + 3NAD^+.$$ (IV)

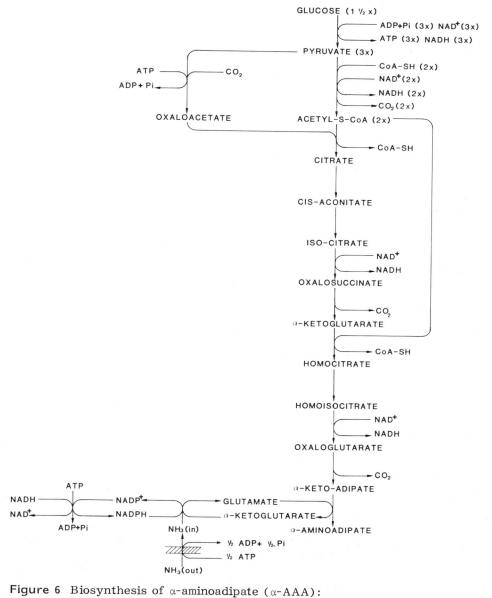

Figure 6 Biosynthesis of α-aminoadipate (α-AAA):

$1\frac{1}{2}$glucose + NH_3 + $1\frac{1}{2}$ADP + $\frac{1}{2}P_i$ + 6NAD$^+$ $\rightarrow$ α-AAA + 3CO$_2$ + $\frac{1}{2}$ATP + 6NADH.

(III)

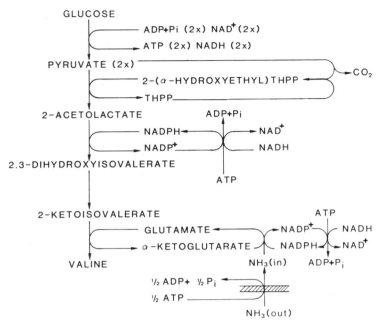

Figure 7 Biosynthesis of valine:

$$\text{glucose} + \tfrac{1}{2}\text{ATP} + \text{NH}_3 \rightarrow \text{valine} + \tfrac{1}{2}\text{ADP} + \tfrac{1}{2}\text{P}_i. \tag{II}$$

Aside from three different routes of cysteine biosynthesis we have considered both the recycling and the secretion of α-aminoadipic acid. The possibility that α-aminoadipic acid may (at least partially) not be reutilized is supported by the discovery of large amounts of 6-oxopiperidine-2-carboxylic acid (cyclic α-aminoadipic acid) in penicillin V fermentation broth (Brundidge et al., 1980). In our own penicillin G fermentations cyclic α-aminoadipic acid accumulates to an equivalent of about 15% of the eventual total penicillin G production in the first part of the fermentation and then remains constant (P.W.M. van Dijck and D. Schipper, 1981, unpublished results). The overall stoichiometry of six possible routes of penicillin biosynthesis is given in Table 2. As the efficiency of the oxidative phosphorylation is not known, two possibilities for ATP production from NADH and FADH$_2$ are taken into account in Table 3. The maximum theoretical conversion yields are shown in columns D and E of Table 3. It can be seen that maximum theoretical conversion yields may vary from 1544 to 638 units of penicillin per milligram of glucose. The actual conversion yield of the best penicillin strains known in the literature is about 200 units of penicillin per milligram of glucose (see Sec. VII.D.3, Table 10). This means that, depending on the actual biosynthetic pathway used, these *P. chrysogenum* strains are producing penicillin at 13−29% of their maximum theoretical conversion yield, a more than twofold difference in efficiency. The degree of improvement still possible is dependent on the amount of glucose necessary for growth and maintenance.

Table 2 Terminal Balance of Six Possible Ways of Penicillin Biosynthesis

α-AAA	Cysteine synthesized from	α-Ketobutyrate transformed into	Equations used	Total equation for penicillin biosynthesis
Recycled	Serine	—	I,II,IV	$1\frac{1}{2}$Glucose + $2NH_3$ + SO_4^{2-} + PAA + 13ATP + 2NADH + FAD → PenG + CO_2 + PAP + 5AMP + $6PP_i$ + 7ADP + $6P_i$ + $2NAD^+$ + $FADH_2$
	Homo-serine	Oxaloacetate	I,II,V	$1\frac{1}{2}$Glucose + $2NH_3$ + SO_4^{2-} + PAA + 16ATP + 3NADH + 2FAD → PenG + CO_2 + PAP + 5AMP + $6PP_i$ + 10ADP + $9P_i$ + $3NAD^+$ + $2FADH_2$
	Isoleucine		I,II,VI	$2\frac{1}{2}$Glucose + $3NH_3$ + SO_4^{2-} + PAA + $17\frac{1}{2}$ATP + 5NADH + FAD → PenG + isoleucine + CO_2 + PAP + 5AMP + $6PP_i$ + $11\frac{1}{2}$ADP + $10\frac{1}{2}P_i$ + $5NAD^+$ + $FADH_2$
Excreted	Serine	—	I,II,III,IV	3Glucose + $3NH_3$ + SO_4^{2-} + PAA + $12\frac{1}{2}$ATP + $4NAD^+$ + FAD → PenG + α-AAA + $4CO_2$ + PAP + 5AMP + $6PP_i$ + $6\frac{1}{2}$ADP + $9\frac{1}{2}P_i$ + 4NADH + $FADH_2$
	Homo-serine	Oxaloacetate	I,II,III,V	3Glucose + $3NH_3$ + SO_4^{2-} + PAA + $15\frac{1}{2}$ATP + $3NAD^+$ + 2FAD → PenG + α-AAA + $4CO_2$ + PAP + 5AMP + $6PP_i$ + $9\frac{1}{2}$ADP + $8\frac{1}{2}P_i$ + 3NADH + $2FADH_2$
	Isoleucine		I,II,III,VI	4Glucose + $4NH_3$ + SO_4^{2-} + PAA + 17ATP + NAD^+ + FAD → PenG + α-AAA + isoleucine + $4CO_2$ + PAP + 5AMP + $6PP_i$ + 11ADP + $10P_i$ + NADH + $FADH_2$

Table 3 Maximum Theoretical Conversion Yield of Six Possible Ways of Penicillin Biosynthesis

Oxidative phosphorylation[a]	Terminal balance after transformation of NADH and FADH to ATP	Moles of glucose needed to synthesize 1 mol of penicillin G[a]	Units of penicillin/mg of glucose[b]
I	$1\frac{1}{2}$ Glucose + 23 ATP → Pen G	2.14	1544
II	$1\frac{1}{2}$ Glucose + 22 ATP → Pen G	2.42	1367
I	$1\frac{1}{2}$ Glucose + 27 ATP → Pen G	2.25	1468
II	$1\frac{1}{2}$ Glucose + 26 ATP → Pen G	2.58	1279
I	$2\frac{1}{2}$ Glucose + $36\frac{1}{2}$ ATP → Pen G	3.51	940
II	$2\frac{1}{2}$ Glucose + $32\frac{1}{2}$ ATP → Pen G	3.85	857
I	3 Glucose + $4\frac{1}{2}$ ATP → Pen G	3.13	1057
II	3 Glucose + $5\frac{1}{2}$ ATP → Pen G	3.40	973
I	3 Glucose + $8\frac{1}{2}$ ATP → Pen G	3.24	1021
II	3 Glucose + $13\frac{1}{2}$ ATP → Pen G	3.56	927
I	4 Glucose + 18 ATP → Pen G	4.50	734
II	4 Glucose + 20 ATP → Pen G	4.83	683

[a]I: NADH → 3ATP, $FADH_2$ → 2ATP, glucose → 36ATP; II: NADH → 2ATP, $FADH_2$ → 1ATP, glucose → 24ATP.
[b]Molecular weight of sodium penicillin G, 356.38; 1 mg sodium penicillin G = 1670 units of sodium penicillin G; molecular weight of glucose, 180.16.

IV. BIOCHEMICAL REGULATION

It has been known for a long time that penicillin production is dependent on the carbon source used (Jarvis and Johnson, 1947; Soltero and Johnson, 1954). Glucose was found to be an excellent growth substrate, but a poor substrate for penicillin production. Lactose, on the other hand, appeared to be an excellent carbon source for penicillin production, but a less efficient growth substrate, suggesting some kind of carbon catabolite regulation. Recently attempts have been made to elucidate the nature of the carbon catabolite regulation in P. chrysogenum (Revilla et al., 1982; Lopez-Niéto et al., 1982). Glucose was added to a lactose batch fermentation and to a resting cell system. Inhibition of penicillin synthesis was observed in the fermentation, while the incorporation of [14C]valine into penicillin was unaffected in the resting cell system, suggesting that repression may be the mechanism involved in carbon catabolite regulation. Unfortunately the rate of penicillin synthesis in the resting cell system was not compared to that in the fermentation to check whether the results of the resting cell studies had any physiological significance. Moreover, no attention was paid to the possibility of oxygen limitation, which may have severely affected the penicillin biosynthetic rate in cases where glucose was added to the fermentation (Secs. VI.E and VII.D.1).

As already discussed in Sec. II on strain selection, regulation in the biosynthesis of the amino acid precursors of penicillin may have played or may still play a role in determining the rate of penicillin biosynthesis once carbon catabolite regulation is eliminated (or diminished). The regulation of the nitrogen metabolism of *P. chrysogenum* was the subject of two articles by a group from the University of Mexico (Sanchez et al., 1982; Lara et al., 1981). In spite of the fact that a low-production strain (NRRL 1951) was used in these studies, the results are interesting because they add some new aspects to the knowledge of the biosynthetic pathway of penicillin and its precursors. In the paper of Sanchez et al., (1982), previous studies on the glutamine synthetase of *Neurospora crassa* were extended to *P. chrysogenum*. By increasing the ammonium concentration in the fermentation broth, the authors inhibited the activity of the glutamine synthetase. As a result, the glutamate pool increased and the glutamine pool and the penicillin production decreased, suggesting that glutamine rather than glutamate is the amino group donor for the amino acid precursors of penicillin synthesis. An alternative explanation may, however, be the involvement of glutamate synthetase in the synthesis of glutamate during penicillin synthesis. In the paper of Lara et al. (1981) the effect of glutamate addition on penicillin production was studied. An increase in penicillin production was observed which was not accompanied by increased penicillin precursor pools. The increase in penicillin synthesis could, however, be correlated with a stimulation of the activity of the dipeptide synthetase. This stimulation could also be achieved by the addition of certain glutamate analogs and could be prevented by the addition of cycloheximide, suggesting that induction or derepression of dipeptide synthetase might be the cause of the stimulation of penicillin production. As cycloheximide has been reported to affect the activities of glutamine synthetase (Legrain et al., 1982) and glutamate dehydrogenase (Wendelberger-Schieweg et al., 1980) additional experiments have to be performed to reveal the exact mechanism by which dipeptide synthetase is stimulated. In addition, preliminary evidence was presented by Lara et al. (1981) to support the theory that the dipeptide synthetase is identical to γ-glutamylcysteine synthetase. Kaszab and Enfors (1981) recently obtained some evidence in favor of this theory too. Little is known about the regulation of the enzymes involved in tripeptide, isopenicillin N, and penicillin G/V synthesis. Although it is known that the two enzymes which are involved in the conversion of isopenicillin N into penicillin G or V, namely, phenyl- (or phenyloxy-)acetic acid-CoA ligase and penicillin acyltransferase are expressed at high levels when the rate of penicillin biosynthesis is high (Brunner et al., 1968; Preuss and Johnson, 1967), nothing is known about the regulation of their synthesis and/or expression. With accumulating penicillin concentrations in the fermentation broth, feed-back inhibition of penicillin on its own biosynthesis may become important (Gordee and Day, 1972). Nestaas and Demain (1981) however presented evidence that the apparent feed-back inhibition observed by Gordee and Day could have been caused by gradual hydrolysis of the added penicillin. By performing resting cell studies Martin et al. (1979) circumvented the hydrolysis problem and found that penicillin biosynthesis was indeed inhibited by exogenously added penicillin G or 6-aminopenicillanic acid.

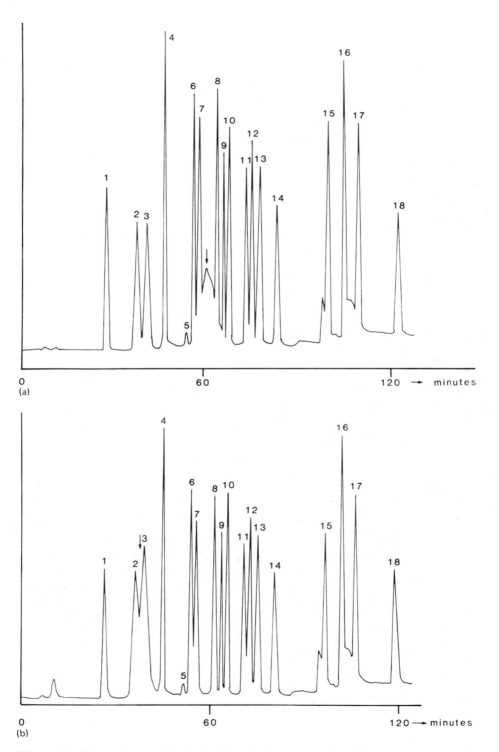

Figure 8 Chromatographic behavior of the LLD-tripeptide on an AS-70 resin column. (a) LLD-tripeptide disulfide in a mixture of amino acids: 1. Asp 2. Thr 3. Ser 4. Glu 5. Pro 6. Gly 7. Ala 8. Val 9. Cys 10. Met 11. lle 12. leu 13. Tyr 14. Phe 15. NH$_3$ 16. lys 17. his 18. Arg. (b) Similar run for the LLD

V. PRODUCT AND INTERMEDIATES OF THE PENICILLIN BIOSYNTHESIS

A. Intermediates

We will deal with the individual steps in the penicillin biosynthetic pathway progressing in order toward the final product penicillin G or V.

1. The Dipeptide

From the three amino acids, first L-α-aminoadipic acid and L-cysteine, or the lyase product of cystathionine (Treichler et al., 1979), are condensed to a dipeptide. To this, L-valine is coupled in a subsequent step to yield the δ-(L-α-aminoadipyl)-L-cysteinyl-D-valine. Lara et al. (1982) have recently published an in vitro assay for this first step catalyzed by the δ-(L-α-aminoadipyl)-L-cysteine synthetase from P. chrysogenum. It is based on the consumption of ATP needed to activate amino acids for nonribosomal peptide synthesis in a dialyzed cell-free system in the presence of both substrates. From the data it can be calculated that the in vitro activity exceeds the overall rate of the in vivo penicillin biosynthesis by several orders of magnitude. Until now, it has not been possible to demonstrate the presence of the dipeptide within the mycelium. There are several possible explanations for this, one of which is that the dipeptide remains enzyme bound. Martin et al. (1979) suggested the thiotemplate mechanism for the formation of the tripeptide. The amino acids are activated by ATP and sequentially bound to thio groups on the multienzyme complex; the product to be released is the completed tripeptide. On the other hand, both Kaszab and Enfors (1981) and Lara et al. (1982) have suggested that the dipeptide synthetase activity is identical to the (γ-glutamyl)-L-cysteine synthetase, the first reaction in the biosynthesis of glutathione. We will discuss the consequences of both these concepts later.

2. The Tripeptide

Minute amounts of the tripeptide were isolated by Arnstein and co-workers (1960a,b) from the mycelium of P. chrysogenum. They purified the compound by paper electrophoresis and showed that it could be hydrolyzed to α-aminoadipic acid, cysteine, and valine. From Cephalosporium acremonium larger quantities could be obtained, which permitted physical analysis (Abraham and Newton, 1965; Loder and Abrahan, 1971). From these analyses it could be concluded that the tripeptide was δ-(L-α-aminoadipyl)-L-cysteinyl-D-valine. By comparing the electrophoretic mobility on paper of the LLL and LLD forms of the tripeptide, Fawcett and Abraham (1976) concluded that in P. chrysogenum the tripeptide was also in the LLD configuration. This was confirmed by chemical means (Adriaens et al., 1975) and by nuclear magnetic resonance spectroscopy (P. W. M. van Dijck, P. A. Deen, A. Kattevilder, and D. Schipper, 1982, unpublished results). Recently large quantities of material have become available from a tripeptide-accumulating mutant of C. acremonium (Shirafuji et al. 1979). The purification of the tripeptide with high-performance liquid chromatography (HPLC) using an ion exchange column has been described by Baldwin and Wan (1981). On an AS70 resin column using the amino acid analyzer in a physiological run the tripeptide elutes between threonine and serine; the disulfide appears between alanine and valine (P. A. Deen and P. W. M. van Dijck, 1982, unpublished data, and Fig. 8). Up to now no proper data

---

Figure 8 (Continued) monomer: the Kontron liquimat III Aminoacid Analyzer was operated with Durrum Pico physiological program.

have become available with respect to the activity of the tripeptide-forming enzyme. Bycroft et al. (1981) reported that in studies with intact mycelium one is hampered by the fact that the dipeptide substrate is externally hydrolyzed before it is taken up by the cell. One way around this problem is to use permeabilized cells (Felix et al., 1980) or to fuse in liposome-entrapped substrates (Makins and Holt, 1982).

3. Isopenicillin N

The formation of the β-lactam and thiazolidine rings yielding isopenicillin N has been well documented in recent years. The information has been obtained predominantly with cell-free systems from *C. acremonium*. O'Sullivan et al. (1979) were the first to show that upon addition of the tripeptide to a cell-free system a compound was formed, which by chemical, chromatographic, and microbiological criteria was identified as isopenicillin N. Demain's group independently confirmed this finding by using a differential bioassay for isopenicillin N and penicillin N (Konomi et al., 1979). Meeschaert et al. (1980) demonstrated the ring closure activity in *P. chrysogenum* cell-free extracts and showed that the LLD tripeptide is converted into isopenicillin N via a monocyclic β-lactam and not via the cyclic dehydrovalinyl intermediate. This finding was recently criticized by Abraham et al. (1982). Whether a hydroxyl intermediate is formed is still uncertain; however, Bahadur et al. (1981c) have demonstrated that this compound does not yield isopenicillin N when added to a cell-free system. The Oxford group (Abraham et al., 1981) has partly purified the isopenicillin N synthetase activity. It is a protein with a molecular weight of about 32,000 which carries a ferro ion as cofactor. The tripeptide must be available in its monomeric form, so bearing a free sulfhydryl group, in order to display activity.

With respect to substrate specificity, it was found that the L-α-aminoadipic acid and L-cysteine moiety are essential in the tripeptide, whereas the D-valine could be replaced by D-isoleucine. Both the D-isoleucine and the L-valine analogs of the tripeptide acted as inhibitors in the ring closure reaction with the normal LLD tripeptide (Abraham et al., 1981). Bahadur et al. (1981b) have studied the ring closure reaction with a variety of valine analogs of the tripeptide. Nuclear magnetic resonance spectroscopy has been used to demonstrate the isopenicillin N formation in the cell-free system using either the unlabeled (Bahadur et al., 1981a), the ^{13}C-labeled (Baldwin et al., 1980), the [2]H-labeled (Baldwin et al., 1981) or the ^{18}O-labeled tripeptide (Adlington et al., 1982). Neuss et al. (1982) demonstrated the applicability of high-performance liquid chromatography (HPLC) in the kinetic approach to the ring closure reaction in a cell-free system. White et al. (1982) have shown that to produce isopenicillin N from the tripeptide, molecular oxygen is needed in a 1:1 stoichiometry.

4. Penicillin G or V

Isopenicillin is converted to the final product penicillin G or V by the action of an acyltransferase. This enzyme has been detected in the mycelium by Pruess and Johnson (1967). Whether or not the exchange of the side chain between different penicillin species (i.e., isopenicillin N) and the hydrolysis of penicillin G or V into 6-aminopenicillanic acid (6-APA) or the reverse reaction is catalyzed by one enzyme has been a matter of debate for quite a while. Spencer and Maung (1970) reported one single protein band on gel electrophoresis displaying both the penicillin acyltransferase, 6-APA-acyltransferase, penicillin acylase, and phenylacetyl-CoA hydrolase activity. All these activities were produced in the same stage of the fermentation and reacted in a identical manner to pH, temperature, and inhibition by N-ethylmaleimide. They also comigrated on DEAE cellulose and Sephadex G-100. That we are dealing with an enzyme complex with multiple activities has been confirmed by O'Sullivan and Abraham (1981). From the experiments of Fawcett et al. (1975), it became clear that isopenicillin N is the substrate for penicillin synthesis in *P. chrysogenum*. Whether or not 6-APA is a transient intermediate of the acyltransferase system has not been resolved satisfactorily. In the acyltransferase reaction the α-aminoadipyl moiety is lost. The presence of large amounts of 6-oxopiperidine-2-carboxylic acid in penicillin G and V fermentation broths (Brundidge et al., 1980; P. W. M. van Dijck and D. Schipper, 1981 unpublished results) suggests the possibility that α-aminoadipic acid is removed from the 6-APA nucleus in the lactam form. Whether this imposes a problem with respect to the recycling of L-α-aminoadipic acid is still an open question.

Phenylacetic acid (PAA) and phenoxyacetic acid (POA) are the side-chain precursors for penicillin G and V, respectively. The side-chain precursor is fed to the fermenter in moderate amounts, since it can be toxic to the cell. They are weak acids which can diffuse freely through the membrane when they are in the nondissociated form. This will cause a rapid dissipation of the proton gradient over the membrane (Bonting and de Pont, 1981). This uncoupling effect on the oxidative phosphorylation explains the toxicity of the precursors. Early methods for assaying phenyl- or phenoxyacetic acid all employ extraction with an organic solvent as a first step. A number of colorimetric methods (Pan, 1955; Birner, 1959; Aly and Faust, 1964) and gas-chromatographic methods (Niedermayer, 1964; Goodwin et al., 1975) have been published. Dunham (1972) has compiled a number of assay methods applicable to phenoxyacetic acid. Nachtmann (1979) has described an analysis by HPLC on a reverse-phase column in which there is no need to extract the side-chain precursor with organic solvents. In one run not only the concentration of precursor is quantified, but also the concentration of penicillin which has been formed and the concentration of a number of penicillin degradation products. The side-chain precursor has to be activated before the transacylation occurs. Brunner et al. (1968) reported the presence of a side-chain CoA ligase in *Penicillium*, an activity which appeared just prior to and during the stage of rapid penicillin production. This enzyme also displayed transacylase activity between CoA derivatives of PAA or POA and 6-APA. The enzyme has been partly purified and characterized (Brunner and Rohr, 1975).

5. *Localization*

In this section we will review the data on the several enzymatic activities of the penicillin biosynthetic pathway with respect to their subcellular localization. In Figure 9 we have depicted a scheme that covers all the activities from the individual amino acids to the secreted penicillin. The data for the tripeptide synthetase are the least controversial. The activity has been reported to be exclusively associated with the particulate fraction of a crude cell homogenate (Fawcett and Abraham, 1975, 1976). Kurylowicz and co-workers (1977, 1979a,b) have shown that the Golgi apparatus is involved in the biosynthesis and secretion of penicillin. Based on these data, we propose that the tripeptide-forming enzyme is bound to the membrane of the Golgi vesicles. Abraham et al. (1981) reported the cyclization activity (isopenicillin

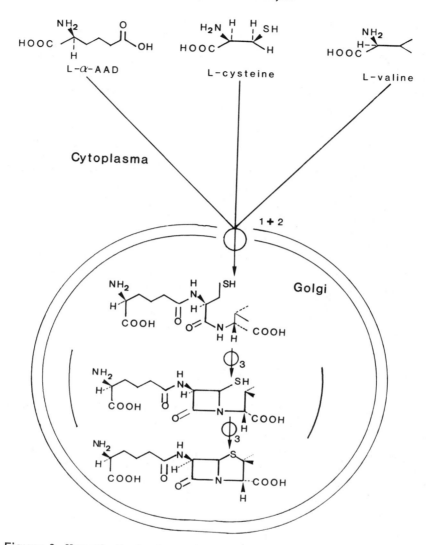

Figure 9 Hypothetical scheme for the localization of the enzymes involved in the biosynthesis of penicillin. (1) δ-(L-α-aminoadipyl)-L-cysteine synthetase, (2) δ-(L-α-aminoadipyl)-L-cysteine-D-valine synthetase, (3) isopenicillin N synthetase, (4) side-chain CoA ligase, and (5) acyltransferase.

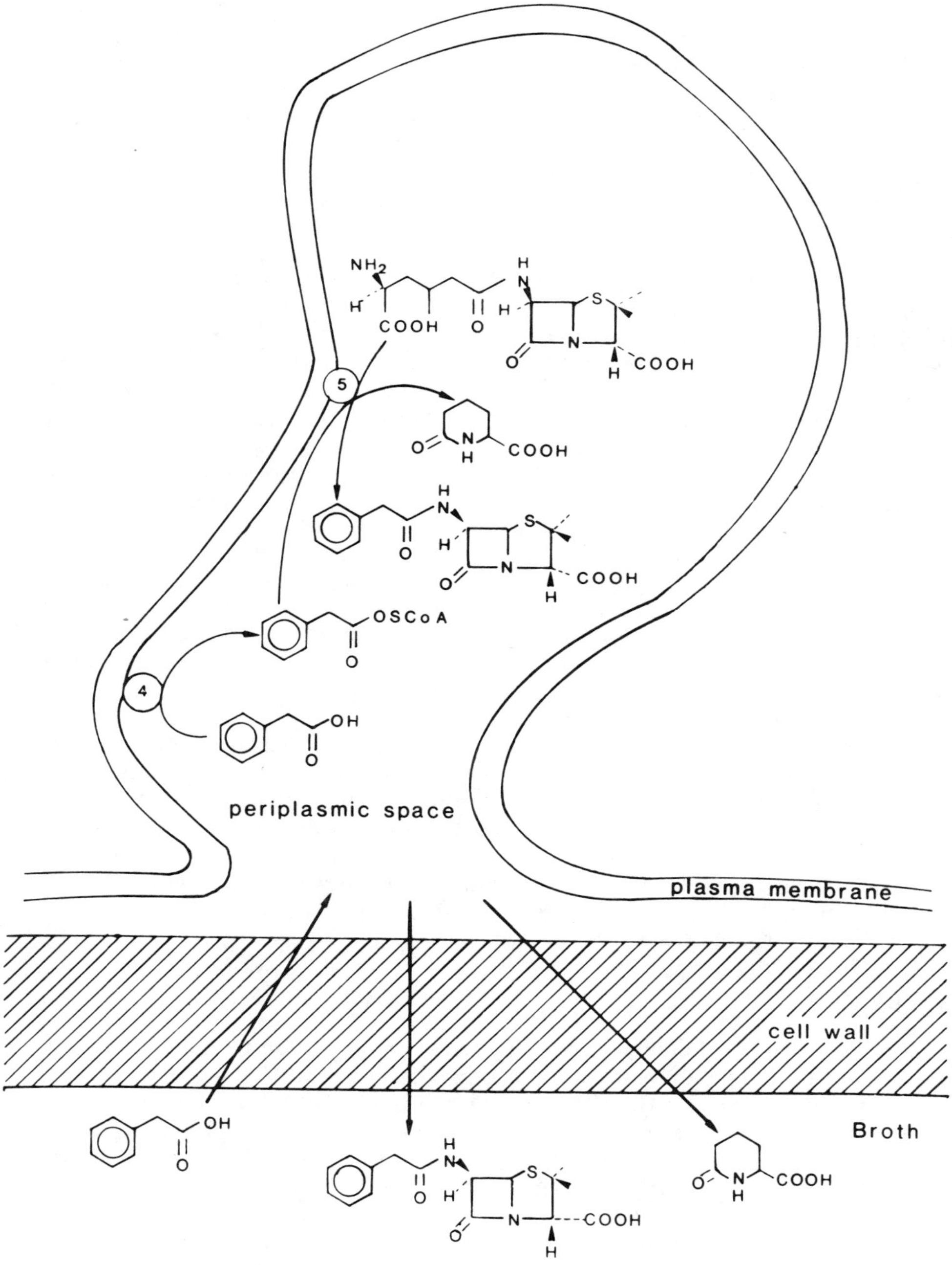
periplasmic space
plasma membrane
cell wall
Broth
OSCoA
COOH

N-synthetase) in *C. acremonium* to be a soluble enzyme, since it does not precipitate upon high-speed centrifugation. Jensen et al. (1982) confirmed this behavior for the enzyme from *Streptomyces clavuligerus*. On the other hand, Sawada et al. (1980) showed that the *C. acremonium* enzyme was stimulated by Triton X-100 and sonication. All data would fit when the cyclization activity is located in the lumen of the Golgi vesicle. Depending on the conditions used to prepare the cell-free extract, the activity might be enhanced by membrane-destabilizing agents so that the enzyme will become accessible to the substrate.

The individual amino acids are synthesized in the cytoplasm. According to Lara et al. (1982), the dipeptide is also formed in the cytoplasm. It has even been proposed that the enzyme is identical to the first enzyme in the biosynthesis of glutathione (Kaszab and Enfors, 1981). On the other hand, recent data of Kurzatkowski (1981) point to an association of the dipeptide synthetase with the same vesicles as the tripeptide synthetase. This leads to a number of possibilities, one of which is that the dipeptide and tripeptide synthetase activities are confined to one multifunctional enzyme. The tripeptide is formed from the three individual activated amino acids bound to thiol groups on the enzyme. Such a thiotemplate mechanism has been suggested by Martin et al. (1979). In addition to this, we propose that the enzyme is an integral membrane protein which takes up the amino acids from the cytoplasm and releases the tripeptide formed in the lumen of the Golgi vesicles, thus avoiding the need for specific permeases.

The isopenicillin N formed within the Golgi interior is transported to the plasma membrane to be secreted. Luengo et al. (1981) have stated that the actual side-chain exchange reaction is confined to the periplasmic space. The periplasmic space is defined here as the space between the cell wall and the plasma membrane caused by large plasma membrane invaginations into the cytoplasma. This implies that the CoA ligase and the transferase complex are located in the periplasmic space probably bound to the plasma membrane. The large penicillin transport vesicles reported by Kurzatkowski (1981) are presumably identical to these plasma membrane invaginations forming sealed vesicles upon cell fractionation (in analogy to microsomes and the endoplasmic reticulum). How the end-product penicillin G or V reaches the broth through the cell wall is still unclear.

B. Product

In the second part of this section some assay methods for penicillin, applicable at various stages of research and development, will be reviewed.

The most classic method to demonstrate penicillin is to make use of the bacteriostatic properties of penicillin. The penicillin-containing systems, agar plates, or paper chromatograms are overlaid with an agar layer containing sensitive gram-positive bacteria which are subsequently incubated for a given period of time. In those regions where penicillin is present, bacterial growth is prevented, resulting in a clear zone in the bacterial layer (Ball and McGonagle (1978). A quantitative test is the Oxford cup test: A series of concentrations of penicillin-containing fluids are put into wells punched in a bacterial layer on agar (Heatley, 1944). It was soon realized that in order to follow the kinetics of penicillin production other assay techniques were needed which could also be used in higher concentration ranges. Various direct spectrophotometric methods have been described which make use of specific ultraviolet absorption maxima. Since they were in vogue in the early days of industrial fermentation, they have already been reviewed extensively in the late 1940s by Colon et al. (1949) and Twigg (1948).

Three types of indirect assay methods have proved to be useful for routine autoanalysis: iodometric techniques and the mercury imidazole and the ferric hydroxamate methods. The iodometric assay method for penicillin was developed by Alicino (1946). The method makes use of the fact that penicilloic acid, which is formed upon hydrolysis of penicillin by alkali or penicillinase, takes up a number of iodine molecules. However, careful blanks have to be made, since many unsaturated organic molecules interact with iodine. Ferrari et al. (1959) modified the method so it could be used in an autoanalyzer system. To increase the color contrast, Thomas (1961) used the 620-nm absorption of the iodine-starch complex to visualize lactam antibiotics on paper chromatograms. The automated method described by Goodall and Davies (1961) also employs the absorption of iodine-starch as chromagen. The iodometric method recently received a new stimulus by the introduction of chloroplatinic acid in the penicillin assay system. The reagent degrades penicillin to penicillenic acid, which in turn causes the disappearance of the claretlike color of the iodine reagent (Ray Chowdhurry and Chacrabarti, 1979). The dye is not very specific, since it had been introduced before by Fowler and Robins (1972) to detect sulfur amino acids and penicillin metabolites on an amino acid analyzer. Ray Chowdhurry et al. (1980) applied the reagent to visualize penicillin produced by colonies on agar as an alternative to the bacterial overlay.

Bundgaard and Ilver (1972) described an assay method for penicillin with mercury and imidazole at neutral pH. Imidazole as a nucleophile attacks the β-lactam bond; both the β-lactam and the thiazolidine ring are opened up and a penicillenic acid mercuric mercaptide is formed that can be measured at 325–345 nm.

Hydroxylamine reacts with penicillin to form a hydroxamic acid. With ferric ions a colored chelate is formed. An assay based on this reaction was described by Boxer and Everett (1949). The complex is more stable when it is extracted with butanol (Henstock, 1949). Niedermayer et al. (1960) adopted the ferrohydroxamate method for an autoanalyzer system. Since hydroxylamine also reacts with other organic compounds bearing a carboxyl group, a penicillinase treated sample had to serve as a blank. Avanzini et al. (1968) showed that this penicillinase blank could be avoided, they obtained blank values by introducing the principle of reagent inversion.

Most research as well as routine laboratories reintroduced the use of direct ultraviolet absorption to quantify penicillins when HPLC was introduced in the mid-1970s. At first, ion exchange columns were used (Blaha et al., 1975; Tsuji and Robertson, 1975), but it soon appeared that penicillins and also other antibiotics could be more conveniently separated on reverse-phase columns. In these columns the stationary phase consists of silanized silica particles; in particular octyl and octadecyl bonded material has been used (Hartmann and Rödiger, 1976; White et al., 1977; Vadino et al., 1979). In recent years comparisons have been made between values obtained by HPLC and the iodometric method (Roksvaag et al., 1979) or the microbiological assay Hekster et al., 1979). Nachtmann and Gstrein (1980) have compared the HPLC data for three penicillins with pharmacopoeia test methods. They concluded that HPLC is highly selective and one of the fastest and most precise methods. Nachtmann's paper, in which in one run complete information could be obtained with respect to the concentration of the side-chain precursor, of penicillin, and of penicillin by-products or degradation products (Nachtmann, 1979), strongly advocates the potential of HPLC in monitoring pencillin fermentations.

Besides the methods described here, numerous other chromatographic, chemical, and physical methods of measuring the concentration of penicillin have been published. Of these the methods employing immobilized penicillinase deserve to be mentioned. Nilsson and Mosbach (1978) and Enfors and Nilsson (1979) introduced this approach in combination with a pH electrode. Owing to the buffering capacity, it was not possible to measure directly in the broth. This problem has now been obviated by using immobilized penicillinase in combination with a thermistor (Mattiasson et al., 1981). Hughes et al.(1976) have compiled a number of assay methods that have not been reviewed here.

VI. FERMENTATION PROCESS PHYSIOLOGY AND TECHNOLOGY

In this section we describe aspects of research on a wide range of important phenomena with respect to the large-scale penicillin production fermentation process. Aspects of fermentation process development are described in Sec. VII. The review of these phenomena is confined to literature concerning research and development work with mutant strains of *P. chrysogenum* (or *P. notatum*).

A. Mycelial Growth

1. Growth Kinetics

Mycelial growth has been reviewed by Righelato (1975, 1979), Bull and Trinci (1977), and van Suijdam and Metz (1981a), and mathematical modeling of spore germination and mycelial growth by Prosser (1979). The increase in biomass is caused by an increase in total hyphal length and the number of growing tips. In batch culture it is difficult to observe exponential biomass increases for more than a few doubling times. The fact that it is possible to obtain steady-state continuous cultures of fungi demonstrates that exponential growth can be maintained. In defining the specific growth rate μ as $(1/x)\,dx/dt$ (with x as the mycelium concentration and t as time), the implicit simplification is made that all of the mycelial mass contributes equally to growth (Righelato, 1979). It is thought that exponential growth of mycelium is maintained by nutrient assimilation and synthesis of macromolecules and cell wall precursors over a large part of the mycelium, while extension only occurs at hyphal tips. The number of tips increases to maintain a constant ratio of tips to hyphal length. The specific growth rate can be described as the ratio of the mean tip extension rate (m/hr) and the hyphal growth unit (m), the mean hyphal length per hyphal tip (Righelato, 1979). According to van Suijdam and Metz (1981a), the hyphal growth unit does not vary with the specific growth rate. This is in agreement with the observation of Trinci (1974), that the hyphal length and number of hyphal tips increase exponentially and approximately proportionally to the specific growth rate. So the hyphal tip extension rate must be proportional to the specific growth rate. However, limited results of Morrison and Righelato (1974) support the hypothesis that branching frequency is the main variant with specific growth rate. Either hyphal tip extension rate or branching frequency may vary with growth rate, depending on the strain and fermentation conditions. The mean length of the main hyphae can be expressed as a function of growth rate (van Suijdam and Metz, 1981a).

Penicillium chrysogenum grown at low specific growth rates exhibits more hyphal vacuolation and other signs of differentiation than do hyphae grown at high specific growth rates (Righelato et al., 1968). The mycelial protein content decreases and the carbohydrate content increases at low specific growth rates, consistent with a decrease in the ratio of cytoplasm to hyphal wall (Righelato, 1979). Submerged conidiation has been observed and was induced at specific growth rates below 0.011/hr (Righelato et al., 1968). The organism could be maintained in a nongrowing state by supplying only the maintenance ration of substrate. The maintenance ration is the glucose requirement of the organism extrapolated to zero growth rate: 0.12 ± 0.03 mmol hexose/g mycelium dry weight per hour (Pirt and Righelato, 1967). When the feed is stopped, the mold autolyses; autolysis is stopped by supplying the maintenance ration of glucose (Righelato et al., 1968). From continuous culture studies the value of the yield of mycelium from hexose, Y_{xs}, is observed to be independent of the specific growth rate in the range of 0.02–0.08/hr: 79 ± 7 (Pirt and Callow, 1960), 81 (Righelato et al., 1968), 87 ± 2 (Mou, 1979), and 87 (Mason and Righelato, 1976) g mycelium dry weight/mol hexose. For the yield of mycelium from oxygen, Y_{xo}, the following is found ($\mu = 0.02$–0.05/hr): 40 (Mason and Righelato, 1976) to 50 (Righelato et al., 1968) g mycelium dry weight/mol O_2. Using elemental balances, as described by Heijnen et al. (1979), 45 g mycelium dry weight/mol O_2 should correspond with 94 g mycelium dry weight/mol hexose (1 mol O_2 = 45 g mycelium dry weight = 1.84 mol C, molecular weight 24.52; 1 mol O_2 = 1.03 mol CO_2 = 1.03 mol C; total 2.87 mol C = 0.48 mol hexose; 45 g mycelium/0.48 mol hexose = 94 g mycelium dry weight/mol hexose). As possible causes of this discrepancy one can mention the formation of (by)products, autolysis of mycelium, and inaccuracy in the estimation of the mycelium dry weight. The following numerical values for the specific parameters for sugar uptake can be mentioned: the maximum specific sugar uptake, $q_{s\ max}$ = 0.001 mol hexose/g mycelium dry weight per hour and the Michaelis constant K_s = 0.0056 mol hexose/kg (Heijnen et al., 1979). From Y_{xs} = 85 and the $q_{s\ max}$ value follows the maximum specific growth rate, μ_{max} = 0.09/hr. Mason and Righelato (1976) examined the effect of several growth-limiting nutrients (sucrose, oxygen, ammonium, nitrate, phosphate, sulfate and magnesium) on the specific growth rate and the sucrose and oxygen utilization rates. The choice of limiting nutrient was seen to have profound effects on energy metabolism. The excess of sucrose over the growth requirement was metabolized to gluconic acid and malic acid. Gluconic acid production was also observed by Mou (1979) in the beginning of a fed-batch process at high glucose concentrations. When growth was oxygen limited, the continuous culture showed a two- to threefold increase in the yield of mycelium from oxygen (Mason and Righelato, 1976). In the presence of excess oxygen the respiration rate is probably controlled by the biosynthetic demand for energy.

2. Mycelial Fragmentation

Van Suijdam and Metz (1981a) and Metz et al. (1981) concluded from continuous culture experiments that breakage of hyphae must occur. Fragmentation of filamentously growing mycelium results in hyphae with a wide range of hyphal lengths and hyphal growth units (van Suijdam and Metz, 1981a; Metz et al., 1981). The effect of shear forces on the hyphal length is small (van Suijdam and Metz, 1981a). Fragmentation may rather be a result of autolysis of hyphal compartments and degradation of the wall of the lysed compartments. Compartments showing advanced signs of autolysis were observed in the middle of intact, apparently growing hyphae (Righelato et al., 1968; Trinci and Righelato, 1970).

3. Growth Determination

Possibilities and difficulties of the evaluation of mycelial growth are described by Calam (1969), and in more general terms by Mor (1982). Though several methods may be used, a reliable evaluation is still impossible. It is only possible to express it by several sets of figures. For practical reasons it is necessary to use a single method to express growth. It is important that when a single index is used its limitations be realized. Moreover, complex media that are often used to grow mycelial cultures contain insoluble material that interferes with the determination of mycelial growth. Attempts must be made to correct for this source of error in growth determinations. Methods for mycelial growth determination include packed cell volume, dry weight, corrected dry weight (crude dry weight less ash), wet weight, optical density, nitrogen content, protein content, nucleic acid content, and carbohydrate content. Dry weight analysis is most commonly used for mycelial growth estimation. Recently Nestaas et al. (1981) described a method for biomass estimation from filtration data. The basic assumptions of the described indirect mycelial growth estimations methods are a constant composition and no autolysis and/or a constant morphology of the mycelium, which in general is not the case.

B. Mycelial Morphology

1. Filamentous and Pelleted Mycelium

The morphology influences mycelial growth and consequently production kinetics by determining the rheology of the culture (see Sec. VI.C) and by causing diffusion limitation of nutrients within mycelial flocs. Hyphal flocs exhibit several morphological forms. Mycelial growth can be filamentous (unstable aregates of filamentously growing hyphae), pelleted, or any combination of these extremes. Several mechanisms can be distinguished for pellet formation (Whitaker and Long, 1973; Metz and Kossen, 1977): agglomeration of spores and hyphae, agglomeration of hyphae, and agglomeration of solid particles and hyphae. Pellet—pellet and pellet—wall (fermenter wall, baffles, cooling coil, stirrer blades, etc.) interactions are considered to be of importance in pellet formation. Metz and Kossen (1977) distinguished three types of pellets: fluffy loose pellets (hairy pellets; these pellets have a compact center and a much looser outer zone), compact smooth pellets (the whole pellet is compact and the outside of the pellet is smooth), and hollow pellets (the center of the pellet is hollow owing to autolysis and the outside is smooth). A survey of pellet parameter data for P. chrysogenum published in the literature has been given by Metz and Kossen (1977). Pellet formation in P. chrysogenum can be regarded as noncoagulating (Takayashi and Yamada, 1959): In principle, one pellet is formed from one spore. However, Trinci (1970) mentions the possibility of agglutination of spores for this genus. Pirt and Callow (1959) observed a link between pellet formation and the formation of short distorted hyphae.

The most important advantage of growth in the pellet form is attributed to a considerable decrease of the broth viscosity in comparison to growth in the filamentous form (Righelato, 1979; Metz et al., 1979). Cultivation in the pellet form may in some cases introduce undesirable limitations. The most important of these is the diffusional limitation of nutrients, that is, oxygen, carbohydrate, and so on, into the pellet. When the radius of the pellet exceeds a critical radius, only a peripheral zone of the pellet will contribute to growth. Also, within unstable aggregates of filamentous growing hyphae nutrient transfer is determined by diffusion until the floc is again disaggregated in a high-shear region (the impeller zone in a stirred vessel). The pellet growth form is more stable and does not easily disaggregate.

2. Morphological Variability

Inoculum Size. Depending on the medium used Camici et al. (1952) found $2 \times 10^{11}-10 \times 10^{11}$ spores per cubic meter as the critical spore inoculum size above which filamentous growth occurred and below which pellets were formed. Calam (1976) and Smith and Calam (1980) estimated this critical level at 5×10^9 spores per cubic meter. This difference can be a consequence of strain attributes, coagulative versus noncoagulative pellet formation, and of a proportional loss of spore inoculum varying with the shape and the scale of the fermentation vessel observed by Camici et al. (1952) and Dion et al. (1954).

Medium. Pirt and Callow (1959) observed that pellet formation occurred when the growth medium contained ammonium sulfate. When ammonium sulfate was substituted by corn-steep liquor, the dispersed form was observed.

Aeration. Oxygen is sparingly soluble in water. The maximum concentrations that can be achieved outside the aggregates of filamentous growing hyphae (flocs) and pellets are low. In the center of flocs and certainly in the center of pellets oxygen depletion can occur very easily and can become the limiting nutrient (Pirt, 1966). In the case of pellets, absence of oxygen in the center results in autolysis of the cells and the formation of a hollow pellet (Camici et al., 1952; Trinci and Righelato, 1970). Phillips (1966) obtained results which indicate that oxygen is supplied to the pellet interior by simple molecular diffusion. He showed that the apparent critical oxygen tension increased rapidly with increasing pellet radius. Modelling the growth of pellets, van Suijdam et al. (1982) introduced an effectiveness factor for the diffusion of oxygen into a pellet. This factor is defined as the ratio of the actual oxygen consumption rate and the maximum oxygen consumption rate, which can be realized in the presence of excess oxygen. Although P. chrysogenum has a high-affinity oxygen uptake mechanism (Mason and Righelato, 1976), much higher dissolved oxygen concentrations are required to avoid oxygen limitation under fermentation conditions. On the hyphal scale van Suijdam and Metz (1981a) could not show any influence for oxygen tensions in the range of $1.6 \times 10^3-40.0 \times 10^3$ Pa upon morphology.

Agitation. In a stirred tank fermenter the pellet size is predictable from the energy input per unit mass (van Suijdam and Metz, 1981b). More agitation power input produces smaller, smoother, and more compact pellets (Dion et al., 1954; Metz, 1976; van Suijdam and Metz, 1981b; Vardar and Lilly, 1982), and hollow smooth pellets develop (Metz, 1976). The hollow pellets break up and pellet fragments remain which tend to disappear completely (Metz, 1976). A model for pellet breakup has been postulated by Bhavaraju and Blanch (1976) and van Suijdam and Metz (1981b). The models are based on the assumption that dynamic turbulent fluid shear stress in a stirred tank fermenter is responsible for the abrasion of particles of about the same density as the suspending fluid.

From an economic point of view for practical applications a sufficiently high pellet density must be attained; this results in a lower broth viscosity for a given biomass concentration. However, the onset of oxygen depletion and absence of oxygen and consequent autolysis of the cells in the center of the pellets (or flocs) is determined by the pellet diameter and the pellet density (and the oxygen demand). Small pellets become more dense with increasing diameter (van Suijdam, 1980) as a consequence of a constant hyphal growth unit (van Suijdam and Metz, 1981a). For larger pellets the growth in the center is hindered by nutrient transfer limitation and in consequence the pellet density is inversely proportional to the pellet diameter (van Suijdam, 1980). So in a stirred tank fermenter small pellets are certainly preferable, although small pellets will still contribute to broth viscosity, and—as also stated by Trinci (1970)—formation of large pellets should be prevented. Although pellet growth can decrease broth viscosity, a yeastlike morphology is even more preferable.

Using Casson's equation for correlation of rheology data (see Sec. VI.C),

$$\tau^{1/2} = \tau_0^{1/2} + K_c\dot{\gamma}^{1/2} \tag{1}$$

the apparent viscosity can be written as

$$\eta_a = \tau/\dot{\gamma} = K_c^2 + \tau_0\dot{\gamma} + 2K_c(\tau_0/\dot{\gamma})^{1/2} \tag{2}$$

With the rheology measurement data of van Suijdam (1980) for various pellet volume fractions and diameters, the apparent viscosity can be calculated from Eq. (2) and the following table can be composed.

ϕ_p [a] (%)	d_p [a] $(10^{-3}m)$	K_c [a] $[(N\ sec/m^2)^{1/2}]$	τ_0 [a] (N/m^2)	$\dot{\gamma}$ (sec^{-1})	η_a $(N\ sec/m^2)$
50.0	1.20	0.422	3.762	1.0[b]	5.58
	0.63	0.601	8.071		11.85
40.0	1.20	0.302	1.739		2.63
	0.63	0.432	2.791		4.42
28.6	1.20	0.210	0.242		0.49
	0.63	0.297	1.250		2.00

[a]Data from van Suijdam (1980).
[b]Assuming the average bulk shear rate; the value of the average bulk shear rate for a vessel of standard geometry is estimated at 0.4–4 liters/sec in the range of large- to small-scale fermenters (van Suijdam, 1980).

d_p pellet diameter
K_c Casson viscosity
$\dot{\gamma}$ shear rate
η_a apparent viscosity
τ_0 yield stress
ϕ_p volume fraction pellets

As can be concluded from this table, the apparent viscosity is inversely proportional to the pellet diameter for different pellet volume fractions.

An increase of energy dissipation results in a decrease in hyphal length (van Suijdam and Metz, 1981a; microscopically observed by Vardar and Lilly, 1982). However, to obtain a substantial decrease in length, an enormous increase in energy input is necessary. Therefore this effect seems to be of little practical value as a method to decrease the viscosity of filamentous mycelial broths (van Suijdam and Metz, 1981a). The considerably lower apparent viscosities found by König et al. (1981b) at higher stirrer speeds for the entire fermentation was caused by the formation of smaller pelletlike agglomerates.

pH. Pirt and Callow (1959) showed a clear influence of pH upon morphology. At pH 6.0 long, thin hyphae were produced and filamentous growth occurred. At pH 7.4, short hyphae containing swollen cells were produced which gave rise to pellet formation. Pirt and Callow suggested that this effect could be due to a change in cell wall structure involving loss of rigidity and a consequent inability to resist the internal osmotic pressure. No influence of pH upon morphology was found when no penicillin G precursor, phenylacetic acid, was present in the broth (van Suijdam and Metz, 1981a)

Growth Rate. Righelato et al. (1968) found the pellet fraction in continuous culture to increase with increasing specific growth rate. This might be caused by the presence of shorter and more branched hyphae at higher specific growth rates, which leads to less interaction between hyphae and more chances for pellet formation.

3. Morphology and Penicillin Production

Dion et al. (1954), Carilli et al. (1961), and Phillips (1966) studied pellet growth with respect to penicillin production. Phillips (1966) observed that when pellets were formed, very high dissolved oxygen levels were necessary to obtain reasonable penicillin productivity. The highest penicillin yields were obtained when the mycelium was in short fragments and agitation was not sufficient to cause excessive mechanical damage (Dion et al., 1954). Calam (1976), Smith and Calam (1980), and Calam and Smith (1981) found filamentous mycelium to produce a higher penicillin titer than mycelium in pellet form. This, however, can also be recognized as a result of the experimental methods applied by these authors (see Sec. VII.A). Carilli et al. (1961) found the strain that grew in pellet form had a higher oxygen uptake rate and produced double the penicillin titer compared to the strain that exhibited the filamentous form. As Whitaker and Long (1973) noticed, this might have been a result of the strain rather than the morphological form. The described effects of morphology upon penicillin production are not homonymous. However, the conclusion can be drawn that the genesis of larger, dense pellets must be prevented and mycelium in the filamentous or small, fluffy loose (hairy pellet form seems preferable (see Sec. VI.B.2) for penicillin production. Vardar and Lilly (1982) observed a good penicillin production in cultures with both filamentous and pelleted ($\phi = 0.2 \times 10^{-3} - 0.5 \times 10^{-3}$ m) mycelium.

König et al. (1981c, 1982a,b) showed that it is possible to produce penicillin in a bubble column fermenter if a pellet suspension with low viscosity is maintained. The volumetric oxygen transfer coefficients in these pelleted mycelium broths were a factor of 4-5 higher than in a filamentous mycelium broth. Mycelium and penicillin concentrations, however, were lower (penicillin, 35% lower) than those attainable in stirred tank reactors, which is probably a consequence of fewer generations and therefore a younger physiological age. The pellet diameter may not exceed a certain critical diameter for a maximum penicillin productivity. A better yield of penicillin seemed to be achievable from substrate than in a stirred tank reactor. If a pelleted mycelium broth is used, broth filtration can be performed very easily.

C. Broth Rheology

Mycelial broths exhibit a non-Newtonian shear thinning or pseudo-plastic flow behavior. The apparent viscosity, defined as the ratio of the shear stress to the shear rate under laminar flow conditions, decreases with increasing shear rate. This leads to relatively low viscosities and turbulent flow in regions of high shear rate (near the impeller in a stirred vessel) and very high viscosities and laminar flow in regions of markedly diminished shear rate (near the wall or cooling coils in a fermenter), where stationary zones can occur. Moreover, a mycelial broth exhibits a yield stress and viscoelastic behavior. The highly viscous mycelial broths cause serious problems in the transfer of momentum, heat, and mass. The importance of mycelial broth rheology in relation to the transport phenomena in fermenters has been extensively reviewed by Blanch and Bhavaraju (1976), Charles (1978), and Metz et al. (1979). The viscosity measurement itself is still a subject of discussion (Charles, 1978; Reuss et al., 1982). The turbine impeller viscosity meter (with a comparatively small range of shear rates) has proven to be useful for rheology measurements on mycelial broths (Roels et al., 1974; Metz et al., 1979). Reuss et al. (1982) used a helical ribbon impeller with a wider range of shear rates and a better mixing capability in the laminar flow region. The use of Casson's equation has been recommended for correlation of rheology data of mycelial broths predicted with turbine impeller systems (Roels et al., 1974; Metz et al., 1979; van Suijdam, 1980). Metz et al. (1979) summarized objections against this theoretical model. Reuss et al. (1982) showed, in addition, that the small range of shear rates of the turbine impeller does not allow the distinction between the equations of Casson, Herschel, and Bulkley and the power-law model. This suggests that the choice for either of these rheology models is arbitrary when the turbine impeller viscosity meter is used. Better rheology measurement systems are required to yield intrinsic parameters (Reuss et al., 1982). Apparent viscosity (viscosity at any specified shear rate) is approximately related to the square of the mycelium concentration (Carilli et al., 1961; Roels et al., 1974; Metz et al., 1979). Yield stress of filamentous mycelium is proportional to mycelium concentrations raised to the 2.5 power (Roels et al., 1974; Metz et al., 1979). A simple correction can be made to account for the presence of mycelial pellets.

Transport phenomena in fermenters influence the fermentation conditions. These conditions influence growth, morphogenesis, and product formation. Use of pellet growth form has the advantage of eliminating the broth viscosity problem to a great extent; however, the transport of nutrients into the pellets can be diffusion limited. The oxygen transfer is markedly influenced by the broth rheology. The volumetric oxygen transfer coefficient, $k_L a$, falls with increasing viscosity (Deindoerfer and Gaden, 1955; Ryu and Humphrey, 1972) and filamentous mycelial broths can quickly become oxygen limited. The poor oxygen transfer is probably a result of low bubble formation rates, increased bubble coalescence rates and decreased broth flow rates around the fermenter (Righelato, 1979; König et al., 1981b), or diffusion limitation into mycelial flocs. As described, the pseudo-plastic flow behavior and the yield stress can cause considerable heterogeneity in the mixing condition in the fermenter. Mycelial hyphae reside for variable periods in the well-aerated, primary flow region and the more stationary, secondary flow region, which can be deficient in oxygen (Metz et al., 1979). Secondary flow regions exchange slowly with the bulk of the fermenter. Moreover, stagnant layers around heat-transfer surfaces (coil, jacket) hinder removal of heat. To overcome mixing problems, high external power inputs are used. An increase in external power input results in smaller flocs and a decrease of secondary flow region size and consequently in an increase of oxygen uptake rate, oxygen exchange, and heat removal, but also in an increased heat generation. Macroscopically, a pelleted morphology seems energetically more efficient than a filamentous morphology (Roels and van Suijdam, 1980).

D. Nutrients for Mycelial Growth and Penicillin Production

1. Nutrient Media and Feeds

The nutrient media used by the penicillin G and V producers are closely guarded trade secrets. Production by submersed fermentation started with the lactose–corn-steep liquor medium (see also Sec. VII.B.1). This medium (A) is summarized in Table 4. The use of additional mineral salts is dependent on the amount of corn-steep solids used. Initially the broth pH was only corrected before sterilization and not controlled during the fermentation. Later pH was controlled by the addition of alkali or ammonia (see Sec. VII.B.2) and acid. Lactose was initially partially and finally completely replaced by a continuous glucose feed during the fermentation (see Sec. VII.B.2). Currently precursor and antifoam are also fed to the broth. Details of the current medium (B) were not published for obvious reasons.

Table 4 Complex Medium Composition

	Medium A	Medium B
Glucose	$0-10$ kg/m^3	$0-10$ kg/m^3 Glucose by continuous feed
Lactose	$20-50$ (or more) kg/m^3	—
Corn-steep liquor solids	$15-50$ kg/m^3	$40-50$ kg/m^3
NaNO$_3$	$0-5$ kg/m^3	
Na$_2$SO$_4$	$0-1$ kg/m^3	
CaCO$_3$	$0-10$ kg/m^3	
KH$_2$PO$_4$	$0-4$ kg/m^3	Identical (?)
MgSO$_4 \cdot$7H$_2$O	$0-0.25$ kg/m^3	
ZnSO$_4 \cdot$7H$_2$O	$0-0.04$ kg/m^3	
MnSO$_4$	$0-0.02$ kg/m^3	
Phenylacetic acid or a derivative	b	By continuous feed
Antifoam	c	Consumable oil by continuous feed
pH	Free after correction	Controlled
References	Moyer and Coghill (1946b), Sylvester and Coghill (1954), Anderson et al. (1956), Perlman (1970), Hockenhull (1963), Calam (1967), El-Marsafy et al. (1977)	Perlman (1970)

[a]Also or besides applied, (NH$_4$)$_2$SO$_4$ or Na$_2$S$_2$O$_3$.
[b]Dependent on the amount of penicillin to produce and the efficiency of precursor utilization.
[c]Dependent on the kind of antifoam.

Already in the 1940s efforts were made to devise chemically defined media. Foster et al. (1943) described a mineral salt solution for a chemically defined medium (Calam and Hockenhull, 1949) for penicillin production in surface culture. The utilization of a chemically defined medium for submersed fermentation (Perlman, 1966) has been reported ever since 1946; its composition is summarized in Table 5. Stone and Farrell (1946) reported that the medium, which they describe, was used in production plants for short periods. The penicillin production was somewhat lower compared to the use of complex media, but the product was purified more easily. Jarvis and Johnson (1947), Gordon et al. (1947), and Koffler et al. (1946, 1947) found lower penicillin titers in a chemically defined medium than when a corn-steep liquor- (CSL) based medium was used. Comparable results in chemically defined and complex media were reported by Soltero and Johnson (1954), Hockenhull (1959), and El-Saied et al. (1976). Calam and Ismail (1980b) found that some fermentation runs with a chemically defined medium were indistinguishable in penicillin production from those with a complex medium. Comparable results were only obtained when a complex medium (medium A in Table 4, with lactose replaced by glucose) was used for the inoculum (see also Hockenhull and Mackenzie, 1968). The pH had to be controlled very carefully (El-Saied et al., 1976). Calam and Ismail (1980b) found the maintenance energy requirements (as indicated by CO_2 production) to be higher and the penicillin productivity decay rate to be equal or less with chemically defined medium than with complex media. A general comparison of chemically defined and complex media is given by Hockenhull (1959). In continuous and fed-batch cultures, CSL appeared to have no specific effect on the synthesis of penicillin (Pirt and Righelato, 1967). An important role of CSL was the provision of substrates that allow a high mycelium concentration to be reached before the growth rate falls below the critical value, and which allow, as a result of a higher specific growth rate, a higher specific production rate (Mou, 1979).

Chemically defined media lack buffering capacity. Before the introduction of pH control, $CaCO_3$ was added to the chemically defined medium to prevent low pH values (Hockenhull, 1959). To increase the buffering capacity KH_2PO_4 was also used (Soltero and Johnson, 1953, 1954; Hosler and Johnson, 1953; Righelato et al., 1968). Good production results were obtained when, in addition, the pH was controlled (Hosler and Johnson, 1953; El-Saied et al., 1976). Phosphate acts as a buffer but is also involved in pH-sensitive precipitation reactions (Hockenhull, 1959), which can, at least partly, be prevented by the use of EDTA (Pirt and Callow, 1960, 1961; Pirt and Righelato, 1967; Righelato et al., 1968). When a high buffering capacity of the medium was not needed because of pH control, $CaCO_3$ was replaced by $CaCl_2$ or $CaSO_4$ for the calcium requirement (Hosler and Johnson, 1953; Pirt and Callow, 1960, 1961; Pirt and Righelato, 1967; Righelato et al., 1968). Greatly differing amounts of Na_2SO_4 and $(NH_4)_2SO_4$ are used in the chemically defined media, which is dependent on the use of, for instance, a sulfuric acid feed. Less or no NH_4NO_3 was used when ammonia was fed to the broth. In some media, thiosulfate, $Na_2S_2O_3$, was used (Verkhovtseva et al., 1970; El-Saied et al., 1976), which was found to stimulate the biosynthesis of penicillin (Hockenhull, 1959), probably by acting as a primary hydrogen acceptor. Other additional compounds supplied in chemically defined media are starch or dextrins (Koffler et al., 1946, 1947; Hockenhull, 1959), citric acid (Hockenhull, 1959), ethylamine (Hockenhull, 1959), L-leucine (Foster, 1949), and Al, Cr (Foster, 1949), Mo, V, Ga, and Sc (Pirt and Callow, 1960) salts.

Table 5

	Medium A	Medium B
Glucose (or sucrose)	$5-15$ kg/m^3	$10-15$ (30) kg/m^3 Glucose by continuous feed
Lactose	$15-60$ kg/m^3	

Acetic acid[a]	$0-6$ kg/m^3
Na_2SO_4	$0-1$ kg/m^3
$(NH_4)_2SO_4$	$0-10$ (18) kg/m^3
NH_4NO_3[b]	$0-8$ kg/m^3
KH_2PO_4	$0.4-8$ kg/m^3
$CaCO_3$	$0-5$ kg/m^3
$MgSO_4 \cdot 7H_2O$	$0.25-0.8$ kg/m^3
$FeSO_4 \cdot 7H_2O$[d]	$0.04-0.4$ kg/m^3
$ZnSO_4 \cdot 7H_2O$	$0.01-0.05$ kg/m^3
$CuSO_4 \cdot 5H_2O$	$0.005-0.01$ kg/m^3
$MnSO_4 \cdot 4H_2O$	$0.01-0.06$ kg/m^3
$CoSO_4 \cdot 7H_2O$	$0-0.005$ kg/m^3
$CaCl_2$	$0-0.05$ kg/m^3

	Medium A	Medium B
Phenylacetic acid or a derivative	See Table 4	By continuous feed
Antifoam	See Table 4	Consumable oil by continuous feed
pH	Free after correction	Controlled
Reference	Koffler et al. (1946, 1947), Stone and Farrell (1946), Gordon et al. (1947), Jarvis and Johnson (1947, 1950), Foster (1949), Hosler and Johnson (1953), Soltero and Johnson (1953, 1954), Hockenhull (1959), Verkhovtseva et al (1970), El-Saied et al. (1976)	Hosler and Johnson (1953), Soltero and Johnson (1954), Pirt and Callow (1960, 1961), Pirt and Righelato (1967), Righelato et al. (1968)

[a]Also applied: NH_4 acetate, lactic acid, or NH_4 lactate.
[b]Also applied: KNO_3.
[c]Also applied: NaH_2PO_4, K_2HPO_4, or $(NH_4)_2HPO_4$.
[d]Also applied: $Fe(NH_4)_2(SO_4)_2$.

Al-Haffar (1970) determined in experiments with a low-production strain, using a systematic variation analysis method, the optimal ionic ratio of K, Ca, and Mg for penicillin production: 30% K, 29% Ca, and 41% Mg. The minimal requirements of S, P, K (40 g/m^3), Mg (8 g/m^3), Fe, and Cu (0.1–0.5 g/m^3) for mycelial growth and penicillin production were determined by Jarvis and Johnson (1950). For penicillin production the requirements for S, P, and Fe are, respectively, 1.5 (100 g/m^3), 2 (200 g/m^3), and 20 (4 g/m^3) times higher. Iron as ferrous ion at a concentration of 6 g/m^3 had no inhibitory effect on penicillin formation. At a concentration of 25 g/m^3, a 5% reduction in penicillin production was observed; at 60 g/m^3, 30% reduction; and at 300 g/m^3, 90% reduction (Pan et al., 1975). A high-yield mutant seemed more sensitive to iron toxicity than the lower-producing progenitors. Sensitivity to high iron concentrations was also observed by Vardar and Lilly (1982). The role of trace elements in the biosynthesis of secondary metabolites was reviewed by Weinberg (1970). As the years elapsed, the amount of nutrients in media and feeds tended to increase for an improved penicillin production by a strain with an increased productivity at an increased mycelium concentration. Critical concentrations of all nutrients essential for producing substrates and enzymes involved in penicillin synthesis are achieved by a correct medium composition and appropriate nutrient feed rates.

2. Alternative Raw Materials

Carbon Source. The carbon source represents a comparatively large part of the overall manufacturing costs for the production of penicillin. Swartz (1979) estimated this portion at 12% and De Flines (1980) probably at even higher amounts (see Sec. VII.F). The cost of the carbon source is therefore important to the overall economics of the process. Alternative fermentation substrates are discussed in general terms by Ratledge (1977). Pentoses, hexoses, mixtures of hexoses, disaccharides, polyols, dextrins, starch hydrolysates, starches, and molasses can be metabolized comparably or less efficiently than glucose (Soltero and Johnson, 1953; Perlman, 1970; Queener and Swartz, 1979; Matelova, 1981). Addition of fatty oils in addition to lactose of a glucose feed was found to increase penicillin titers (Pan et al., 1959; Lurie et al., 1979; see Sec. VII.B.2). Ethanol was employed as a carbon feed source and therefore as the major carbon and energy source (Matelova, 1976; Squibb and Sons, U.S. Patent No. 4164445, 1979; Matelova, 1981). Acetic acid was used as a carbon feed source together with glucose in a fermentation with a low-production strain (Jensen et al. 1981). Acetic acid is quickly metabolized by *P. chrysogenum* (Jarvis and Johnson, 1947; Hockenhull et al., 1954) at pH 6–7 (Hockenhull et al., 1954), the TCA cycle being the main pathway for acetate oxidation (Yall, 1955). Acetate was also found to be used for penicillin production (Martin et al., 1953; Tome et al., 1953). When 20% of the carbon source in the feed was acetic acid, penicillin (V) titers were about 25% higher than in fermentations where glucose was the only carbon source in the feed (Jensen et al., 1981). This might be explained by an increased availability of precursors of cysteine and valine for penicillin biosynthesis (Jensen et al., 1981), which might be related to the effects of fatty oils (Hockenhull, 1963).

Nitrogen Source. Because of the variability of composition of corn-steep liquor (El-Marsafy et al., 1975) substitutes were sought. Cottonseed meal (Foster et al., 1946), Pharmamedia (Queener and Swartz, 1979; Matelova, 1981), different kinds of brans with a considerable amount of phytic acid, phosphorus (El-Saied et al., 1977), other raw materials (Matelova, 1981), and *Penicillium* mycelium (Ghosh and Ganguli, 1961; Pathak, 1967; Orlov et al., 1970) can be regarded as usable substitutes. Inorganic or defined organic nitrogen sources in chemically defined media (see Sec. VI.D.1) are also usable substitutes.

Precursor Source. Szarka (1981) found that some 1-phenyl-n-alkanes are suitable carbon, energy, and precursor sources for penicillin G fermentation. Using 1-phenyl-n-decane and 1-phenyl-n-dodecane 70–80% was converted to phenylacetic acid and the produced penicillins consisted for more than 90% of benzylpenicillin; the penicillin potency being comparable to the fermentation with phenylacetic acid as precursor.

A disadvantage of substrates with a higher degree of reduction than glucose, such as ethanol, fatty oils, and 1-phenyl-n-alkanes, is a higher oxygen demand and a considerably higher heat production (Roels, 1980). Oxygen demand and heat generation by the organism increase with the degree of reduction on the substrate. The use of a substrate with a degree of reduction higher than glucose as a minor or major source of carbon and energy beside glucose has the consequence that the aeration and cooling capacity limits of the fermenter are reached at a lower total carbon dosage than when glucose is the only carbon and energy source. The use of fatty oils as consumable antifoam was found to affect oxygen transfer and to cause a decrease in dissolved oxygen tension. The metabolism of the consumable antifoam resulted in an increased growth rate and a decreased penicillin production rate (Vardar and Lilly, 1982).

E. Conditions for Mycelial Growth and Penicillin Production

Temperature. Owen and Johnson (1955) showed that more penicillin was produced by fermentations which were started at 30°C and transferred to 20°C or 25°C about 2 days after inoculation than by control fermentations kept at 25°C for the entire process. Optimal temperature profiles for batch fermentation processes were determined by Constantinides et al. (1970b) and Constantinides and Rai (1974). The optimal temperature profile, which was found with two models, starts at a high temperature, favoring faster growth, and than drops to a lower temperature, which favors a high level of cell concentration and a low rate of penicillin degradation (see Sec. VI.G). At the end of the cycle, when the cells have reached their maximum concentration, the optimum temperature shifts to a level which favors penicillin formation only. When a constant temperature is used throughout the fermentation period, McCann and Calam (1972) found a temperature optimum of 25–27°C. However, a fermentation at 22°C had a penicillin titer 50% higher than one at 30°C.

pH. For both mycelial growth and penicillin biosynthesis a pH range of 6.8–7.4 was usually found to be optimal (Brown and Peterson, 1950; Pirt and Callow, 1960; Pan et al., 1972; Pirt and Mancini, 1975). Lurie and Levitov (1976) found the optimum pH for their strain to be around 6.5. As can be found in the recent literature, nowadays the pH is maintained at a value in the range 6.2–6.8. pH values above 7.5 lead to rapid penicillin degradation (Hockenhull, 1963; see Sec. VI.G).

Dissolved Oxygen. The effect of dissolved oxygen tension on penicillin production was recently reported by Fish et al. (1981) and Vardar and Lilly (1982). At 30% air saturation the dissolved oxygen tension in the broth was found to be critical for penicillin production. Below this value a sharp decrease in specific penicillin production rate was observed, which could be considered a physiological effect (Vardar and Lilly, 1982). Penicillin synthesis was impaired irreversibly below 10% air saturation (Vardar and Lilly, 1982) or 5-10% air saturation (Squires 1972). König et al. (1981b) found 8-10% air saturation critical for penicillin production, and Bernard and Cooney (1981) 10%, which might be the point of irreversible productivity damage. Giona et al. (1976a,b) correlated dissolved oxygen concentration values with the specific penicillin production rate q_p. Below C_{oL} (dissolved oxygen concentration) = 0.022 mole/m^3 (broth) irreversible damage to the productivity of the mycelium resulted (Giona et al., 1976a), a value in agreement with an earlier observation (see Sec. VII.B.1). At higher C_{oL} values q_p was found to increase (Giona et al., 1976a,b) and the total penicillin production increased (Giona et al., 1976b). Oxygen uptake rate was found to be affected significantly at a dissolved oxygen tension below 7% air saturation (Vardar and Lilly, 1982). The critical dissolved oxygen tension for oxygen uptake was determined earlier at 0.013 mol/m^3 (broth) (Phillips and Johnson, 1961). Vardar and Lilly (1982) noted that the critical dissolved oxygen tension value for penicillin production and oxygen uptake can be recognized as two distinct parameters.

Limitations in mass and momentum transfer coupled with a high hydrostatic pressure create significant spatial variations in dissolved gas concentrations in large-scale fermenters. These effects have to be accounted for during scale-up. The heterogenous conditions in a large-scale fermenter were simulated by sinusoidally cycling the head pressure of a small-scale fermenter at a frequency in the range of the mixing times of a large-scale fermenter (Vardar and Lilly, 1982). When the dissolved oxygen concentration was cycled around the critical value for penicillin production, a considerable decrease in the specific penicillin production rate was observed. The effect did not appear to be an irreversible change in metabolic activity, and it was not transient, which suggests that the release from inhibition may be slower than inhibition itself.

Carbon Dioxide. The presence of about 1% carbon dioxide in the air seems essential for germination of spores (Nyiri, 1967). The absence of carbon dioxide causes a much slower germination of the inoculated spores (Nyiri, 1967; van Suijdam and Metz, 1981a), as does the presence of a high amount of carbon dioxide (Nyiri, 1967). When carbon dioxide is absent, the lag phase is about twice as long compared to when carbon dioxide is present (van Suijdam and Metz, 1981a).

Foster (1949) found that some carbon dioxide in the atmosphere is essential for penicillin formation and that excess has an inhibitory effect. The experiments of Nyiri and Lengyel (1965) and Lengyel and Nyiri (1965) pointed out that the presence of 4% or more carbon dioxide in the effluent gas adversely influenced the respiration and the rates of sugar uptake and penicillin biosynthesis, even though the dissolved oxygen concentration was far above its critical value (0.022 mol/m^3). Pirt and Mancini (1975) also showed that carbon dioxide strongly inhibits penicillin production. The specific penicillin production rate was decreased by about 34% at a carbon dioxide tension of 5×10^3 Pa in the effluent gas, 50% at 8×10^3 Pa, and 63% at values higher than 10×10^3 Pa when the rate at 0.7×10^3 Pa was taken as the 100% value.

In conclusion, we can state that aeration is necessary to a biosynthetic process such as penicillin formation in two important ways: to maintain the dissolved oxygen level above the critical value and to assure ventilation by which the carbon dioxide level may be kept below its critical inhibitory value (Nyiri and Lengyel, 1965; Fox, 1978).

Agitation. At a comparatively high stirrer speed, König et al. (1981a,b) observed a shorter penicillin production phase, great cell growth, and a higher fraction of the carbon source converted into carbon dioxide. Vardar and Lilly (1982) found a severe decrease in the specific penicillin production rate at higher stirrer speeds. They concluded that the cell metabolism might have shifted toward the production of cell constituents as a reaction to intense agitation. This may occur by strengthening the cell wall or replacing the leaking cell constituents or some repair process. In reactors with a draught tube and propeller, a lower penicillin productivity was attained than in those with turbine stirrers (König et al., 1981b). This was probably due to a much higher specific power input and oxygen transfer rate attained by the turbine stirrer than by the propeller and to the higher shear stress produced in the immediate vicinity of the high-speed propeller.

F. Penicillin Production Kinetics

1. Continuous Culture

The application of continuous flow cultures in penicillin production kinetics research has been described by Righelato and Pirt (1967), Pirt (1968), and Righelato and Elsworth (1970). Penicillin production kinetics have been reviewed by Pirt (1969). Pirt and Righelato (1967) found the specific penicillin production rate q_p to be independent of the specific growth rate μ over the range of 0.014–0.086/hr. With glucose fed at the maintenance rate (0.12 ± 0.03 mmol hexose/g mycelium dry weight per hour; see Sec. VI.A.1), the mycelial dry weight remained constant, but the q_p fell approximately linearly to zero with a rate inversely proportional to the previous growth rate. The q_p decay was found to be prevented at a critical specific growth rate between 0.009 and 0.014 hr^{-1} (0.21–0.31 mmol hexose/g mycelium dry weight per hour). So penicillin formation appears to be growth associated. The results of Pirt and Righelato (1967) are, in general terms, similar to those of Ryu and Humphrey (1972, 1973) and Ryu and Hospodka (1980). The latter, however, found a more stable, proportionally increasing q_p with μ up to μ = 0.014 hr^{-1} (0.30 mmol hexose/g mycelium dry weight per hour; 1.4 mmol O_2/g mycelium dry weight per hour; Ryu and Humphrey, 1972), or μ = 0.015 hr^{-1} (0.33 mmol hexose/g mycelium dry weight per hour; 1.6 mmol O_2/g mycelium dry weight per hour; Ryu and Hospodka, 1980; see Sec. VII.D.2). For μ above 0.014 or 0.015 hr^{-1}, q_p appeared to be independent of μ. Young and Koplove (1972) found a comparable relation.

Degeneration of the strain's productivity (q_p) was generally observed in continuous culture. A marked correlation between pH and q_p suggested that the stability is higher at pH values above 7 than it is at pH 6.6 (Pirt and Callow, 1961). Pirt and Callow (1960) observed that in continuous culture q_p remained at its maximum value for about 1000 hr at μ = 0.04–0.08/hr. In phosphate-limited cultures the strain's productivity seemed to be comparable to that in carbon-limited cultures (Mason and Righelato, 1976).

2. Fed-Batch Culture

Aspects of fed-batch culture are described in more detail in Sec. VII (see Secs. VII.D and VII.B.2). The specific production rate q_p in batch culture is usually higher than that observed in a chemostat (Pirt and Righelato, 1967; Wright and Calam, 1968). Young and Koplove (1972) described that batch culture and continuous culture q_p versus μ (specific growth rate) relations were in agreement. However, for continuous cultures they only presented data for a small range of μ values. Under optimum conditions the fed-batch quasi-steady-state q_p was found to be slightly higher than the continuous culture steady-state q_p (Court and Pirt, 1977), although q_p did not reach a quasi-steady state in fed-batch culture (Court and Pirt, 1976). The q_p decay rate after a maximum q_p was reached in fed-batch culture appeared to be independent of the initial growth rate (Mou, 1979). The use of a fed-batch culture prolonged the penicillin fermentation, provided that the extent of growth rate changes was controlled within certain limits (Court and Pirt, 1976). Pilát (1979, 1980) studied penicillin production kinetics in repeated fed-batch culture as a function of the sugar (sucrose) feed rate to determine the optimum feed rate. In fed-batch culture the maintenance demand for sugar varied with time, as it was apparently influenced by the mycelium growth history (Mou, 1979). The minimum value was approximately 0.04 mmol hexose/g mycelium dry weight per hour. Mou (1979) explained the higher q_p in (fed-) batch culture by the dilution effect in continuous culture. In continuous culture, owing to the broad distribution of growth activity, only a fraction of the total population contributes to the production. Although the producing cells may have comparable productivity under both culture conditions, the apparent q_p will be lower in a continuous culture than in a batch culture. This effect can be regarded as a major drawback of a continuous process for penicillin production, because substrates are used to support a heterogeneous population and producing cells are continuously washed out of the reactor (see also Sec. VII.F). The latter effect will also manifest itself in a repeated fed-batch culture, where producing cells are withdrawn at intervals.

Degeneration of the strain's productivity was also observed in fed-batch culture after 300–400 hr (Court and Pirt, 1977). Ammonium- or phosphate-limited or ammonium and sucrose colimited fed-batch culture resulted in a significant reduction of q_p (Court and Pirt, 1977, 1981).

G. Penicillin Degradation

Nestaas and Demain (1981) recently reviewed the stability of penicillin (G) during production (see also Sec. IV). They calculated the first-order penicillin hydrolysis rate constant as a function of pH and temperature from the data of Benedict et al. (1946). The highest stability was found at pH 6.0 and low temperature. Gordee and Day (1972) observed that the net accumulation decreased with increasing amounts of penicillin added to a penicillin culture. They concluded that penicillin was regulating its own synthesis. Nestaas and Demain (1981), using the calculated penicillin hydrolysis rate constants and pH and temperature data, showed that the apparent decrease of net accumulation, as also observed by Demain (1957), can be explained by gradual hydrolysis of the added penicillin. They emphasized that no clear evidence has been presented in the literature that penicillin production is limited by feed-back regulation. The gradual decrease in product accumulation might therefore be explained by penicillin hydrolysis combined with a mycelium activity decrease, as measured by Nestaas and Wang (1981c). The mycelium activity was measured by calculating the hyphae density (dry weight hyphae/volume hyphae) from filtration data. With a decrease of hyphae density due to a loss of protein, respiratory activity and penicillin synthesis decreased linearly.

A significant amount of the penicillin produced hydrolyzes during the course of the fermentation to yield penicilloic acid. From their penicillin fermentation modeling study using a comparatively low hydrolysis constant, Heijnen et al. (1979) found that the penicillin hydrolysis might amount to 10–20% of the total penicillin production. Bundgaard and Hansen (1981) have shown that the catalytic effect of phosphate on the rate of degradation of penicillin G to benzylpenicilloic acid is due to a nucleophilic reaction mechanism involving formation of a penicilloyl phosphate intermediate. Hydrolysis of bensylpenicillin was also found to be catalyzed by zinc ions (Schwartz, 1982).

H. Penicillin Production with Immobilized Mycelium

Comparatively little research has been carried out on penicillin fermentation with immobilized mycelium. Suzuki and Karube (1978) and Morikawa et al. (1979) immobilized mycelium in collagen, which was cross-linked with glutardialdehyde, in calcium alginate or in polyacrylamide. The productivity of mycelium immobilized in collagen was nil, which is probably due to the toxicity of glutardialdehyde. The polyacrylamide monomers and the cross-linking agent which was used also appeared to be toxic, although some penicillin G productivity remained. Calcium alginate disintegrated in the phosphate buffer which was used. Polyacrylamide proved to be the best matrix. In shake flasks the penicillin production rate of polyacrylamide-immobilized mycelium was initially found to be 17% of that of free mycelium. The production rate of the immobilized mycelium decreased after repeated use. Kurzatkowski et al. (1982) found the penicillin G production activity of calcium alginate immobilized vesicles to be 44% in comparison to native vesicles (Vandamme, 1983; Karube et al., 1984).

Gbewonyo and Wang (1979) immobilized mycelium by adsorption of spores on microscopic beads. They investigated adsorption on anion exchangers, cation exchangers, glass, active carbon, magnesium/aluminum silicates, and several synthetic polymers. Adsorption was obtained with anion exchangers (Amberlite) and magnesium/aluminum silicates (Celite filter aid) and carboxylized polyethylene. In submerged culture mycelium was successfully grown on porous Celite beads after entrapping spores into the beads. In a bubble column starting with about 3000 spores per bead, 6 kg of penicillin per cubic meter were produced by 30 kg of mycelium dry weight per cubic meter in 140 hr (Gbewonyo and Wang, 1981). With this spore-entrapping immobilization technique, artificial pellets are grown with the advantages but also the disadvantages of natural pellets (see Sec. VI.B). However, these pellets do not easily break down, but hyphae growing out of the beads can be broken off and can continue to grow as nonimmobilized mycelium.

VII. LARGE-SCALE FERMENTATION PROCESS

As already noted by Queener and Swartz (1979), penicillin-manufacturing firms keep the details of their processes secret. Improved technology provides a competitive advantage. Patenting process technology is not always deemed to be adequate protection for obvious reasons. The manufacture of penicillins can therefore only be described by a general discussion of the kinds of technology available and described in the literature.

A flow sheet of the penicillin manufacturing processes used by Gist-Brocades is presented in Figure 10. In general, four processes can be distinguished (Hersbach, 1982): fermentation, product recovery (see Sec. VIII), production of semisynthetic penicillins (and cephalosporins), and patent medicine production (see Sec. IX).

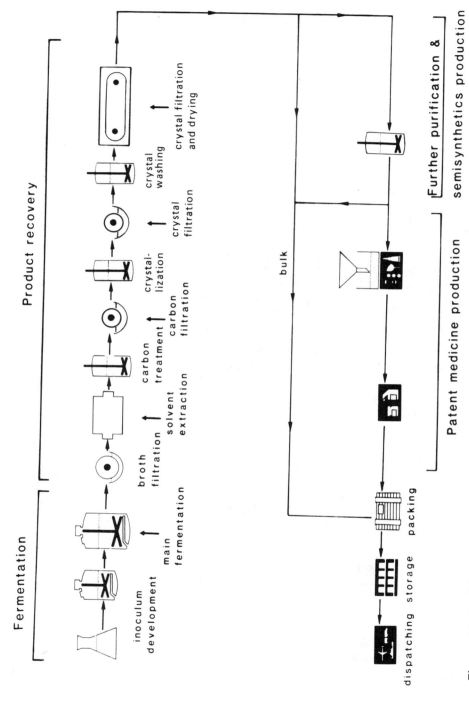

Figure 10 Penicillin manufacturing processes of Gist-Brocades.

A. Inoculum Development Process

From a master stock of the production organism, vegetative mycelium is prepared in a series of inoculum development steps (for detailed inoculation techniques the reader is referred to Meyrath and Suchanek, 1972). The main purpose of the inoculum development process is to increase the concentration of fungal mycelium. The subsequent inoculum development stages are carried out in shake flasks and ultimately in an agitated vessel. The medium contains sufficient fermentable carbohydrate, an organic nitrogen source, calcium carbonate as a buffer to prevent low pH values, and, as far as required, other inorganic (nitrogen, phosphorus, sulfur, potassium, and metal) salts (Queener and Swartz, 1979; see also medium A, Table 4, where generally lactose is replaced by glucose for inoculum cultures).

Inocula sown with spores above or below a critical level show differences in morphology and productivity a few hours after swelling and germination. At this time patterns of productivity and morphology appear (Calam and Smith, 1981). Inoculum quality and quantity have an important effect on the production of penicillin (Calam, 1976; Smith and Calam, 1980). An unexpectedly sudden breakdown of productivity in the main fermentation can probably be ascribed to the development of an ineffective inoculum culture (Calam and Ismail, 1980a). Above 5×10^9 spores/m^3, more filamentous mycelium was formed, enzyme (glucose 6-phosphate dehydrogenase, isocitrate dehydrogenase, and aldolase) levels were high, and penicillin production was good (Smith and Calam, 1980; Calam, 1982). Below this spore concentration dense pellets were formed, utilization or glucose was less efficient, and enzyme levels were lower, as was the penicillin production. The variation in enzyme levels persisted in the production stage. Although these results can be a consequence of morphology, they are also a result of the amount of mycelium inoculated. This amount increases with spore concentration, as the enzyme levels still probably do. Above the critical spore concentration an increasing amount of mycelium is produced in the seed stage and inoculated into the production stage. The total penicillin production remains the same, but the production maximum is reached sooner, which is probably a result of an increased consumption rate of the lactose, used in these experiments (Smith and Calam, 1980). The optimum inoculum size will be a function of the morphological characteristics of the strain, as well as of the design and operation of the fermenter. Vardar and Lilly (1982) found that a 10% inoculum with 20 kg mycelium dry weight/m^3 is required to result in a high specific penicillin production rate in the main fermentation.

Mycelium pellets can be obtained directly from spores as described above (and in Sec. VI.B) or from a preculture of filamentous mycelium (van Suijdam et al., 1980). Of importance are medium composition, inoculum concentration, polymer additives (affecting electrostatic forces among spores), and shear forces in the fermenter. Mild agitation and shear stresses favor the development of pellets.

During the subsequent inoculum development steps, samples are withdrawn periodically and examined for growth and morphology, nutrients consumed, and contamination to estimate the moment for transferring the broth to the next inoculum development step or to the production stage when a desired criterion is reached.

B. History of Production Fermentation Process Development

1. Batch Fermentation

Submerged penicillin fermentation has a long history of development (see Sec. I) since the first trials during the Second World War (Coghill, 1944). The start of the penicillin production history is reviewed by Sylvester and Coghill (1954). The first penicillin titers (Perlman, 1974) were of the order of 2 units/ml. The International Penicillin Master Standard (a pure crystalline sodium salt of penicillin G): 1 IU is equivalent to 0.6 g, which is approximately equivalent to 1 Oxford unit (Heatley, 1949). For the submerged process a lactose–corn steep liquor medium was developed (Sylvester and Coghill, 1954; Hockenhull, 1959, 1963; Perlman, 1970; see Sec. VI.D.1). Lactose was found to be a penicillin production-improving carbon source (Moyer and Coghill, 1946a,b). Lactose is a slow enzymically hydrolyzed carbohydrate. The slow rate of enzymic hydrolysis prevents accumulation of hexose (Matelova, 1976), which can cause inhibition of the penicillin synthesis (see Sec. IV). Corn-steep liquor was to markedly improve the yield of penicillin (Moyer and Coghill, 1946a,b; Bowden and Peterson, 1946). The value of corn-steep liquor was thought to lie in its balance of materials (amino acids, polypeptide, lactic acid, minerals, etc.) generally favorable for penicillin synthesis (Liggett and Koffler, 1948). The corn-steep liquor effect could, at least partly, be explained by the presence of compounds such as phenylalanine. These could be broken down to phenylacetic acid (Smith and Bide, 1948; Mead and Stack, 1948), which is used as a penicillin side-chain precursor, resulting in the natural synthesis of penicillin G. When phenylacetic acid or a derivative was added to the medium ($0.2-0.8$ kg/m^3, dependent on pH) the production of penicillin G increased greatly and that of other penicillins decreased (Moyer and Coghill, 1947). Early work on side-chain precursors has been reviewed by Behrens (1949), Johnson, (1952), Demain (1959, 1966), and Hockenhull (1959, 1963). To prevent β-oxidation of the precursor by the fungus, the presence of an interrupting group in the carbon chain is advantageous. β-Substituted acetic acids or compounds unsaturated in the β–γ positions are effective.

After the glucose and corn-steep liquor carbon compounds are metabolized and rapid mycelium development has occurred, lactose utilization begins (Sylvester and Coghill, 1954) and slow mycelial growth takes place. The corn-steep liquor nitrogen compounds are deaminated and ammonia nitrogen is formed. The nitrate is utilized when organic nitrogen is no longer available (Koffler et al., 1945). In the early submerged fermentation the pH was not controlled. Owing to nutrient utilization, the pH increased first, dropped slightly when lactose was utilized, and increased when all the carbon compounds were exhausted (Koffler et al., 1945). For a good penicillin yield the pH should stay between 6.8 and 7.4 (Brown and Peterson, 1950). During the rapid growth phase, the rate of antibiotic production was comparatively low. The maximum penicillin synthesis rate occurred during the lactose utilization phase, the slow growth phase. After the lactose was exhausted autolysis began and the onset of a decrease in penicillin concentration was observed. The fermentations were harvested before this point (Sylvester and Coghill, 1954).

Oxygen transfer was recognized early on as a very important factor for penicillin fermentation (Brown and Peterson, 1950; Bartholomew et al., 1950; Calam et al., 1951; Johnson, 1952; Donovick, 1960; Bartholomew, 1960). The oxygen uptake rate should not exceed the obtainable oxygen transfer rate in the fermenter being used. The oxygen uptake rate reached values of 5-6 mol/m³ (broth) per hour (Koffler et al., 1945; Johnson, 1946) or maximal 20-30 mol/m³ per hour (Finn, 1954). The carbon dioxide production rate reached values of about 7 mol/m³ per hour and, when well aerated, about 27 mol/m³ per hour (Johnson, 1946). The critical dissolved oxygen concentration could be estimated at 0.022 mol/m³ (Rolinson, 1952; Finn, 1954) or an oxygen tension of 2×10^3 Pa (Finn, 1954). Scale-up criteria were formulated on the basis of oxygen transfer and total power input (Karow et al., 1953; Bartholomew, 1960).

2. Fed-Batch Fermentation

The history of development of a fed-batch fermentation process for penicillin (see Table 6) has been extensively reviewed by Whitaker (1980). Improved penicillin yields were initially obtained by daily or more frequent additions of carbon compounds (Johnson, 1952). Hexose monomers (glucose) or carbohydrate compounds which are readily hydrolyzed to hexose monomers were found to be the best carbon sources. Easily fermentable carbon sources such as glucose or sucrose, however, have to be added continuously in controlled amounts during the fermentation to establish optimal conditions for the formation of both biomass and penicillin and to prevent accumulation of hexose in the broth (Matelova, 1976; Itsygin et al., 1977). Because of the risks associated with over- and underfeeding, early glucose feeding was initially not generally practiced (Hockenhull and Mackenzie, 1968). An amount of lactose was still poured in the medium, and after the lactose was exhausted a glucose or sucrose solution was continuously fed to the broth (McCann and Calam, 1972). Probably the concentration of lactose present in the medium induces sufficient β-galactosidase activity to allow a comparatively high growth rate early in the fermentation. As carbohydrate concentrations fall, low levels of β-galactosidase are produced, which presumably limit growth to a rate appropriate for the latter stage of the fermentation (Queener and Swartz, 1979). In general, lactose is now completely replaced by a glucose feed (Verkhovtseva et al., 1975), which was originally based on the lactose consumption data (see Sec. VII.B.1). The glucose feed is started after medium carbon sources (glucose, corn-steep liquor carbon, etc.) are consumed in the first fermentation phase (Verkhovtseva et al., 1975), the rapid growth phase. The lower limit of the sugar feed rate is determined by the amount needed to prevent autolysis, and the upper limit by oxygen limitation (Hosler and Johnson, 1953). With the glucose or sucrose feed the fermentation time could be extended considerably. Fatty oils containing a high content of oleic acid and linoleic acid, such as lard, soybean, and corn oil, were very efficiently used as carbon and energy sources when fed to the fermentation (Pan et al., 1959) as an antifoam. Broth withdrawals and complete medium additions have not been successful (see Table 6).

Ammonia has been fed both as a nitrogen source and to control the pH (Wright and Calam, 1968; Hockenhull and Mackenzie, 1968), as penicillin production was found to depend greatly on the pH value of the broth (Lurie and Levitov, 1976, optimum pH 6.5). It was found essential to maintain the residual concentration of ammonia at about 0.3-0.4 kg/m³ to obtain a high specific penicillin production rate and a continued synthesis of penicillin and to prevent nitrogen/carbon colimitation (Lurie et al., 1976; Court and Pirt, 1981) Lurie et al. (1976) reached the desired ammonium concentration by feeding ammonium sulfate to the fermentation. Ammonium sulfate provides nitrogen and sulfur in the proportions necessary for penicillin synthesis.

Table 6 Fed Batch Experiments to Increase Penicillin Yields

Treatment(s)	Highest penicillin titer		Year
	(units/ml)	(mol/m^3)[a]	
Glucose, lactose, corn starch, or corn-steep liquor	220	0.37	1946
Phenylacetic acid	—	—	1948
Glucose additions or withdrawals and additions of media	200	0.34	1948
Phenylacetic acid	1823	3.07	1950
Lactose and phenylacetic acid	2195	3.70	1950
Withdrawals and additions of media	70	0.12	1952
Glucose to shake flasks, continuous additions of glucose to shake flasks, glucose additions to fermenter	1600	2.69	1952
Withdrawals and additions of media	—	—	1952
Limiting substrate feed	1000	1.68	1953
Continuous additions of glucose, lactose, or sucrose	1520	2.56	1953
Glucose and ammonia additions	1774	2.99	1953
Phenylacetic acid	1850	3.11	1953
Continuous glucose feed	2225	3.75	1954
Lard oil additions	2200	3.70	1956
Continuous additions of glucose, sucrose, or molasses at variable feed rates	—	—	1958
Additions of fatty oils	4000	6.73	1959
Continuous sucrose additions	7000	11.79	1960
Glucose and precursors	4700	7.91	1961
Withdrawals and additions of media	3640	6.13	1961
Glucose and ammonia	7500	12.63	1968
Sucrose, ammonium nitrate, arachis oil, and phenoxyacetic acid	11,240	18.92	1972
Sucrose and phenylacetic acid	16,150	27.19	1972
Varying rates of continuous glucose feed	—	—	1972
Glucose feed	—	—	1974
Varying rates of continuous glucose feed	—	—	1975

[a]1 unit/ml $\equiv$ 1.684 $\times$ 10^{-3} mol/m^3.
Source: Data adapted from Whitaker (1980).

The residual concentration of the penicillin side chain precursor as phenyl-acetic acid (PAA) in the broth must be controlled within $0.1\,kg\,PAA/m^3$ (Court and Pirt, 1981) and $1\,kg\,PAA/m^3$ (Johnson, 1952; Chaturbhuj, 1961). Low con-centrations will result in the production of penicillins other than penicillin G. High concentrations can cause toxicity and hydroxylation by the produc-ing organism. The use of less toxic derivatives of phenylacetic acid or phe-oxyacetic acid has been proposed (Demain, 1966, 1974). Continuous feeding techniques and periodic analyses (see Sec. V) can be used to control the pre-cursor concentration and to ensure more efficient precursor utilization (Cha-turbhuj et al., 1961; McCann and Calam, 1972; Perlman, 1970, phenylacetic acid efficiency greater than 90%) and make the use of less toxic precursors super-fluous.

The penicillin production improvement due to the development of a fed-batch process, as shown in Table 6, must be regarded as a result of both fermentation process and strain improvement. During the last decade little has been published on penicillin production yields. Queener and Swartz (1979) and Swartz (1979) gave a survey of the reported penicillin titers on a production scale with strains distributed by Panlabs, Inc., to their clients (Table 7). These titer increases are also a combination of both fermentation process and strain improvement. Improvement of equipment and the fermen-tation process have played an important role in increasing the penicillin titers. The improvement of the specific production rates of the strains was estimated to be 20-fold (van der Waard, 1976). The present maximum specific penicillin G production rate is said to be about 8–9 units/mg (mycelium dry weight) per hour (Lurie et al., 1976, 9.0 units/mg per hour; Heijnen et al., 1979, for modeling study 8.0 units/mg per hour; Mou, 1979, 8 units/mg per hour; Vardar and Lilly, 1982, 8.2 units/mg per hour).

C. Fermentation Process Equipment

Each penicillin manufacturer's plant has been through a series of independent developments. Modern fermenters are still stirred vessels, mostly made of stainless steel and equipped with one or more turbine stirrers and baffles. To provide good dispersion, the air inlet (ring or fishtail sparger) is con-structed immediately below the bottom turbine. The vessel is provided with a cooling jacket or coil, depending on the scale. The fermentation volume has increased considerably over the last three decades. In the 1950s the maximum total volume amounted to $20\,m^3$ (Sylvester and Coghill, 1954; Vinze and Ghosh, 1959). Nowadays fermenters of $150\,m^3$ (van der Waard, 1976) and $200\,m^3$ (Queener and Swartz, 1979) are used. The filling can be up to about 80% of the total volume and is dependent on the hold-up and foam char-acteristics of the broth (König et al., 1979; Viesturs et al., 1982). The ini-tial broth volume is determined by the feeds and withdrawals during the fer-mentation process. The seed fermenter volume is of the order of 10% of the main fermenter volume (Sylvester and Coghill, 1954; Queener and Swartz, 1979). The seed and main fermenter are connected by piping. Aiba and Okabe (1977) and Young (1979) emphasized the significance of a total environ-mental approach to equipment optimization, process optimization, and scale-up.

The fermenters must be kept free of contamination. Contamination re-sults in competition for nutrient consumption; in addition, contamination with a β-lactamase-producing organism results in degradation of the produced peni-cillin. Compressed air is decontaminated by filtration. The filters are steril-

Table 7 Results of the Panlabs Inc. Strain Development Program

Strain	Year	Penicillin G[a]			Penicillin V			Fermentation cycle (hr)
		units/ml	Na salt kg/m³	mol/m³	units/ml	Na salt kg/m³	mol/m³	
P-1	1973	—	—	—	13,600	8.5	22.8	130
P-3	1973	17,600	10.6	29.6	16,400	10.3	27.7	182/180
P-4		17,800	10.7	29.9	20,000	12.5	33.7	182
P-5	1974	23,100	13.9	38.9	24,600	15.4	41.4	182
P-7		25,600	15.3	43.1	27,300	17.1	46.0	190
P-8	1975	29,500	17.7	49.6	32,400	20.3	54.5	203
P-10 (pilot scale)	1976	—	—	—	33,600	21.0	56.5	185
P-12 (pilot scale)		—	—	—	34,900	21.9	58.8	185
P-13		42,100	25.3	70.9	40,700	25.5	68.5	185
P-15		46,700	28.0	78.7	—	—	—	183

[a] Production scale, except as noted. Harvest volumes around 110% of starting volumes. Production medium and feeds changed along with cultures.

[b] 1 unit of penicillin G or V/ml $\equiv 1.684 \times 10^{-3}$ mol/m³, 1 unit of penicillin G/ml $\equiv 6.000 \times 10^{-4}$ kg of sodium penicillin G/m³, 1 unit of penicillin V/ml $\equiv 6.269 \times 10^{-4}$ kg of sodium penicillin V/m³.

Source: Data adapted from Swartz (1979).

ized before each fermentation run. The medium ingredients can be mixed in a comparatively small tank with sufficient water to make an easily pumpable solution, and sterilized continuously by heat (at about 140°C) or filtration, or heat-sterilized batchwise (at about 120°C) and subsequently pumped into the presterilized fermenter. The medium ingredients can also be mixed with water and heat-sterilized in the fermenter. The feeds (glucose, ammonia, ammonium sulfate, sulfuric acid, and phenylacetic or phenoxyacetic acid, etc.) can be dosed aseptically after batch heat sterilization in a holding tank or after continuous sterilization by heat or filtration. The principles of continuous sterilization have recently been described by Ashley (1982). The fermentation equipment, piping and filters included, must be carefully sterilized before each run. During the fermentation contamination is prevented by maintaining a head pressure and applying steam seals. Some of the operating variables are summarized in Table 8. An example of ammonia, sulfuric acid, and precursor feed concentrations are given by Heijnen et al. (1979). The harvesting point can be determined (van der Waard, 1976) by a decrease in production per unit volume and per unit time (total fermenter occupation time being the time needed for filling, sterilization, fermentation, emptying, etc.). Sittig (1982) recently concluded that the stirred tank reactor remains the type of fermenter most commonly used and that the trend toward higher power inputs per unit volume seems to have stopped.

D. Present Production Fermentation Process

Penicillin fermentation has developed to a fed-batch process (see Sec. VII.B.2) or a repeated fed-batch process (a portion of the broth is withdrawn at intervals). The theory of fed-batch fermentations has been clarified by Pirt (1974), Dunn and Mor (1975), Lim et al. (1977), and Keller and Dunn (1978). The relationship between the nutrient feed and growth for such fermentations was described by Yamané and Hirano (1977) and Jones and Anthony (1977). Optimization of the nutrient feed for fed-batch fermentations in general has been described by Yamané et al. (1977) and Choi and Park (1981), and that for penicillin fermentation by Fishman and Biryukov (1974). There is some evidence that penicillin is produced at maximum rates only transiently during dynamic shift conditions (Pirt, 1974); therefore penicillin fermentation is a fed-batch process.

The total penicillin production P in a fermenter can be expressed by

$$P = \int_{t_{inoculation}}^{t_{harvest}} V(t) \cdot q_p(t) \cdot C_x(t) \cdot dt \tag{3}$$

where V is the broth volume, q_p the specific penicillin production rate $1/C_x \cdot dC_p/dt$, C_p the penicillin concentration, C_x the mycelium concentration, and t time. The broth volume is variable during the course of the fermentation process. Various feeds cause an increasing total broth volume. When fractions of broth are withdrawn at certain points of time, the broth volume is a discontinuous function of time. The object is to optimize P within the constraints of the equipment and overall economics (Queener and Swartz, 1979). An algorithm for fermentation optimization including product purification was constructed by Okabe and Aiba (1975).

Table 8 Operating Variables

Variable	Value	Reference
Fermenter volume	150–200 m3	van der Waard (1976), Queener and Swartz (1979)
Filling	Up to about 80%	
Mechanical power input[a]	About 3–4 kW/m3	Sittig and Heine (1977), Sittig (1982)
Air flow rate	30–60 N m3/m3 broth per hour	Giona et al. (1976a), Sittig (1982), Queener and Swartz (1979)
Air pressure (absolute	~300 kPa	Sittig and Heine (1977)
Fermenter head pressure (absolute)	135–170 kPa	Sylvester and Coghill (1954)
$k_L a$	~200 hr^{-1}	Sittig and Heine (1977)
Oxygen uptake rate	25–50 mol/m3 broth per hour	Perlman (1974)
Initial mycelium concentration	1–2 kg dry weight/m3	Heijnen et al. (1979), Queener and Swartz (1979)
Broth temperature	~25°C[b]	Sittig (1982) (see Sec. VI.E)
Broth pH	~6.5[b]	(see Sec. VI.E)
Glucose (equivalent) concentration in feed	500 kg/m3 solution	Heijnen et al. (1979)
Glucose (equivalent) feed rate	1.0–2.5 kg/m3 broth per hour	Giona et al. (1976a)
Glucose (equivalent) optimum feed rate	1.8 kg/m3 broth per hour	Giona et al. (1976a)
Duration of fermentation	180–220 hr	Giona et al. (1976a), De Flines (1980), Sittig (1982)

[a]Dependent on the fermenter dimensions one or more turbine stirrers.
[b]Strongly dependent on the strain.

1. *Productivity and Oxygen Transfer*

Since the rate of penicillin production depends upon biomass concentration, it is desirable to have a high biomass concentration in the vessel. The medium contains carbon compounds and other nutrients and they allow the organism to grow initially near the maximum specific growth rate. This rapid growth causes a considerable increase in the initial oxygen uptake rate (Heijnen et al., 1979, the maximum specific oxygen uptake rate amounts to 3.9 mmol O_2/g mycelium dry weight per hour; Phillips and Johnson, 1961) and the carbon dioxide evolution rate (Mou, 1979). For a given oxygen capacity, there is an optimum initial sugar (or total carbon) concentration (Ryu and Humphrey, 1972; Calam and Russel, 1973, Hegewald et al., 1981). Care must be taken to prevent excessive pH changes (Queener and Swartz, 1979). Once a cell concentration sufficient to support a satisfactory penicillin production has been obtained, growth must continue at a certain minimum rate (Pirt and Righelato, 1967; Young and Koplove, 1972; Ryu and Humphrey, 1972, 1973; Ryu and Hospodka, 1980) to maintain a high q_p (specific penicillin production 0.015 hr^{-1} rate) value. The mycelial mass achieved during the rapid growth phase must be limited to allow for the additional mycelium mass during the slow growth phase of the fermentation. The optimal initial sugar concentration is therefore dependent upon the fermenter aeration capacity and the glucose feeding rate (Bajpai and Reuss, 1980, 1981). A minimum oxygen supply is required to support maximum q_p values. Since both oxygen demand and apparent viscosity increase with increasing mycelium concentration (see Sec. VI.C) and since the oxygen transfer coefficient (k_La) decreases exponentially with increasing apparent viscosity, beyond a certain mycelium concentration the oxygen supply in a particular fermenter will be insufficient to support maximal q_p for a sufficient duration, and the overall rate of penicillin synthesis will decrease (Ryu and Hospodka, 1980).

Small variations in k_La would dramatically influence the penicillin productivity in the region of low values of k_La, which is the range of practical interest (Bajpai and Reuss, 1980). At low k_La values the effect of the mechanical power inpur on the penicillin production was found to be more profound (Takei et al., 1975). It is therefore extremely important to maintain the optimum specific growth rate during the slow growth phase of the fermentation. Giona et al. (1976a,b) correlated dissolved oxygen concentration values (C_{oL}) with q_p in an industrial repeated glucose fed-batch fermentation (see Sec. VI.E). Within the range of variables tested glucose feed rates were shown to have the most profound effect on overall penicillin production (Giona et al., 1976a). Increasing sugar feeds decreased the C_{oL} value (Giona et al., 1976b).

Increasing the maximum oxygen transfer rate of the fermenter is an obvious possibility to improve penicillin production by increasing the mycelial mass which can be supported in a particular physiological state. This can be achieved by increasing the agitation power, the aeration rate, and the head pressure of the fermenter (increasing the head pressure, however, decreases the ventilation and can result in inhibiting carbon dioxide concentrations; see Sec. VI.E) and by selecting strains (see Sec. II) and fermentation conditions which decrease broth viscosity. Heat transfer can also limit the penicillin production in fermenters. The temperature for obtaining maximal q_p values can only be obtained by removal of the metabolic heat generated by the mycelium and the mechanical heat generated by agitation and aeration. So when the agitation and aeration power are increased in order to increase the mycelial mass, more heat removal will be necessary.

2. Productivity and Nutrient Feeding

Glucose feeding may be commenced during the rapid growth phase in order to control the growth in the initial period of the fermentation. The high specific growth rate needed to obtain the desired mycelial mass at the end of the rapid growth phase must be limited in order to prevent the oxygen uptake exceeding the oxygen transfer and to prevent the fed sugar from being wasted. Otherwise high q_p (specific penicillin production rate) values are not achieved. Mou (1979) reported the production of appreciable amounts of gluconic acid when glucose concentrations were high. Wright and Calam (1968) and Mou (1979) experimentally illustrated increasing q_p values during the rapid growth phase. At the end of the rapid growth phase the specific carbon feed is diminished in order to sustain a satisfactory q_p for the rest of the fermentation cycle.

Control algorithms for the carbon feed are summarized in Table 9. pH is not a very sensitive indicator of the process during penicillin production. The control of carbon feeding via the pH value is insufficient to guarantee a stable fermentation. If the process becomes unbalanced, it is not always possible to restore the productivity by changing the carbon feed. Dissolved oxygen is more responsive to changes in values of q_p than pH (Squires, 1972). The control of carbon feeding via the dissolved oxygen value essentially increases the reliability of the fermentation. However, the control is largely dependent on the availability of good and reliable electrodes. As penicillin formation is growth related, control of the specific growth rate by the carbon feed seems most attractive. The upper limit of the specific growth rate is determined by several fermenter constraints. The specific growth rate varies inversely with the mycelial mass at the end of the rapid growth phase (see Sec. VII.D.1). Ryu and Hospodka (1980) reported the minimum μ (specific growth rate) for a maximum q_p to be 0.015 hr^{-1}. The specific oxygen uptake rate q_0 required to maintain this μ was found to be 1.2 mmol oxygen/g mycelium dry weight per hour for maintenance and growth and 1.6 mmol oxygen/g mycelium dry weight per hour for maintenance, growth, and penicillin production. This is in agreement with the critical oxygen tension parameters for

Table 9 Control Algorithms for Carbon Feeding

Variable to control with the carbon feed	Reference
Dissolved oxygen (with control of aeration and agitation)	Squires (1972), Giona et al. (1976a,b)
pH (in a poorly buffered medium)	Shu (1972), Squires (1972), Pan et al. (1972), Jensen et al. (1981)
Carbon dioxide evolution	Itsygin et al. (1976)
Growth curve	Hockenhull and MacKenzie (1968)
Specific growth rate (specific carbon feed)	Calam and Russel (1973), Ryu and Humphrey (1972,1973), Ryu and Hospodka (1977,1980), Mou (1979)

oxygen uptake and penicillin production (see Sec. VI.E). So penicillin synthesis commences only when the needs for maintenance and growth are satisfied (see also Vardar and Lilly, 1982).

With the growth rate controlled by the sugar feed, it is desirable to prevent colimitation effects of other nutrients. All noncarbon nutrients must be available in concentrations sufficient to maintain their μ-dependent specific uptake rates. The specific nutrient uptake rates have been precisely determined by Ryu and Hospodka (1980) in continuous culture and may be used for optimum nutrient feed rates calculations:

Nutrients and precursor	Specific uptake rates (at $\mu = 0.015$ hr^{-1}; in mmol/g mycelium dry weight per hour)
Carbon (hexose)	0.33
Oxygen	1.6[a]
Nitrogen (NH$_3$)	0.12
Phosphorus (PO$_4^{3-}$)	0.006
Sulfur (SO$_4^{2-}$)	0.029
Precursor (phenylacetic acid)	0.013

[a]For maintenance, growth, and penicillin production.

Lee et al. (1978) determined specific hexose and NH$_3$ uptake rates in fed-batch culture for glucose and lactose as the major carbon source. The optimum specific uptake rates were found to vary from strain to strain. A typical concentration profile of a controlled fed-batch penicillin fermentation is shown in Figure 11. At the end of the fermentation oxygen becomes colimiting with

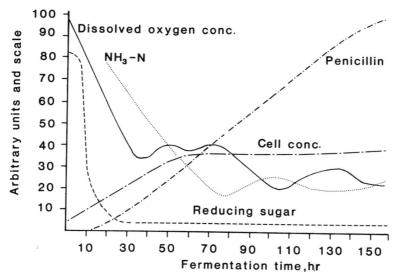

Figure 11 Typical concentration profile of fed-batch penicillin fermentation. (From Ryu and Hospodka, 1980).

Table 10   Yield of Penicillin from Glucose

	Reference	Units pen G or V per mg glucose[a,b]	Mole pen G or V per mole glucose[a,b]
maximum yield value model calculation	Stouthamer (1977)	2198	0.667
	Cooney and Acevedo (1977)	1800	0.56
		1100	0.33
	Cooney (1979)	1978	0.60
	Heynen et al. (1979)	2080/1517	0.631/0.46
	Section III	683−1541	0.207−0.467
real yield value — model calculation	Giona et al. (1976b)	134	0.041
		143	0.043
		146	0.044
	Cooney and Acevedo (1977)	200	0.061
		183	0.056
	Heijnen et al. (1979)	105	0.032
	Bajpai and Reuss (1981)	95	0.029
experimental	MacCann and Calam (1972)	123	0.037
	Pilat (1979)	100−113	0.030−0.034
	Cooney (1979)	88	0.027
		80	0.024
		160	0.049
		192	0.058
		143	0.043
		192	0.058
	Mou (1979)	122	0.037
	König et al. (1981b)	212	0.064

[a]1 unit pen G/mg glucose $\equiv$ 3.033 $\times$ 10^{-4} mole pen G/mole glucose $\equiv$ 6.00; 10^{-4} g pen G-Na/g glucose.

[b]1 unit pen V/mg glucose $\equiv$ 3.033 $\times$ 10^{-4} mole pen V/mole glucose $\equiv$ 6.269; 10^{-4} g pen V-Na/g glucose.

[c]Panlabs, Inc. strains; data modified by Cooney (1979).

g pen G-Na per g glucose[a]	g pen V-Na per g glucose[b]	Remarks
1.32	—	Stoichiometry
1.1	—	Biochemical pathway α-aminoadipic acid recycled
0.66	—	α-aminoadipic acid discarded
1.19	—	Stoichiometry
1.25/0.91	—	Literature
0.410−0.924	—	
0.080	—	140 hr
0.086	—	180 hr
0.088	—	220 hr
0.12	—	200 hr α-aminoadipic acid recycled
0.11	—	α-aminoadipic acid discarded
0.063	—	135 hr
0.057	—	150 hr
		Simulated from data reported by Mou (1979)
—	0.077	137 hr, 18 liters; 11,240 units penicillin V/ml; 1562 g sucrose equivalent; oil and CSL carbon components not included
0.060−0.068	—	Carbon source: sucrose
0.053	—	114 hr
		Panlabs, Inc. strain P-2[c]
0.048	—	P-2
0.096	—	P-7
0.115	—	P-11
0.086	—	P-13
0.115	—	P-15
0.073	—	160 hr
—	0.133	110 hr; only consumed lactose accounted for

carbon. To keep carbon the primary colimiting substrate, the carbon feed rate must be reduced; consequently μ will decrease. So oxygen transfer rates may limit μ to values below a strain's optimum for maximum fermentation productivity. Deviations from the ideal conditions are sometimes dictated by the limitations of existing production equipment (Queener and Swartz 1979) and by the optimum substrate conversion efficiency to penicillin and overall penicillin productivity (Mou, 1979).

The duration of the repeated fed-batch fermentation is limited by the oxygen transfer capacity of the fermenter, the exhaustion of nutrients, which cannot be fed conveniently or economically, the buildup of toxic and inhibiting compounds, and/or the production and selection of mutants with a lower productivity (Queener and Swartz, 1979).

3. Yield of Penicillin from Glucose

Reported yields of penicillin from glucose (or glucose equivalents) are summarized in Table 10. The maximum yield values are calculated with only simple stoichiometric relations or with the aid of biochemical pathways (see Sec. III). The maximum yield value tends to decrease as more biochemical knowledge and assumptions are introduced. The real yield value, which is or can be obtained during a fermentation, is much lower and is dependent on the carbon distribution for penicillin production, and mycelium growth and maintenance. The real yield value increases with fermentation time up to a certain point, after which it decreases again (Heijnen et al., 1979; Mou, 1979). It is very dependent on the carbon sources accounted for; broth withdrawals cause lower values (Mou, 1979). Carbon distribution values are given in Table 11. The distribution is determined by the penicillin productivity of the strain and its maintenance requirements. In general, 65% of the consumed carbon is used for maintenance, 25% for mycelial growth, and only 10% for penicillin production. Distribution data of Cooney (1979) are not in agreement with the general distribution, which is probably a result of calculating the main-

Table 11 Carbon Distribution for Penicillin, Mycelium, and Maintenance

Reference	Percentage of consumed carbon for			Remarks
	Penicillin	Mycelium	Maintenance	
Stouthamer (1977)	6	25	69	140 hr
Cooney and Acevedo (1977)	11	28	61	200 hr; α-aminoadipic acid recycled
	16	27	57	α-aminoadipic acid discarded
Mou (1979)	6	29	65	142.5 hr
Cooney (1979)	6	70	26	126 hr
Heijnen et al. (1979)	10	20	70	200 hr

tenance requirements as the difference between consumed carbon and carbon used for growth and penicillin production. An error in the consumed carbon calculation will be expressed in the maintenance requirement. An independent maintenance calculation is more reliable. Variation of the maintenance requirement has a great effect on mycelium and penicillin production (Heijnen et al., 1979; Mou, 1979). Lowering the maintenance factor results in an increased penicillin yield and a higher mycelial dry weight; an increased maintenance factor has the opposite effect.

4. Fermentation Modeling

Several attempts have been made to model the fed-batch penicillin fermentation to study the process and optimize operating conditions (Biryukov and Kantere, 1982). Fermentation modeling has been reviewed by Roels and Kossen (1978) and Roels (1982). Shu (1961, 1972), Fishman and Biryukov (1974), and Calam and Ismail (1980a,b) assumed a relation between mycelial age and mycelial productivity, resulting in a culture age model. This model is based on the view that the specific rate of penicillin production is independent of the growth rate. However, Calam (1979) concluded that because penicillin formation can be related to the culture age, it is related to growth. In the culture age model the product activity was assumed to decrease at high mycelial age, which was explained by a decreasing concentration of active cells by Calam (1982). With the culture age model the decreasing slope of the penicillin production curve in the last part of the fermentation could be modeled. This might, however, also be explained from dilution due to increasing broth volume and hydrolysis of penicillin (Heijnen et al., 1979).

Formerly penicillin fermentation was often classified in the group of product formation processes of the non-growth-associated type (Gaden, 1959). The penicillin fermentation process was assumed to consist of two phases: a phase of rapid growth with no product formation (tropophase) and a phase with no growth in which the product is formed (idiophase). This picture seems to originate from a lack of insight in the use of carbon for maintenance requirements (Stouthamer, 1977). However, the apparent separation between production and growth phases does not necessarily mean that penicillin production is of the non-growth-associated type. Heijnen et al. (1979) assumed in their penicillin fermentation modeling study—applying balancing methods— a direct coupling between specific growth rate and specific penicillin formation rate. The results commonly observed could be explained perfectly by their model and were seemingly in agreement with the tropophase/idiophase description. The simulation results also showed an apparent lag phase in the penicillin production curve.

Heijnen et al. (1979) concluded that growth-coupled penicillin production provides an adequate description of most of the observed phenomena in the penicillin fermentation process and that there appears to be no evidence necessitating the introduction of an age dependency or a time delay in modeling the penicillin production kinetics. Using a growth-coupled penicillin formation relation, a lag phase (of about 20 hr) was introduced in the penicillin production model by Constantinides et al. (1970a) and Calam and Russel (1973). Bajpai and Reuss (1980, 1981) introduced substrate inhibition kinetics, although the exact nature of the biochemical mechanism of product formation regulation (catabolite repression/inhibition) is unknown (Calam, 1979; Demain et al., 1979; see Sec. IV). They clearly showed the influence of substrate inhibition on the $\mu-q_p$ relation. To verify their model, Bajpai and Reuss (1980, 1981) used data sets reported by Pirt and Righelato (1967) and Mou (1979). This mechanistic model has been found to predict the time course of fed-batch penicillin fermentations. However, the simulated penicillin production curve showed divergence from the data determined by Mou (1979). Bajpai and Reuss (1980, 1981) used Contois kinetics for the specific growth rate. Limitation of growth and penicillin production by dissolved oxygen is introduced with a simple model. Heijnen et al. (1979) modeled the substrate uptake rate with a Monod-type relation.

Bajpai and Reuss (1981) described two different carbon feeding strategies: constant carbon feed rate and controlled carbon feeding so as to reproduce a predicted growth pattern. They concluded that the maximum predicted productivities are more or less independent of the feeding scheme strategy. This conclusion is not in agreement with the modeling study results of Heijnen et al. (1979). These authors showed that increasing the glucose feed rate with time will result in a higher penicillin productivity with a higher efficiency as compared with other possible schemes such as a constant feed rate or decreasing the feed rate with time. With the latter schemes the large amount of biomass formed cannot be kept at the specific growth rate necessary for optimal penicillin productivity. The limitations to the feed rate scheme set by the capacity of the equipment with respect to the feed supply or the removal of heat and carbon dioxide can also be accounted for (Humphrey and Jefferis, 1973). Hegewald et al. (1981) proposed a model whose kinetic equations were very similar to those of the model described by Heijnen et al. (1979). Hegewald et al. (1981) demonstrated that high fermentation productivity is connected with a low stability of the fermentation. The duration, strength, and the beginning of a disturbance of the oxygen transfer coefficient determine the production failure. The sensitivity of the process to such disturbances increases with the penicillin productivity at the production onset.

The nitrogen demand for the penicillin production accounts for 20% of the total nitrogen demand during the fed-batch penicillin fermentation (Stouthamer, 1977; Heijnen et al., 1979). From fermentation modeling with the macroscopic electric charge balance taking only biomass production into account, Heijnen (1981) concluded that it seems impossible to obtain both constant pH control and constant dissolved NH_3 concentration control through the use of ammonia feed only. Whenever a constant pH and a constant ammonium ion concentration is desired, a sulfuric acid feed or an ammonium sulfate feed is needed. The sulfur demand for penicillin production accounts for 80% of the total sulfur demand of the fermentation (Stouthamer, 1977; Heijnen et al., 1979). This stresses the importance of the sulfur metabolism in the production of penicillin (see Sec. III).

The development of a model capable of predicting the fermentation pattern for different sets of operating conditions is important for understanding the fed-batch penicillin fermentation process. By carrying out suitable experiments to determine the basic model parameters, which can be verified by subsequent experiments, it is possible to simulate the process for a whole set of fermentation conditions. Validity of the model predictions in the regions must be verified by further experiments. The different process operation strategies can first be simulated so as to reach a better and more suitable basis for experimentation.

E. Fermentation Process Control

Recently Hatch (1982) reviewed computer fermentation process analysis and control. Instrumentation for fermentation process control has been reviewed by Fleischaker et al. (1981) and Mor (1982). A mass spectrometer is suitable for on-line fermentation effluent gas analysis (Buckland and Fastert, 1982). For the on-line measurement of dissolves nutrients, development of sensors with specific membranes is important (Yamané et al. 1981; dissolved oxygen: Lee and Tsao, 1979). Microbial sensors can also be applied (Hikuma et al., 1981; penicillin: Enfors and Nilsson, 1979). However, in general, these sensors are not sterilizable and have to be used off-line or semi-on-line. This requires the development of an automatic broth sampling system and broth transport system.

Using a physiological approach, Yough and Koplove (1972), Ryu and Humphrey (1972, 1973), Ryu and Hospodka (1977, 1980), and Mou (1979) described principles for controlling the mycelial growth and productivity by controlling the sugar feed and other nutrient feeds (see Sec. VII.D.2). Mou (1979), Cooney and Mou (1982), and Mou and Cooney (1983a,b) used a highly instrumented, computer-coupled fermenter to control the biomass growth rate profile during penicillin fermentation. In the rapid growth phase a rapid mycelium accumulation is attained by controlling the growth rate at a high value, and in the slow growth phase the growth rate is controlled at a lower value in order to permit penicillin production. Mou (1979), Cooney and Mou (1982), and Mou and Cooney (1983a,b) have shown that it is possible to calculate the growth rates using an empirical correlation for CO_2 production data in the rapid growth phase (accuracy ± 1 kg mycelium dry weight/m^3 and $\mu \pm 0.01/$ hr) and a carbon balancing method in the slow growth phase (accuracy $\mu \pm 0.002/$ hr) both in chemically defined and complex medium. Calam and Ismail (1980a, b) determined the mycelial growth from the CO_2 evolution as suggested by Humphrey (1977). Based on the instantaneous growth rate data generated by computer, a strategy for glucose feeding was formulated by Mou (1979), Cooney and Mou (1982), and Mou and Cooney (1983a,b) to control the growth rate at predicted values. Growth rates during the phases of the fermentation process were controlled by regulating the sugar feed rate to the fermenter using a proportional control. Since the rates of sugar feeding during the process may be quite different, a gradual change according to a predicted pattern was enforced after achieving a critical biomass concentration. A feed-forward modification of feed-back control was applied. Physical limitations of the equipment can be taken into account. For computer coupling, it is essential to have a model available which is capable of predicting the fermentation pattern for different strategies with different sets of operating conditions (see Sec. VII.D.4).

Recently the Eli Lilly fermentation plant computer system has been described by Alford (1982), and the development of the NOVO system by Falch et al. (1982).

F. Fermentation Process Economics

Analyses of the manufacturing cost of penicillin G as the percentage of the total cost) can be summarized as follows

	Swartz (1979)	de Flines (1980)
Raw materials	28% (glucose, 12%; sodium phenylacetate, 11%)	44%
Utilities	24% (electricity, 9%)	8%
Fixed costs (including plant overhead)	27%	26%
Separation and purification	21%	22%
Total	100%	100%

As can be concluded from these figures, the contributions of raw materials and utilities to the total manufacturing cost are different in the United States (Eli Lilly; Swartz, 1979) and in the Netherlands (Gist-Brocades; de Flines, 1980). This is probably due to differences in the price level of these variable costs. From the summarized analyses it is clear that penicillin production is fermentation cost intensive. Fermentation constitutes almost 80% of the

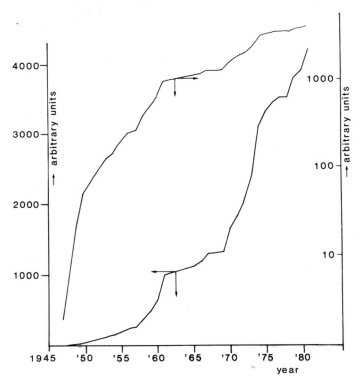

Figure 12 Mean penicillin G productivity of Gist-Brocades (fermentation production per unit volume and per unit time).

total manufacturing cost of penicillin G. Raw material costs and product yields are very significant in penicillin production (Bungay, 1963). Mixing costs are also considerable in antibiotic fermentations (Oldshue et al., 1978). In the fed-batch penicillin fermentation a significant cost price decrease and fermenter productivity increase can be obtained by an increase in the specific penicillin production rate through improvement of nutrient availability control, strain improvement, a cell mass increase as far as allowed by equipment constraints, and the use of cheaper raw materials (see Sec. VI.D.2). On the other hand, continuous operation (see Sec. VI.F.1) or a recycle of killed cells does not lead to productivity improvement (Swartz, 1979). An economic analysis (with quantification of the expected annual savings) can provide a guide to the potential of approaches to fermentation process improvement and will determine their priorities. Although a significant improvement in penicillin production yield has been obtained during the last 40 years, productivity can probably still be increased, since about 10% of the total sugar feed is used for penicillin production (see Sec. VII.D.3; Hersbach, 1982). The present and near future penicillin fermentation level has been estimated at about 40–50 kg of sodium penicillin G per cubic meter of fermentation broth (or 67,000–83,000 units of penicillin G per milliliter of broth or 112–140 mol of penicillin G per cubic meter of broth) by Bérdy (1980).

An example of the development of the mean penicillin productivity (penicillin fermentation production per unit of volume and per unit time) is shown in Figure 12. The increase in productivity is regarded as due to the interaction of results obtained in the field of strains, inoculum, medium, equipment, feed controls, and control of physical parameters. Although up to now productivity is still significantly increasing, improvements are becoming more difficult to achieve.

VIII. PRODUCT RECOVERY PROCESS

The basic design of a penicillin separation and purification plant has changed very little over the last 30 years (Sylvester and Coghill, 1954). Scale has increased, automatic controls have been added, and facilities have been adapted to higher potency penicillin broths. The purification process as used by Gist-Brocades is outlined in Figure 13, which can be regarded as a detail of Figure 10. For a detailed description of penicillin recovery we refer the reader to Queener and Swartz (1979). The duration of the penicillin recovery cycle from one fermenter is about 15 hr (de Flines, 1980).

A. Broth Filtration

The mycelium is usually separated from the penicillin-containing broth on a rotary vacuum filter. The mycelium is also washed on the filter. Filter aids or precoats may be necessary. Formerly a second filtration step was sometimes necessary to clarify the filtrate from proteinaceous material which gave rise to emulsion difficulties during extraction (Queener and Swartz, 1979). Present extractors and broth types make it possible to use only one filtration step and will achieve good separation during extraction. The composition of the fermentation media has been shown to exert a substantial influence on filtration (Zhukovskaya and Gorskaya, 1974). Filtration is impeded by protein-containing constituents of the media. The rate of filtration was found to change with the duration of the fermentation. Mycelial lysis causes a sharp decrease

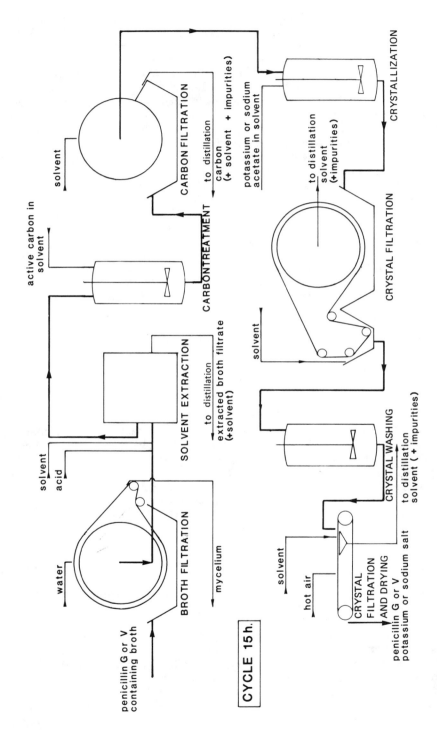

Figure 13 Penicillin purification process of Gist-Brocades.

in filterability. Whole broth extraction can be recognized as an alternative to filtration (Podbielniak et al., 1970); nevertheless, extraction of mycelium components cannot be prevented under these conditions. These components will place an extra load on the subsequent purification steps. In addition, the remaining broth has implications for waste treatment. When the filtration step is incorporated in the recovery process, the mycelium can be sold as a protein-rich, penicillin-free cattle fodder additive.

B. Solvent Extraction

The conditions (temperature, pH, sterility) of the penicillin-containing filtrate have to be controlled very carefully in order to minimize chemical and enzymatic degradation of the penicillin in the pumps, collecting vessel, and extractor. Penicillin is extracted in the acid form (Souders et al., 1970; Podbielniak et al., 1970) into amyl acetate or butyl acetate in a continuous countercurrent multistage centrifugal extractor at pH 2.5–3.0 and at 0–3°C (Queener and Swartz, 1979). Emulsion formation can be reduced by using demulsifying agents in order to obtain a high separation efficiency. The penicillin-rich filtrate is mixed with dilute sulfuric or phosphoric acid and a demulsifier. The mixture and the organic solvent are fed into the extractor (Podbielniak et al., 1970). The flows are in the proportion of 4 or more to 1. In the water phase residual penicillin is measured and controlled at a low level. In order to minimize degradation of penicillin under the acid conditions necessary for efficient extraction, short residence times are essential. These and other operational parameters are kept optimal by controlling the liquid flow pattern (Todd and Davies, 1973).

C. Carbon Treatment

To remove pigments and other impurities, the penicillin-containing solvent is treated with active charcoal (Queener and Swartz, 1979). The penicillin extract is therefore mixed with charcoal in the solvent. The charcoal is separated from the extract on a precoated rotary vacuum filter and washed with the solvent.

D. Crystallization

Through addition of potassium or sodium acetate in the solvent, the penicillin G or V can be crystallized as potassium or sodium salt from the solvent phase. Critical parameters include potassium or sodium concentration, pH value, penicillin concentration, and temperature (Queener and Swartz, 1979). The crystallization yield and the mean crystal size have to be carefully controlled with the aid of these parameters. The penicillin potassium or sodium salt crystals are collected on a rotary vacuum filter.

E. Crystal Cleaning and Drying

The penicillin salt crystals are washed and predried with anhydrous isopropyl alcohol, butyl alcohol, or other volatile solvents (Queener and Swartz, 1979) to remove residual impurities (pigments, unused potassium or sodium acetate, etc.). The crystal slurry is then dosed to a large horizontal vacuum belt filter, washed with fresh anhydrous solvent, and dried with warm air (Queener and Swartz, 1979). At the end of the belt the penicillin salt, at 99.5% purity (de Flines, 1980), is collected in a container. This crude penicillin salt is sold for about $35 per kilogram (de Flines, 1980).

F. Penicillin Losses and Solvent Recovery

In the penicillin purification process several causes of penicillin loss can be mentioned (see Fig. 13): (1) Filtered mycelium, (2) extracted broth filtrate, (3) filtered charcoal, (4) spent solvents, and (5) chemical and enzymatic breakdown during the purification steps. In an optimized (Okabe and Aiba, 1974, 1975), automated computer-controlled (Berkovitch, 1976) purification plant these losses can be minimized. The development of Gist-Brocades' penicillin recovery yield is shown in Figure 14. As can be seen, the recovery yield has increased as a consequence of steadily decreasing losses. The streams containing spent solvents (see Fig. 13) are processed for solvent purification by washing, distillation, and drying. The recovered solvents are reused in the penicillin isolation and purification process.

G. Further Processing

Crystalline penicillin G or V potassium or sodium salt is further processed to pharmaceutical grade. It is also used as an intermediate for the production of semisynthetic penicillins and cephalosporins (see Sec. IX).

IX. MODIFICATION AND MODE OF ACTION

Penicillin G is produced in bulk quantities not only to be sold as the pharmaceutical itself, but mainly to be used as a starting material for the synthesis of other β-lactam antibiotics. The so-called semisynthetic penicillins and a number of cephalosporins are produced from penicillin G. The semisynthetic penicillins are produced with 6-aminopenicilanic acid as the starting material. This is prepared by chemical or enzymatic deacylation of penicillin G. Some

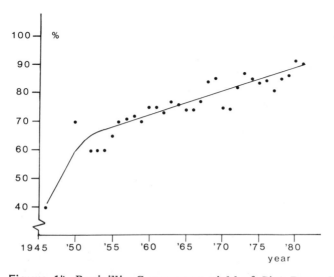

Figure 14 Penicillin G recovery yield of Gist-Brocades.

cephalosporins are produced with penicillin G itself as the starting material. Why these modified antibiotics have been developed will become apparent below in the section dealing with the mode of action of penicillin. We will also describe the use of penicillin G and 6-aminopenicillanic acid as "substrates" for semisynthetic β-lactam antibiotics.

A. Mode of Action

Since its discovery it has been known that penicillin has a lethal effect on bacteria. The mechanism of this lethal effect has been a subject of research for many years.

Park and Johnson (1949) found that uridine nucleotides accumulated in the cytoplasm of penicillin-treated bacteria. They were of a composition identical to those in the cell wall (Park and Strominger, 1957). In an isotonic medium penicillin induced the formation of spheroplasts rather than leading to lysis (Lederberg, 1956). From these results it was concluded that the primary target of penicillin is the bacterial cell wall. The unique structural element of the bacterial cell wall is peptidoglycan. In gram-positive bacteria this compromises some 50% of the cell wall. It also is found in gram-negative bacteria in smaller amounts. The peptidoglycan polymer consists of glycan chains of repeating disaccharide units in a β 1 → 4 linkage. The disaccharide is composed of N-acetyl glucosamine and the lactate derivative N-acetylmuraminic acid. The polymer can be more than 200 disaccharide units in length. In the first step in the biosynthesis of peptidoglycan (Blumberg and Strominger, 1974), a pentapeptide derivative of UDP-N-acetyl-muraminic acid is formed. The pentapeptide attached to the lactyl moiety is composed of three L and D amino acids, which are characteristic for the organism, and two terminal D-alanines. The third amino acid has an unsubstituted amino group (lysine, diaminopimelinic acid). In the next sequence of reactions the N-acetyl glucosamine is coupled to the N-acetylmuraminic acid pentapeptide. The product is structurally linked to phospholipids through the pentapeptide. At this stage the second and third amino acids may become carboxylated or, as in the case of *Staphylococcus aureus,* a pentaglycine is attached to lysine. In the last stage the disaccharide is inserted into peptidoglycan. During this final step of peptidoglycan synthesis a bridge is formed between the carboxyl group of the D-alanine at position 4 and a free amino group of the third residue of an adjacent glycan chain. In the case of the pentaglycine it will be the terminal amino group of the glycine. In the transpeptidase reaction the terminal D-alanine is eliminated; the reaction yields the final cross-linked peptidoglycan complex. The biosynthesis is summarized in Figure 15. In gram-positive bacteria the complex is cross-linked in three dimensions to form concentric sheets. Gram-negative bacteria have a more restricted complex, in that a two-dimensional monolayer is formed. Wise and Park (1965) and Tipper and Strominger (1965) have shown that penicillin works at the level of transpeptidase activity. Based on the structural similarities between penicillin and the D-alanyl-D-alanine portion of nascent peptidoglycan, Tipper and Strominger (1965) postulated that penicillin acts as a substrate analog for transpeptidase TPase. Instead of a transient acyl enzyme intermediate, a covalent stoichiometric penicilloyl enzyme complex is formed, preventing proper transpeptidation activity. As a consequence, a fragile cell wall is formed, leading to cell lysis in a medium that is not kept isotonic.

Penicillin and other β-lactam antibiotics bind to a special class of proteins that are embedded exclusively in the bacterial cytoplasmic membrane, the

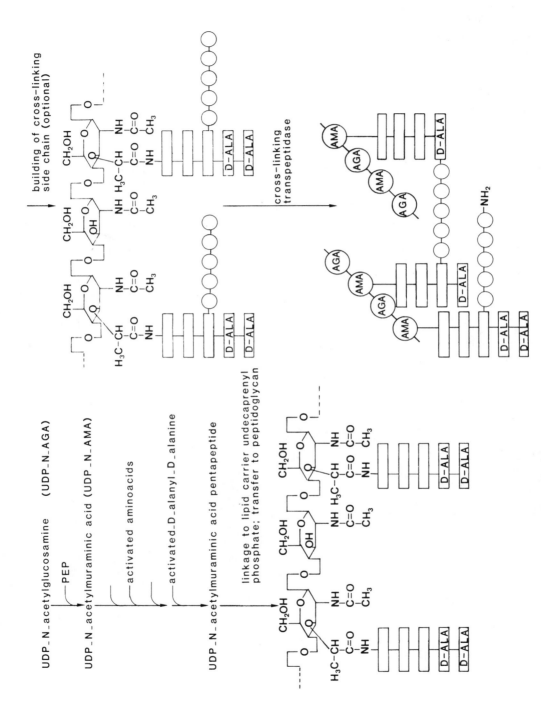

Figure 15 Biosynthesis of the bacterial cell wall.

penicillin-binding proteins (PBPs). In bacteria six to nine classes of binding proteins have been detected by electrophoresis on SDS polyacrylamide gels after labeling the cells with radioactive β-lactams. It is not fully clear what specific enzymatic functions can be attributed to all these PBPs. Upon solubilization of the PBPs from the plasma membrane the peptidoglycan polymerase and ATPase activity is lost; only a D-alanine carboxypeptidase (CPase) activity is retained which is associated with PBPs 5 and 6 (Yocum et al., 1980). These are the major PBPs which in rod-shaped bacteria account for 70–90% of the β-lactams bound to the membrane. CPases remove the terminal D-alanide from the pentapeptide, thus restricting cross-linking of the peptidoglycan complex (Izaki et al., 1968). Whether or not this is the in vivo function of PBPs 5 and 6 has been a matter of debate. The findings that negative mutants of *Escherichia coli* grow normally and possess normally cross-linked peptidoglycan (Suzuki et al., 1977) and that the *Bacillus subtilis* CPase can be inhibited for 95% by penicillin without any effect on the growth (Blumberg and Strominger, 1974) indicate that CPase and TPase in vivo are not identical. Both the identity of the in vivo TPase and the in vivo function of CPase are not yet known. A number of carboxypeptidases have been isolated and (partially) sequenced.

Penicillinase not only binds penicillin, but also rapidly releases the formed penicilloic acid. The enzyme was first discovered by Abraham and Chain (1940) in *E. coli* membranes. Over the years the ability to develop penicillinases and β-lactamases in general has resulted in certain bacterial strains becoming resistant to a number of β-lactam antibiotics. β-lactamases may be acquired by sensitive strains from resistant bacteria as either an extrachromosomal plasmid or transposon or are mediated chromosomally. Some PBPs, in particular CPases, display weak lactamase activity (Tamura et al., 1976; Kozarich and Strominger 1978). Structural analysis of CPases and lactamases in the N-terminal portion of the proteins (Waxman et al., 1982) as well as in the active center has revealed a statistically significant degree of homology. Both in the CPases, reacted with substrate or penicillin, and in the lactamase, reacted with the active site-directed inhibitor 6-bromopenicillanic acid, a serine residue becomes acylated at position 36 (Yocum et al., 1979; Knott-Hunziker et al., 1979; Waxman and Strominger, 1980; Cohen and Pratt, 1980; Fisher et al., 1980).

It has been postulated that β-lactamases have evolved from the cell wall enzymes or, at least, have a common evolutionary origin (Yocum et al., 1980). There is a wide variety of β-lactamases, some constitutive and some inducible. The inducer capacity of a given β-lactam antibiotic is an important parameter to consider in treating an infectious disease.

Two aspects still need to be discussed here. The first is the question of whether β-lactam antibiotics have only one site of action. There is recent evidence that not only the wall is damaged, but that also phospholipid metabolism is affected. Rozgonyi et al. (1981) noted alterations both in the level and in the composition of the phospholipids in methicillin-treated S. sureus cells. Especially the increased level of lysophosphatidyl glycerol may contribute to cell lysis. It is not known whether these effects on the membrane level are primary or secondary effects and whether PBPs are involved in phospholipid metabolism. The last point to be mentioned is the fact that β-lactam antibiotics are especially powerful therapeutic agents, since they act on the peptidoglycan biosynthesis, which is unique for bacteria. Nevertheless, treatment with penicillins and other β-lactam antibiotics can produce an adverse reaction, anaphylaxis. This antigenic and allergic reaction is probably caused by polymers. These are formed from the β-lactam antibiotics upon cleavage of the β-lactam ring (Munro et al., 1976; Ahlstedt et al., 1976, 1977).

The pharmaceutical companies are still developing new semisynthetic penicillins and β-lactam antibiotics. They do this for several reasons. One obvious reason stems from the desire to have β-lactam antibiotics that can be taken orally; they have to be resistant to the acid environment of the stomach. Furthermore, there are structural modifications to be made to enable the antibiotic to pass the intestinal barrier. The classic example of an acid-stable penicillin is penicillin V, with phenoxyacetic acid as a side chain. The other reasons are related to the mode of action and the bacterial defense system. Different β-lactam antibiotics interact with other subsets of PBPs and therefore affect other cell wall enzymes. Generations of β-lactam antibiotics which kill a wide variety of both gram-positive and gram-negative bacteria (broad-spectrum antibiotics) have been developed along these lines. A major goal is the creation of β-lactam antibiotics with side-chain substitutions resulting in a low affinity for certain classes of lactamases. As mentioned before, the capacity to induce certain lactamases is important and related to the nature of the chemical substitutions which have been made in the antibiotic. Lactamase inhibitors are a special line of research, but apart from those derived by chemical synthesis from 6-aminopenicillanic acid, they will be dealt with in other chapters in this volume. Bacteria may have become insensitive to an old generation of antibiotics as a consequence of structural alterations in the active center of the cell wall enzymes; new β-lactam antibiotics will have to deal with these. Some bacterial resistance patterns are associated with morphological changes in the bacterial cell wall. It has been suggested that the resistance is caused by the inability of the β-lactam antibiotic to penetrate the cell wall to reach its target site. Unfortunately we do not yet know how β-lactam antibiotics reach their targets and exactly which target is decisive in their mode of action.

Detailed knowledge of the primary targets in the mode of action and insight into bacterial resistance mechanisms at a molecular level can lead to the development of new generations of β-lactam antibiotics. The old generations of β-lactam antibiotics can also be used successfully when administered together with lactamase inhibitors or cell wall-destabilizing agents. In any case, the need of penicillin G (and 6-aminopenicillanic acid) as basic substrates in the synthesis of semisynthetic penicillins and cephalosporins will remain unchanged.

B. Modification

The production costs of penicillin G, as elaborated in the previous sections, can be kept low by strain improvement, a sophisticated fermentation regime, and relatively cheap processes for separating the product from all other constituents in the broth. Part of penicillin G is used as the pharmaceutical, but the bulk of the world production of penicillin G is converted into other β-lactam antibiotics, the so-called semisynthetic β-lactam antibiotics. Here we will discuss the preparation of semisynthetic β-lactam antibiotics, with an emphasis on the production of bulk intermediates.

1. Semisynthetic Penicillins

The semisynthetic penicillins all contain the 6-aminopenicillanic acid (6-APA) nucleus. 6-APA can be obtained by fermentation in the absence of precursor (Batchelor et al., 1959), but the yield is low and the product contaminated with natural penicillins with a variety of lipidic side chains. 6-APA is commercially obtained by removal of the phenylacetate side chain from penicillin G. This can be accomplished by chemical means or by using acylases of microbial origin. The chemical route involves protection of the 3-carboxyl group, selective cleavage of the amide bond, and removal of the protecting group. Weissenburger and van der Hoeven (1970) have developed a method using silyl protection in which the various reaction steps could be carried out in one vessel without isolating intermediate products. The procedure in its present state (Fig. 16) yields 6-APA with an overall recovery of 90-95%. Sakaguchi and Murao (1950) were the first to report a penicillin amidase in *P. chrysogenum* yielding 6-APA. However, Rolinson et al. (1960) found that the enzyme preferentially hydrolyzed the natural penicillins; with penicillium G no activity was found. Other actinomycetes and filamentous fungi were also shown to possess a stable extracellular enzyme with low activity toward penicillin G. Among representatives of the bacterial genera *Escherichia* and *Alcaligenes*, a membrane-bound enzyme was found with high specificity for penicillin G. It also demonstrated acylase activity (Rolinson et al. (1960). Acylases useful in the preparation of 6-APA can be obtained from a variety of microbial sources (Sebek, 1980; Vandamme, 1980). Several types of enzymatic deacylation processes are used. The penicillin can be brought in contact with intact bacterial cells or with isolated enzymes. In recent years intact cells or isolated enzymes immobilized on a support have been used. Queener and Swartz (1979) have given typical examples of all four variations. The yield of 6-APA in the enzymatic procedures is usually somewhat lower than that of the chemical process; however, immobilized cells or enzymes have the advantage that they can be reused (batch process) or used continuously (column process). Furthermore, the rising prices of chemicals and energy will work in favor of the enzymatic approach. A serious problem is the drop in the local pH upon release of the side chain; at acidic pH 6-APA is unstable. Apparently this problem has been dealt with since then. At present, the majority of the bulk producers employ an enzymatic process.

New penicillins can be made from 6-APA by chemical acylation with the appropriate acid chlorides (Kaiser and Kukolja, 1972; Queener and Swartz, 1979). Well-known examples of these semisynthetic penicillins are propicillin, methicillin, ampicillin, amoxycillin, carbenicillin, oxacillin, cloxacillin, dicloxacillin, and flucloxacillin (Fig. 17). A classification of the semisynthetic

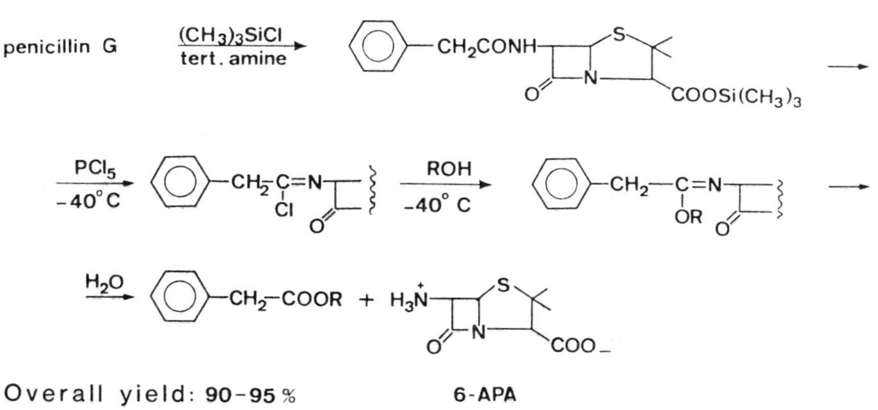

Figure 16 Chemical synthesis of 6-APA from penicillin G.

penicillins according to their clinical usefulness and pharmacokinetic properties has been given by Noguchi and Mitsuhashi (1981). Rolinson (1979) has published a pedigree of 6-APA-derived penicillins. Besides N-acylated 6-β-penicillins, other penicillins have been prepared with substitutions at other positions or other types of bond. None of them became commercially interesting products, except for 6-β-amidinopenicillins (Lund, 1976). Since the bacterial acylases can display both hydrolytic as well synthetic activity, depending on the conditions, semisynthetic penicillins also may be prepared enzymatically (Shimizu et al., 1975a,b); Vandamme, 1981; Kasche and Galunski, 1982; McDougall et al., 1982). At present the enzymatic approach cannot compete with the chemical process (Vandamme, 1980, 1983; Savidge, 1984).

2. Semisynthetic 7-Acylcephalosporins

A new area for the use of penicillin G in semisynthesis has been found in the preparation of cephalosporins. The starting products in the semisynthesis of cephalosporins are 7-aminocephalosporanic acid (7-ACA) and 7-aminodeacetoxycephalosporanic acid (7-ADCA). 7-ACA is produced by chemical de-

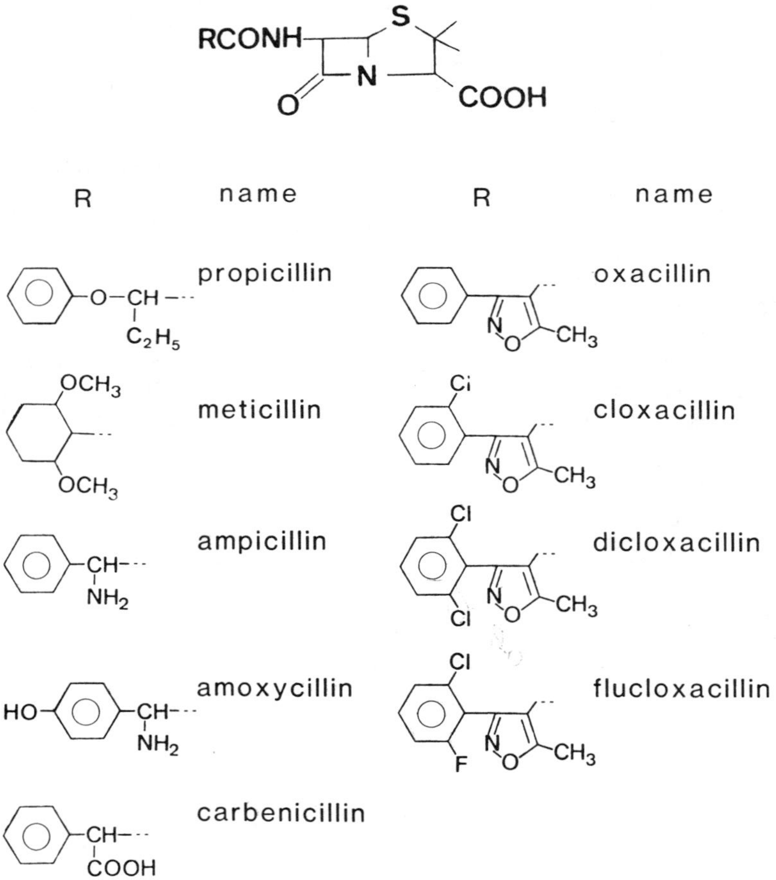

Figure 17 Chemical structures of some semisynthetic penicillins.

acylation of cephalosporin C (Fig. 18). Direct enzymatic deacylation is difficult possibly owing to the D configuration of the α-aminoadipyl side chain of cephalosporin C (Vandamme, 1980, 1981). Recently D-amino acid oxidases of bacterial origin have been isolated which can convert cephalosporin C into a product susceptible to enzymatic deacylation (Toyo Jozo, Japanese Patent No. 117205). Several cephalosporins are prepared by chemical acylation of 7-ACA. 7-ADCA can be prepared from cephalosporin C by hydrogenation into deacetoxycephalosporin C and subsequent deacylation (Stedman et al., 1964). Subsequently mutants have been isolated that are deficient in the final stage of cephalosporin C biosynthesis. They secrete deacetoxycephalosporin C instead of cephalosporin C (Liersch et al., 1976). Deacetoxycephalosporin C can be converted into 7-ADCA, but this is not economical. In the early 1960s it was demonstrated that the thiazolidine ring of penicillin could be expanded, albeit in low yield, into the dihydrothiazine ring of cephalosporin (Morin et al., 1963). The production of 7-ADCA from penicillin G became preferable to other methods, when an efficient procedure with high yields (>70%) was found (de Koning et al., 1975). It involved three steps: oxidation of penicillin G to its sulfoxide, conversion of the sulfoxide to the corresponding deacetoxycephalosporanic acid with transient protection of the carboxyl group, and subsequent deacylation (Fig. 19). From 7-ADCA, medically important deacetoxycephalosporins with various side chains can be prepared. Examples are cephalexin, cephradin, and cephadroxil (Fig. 20).

3. Semisynthetic Lactamase Inhibitors

A number of semisynthetic penicillins such as oxacillin, cloxacillin, dicloxacillin, and flucloxacillin have been selected, since they display lactamase inhibi-

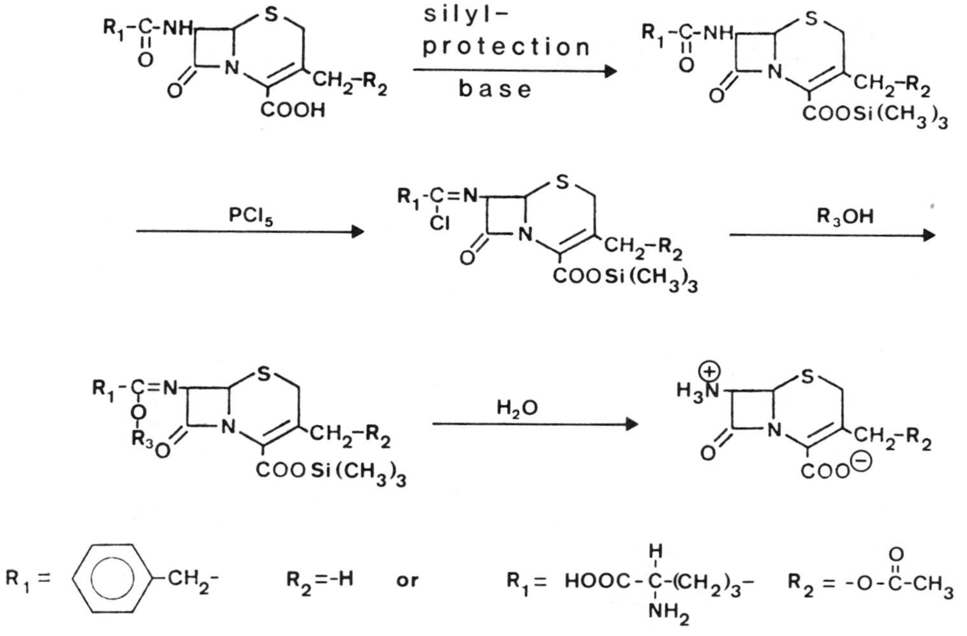

Figure 18 Chemical deacylation of cephalosporin C.

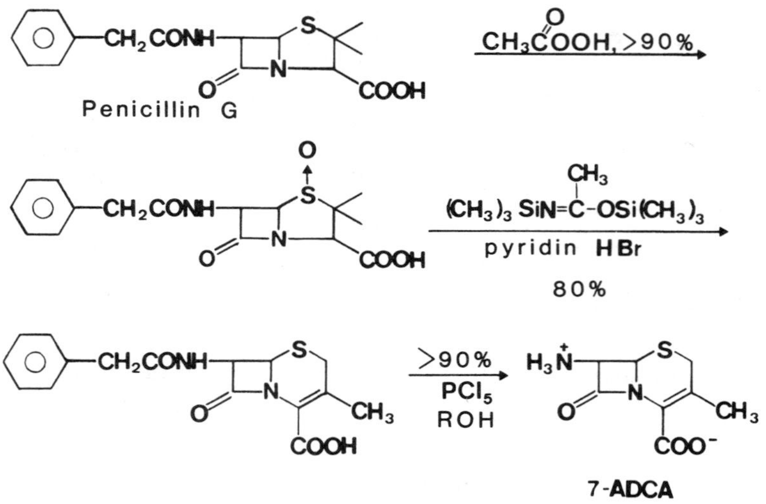
Penicillin G
7-ADCA

Figure 19 Conversion of penicillin G into 7-ADCA.

tory properties. A special class of β-lactam compounds are those with no or only weak antibiotic function, but which are potent inhibitors of β-lactamases. Clavulanic acid and olivanic acid belong to this class; they are produced by fermentation (see Chaps. 6 and 7 in this volume). Some lactamase inhibitors are prepared by chemical manipulation of 6-APA. 6-Bromopeni-

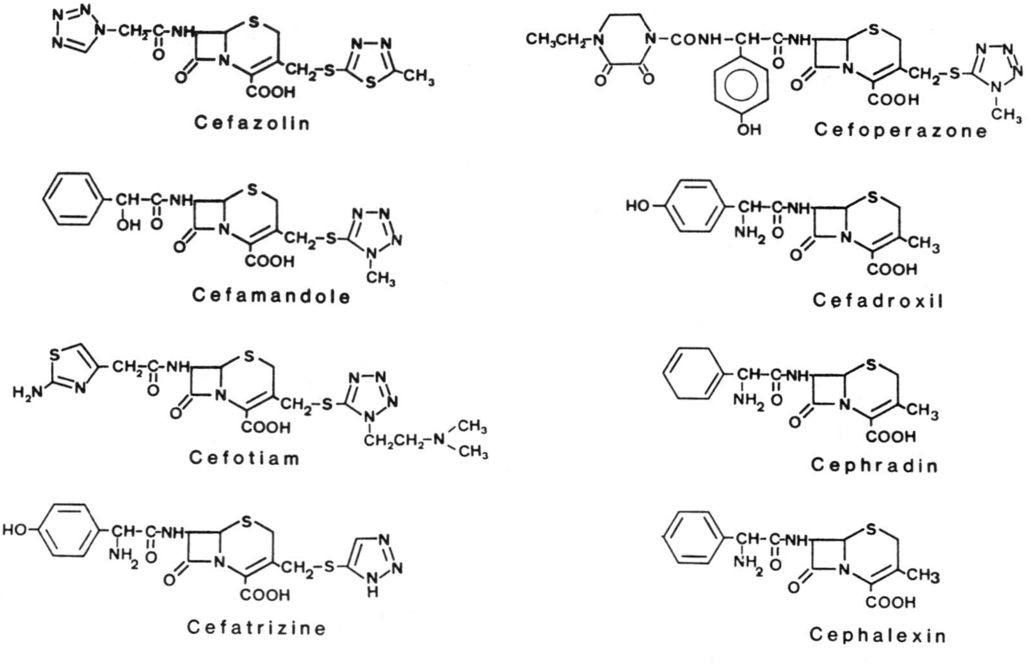
Cefazolin
Cefoperazone
Cefamandole
Cefadroxil
Cefotiam
Cephradin
Cefatrizine
Cephalexin

Figure 20 Chemical structures of some semisynthetic cephalosporins.

cillanic acid has already been mentioned as an active site-directed inhibitor (suicide-killer) of β-lactamase. A recent review on the lactamase inhibitory properties of this compound, other halogenated penicillanic acids, and penicillanic acid sulfone has been given by Cole (1981).

4. Semisynthetic 3'-Substituted Cephalosporins

In penicillins only modifications at position 6 have led to useful pharmaceutical compounds; in cephalosporins both the 7-position as well as the 3'-position can be modified. Examples of this type of cephalosporin are cefazolin, cefatrizine, cefoperazone, cefamandole, and cefotiam (Fig. 20). Thus far such compounds have been produced in general by chemical substitution of the naturally occurring acetoxy group, carried out before or after acylation of the amino group at position 7. In recent years considerable research efforts (Gist-Brocades) have culminated in an alternative preparation method for these structurally modified cephalosporins, whereby the involved 3'-substituted 7-aminocephalosporanic acid intermediates, as depicted in Figure 21, can be obtained from benzylcarbonamidodesacetoxycephalosporin, which in turn is produced from penicillin G by oxidation and ring expansion, as shown in Figure 19. By a series of inventive interventions, whereby the formerly prevailing necessity of intermediate isolation has been considerably reduced, this alternative multistep approach now appears to have reached a competitive level.

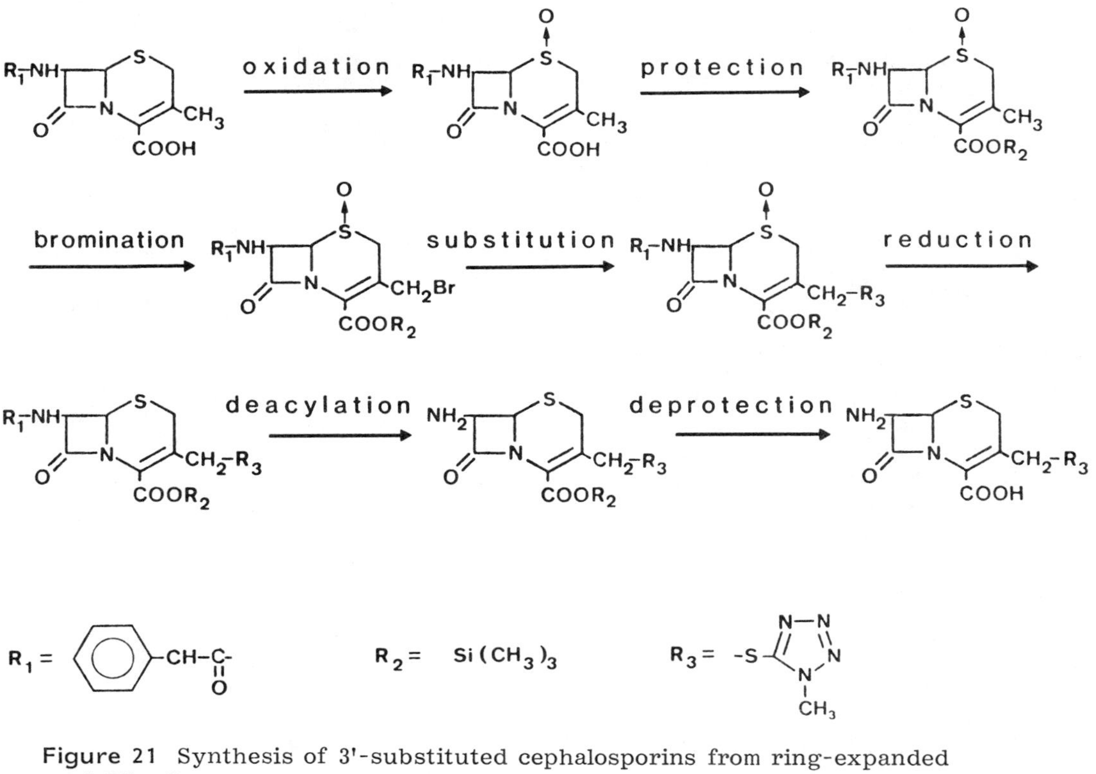

Figure 21 Synthesis of 3'-substituted cephalosporins from ring-expanded penicillin G.

X. FUTURE PROSPECTS

Important phenomena in the penicillin fermentation, which for economic reasons is a fed-batch process, include medium composition and nutrient and precursor feeds (faw materials); mass, heat, and momentum transfer; and morphology of the mold. Mass and heat transfer in the fermentation broth are enhanced by decreasing the broth viscosity or increasing the total mechanical power input. The latter method has several disadvantages. A lower broth viscosity can be obtained by favoring a pelletous mrophology, by mycelium immobilization, and by screening strains with better morphological characteristics and an unchanged or improved penicillin productivity. The introduction of molecular genetics in order to increase rate-limiting enzyme levels by means of gene amplification or enhancement of transcription may increasingly contribute to strain selection programs. As a detailed knowledge of the penicillin biosynthetic pathway and well-characterized mutants are required for the application of these techniques, these research areas will expand. Until now no commercial-scale process with immobilized mycelium has been developed and it is doubtful whether such a development will be successful.

Raw material costs, which may constitute up to 50% of the total manufacturing cost of penicillin G, and production yields are very significant in penicillin production. Since only 10% of the consumed carbon compounds is used for penicillin production (about 65% for maintenance requirements and 25% for mycelial growth), the carbon yield of penicillin can be theoretically increased. This will lead to an important reduction in production costs. In the near future production of 40–50 kg of sodium penicillin G per cubic meter of fermentation broth (about 70,000–80,000 units of penicillin G per milliliter of broth) must be attainable. Fermentation process computer control strategy development can be regarded as important for the near future, preceded by the development of sterilizable sensors. It is to be expected that both pilot-plant and production-scale fermentations will become progressively automated in order to optimize the penicillin production process. The separation and purification process constitutes a minor part of the penicillin manufacturing cost (about 20%). The recovery yield nowadays reaches 90%. No large cost decrease may be expected from recovery process improvement.

It is worth noting that during the last decade major producers like Squibb and Eli Lilly decided to stop their penicillin G fermentation and have turned to the screening and production of new β-lactam antibiotics. The commercial success of this new approach is not yet clear. It is difficult therefore to evaluate what the consequences will be for the remaining old major penicillin producers and the newcomers.

The development of new generations of β-lactam antibiotics will depend more and more on the growing knowledge of the working mechanism, the primary targets, and the pharmacokinetics of penicillin.

ACKNOWLEDGMENTS

We wish to acknowledge the opportunity given to us by the Board of Directors of Gist-Brocades to write this review. Furthermore, we thank all our colleagues for support and stimulating and critical discussions, particularly A. J. P. M. van Asten, Dr. M. J. Bull, J. J. Heijnen, Dr. B. E. Jones, Dr. A. de Leeuw, H. D. Meijer, H. L. Nijenhuis, Dr. J. C. van Suijdam, Dr. P. J. Strijkert, Dr. J. Verweij and J. van Wingerde.

Thanks also to Corry Brunet de Rochebrune, Helma van Dijk, and Ineke Taal, but particularly Henny Rodrigues de Miranda for typing the manuscript.

REFERENCES

Aberhart, D. J. (1977). Biosynthesis of β-lactam antibiotics. *Tetrahedron* *33*:1545–1559.

Abraham, E. P., and Chain, E. (1940). An enzyme from bacteria able to destroy penicillin. *Nature 146*:837.

Abraham, E. P., and Newton, G. G. F. (1965). The cephalosporins. *Adv. Chemother.* *2*:23–90.

Abraham, E. P., Chain, E., Fletcher, C. M., Gardner, A. D., Heatley, N. G., Jennings, M. A., and Florey, H. W. (1941). Further observations on penicillin. *Lancet 2*:177–188.

Abraham, E. P., Baker, W., Chain, E., and Robinson, R. (1943). Pen 103, October 23, 1943.

Abraham, E. P., Huddlestone, J. A., Jayatilake, G. S., O'Sullivan, J., and White, R. L. (1981). Conversion of δ (L-α-Aminoadipyl)-L-cysteinyl-D-valine to isopenicillin N in cell-free extracts of *Cephalosporium acremonium*. In *Recent Advances in the Chemistry of β-Lactam Antibiotics*. G. I. Gregory (Ed.). Royal Society of Chemistry, London, pp. 125–134.

Abraham, E. P., Adlington, R. M., Baldwin, J. E., Crimmin, M. J., Field, L. D., Jayatilake, G. S., and White, R. L. (1982). Monocyclic β-lactam tripeptide, 1-(D-carboxy-2-methylpropyl)-3-L(δ-L-2-aminoadipamido)-4-mercapto-azetidin-2-one, a putative intermediate in penicillin biosynthesis. *J. Chem. Soc. Chem. Commun.* 1130–1132.

Adlington, R. M., Aplin, R. T., Baldwin, J. E., Field, L. D., John, E.-M. M., Abraham, E. P., and White, R. L. (1982). Conversion of $^{17}O/^{18}O$-labeled δ(L-α-aminoadipyl-L-cysteinyl-D-valine into $^{17}O/^{18}O$-labeled isopenicillin N in a cell-free extract of *Cephalosporium acremonium*. *J. Chem. Soc. Chem. Commun.* 137–139.

Adriaens, P., Meeschaert, B., Wuyts, W., Vanderhaeghe, H., and Eyssen, H. (1975). Presence of δ(L-α-aminoadipyl)-L-cysteinyl-D-valine in fermentations of *Penicillium chrysogenum*. *Antimicrob. Agents Chemother.* *8*:638–642.

Aiba, S., and Okabe, M. (1977). A complementary approach to scale-up. Simulation and optimization of microbial processes. *Adv. Biochem. Eng.* *7*:111–130.

Alford, J. S. (1982). Evolution of the fermentation computer system at Eli Lilly & Co. *3rd International Conference on Computer Applications in Fermentation Technology* 1981, Manchester, England, pp. 67–74.

Al-Haffar, S. (1970). Optimum cationique pour la production de pénicilline. *Ann. Physiol. Veg. Univ. Bruxelles 15*:101–114.

Alicino, J. F. (1946). Iodometric method for the assay of penicillin preparations. *Ind. Eng. Chem. Anal. 18*:619–620.

Aly, O. M., and Faust, S. D. (1964). Determination of phenoxyacetic acids with J and phenyl J acids. *Anal. Chem. 36*:2200–2201.

Ahlstedt, S., Kristoffersson, A., Svärd, P. O., Thor, L., and Örtengren, B. (1976). Ampicillin polymers as elicitors of passive cutaneous anaphylaxis. *Int. Arch. Allergy Appl. Immunol. 51*:131–139.

Ahlstedt, S., Kristoffersson, A., Svärd, P. O., and Strannegard, Ö. (1977). Immunological properties of ampicillin polymers. *Int. Arch. Allergy Appl. Immunol. 53*:247–253.

Anderson, R. E., Törnqvist, E. G. M., and Peterson, W. H. (1956). Effect of oil in pilot plant fermentations. *Agric. Food Chem.* 4:556–559.

Arnstein, H. V. R., and Morris, D. (1960a). The structure of a peptide containing α-aminoadipic acid, cystine and valine, present in the mycelium of *Penicillium chrysogenum*. *Biochem. J.* 76:357–361.

Arnstein, H. V. R., Hartman, M., Morris, D., and Toms, E. J. (1960b). Sulphur containing amino acids and peptides in the mycelium of *Penicillium chrysogenum*. *Biochem. J.* 76353–357.

Ashley, M. H. J. (1982). Continuous sterilization of media. *Chem. Eng.* 377:54–58.

Avanzini, E., Magnanelli, D., and Cerrone, G. (1968). An automated assay for penicillin in fermentation media using hydroxyl amine reagent but avoiding penicillinase. *Ann. N.Y. Acad. Sci.* 153:534–540.

Backus, M. P., and Stauffer, J. F. (1955). The production and selection of a family of strains in *Penicillium chrysogenum*. *Mycologia* 47:429–463.

Bahadur, G., Baldwin, J. E., Field, L. D., Lektonen, E. M. M., Usher, J. J., Vallejo, C. A., Abraham, E. P., and White, R. L. (1981a). Direct 1H-N.M.R. observation of the cell-free conversion of δ(L-α-aminoadipyl)-L-cysteinyl-D-valine and δ(L-α-aminoadipyl)-L-cysteinyl-D-(−)-isoleucine into penicillins. *J. Chem. Soc. Chem. Commun.* 917–919.

Bahadur, G. A., Baldwin, J. E., Usher, J. J., Abraham, E. P., Jayatilake, G. S., and White, R. L. (1981b). Cell-free biosynthesis of penicillins. Conversion of peptides into new β-lactam antibiotics. *J. Am. Chem. Soc.* 103:7650–7651.

Bahadur, G., Baldwin, J. E., Wan, T., Jung, M., Abraham, E. P., Huddleston, J. A., and White, R. L. (1981c). On the proposed intermediary of β-hydroxyvaline- and thiazepinone-containing peptides in penicillin biosynthesis. *J. Chem. Soc. Chem. Commun.* 1146–1147.

Bajpai, R. K., and Reuss, M. (1980). A mechanistic model for penicillin production. *J. Chem. Technol. Biotechnol.* 30:332–344.

Bajpai, R. K., and Reuss, M. (1981). Evaluation of feeding strategies in carbon-regulated secondary metabolite production through mathematical modeling. *Biotechnol. Bioeng.* 23:717–738.

Baldwin, J. E., and Wan, T. S. (1981). Penicillin biosynthesis. Retention of configuration at C-3 of valine during its incorporation into the Arnstein tripeptide. *Tetrahedron* 37:1589–1595.

Baldwin, J. E., Johnson, B. L., Usher, J. J., Abraham, E. P., Huddleston, J. A., and White, R. L. (1980). Direct N.M.R. observation of cell-free conversion of (L-α-amino-δ-adipyl)-L-cysteinyl-D-valine into isopenicillin N. *J. Chem. Soc. Chem. Commun.* 1271–1273.

Baldwin, J. E., Jung, M., Usher, J. J., Abraham, E. P., Huddleston, J. A., and White, R. L. (1981). Penicillin biosynthesis: Conversion of deuterated (L-α-amino-δ-adipyl)-L-cysteinyl-D-valine into isopenicillin N by a cell-free extract of *Cephalosporium*. *J. Chem. Soc. Chem. Commun.* 246–247.

Ball, C. (1973). The genetics of *Penicillium chrysogenum*. *Prog. Ind. Microbiol.* 12:47–72.

Ball, C. (1978). Genetics in the development of the penicillin process. In *Antibiotics and other Secondary Metabolites. Biosynthesis and Production.* R. Hütter, T. Leisinger, J. Nüesch, and W. Wehrli (Eds.). Academic, New York, pp. 165–176.

Ball, C. (1982). Genetic approaches to overproduction of β-lactam antibiotics in eukaryotes. In *Overproduction of Microbial Products.* V. Krumphanzl, B. Sikyta, and Z. Vanek (Eds.). Academic, New York, pp. 515–534.

Ball, C., and McGonagle, M. P. (1978). Development and evaluation of a potency index screen for detecting mutants of *Penicillium chrysogenum* having increased penicillin yield. *J. Appl. Bacteriol. 45*:67–74.

Bartholomew, W. H. (1960). Scale-up of submerged fermentations. *Adv. Appl. Microbiol. 2*:289–300.

Bartholomew, W. H., Karow, E. O., Sfat, M. R., and Wilhelm, R. H. (1950). Effect of air flow and agitation rates upon fermentation of *Penicillium chrysogenum* and *Streptomyces griseus*. *Ind. Eng. Chem. 42*:1810–1815.

Batchelor, F. R., Doyle, F. P., Nayler, J. H. C., and Rolinson, G. N. (1959). Synthesis of penicillin: 6-Amino penicillanic acid in penicillin fermentations. *Nature 183*:257–258.

Behrens, O. K. (1949). Biosynthesis of penicillins. In *The Chemistry of Penicillins*. H. T. Clarke, J. R. Johnson, and R. Robinson (Eds.). Princeton University Press, Princeton, N.J., pp. 657–679.

Benedict, R. G., Schmidt, W. H., and Coghill, R. W. (1946). The stability of penicillin in aqueous solution. *J. Bacteriol. 51*:291–292.

Bérdy, J. (1980). Recent advances in and prospects of antibiotic research. *Proc. Biochem. Oct/Nov*:28–35.

Berkovitch, J. (1976). Painless computer control at penicillin plant. *Proc. Eng. Oct*:95.

Bernard, A., and Cooney, C. L. (1981). Studies on oxygen limitations in the penicillin fermentation. 2nd European Congress of Biotechnology, Eastbourne, England. Abstracts of Communications. Soc. Chem. Ind., London, England, p. 21.

Bhavaraju, S. M., and Blanch, H. W. (1976). A model for pellet breakup in fungal fermentations. *J. Ferment. Technol. 54*:466–468.

Birner, J. (1959). Colorimetric estimation of phenoxy methyl-penicillin (penicillin V) and phenoxyacetic acid in samples from penicillin fermentations. *Anal. Chem. 31*:271–273.

Biryukov, V. V., and Kantere, V. M. (1982). Optimization of secondary metabolite production using on-line real-time computer. In *Overproduction of Microbial Products*. V. Krumphanzl, B. Sikyta, and Z. Vanek (Eds.). Academic, New York, pp. 687–701.

Blaha, J. M., Knevel, A. M., and Hem, S. L. (1975). High pressure liquid chromatographic analysis of penicillin G potassium and its degradation products. *J. Pharm. Sci. 64*:1384–1386.

Blanch, H. W., and Bhavaraju, S. M. (1976). Non-Newtonian fermentation broths: Rheology and mass transfer. *Biotechnol. Bioeng. 18*:745–790.

Blumberg, P. M., and Strominger, J. L. (1974). Interaction of penicillin with the bacterial cell wall: Penicillin binding proteins and penicillin sensitive enzymes. *Bacteriol. Rev. 38*:291–335.

Bonting, S. L., and Pont, J. J. H. H. M. de. (1981). *New Comprehensive Biochemistry*, Vol. 2. Membrane transport. Elsevier/North-Holland, Amsterdam.

Bowden, J. P., and Peterson, W. H. (1946). The role of corn-steep liquor in production of penicillin. *Arch. Biochem. 9*:387–399.

Boxer, G., and Everett, P. (1949). Colorimetric determination of benzylpenicillin. *Anal. Chem. 21*:670–673.

Brown, W. E., and Peterson, W. H. (1950). Factors affecting production of penicillin in semi-pilot plant equipment. *Ind. Eng. Chem. 42*:1769–1774.

Brundidge, S. P., Gaeta, F. C. A., Hook, D. J., Sapino, C., Elander, R. P., and Morin, R. B. (1980). Association of δ-oxo-piperidine-2-carboxylic acid with penicillin V production in *Penicillium chrysogenum* fermentations. *J. Antibiot.* 33:1348–1351.

Brunner, R., and Rohr, M. (1975). Phenacyl: Coenzyme A ligase. *Methods Enzymol.* 43:476–481.

Brunner, R., Rohr, M., and Zinner, M. (1968). Zur Biosynthese des Penicillins. *Hoppe Seyler's Z. Physiol. Chem.* 349:95–103.

Buckland, R. C., and Fastert, H. (1982). Analysis of fermentation exhaust gas using a mass spectrometer. 3rd International Conference on Computer Applications in Fermentation Technology, 1981, Manchester, England.

Bull, A. T., and Trinci, A. P. J. (1977). The physiology and metabolic control of fungal growth. *Adv. Microbiol. Physid.* 15:1–84.

Bundgaard, H., and Hansen, J. (1981). Nucleophilic phosphate-catalyzed degradation of a penicilloyl phosphate intermediate and transformation of ampicillin to a piperazinedione. *Int. J. Pharm.* 9:273–283.

Bundgaard, H., and Ilver, K. (1972). A new spectrophotometric method for the determination of penicillins. *J. Pharm. Pharmacol.* 24:790–794.

Bungay, H. R. (1963). Economic definition of continious fermentation goals. *Biotechnol. Bioeng.* 5:1–7.

Bycroft, B. W., Taylor, P. M., and Corbett, K. (1981). The role of cysteinyl-valine peptides as precursors of penicillin G in *Penicillium chrysogenum*. In *Recent Advances in the Chemistry of β-Lactam Antibiotics*. G. I. Gregory (Ed.). Royal Society of Chemistry, London, pp. 135–141.

Calam, C. T. (1967). Media for industrial fermentations. *Proc. Biochem. June*:19–22,46.

Calam, C. T. (1969). The evaluation of mycelial growth. In *Methods in Microbiology*, Vol. 1. J. R. Norris and D. W. Ribbons (Eds.). Academic, New York, pp. 567–591.

Calam, C. T. (1976). Starting investigational and production cultures. *Proc. Biochem. April*:7–12.

Calam, C. T. (1979). Secondary metabolism as an expression of microbial growth and development. *Folia Microbiol.* 24:276–285.

Calam, C. T. (1982). Factors governing the production of penicillin by *Penicillium chrysogenum*. In *Overproduction of Microbial Products*. V. Krumphanzl, B. Sikyta, and Z. Vanek (Eds.). Academic, New York, pp. 89–95.

Calam, C. T., and Hockenhull, D. J. D. (1949). The production of penicillin in surface culture using chemically defined media. *J. Gen. Microbiol.* 3:19–31.

Calam, C. T., and Ismail, B. A.-K. (1980a). Calculation of growth from carbor dioxide output and problems with the early growth stages in penicillin production. In *7th Symposium on Continuous Cultivation of Microorganisms*, Prague, 1978, B. Sikyta, Z. Fencl, and V. Polacek (Eds.). Czechoslovak Academy of Sciences, Prague, pp. 745–751.

Calam, C. T., and Ismail, B. A.-K. (1980b). Investigation of factors in the optimization of penicillin production. *J. Chem. Technol. Biotechnol.* 30:249–262.

Calam, C. T., and Russel, D. W. (1973). Microbial aspects of fermentation process development. *J. Appl. Chem. Biotechnol.* 23:225–237.

Calam, C. T., and Smith, G. M. (1981). Regulation of the biochemistry and morphology of *Penicillin chrysogenum*, in relation to initial growth. *FEMS Microbiol. Lett.* 10:231–234.

Calam, C. T., Driver, N., and Bowers, R. H. (1951). Studies in the production of penicillin, respiration and growth of *Penicillium chrysogenum* in submerged culture in relation to agitation and oxygen transfer. *J. Appl. Chem. (London)* 1:209–216.

Camici, L., Sermonti, G., and Chain, E. B. (1952). Observation on *Penicillium chrysogenum* in submerged culture. *Bull. WHO* 6:265–275.

Carilli, A., Chain, E. B., Gualandi, G., and Morisi, G. (1961). Aeration studies III. Continuous measurement of dissolved oxygen during fermentation in large fermentors. *Sci. Rep. Ist. Super. Sanita* 1:177–189.

Chain, E. (1948). The chemistry of penicillin. *Annu. Rev. Biochem.* 17:657–704.

Chain, E. (1980). A short history of the penicillin discovery from Fleming's early observations in 1929 to the present time. In *The History of Antibiotics. A Symposium.* J. Parascandola (Ed.). American Institute of the History of Pharmacy, Madison, Wisconsin, pp. 15–29.

Chain, E., Florey, H. W., Gardner, A. D., Heatley, N. G., Jennings, M. A., Ewing, J. O. M., and Sanders, A. G. (1940). Penicillin as a chemotherapeutic agent. *Lancet* 2:226–228.

Chang, L. T., and Elander, R. P. (1979). Rational selection for improved cephalosporin C productivity in strains of *Acremonium chrysogenum* Gamus. *Dev. Ind. Microbiol.* 20:367–379.

Charles, M. (1978). Technical aspects of the rheological properties of microbial cultures. *Adv. Biochem. Eng.* 8:1–62.

Chaturbhuj, K., Gopalkrishnan, K. S., and Ghosh, D. (1961). Studies on the feed rate for precursor and sugar in penicillin fermentation. *Hind. Antibiot. Bull.* 3:144–151.

Choi, C. Y., and Park, S. Y. (1981). The parametric sensitivity of the optimal fed-batch fermentation policy. *J. Ferment. Technol.* 59:65–71.

Clutterbuck, P. W., Lovell, R., and Raistrick, H. (1932). The formation from glucose by members of the *Penicillium chrysogenum* series of a pigment, an alkali-soluble protein and penicillin—The antibacterial substance of Fleming. *Biochem. J.* 26:1907–1918.

Coghill, R. D. (1944). Penicillin. Science's Cinderella. *Chem. Eng. News* 22:588–593.

Cohen, S. A., and Pratt, R. F. (1980). Inactivation of *Bacillus cereus* β-lactamase I by 6β-bromopenicillanic acid: mechanism. *Biochemistry* 19:3996–4003.

Cole, M. (1981). Inhibitors of bacterial β-lactamases. *Drugs Future* 6:697–727.

Colon, A. A., Herpich, G. E., Neuss, J. D., and Frediani, H. A. (1949). The assay of relatively pure benzylpenicillin by ultraviolet absorption. *J. Am. Pharm. Assoc.* 38:138–142.

Constantinides, A., and Rai, V. R. (1974). Application of the continuous maximum principle to fermentation processes. *Biotechnol. Bioeng. Symp.* 4:663–680.

Constantinides, A., Spencer, J. L., and Gaden, E. L., Jr. (1970a). Optimization of batch fermentation processes. I. Development of mathematical models for batch penicillin fermentations. *Biotechnol. Bioeng.* 12:803–830.

Constantinides, A., Spencer, J. L., and Gaden, E. L., Jr. (1970b). Optimization of batch fermentation processes. II. Optimum temperature profiles for batch penicillin fermentations. *Biotechnol. Bioeng.* 12:1081–1098.

Cooney, C. L. (1979). Conversion yields in penicillin production: Theory vs practice. *Proc. Biochem.* May:31–33.

Cooney, C. L., and Acevedo, F. (1977). Theoretical conversion yields for penicillin synthesis. *Biotechnol. Bioeng.* *19*:1449–1462.

Cooney, C. L., and Mou, D.-G. (1982). Application of computer monitoring and control to the penicillin fermentation. 3rd International Conference on Computer Applications in Fermentation Technology, Manchester, England, 1981, pp. 217–225.

Court, J. R., and Pirt, S. J. (1976). The application of fed batch culture to the penicillin fermentation. *5th International Fermentation Symposium*. Berlin 1976. H. Dellweg (Ed.). Verlag Versuchs, Berlin, p. 127.

Court, J. R., and Pirt, S. J. (1977). Fed batch culture of *Penicillium chrysogenum* for penicillin production. *Proc. Soc. Gen. Microbiol.* *4*:139–140.

Court, J. R., and Pirt, S. J. (1981). Carbon and nitrogen-limited growth of *Penicillium chrysogenum* in fed batch culture: The optimal ammonium ion concentration for penicillin production. *J. Chem. Technol. Biotechnol.* *31*:235–240.

Danielsson, B., and Mosbach, K. (1982). The prospects for enzyme-coupled probes in fermentation. Paper presented at the 3rd International Conference on Computer Application in Fermentation Technology, Manchester, England, 1981, pp. 137–145.

Deindoerfer, F. H., and Gaden, E. L., Jr. (1955). Effects of liquid physical properties on oxygen transfer in penicillin fermentation. *Appl. Microbiol.* *3*:253–257.

Demain, A. L. (1957). Stability of benzylpenicillin during biosynthesis. *Antibiot. Chemother.* *7*:361–362.

Demain, A. L. (1959). The mechanism of penicillin biosynthesis. *Adv. Appl. Microbiol.* *1*:23–47.

Demain, A. L. (1966). Biosynthesis of penicillins and cephalosporins. In *Biosynthesis of Antibiotics*. J. F. Snell (Ed.). Academic, New York, pp. 29–94.

Demain, A. L. (1974). Biochemistry of penicillin and cephalosporin fermentations. *Lloydia* 37147–167.

Demain, A. L. (1981). Industrial microbiology. *Science* *214*:987–995.

Demain, A. L., Kennel, Y. M., and Aharonowitz, Y. (1979). Carbon catabolite regulation of secondary metabolism. In *Microbial Technology: Current State, Future Prospects*. A. T. Bull, D. C. Ellwood, and C. Ratledge (Eds.). Cambridge University Press, Cambridge, pp. 163–186.

Dion, W. M., Carilli, A., Sermonti, G., and Chain, E. B. (1954). The effect of mechanical agitation on the morphology of *Penicillium chrysogenum* Thom in stirred fermentors. *Rend. Ist. Super. Sanita* *17*:187–205.

Ditchburn, P., Giddings, B., and MacDonald, K. D. (1974). Rapid screening for the isolation of mutants of *Aspergillus nidulans* with increased penicillin yields. *J. Appl. Bacteriol.* *37*:515–523.

Donovick, R. (1960). Some considerations of bioengineering from a microscopic viewpoint. *Appl. Microbiol.* *8*:117–122.

Dynham, J. M. (1972). Potassium phenoxymethyl penicillin. In *Analytical Profiles of Drug Substances*, Vol. 1. K. Florey (Ed.). Academic, New York, pp. 249–300.

Dunn, I. J., and Mor, J. R. (1975). Variable-volume continuous cultivation. *Biotechnol. Bioeng.* *17*:1805–1822.

Elander, R. P. (1967a). Enhanced penicillin biosynthesis in mutant and recombinant strains of *Penicillium chrysogenum*. *Abh. Dtsch. Akad. Wissensch. Berlin* *2*:403–423.

Elander, R. P. (1967b). Enhanced penicillin synthesis in mutant and recombinant strains of *Penicillium chrysogenum*. In *Induced Mutations and Their*

Utilization. H. Slubbe (Ed.). Akademie Verlag, Berlin, pp. 403–423.

Elander, R. P. (1980). New genetic approaches to industrially important fungi. *Biotechnol. Bioeng.* 22:49–61.

Elema, B. (1970). One hundred years of yeast research. Koninklijke Nederlandse Gist en Spiritusfabrick NV, Delft.

El-Marsafy, M., Abdel-Akher, M., and El-Saied, H. (1975). Evaluation of various brands of corn-steep liquor for penicillin production. *Stärke* 27:91–93.

El-Marsafy, M., Abdel-Akher, M., and El-Saied, H. (1977). Effects of media composition on the penicillin production. *Zentralbl. Bakteriol. II Abt.* 132:117–122.

El-Saied, H. M., El-Marsafy, M. K., and Darwish, M. M. (1976). Chemically defined medium for penicillin production. *Stärke 28*:282–284.

El-Saied, H. M., El-Din, S. B., and Akher, M. A. (1977). Replacement of corn-steep liquor in penicillin fermentation. *Proc. Biochem.* Oct.:31–32, 36.

Enfors, S. O., and Nilsson, H. (1979). Design and response characteristics of an enzyme electrode for measurement of penicillin in fermentation broth. *Enzyme Microb. Technol.* 1:260–264.

Erickson, R. C., and Dean, D. (1966). Acylation of 6-aminopenicillanic acid in *Penicillium chrysogenum*. *Appl. Microbiol.* 14:1047–1048.

Falch, E. A., Hjortkjaer, P., and Pedersen, P. (1982). Computer applications in industry: Justification and planning. Paper presented at the 3rd International Conference on Computer Applications in Fermentation Technology, Manchester, England, 1981, pp. 207–215.

Fawcett, P. A., and Abraham, E. P. (1975). δ(L-α-Aminoadipyl)cysteinyl-valine synthetase. *Methods Enzymol.* 43:471–473.

Fawcett, P. A., and Abraham, E. P. (1976). Biosynthesis of penicillin and cephalosporins. *Biosynthesis* 4:248–265.

Fawcett, P. A., Usher, J. J., and Abraham, E. P. (1975). Behavior of tritrium-labeled isopenicillin-N and 6-aminopenicillanic acid as potential penicillin precursors in an extract of *Penicillium chrysogenum*. *Biochem. J.* 151:741–746.

Felix, H., Nüesch, J., and Wehrli, W. (1980). A convenient method for permeabilizing the fungus *Cephalosporium acremonium*. *Anal. Biochem.* 103:81–86.

Ferrari, A., Russo-Alesi, F. M., and Kelly, J. M. (1959). A completely automated system for the chemical determination of streptomycin and penicillin in fermentation media. *Anal. Chem.* 31:1710–1717.

Finn, R. K. (1954). Agitation-aeration in the laboratory and in industry. *Bacteriol. Rev.* 18:254–274.

Fish, N. M., Vardar, F., and Lilly, M. D. (1981). Effect of dissolved gas concentrations on microbial product formation. 2nd European Congress of Biotechnology, Eastbourne, England. Abstracts of communications, Soc. Chem. Ind., London, England, p. 77.

Fisher, J., Belasco, J. G., Khosla, S., and Knowles, J. R. (1980). β-Lactamase proceeds via an acyl-enzyme intermediate: Interaction of the *Escherichia coli* RTEM enzyme with cefoxitin. *Biochemistry* 19:2895–2901.

Fishman, V. M., and Biryukov, V. V. (1974). Kinetic model of secondary metabolite production and its use in computation of optimal conditions. *Biotechnol. Bioeng. Symp.* 4:647–662.

Fleischaker, R. J., Weaver, J. C., and Sinskey, A. J. (1981). Instrumentation for process control in cell culture. *Adv. Appl. Microbiol* 27:137–167.

Fleming, A. (1929). On the antibacterial action of cultures of a *Penicillium*, with special reference to their use in the isolation of *B. influenza*. *Br. J. Exp. Pathol. 10*:226–236.

Flines, J. de (1980). Biotechnology—Its past, present and future. In *Biotechnology a Hidden Past, a Shining Future*. A. Verbraeck (Ed.). 13th International TNO Conference, Rotterdam, The Netherlands. Netherlands Central Organization for Applied Scientific Research TNO, The Hague, pp. 12–17.

Foster, J. W. (1949). *Chemical Activities of Fungi*. Academic, New York, pp. 587–593.

Foster, J. W., Woodruff, H. B., and McDaniel, L. E. (1943). Microbiological aspects of penicillin. III. Production of penicillin in surface cultures of *Penicillium notatum*. *J. Bacteriol. 46*:421–433.

Foster, J. W., Woodruff, H. B., Perlman, D., McDaniel, L. E., Wilker, B. L. and Hendlin, D. (1946). Microbiological aspects of penicillin. IX. Cottonseed meal as a substitute for corn steep liquor in penicillin production. *J. Bacteriol. 51*:695–698.

Fowler, B., and Robins, A. J. (1972). Methods for the quantitative analysis of sulphur containing compounds in physiological fluids. *J. Chromatogr. 72*:105–111.

Fox, R. I. (1978). The applicability of published scale-up criteria to commercial fermentation process. 1st European Congress on Biotechnology, Interlaken, Switzerland. Preprints, part 1, DECHEMA, Frankfurt, FRG, pp. 80–83.

Fuska, J., and Welwardova, F. (1969). Selection of productive strains of *Penicillium chrysogenum*. *Biologia 24*:691–698.

Gaden, E. L. (1959). Fermentation process kinetics. *J. Biochem. Microbiol. Technol. Eng. 1*:413–429.

Gbewonyo, K., and Wang, D. I. C. (1979). Growth of mycelial microorganisms on microscopic beads. Paper presented at the 178th Annual Meeting of the American Chemical Society, Washington, D.C., September.

Gbewonyo, K., and Wang, D. I. C. (1981). Enhanced performance of the penicillin fermentation using microbeads. 182nd ACS national meeting. New York. American Chemical Society, New York, N.Y., Micr. 52.

Ghosh, D., and Sanguli, B. N. (1961). Production of penicillin with waste mycelium of *Penicillium chrysogenum* as the sole source of nitrogen. *Appl. Microbiol. 2*:252–255.

Giona, A. R., Marelli, L., Toro, L., and De Santis, R. (1976a). Kinetic analysis of penicillin production by semicontinuous fermentors. *Biotechnol. Bioeng. 18*:473–492.

Giona, A. R., De Santis, R., Marelli, L., and Toro, L. (1976b). The influence of oxygen concentrations and of specific rate of growth on the kinetics of penicillin production. *Biotechnol. Bioeng. 18*:493–512.

Goodall, R. R., and Davies, R. (1961). Automatic determination of penicillin in fermentation broth. An improved iodometric assay. *Analist 86*:326–335.

Goodwin, B. L., Ruthven, C. R. J., and Sandler, M. (1975). Gas chromatographic assay of phenylacetic acid in biological fluids. *Clin. Chim. Acta 62*443–446.

Gordee, E. Z., and Day, L. E. (1972). Effect of exogenous penicillin on penicillin biosynthesis. *Antimicrob. Agents Chemother. 1*:315–322.

Gordon, J. J., Grenfell, E., Knowles, E., Legge, B. J., McAllister, R. C. A., and White, T. (1947). Methods of penicillin production in submerged culture on a pilot-plant scale. *J. Gen. Microbiol. 1*:187–202.

Goulden, S. A., and Chattaway, F. W. (1969). End product control of aceto-hydroxy acid synthetase by valine in *Penicillium chrysogenum* Q-176 and a high penicillin-yielding mutant. *J. Gen. Microbiol.* 59:111–118.

Hartmann, V., and Rödiger, M. (1976). Anwendung der Hochdruck Flüssig-keits Chromatographie zur Analyse von Penicillinen und Cephalosporinen. *Chromatographia* 9:266–272.

Hatch, R. T. (1982). Computer applications for analysis and control of fer-mentation. *Ann. Rep. Ferment. Proc.* 5:291–311.

Heatley, N. G. (1944). A method for the assay of penicillin. *Biochem. J.* 38: 61–65.

Heatley, N. G. (1949). The assay of antibiotics. In *Antibiotics*, Vol. 1. H. W. Florey, E. Chain, N. G. Heatley, M. A. Jennings, A. G. Sanders, E. P. Abraham, and M. E. Florey (Eds.). Oxford University Press, London, pp. 110–199.

Hegewald, E., Wolleschensky, B., Guthke, R., Neubert, M., and Knorre, W. A. (1981). Instabilities of product formation in a fed-batch culture of *Penicillium chrysogenum*. *Biotechnol. Bioeng.* 23:1563–1572.

Heijnen, J. J. (1981). Application of the macroscopic electric charge balance in fermentation modeling. *Biotechnol. Bioeng.* 23:1133–1144.

Heijnen, J. J., Roels, J. A., and Stouthamer, A. H. (1979). Application of balancing methods in modeling the penicillin fermentation. *Biotechnol. Bioeng.* 21:2175–2201.

Hekster, Y. A., Baars, A. M., Vree, T. B., Klingeren, B. van, and Rutgers, A. (1979). *Pharm. Weekbl.* 1:695–700.

Helfand, W. H., Woodruff, H. B., Coleman, K. M. H., and Cowen, D. L. (1980). Wartime industrial development of penicillin in the United States. In *The History of Antibiotics. A Symposium*. J. Parascandola (Ed.). American Institute of the History of Pharmacy, Madison, Wisconsin, pp. 31–56.

Henstock, H. (1949). Hydroxylamine assay of penicillin. *Nature* 164:139–140.

Hersbach, G. J. M. (1982). Penicillin production. Paper presented at the Meeting of the Netherlands Society for Microbiology, Section for Techni-cal Microbiology, and the Netherlands Biotechnological Society, March 17th, Wageningen, The Netherlands.

Hikuma, M., Yasuda, T., Karube, I., and Suzuki, S. (1981). Application of microbial sensors to the fermentation process. *Biochem. Eng.* 2:307–319.

Hockenhull, D. J. D. (1959). The influence of medium constituents on the bio-synthesis of penicillin. *Progr. Ind. Microbiol.* 1:3–27.

Hockenhull, D. J. D. (1963). Antibiotics. In *Biochemistry of Industrial Micro-organisms*. C. Rainbow and A. M. Rose (Eds.). Academic, New York, pp. 227–299.

Hockenhull, D. J. D., and Mackenzie, R. M. (1968). Present nutrient feeds for penicillin fermentation on defined media. *Chem. Ind.* xx:607–610.

Hockenhull, D. J. D., Herbert, M., Walker, A. D., Wilkin, G. D., and Winder, F. G. (1954). Organic acid metabolism of *Penicillium chryso-genum*. 1. Lactate and acetate. *Biochem. J.* 56:73–82.

Hoogerheide, J. C. (1980). The penicillin legend remembered. *Chim. Ind.* (Milan) 62:440–445.

Hosler, P., and Johnson, M. J. (1953). Penicillin from chemically defined media. *Ind. Eng. Chem.* 45:871–874.

Hughes, D. W., Vilim, A., and Wilson, W. L. (1976). Chemical and physical analysis of antibiotics. Part III. *Can. J. Pharm. Sci.* 11:97–108.

Hult, K., Veide, A., and Gatenbeck, S. (1980). The distribution of the

NADPH regenerating mannitol cycle among fungal species. *Arch. Microbiol.* *128*:253–255.

Humphrey, A. E. (1977). The use of computers in fermentation systems. *Proc. Biochem. March*:19–25.

Humphrey, A. E., and Jefferis, R. R., III. (1973). Optimization of batch fermentation processes. 4th GIAM Meeting Global Impacts of Applied Microbiology, Sao Paulo, Brazil. J. S. Furtada (Ed.)., pp. 767–774.

Ichikawa, T., Date, M., Ishikura, T., and Ozaki, A. (1971). Improvement of kasugamycin producing strain by the agar piece method and the prototroph method. *Folia Microbiol. 16*:218.

Ingram, G. I. C., Foxell, A. W. H., and Armitage, P. (1953). Nature of the dose response curve in penicillin plate assay. *Antibiot. Chemother.* *3*:1247–1257.

Itsygin, S. B., Biryukov, V. V., and Lurie, L. M. (1976). Respiration rate as a parameter for the process control of penicillin biosynthesis. *Pharm. Chem. J. 10*:119–123.

Itsygin, S. B., Biryukov, V. V., Lurie, L. M., and Berezovskaya, A. I. (1977). Studies on effect of microconcentrations of glucose on biosynthesis of penicillin. *Antibiotiki (Moscow) 22*:581–587.

Izaki, K., Matsuhashi, M., and Strominger, J. L. (1968). Biosynthesis of the peptidoglycan of bacterial cell walls XIII. *J. Biol. Chem. 243*:3180-3192.

Jarvis, G. F., and Johnson, M. J. (1947). The role of the constituents of synthetic media for penicillin production. *J. Am. Chem. Soc. 69*:3010–3017.

Jarvis, F. G., and Johnson, M. J. (1950). The mineral nutrition of *Penicillium chrysogenum* Q176. *J. Bacteriol. 59*:51–60.

Jensen, E. B., Nielsen, R., and Emborg, C. (1981). The influence of acetic acid on penicillin production. *Eur. J. Appl. Microbiol. Biotechnol. 13*: 29–33.

Jensen, S. E., Westlake, D. W. S., and Wolfe, S. (1982). Cyclization of δ(L-α-aminoadipyl)L-cysteinyl-D-valine to penicillins by cell free extracts of *Streptomyces clavuligerus*. *J. Antibiot. 35*:483–490.

Johnson, M. J. (1946). Metabolism of penicillin producing molds. *Ann. N.Y. Acad. Sci. 48*:57–66.

Johnson, M. J. (1952). Recent advances in penicillin fermentation. *Bull. WHO 6*:99–121.

Jones, R. C., and Anthony, R. M. (1977). The relationship between nutrient feed rate and specific growth rate in fed batch cultures. *Eur. J. Appl. Microbiol. 4*:87–92.

Kaiser, G. V., and Kukolja, S. (1972). Modifications of the β-lactam system. In *Cephalosporins and Penicillins: Chemistry and Biology*. E. H. Flynn (Ed.). Academic, New York, pp. 74–133.

Karow, E. O., Bartholomew, W. H., and Sfat, M. R. (1953). Oxygen transfer and agitation in submerged fermentations. *J. Agric. Food Chem. 1*:302–306

Karube, I., Suzuki, S., and Vandamme, E.J. (1984). This volume, pp. 761-780.

Kasche, V., and Galunsky, B. (1982). Ionic strength and pH effects in the kinetically controlled synthesis of benzylpenicillin by nucleophilic deacylation of free and immobilized phenyl-acetyl-penicillin amidase with 6-amino penicillanic acid. *Biochem. Biophys. Res. Commun. 104*:1215–1222.

Kaszab, I., and Enfors, S. O. (1981). The γ-glutamyl-cysteine synthetase: A possible connecting point of primary and secondary metabolism of *Penicillium chrysogenum*. *2nd European Congress of Biotechnology*, Eastbourne, England. Abstracts of communications, Soc. Chem. Ind., London, England, p. 77.

Keller, R., and Dunn, I. J. (1978). Fed-batch microbial culture: Models, errors and applications. *J. Appl. Chem. Biotechnol.* 28:508–514.

Kiese, M. (1943). Chemische Therapie mit antibakteriellen Stoffen aus niederen Pilzen und Bakterien. *Klin. Wochenschr.* 22:505.

Knott-Hunziker, V., Walek, S. G., Orlek, B. S., and Sammes, P. G. (1979). Penicillinase active sites: Labelling of serine 44 in β-lactamase I by 6β-bromo-penicillanic acid. *FEBS Lett.* 99:59–61.

Koffler, H., Emerson, R. L., Perlman, D., and Burris, R. H. (1945). Chemical changes in submerged penicillin fermentations. *J. Bacteriol.* 50: 516–548.

Koffler, H., Knight, S. G., and Frazier, W. C. (1947). The effect of certain mineral elements on the production of penicillin in shake flasks. *J. Bacteriol.* 53:115–123.

Koffler, H., Knight, S. G., Frazier, W. C., and Burris, R. H. (1946). Metabolic changes in submerged penicillin fermentations on synthetic media. *J. Bacteriol.* 51:385–392.

König, B., Kalischewski, K., and Schügerl, K. (1979). Foam behavior of biological media. III. *Penicillium chrysogenum* cultivation foam. *Eur. J. Appl. Microbiol. Biotechnol.* 7:251–258.

König, B., Seewald, C., and Schügerl, K. (1981a). Untersuchungen zur Penicillin-Produktion unter verfahrenstechnischen Gesichtspunkten. *Chem. Ing. Techn.* 53:56–57.

König, B., Seewald, C., and Schügerl, K. (1981b). Process engineering investigations of penicillin production. *Eur. J. Appl. Microbiol. Biotechnol.* 12:205–211.

König, B., Seewald, C., and Schügerl, K. (1981c). Strategien zur Penicillin-Fermentation. *Chem. Ing. Techn.* 53:391.

König, B., Schügerl, K., and Seewald, C. (1982a). Strategies for penicillin fermentation in tower-loop reactors. *Biotechnol. Bioeng.* 24:259–280.

König, B., Seewald, C., and Schügerl, K. (1982b). Penicillin production in a bubble column air lift loop reactor. In *Advances in Biotechnology*, Vol. 1, Scientific and Engineering Principles. Pergamon Press, New York, pp. 573–579.

Koning, J. J. de, Kooreman, H. J., Tan, H. S., and Verweij, J. (1975). One step, high yield conversion of penicillin sulfoxides to deacetoxy-cephalosporins. *J. Org. Chem.* 40:1346–1347.

Konomi, T., Herchen, S., Baldwin, J. E., Yoshida, M., Hunt, N. A., and Demain, A. L. (1979). Cell-free conversion of δ(L-α-aminoadipyl)L-cysteinyl-D-valine into an antibiotic with the properties of isopenicillin N in *Cephalosporium acremonium*. *Biochem. J.* 184:427–430.

Kozarich, J. W., and Strominger, J. L. (1978). A membrane enzyme from *Staphylococcus aureus* which catalyzes transpeptidase carboxypeptidase and penicillinase activities. *J. Biol. Chem.* 253:1272–1278.

Kurylowicz, W. (1977). The site of antibiotic accumulation in streptomycytes and *Penicillium chrysogenum*. *Acta Microbiol. Acad. Sci. Hung.* 24: 263–271

Kurylowicz, W., Kurzatkowski, W., Woznicka, W., Polowniak-Pracka, H., and Paszkiewica, A. (1979a). The ultrastructure of *Penicillium chrysogenum* in the course of benzyl-penicillin biosynthesis. *Zentralbl. Bakteriol.* II Abt. 134:706–720.

Kurylowicz, W., Kurzatkowski, W., Woznicka, W., Polowniak-Pracka, H., and Paszkiewicz, A. (1979b). The site of benzyl penicillin accumulation in *Penicillium chrysogenum*. *Zentralbl. Bakteriol.* II Abt. 134:721–732.

Kurzatkowski, W. (1981). Localization of some steps of penicillin G biosynthesis in subcellular structures of *Penicillium chrysogenum* PQ-96. *Med. Dosw. Microbiol. 33*:15–29.

Kurzatkowski, W., Kurylowicz, W., and Paszkiewicz, A. (1982). Penicillin G production by immobilized fungal vesicles. *Europ. J. Appl. Microb. Biotechnol. 15*:211–213.

Lara, F., Mateos, R. C., Vásquez, G., and Sanchez, S. (1982). Induction of penicillin biosynthesis by L-glutamate in *Penicillium chrysogenum*. *Biochem. Biophys. Res. Commun. 105*:172–178.

Lederberg, J. (1956). Bacterial protoplasts induced by penicillin. *Proc. Nat. Acad. Sci. U.S.A. 42*:574–577.

Lee, J. S., Shin, K. C., Yang, H. S., and Ryu, D. D. Y. (1978). Effects of carbon sources and other process variables in fed-batch fermentation of penicillin. *Korean J. Microbiol. 16*:21-29.

Lee, Y. H., and Tsao, G. T. (1979). Dissolved oxygen electrodes. *Adv. Biochem. Eng. 13*:35–86.

Legrain, C., Vissers, S., Dubois, E., Legrain, M., and Wiame, J. M. (1982). Regulation of glutamine synthetase from *Saccharomyces cerevisiae* by repression, inactivation and proteolysis. *Eur. J. Biochem. 123*:611–616.

Lengyel, Z. L., and Nyiri, L. (1965). The inhibitory effect of CO_2 on the penicillin biosynthesis. *Antibiot. Adv. Res. Prod. Clin. Use*, Proc. Congr. on Antibiotics, Prague, Czechoslavakia, pp. 733–735.

Liersch, M., Nuesch, J., and Treichler, H. J. (1976). Final steps in the biosynthesis of cephalosporin C. In *2nd International Symposium on the Genetics of Industrial Microorganisms*, 1974. K. D. MacDonald (Ed.). Academic, London, England, pp. 179–195.

Ligget, R. W., and Koffler, H. (1948). Corn steep liquor in microbiology. *Bacteriol. Rev. 12*:297–311.

Lim, H. C., Chen, B. J., and Creagan, C. C. (1977). An analysis of extended and exponentially-fed-batch cultures. *Biotechnol. Bioeng. 19*:425–433.

Loder, P. B., and Abraham, E. P. (1971). Isolation and nature of intracellular peptides from a cephalosporin C producing *Cephalosporium* sp. *Biochem. J. 123*:471–476.

López-Nieto, M. J., Revilla, G., and Martin, J. F. (1982). Biosynthesis of the tripeptide δ(L-α-aminoadipyl)L-cysteinyl-D-valine in *Penicillium chrysogenum* and its control by glucose. 4th International Symposium on the Genetics of Industrial Microorganisms, Kyoto, Japan. Y. Ikeda and T. Beppu (Eds.). p. 111.

Luengo, J. M., Revilla, G., Lopez, M. J., Villanueva, J. R., and Martin, J. F. (1979a). Penicillin production by mutants of *Penicillium chrysogenum* resistant to polyene macrolide antibiotics. *Biotechnol. Lett. 1*: 233–238.

Luengo, J. M., Revilla, G., Villanueva, J. R., and Martin, J. F. (1979b). Lysine regulation of Penicillin biosynthesis in low-producing and industrial strains of *Penicillium chrysogenum*. *J. Gen. Microbiol. 115*:207–211.

Luengo, J. M., Revilla, G., Lopez-Nieto, M. J., and Martin, J. F. (1981). Biosynthesis and excretion of penicillin by *Penicillium chrysogenum*. Paper presented at the *2nd Eur. Congri. Biotechnol.*, Eastbourne, England.

Lund, F. J. (1976). 6β-Aminodinopenicillanic acids—Synthesis and antibacterial properties. In *Recent Advances in the Chemistry of β-Lactam Antibiotics*. J. Celks (Ed.). Royal Society of Chemistry, London, pp. 25–45.

Lurie, L. M., and Levitov, M. M. (1967). pH value in the biosynthesis of penicillin. *Antibiotiki* 12:395–400.

Lurie, L. M., Verkhovtseva, T. P., Orlova, A. I., and Levitov, M. M. (1976). Technology of drug manufacture. Nitrogen nutrition as a factor in the intensification of penicillin synthesis. *Pharm. Chem. J.* 10:218–222.

Lurie, L. M., Stepanova, N. E., Bartoshevich, Yu. E., and Levitov, M. M. (1979). Study on physiological role of oils in penicillin biosynthesis. *Antibiotiki (Moscow)* 24:86–92.

McCann, E. P., and Calam, C. T. (1972). The metabolism of *Penicillium chrysogenum* and the production of penicillin using a high yielding strain at different temperatures. *J. Appl. Chem. Biotechnol.* 22:1201–1208.

McDougall, B., Dunnill, P., and Lilly, M. D. (1982). Enzymatic acylation of 6-aminopenicillanic acid. *Enzyme Microb. Technol.* 4:114–115.

Makins, J. F., and Holt, G. (1982). Liposome protoplast interactions. *J. Chem. Technol. Biotechnol.* 32:239–250.

Martin, E., Berky, J., Godzesky, C., Miller, P., Tome, J., and Stone, R. W. (1953). Biosynthesis of penicillin in the presence of C14. *J. Biol. Chem.* 203:239–250.

Martin, J. F., Luengo, J. M., Revilla, G., and Villanueva, J. R. (1979). Biochemical genetics of the β-lactam antibiotic biosynthesis. In *Genetics of Industrial Microorganisms*. O. K. Sebek and A. I. Laskin (Eds.). American Society for Microbiology, Washington, D.C., pp. 83–89.

Mason, H. R. S., and Righelato, R. C. (1976). Energetics of fungal growth: The effects of growth-limiting substrate on respiration of *Penicillium chrysogenum*. *J. Appl. Chem. Biotechnol.* 26:145–152.

Masurekar, P. S., and Demain, A. L. (1974). Impaired production in lysine regulatory mutants of *Penicillium chrysogenum*. *Antimicrob. Agents Chemother.* 6:366–368.

Matelová, V. (1976). Utilization of carbon sources during penicillin biosynthesis. *Folia Microbiol.* 21:208.

Matelová, V. (1981). Less customary raw materials for the production of penicillin. *Kvasny Prum.* 27:21–22.

Mattiasson, B., Danielsson, B., Winquist, F., Milsson, H., and Mosbach, K. (1981). Enzyme thermistor analysis of penicillin in standard solutions and in fermentation broth. *Appl. Env. Microbiol.* 41:903–908.

Mead, T. H., and Stack, M. V. (1948). Penicillin precursors in corn-steep liquor. *Biochem. J.* 42:58.

Meesschaert, B., Adriaens, P., and Eyssen, H. (1980). Studies on the biosynthesis of isopenicillin N with a cell-free preparation of *Penicillium chrysogenum*. *J. Antibiot.* 33:722–730.

Metz, B. (1976). From pulp to pellet: An engineering study on the morphology of moulds. Ph.D. thesis, Delft University of Technology, Delft, The Netherlands.

Metz, B., and Kossen, N. W. F. (1977). The growth of molds in the form of pellets. A literature review. *Biotechnol. Bioeng.* 19:781–799.

Metz, B., Kossen, N. W. F., and Suijdam, J. C. van. (1979). The rheology of mould suspensions. In *Advances in Biochemical Engineering*, Vol. 11. T. K. Ghose, A. Fiechter, and N. Blakebrough (Eds.). Springer Verlag, Berlin, pp. 103–156.

Metz, B., Bruyn, E. W. de, and Suijdam, J. C. van. (1981). Method for quantitative representation of the morphology of molds. *Biotechnol. Bioeng.* 23:149–162.

Meyrath, J., and Suchanek, J. (1972). Inoculation techniques effects due to quality and quantity of inoculum. *Methods Microbiol.* 7B:159–209.

Mor, J.-R. (1982). A review of instrumental analysis in fermentation technology. Paper presented at *3rd International Conference on Computer Applications in Fermentation Technology*, Manchester, England, 1981, pp. 108–118.

Morikawa, Y., Karube, I., and Suzuki, S. (1979). Penicillin G production by immobilized whole cells of *Penicillium chrysogenum*. *Biotechnol. Bioeng.* 21:261–270.

Morin, R. B., Jackson, B. G., Mueller, R. A., Lavagnino, E. R., Scanlon, W. B., and Andrews, S. L. (1963). Chemistry of cephalosporin antibiotics III. Chemical correlations of penicillin and cephalosporin antibiotics. *J. Am. Chem. Soc.* 85:1896–1897.

Morrison, K. B., and Righelato, R. C. (1974). The relationship between hyphal branching specific growth rate and colony radial growth rate in *Penicillium chrysogenum*. *J. Gen. Microbiol.* 81:517–520.

Mou, D.-G. (1979). Toward an optimum penicillin fermentation by monitoring and controlling growth through computer-aided mass balancing. Ph.D. thesis, Massachusetts Institute of Technology, Cambridge, Massachusetts.

Mou, D.-G., and Cooney, C. L. (1983a). Growth monitoring and control through computer aided on-line mass balancing in a fed-batch penicillin fermentation. *Biotechnol. Bioeng.* 25:225–255.

Mou, D.-G., and Cooney, C. L. (1983b). Growth monitoring and control in a complex medium: A case study employing fed-batch penicillin fermentation and computer-aided on-line mass balancing. *Biotechnol. Bioeng.* 25:257–269.

Moyer, A. J., and Coghill, R. D. (1946a). Penicillin VIII. Production of penicillin in surface cultures. *J. Bacteriol.* 51:57–78.

Moyer, A. J., and Coghill, R. D. (1946b). Penicillin IX. The laboratory scale production of penicillin in submerged cultures by *Penicillium notatum* Westling (NRRL 832). *J. Bacteriol.* 51:79–93.

Moyer, A. J., and Coghill, R. D. (1947). Penicillin X. The effect of phenylacetic acid on penicillin production. *J. Bacteriol.* 53:329–341.

Munro, A. C., Dewdney, J. M., Smith, H., and Wheeler, A. W. (1976). Antigenic properties of polymers formed by β-lactam antibiotics. *Int. Arch. Allergy Appl. Immunol.* 50:192–205.

Nachtmann, F. (1979). Automated high-performance liquid chromatography as a means of monitoring the production of penicillins and 6-aminopenicillanic acid. *Chromatographia* 12:380–385.

Nachtmann, F., and Gstrein, K. (1980). Comparison of HPLC and some official test methods for different penicillins. *Int. J. Pharm.* 7:55–62.

Nestaas, E., and Demain, A. L. (1981). Influence of penicillin instability on interpretation of feedback regulation experiments. *Eur. J. Appl. Microbiol. Biotechnol.* 12:170–172.

Nestaas, E., and Wang, D. I. C. (1981). A new sensor, the "filtration probe" for quantitative characterisation of the penicillin fermentation. I. Mycelial morphology and culture activity. *Biotechnol. Bioeng.* 23:2803–2813.

Nestaas, E., Wang, D. I. C., Suzuki, H., and Evans, L. B. (1981). A new sensor, the "filtration probe" for quantitative characterisation of the penicillin fermentation. II. The monitor of mycelial growth. *Biotechnol. Bioeng.* 23:2815–2824.

Neuss, N., Berry, D. M., Kupka, J., Demain, A. L., Queener, S. W., Duckworth, D. C., and Huckstep, L. L. (1982). High performance liquid chromatography (HPLC) of natural products. V. The use of HPLC in the cell-free biosynthetic conversion of α-aminoadipyl-cysteinyl-valine (LLD) into isopenicillin N. *J. Antibiot.* 35:580–584.

Niedermayer, A. O. (1964). Determination of phenylacetic acid in penicillin fermentation media by means of gas chromatography. *Anal. Chem. 36:* 938–939.

Niedermayer, A. O., Russo-Alesi, F. M., Lendzian, C. A., and Kelly, J. M. (1960). Automated system for the continuous determination of penicillin in fermentation media using hydroxylamine reagent. *Anal. Chem. 32:* 664–666.

Nilsson, H., Mosbach, K., Enfors, S. O., and Molin, N. (1978). An enzyme electrode for measurement of penicillin in fermentation broth: A step toward the application of enzyme electrodes in fermentation control. *Biotechnol. Bioeng.* 20:527–539.

Noguchi, H., and Mitsuhashi, S. (1981). Structure–activity relationship. In *Beta-Lactam Antibiotics.* S. Mitsuhashi (Ed.). Japan Science Society Press, Tokyo, pp. 59–81.

Nyiri, L. (1967). Effect of CO_2 on the germination of *Penicillium chrysogenum* spores. *Z. Allg. Mikrobiol.* 7:107–111.

Nyiri, L., and Leugyel, Z. L. (1965). Studies on automatically aerated biosynthetic processes. I. The effect of agitation and CO_2 on penicillin formation in automatically aerated liquid cultures. *Biotechnol. Bioeng.* 7:343–354.

Okabe, M., and Aiba, S. (1974). Simulation and optimization of cultured broth filtration. *J. Ferment. Technol.* 52:759–777.

Okabe, M., and Aiba, S. (1975). Optimization of a fermentation plant—example of antibiotic production. *J. Ferment. Technol.* 53:730–743.

Oldshue, J. Y., Coyle, C. K., Zemke, H., and Bruegger, K. (1978). Fluid mixing variables in the optimization of fermentation production. *Oric, Biochem. Nov.*:16–18, 24.

O'Sullivan, C. Y., and Pirt, S. J. (1973). Penicillin production by lysine auxotrophs of *Penicillium chrysogenum. J. Gen. Microbiol.* 76:65–75.

O'Sullivan, J., and Abraham, E. P. (1981). Biosynthesis of β-lactam antibiotics. In *Antibiotics*, Vol. IV. J. W. Corcoran (Ed.). Springer Verlag, Berlin, pp. 101–122.

O'Sullivan, J., Bleaney, R. C., Huddleston, J. A., and Abraham, E. P. (1979). Incorporation of ^{3}H from δ(L-α-amino-[4.5-^{3}H]-adipyl-L-cysteinyl-D-[4.4-^{3}H]-valine into isopenicillin-N. *Biochem. J.* 184:421–426.

Orlov, I. A., Berezovskaja, A. I., Sokolovskaja, A. E., and Beljanina, V. F. (1970). Replacement of corn steep liquor by mycelium in penicillin fermentation. *Antibiotiki (Moscow)* 15:1067–1068.

Owen, S. P., and Johnson, M. J. (1955). The effect of temperature changes on the production of penicillin by *Penicillium chrysogenum* W49-133. *Appl. Microbiol.* 3:375–379.

Pan, C. H., Hepler, L., and Elander, R. P. (1972). Control of pH and carbohydrate addition in the penicillin fermentation. *Dev. Ind. Microbiol.* 13:103–112.

Pan, C. H., Hepler, L., and Elander, R. P. (1975). The effect of iron on a high-yielding industrial strain of *Penicillium chrysogenum* and production levels of penicillin G. *J. Ferment. Technol.* 53:854–861.

Pan, S. C. (1955). Colorimetric determination of O-hydroxy-phenylacetic acid in samples from penicillin fermentations. *Anal. Chem.* 27:65–67.

Pan, S. C., Bonanno, S., and Wagman, G. H. (1959). Efficient utilization of fatty oils as energy source in penicillin fermentation. *Appl. Microbiol.* 7:176-180.

Park, J. T., and Johnson, M. J. (1949). Accumulation of labile phosphate in *Staphylococcus aureus* grown in the presence of penicillin. *J. Biol. Chem.* 179:585-592.

Park, J. T., and Strominger, J. L. (1957). Mode of action of penicillin. Biochemical basis for the mechanism of action of penicillin and for its selective toxicity. *Science* 125:99-101.

Pathak, S. G. (1967). Use of dried mycelium of *Penicillium chrysogenum* as the only constituent of seed medium in penicillin fermentation. *Can. J. Microbiol.* 13:730-731.

Perlman, D. (1966). Chemically defined media for antibiotic production. *Ann. N.Y. Acad. Sci.* 139:258-269.

Perlman, D. (1970). The evolution of penicillin manufacturing processes. In *The History of Penicillin Production.* Chem. Eng. Prog. Symp. Ser. 100 Vol. 66. American Institute of Chemical Engineers, New York, N.Y., pp. 23-30.

Perlman, D. (1974). Evolution of the antibiotics industry 1940-1975. *ASM News* 40:910-916.

Perquin, L. H. C. (1938). Bijdkage tot de kennis dek oxydatieve dissimilatie van *Aspergillus niger* van Tieghem. Ph.D. thesis, Delft University of Technology, Delft, The Netherlands.

Phillips, D. H. (1966). Oxygen transfer into mycelial pellets. *Biotechnol. Bioeng.* 8:456-460.

Phillips, D. H., and Johnson, M. V. (1961). Aeration in fermentations. *J. Biochem. Microbiol. Technol. Eng.* 3:277-309.

Pilát, P. (1979). Die Aussichten der halbkontinuierlichen Biosynthese von Penicillin. *Abhand. Dtsch. Akad. Wissen.* 3N:53-57.

Pilát, P. (1980). The effect of sucrose feeding on semi-continuous penicillin biosynthesis. In *7th Symposium on Continuous Cultivation of Microorganisms, Prague, 1978.* B. Sikyta, Z. Fencl, and V. Polacek (Eds.). pp. 753-758.

Pirt, S. J. (1966). A theory of the mode of growth of fungi in the form of pellets in submerged culture. *Proc. R. Soc. London Ser. B* 166:369-373.

Pirt, S. J. (1968). Application of continuous-flow methods to research and development in the penicillin fermentation. *Chem. Ind. May*:601-603.

Pirt, S. J. (1969). Microbial growth and product formation. In *Microbial Growth. 19th Symposium of the Society to General Microbiology.* P. M. Meadow and S. J. Pirt (Eds.). Cambridge University Press, London, pp. 199-221.

Pirt, S. J. (1974). The theory of fed-batch culture with reference to the penicillin fermentation. *J. Appl. Chem. Biotechnol.* 24:415-424.

Pirt, S. J., and Callow, D. S. (1959). Continuous-flow culture of the filamentous mould *Penicillium chrysogenum* and the control of its morphology. *Nature* 134:307-310.

Pirt, S. J., and Callow, D. S. (1960). Studies of the growth of *Penicillium chrysogenum* in continuous flow culture with reference to penicillin production. *J. Appl. Bacteriol.* 23:87-98.

Pirt, S. J., and Callow, D. S. (1961). The production of penicillin by continuous flow fermentation. *Sci. Rep. Ist. Super. Sanita* 1:250-259.

Pirt, S. J., and Mancini, B. (1975). Inhibition of penicillin production by carbon dioxide. *J. Appl. Chem. Biotechnol.* 25:781-783.

Pirt, S. J., and Righelato, R. C. (1967). Effect of growth rate on the synthesis of penicillin by *Penicillium chrysogenum* in batch and chemostat cultures. *Appl. Microbiol. 15*:1284–1250.

Podbielniak, W. J., Kaiser, H. R., and Ziegenhorn, G. J. (1970). Centrifugal solvent extraction. In *The History of Penicillin Production*. Chem. Eng. Prog. Symp. Ser. 100. Vol. 66. American Institute of Chemical Engineers, New York, pp. 45–50.

Pontecorvo, G., and Sermont, G. (1953). Recombination without sexual reproduction in *Penicillium chrysogenum*. *Nature 172*:126–127.

Prosser, J. I. (1979). Mathematical modelling of mycelial growth. In *Fungal Walls and Hyphal Growth*. J. H. Burnett and A. P. J. Trinci (Eds.). Cambridge University Press, London, pp. 385–401.

Pruess, D. L., and Johnson, M. J. (1967). Penicillin acyltransferase in *Penicillium chrysogenum*. *J. Bacteriol. 94*:1501–1508.

Queener, S., and Swartz, R. (1979). Penicillins: Biosynthetic and semisynthetic. In *Economic microbiology*, Vol. 3. A. H. Rose (Ed.). Academic, New York, pp. 35–122.

Queener, S., and Neuss, N. (1982). The biosynthesis of β-lactam antibiotics. In *The Chemistry and Biology of β-Lactam Antibiotics*, Vol. 3. R. B. Morin and M. Gorman (Eds.). Academic, New York, pp. 1–810.

Ratledge, C. (1977). Fermentation substrates. *Ann. Rep. Ferment. Proc. 1*: 49–71.

Ratzkin, B. (1981). Creation of a gene library from *Penicillium chrysogenum*: First steps in the application of recombinant DNA techniques to β-lactam antibiotic fermentations. 182nd ACS National Meeting, New York. American Chemical Society, New York, Micr. 43.

Ray Chowdhurry, M. K., and Chakrabarti, P. (1979). A new method of estimation of β-lactam antibiotics. *Anal. Biochem. 95*:413–415.

Ray Chowdhurry, M. K., Goswani, R., and Chakrabarti, P. (1980). A new rapid agar cup assay of β-lactam antibiotics in the presence of a chemical reagent. *Anal. Biochem. 108*:126–128.

Reuss, M., Debus, D., and Zoll, G. (1982). Rheological properties of fermentation fluids. *Chem. Eng. June*:233–236.

Revilla, G., Luengo, J. M., Villanueva, J. R., and Martin, J. F. (1982). Carbon catabolite repression of penicillin biosynthesis. In *Advances in Biotechnology*, Vol. 3. M. Moo-Young (Ed.), Pergamon, New York, pp. 155–160.

Righelato, R. C. (1975). Growth kinetics of mycelial fungi. In *Filamentous Fungi*. J. E. Smith and D. R. Berry (Eds.). Edward Arnold, London, pp. 79–103.

Righelato, R. C. (1979). The kinetics of mycelial growth. *Fungal Walls and Hyphal Growth*. J. H. Burnett and A. P. J. Prince (Eds.). Cambridge University Press, London, pp. 385–401.

Righelato, R. C., and Elsworth, R. (1970). Industrial applications of continuous culture: Pharmaceutical products and other products and processes. *Adv. Appl. Microbiol. 13*:399–417.

Righelato, R. C., and Pirt, S. J. (1967). Improved control of organism continuous cultures of filamentous micro-organisms. *J. Appl. Bacteriol. 30*:246–250.

Righelato, R. C., Trinci, A. P. J., Pirt, S. J., and Peat, A. (1968). The influence of maintenance energy and growth rate on the metabolic activity. Morphology and conidiation of *Penicillium chrysogenum*. *J. Gen. Microbiol. 50*:399–412.

Roels, J. A. (1980). Application of macroscopic principles to microbial metabolism. *Biotechnol. Bioeng. 22*:2457–2514.

Roels, J. A. (1982). Kinetic models in bioengineering. Applications, Prospects and problems. 3rd International Conference on Computer Applications in Fermentation Technology. Manchester, England, 1981, pp. 37–46.

Roels, J. A., and Kossen, N. W. F. (1978). On the modelling of microbial metabolism. In *Progress in Industrial Microbiology*, Vol. 14. M. J. Bull (Ed.). Elsevier, Amsterdam, pp. 95–203.

Roels, J. A., and Suijdam, J. C. van (1980). Energetic efficiency of a microbial process with an external power input: Thermodynamic approach. *Biotechnol. Bioeng. 22*:463–471.

Roels, J. A., Berg, J. van den, and Voncken, R. M. (1974). The rheology of mycelial broths. *Biotechnol. Bioeng. 16*:181–208.

Roksvaag, T. O., Brummenaes, H. I., and Waaler, T. (1979). Quantitative determination of benzylpenicillin sodium in the presence of its degradation products. A comparison between the iodometric and the HPLC method. *Pharm. Acta Helv. 54*:180–185.

Rolinson, G. N. (1952). Respiration of *Penicillium chrysogenum* in penicillin fermentations. *J. Gen. Microbiol. 6*:336–343.

Rolinson, G. N. (1979). 6-APA and the development of the β-lactam antibiotics. *J. Antimicrob. Chemother. 5*:7–14.

Rolinson, G. N., Batchelor, F. R., Butterworth, D., Cameron-Wood, J., Cole, M., Eustace, G. C., Hart, M. V., Richards, M., and Chain, E. B. (1960). Formation of 6-aminopenicillanic acid from penicillin by enzymatic hydrolysis. *Nature 187*:236–237.

Rozgonyi, F., Kiss, J., Biacs, P., and Váczi, L. (1981). May Phospholipid synthesis by a site of action of β-lactam antibiotics in *Staphylococcus aureus*. In *Staphylococci and Staphylococcal infections*. J. Jeljaszewicz (Ed.). Gustav Fischer Verlag, Stuttgart, pp. 741–747.

Ryu, D. D. Y., and Hospodka, J. (1977). Quantitative physiology in penicillin fermentation as a directed fermentation. 174th ACS National Meeting, Chicago, Ill., Port City Press, Baltimore, Md., Micr. 8.

Ryu, D. D. Y., and Hospodka, J. (1980). Quantitative physiology of *Penicillium chrysogenum* in penicillin fermentation. *Biotechnol. Bioeng. 22*: 289–298.

Ryu, D. D. Y., and Humphrey, A. E. (1972). A reassessment of oxygen-transfer rates in antibiotics fermentations. *J. Ferment. Technol. 50*: 424–431.

Ryu, D. D. Y., and Humphrey, A. E. (1973). Examples of computer-aided fermentation systems. *J. Appl. Chem. Biotechnol. 23*:283–295.

Sakaguchi, K., and Murao, S. (1950). A preliminary report on a new enzyme "penicillin-amidase". *J. Agric. Chem. Soc. Jpn. 23*:411.

Sánchez, S., Paniagua, L., Mateos, R. C., Lara, F., and Mara, J. (1982). Nitrogen regulation of penicillin G biosynthesis in *Penicillium chrysogenum*. In *Advances in Biotechnology*, Vol. 3. M. Moo-Young (Ed.). Pergamon, pp. 147–154.

Savidge, T.A. (1984). This volume, pp. 171-224.

Sawada, Y., Baldwin, J. E., Singh, P. D., Solomon, N. A., and Demain, A. L. (1980). Cell-free cyclization of δ(L-α-aminoadipyl)-L-cysteinyl-D-valine to isopenicillin N. *Antimicrob. Agents Chemother. 18*:465–470.

Schwartz, M. A. (1982). Catalysis of hydrolysis and aminolysis of benzylpenicillin mediated by a ternary complex with zinc ion and tris(hydroxymethyl)aminomethane. *Bioorg. Chem. 11*:4–18.

Sebek, O. K. (1980). Microbial transformations of antibiotics. In *Economic Microbiology*, Vol. 5. A. H. Rose (Ed.). Academic, New York, pp. 575–612.

Segel, I. H., and Johnson, M. J. (1961). Accumulation of intracellular inorganic sulfate by *Penicillium chrysogenum*. *J. Bacteriol. 81*:91–98.

Selwyn, S. (1980). The discovery and evolution of the penicillins and cephalosporins. In *The Beta-Lactam Antibiotics: Penicillins and Cephalosporins in Perspective*. S. Selwyn (Ed.). Hodder and Stoughton, London, pp. 1–48.

Shimizu, M., Masuike, T., Fujita, H., Kimura, K., Okachi, R., and Nara, T. (1975a) Enzymatic synthesis of cephalosporins I. Search for microorganisms producing cephalosporin acylase and enzymatic synthetis of cephalosporins. *Agric. Biol. Chem. 39*:1225–1232.

Shimuzu, M., Okachi, R., Kimura, K., and Nara, T. (1975b). Enzymic synthesis of cephalosporins III. Purification and properties of penicillin acylase from *Kluyvera citrophila*. *Agric. Biol. Chem. 39*:1655–1661.

Shirafuji, H., Fujisawa, Y., Kida, M., Kanzaki, T., and Yoneda, M. (1979). Accumulation of tripeptide derivatives by mutants of *Cephalosporium acremonium*. *Agric. Biol. Chem. 43*:155–160.

Shu, P. (1961). Mathematical models for the product accumulation in microbiological processes. *J. Biochem. Microbiol. Technol. Eng. 3*:95–109.

Shu, P. (1972). Characteristics of a fermentation process and the control of inter-related variables. In *Fermentation Technology Today*. 4th International Fermentation Symposium, Kyoto, Japan, Proceedings, Soc. Ferm. Technol., Osaka, Japan, pp. 183–186.

Sittig, W. (1982). The present state of fermentation reactors. *J. Chem. Technol. Biotechnol. 32*:47–58.

Sittig, W., and Heine, H. W. (1977). Erfahrungen mit großtechnisch eingesetzten Bioreaktoren. *Chem. Ing. Techn. 49*:595–605.

Smith, E. L., and Bide, A. E. (1948). Biosynthesis of the penicillins. *Biochem. J. 42*:17.

Smith, G. M., and Calam, C. T. (1980). Variations in inocula and their influence on the productivity of antibiotic fermentations. *Biotechnol. Lett. 2*:261–266.

Soltero, F. V., and Johnson, M. J. (1953). The effect of the carbohydrate nutrition on penicillin production by *Penicillium chrysogenum* O-176. *Appl. Microbiol. 1*:52–57.

Soltero, F. V., and Johnson, M. J. (1954). Continuous addition of glucose for evaluation of penicillin-producing cultures. *Appl. Microbiol. 2*:41–44.

Souders, M., Pierotti, G. J., and Dunn, C. L. (1970). The recovery of penicillin by extraction with a pH gradient. In *The History of Penicillin Production*. Chem. Eng. Prog. Symp. Ser. 100, Vol. 66. American Institute of Chemical Engineers, New York, N.Y., pp. 37–42.

Spencer, B., and Maung, C. (1970). Multiple activities of penicillin acyltransferase of *Penicillium chrysogenum*. *Biochem. J. 118*:29p–30p.

Squires, R. W. (1972). Regulation of the penicillin fermentation by means of a submerged oxygen-sensitive electrode. *Dev. Ind. Microbiol. 13*:128–135.

Stedman, R. J., Swered, K. H., and Hoover, J. R. E. (1964). 7-Aminodeacetoxy cephalosporanic acid and its derivatives. *J. Med. Chem. 7*:117–119.

Stone, R. W., and Farrell, M. A. (1946). Synthetic media for penicillin production. *Science 104*:445–446.

Stouthamer, A. H. (1977). Penicilline-Na ongeveer 50 jaar onderzoek. *Versl. Afd. Natuurk. Koninkl. Akad. Wetensch.*, Amsterdam, *86*:134–138.

Suzuki, H., Nishimura, Y., and Hirota, Y. (1977). On the process of cellular division in *Escherichia coli*: A series of mutants of *E. coli* altered in the penicillin-binding proteins. *Proc. Nat. Acad. Sci. U.S.A.* 75:664–668.

Suzuki, S., and Karube, I. (1978). Production of antibiotics and enzymes by immobilized whole cells. American Chemical Society, Symposium series *106*:59–72.

Swartz, R. W. (1979). The use of economic analysis of penicillin G manufacturing costs in establishing priorities for fermentation process improvement. In *Annual Reports on Fermentation Processes*, Vol. 3. D. Perlman and G. T. Tsao (Eds.). Academic, New York, pp. 75–110.

Sylvester, J. C., and Coghill, R. D. (1954). The penicillin fermentation. In *Industrial Fermentations*, Vol. 2. L. A. Underkofler and R. J. Hickey (Eds.). Chemical Publishing, New York, pp. 219–263.

Szarka, L. J. (1981). The use of different 1-phenyl-n-alkanes for benzylpenicillin biosynthesis by *Penicillium chrysogenum*. In *Advances Biotechnology*, Vol. 3. M. Moo-Young (Ed.). Pergamon, New York, pp. 167–173.

Takayashi, J., and Yamada, K. (1959). Studies on the effects of some physical conditions on the submerged mold culture. Part II. On the two types of pellet formation in the shaking culture. *J. Agric. Chem. Soc. Jpn.* 33:707–710.

Takei, H., Mizusawa, K., and Yoshida, F. (1975). Effect of initial aeration and agitation conditions on production of protease and penicillin. *J. Ferment. Technol.* 53:151–158.

Tamura, T., Imal, Y., and Strominger, J. L. (1976). Purification to homogeneity and properties of two D-alanine carboxypeptidases I from *Escherichia coli*. *J. Biol. Chem.* 251:414–423.

Tardew, P. L., and Johnson, M. J. (1958). Sulfate utilization by penicillin producing mutants of *Penicillium chrysogenum*. *J. Bacteriol.* 76:400–405.

Thomas, R. (1961). Colorimetric detection of penicillins and cephalosporins on paper. *Nature 191*:1161–1163.

Tien, W. (1981). Isolation of *Penicillium chrysogenum* mutants by mutation and selection technique. *Proc. Natl. Sci. Counc. R.O.C. (A)* 5:256–261.

Tipper, D. J., and Strominger, J. L. (1965). Mechanism of action of penicillins: A proposal made on their structural similarity to acyl-D-alanyl D-alanine. *Proc. Nat. Acad. Sci. U.S.A.* 54:1133–1141.

Todd, D. B., and Davies, G. R. (1973). Centrifugal pharmaceutical extractions. *Filtr. Sep.* :663–666.

Tome, J., Zook, H. D., Wagner, R. B., and Stone, R. W. (1953). Degradation of radio-active penicillin G. *J. Biol. Chem.* 203:251–-255.

Treichler, H. J., Liersch, M., Nüesch, J., and Döbeli, H. (1979). Role of sulphur metabolism in cephalosporin C and penicillin biosynthesis. In *Genetics of Industrial Microorganisms*. O. K. Sebek and A. L. Laskin (Eds.). American Society for Microbiology, Washington, D. C., pp. 97–104.

Trilli, A., Michelini, V., Mantovani, V., and Pirt, S. J. (1978). Development of the agar disk method for the rapid selection of cephalosporin producers with improved yields. *Antimicrob. Agents Chemother.* 13:7–13.

Trinci, A. P. J. (1970). Kinetics of the growth of mycelial pellets of *Aspergillus nidulans*. *Arch. Microbiol.* 73:353–367.

Trinci, A. P. J. (1974). A study of the kinetics of hyphal extension and branch initiation of fungal mycelia. *J. Gen. Microbiol.* 81:225–236.

Trinci, A. P. J., and Righelato, R. C. (1970). Changes in the constituents and ultrastructure of hyphal compartments during autolysis of glucose-starved *Penicillium chrysogenum*. *J. Gen. Microbiol. 60*:239–249.

Tsuji, K., and Robertson, J. H. (1975). High performance liquid chromatographic analysis of ampicillin. *J. Pharm. Sci. 64*:1542–1545.

Twigg, G. H. (1948). The spectroscopic estimation of penicillin. *Analyst 73*: 211–214.

Vadino, W. A., Sugita, E. T., Schnaare, R. L., Ando, H. Y., and Nibergall, P. J. (1979). Separation of penicillin G potassium and its degradation products using high pressure liquid chromatography. *J. Pharm. Sci. 68*:1316–1318.

van Suijdam, J. C. (1980). Mycelial pellet suspensions. Biotechnological aspects. Ph.D. thesis, Delft University of Technology, Delft, The Netherlands.

van Suijdam, J. C., and Metz, B. (1981a). Influence of engineering variables upon the morphology of filamentous molds. *Biotechnol. Bioeng. 23*:111–148.

van Suijdam, J. C., and Metz, B. (1981b). Fungal pellet breakup as a function of shear in a fermentor. *J. Ferment. Technol. 59*:329–333.

van Suijdam, J. C., Kossen, N. W. F., and Paul, P. G. (1980). An inoculum technique for the production of fungal pellets. *Eur. J. Appl. Microbiol. Biotechnol. 10*:211–221.

van Suijdam, J. C., Hols, H., and Kossen, N. W. F. (1982). Unstructured model for growth of mycelial pellets in submerged cultures. *Biotechnol. Bioeng. 24*:177–191.

Vandamme, E. J. (1980). Penicillin acylases and beta-lactamases. In *Econimic Microbiology*, Vol. 5. A. H. Rose (Ed.). Academic, New York, pp. 467–522.

Vandamme, E. J. (1981). Use of microbial enzyme and cell preparations to synthesize oligo-peptide antibiotics. *J. Chem. Tech. Biotechnol. 31*: 637–659.

Vandamme, E. J. 1983. Peptide antibiotic production through immobilized biocatalyst technology. *Enz. Microbial. Technol. 5*:403–416.

Vardar, F., and Lilly, M. D. (1982). Effect of cycling dissolved oxygen concentrations on product formation in penicillin fermentation. *Eur. J. Appl. Microbiol. Biotechnol. 14*:203–211.

Verkhovtseva, T. P., Lurie, L. M., and Levitov, M. M. (1970). Nature of penicillin receptor sides and certain aspects of metabolism in penicillin resistant bacteria. *Antibiotiki 15*:876–879.

Verkhovtseva, T. P., Lurie, L. M., Itsygin, S. B., Stepanova, N. E., and Levitov, M. M. (1975). Carbon metabolism under conditions of controlled penicillin biosynthesis. *Antibiotiki (Moscow) 20*:102–106.

Viesturs, U. E., Kristapsons, M. Z., and Levitans, E. S. (1982). Foam in microbiological processes. *Adv. Biochem. Eng. 21*:169–224.

Vinze, V. L., and Ghosh, D. (1959). Respiratory metabolism of *Penicillium chrysogenum* during commercial penicillin fermentation. *J. Sci. Ind. Res. 18C*:73–76.

Waard, W. F. van der (1976). De bereiding van penicilline; een voorbeeld van het gebruik van micro-organismen in de proces-industrie. In *De microbiologie drie eeuwen na Antonie van Leeuwenhoek*. F. Wensinck (Ed.). Pudoc, Centrum voor Landbouwpublicatie en Landbouwdocumentatie, Wageningen, pp. 162–179.

Waxman, D. J., and Strominger, J. L. (1980). Sequence of active site peptides from the penicillin sensitive D-alanine carboxy peptidase of *Bacillus subtilus*. *J. Biol. Chem. 255*:3964–3976.

Waxman, D. J., Amanuma, H., and Strominger, J. L. (1982). Amino acid sequence homologies between *Escherichia coli* penicillin-binding protein 5 and class A β-lactamases. *FEBS Lett.* *139*:159–163.

Weinberg, E. D. (1970). Biosynthesis of secondary metabolites: Roles of trace metals. *Adv. Microb. Physiol.* *4*:1–44.

Weissenburger, H. W. O., and Hoeven, M. B. van der. (1970). An efficient non-enzymatic conversion of benzylpenicillin to 6-aminopenicillanic acid. *Rec. Trav. Chim. Pays-Bas* *89*:1081–1084.

Wendelberger-Schieweg, G., Hütterman, A., and Haugli, F. B. (1980). Multiple sites of action of cycloheximide in addition to inhibition of protein synthesis in *Physarum polycephalum*. *Arch. Microbiol.* *126*:109–115.

Whitaker, A. (1980). Fed-batch culture. *Proc. Biochem.* May:10–15, 32.

Whitaker, A., and Long, P. A. (1973). Fungal Pelleting. *Proc. Biochem.* Nov.:27–31.

White, E. R., Carroll, M. A., and Zarembo, J. E. (1977). Reverse phase high speed liquid chromatography of antibiotics. III. Use of higher efficiency small particle columns. *J. Antibiot.* *30*:811–818.

White, R. L., John, E.-M. M., Baldwin, J. E., and Abraham, E. P. (1982). Stoichiometry of oxygen consumption in the biosynthesis of sipenicillin-N from a tripeptide. *Biochem. J.* *203*:791–793.

Wise, E. M., and Park, J. T. (1965). Penicillin: Its basic site of action as an inhibitor of a peptide cross linking reaction in cell wall mucopeptide synthesis. *Proc. Nat. Acad. Sci. U.S.A.* *54*:75–81.

Wright, D. G., and Calam, C. T. (1968). Importance of the introductory phase in penicillin production, using continuous flow culture. *Chem. Ind.* 1274–1275.

Yall, I. (1955). Pathways of acetate oxidation in *Penicillium chrysogenum* Q-176. Ph.D. thesis, Purdue University, Lafayette, Indiana.

Yamané, T., and Hirano, S. (1977). Semi-batch culture of microorganisms with constant feed of substrate. A mathematical simulation. *J. Ferment. Technol.* *55*:156–165.

Yamané, T., Kume, T., Sada, E., and Takamatsu, T. (1977). A simple optimization technique for fed-batch culture. *J. Ferment. Technol.* *55*:587–598.

Yamané, T., Matsuda, M., and Sada, E. (1981). Application of porous Teflon tubing method to automatic fed-batch culture of micro-organisms. I. Mass transfer through porous Teflon tubing. *Biotechnol. Bioeng.* *23*:2493–2507.

Yocum, R. R., Waxman, D. J., Rasmussen, J. R., and Strominger, J. L. (1979). Mechanism of penicillin action: Penicillin and substrate bind covalently to the same active site serine in two bacterial D-alanine carboxypeptidases. *Proc. Nat. Acad. Sci. U.S.A.* *76*:2730–2734.

Yocum, R. R., Waxman, D. J., and Strominger, J. L. (1980). Interaction of penicillin with its receptors in bacterial membranes. *Trends Biochem. Sci.* *5*:97–101.

Young, T. B. (1979). Fermentation scale up: Industrial experience with a total environmental approach. *Ann. N.Y. Acad. Sci.* *326*:165–180.

Young, T. B., and Koplove, H. M. (1972). A systems approach to design and control of antibiotic fermentations. In *Fermentation Technology Today*. 4th International Fermentation Symposium, Kyoto, Japan. Proceedings, Soc. Ferm. Technol., Osaka, Japan, pp. 163–166.

Zhukovkaya, S. A., and Gorskaya, S. V. (1974). The effect of fermentation conditions on the filtration characteristics of fermented broths. *Biotechnol. Bioeng. Symp.* *4*:935–952.

4

CEPHALOSPORIN C FERMENTATION: BIOCHEMICAL AND REGULATORY ASPECTS OF SULFUR METABOLISM

STEPHEN W. QUEENER, SARAH WILKERSON, DONALD R. TUNIN, JAMES P. McDERMOTT, JERRY L. CHAPMAN, CLAUDE NASH, CHESTER PLATT, AND JANET WESTPHELING* *Eli Lilly and Company, Indianapolis, Indiana*

---

*Present affiliation: Biology Laboratory, Harvard University, Cambridge, Massachusetts.

I. INTRODUCTION

Since cysteine is a precursor of cephalosporin C, regulation of the conversion of methionine to cysteine and of sulfate to cysteine is an important consideration in the economics of cephalosporin C fermentation. To understand the regulatory mechanisms governing the flow of metabolites through these pathways, one must first understand the pathways themselves. The pathways by which cysteine and methionine are synthesized and interconverted have been elucidated by over 30 years of study involving many different microorganisms. Figure 1 depicts the present state of knowledge of this complex set of interconnecting pathways as they exist in yeast and fungi.

Five enzymes [in Fig. 1: (1) sulfate permease, (2) ATP:sulfate adenyltransferase (ATP sulfurylase), (3) ATP:adenylsulfate 5'-phosphotransferase (APS kinase), (4) adenosine 3'-phosphate 5'-sulfatophosphate reductase (PAPS reductase), and (5) sulfite reductase] function to convert exogenous sulfate to endogenous sulfide. Endogenous sulfide can be converted to homocysteine by two pathways.

In the first pathway from sulfide to homocysteine, O-acetylserine (OAS) is produced from acetyl coenzyme A (acetyl CoA) and serine by serinetransacetylase (enzyme 6 in Fig. 1). The OAS condenses with sulfide to produce cysteine in a reaction catalyzed by OAS-sulfhydrylase (enzyme 7 in Fig. 1).* Cysteine then condenses with O-acetylhomoserine to produce cystathionine. This reaction is catalyzed by cystathionine γ-synthase (enzyme 10 in Fig. 1). Cystathionine is converted to homocysteine by action of β-cystathionase (enzyme 11 in Fig. 1).

The enzyme homoserine transacetylase (enzyme 8 in Fig. 1) plays a key role in homocysteine synthesis since it is required to produce O-acetylhomoserine from acetyl-CoA and homoserine. O-Acetylhomoserine is required in both pathways of homocysteine biosynthesis. It is the substrate for enzyme 10 in the first pathway, and it is the substrate for O-acetylhomoserine sulfhydrylase (also called homocysteine synthase, enzyme 9 in Fig. 1) in the second pathway. In this second pathway from sulfide to homocysteine, enzyme 9 catalyzes the condensation of hydrogen sulfide with O-acetylhomoserine.

Homocysteine is methylated to form methionine by a very complex enzyme, homosteine methylase (enzyme 12 in Fig. 1). This methylase utilizes

---

*Some years ago, thisulfate and S-sulfocysteine were considered as possible intermediates in the conversion of sulfide to cysteine in fungi. Thiosulfate and S-sulfocysteine can serve as sulfur sources for cephalosporin C synthesis, and the detection of S-sulfocysteine in *Cephalosporium acremonium* cells has been reported. An enzymic reduction of NADPH which was dependent on glutathionine and S-sulfocysteine was also reported (Suzuki et al., 1980). However, these data do not mean that thiosulfate and S-sulfocysteine are intermediates in the conversion of sulfide to cysteine as was once suggested for *Aspergillus nidulans* (Nakamura and Sato, 1963). Recent studies indicate that thiosulfate and S-sulfocysteine are not on the physiologically significant route of cysteine synthesis from sulfate in *A. nidulans* (Stepien et al., 1975). Woodin and Segel (1968) observed cell-free conversion of S-sulfocysteine to cysteine and sulfite by crude extracts of *Penicillium chrysogenum*, but noted that the results could be explained by a series of nonenzymatic exchange reactions involving glutathionine. Reduction of oxidized glutathione (G-S-S-G) by glutathionine reductase [NAD(P)H dependent] would drive the equilibrium of the chemical reaction toward cysteine.

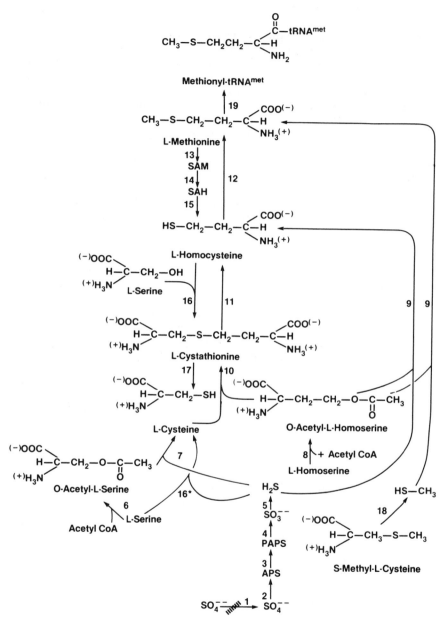

Figure 1 Sulfur metabolism pathways in yeast and fungi, including *C. acremonium*. In yeast the major route from sulfate to methionine involves enzymes 1–5, 8, 9, and 12, and in some fungi, for example, *Neurospora crassa*, the major pathway involves enzymes 1–5, 7, 10, 11, and 12. In *A. nidulans* and *C. acremonium* both pathways probably function significantly. In wild-type strains the O-acetylserine sulfhydrylase (enzyme 7) pathway appears to dominate. In some mutants (see text) the homocysteine synthase (enzyme 9) pathway dominates. Enzyme 16 catalyzes the formation of L-cystathionine from L-serine and L-homocysteine and the formation of cysteine from L-serine and H_2S. The former reaction is significant in vivo.

N^5-methyltetrahydrofolate as a methyl donor and is activated by catalytic amounts of S-adenosylmethionine.

Cystathionine β-synthase*, and γ-cystathionase (enzymes 16* and 17 in Fig. 1) function to convert homocysteine to cysteine. These enzymes can function using homocysteine produced from sulfide via enzymes 8 and 9 or from homocysteine produced from methionine via the sequential action of three enzymes: S-adenosylmethionine synthetase, the set of enzymes known as S-adenosylmethionine methyltransferases, and S-adenosylhomocysteine hydrolase (enzymes 13–15 in Fig. 1).

Due to regulatory mechanisms, however, enzymes 16 and 17 do not degrade homocysteine produced by the action of enzymes 10 and 11. Enzymes 16 and 17 are not produced under conditions which allow function of enzymes 10 and 11 and vice versa. Enzymes 16 and 17 are absent in most bacteria.† They exist in all yeast and fungi studied to date.

At least two different regulatory mechanisms in which methionine controls the conversion of sulfate to cysteine have been elucidated in yeast and fungi. In this discussion these mechanisms are referred to as methionyl-tRNA-mediated repression (MtR-mr) and reverse trans-sulfuration metabolite-mediated repression (RTM-mr). In a third mechanism (demonstrated in the bacterium *Salmonella typhimurium*), O-acetylserine induced the enzymes which convert sulfate to cysteine. In this chapter this mechanism is termed O-acetylserine-mediated induction (OAS-mi).

Microbiological and biochemical analyses of mutants of *Cephalosporium acremonium* with altered sulfur metabolism indicate that these three regulatory mechanism probably occur in wild-type *C. acremonium*. The basis of this conclusion is the observation that the biochemical characteristics of each of the *C. acremonium* mutants can be explained via the regulatory mechanisms previously established in *Saccharomyces cerevisiae*, *Aspergillus nidulans*, and *S. typhimurium*. A description of each regulatory mechanism together with the results obtained for each *C. acremonium* mutant is presented.

II. MATERIALS AND METHODS

A. Strains and Media

The lineage M8650 → M8650-1 → M8650-5 (intermediate steps not indicated) was obtained by selection for improved cephalosporin C production on a

---

*Enzyme 16 can also catalyze the condensation of hydrogen sulfide with L-serine to form cysteine. When catalysis of the formation of L-cystathionine from L-serine and homocysteine is being measured, enzyme 16 is referred to as cystathionine-β-synthase. When synthesis of cysteine from L-serine and hydrogen sulfide is measured, enzyme 16 is referred to as L-serine sulfhydrylase. The cystathionine-β-synthase activity is physiologically significant. The L-serine sulfhydrylase activity is probably not significant in vivo. Under normal conditions the intracellular concentration of hydrogen sulfide is below the K_m (H_2S) of the enzyme.

†Recently γ-cystathionase (also called cystathionine γ-lyase) has been detected in extracts of the cephamycin C producer *Streptomyces lactamdurans*. This is the only known report of the presence of this reverse trans-sulfuration pathway enzyme in a prokaryote (Kern and Inamine, 1981).

complex medium containing high concentrations of methionine (10 g/liter). No sulfate salts were present in this medium. The lineage M8650 → 92G-AD-11 (intermediate steps not indicated) was obtained by selections on a complex medium containing methionine (6 g/liter) and sulfate (16 g/liter).

Strain ACI-47 was present in a population of 10^5 N-methyl-N'-nitro-N-nitrosoguanidine- (NTG) treated spores of strain M8650-5. Strain BAH-16 was present in a population of 10^5 ultraviolet-treated arthrospores of strain 92G-AD-11. Strain 394-4 was present in a population of NTG-treated protoplasts of strain 92G-AD-11. Each strain was isolated as a large colony on solid agar medium containing sufficient selenomethionine to inhibit the growth and cell wall regeneration of the parent cell population.

The minimal medium (liquid or solid) used in experiments was Czapek Dox, and pH was adjusted to 7.3. For solid medium 1.5 g/liter Bacto agar was added.

B. Residual Sulfate and Methionine Concentrations in Fermentation Broths

Cephalosporin C fermentations were carried out in stirred fermenters containing 100 liters of a basal complex medium [e.g., lard oil, 10%; beet molasses, 3.9%; peanut meal, 3.9%; fish meal, 3.6%; $CaCO_3$, 0.65%; and $Ca(OH)_2$, 0.1%] to which various amounts (see text) of DL-methionine or ammonium sulfate were added. Whole broth samples were removed throughout the fermentation. Cells were separated from the fermentation broth by centrifugation. Aliquots of the clear amber supernatants were analyzed for sulfate and methionine by published procedures (Garrido, 1964; Gehrke and Neuner, 1974).

C. Assay of O-Acetylserine Sulfhydrylase

The method of Kredich and Becker (1971) was used, except that 0.01 M 2-mercaptoethanol was added to the reaction mixture. Cells for enzyme extraction were grown for 48, 72, and 96 hr in the following medium: corn meal, 2.0%; soy flour, 1.5%; $CaCO_3$, 0.3%; $(NH_4)_2SO_4$, 0.1%; methyl oleate, 2.0%; with or without DL-methionine at 0.5%. Cells (70 g damp) were removed from the whole broth by centrifugation (20 min, 5000 rpm, 4°C). Cells were added to 145 ml of 10 mM tris buffer at pH 7.8. The suspension was milled for 3 min (Braun homogenizer) to break cells. Unbroken cells (∿30%) and cell debris were removed by centrifugation (13,000 g, 60 min, 4°C). Supernatant was passed through a 0.8-μm Millipore filter. To 170 ml of filtrate, 41 g of $(NH_4)_2SO_4$ (enzyme grade) were added (40% saturation). Precipitated protein was removed by centrifugation. The resulting supernatant (180 ml) was adjusted to 60% saturation by the addition of 38 g of $(NH_4)_2SO_4$. Precipitated protein was recovered by centrifugation (10 min, 13,000 g, 4°C) and supernatant was discarded. Protein pellet was dissolved in 20 ml of 10 mM tris buffer pH 7.8. The enzyme solutions were dialyzed overnight at 4°C against three changes (500 ml each) of tris buffer at pH 7.8. Enzyme solutions were stored at 4°C prior to assay for O-acetylserine sulfhydrylase.

D. Sulfate Permease and Methionine Permease Assays

Cells were grown in the vegetative medium (100 ml per 500-ml flask) described in the section above for 72 hr at 250 rpm (except for strain ACI-

47, which required 96 hr). A 5-ml aliquot of the vegetative culture served as inoculum for basal complex medium +1% DL-methionine or 1% ammonium sulfate. Cells were harvested after 96 hr of fermentation (250 rpm, 25°C). Cells were centrifuged in 15-ml tubes for 10 min at 2600 rpm. Cells from six tubes were placed into a standard mixing vessel and suspended in distilled ice cold (4°C) water (to a total volume of ~ 50 ml) with an automatic stirrer (Tri R Stir R). Cells were centrifuged as above and the wash procedure was repeated. Twice-washed cells were diluted to 0.5% (vol/vol) in 0.1 M potassium phosphate buffer at pH 6.0 for starvation. Cells were starved in a 50-ml Erlenmeyer flask containing 10 ml of 0.5% cell suspension at 25°C and 250 rpm. At timed intervals 1-ml aliquots of cell suspension were withdrawn for assay.

In the assay for methionine uptake, 1 ml of cell suspension was added to 9 ml of 0.1 M potassium phosphate of pH 6.0, and the reaction was initiated by simultaneous addition of 1 ml of 10 mM DL-methionine and 0.1 ml of 0.35 mM [14C]DL-methionine (total radioactivity, 1.0 μCi). The total volume was 11.1 ml. At timed intervals 1.0-ml aliquots of reaction mixture were removed and placed in a Millipore suction–filtration unit (glass) fitted with a prefilter pad to trap cells. The uptake reaction was quenched by immediately adding 10 ml of 1% ice-cold DL-methionine. A second 10 ml 1% DL-methionine wash was used to rinse the sides of the filtration column. The pad and cells were placed face down in a scintillation vial containing 10 ml of Aquasol II. Radioactivity was measured in a Nuclear-Chicago Isocap/300 Liquid Scintillation System.

For assay of sulfate ion, the assay was initiated by the simultaneous addition of 1.0 ml of 10 mM Na_2SO_4 and 0.1 ml of 0.135 mM [35S]Na_2SO_4 (total radioactivity, 1.0 μCi).

E. Cephalosporin C Assay

Cells were removed from broth by filtration and filtrates were assayed for cephalosporin C as described previously (Redstone, 1971). Antibiotic concentrations are expressed in arbitrary units.

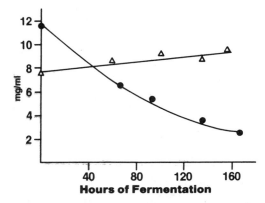

Figure 2 Residual sulfate (△) and residual methionine (●) in the fermentation broth during the production of cephalosporin C by strain M8650-5.

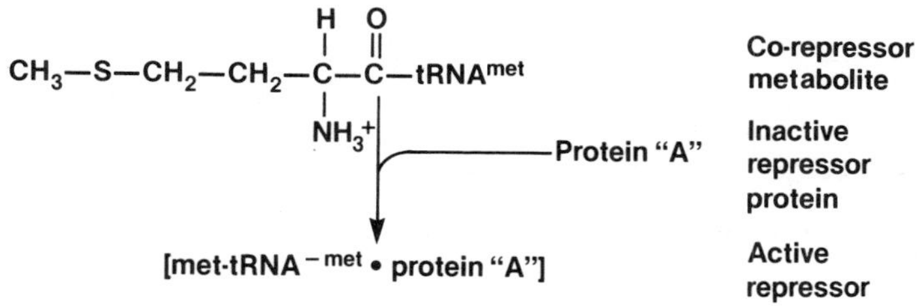

$$CH_3-S-CH_2-CH_2-\overset{\overset{\displaystyle H}{|}}{\underset{\underset{\displaystyle NH_3^+}{|}}{C}}-\overset{\overset{\displaystyle O}{||}}{C}-tRNA^{met}$$

Co-repressor
metabolite

———— Protein "A"

Inactive
repressor
protein

[met·tRNA $^{-met}$ • protein "A"]

Active
repressor

Figure 3 The methionyl-tRNA-mediated repression system. Active repressor inhibits synthesis of some enzymes involved in converting sulfate to methionine. Active repressor is formed when the co-repressor metabolite (charged methionyl-tRNA) binds to repressor protein (protein A).

III. RESULTS AND DISCUSSION

A. Strains with Wild-Type Regulation of Sulfate Assimilation

Cephalosporium acremonium strains M8650-3 and M8650-5 are believed to regulate the conversion of sulfate to cysteine and methionine by the same mechanisms employed in the wild-type strain CMI 49,137. All strains in the CMI 49,137 → M8650-5 lineage were screened in media containing significant concentrations of methionine. For example, strains in the lineage M8650-1 to M8650-5 were obtained by screening for maximal cephalosporin C production on media containing 1% or more methionine. No sulfate salts were added. Under these conditions cells deregulated for the production of cysteine from sulfate would not be expected to produce more cephalosporin C and would be eliminated from the screen. When strain M8650-5 was grown in basal complex medium supplemented with 1.2% DL-methionine and 0.7% sulfate (added as ammonium and magnesium salts), the strain was unable to utilize sulfate; however, methionine was used readily (Fig. 2). A similar inability to utilize sulfate under these conditions is observed by all strains of the CMI 49,137 → M8650-5 lineage. In the absence of methionine all these strains utilize sulfate readily.

B. Strains Without Methionyl-tRNA-Mediated Repression

In yeast, a system for repressing enzymes involved in the biosynthesis of methionine becomes active when repressor protein (referred to in this chapter as "protein A") binds to a co-repressor, charged methionyl-tRNA (Fig. 3) (Cherest et al., 1971; Surdin-Kerjan et al., 1973). In this discussion this system is referred to as methionyl-tRNA-mediated repression (MtR-mr). MtR-mr is activated by growth of yeast on methionine-rich media, and sulfate utilization for the formation of cysteine and methionine is drastically curtailed by a strong repression of enzymes 1, 2, and 5 (Fig. 4).*
The synthesis of other enzymes involved in methionine biosynthesis—for example, 8 and 9—is controlled by the same mechanism (Fig. 5) (Surdin-Kerjan et al., 1976).

---

*Assays of enzymes 3 and 4 (Fig. 1) from yeast grown in media with and without methionine were not reported. Synthesis of these enzymes may also be inhibited in media containing methionine.

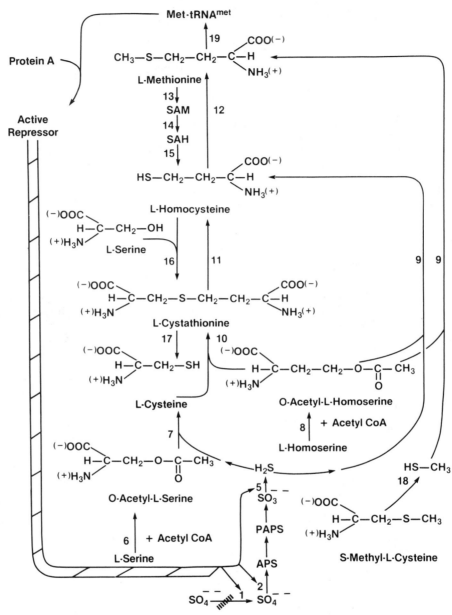

Figure 4 Enzymes of cysteine biosynthesis which have been shown to be repressed via the action of the methionyl-tRNA[met]-activated repressor in *S. cerevisiae*.

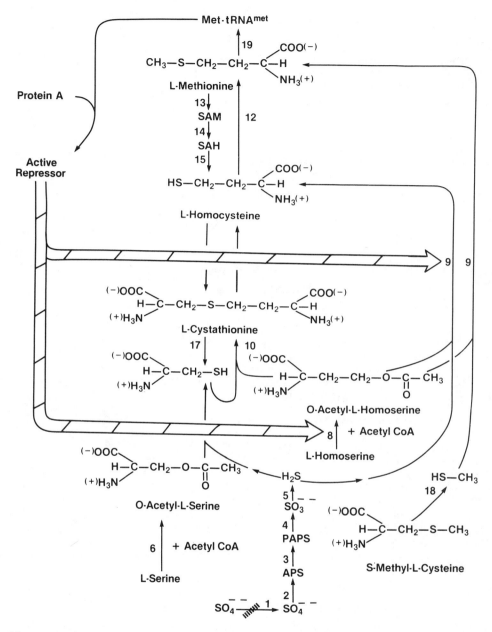

Figure 5 Enzymes of methionine biosynthesis repressed by the methionyl-tRNA^{met}–protein A complex in *S. cerevisiae*.

Analogs of methionine such as ethionine, selenomethionine, and trifluoromethionine (Fig. 6) can be charged onto the tRNA which normally accepts methionine. Hence addition of these analogs to the growth medium leads to repression of methionine biosynthesis. These analogs are incorporated into proteins in place of methionine. Such proteins and enzymes

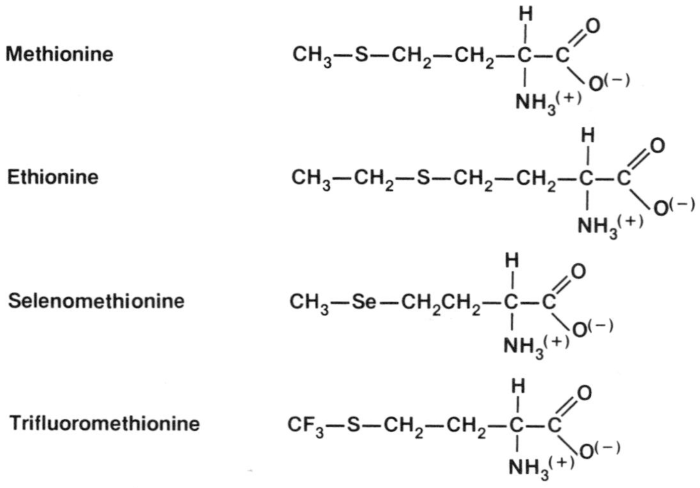

Methionine
$$CH_3-S-CH_2-CH_2-\underset{\underset{NH_3^{(+)}}{|}}{\overset{\overset{H}{|}}{C}}-C\underset{O^{(-)}}{\overset{O}{\diagup}}$$

Ethionine
$$CH_3-CH_2-S-CH_2-CH_2-\underset{\underset{NH_3^{(+)}}{|}}{\overset{\overset{H}{|}}{C}}-C\underset{O^{(-)}}{\overset{O}{\diagup}}$$

Selenomethionine
$$CH_3-Se-CH_2CH_2-\underset{\underset{NH_3^{(+)}}{|}}{\overset{\overset{H}{|}}{C}}-C\underset{O^{(-)}}{\overset{O}{\diagup}}$$

Trifluoromethionine
$$CF_3-S-CH_2-CH_2-\underset{\underset{NH_3^{(+)}}{|}}{\overset{\overset{H}{|}}{C}}-C\underset{O^{(-)}}{\overset{O}{\diagup}}$$

Figure 6 Structure of methionine and analogs often used to select deregulated mutants in bacteria and fungi.

do not function at 100% efficiency. The analogs also interfere with transmethylation reactions required for the synthesis of functional RNA and many other cellular constituents.

Because of the toxic effects of these analogs and their ability to repress endogenous methionine biosynthesis, low concentrations of the analogs strongly inhibit the growth of microorganisms.

1. met-tRNA Synthetase-Negative Mutants

Two types of mutation destroy MtR-mr in yeast. First, mutations can drastically increase the K_m of yeast met-tRNA synthetase (enzyme 19, in Fig. 1) for its substrate, methionine (Cherest et al., 1971). At any given concentration of methionine in the medium, strains carrying these mutations have been shown to have less intracellular charged methionyl-tRNA than wild-type strains. Hence in yeast strains carrying these mutations, the synthesis of enzymes 1, 2, and 5 is less responsive to the levels of methionine in the growth medium (Surdin-Kerjan et al., 1976, 1973). Other enzymes involved in methionine biosynthesis, for example, 8 and 9, are also less responsive in these mutants (Surdin-Kerjan et al., 1976).

These yeast mutants with altered methionyl-tRNA synthetases are methionine auxotrophs. They require high concentrations of methionine to support their growth. They are also temperature sensitive. The degree of regulation of methionine biosynthetic enzymes that occurs in these mutants can be directly correlated with the amounts of charged methionyl-tRNA present intracellularly in these mutants (Surdin-Kerjan et al., 1976; Cherest et al., 1975).

2. "Protein A"-Negative Mutants

A second kind of mutation can destroy the charged methionyl-tRNA-mediated regulatory system. These mutations are believed to affect the repressor protein, protein A, which binds charged methionyl-tRNA (Cherest et al.,

1971). Yeast mutants carrying such mutations* can no longer efficiently repress the formation of methionine biosynthetic enzymes in response to the presence of methionine or analogs of methionine in the growth medium. They are not methionine auxotrophs, as they do not lack any of the enzymes in Figure 1 (Cherest et al., 1971, 1973b).

These mutants efficiently utilize sulfate for methionine biosynthesis, even in the presence of methionine or its analogs. As a consequence, they overproduce methionine and are tolerant to levels of methionine analogs that would inhibit wild type (Cherest et al., 1971, 1973b; Masselot and deRobichon-Szulmajster, 1972).

The existence of MtR-mr in *C. acremonium* should mean that one could obtain *C. acremonium* mutants analogous to yeast protein A-negative mutants. By analogy to these yeast mutants, the protein A-negative mutant derived from wild-type *C. acremonium* should exhibit three easily detectable phenotypic properties: (1) ability to utilize sulfate in the presence of methionine, (2) tolerance to ethionine at concentrations that inhibit MtR-mr-positive (wild-type) strains, and (3) absence of nutritional requirements for methionine.†

The *C. acremonium* mutant 92G-AD-11 is able to utilize sulfate even in the presence of significant concentrations of methionine. The relative abilities of strains 92G-AD-11 and M8650-5 to assimilate sulfate were measured by growing each strain in the fermentation medium in which it was selected. Each medium was modified so that the initial concentrations of sulfate and methionine were 0.8 and 1.0%, respectively. Inocula of identical size for each strain were transferred to the respective media and the residual sulfate and methionine levels in the fermentation broth were measured throughout the fermentation. The residual concentration of sulfate in such complex media as a function of fermentation time is shown in Figure 7a for strain M8650-5. This strain is believed to have wild-type sulfur regulation. The same profile is also shown for strain 92G-AD-11. The ability of strain 92G-AD-11 to utilize sulfate has a sparing effect on methionine utilization (Fig. 7b).

The deregulated mutant of *C. acremonium*, 92G-AD-11, is resistant to ethionine. An increasing concentration of ethionine in agar medium completely shuts off the growth of the regulated strain, M8650-5. At the same concentrations of ethionine the deregulated strain, 92G-AD-11, is able to grow (Fig. 8).

---

*At least three genes, ETH 2, ETH 3, and ETH 10, plus the presence of charged methionyl-tRNA are required for MtR-mr (Cherest et al., 1973b; Masselot and deRobichon-Szulmajster, 1972). ETH 2, 3, and 10 are not involved in methionyl-tRNA synthesis; they are therefore assumed to be involved in the synthesis of repressor protein. The repressor protein, designated "protein A" in this chapter, may have nonidentical subunits. Post-transcriptional modification may be required to enable the protein to bind methionyl-tRNA.
†Two mutant types with deregulated sulfate assimilation exhibit nutritional requirements for methionine. Mutants of yeast with altered methionyl-tRNA synthetase are methionine auxotrophs. Some mutants which are blocked in reverse trans-sulfuration cannot utilize methionine as their sole source of sulfur.

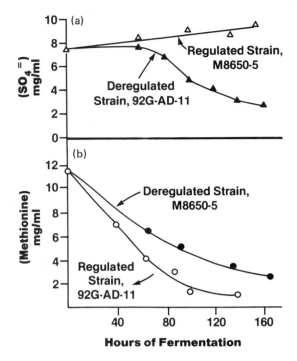

Figure 7 Residual sulfate (a) and methionine (b) in the fermentation broth during the production of cephalosporin C by strains M8650-5 and 92G-AD-11. Basal complex media are supplemented with 1.2% methionine and 0.5% sulfate.

Mutant 92G-AD-11 is prototrophic, as shown by its ability to grow on Czapek Dox medium.

All strains of the CMI 49,137 → M8650-5 lineage are (1) sensitive to ethionine at concentrations tolerated by strain 92G-AD-11, (2) unable to utilize sulfate in the presence of methionine in complex medium,* and (3) prototrophic.

In summary, the phenotypes of deregulated sulfate assimilation, ethionine resistance, and prototrophy are shared by the protein A-negative mutants of yeast and the *C. acremonium* strain 92G-AD-11. In the absence of further information, we assume that the genotype of 92G-AD-11 is protein A-negative and that therefore this strain is deficient in MtR-mr.

Loss of MtR-mr in 92G-AD-11 should mean that the levels of enzymes involved in the biosynthesis of methionine from sulfate should be less responsive to the level of methionine in the medium. Direct evidence for this is provided by an experiment in which the maximum levels of O-acetylserine sulfhydrylase (enzyme 7, Fig. 1) during exponential growth phase were

---

*Mutant 92G-AD-11 and all strains of the CMI 49,137 → M8650-5 lineage rapidly deplete sulfate from the basal complex media that contains no added methionine.

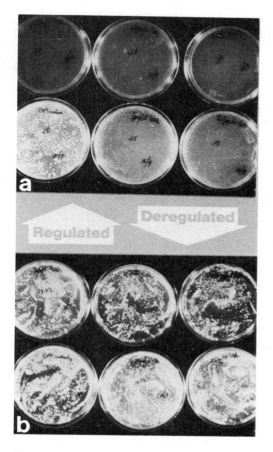

Figure 8 Comparison of tolerance of strains M8650-5 and 92G-AD-11 to ethionine. (a) Six plates of Czapek Dox media inoculated with 10^5 arthrospores of strain M8650-5. Plates contain varying concentrations of ethionine: bottom row of three plates, 0, 1, and 2.5 µg/ml; top row of three plates, 10, 100, and 250 µg/ml. (b) Six plates of Czapek Dox media inoculated with 10^5 arthrospores of strain 92G-AD-11. Plates contain varying concentrations of ethionine: bottom row of three plates, 0, 50, and 100 µg/ml; top row of three plates, 250, 500, and 1000 µg/ml.

determined for strains M8650-3 (a strain of the CMI 49,137 → M8650-5 lineage) and 92G-AD-11. The maximum level obtained in strain 92G-AD-11 was six times that obtained in strain M8650-3* (Table 1).

---

*The repressed (methionine added to medium) and derepressed (no methionine added) levels of O-acetylserine sulfhydrylase in strain M8650-3 are similar. Apparently even the low levels of methionine released by hydrolysis of soy flour are sufficient to cause almost complete repression of O-acetylserine sulfyhydrylase in strains of the lineage CMI 49,137 → M8650-5.

Table 1 O-Acetyl-L-Serine Sulfhydrylase Levels in Strains M8650-3 (Regulated Strain) and 92G-AD-11 (Deregulated Strain)[a]

Regulated strain (M8650-3)		Deregulated strain (92G-AD-11)	
Without methionine	With methionine	Without methionine	With methionine
1.4	1.2	9.5	5.3

[a]Data expressed in nanomoles per minute per milligram of protein.

C. Strains Without Reverse Trans-Sulfuration Metabolite-Mediated Repression

A second repression mechanism for controlling the rate of conversion of sulfate to cysteine and methionine is believed to exist in microorganisms. In this report the second mechanism is referred to as "reverse trans-sulfuration metabolite-mediated repression" (RTM-mr). The model of RTM-mr is as follows: A regulatory protein, which we designate "protein C," will interact with a metabolite involved in reverse trans-sulfuration (steps 13-17, Fig. 1) when the concentration of that metabolite is sufficiently high. The complex formed by this interaction triggers a repression of synthesis of enzymes involves in the uptake and conversion of sulfate to methionine (enzymes 1-5, 7-12 in Fig. 1).

In *S. cerevisiae* S-adenosylmethionine (SAM) appears to act as the corepressor metabolite in the RTM-mr. Addition of SAM to wild-type *S. cerevisiae* results in high intracellular concentration of SAM, but the methionine pool remains low. The levels of three methionine biosynthetic enzymes have been shown to be repressed by the addition of SAM to cultures of *S. cerevisiae*: adenosine triphosphate sulfurylase (enzyme 2, Fig. 1), sulfite reductase (enzyme 3), and homocysteine synthetase (enzyme 9) (Cherest et al., 1973a). Mutants of *S. cerevisiae* with altered methionyl-tRNA synthetase (high K_m for methionine), when grown in the presence of SAM, exhibited repressed levels of homocysteine synthetase (enzyme 9, Fig. 1) and ATP sulfurylase (enzyme 2). The presence of SAM in the medium caused elevation of the intracellular SAM level, but the levels of methionine and methionyl-tRNA[met] did not increase (Cherest et al., 1975).

The kinetics of derepression of homocysteine synthetase (HCS) have been studied in the absence and in the presence of cycloheximide (inhibitor of protein synthesis) and lomofungin (inhibitor of mRNA synthesis). Wild-type cells grown in the presence of methionine and transferred to methionine-free medium exhibit repressed levels of HCS for 40 min; then the rate of HCS synthesis accelerated to attain a maximum rate of synthesis after 30 min. If lomofungin is added during the lag period, no HCS is synthesized. When lomofungin is added after HCS synthesis is maximal, synthesis of HCS continues unabated for about 20 min. These results suggest that the MtR-mr acts at the level of transcription. When cells are grown in the presence of SAM and in the absence of lomofungin, HCS synthesis follows the same lag phase and acceleration phase after transfer to SAM-free medium. The same maximal rate of HCS synthesis is attained. However, when lomofungin is added during the lag phase, HCS synthesis is not affected. HCS synthesis accelerates normally to maximal rate. This suggests that SAM-mediated repression of methionine biosynthetic enzymes occurs at a posttranscriptional level. As expected, cycloheximide completely inhibited HCS synthesis almost immediately after its addition during the period of maximal HCS synthesis. This occurred for cells grown in the presence of methionine or SAM (Surdin-Kerjan et al., 1976).

Belief in the existence of RTM-mr is to a large degree based on the properties of mutants. The model of RTM-mr, as outlined above, predicts that this repression mechanism could be blocked either by preventing the conversion of methionine to the corepressor metabolite (defective enzyme at steps 13-17) or by loss of the ability to produce "protein C." Mutants blocked in reverse trans-sulfuration (steps 13-17 in Fig. 1) are referred to as mec mutants by investigators of *A. nidulans* (mec stands for *methionine catabolism*). In this report we adopt this nomenclature to refer to reverse trans-sulfuration mutants of other fungi and yeast. In addition, mutants of *S. typhimurium* blocked in step 13 are discussed under this classification. Mutants believed to lack functional "protein C" are referred to as protein C-negative mutants.

1. *"Methionine Catabolism" Mutants*

If S-adenosylmethionine (SAM) were the co-repressor metabolite in RTM-mr, then this regulatory mechanism would be inoperative in mutants deficient in methionine adenosyltransferase (enzyme 13, Fig. 1). Such mutants would therefore be expected to produce higher levels of methionine biosynthetic enzymes than the corresponding wild type. As a consequence of the increased levels of the methionine biosynthetic enzymes, methionine should be overproduced. As a consequence of overproduction of methionine, these mutants would be expected to have high intracellular levels of methionine. The high intracellular levels should dilute the effect of toxic analogs of methionine, such as ethionine and selenomethionine. Mutants deficient in methionine adenosyltransferase (MAT) have been isolated in *S. typhimurium* (Hobson, 1976; Lawrence et al., 1968; Hobson and Smith, 1973), *A. nidulans* (Pieniazek et al., 1973; Paszewski and Grabski, 1974), *Corynebacterium glutamicum* (Kase and Nakayama, 1975), *Candida petrophilum* (Komatsu et al., 1974), and *S. cerevisiae* (Mertz and Spence, 1972). These MAT-deficient strains are all resistant to methionine analogs. Resistant strains exhibited elevated levels of methionine biosynthetic enzymes and methionine pools. MAT-deficient strains of the *C. glutamicum* cannot grow in minimal medium in which methionine is the only source of sulfur; when the minimal medium is supplemented with SAM, they grow normally (Kase and Nakayama, 1975). Growth of MAT-deficient strains (mecC mutants) of *A. nidulans* is inhibited by selenate. This inhibition is not reversed by methionine, but is reversed by homocysteine, cystathionine, or cysteine. These results have been explained by suggesting that cysteine is the co-repressor metabolite in RTM-mr in *A. nidulans*. Following this hypothesis, it is reasoned that as long as the cell has a sufficient intracellular pool of cysteine to activate the RTM-mr, sulfate permease is repressed, thereby limiting the access of selenate to the cell. The cysteine can be provided without the introduction of selenate by the direct uptake of cysteine or the degradation of exogenously supplied methionine. Enzymatic deficiencies in reverse trans-sulfuration would prevent transformation of methionine to cysteine. Hence mutants blocked in reverse trans-sulfuration would be sensitive to selenate toxicity in the absence or presence of methionine, whereas, the parental wild type would be sensitive to selenate in the absence of methionine, but tolerant to selenate as long as sufficient exogenous methionine were present. The hypothesis is supported by the phenotypes of all three types of reverse trans-sulfuration mutation which have been isolated in *A. nidulans*. The mecA mutants (deficient in cystathionine β-synthase, enzyme 16, Fig. 1),

the *mecB* mutants (deficient in γ-cystathionase, enzyme 17, Fig. 1), and
mecC mutants (deficient in methionine adenosyltransferase, enzyme 13, Fig.
1). Each exhibit sensitivity to selenate which cannot be reversed by methionine.
From the examination of the literature above, we see that the following characteristics have been documented for fungal mec mutants and/or bacterial mutants blocked in the conversion of methionine to S-adenosylmethionine (same
as the first step in the fungal reverse trans-sulfuration pathway):

1. Resistant to analogs of methionine.
2. Exhibit derepressed levels of methionine biosynthetic enzymes.
3. Exhibit increased pool levels of methionine.
4. Are sensitive to selenate. Sensitivity to selenate is not reversed
   by methionine, although methionine does reverse selenate toxicity
   in wild type.
5. Unable to grow in minimal medium supplemented with only methionine
   (demonstrated only for *C. glutamicum* SAM synthetase mutants).

The existence of RTM-mr in *C. acremonium* should mean that one could
obtain mutants analogous to the mec mutants obtained in other microorganisms.
The *C. acremonium* mutant strain ACI-47 (derived from strain M8650-5) has
the characteristics of a mec mutant.

Strain ACI-47 has the nutritional characteristics of a strain blocked at
step 17 (Fig. 1) that is, deficient in the enzyme γ-cystathionase. The mutant
grows very poorly relative to its parent on liquid minimal medium which contains methionine, S-adenosylmethionine, homocysteine, or cystathionine as
sole sources of sulfur. It should be noted that these results are only obtained
in liquid medium (see Sec. II). Agar contains significant quantities of covalently bound organic sulfur in a form which supports the growth of this strain.
On liquid minimal medium supplemented with cysteine, the ACI-47 strain grows
at essentially the same rate as its parent. It must be noted, however, that
no assays of enzymes 13−17 have been conducted on extracts prepared from
strain ACI-47. This strain has never been observed to revert as judged by
any of its distinctive phenotype properties. It is therefore distinctly possible
that this mutant represents a deletion mutant which is deficient in the synthesis of two or more of the enzymes of reverse trans-sulfuration.

The existence of methionine permease in strain ACI-47 was established
by following uptake of [^{14}C]DL-methionine and by disappearance of methionine
from fermentation broth. The rate of disappearance was the same as for its
parent strain M8650-5 (Fig. 9). Thus strain ACI-47 is different from other
selenomethionine-resistant mutants previously reported (Nüesch et al., 1976).
Methionine does not repress sulfate permease in ACI-47. Methionine does repress sulfate permease in M8650-5 (the parent of ACI-47; Fig. 10).

Like MAT-deficient strains in bacteria and yeast, the ACI-47 strain of
C. acremonium is highly resistant to analogs of methionine such as selenomethionine and ethionine. The parent of strain ACI-47, strain M8650-5, appears to possess all the wild-type mechanisms governing conversion of sulfate
to methionine. Strain M8650-5 is highly sensitive to methionine analogs, ethionine and selenomethionine (e.g., 1 μg/ml), in minimal medium.

Like the mec strains of *A. nidulans*, the growth of ACI-47 on minimal
medium (Czapek Dox) is inhibited by selenate (500 μg/ml). Exogenous methionine (100 μg/ml) does not reverse this selenate toxicity. Methionine (100
μg/ml) does reverse the selenate (500 μg/ml) toxicity for the parent of ACI-
47, strain M8650-5. A very modest reduction in selenate toxicity was achieved
by addition of cysteine (150 μg/ml) to M8650-5. For strain ACI-47 addition

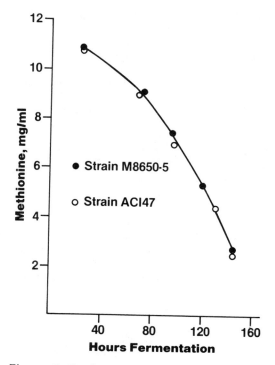

Figure 9 Residual methionine in fermentation broth of strain ACI-47 (mutant type, deregulated) and parental strain, M8650-5 (wild type, regulated). The basal complex medium was supplemented with 1.2% methionine.

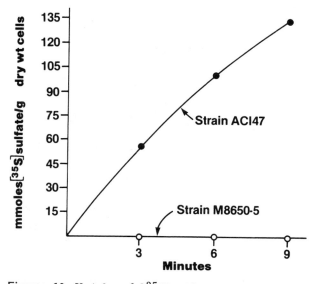

Figure 10 Uptake of [^{35}S]sulfate by *C. acremonium* cells grown in the presence of DL-methionine. Freshly washed cells of strains ACI-47 (deregulated) and M8650-5 (regulated) were assayed prior to starvation, as described in Sec. II.

of cysteine did not reduce selenate toxicity. Increasing the cysteine concentration did not increase the protection from selenate toxicity. These results are consistent with the existence of an RTM-mr of sulfur metabolism in *C. acremonium*. The poor reversal of selenate toxicity by cysteine may reflect the relatively poor uptake of cysteine by *C. acremonium*, but, more likely, it may mean that SAM or homocysteine rather than cysteine is the co-repressor of the RTM-mr in *C. acremonium*.

The production of cephalosporin C by strain ACI-47 has been compared to that of its parent in complex media in which the major sulfur source was methionine, sulfate, or a mixture of methionine and sulfate.

In experiments in 150-liter fermenters, strain ACI-47 produced nearly as much cephalosporin C on complex medium with 1% ammonium sulfate as the sulfur source as its parent, M8650-5, did on basal complex medium with 1% DL-methionine as the sulfur source (Table 2). In experiments in which sulfate replaced DL-methionine, M8650-5 produced 6.6 units/ml as compared to 12 units/ml with methionine as fulfur source.

In shake flasks, numerous fermentations of strain M8650-5 were run in basal complex medium supplemented with 1.0% methionine or ammonium sulfate. The best potencies in the respective media were 10 ± 0.2 and 6.4 ± 0.3 units/ml. In four shake flask experiments, the ACI-47 strain produced $1.96 \pm .11$

Table 2 Cephalosporin C Concentrations in Cell-Free Broth

Experiment	Peak cephalosporin C potency (units/ml hr)	
	M8650-5/1.09% DL-methionine	ACI-47/1.09% $(NH_4)_2SO_4$
DCED-8	12.93 (119)	—
	12.04 (107)	—
DCED-9	—	12.61 (131)
	—	11.11 (131)
DCED-10	11.70 (107)	12.07 (119)
DCED-11	11.45 (107)	—
	11.26 (119)	—
DCED-12	11.92 (103)	11.67 (107)
	—	10.03 (131)
DCED-14	—	10.22 (115)
DCED-15	11.98 (115)	—
	10.99 (115)	—
	11.30 (124)	—
DCED-17	11.50 (103)	—
DCED-18	—	10.29 (139)
DCED-19	9.98 (123)	—
	10.21 (114)	—
	11.44 ± 0.77 (113 ± 7 hr) (100%)	11.14 ± 0.93 (127 ± 10 hr) (98%)

Table 3 Relative Ability of Strains ACI-47 and M8650-5 to Produce Cephalosporin C (CPC) from Sulfate and Methionine in 500-ml Shake Flasks[a]

Experiment number	Strain M8650-5 (CPC, units/ml)		Strain ACI-47 (CPC, units/ml)	
	BCF5[b] + 1% DL-methionine medium	BCF5 + 1% $(NH_4)_2SO_4$ medium	BCF5 + 1% DL-methionine medium	BCF5 + 1% $(NH_4)_2SO_4$ medium
18	9.78 10.23	5.78 6.73		
54			1.90 2.15	
62	9.00 10.07			
68		5.96 6.25		
71				8.38 8.41
72				8.94 8.26
77				8.09 8.03
92				7.49 8.26
96	10.01 9.89			
99			1.95 1.85	
Average	9.83 ± 0.40	6.18 ± 0.36	1.96 ± 0.11	8.24 ± 0.39

1. $\dfrac{\text{CPC-SO}_4,\ \text{M8650-5}}{\text{CPC-met},\ \text{M8650-5}} = \dfrac{6.18}{9.83} = 63\%$

2. $\dfrac{\text{CPC-SO}_4,\ \text{ACI-47}}{\text{CPC-SO}_4,\ \text{M8650-5}} = \dfrac{8.24}{6.18} = 133\%$

3. $\dfrac{\text{CPC-SO}_4,\ \text{ACI-47}}{\text{CPC-met},\ \text{M8650-5}} = \dfrac{8.24}{9.83} = 84\%$

4. $\dfrac{\text{CPC-met},\ \text{ACI-47}}{\text{CPC-SO}_4,\ \text{ACI-47}} = \dfrac{1.96}{8.24} = 24\%$

5. $\dfrac{\text{CPC-met},\ \text{ACI-47}}{\text{CPC-met},\ \text{M8650-5}} = \dfrac{1.96}{9.83} = 20\%$

[a]All fermentations were carried out in 500-ml shake flasks with 30 ml of basal complex media shaken at 285 rpm, 2-in. throw, 25°C. Media contain sulfate or DL-methionine, as listed above.
[b]BCF5 = Basal Complex Fermentation medium #5 (see B, p. 145).

units/ml cephalosporin C on the methionine-based medium, and $8.24 \pm .39$ units/ml on the sulfate-based medium (Table 3).

These cephalosporin C synthesis data can all be explained by assuming that strain ACI-47 is a mec mutant with an intact methionyl-tRNA-mediated regulatory mechanism. Such a strain would be expected to produce very little cephalosporin C on a complex medium in which methionine was the major sulfur source. In this medium the reverse trans-sulfuration pathway would be required to supply the bulk of methionine for cephalosporin C synthesis. In complex medium containing both sulfate and methionine, a mec strain which possessed the methionyl-tRNA mediated regulatory mechanism would be expected to produce very little cephalosporin C from sulfate. Sulfur for cephalosporin C biosynthesis could not be derived efficiently from sulfate, because the exogenous methionine would create a large intracellular methionine pool and a methionyl-tRNA pool and therefore activate the MtR-mr of enzymes 1-5 and 8-11 (Fig. 1). In complex medium containing sulfate as the major sulfate course, a mec strain which possessed the methionyl-tRNA-mediated mechanism might be expected to produce only modestly more cephalosporin C than its parent.* For example, suppose that SAM served as a co-repressor in a mechanism that repressed the synthesis of enzymes 1-5. In the parent in sulfate-based medium, SAM formed from endogenous methionine could trigger repression of enzymes 1-5. In a mutant impaired in SAM synthesis, the level of SAM would be expected to be lower, and therefore the repression mechanism would not be triggered at all. In a sulfate-based medium, reverse transsulfuration would not be expected to be an important source of sulfur for cephalosporin C synthesis either in the parent wild type or in the mutant.

2. mec/Protein A-Negative Double Mutants

Evidence suggesting that strain 92G-AD-11 lacks MtR-mr and that strain ACI-47 lacks RTM-mr has been presented. The existence of both of these mutants suggests both MtR-mr and RTM-mr operate in wild-type *C. acremonium*. Loss of only one regulatory mechanism in 92G-AD-11 and ACI-47 should mean that strain 92G-AD-11 should exhibit properties which reflect the presence of RTM-mr, and strain ACI-47 should exhibit properties which reflect the presence of MtR-mr.

Several observations indicate that some regulation of sulfate assimilation still exists in strain ACI-47. Although ACI-47 produces more cephalosporin C on sulfate-based medium than its parent, this yield is somewhat less than that of the parent strain on methionine-based medium (see Table 2). In

---

*A *C. acremonium* mec mutant, 8650/113, which lacks γ-cystathionase activity has been reported (Treichler et al., 1979). This mutant exhibits *reduced* ability to produce cephalosporin C from sulfate. Absence of γ-cystationase activity was demonstrated by enzyme assay. The mutation in 8650/113 was reversible. The phenotype of 8650/113 was explained by postulating that γ-cystathionase acted as an inducer for the enzymes which convert α-aminoadipic acid, cysteine, valine, and acetyl CoA into cephalosporin C (Treichler et al., 1979). If one uses the "Treichler model" to interpret the genotype of strain ACI-47, then one would suggest that ACI-47 may carry a deletion which has destroyed cystathionase activity and left the inducer function of mecB gene product.

medium containing both methionine and sulfate, strain ACI-47 depleted sulfate from the medium at a rate only equivalent to that of the fully regulated strain M8650-5 (Fig. 9). On sulfate-based medium the optimal yield of cephalosporin C from ACI-47 is considerably less than that of 92G-AD-11. The rate of sulfate utilization with strain ACI-47 in a medium containing both methionine and sulfate was significantly less than with strain 92G-AD-11.

There are several lines of evidence indicating that some regulation of methionine biosynthesis still occurs in the strain 92G-AD-11. For example, when the levels of O-acetylserine sulfhydrylase are determined for cells grown in the presence and absence of methionine, significant (44%) repression of O-acetylserine sulfhydrylase due to the presence of methionine in the medium can be seen to occur for this strain (Table 1).

A second observation suggests that a regulatory mechanism controlling sulfate utilization still exists in the deregulated strain 92G-AD-11. During log phase growth in the presence of methionine there is virtually no utilization of sulfate (Fig. 7a). The onset of sulfate utilization at 60–80 hr appears to correspond to the heavy draw on the intracellular cysteine pool caused by cephalosporin C synthesis. Cephalosporin C synthesis achieves maximal rate between 60 and 80 hr.

A third observation suggests the existence of a regulatory mechanism in the partially deregulated 92G-AD-11 strain. Although this strain is highly resistant to ethionine (Fig. 8), it is rather sensitive to the toxic effects of selenomethionine; 2 ppm selenomethionine in Czapek Dox minimal medium strongly inhibit the growth of strain 92G-AD-11. Mutants of 92G-AD-11 resistant to various levels of selenomethionine arise spontaneously at a frequency of 3×10^{-5} when protoplasts of the strain are regenerated in the presence of selenomethionine. Treatment of protoplasts with NTG prior to regeneration can increase the frequency to approximately 10^{-3}. Some selenomethionine-resistant mutants of C. acremonium are mutants which cannot efficiently take up methionine (Nüesch et al., 1976). However, some of the selenomethionine-resistant mutants derived from 92G-AD-11 (e.g., strain BAH-16) retained the ability to take up methionine.

Mutant BAH-16 appears to be deficient in both MtR-mr and RTM-mr. Addition of methionine to sulfate-based medium inhibited cephalosporin C synthesis in strain ACI-47 (presumably because of MtR-mr). Addition of methionine to sulfate-based medium modestly stimulated cephalosporin C synthesis by stain BAH-16. A strain lacking both MtR-mr and RTM-mr would be expected to produce more cysteine and methionine from sulfate than one lacking only MtR-mr. Increased cytoplasmic cysteine, as a precursor of cephalosporin C, and methionine, as a regulator of cephalosporin C biosynthesis (Martin and Demain, 1980), should improve cephalosporin C production. As expected, strain BAH-16 produced more cephalosporin C on sulfate-based medium than its parent 92G-AD-11 (Table 4).

3. Protein A-Negative/Protein C-Negative Double Mutants

A second class of selenomethionine-resistant mutants which can take up methionine has been derived from strain 92G-AD-11. This class can be distinguished from the BAH-16 type by the ability to produce cephalosporin C on methionine-based medium. The selenomethionine-resistant class represented by strain 394-4 produced as much cephalosporin C as its selenomethionine-sensitive parent 92G-AD-11 in methionine-based medium, whereas the class represented by strain BAH-16 produced poorly on this medium. This characteristic of the 394-4 class is consistent with a genotype which prevents synthesis of a

Table 4 Cephalosporin (units/ml) Produced by Strain 92G AD-11 on Media with Different Sulfur Sources in Shake Flasks[a]

Strain	Major sulfur source		
	Methionine[b]	Sulfate[c]	Methionine + sulfate[d]
92G-AD-11	8.85	8.35 ± 0.31	9.68
BAH-16	2.54 ± 0.26	9.84 ± 0.56	10.36

[a]With 30 ml of medium per 500-ml flask.
[b]A complex basal medium with methionine raised from 0.6 to 2.0% and with $(NH_4)_2SO_4$ replaced with 0.75% NH_4Cl.
[c]Complex basal medium without methionine and with $(NH_4)_2SO_4$ reduced from 1.3 to 1.0%.
[d]Complex basal medium.

functional regulatory protein, protein C. The ability of strain 394-4 to produce cephalosporin C from sulfate is improved over that of strain 92G-AD-11, and its ability to produce cephalosporin C on media containing both methionine and sulfate is improved over strain BAH-16.

We note with interest a recent report on a selenomethionine resistant mutant of *Cephalosporium acremonium* (Matsumura et al., 1982). This mutant, SMR-13, like our mutant 394-4, showed substantial improvement over parent for cephalosporin C synthesis with sulfate as sulfur source. Like mutant 394-4, SMR-13 retained its ability to produce cephalosporin C with methionine as sulfur source. Like mutant IS-5 (see section D, below) reported by Komatsu et al. (1975), SMR-13 produced cephalosporin C maximally on very low amounts of methionine (0.5 g/liter) and antibiotic production was inhibited at high methionine concentrations (4 g/liter). Inhibition of cephalosporin C synthesis on 4 g/liter methionine has not been observed for mutant 394-4.

D. Strains Without O-acetylserine-Mediated Induction

In *S. typhimurium* it has been shown that induction of enzymes of sulfate assimilation (enzymes 2–5,* Fig. 1) and the induction of O-acetylserine sulfhydrylase (enzyme 7, Fig. 1) require the presence of an intracellular coinducer metabolite, O-acetylserine (Kredich, 1971). The cysB mutants (cysteine auxotrophs) are believed to be deficient in a regulatory protein which acts to induce the enzymes of sulfate assimilation only when O-acetylserine is bound to that regulatory protein. Mutants which no longer require O-acetylserine for induction of sulfate assimilation enzymes have been isolated (Kredich, 1971). The mutation corresponding to this phenotype also maps at the cysB locus. These mutants are believed to have an altered cysB allele that codes for an altered regulatory protein which no longer requires binding of O-acetylserine to activate its regulatory function.

An O-acetylserine induction of enzymes 2–5* (Fig. 1) may also exist in fungi. In fungi, however, enzymes 8, 9, 16, and 17 may also be induced by O-acetylserine.

---

*Sulfate permease (enzyme 1, Fig. 1) is probably also regulated by this mechanism; however, this enzyme was not assayed by Kredich (1971).

If O-acetylserine-mediated induction (OAS-mi) occurs in fungi, mutations which block the conversion of O-acetylserine to cysteine (deficient in enzyme 7, Fig. 1) would be expected to cause a significant accumulation of O-acetylserine and therefore cause a dramatic induction of sulfate assimilation enzymes. Mutants deficient in O-acetylserine sulfhydrylase have been isolated for both *C. acremonium* and *A. nidulans*. These mutants are *not* cysteine auxotrophs. They are prototrophs because the block in cysteine formation from O-acetylserine is compensated for by increased cysteine formation from O-acetylhomoserine via enzymes 8, 9, 16, and 17 (Fig. 1). O-Acetylserine sulfhydrylase-negative mutants of *A. nidulans* have been shown to produce increased levels of enzyme 9. A number of O-acetylserine sulfhydrylase-negative mutants of *C. acremonium* were shown to produce more cephalosporin C from sulfate than their O-acetylserine sulfhydrylase-proficient parents. The amount of cephalosporin C produced from sulfate in the mutants were equivalent to the cephalosporin C produced from methionine in their parents (Treichler et al., 1979).

If the OAS-mi of enzymes 1–5, 8, 9, 16, and 17 does occur in *C. acremonium*, then it should be possible to isolate mutants that produce an altered cysB protein which requires less co-inducer, O-acetylserine, to be activated than does the normal cysB protein. Such mutants would be expected to produce more cysteine and cephalosporin C from sulfate than their parent. Such mutants would be O-acetylserine sulfhydrylase proficient. Komatsu et al. (1975) have isolated mutant IS-5, which exhibits these two properties. It cannot be a mec mutant, because it was observed to form more internal cysteine from methionine than its parent (Komatsu and Kodaira, 1977).

Supplementation of sulfate-based medium with methionine will improve the yield of cephalosporin C for O-acetylserine sulfhydrylase-negative mutants, the IS-5 mutant, and their parents. However, the optimal amount of methionine supplement is low (e.g., 0.2%) for the O-acetylserine sulfhydrylase-negative mutants and the IS-5 mutant. The optimal supplement of methionine is high (e.g., 0.75%) for their parents. Minor increases in methionine level over the level optimal for cephalosporin C synthesis causes a dramatic reduction in cephalosporin C synthesis for O-acetylserine sulfhydrylase-negative mutants and mutant IS-5. Minor increases in methionine concentration over the optimal level for the parents of these mutants have little or no effect on cephalosporin C synthesis. Because of these characteristics, the O-acetylserine sulfhydrylase-negative mutants and the IS-5 mutant are said to be "methionine sensitive."

Komatsu and Kodaira (1977) have suggested that methionine sensitivity might be caused by an excessively high intracellular pool of cysteine brought about by increased concentrations of enzyme 16 (Fig. 1). This hypothesis was consistent with the observed intracellular concentrations of enzyme 16 and cysteine, and extracellular concentrations of cephalosporin C obtained from cells taken from 12-hr replacement cultures* containing different sulfur sources (Table 5). Too much internal cysteine might competitively inhibit the condensation of L-α-aminoadipyl-L-cysteine with L-valine to form L-α-aminoadipyl-L-cysteine-D-valine (step 3 in the biosynthesis of cephalosporin C; Fig. 11.†

---

*Sulfur-starved cells were transferred to a basal glucose/sucrose–ammonium chloride medium containing supplements of various sulfur sources and incubated 12 hr.

†For a discussion of the biosynthesis of cephalosporin C, see Queener and Neuss (1982).

Table 5 Intracellular Concentrations of Cysteine and Enzyme 16 and Extracellular Levels of Cephalosporin C (CPC) for Mutant IS-5 and Its Parent Incubated in the Presence of Sulfate, Sulfate + Norleucine, or L-Methionine

Strain	NaSO$_4$ (0.5%)			NaSO$_4$ (0.5%) 0.25% DL-norleucine			NaSO$_4$ (0.5%) 0.25% DL-methionine		
	Cysteine[a]	Enzyme 16[b]	CPC[c]	Cysteine	Enzyme 16	CPC	Cysteine	Enzyme 16	CPC
IS-5	Trace	1.36	240	4.49	d	110	6.34	1.43	130
Parent (N16)	0	0.70	150	2.19	d	230	3.46	0.61	220

[a]Formed from H$_2$S and L-serine and measured in nanomoles per minute per milligram of protein.
[b]Micromoles per gram dry weight of cells.
[c]Cephalosporin C, in micrograms per milliliter of culture broth.
[d]Not measured.
Source: Adopted from Komatsu and Kodaira, 1977.

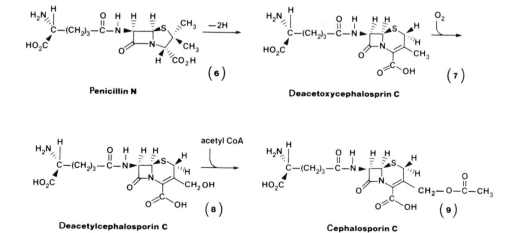

Figure 11 Proposed biosynthetic pathway to cephalosporin C (see Queener and Neuss, 1982).

Table 6 Suggested Genotypes for Mutants of *C. acremonium* with Altered Regulatory Mechanisms

Mutant designation[a]	Methionyl-tRNA^{met} mediated repression[b]		Reverse trans-sulfuration metabolite-mediated repression[c]		O-Acetylserine-mediated induction[d]	
	Corepressor	Regulatory protein	Corepressor	Regulatory protein	Coinducer	Regulatory protein
	Methionyl-tRNA^{met}	Protein A	S-Adenosylmethionine, homocysteine, or cysteine (unknown at present)	Protein C	O-Acetylserine	"cysB protein"
M8650-3, M8650-5	Normal	Normal	Normal	Normal	Normal	Normal
ACI-47	Normal	Normal	Defective, corepressor metabolite cannot be formed	Normal	Normal	Normal
92G-AD-11	Normal	Defective, protein does not bind corepressor	Normal	Normal	Normal	Normal

BAH-16	Normal	Defective, protein does not bind corepressor	Defective, corepressor metabolite cannot be formed	Normal	Normal	Normal
394-4	Normal	Defective protein does not bind corepressor	Normal	Defective, protein does not bind corepressor	Normal	Normal
OASS-7	Normal	Normal	Normal	Normal	O-Acetylserine is accumulated owing to block at cysteine synthetase	Normal
IS-5	Normal	Normal	Normal	Normal	Normal	Altered protein, does not require O-acetylserine to induce

aControl effect by normal mechanism (theoretical).
bEnzymes 1–5, 7, 8, 9, 10, 11, and 12 are repressed.
cEnzymes 1–5, 7, 8, 9, 10, 11, and 12 are repressed.
dEnzymes 1–5, 7, 8, 9, 16, and 17 are induced.

The ratio of the concentration of enzyme 16 in IS-5 to the concentration of enzyme 16 in N16 was at least 2. This increased concentration of enzyme 16 in IS-5 is consistent with a genotype in which cysB protein is altered so that it no longer requires the presence of O-acetylserine to induce the enzymes 1–5, 8, 9, 16, and 17. The similarity of concentration of enzyme 16 for cells incubated in the presence of sulfate or methionine suggests that enzyme 16 is not under control by MtR-mr or RTM-mr. Lack of control of enzyme 16 by MtR-mr or RTM-mr would suggest that the primary function of enzyme 16 in *C. acremonium* is catalysis of cystathionine synthesis from homocysteine and serine.

IV. SUMMARY

The properties of seven mutants (92G-AD-11, ACI-47, M8650-5, BAH-16, 394-4, 8650/OAS⁻-7, and IS-5) are explained in terms of regulatory mechanisms which are currently believed to operate in controlling sulfur metabolism in bacteria and fungi. The genotypes suggested for these mutants are summarized in Table 6.

ACKNOWLEDGMENTS

The primary author (SWQ) wishes to acknowledge J. Chapman, J. McDermott, C. Nash, C. Platt, D. Tunin, S. Wilkerson, and J. Westpheling for the microbiological and biochemical experiments upon which this discussion is based. Dr. Lawrence E. Day deserves mention for encouragement and suggestions during the preparation of the manuscript. The author is grateful for the expert secretarial assistance of Mrs. Jo Ann Hamilton.

REFERENCES

Cherest, H., Surdin-Kerjan, Y., and deRobichon-Szulmajster, H. (1971). Methionine-mediated repression in *Saccharomyces cerevisiae*: A pleiotropic regulatory system involving methionyl transfer ribonucleic acid and the product of the gene eth 2. *J. Bacteriol. 106*:758–777.

Cherest, H., Surdin-Kerjan, Y., Antoniewski, J., and deRobichon-Szulmajster, H. (1973a). S-Adenosylmethionine-mediated repression of methionine biosynthetic enzymes in *Saccharomyces cerevisiae*. *J. Bacteriol. 114*: 928–983.

Cherest, H., Surdin-Kerjan, Y., Antoniewski, J., and deRobichon-Szulmajster, H. (1973b). Effects of regulatory mutations upon methionine biosynthesis in *Saccharomyces cerevisiae*: Loci eth 2-eth 3-eth 10. *J. Bacteriol. 115*:1084–1093.

Cherest, H., Surdin-Kerjan, Y., and deRobichon-Szulmajster, H. (1975). Methionine- and S-adenosylmethionine-mediated repression in a methionyl-transfer ribonucleic acid synthetase mutant of *Saccharomyces cerevisiae*. *J. Bacteriol. 123428*–435.

Garrido, M. L. (1964). Determination of sulfur in plant material. *Analyst 89*:61.

Gehrke, W. C., and Neuner, R. E. (1974). Automated chemical determination of methionine. *J. Assoc. Off. Anal. Chem. 57*:682.

Hobson, A. C. (1974). The regulation of methionine and S-adenosylmethionine biosynthesis and utilization in mutants of *Salmonella typhimurium* with defects in S-adenosylmethionine synthesis. *Mol. Gen. Genet. 131*:263–273.

Hobson, A. C. (1976). The synthesis of S-adenosylmethionine by mutants with defects in S-adenosylmethionine synthetase. *Mol. Gen. Genet. 144*: 87–95.

Hobson, A. C., and Smith, D. A. (1973). S-Adenosylmethionine synthetase in methionine regulatory mutants of *S. typhimurium. Mol. Gen. Genet. 126*:7–18.

Kase, H., and Nakayama, K. (1975). Isolation and characterization of S-adenosylmethionine requiring mutants and role of S-adenosylmethionine in the regulation of methionine biosynthesis in *Corynebacterium glutamicum. Agric. Biol. Chem. 39*:161–163.

Kern, B. A., and Inamine, E. (1981). Cystathionine n-lyase activity in the cephamycin C producer *Streptomyces lactamdurans. J. Antibiot. 34*: 583–589.

Komatsu, K., and Kodaira, R. (1977). Sulfur metabolism of a mutant of *Cephalosporin acremonium* with enhanced potential to utilize sulfate for cephalosporin C production. *J. Antibiot. 30*:226–233.

Komatsu, K., Yamada, T., and Kodaira, R. (1974). Isolation and characteristics of pool methionine-rich mutants of *Candida* sp. *J. Ferment. Tech. 52*:93–99.

Komatsu, K., Mizumo, M., and Kodaira, R. (1975). Effect of methionine on cephalosporin C and penicillin N production by a mutant of *Cephalosporium acremonium. J. Antibiot. 28*:881–887.

Kredich, N. M. (1971). Regulation of L-cysteine biosynthesis in *Salmonella typhimurium. J. Biol. Chem. 246*:3474–3484.

Kredich, N. M., and Becker, M. A. (1971). Cysteine biosynthesis: Serine transacetylase and O-acetylserine sulfhydrylase. In *Methods in Enzymology*, Vol. 17-B. H. Tabor and C. W. Tabor (Eds.). Academic, New York, pp. 459–463.

Lawrence, D. A., Smith, D. A., and Rowbury, R. J. (1968). Regulation of methionine synthesis in *Salmonella typhimurium* mutants resistant to inhibition by analogs of methionine. *Genetics 58*:473–492.

Martin, J. F., and Demain, A. L. (1980). Control of antibiotic biosynthesis. *Microbiol. Rev. 44*:232–251.

Masselot, M., and deRobichon-Szulmajster, H. (1972). No nonsense mutation in the regulatory gene ETH 2 involved in methionine biosynthesis in *Saccharomyces cerevisiae. Genetics 71*:535–550.

Mertz, J. E., and Spence, K. D. (1972). Methionine adenosyltransferase and ethionine resistance in *S. cerevisiae. J. Bacteriol. 111*:778–783.

Nakamura, T., and Sato, R. (1963). Synthesis from sulfate and accumulation of S-sulfocysteine by a mutant strain of *Aspergillus nidulans. Biochem. J. 86*:328–335.

Nüesch, J., Hinnen, A., Liersch, M., and Treichler, H. J. (1976). A biochemical and genetical approach to the biosynthesis of cephalosporin C. In *Second International Symposium on the Genetics of Industrial Microorganisms*. K. D. Macdonald (Ed.). Academic, New York, pp. 451–472.

Paszewski, A., and Grabski, J. (1963). Studies on γ-cystathionase and O-acetylhomoserine sulfhydrylase as enzymes of alternative methionine biosynthetic pathways in *Aspergillus nidulans. Acta Biochim. Pol. 20*:159–167.

Paszewski, A., and Grabski, J. (1974). Regulation of S-amino acids biosynthesis in *Aspergillus nidulans*. *Mol. Gen. Genet.* *132*:307–320.

Matsumara, M., Yoshida, T., and Taguchi, H. (1982). Synthesis of Cephalosporin C by a methionine analog resistant mutant of *Cephalosporium acremonium*. *Eur. J. Appl. Microbiol. Biotechnol.* *16*:114–118.

Pieniazek, N. J., Kowalska, I. M., and Stepien, P. P. (1973). Deficiency in methionine adenosyltransferase resulting in limited repressibility of methionine biosynthetic enzymes in *Aspergillus nidulans*. *Mol. Gen. Genet.* *126*:367–374.

Pieniazek, N. J., Bal, J., Blabin, E., and Stepien, P. P. (1974). An *Aspergillus nidulans* mutant lacking serine transacetylase: Evidence for two pathways of cysteine biosynthesis. *Mol. Gen. Genet.* *132*:363–366.

Queener, S. W., and Ellis, L. F. (1975). Differentiation of mutants of *Cephalosporium acremonium* in complex medium: The formation of unicellular arthrospores and their germination. *Can. J. Microbiol.* *21*:1981–1996.

Queener, S. W., and Neuss, N. (1982). Biosynthesis of β-lactam antibiotics. In β-*Lactam Antibiotics*. R. Morin and M. Gorman (Eds.). Academic, New York, pp. 1–81.

Queener, S. W., Capone, J. J., Radue, A. B., and Nagarajan, N. (1974). Synthesis of deacetoxycephalosporin C by a mutant of *Cephalosporium acremonium*. *Antimicrob. Agents Chemother.* *6*:334–337.

Queener, S. W., McDermott, J., and Radue, A. B. (1975). Glutamate dehydrogenase specific activity and cephalosporin C synthesis in the M8650 series of *Cephalosporium acremonium*. *Antimicrob. Agents Chemother.* *7*:646–651.

Redstone, M. (1971). Prog. Abstr. Intersci, Conf. Antimicrob. Ag. Chemother., 11th, Atlantic City, N.J., Abstr. 73, p. 37.

Stepien, P. P., Pieniazek, N. J., Bal, J., and Morzycka, E. (1975). Cysteine biosynthesis in *Aspergillus nidulans*. *Acta Microbiol. Pol.* *7*:201–210.

Surdin-Kerjan, Y., Cherest, H., and deRobichon-Szulmajster, H. (1973). Relationship between methionyl transfer ribonucleic acid cellular content and the synthesis of methionine enzymes in *Saccharomyces cerevisiae*. *J. Bacteriol.* *113*:1156–1160.

Surdin-Kerjan, Y., Cherest, H., and deRobichon-Szulmajster, H. (1976). Regulation of methionine synthesis in *Saccharomyces cerevisiae* operates through independent signals: Methionyl-tRNA^{met} and S-adenosylmethionine. *Acta Microbiol. Acad. Sci. Hung.* *23*:109–120.

Suzuki, M., Fujisawa, Y., and Uchida, M. (1980). S-Sulfocysteine as a source of the sulfur atom of cephalosporin C. *Agric. Biol. Chem.* *44*:1995–1997.

Treichler, H. J., Liersch, M., Nüesch, J., and Bobeei, H. (1979). Role of sulfur metabolism in cephalosporin C and penicillin N biosynthesis. In *Third International Symposium on the Genetics of Industrial Microorganisms*. O. K. Sebek and A. I. Laskin (Eds.). American Society for Microbiology, Washington, D.C., pp. 97–104.

Woodin, T. S., and Segel, I. H. (1968). Glutathione reductase-dependent metabolism of cysteine-S-sulfate by *Penicillium chrysogenum*. *Biochim. Biophys. Acta* *167*:78–88.

5

ENZYMATIC CONVERSIONS USED IN THE PRODUCTION OF PENICILLINS AND CEPHALOSPORINS

T. A. SAVIDGE *Beecham Pharmaceuticals, Worthing, Sussex, England*

I. INTRODUCTION

The isolation and characterization of the penicillin nucleus, 6-aminopenicil-
lanic acid (6-APA), by Batchelor et al. (1959) and the development of "semisynthetic penicillins" made by attaching different side chains to 6-APA led
to a search for a production method based on the deacylation of naturally produced penicillin G or V. The initial method for making 6-APA by direct fermentation, (essentially carried out as a penicillin G or V fermentation, but
without the addition of side-chain precursor) was very inefficient, because
6-APA broth titers were low and the procedure for its recovery from broth
was extremely complex and costly. At that time no chemical deacylation process was available which did not also open the labile β-lactam ring of the penicillin to form unwanted penicilloic acid, and an enzymatic process offered the
prospect of the mild reaction conditions necessary to avoid this.

Subsequently, four pharmaceutical companies almost simultaneously reported on finding microbial enzymes capable of deacylating either penicillin
G or V: Bayer A. G. with *Escherichia coli* (Kaufman and Bauer, 1960), Beecham with *Streptomyces lavendulae* (Rolinson et al., 1960), Bristol-Myers
with *Alcaligenes faecalis* (Claridge et al., 1960), and Pfizer with *Proteus*

rettgeri, (Huang et al., 1960). As screening programs continued, it became evident that this "penicillin acylase" activity was present in a wide range of bacteria and fungi, the preferred substrate for bacterial enzyme was generally penicillin G, while fungal enzymes preferentially hydrolyzed penicillin V.

Although an efficient chemical deacylation process was ultimately developed and operated on a commercial scale (reviewed by Huber et al., 1972), the enzymatic process is still used extensively. The relative merits of either process will to a large extent depend upon the detailed expertise and efficiency developed within individual pharmaceutical companies, but there now seems to be a preference for the enzymatic route in view of the high energy and solvent costs together with some environmental problems associated with the "hard technology" of the chemical route (Dunnill, 1980). A further factor in favor of the enzymatic route has been due to the development during the 1970s of immobilized enzyme processes which were more efficient both in terms of reduced enzyme costs as a result of reusing the immobilized enzyme and improved deacylation yields. Such processes have now largely supplanted the whole microbial cell processes used originally. Indeed, 6-APA production was one of the first successful commercial applications of immobilized enzyme technology and the various processes that have emerged use many of the different types of immobilization methods, including immobilization of whole cells (Vandamme, 1981b; Karube et al., 1984).

Some idea of the size of the 6-APA market may be gained from published estimates from various sources which indicate a total worldwide production in the region of 2500 tons/per year at a current cost of around $75,000 per ton, depending upon market availability. Most of this 6-APA is not sold as such, but is used directly by the producing companies for the manufacture of semisynthetic penicillins.

The most widely used source of acylase is *E. coli*, the substrate for which is penicillin G. Some companies, however, prefer to use penicillin V, and processes using penicillin acylase of fungal or bacterial origin have been developed, although the extent to which they are operated commercially is uncertain.

Early in the development of penicillin acylases it was realized that they could be used in the reverse direction to synthesize semisynthetic penicillins. Until recently, such processes were relatively inefficient and were not competitive with the chemical routes, but recent reports of processes based on the use of α-amino acid ester hydrolases seem more encouraging.

The later but parallel development of semisynthetic cephalosporins provided further opportunity for enzymatic conversions, and penicillin acylase has been used both to produce intermediates by deacylating the compounds derived by ring expansion of penicillins to cephalosporins and to synthesize some semisynthetic cephalosporins.

While penicillin acylases occupy a central role in the interconversions of β-lactam antibiotics, some other enzymatic reactions have been described, for example, the use of an acylesterase to produce the intermediate deacyl-7-aminocephalosporanic acid.

Enzyme processes are used not only in the production of penicillins and cephalosporins, but also in the manufacture of some of the side-chain acids which are attached to either the penicillin of cephalosporin nucleus to form the corresponding semisynthetic antibiotic. These processes are concerned with the optical resolution of racemic mixtures of the side chains, since the antibacterial activity of both penicillins and cephalosporins having a side chain with an α-amino substituent is greater when the D($-$) isomer is used.

II. PENICILLIN ACYLASES

A. Classification

The official nomenclature for penicillin acylases, irrespective of the preferred substrate, is penicillin amidohydrolase (E.C.3.5.1.11), with the trivial names "penicillin acylase" or "penicillin amidase." Penicillin acylase is preferred, in that it more accurately describes the reaction, namely, the removal of N-acyl groups from N-acyl 6-APA derivatives.

The reactions, which can proceed in either direction, are shown in Figure 1. The inclusion of the semisynthetic penicillin ampicillin stems from a later discovery of a class of enzymes for which preferred substrates are ampicillin or amoxycillin. Some of these enzymes are completely inactive in the hydrolysis of penicillins G or V, and Takahashi et al. (1974) have classified such enzymes as α-amino acid ester hydrolases (see Sec. VI.C).

Enzymes for which the preferred substrate is penicillin V are generally found among the fungi, and this has led in the past to a classification of the enzymes on the basis of their origin, that is, bacterial and fungal acylases for which preferred substrates are penicillins G and V, respectively. This classification is insufficiently rigid, however, and therefore one based on the preferred substrate is now recommended: for example, benzylpenicillin (penicillin G) acylase, phenoxymethylpenicillin (penicillin V) acylase, and D(−)-α-aminobenzylpenicillin (ampicillin) acylase. These enzymes can also be used in certain bioconversions of cephalosporins, as discussed in Sec. VII, and can also hydrolyze a variety of acylamino acids, amides, and esters.

B. Assay Methods

1. Penicillin G and Penicillin V Acylase

The most reliable method for assaying penicillin acylase is to determine the end product 6-APA which is produced when the enzyme is incubated at a suitable pH in the presence of an appropriate penicillin.

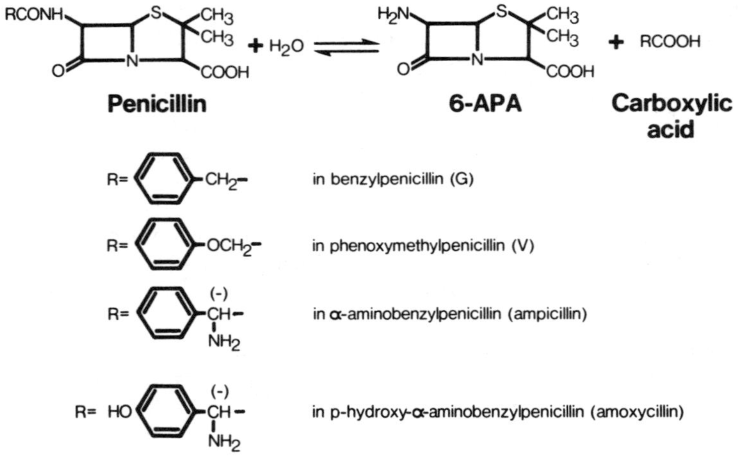

Figure 1 Reactions of penicillin acylases.

For a program in which microorganisms are being screened for the presence of penicillin acylase activity, a culture filtrate or suspension of microbial cells in buffer is incubated with substrate penicillin solution for a few hours at 30-37°C and pH 8.0, with gentle shaking if cells are present. Choice of substrate and exact reaction conditions must be determined experimentally. Cell-free samples from the reaction mixtures are then examined for 6-APA content by biochromatography, which, although time-consuming, is extremely sensitive and therefore suitable for detecting the low level of enzyme which may be produced in initial culture screens. In this procedure, loaded paper chromatography strips are chromatographed overnight in a suitable solvent system, and after drying, duplicate strips, one of which has been sprayed with phenylacetylchloride, are placed on agar plates seeded with *Bacillus subtilis*. Phenylacetylation converts any 6-APA present on the strip to antibacterially active penicillin G, which produces a zone of inhibition on the assay plate. Quantification of 6-APA concentration is made by the usual bioassay procedures and the method is capable of detecting 6-APA unequivocally down to a concentration of 1 µg/ml. Full experimental details of this method are described by Cole et al. (1975). This rather laborious biochromatographic technique for detecting 6-APA was replaced by thin-layer chromatography, in order to facilitate screening by Nara et al. (1971) and Lowe et al. (1981), who detected unreacted substrate (penicillin V), phenoxyacetic acid, and penicilloic acid by short-wave ultraviolet and 6-APA by spraying with ninhydrin.

Once a suitable culture which produces reasonable quantities of enzyme has been selected, several alternative assay methods are available. These may vary in the specificity of 6-APA determination, and their suitability will depend upon the care taken to ensure that the selected organism does not also produce β-lactamase, which opens the β-lactam ring of penicillins to form penicilloic acid (for a recent review of these enzymes see Vandamme, 1981a). Penicilloic acid may also be formed as a result of alkaline hydrolysis of penicillin and care must therefore be exercised in devising the conditions for the enzyme reaction.

The colorimetric hydroxylamine method for determining penicillins has been used extensively (Batchelor et al., 1961), although the method is somewhat time-consuming in that after enzymation, unreacted penicillin substrate must be removed by solvent extraction at pH 2. The 6-APA which remains is determined by reacting with hydroxylamine and measuring absorbance at 490 nm. The method cannot be used if ampicillin is the substrate, since it is not solvent extractable. Penicilloic acid does not react with hydroxylamine. For complete experimental details see Cole et al. (1975).

Another colorimetric method which does not necessitate removal of unreacted substrate penicillin is that based upon the reaction of the 6-APA amino group with p-dimethylaminobenzaldehyde, as first described by Bomstein and Evans (1965). This was successfully applied to the assay of penicillin acylase by Balasingham et al. (1972). Since desacetylpenicilloic acid will also react, although at a slower rate, care must be taken to ensure that no β-lactamase activity is present. The method is particularly amenable to automation by autoanalysis, and therefore especially suitable for use in culture development programs. A similar method for the spectrophotometric determination of 6-APA in the presence of penicillins using D-glucosamine as the reagent has been described (Shaikh et al., 1973). High-pressure liquid chromatography can also be used for the specific measurement not only of 6-APA, but also of unreacted substrate, side-chain acid, and penicilloic acid (Lowe et al.,

1981), but the time involved for each assay may preclude its use for routine culture work.

As an alternative to assays based upon the determination of 6-APA, the determination of end-product carboxylic acid has been used. Niedermayer (1964) and Chiang and Bennett (1967) determined phenylacetic acid using gas chromatography, and several workers have described assays based upon titration of side chain acid using pH stats; for experimental details see Lagerlöf et al. (1976).

A number of assays for penicillin acylase using model substrates have also been developed. All have been used only with the enzyme derived from *E. coli* and therefore their wider applicability is not known. Bauer et al. (1971) described an assay based upon the ability of the enzyme to hydrolyze phenylacetyl-L-asparagine to L-asparagine and phenylacetic acid. By concurrently reacting the asparagine to aspartic acid and ammonia with an excess of L-asparaginase, the assay of penicillin acylase was based upon quantitative determination of released ammonia using Nessler's reagent.

Walton (1964) based an assay on the hydrolysis of phenylacetyl-p-nitroaniline to release p-nitroaniline which was determined colorimetrically, but application of the method was limited by the poor solubility of the substrate in aqueous systems. Kutzbach and Rauenbusch (1974) overcame this limitation by introducing a carboxylic group into the meta position of the anilide group. This compound, 6-nitro-3-phenylacetamido-benzoic acid, has been used very successfully as a substrate for the assay of enzyme solutions and to visualize the acylase band in gel electrophoresis of enzyme extracts; its use for the assay of fermentation broths may be limited owing to interference by other broth constituents.

Units of enzyme activity are usually expressed as μmoles of 6-APA per minute under the conditions of the test, generally pH 7.8 or 8.0 and 37 or 40°C.

2. α-Amino Acid Ester Hydrolase

The assay method described by Takahashi et al. (1974) was based upon titration of the D(−)-aminophenylacetic acid released upon incubation of the corresponding methyl ester with enzyme at 27°C and pH 6.0.

C. Selection of Penicillin Acylase-Producing Microorganisms

1. Initial Screening

Random screening has proved to be the most productive method for selecting microorganisms expressing penicillin acylase. Attempts have been made to correlate this activity with penicillin resistance. Holt and Stewart (1964) found that 40% of 310 clinical isolates of *E. coli* produced penicillin acylase, but this result was not confirmed by the more extensive investigations carried out by Cole and Sutherland (1966), who concluded that penicillin acylase was not a significant factor in determining resistance to penicillin. Moreover, the conditions prevailing in sensitivity tests (alkaline pH, 37°C, and stationary culture) are not conducive to the production of penicillin acylase. A similar conclusion (8% acylase positive isolates) was reported to Rozansky et al (1969).

A selective isolation procedure based on growth on a mineral medium containing benzylpenicillin or phenylacetic acid as sole carbon source was proposed by Kameda et al. in 1961; they also correlated hydrolysis of acyl-DL-amino acids with the presence of penicillin acylase.

Batchelor et al. (1961) screened 215 species of fungi, yeasts, and actino-mycetes and selected 38 phenoxymethylpenicillin acylase-producing strains after growth on corn-steep liquor and related penicillin fermentation broths, while Huang et al. (1963) found 60 benzylpenicillin acylase producers from 392 isolates screened on media based on either corn-steep liquor, yeast extract, or casamino acids. In searching for a microbial method for the production of ampicillin, Nara et al. (1971) first screened 251 species of bacteria, 229 species of actinomycetes, 2 species of yeast, and 37 species of basidiomycetes using thin-layer chromatography of culture filtrates to which either penicillin G or V had been added, and selected only 9 bacteria and 5 actinomycetes the basis of 6-APA production. Vandamme and Voets (1973) isolated acylase-producing microorganisms from soil using a mineral medium to which N-acetylglycine, N-glycylglycine, or phenylacetamide was added as sole carbon source.

More recently, Diers and Emborg (1979) isolated phenoxymethylpenicillin acylase-producing bacteria by enrichment culture of soil in a mineral salts medium to which phenoxyacetic acid and penicillin V were added as carbon and nitrogen sources, while Lowe et al. (1981) screened 2000 bacterial cultures from random soil suspensions and found that over 10% possessed some penicillin V acylase activity. The screening procedure used by the latter authors was shaken-flask incubation in tryptone soya broth either with or both with and without phenoxyacetic acid (0.1%) as inducer. Samples of whole broth were tested for acylase activity by shaking with penicillin V solution (3%) buffered to either pH 5.5 or 7.5 for 2 hr at 30°C. Samples of reaction mixture were assayed by the thin-layer chromatographic procedure described in Sec. II.B.1. Cultures with high β-lactamase activity (determined by presence of penicilloic acid zones) were discarded. About 2% of the screened cultures selected on the basis of good acylase activity were examined to determine the optimum pH of enzyme activity and optimum medium composition. The three highest yielding strains were all identified as species of *Pseudomonas*. Enzyme from these cultures was extracted from disrupted cells, and both the soluble and cell-bound enzyme examined to determine optimum temperature, pH, and substrate inhibition characteristics. This led to the elimination of two isolates on the basis of low temperature optima and poor substrate tolerance, leaving one isolate, *Pseudomonas acidovorans,* which was acceptable on all counts. The activity of this enzyme on penicillin G was 10% of that on penicillin V; penicillins F and dihydro F and K were not split. Penicillin K (n-pentylpenicillin) strongly inhibited the deacylation of penicillin V. The enzyme may be distinguished from the ampicillin acylases of *Pseudomonas* (Sec. VI.B) on the basis of substrate profiles for hydrolysis and by the fact that no synthetic activity could be detected using various side chains either as free acids or methyl esters.

From such screening programs commercial companies selected their preferred organism which was then developed to provide processes suitable for producing 6-APA on an industrial scale, as described in Secs. III and IV. It is apparent from these studies, therefore, that although penicillin acylase activity is not widespread among microorganisms, it can be detected fairly readily. For detailed surveys of strains reported to produce penicillin acylase activity see Vandamme and Voets (1974) and Vandamme (1981a).

The organisms selected by the various companies making 6-APA and semi-synthetic penicillins industrially may be determined from the patent literature (Table 1), although, of course, this is no guarantee that these organisms are actually used on an industrial scale—since this information is not generally made available.

Table 1 Some Examples of Penicillin Acylase-Producing Microorganisms Described in the Patent Literature

Company	Patent reference	Organism	Enzyme type[a]	Intra- or extracellular
Bayer A. G.	BP 893418 (1962)	*Escherichia coli* ATCC 11105 and 9637	I	Intra
Beecham	BP 892144 (1962)	*Streptomyces lavendulae*	II	Extra
Beecham	BP 1009028 (1965)	*Achromobacter* sp.	III	Intra
Squibb	USP 3121667 (1964)	*Bacillus megaterium*	I	Extra
Pfizer	USP 3088880 (1963)	*Proteus rettgeri*	I	Intra
Eli Lilly	USP 3150059 (1964)	*Actinoplanes* sp.	II	Extra
Biochemie	BP 961517 (1964)	*Fusarium* sp.	II	Intra
Biochemie	DT 2503584 (1974)	*Bovista plumbea*	II	Intra
Kyowa Hakko	DT 1967074 (1977)	*Kluyvera citrophila*	I	Intra
Kyowa Hakko	DT 2050983 (1971)	*Pseudomonas melanogenum*	III[b]	Intra
Takeda	USP 374962 (1973)	Species of *Mycoplana*, *Protaminobacter*, *Acetobacter*, *Aeromonas*, *Xanthomonas*	III[b]	Intra
Novo	BP 2021119 (1979)	Gram-negative bacillus	II	Intra

[a]Type I, benzylpenicillin (penicillin G) acylase; type II, phenoxymethylpenicillin (penicillin V) acylase; type III, D(−)-α-aminobenzylpenicillin (ampicillin) acylase.

[b]These enzymes have been used only for the synthesis of semisynthetic penicillins (ampicillin and amoxycillin).

2. Strain Improvement

Having selected the strain to be used for penicillin acylase production, it is likely that most industrial companies will have endeavored to improve the strain by culture development. If the strain also produced β-lactamase, then the first task may well be to select β-lactamase-negative mutants. Improvement of acylase production has in the past been achieved by conventional mutation and selection for higher-yielding strains by productivity testing. Thus in the author's laboratory, the productivity of a selected strain of E. coli was increased significantly by mutation, using for example, irradiation with ultraviolet light or treatment with N-methyl-N'-nitro-N-nitrosoguanidine (NTG) using the method of Adelberg et al. (1965) and screening randomly selected isolates for penicillin acylase activity by assaying the cells obtained following fermentation in 250-ml shaken flasks. Recently Son et al. (1982) have described the mutation and screening methods used to isolate more productive mutants of the benzylpenicillin acylase-producing strain, *Bacillus megaterium* (ATCC 14945). Bacterial cells at logarithmic growth phase were diluted to 10^6-10^8 cells/ml and treated with UV light or with ethylmethanesulfonate (EMS) followed by NTG to achieve a kill of 99.9%. However, it was found that milder treatment with UV light (3–4 minutes exposure with a 15 W lamp at a distance of 27 cm) was most effective in inducing mutation at the acylase gene locus.

Mutants were selected which not only produced enzyme in greater quantity, (e.g., threefold) but which also were partially constitutive and were capable of some production in the absence of the inducer phenylacetic acid (Sec. III.A) although maximum productivity was achieved in the presence of inducer. Similarly, Vojtisek et al. (1973) reported isolation of constitutive mutants of E. coli and mutants that were more resistant to catabolite repression of enzyme synthesis by either glucose or acetate.

While the more traditional methods of mutation and selection have been used successfully in the past, current programs to improve yield are likely to involve some aspects of modern recombinant DNA technology. An account of the application of these techniques to improving the penicillin acylase activity of E. coli has been given by Mayer et al. (1979, 1980), who cloned the acylase gene from E. coli ATCC 11105 into a non-acylase-producing strain, E. coli 5K, using a cosmid vector system.

Cosmids are plasmids carrying the λ bacteriophage cos site which allows the plasmid, including large foreign DNA fragments which have been ligated into suitable restriction sites, to be taken up into bacteriophage particles. These particles containing hybrid molecules are adsorbed to E. coli recipients and the hybrid plasmids introduced with high efficiency into the recipient bacteria. Since the original cosmids do not meet packaging requirements, they cannot form infective particles and therefore this provides a direct physical selection for hybrids. Another advantage of this system is that large fragments can be cloned, thus reducing the number of hybrid clones required to be screened before the desired gene is selected.

The gene library was established after restriction endonuclease (Hind III) fragmentation of E. coli ATCC 11105 total chromosome and the cosmid pJC720. After ligation and in vitro packaging, 10,000 clones were screened initially for penicillin acylase using a plate overlay test (soft agar containing *Serratia marcescens* and penicillin G) from which one positive clone (pHM5) was obtained. This strain, which was inducible and produced the same level of penicillin acylase as ATCC 11105, was then cloned onto the multicopy plasmid pBR322, from which the β-lactamase gene had been destroyed by replace-

Table 2 Penicillin G Acylase Levels of Hybrid Strains[a]

Strain	Specific activity (units/g bacteria wet weight at 37°C and pH 7.5)	
	Uninduced	Induced
E. coli ATCC 11105	20	120
E. coli 5K/pHM6 (cloned on pBR322)	560	700
E. coli 5K/pHM11 (cloned on pOP203-3)	Not tested	280
E. coli 5K/pHM12 (U.V. mutant of pHM6)	1050[b]	Not tested

[a]Fermentation conditions: For constitutive strains a medium of yeast extract (1%), tryptone (1%), beef extract (0.5%), NaCl (0.25%), and fructose (1%); for inducible strains the fructose was omitted and sodium phenylacetate (0.1%) added. Fermentation was carried out in a 10-liter reactor at 27°C, pH 7.0, agitated at 300 rpm, with an airflow of 0.4 vol/vol per minute.
[b]Activity determined at 45°C.
Source: Data taken from Mayer et al. (1979) and Stoppok et al. (1980).

ment of the small fragment produced by treatment with other restriction endo-nucleases (Eco R1 and Pst 1) with fragments from the hybrid clone, pHM5. Clones were selected by virtue of the remaining tetracycline resistance region on the pBR322 plasmid. The gene was also subcloned onto the multicopy plasmid pOP203-3 in an attempt to bring it under the influence of the active lac promoter. Screening for acylase-positive strains using the plate overlay techniques led to the isolation of several higher-yielding strains which had the added advantage of being constitutive (Table 2).

The stability of these plasmids in the host bacteria was satisfactory, as indicated by the fact that cell paste lost no enzyme activity during 15 days storage at 4°C. Mayer et al. (1979, 1980) have stated their intention to further subclone the acylase gene in order to bring it under the control of a strong promoter and predict that this will lead to enzyme levels far in excess of that already achieved. Progress in this direction has been achieved by ultra-violet mutation of strain pHM6, which led to strain pHM12 and a further doubling of specific enzyme activity (Table 2).

III. PRODUCTION OF PENICILLIN G ACYLASE

A. Fermentation

The most widely used organism is *E. coli* and a typical process has been described by Savidge and Cole (1975). A cell suspension prepared from an agar slant culture was used to inoculate flasks containing sterile corn-steep liquor (CSL) medium (typically 1.2% on dry solids basis adjusted to pH 7.0 with NaOH). After incubation on a rotary shaker at 24°C for 24 hr, these seed flasks were used either to inoculate final-stage production medium directly or, depending upon the size of the production fermenter, to initiate a second seed fermentation in a stirred fermenter. An inoculum level to the final production stage of around 0.5% vol/vol was required. As first noted by Kaufmann and Bauer (1960), a most important constituent of the medium is phenylacetate, which is metabolized during the course of the fermentation to induce enzyme synthesis, increasing its production by 5- to 10-fold. It is probable that most strains in current industrial use have relaxed regulatory requirements as a result of mutation, but the only reported examples of fully constitutive strains are those obtained by genetic engineering. Because of carbon catabolite repression of enzyme synthesis, the medium must be substantially free from fermentable carbohydrate. Typically, components for industrial-scale (e.g., 30,000 liters) fermentations are selected from complex organic sources, for example, corn-steep liquor or yeast extract. A typical medium as described by Savidge and Cole (1975) is CSL (1.2% on a dry solids basis), ammonium sulfate (0.1%), and either sodium, potassium, or ammonium phenyl-acetate (0.1%). The pH was adjusted to 6.5 with either KOH or NaOH before sterilization. The aeration and agitation rates during fermentation must be optimized for individual fermenters and processes. However, in general, rather low aeration rates seem preferable, but this does appear to depend upon culture and medium. Thus Cole (1969a) reported an aeration rate of 0.4 vol/vol per minute, while Kaufmann and Bauer (1980) used 0.9 vol/vol per minute. A more detailed study of the effect of dissolved oxygen was carried out by Vojtisek and Slezak (1975a), who showed that enzyme synthesis was completely inhibited at elevated dissolved oxygen concentration. Incubation temperature must also be optimized, although it is preferable to incubate at below the optimum temperature for growth (at least during the enzyme production phase). Thus a temperature of 24°C was used by Carrington et al. (1966), and one of 31°C by Kaufmann and Bauer (1960). No enzyme synthesis occurred at 37°C (Levitov et al., 1967). Klein and Wagner (1980) showed that enzyme activity could be increased by shifting incubation temperature from 27 to 24°C when the cell concentration reached 1.8 g of dry cells per liter with a dissolved oxygen concentration below 2%. The physiological age of the cells at the time of the temperature shift was important (Table 3). The time course of growth and enzyme activity of a typical fermentation is given in Fig. 2. Incubation is continued until the pH of the medium increases to 8.0, generally after around 20 hr, depending upon the temperature and the aeration rate. Enzyme levels were doubled by slowly adding ammonium phenyl-acetate (total addition, 0.35%) during the course of fermentation (Carrington et al., 1966).

The process is amenable to continuous culture; for example, Sikyta and Slezak (1964) reported that *E. coli* ATCC 9637 could be grown continuously at a dilution rate of 0.5 hr^{-1} in a CSL peptone medium to produce cells having the same acylase activity as in batch culture, but the concentration of ammonium phenylacetate inducer was 10 times that required in a batch culture. The overall productivity of this continuous culture was seven times that of a batch culture.

The effect of phenylacetate has been studied in greater detail by Levitov et al. (1967), who found that in a corn-steep-liquor-based medium, enzyme induction occurred within 30 min of adding phenylacetate; simultaneous addition of glucose totally repressed enzyme synthesis. A range of phenylacetic acid derivatives and structurally related compounds were tested as inducers, but all were inferior to phenylacetate. Similar results were obtained by Vojtisek and Slezak (1975a), who also showed maximum specific enzyme synthesis to occur in a minimal medium with phenylacetate as the sole carbon source. Enzyme synthesis was greatest during exponential growth; no synthesis occurred when the cells ceased growing. The induction time was found to be equivalent to two generation times. The inferiority as inducers of compounds structurally related to phenylacetic acid was also confirmed; only phenoxyacetic acid gave significant induction, but then only about 25% of that with phe-

Table 3 Effect of Temperature Shift on Penicillin G-Acylase Activity at Different Growth Phases[a]

Temperature shift (°C)	Cell density (OD$_{546 \text{ nm}}$)	Specific activity (unit/g wet weight)
27-24	5.5	44
27-24	6.0[b]	69
27-24	6.5	41
27-24	7.5	30
24-27	6.0	36

[a]Fermentation conditions: *E. coli* ATCC 11105; medium, yeast extract (1%), tryptone (1%), beef extract (0.5%), NaCl (0.25%), sodium phenylacetate (0.1%); pH controlled at 7.0; 10-liter fermenter at 470 rpm; airflow, 1 vol/vol per minute; incubation time, 12 hr.
[b]Equivalent to 1.8 g dry cells/liter.
Source: Data taken from Klein and Wagner (1980).

nylacetic acid. Further study of glucose repression in the synthetic medium (Vojtisek and Slezak, 1975b) showed absolute enzyme repression by glucose to be independent of glucose concentration within the range of 0.1-1.0%. The growth diauxie seen in media containing glucose and phenylacetate was overcome completely by adding cyclic adenosine 5'-monophosphate (cAMP), phenylacetate utilization was restored, and enzyme synthesis was totally derepressed. Acetate instead of glucose caused only partial repression, while lactate served as a "catabolically neutral" carbon source for maximal enzyme production.

This acceptance of lactate was not confirmed by Klein and Wagner (1980), who found lactate as well as glucose and fructose to strongly repress acylase production with the inducible strain of *E. coli*, ATCC 11105. With the constitutive hybrid strains, however, these carbon sources increased enzyme levels (Table 4).

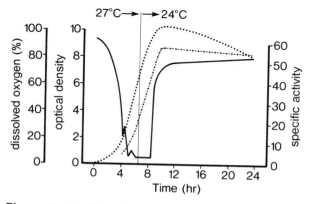

Figure 2 Profiles for 10-liter penicillin G acylase fermentation of *E. coli* ATCC 11105 (—), dissolved oxygen; ••••, optical density; •-•-•, specific activity). Data from Klein and Wagner (1980).

Table 4 Induction and Repression of Penicillin G Acylase in *E. coli*[a]

Additional carbon source (10 g/liter)	*E. coli* ATCC 11105		*E. coli* 5K(pHM12)	
	Phenylacetate (1 g/liter)	Specific activity (unit/g wet weight)	Phenylacetate (1 g/liter)	Specific activity (unit/g wet weight)
None	No	1	No	46
None	Yes	23	Yes	18
Glucose	No	1	No	11
Glucose	Yes	1	Not tested	Not tested
Lactose	No	0	No	72
Lactose	Yes	3	Not tested	Not tested
Fructose	Yes	1	No	85

[a]Fermentation conditions: Shaken flasks of medium described in Table 3; incubated for 25 hr at 27°C.
Source: Data from Klein and Wagner (1980).

Further study of the effect of cAMP by Gang and Shaikh (1976) showed stimulation of acylase synthesis in *E. coli* cells not repressed with glucose in the presence of added cAMP (Fig. 3). From studies with inhibitors, it was concluded that cAMP acted at the transcriptional level. No cAMP stimulation occurred in the absence of inducer.

The effect of glucose on phenylacetate uptake was further studied by Golub and Bel'Kind (1977), who found that glucose repressed synthesis of components of the phenylacetic acid transport system and inhibited their activity. Thus *E. coli* growing on glucose did not take up phenylacetic acid (PAA) and hence did not synthesize acylase.

Another source of benzylpenicillin acylase which has been investigated and reported on in some detail is the extracellular enzyme produced by *Bacillus megaterium* (Chiang and Bennett, 1967; Ryu et al., 1972). The enzyme was readily produced by fermentation on a medium consisting of enzyme-hydrolyzed casein (3%), glucose (0.5%) adjusted to pH 7.0. After an initial growth period of 8 hr at 30°C, enzyme production was induced by the addition of phenylacetic acid (0.15%) and incubation continued for a further 70 hr. Optimum fermentation conditions for the partially constitutive mutant isolated by Son et al. (1982), (Sec. II.C) have been established. Highest productivity, (1.4 unit/ml after 20 hr fermentation) was obtained on a medium consisting of soytone (2.5%), yeast extract (0.5%), glucose (1.0%) to which additions of phenylacetic acid (each of 0.3%) were made after 8 and 12 hr incubation at 30°C; the pH was controlled at 8.5 after the first PAA addition.

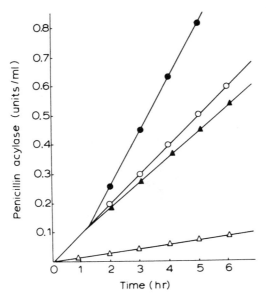

Figure 3 Effect of glucose and cAMP on penicillin acylase production [○——○, synthetic medium + phenylacetic acid (PAA) (3.6 mM); △——△, synthetic medium + PAA (3.6 mM) + glucose (10 mM); ▲——▲, synthetic medium + PAA (3.6 mM) + glucose (10 mM) + cAMP (2.5 mM); ●——●, synthetic medium + PAA (3.6 mM) + cAMP (2.5 mM)]. Data from Gang and Shaikh (1976).

B. Cell Harvest and Enzyme Purification

After fermentation the cells are harvested by centrifugation or filtration. On the industrial scale a self-cleaning centrifugal separator may be used to produce a slurry at a cell concentration around 10 times that of the fermenter broth. If whole cells are to be used for subsequent enzymatic reactions, care must be taken to select centrifuge operating conditions that do not cause cell disruption. Precoated rotary vacuum filtration has also been used for cell recovery, although the presence of an abrasive filter aid may impose constraints on any subsequent processing to extract enzyme.

Originally, whole *E. coli* cells were used for the production of 6-APA by the enzymatic hydrolysis of penicillin G, but during the last decade the whole-cell process has been superceded by immobilized enzymes. Such processes necessitate extraction of the enzyme from the cells followed by some degree of purification. Examples of all these stages are described below. Reports of immobilizing whole cells rather than enzyme have appeared in the literature, although it seems unlikely that such processes have yet gained any widespread acceptance on the industrial scale. Examples of whole cell immobilization are given in Sec. III.E.

The intracellular enzyme is released by cell disruption, which on the industrial scale is most conveniently carried out by repeated passage of cell slurry through a high-pressure homogenizer. A machine commonly used for this purpose is the Manton Gaulin homogenizer; the kinetics of cell disruption on this machine have been investigated (Hetherington et al., 1971). As mentioned above, care must be taken during cell harvest on a desludging centrifuge to prevent cell disruption. However, when enzyme is to be extracted, this has been turned to advantage by Delin et al. (1972), who operated a De Laval separator at a centrifugal pressure of around 1000 psi (gauge) and periodically discharged the slurry through narrow peripheral slits; 28% of the enzyme was solubilized.

Several methods for purifying the enzyme have been reported. Balasingham et al. (1972) described a procedure in which, after cell debris and precipitated nucleic acids were removed, enzyme was purified 10-fold by fractional precipitation with ammonium sulfate and polyethylene glycol. A further sixfold purification was achieved by diethylaminoethyl-(DEAE) cellulose column chromatography to provide enzyme with a specific activity of 6.35 units/mg. The overall yield from disrupted suspension was 25%.

More recently Lagerlöf et al. (1976) described a purification process in which disrupted *E. coli* fermentation broth was clarified by centrifugation and Seitz filtration and then concentrated on a thin-film evaporator. The enzyme was precipitated by the addition of 20% aqueous tannin and recovered from the tannin–acylase complex by homogenizing in cold acetone from which the enzyme was precipitated by further acetone addition. The precipitate was dissolved in water and then further purified by adsorption to carboxymethyl cellulose, followed by elution with phosphate buffer (pH 8.0). The results of this procedure are shown in Table 5.

Kutzbach and Rauenbusch (1974) succeeded in crystallizing enzyme from enzyme solution purified 50-fold by ammonium sulfate and column chromatography on sulfoethyl-(SE) and DEAE-Sephadex, as shown in Table 6.

C. Properties

Kutzbach and Rauenbusch (1974) showed the crystalline enzyme from *E. coli* forms regular rectangular plates (150 × 80 μm) with a specific activity of

Table 5 Purification of Benzylpenicillin Acylase via a Tannin-Acylase Complex

Material	Volume (liter)	Total enzyme activity units	Specific activity (A/E)[a]	Yield (%)
Disrupted cell suspension	4000	51.2×10^6	—	100
Concentrated cell extract	290	43.5×10^6	0.77	85
Crude acylase solution	170	31.4×10^6	2.57	61
Purified acylase solution	86	26.7×10^6	10.80	52

[a]Specific activity is expressed as A/E, where A stands for the activity in unit/ml, and E for the optical density at 280 nm.
Source: Data from Lagerlöf et al. (1976).

48 units/mg protein. The molecular weight by thin-layer gel filtration was 70,000; electrophoresis in the presence of sodium dodecylsulfate (1%) revealed two species with molecular weights of 71,000 and 20,500. The isoelectric point was pH 6.8, and the optimum pH for the hydrolysis of penicillin G, was 8.1. The K_m value for the pure enzyme was 0.02 mM, as compared to the 30.0 mM determined by Cole (1969a) for cell-bound enzyme. The enzyme is inhibited by both reaction products and by the substrate penicillin G: K_i values for 6-APA (noncompetitive inhibition) and phenylacetic acid (competitive inhibition) of 71 and 4.8 mM, respectively, and a K_s value of 270 mM were determined by Balasingham et al. (1972). The enzyme from *B. megaterium* is less susceptible to product inhibition and is not at all inhibited by substrate. Kinetics and substrate profiles for both the *E. coli* and *B. megaterium* enzymes were compared in the review by Savidge and Cole (1975). The hydrolysis by *E. coli* enzyme of compounds other than penicillins was reviewed by Cole (1969b).

D. Enzyme Immobilization

There is extensive literature on the immobilization of benzylpenicillin acylase, and the methods used span all four immobilization categories: adsorption, adsorption and cross-linking, covalent attachment, and physical entrapment. Both soluble and insoluble carriers have been used. The deacylation of penicillin G to produce 6-APA represents one of the first successful applications of immobilized enzyme technology to an industrial-scale process. Examples to illustrate each of the immobilization methods are given in Table 7. For a more extensive review of immobilization methods, see Abbot (1976) and Vandamme (1981b). A detailed account of an industrial-scale process for 6-APA production using immobilized enzyme has been described by Lagerlöf et al. (1976). Covalent attachment to cyanogen bromide-activated Sephadex G-200 (Axen et al., 1967) was the immobilization method selected from a number of carriers investigated on the basis of specific activity, coupling efficiency, filtration properties, stability, and cost. The procedure used was activation of the Sephadex G-200 by overnight swelling in deionized water, followed by addition of cyanogen bromide. After filtering off and washing, the activated polymer was stirred overnight with a buffered solution of purified acylase (pH 8.75 and 4°C). The immobilized enzyme preparation was filtered off and washed with deionized water. The specific activity was 225 units/g wet weight and the coupling efficiency was 47%.

Table 6 Purification and Crystallization of Benzylpenicillin Acylase

Step	Volume (ml)	Protein concentration[a] (mg/ml)	Total activity[b] (units)	Specific activity[b] (unit/mg protein)	Yield (%)
1. Disrupted sludge	4000	—	9200	—	100
2. pH 5.0 supernatant	2340	23.2	6640	0.12	72
3. 40–60% Ammonium sulfate precip.	440	52.5	5980	0.26	65
4. Pooled fraction SE-Sephadex	670	1.6	3380	3.30	37
5. Pooled fraction DEAE-Sephadex	320	1.7	3220	6.00	35
6. Crystal suspension	5.4	35.2	2260	11.90[c]	25

[a]Biuret test.

[b]Tested with NIPAB as substrate (Sec. II.B.1).

[c]Equivalent to approximately 46 units/mg when assayed with penicillin G at pH 7.8 and 37°C.

Source: Data from Kutzbach and Rauenbusch (1974).

Table 7 Examples of Methods for Immobilization of Benzylpenicillin Acylase and Use for 6-APA Production

	Immobilization category	Immobilization method
Water-insoluble carriers	Adsorption	Adsorption of clarified fermentation broth onto bentonite and diatomaceous earth at pH 6.2–6.3 for 1 hr
	Adsorption of *modified* enzyme	Enzyme modified to improve adsorption to basic carrier (DEAE-Sephadex) by succinaylating with succinic anhydride
	Adsorption and cross-linking	Adsorption to polymethacrylate resin (Amberlike XAD-7) at pH 5.0 and cross-linking with glutaraldehyde
	Covalent attachment	1. Covalent coupling to cyanogen bromide-activated Sephadex G200
		2. Covalent coupling via anhydride groups to a copolymer of acrylamide, maleic anhydride, and N,N'-methylene bis-acrylamide
	Physical entrapment	Entrapment in cellulose triacetate fibers
Water-soluble carriers	Covalent attachment	1. Covalent coupling to cyanogen bromide-activated sucrose/epichlorhydrin co-polymer to form a water-soluble immobilized enzyme
		2. Covalent coupling via anhydride groups to ethylene/maleic anhydride copolymers modified to form a hydrophobic enzyme–polymer conjugate by reacting with n-octadecylamine

A process for the physical entrapment of enzyme within cellulose triacetate fibers has been described by Dinelli (1972) and Marconi et al. (1973). Partially purified enzyme solution from *E. coli* in phosphate buffer at pH 8.0 and containing glycerol, 30% vol/vol, was added dropwise to a solution of cellulose triacetate dissolved in methylene chloride to form an emulsion. This emulsion was stirred for 30 min, allowed to stand for a further 30 min, and

Enzyme source	Reactor system for using immobilized enzyme	References
Bacillus megaterium	1. Batchwise in stirred tank using using NH₄OH solution to control pH at 8.0–8.2	Squibb (1969)
	2. Single-stage continuous stirred tank reactor with enzyme retention by ultrafiltration	Ryu et al. (1972)
Bacillus circulans		Otsuka (1972a,b)
Escherichia coli	Batchwise in stirred tank using NaOH to control pH at 7.8–8.0; good mechanical strength of carrier claimed as advantage	Beecham (1975)
E. coli	1. Batchwise in stirred tank or recirculation reactor pH controlled at 7.8 with NaOH	1. Lagerlöf et al. (1976)
E. coli	2. Batchwise in stirred tank using triethylamine to control pH at 7.8–8.0	2. Bayer A. G. (1973a,b)
E. coli	Batchwise in recirculation reactor; pH controlled at 8.0 with NaOH	Snam Proghetti (1974), Marconi et al. (1973)
E. coli	1. Batchwise in stirred tank using NaOH to control pH at 7.8; 6-APA liquors separated from immobilized enzyme by ultrafiltration	Beecham (1976)
E. coli	2. Batchwise in a tank which is vigorously stirred in order to disperse the immobilized enzyme throughout the aqueous phase; enzyme separated by flotation or centrifugation	Smith (1976)

then extruded through a spinneret into a coagulating bath of toluene. The enzyme remained entrapped as an aqueous solution within the cavities formed within the fibers. The fibers were vacuum dried to eliminate solvents, wound into skeins, and stored at 5°C. Although the organic solvents did not denature the enzyme, the acylase fibers exhibited less activity than the soluble enzyme owing to diffusional limitations caused by the fiber matrix.

E. Whole Cell Immobilization

In several enzymatic processes, avoiding the necessity for extracting, purifying, and immobilizing the enzyme by immobilizing whole cells has proved
to be very effective; for general reviews see Abbott (1978) and Vandamme
(1981b). The same approach has been applied to penicillin G acylase and a
number of reports on the immobilization of E. coli cells have appeared, the
first being by Sato et al. (1976), who successfully entrapped cells within a
polyacrylamide gel and used the preparation to continuously produce 6-APA.
One of the disadvantages of such a process is due to the low specific activity
of these preparations compared with immobilized enzymes, but two approaches
for overcoming this have been proposed by Klein and Wagner (1980): (1) increasing the specific activity of the cells and (2) selecting the most appropriate immobilization technique. The first approach was achieved by developing new hybrid strains as described in Sec. II.C.2 and by optimizing fermentation conditions to maximize specific activity as described in Sec. III.A. The
selection of the optimum immobilization was made by comparing five different
methods for polymer entrapment (for a review of the various methods of whole
cell immobilization, see Klein and Wagner, 1978a) on the basis of specific activity, efficiency of immobilization, reaction kinetics, mechanical strength,
and enzymatic stability. Cells were either E. coli ATCC 11105 or the new
hybrid strain K5/pHM 12. The immobilization methods evaluated were

1. Ionotropic gelation of alginates with Ca^{2+} (Klein and Wagner, 1978b)
2. Cross-linking polymerization of methacrylamide (Klein et al., 1979)
3. Polycondensation of epoxides (Klein and Eng, 1979)
4. Polycondensation of polyurethane (Klein and Kluge, 1981)
5. A simple precipitation process using the synthetic polymer "Eudragit"
   (Vorlop, 1978)

Specific activities ranged between 0.5 and 4.8 units/g, with efficiencies
between 27 and 65%. While part of the loss was undoubtedly due to enzyme
inactivation, a considerable diffusional limitation may have existed. This was
confirmed by observing the decrease in yield which occurred when either particle size or cell loading was increased. Previously, Klein and Eng (1979)
had observed that increasing the porosity of the epoxide resin increased
yields. When the highly active cells obtained from fermentation of the high-
yielding hybrid strain were immobilized in epoxy resin, yields were only 10% at
all particle diameters above 0.1 mm, compared to 20–30% for ATC 11105. It would
seem, therefore, that in order to realize the potential of the highly active
cells in immobilized cell systems, a combination of small particle size and high
bead porosity is required.

It was concluded from this study that the epoxy bead preparation was
preferable, since it combined high cell loading capacity, which led to high
specific activity, and immobilization yield together with excellent mechanical
properties, all of the other preparations being deficient in one or more of
the above factors. Preliminary work established that the immobilized cells
were viable and that the preparations could be reincubated to increase catalytic activity. This aspect has been more extensively studied by Morikawa
et al. (1980), using Kluyvera citrophila cells, who showed that the acylase
activity was increased when cells immobilized in polyacrylamide gel were continuously cultivated in an aerated fermenter at 30°C for 24–40 hr in various
media, the optimal composition being peptone, NaCl, and sodium phenylacetate
(0.05%) as inducer. Acylase activity (determined on the basis of ampicillin

synthesized from 6-APA and D(−)-phenylglycine methyl ester) was up to 2.7 times higher in the cultivated immobilized cells. The activity was further increased (up to 4.4 times) when the cell preparations were treated with alkali (1 hr at pH 9.0, 40°C with aeration) to eliminate β-lactamase activity.

For the optimum epoxy resin cell immobilization developed by Klein and Wagner (1980), beads of sufficient porosity and size were prepared by mixing together epoxy precursor, cells, and curing agent for 36 hr, and then adding sodium alginate solution. This suspension was periodically injected into calcium chloride solution to give stabilized drops by cross-linking the polyelectrolyte at the drop surface. After drying the particles, the alginate was dissolved away with phosphate buffer to give rigid, open-pored beads. The operational stability of these beads was good, for example, less than 10% loss of activity during 30 days of continuous operation of a fluidized bed reactor. Similarly, Sato et al. (1976) produced 6-APA continuously from a column of polyacrylamide-entrapped cells and calculated the operational half-life at 30°C to be 42 days. A buffered solution of penicillin G (50 mM) was flowed through a column and the eluate collected, from which 6-APA was recovered in 78% yield. Bacterial cells containing penicillin V acylase have also been immobilized as described in Sec. V.D.2.

IV. PRODUCTION OF 6-APA USING PENICILLIN G ACYLASE

The process involves an enzymation stage in which the penicillin is deacylated followed by 6-APA extraction to provide crystalline material; the other product of the reaction, phenylacetic acid, may also be recovered, either as an aqueous extract or as a solid, for reuse as a precursor in the penicillin G fermentation process.

The overall 6-APA activity yield, that is, the weights of both penicillin G and 6-APA corrected for purity, quoted by Lagerlöf et al. (1976) was 90%. Similar yields have been quoted for the fiber-entrapped acylase system (Snam Proghetti, S.P.A., 1974). Bearing in mind the 6-APA which is left in the crystallization mother liquors (about 4%), it is evident that the enzymatic process for 6-APA production can be extremely efficient.

A. Enzymation of Penicillin G

As shown in Figure 4, various types of reactors have been used to carry out the deacylation reaction. In some cases, the conventional stirred tank reactors used for the original cell-bound enzymation process will have been used, with perhaps minor modifications (Fig. 4a) to permit

1. Retention of the immobilized enzyme preparation after the enzymation has been completed on, for example, simple wire mesh screens
2. Modified (less shear intensive) agitation to avoid mechanically damaging the enzyme carrier
3. More efficient means of alkali distribution to avoid excessive alkaline degradation of enzyme and penicillin which might otherwise occur with less vigorous agitation

This type of reactor is especially suitable for immobilized enzyme preparations which are in the form of discrete particles, for example, bead polymers. It is evident that optimum agitation conditions will be a compromise between the second and third requirements above; if satisfactory agitation

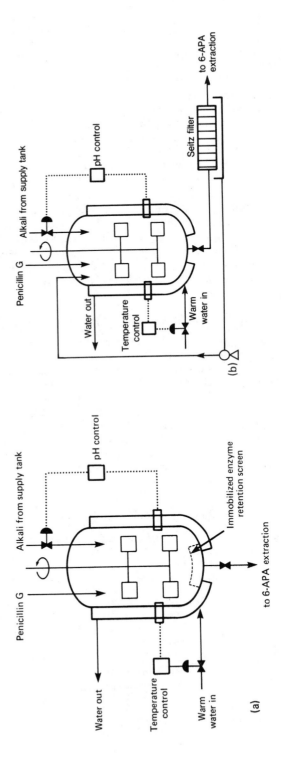

(a)

(b)

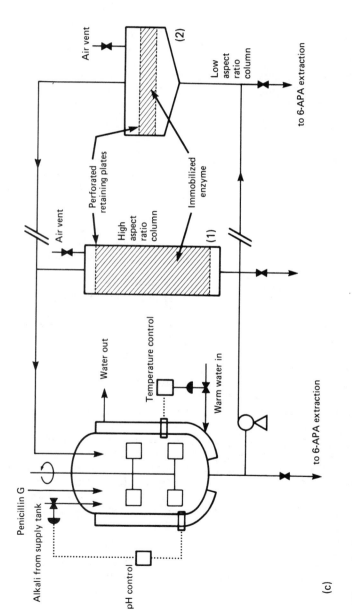

Figure 4 Reactors for enzymatic deacylation of penicillins and cephalosporins.

conditions in this type of reactor cannot be achieved because of enzyme carrier limitations, then a recycle reactor may be preferable. This permits pH adjustment to be separated from the immobilized enzyme (Fig. 4c1). The residence time in the enzyme column must be minimized (e.g., <1 min) in order to prevent an excessive pH fall; slightly buffering the penicillin G solution may also be advantageous.

This type of reactor was also used for the fiber-entrapped enzyme described below. A similar system may be used for very small particles, but in order to minimize residence time and avoid an excessive pressure drop, a column with a low aspect ratio may be preferable (Fig. 4c2).

Use of both batch and recirculation processes have been described by Lagerlöf et al., (1976). In the batch process, 100 kg of penicillin G (3.3%) were hydrolyzed with 16.5 kg wet weight of the Sephadex immobilized enzyme preparation described above (total activity, 3.7×10^6 units); the reaction temperature was controlled at 35°C and the pH at 7.8 by addition of sodium hydroxide solution. After enzymation the immobilized enzyme was recovered on a Seitz filter press, collected in the trough beneath the filter, and returned to the reactor for reuse (Fig. 4b). This procedure was obviously fairly laborious and likely to result in some physical losses, as well as increase the probability of microbial contamination. Nevertheless, it was found that if this operation was carefully controlled, the enzyme preparation could be reused more than 100 times without the need for fresh enzyme addition.

The very high stability of this preparation was demonstrated in an earlier publication by these authors (Ekstrom et al., 1974) which showed a gradual rise in the time taken for the reaction to reach 50% completion: from 50 to 60 min through 60 reuses. A total of 30 reuses were required to offset the cost of immobilizing the enzyme. After 60 reuses a rapid rise in reaction time was attributed to microbial contamination.

To avoid handling the enzyme, a recirculation process has been developed on a pilot plant scale (Fig. 4c2). The enzyme preparation (500 g) was retained and enclosed in the funnel by means of perforated disks. Water (90 liters) and monosodium phosphate (600 g) were added to a vessel in series with the column. The pH was adjusted to 7.8 with sodium hydroxide (10%), and the temperature to 37°C. Potassium penicillin G (3 kg) was added and the reaction started by pumping this solution through the filter bed. The pH was maintained at 7.8 by automatic addition of sodium hydroxide to the vessel, which was vigorously agitated to minimize alkaline degradation of the β-lactam nucleus. The reaction was continued until the requirement for sodium hydroxide to maintain pH ceased.

For 6-APA production using the fiber-entrapped enzyme (Marconi et al., 1973), the acylase fibers (1 kg at 167 units/g cellulose triacetate) were packed in a jacketed column reactor of high aspect ratio, through which flowed 5.5 liters of a solution of penicillin G (e.g. 12%) in phosphate buffer at pH 8.0 (Fig. 4c1). Temperature was maintained at 37°C and pH at 8.0 by automatic addition of sodium hydroxide to the agitated vessel in series with the column. Some chemical decomposition of the labile β-lactam nucleus occurred during the reaction, since at the end the conversion efficiency was 88%, with 3% unconverted penicillin G. Reducing the reaction time from 310 min to around 200 min by increasing the flow rate through the column from 50 to 260 liters/hr reduced degradation and improved conversion efficiency to over 90%. A further advantage of a high flow rate was that it overcame diffusional limitations, the mechanical strength of the fibers being sufficient to withstand this. Nevertheless, it is likely that 12% penicillin G was too high a concentration, despite the high enzyme loading; a more cost-effective process is likely to be achieved at a penicillin G concentration of around 6%, with an enzyme loading of around 50 units/g penicillin and a reaction time of about 6 hr.

Some preliminary results indicated the feasibility of a continuous process in which activity of the acylase fibers was sufficient to permit about 93% conversion of a well-buffered solution of 4% penicillin G (pH 8.5) in a single pass through the column.

From a consideration of the kinetics of the degradation of penicillin G, 6-APA, and enzyme, Ho and Humphrey (1970) determined the optimum process conditions for maximizing 6-APA yield and minimizing enzyme loss by inactivation. This was achieved by profiling both temperature and pH during the course of a 3-hr enzymation; this was particularly beneficial when the enzyme cost was very high.

B. Extraction of 6-APA

The 6-APA may be extracted from the cooled enzymation liquors by the procedure described by Lagerlöf et al. (1976). The liquors were adjusted to pH 2.5 with hydrochloric acid and methyl isobutyl ketone (0.5 vol) added to extract the phenylacetic acid and residual penicillin G. After separating the phases, the aqueous phase was adjusted to pH 7.8 with sodium hydroxide and evaporated under vacuum (conditions must be suitably mild to avoid degradation of the labile β-lactam ring, e.g., <30°) to a 6-APA concentration of 12–15%. The 6-APA was then precipitated by adjusting the pH to the isoelectric point of 4.3. The crystals were washed successively with water and acetone and dried under vacuum.

If all stages of this extraction procedure are well controlled (e.g., residence times and temperature of process streams kept low, pH adjustments carried out carefully), then a yield of about 94% can be achieved (about 4% 6-APA remains in the mother liquors). This in combination with an enzymation stage efficiency of 95–96% results in the overall 6-APA yield of 90% quoted above.

V. PRODUCTION OF PENICILLIN V ACYLASE

Since some manufacturers prefer to make penicillin V rather than penicillin G, there is a requirement for acylases suitable for use with penicillin V. These have been found to be fairly widely distributed among fungi, for example, the genera *Fusaria* and *Penicillia* and the basidiomycetes *Pleurotus* and *Bovista*. The incidence among bacteria was thought to be lower, but the recent work of Lowe et al. (1981) clearly indicates a much wider distribution among soil bacteria, as described in Sec. II.C.1. For general reviews of these enzymes see Vandamme and Voets, (1974), Vandamme, (1981a), and Vanderhaeghe (1975).

As an example of a process suitable for the industrial scale, production of the enzyme from the basidiomycete *Bovista plumbea* NRRL 3501* (Brandl and Knauseder, 1975; Schneider and Roehr, 1976) will be described. Mention will also be made of enzymes of bacterial origin.

---

*Strain reclassified as *Pleurotus ostreatus* by Stoppok (1981).

A. Enzyme Preparation from *Bovista plumbea**

The mycelium from a slant culture of the organism grown on Sabouraud agar for 14 days at 24°C was suspended in a sterile medium (5 ml) consisting of yeast extract (equivalent to 0.1% nitrogen), glucose (5.0%), KH_2PO_4 (0.1%), $MgSO_4 \cdot 7H_2O$) (0.05%), $Ca(NO_3)_2$ (0.05%), NaCl (0.01%), and $FeSO_4 \cdot 7H_2O$ (0.005%), adjusted to pH 6.0, and the mycelium was broken up with a glass rod. This fragmented mycelium was used to initiate a series of three successive seed stages in Erlenmeyer flasks. The medium for each was as above, plus sperm oil (0.6%), and incubation was for 96 hr at 24°C on a rotary shaker. The mycelium after each stage, described as "globular," was aseptically comminuted before aliquots were used to inoculate the next stage. Mycelium from the third seed stage was used to inoculate a 5-liter enzyme production fermentation using the same medium; incubation was for 96 hr at 24°C with aeration and agitation. The enzyme is mainly extracellular, and therefore after the broth was clarified the filtrate was cooled and adjusted to pH 4.5. The enzyme was partially purified by adsorption onto bentonite, which, after filtering off, was washed with physiological saline adjusted to pH 10.0. Enzyme was twice eluted from the bentonite, first with 20 vol and then with 10 vol of the pH 10 saline. The combined eluates were adjusted to pH 7.5 and carefully concentrated under vacuum (around 10°C) to 10% of the original volume to provide an enzyme solution suitable for immobilization.

Enzyme may also be extracted from the mycelium (Schneider and Roehr, 1976) by homogenization of a buffered suspension of acetone-dried mycelium by stirring with an Ultra-Turrax high-efficiency stirrer followed by disruption in a Manton Gaulin homogenizer. This enzyme was purified 220-fold by gel filtration (Sephadex G-25) and successive chromatography on DEAE-cellulose, Bio Gel P-200, and hydroxyapatite.

B. Properties of the *Bovista* Enzyme and Comparison with the *E. coli* Enzyme

The molecular weight of the *Bovista* enzyme was determined by Schneider and Roehr (1976) to be 88,000 ± 5000. The K_m for the best substrate, penicillin V, was a satisfactory 1.67 mM. No substrate inhibition was observed up to the maximum concentration tested (50 mM), compared to inhibition by both 6-APA and phenylacetic acid for the *E. coli* benzylpenicillin acylase.

The substrate specificity and kinetics of the cell-bound enzymes *B. plumbea* and the genetically engineered *E. coli* 5K/pHM 12 have been compared by Stoppok et al. (1980). Although, as shown in Table 8, higher values of penicillin V acylase were produced by the genetically engineered *E. coli* strain, it was concluded that the *Bovista* enzyme was preferable for industrial use, because more favorable inhibition characteristics would permit use of a higher concentration of penicillin V (Table 9).

The substrate specificity of the *Bovista* enzyme has been determined by Schneider and Roehr (1976), and that for the enzymes from *Penicillium chrysogenum* and *Fusarium avenaceum* by Vanderhaeghe (1975). It is evident from the latter review and from Batchelor et al. (1961) that some penicillin V acylases seem to have more activity against the natural aliphatic penicillins (e.g., penicillins F and dihydro F and K) than penicillin G acylases (although *P. acidovorans* is an exception). This could be advantageous in industrial processes, depending upon the extent to which these penicillins are present in "technical-grade" penicillin G used for 6-APA manufacture.

---

*Strain reclassified as *Pleurotus ostreatus* by Stoppok (1981).

Table 8 Substrate Specificity of *Bovista plumbea* NRRL 3824 and *E. coli* 5K/pHM 12

| Substrate | Specific acylase activity (units/g dry weight) | |
	B. plumbea	*E. coli*
Penicillin G	3.5	1050
Penicillin V	70	105
p-OH-penicillin V	8.4	0
Ampicillin	0	330

Source: Data from Stoppok et al. (1980).

C. Immobilization and Use of Immobilized Enzyme for 6-APA Production

While whole, washed mycelium may be used to produce 6-APA, a more efficient process has been described by Brandl and Knauseder, (1975). The bentonite-purified enzyme was immobilized by entrapment in cellulose triacetate fibers via the procedure described in Sec. III.D. The fibers were used in the same type of reactor as for penicillin G deacylation (Sec. IV and Fig. 4c1), except that the penicillin V concentration was 8%, the temperature was 32°C, and the pH was controlled at 7.5 with 10% ammonia. The reaction mixture (600 ml) was pumped through the column of enzyme fibers at a rate of 10–30 ml/min. After 3 hr the penicillin V conversion to 6-APA was 97%. Crystalline 6-APA was extracted in an overall yield of 91.5%. This immobilized enzyme was so stable that 100 reuses were achieved without noticeable loss of enzyme activity.

Table 9 Kinetic Constants of Whole Cell Acylase of *Bovista Plumbea* NRRL 3824 and *E. coli* 5K/pHM 12[a]

| Substrate | *E. coli* 5K (pHM 12) | | *B. plumbea* NRRL 3824 |
	Penicillin G	Penicillin V	Penicillin V
pH optimum	7.8	8.1	7.0
K_M	9–11	5–9	5–10
K_S	1570	170–180	1500–1700
K_i 6-APA	131	43–52	125
K_i acyl group	130	4–5	240

[a]Concentrations expressed as mM.
Source: Data from Stoppok et al. (1980).

D. Penicillin V Acylase of Bacterial Origin

As described in Sec. II.C.1, Diers and Emborg (1979) isolated a gram-negative bacterial strain producing a penicillin V acylase. This strain was stated to have the advantage over previously described enzymes of bacterial origin, for example, from *Erwinia aroideae* (Vandamme and Voets, 1975), being of greater activity at higher pH values and giving in better conversion due to a more favorable equilibrium. The bacterial strain had been deposited without restriction (NRRL 11240).

1. *Fermentation*

Fermentation was carried out in a 550-liter vessel containing 400 liter of a medium consisting of yeast extract (1.0%), corn-steep liquor (4.0%), K_2HPO_4 (0.025%), $MgSO_4 \cdot 7H_2O$ (0.025%), and acetic acid (0.25%) at 30°C with an airflow of 160 liters/min and agitated at 251 rpm for 18 hr; the pH was maintained at 7.0 with acetic acid. The activity was 1.2 units/ml (determined at 40°C and pH 7.5). The intracellular enzyme was extracted from cells by Manton Gaulin homogenization, and the disrupted sludge heated for 90 min at 50°C in order to destroy coproduced β-lactamase. The enzyme was purified by ammonium sulfate precipitation and ultrafiltration, and the enzyme used to split penicillin V (3%) in 90% efficiency.

2. *Whole Cell Immobilization and Use for 6-APA Production*

Since the enzyme was inhibited by p-chloromercuribenzoate and sodium tetrathionate and the lost activity could be partially restored by treating with cysteine, it was concluded that the enzyme contained a thiol group in or near the active center. It was reported (Gestrelius, 1979) that such enzymes could not be efficiently immobilized via procedures involving glutaraldehyde, because of inactivation, and therefore a modified method was developed. Treating with a branched polyethylene imine having 20% primary amine groups prior to the addition of glutaraldehyde increased the coupling efficiency from <10% to around 70%. In a typical example, a 10% solution of polyethylene imine was added with stirring to whole fermentation broth to a final concentration of 1% followed by addition of glutaraldehyde solution to a concentration of 0.5%. After further stirring the immobilized cells were filtered off, washed with phosphate buffer containing β-mercaptoethanol, and air dried. The activity of the particles (200–700 μm in diameter) was around 100 units/g dry weight. The immobilized cell mass could be extruded prior to drying in order to control the size of the immobilized cell particles. The dried cell particles (5–10%) so produced were reswollen in the mercaptoethanol phosphate buffer and used to deacylate buffered solutions of penicillin V (3%). Optimum conditions were pH 7.5 (controlled with NaOH) and 40°C and the reactor was either a stirred tank (Fig. 4a), where up to 20 reuses were achieved without any substantial decrease in the initial conversion efficiency of 96%, or a recirculation column reactor (Fig. 4c1). For this, the reswollen cells were packed in a column and used to deacylate 4.3% solutions of penicillin V; at flowrates of 13 to 26 bed vol/hour, conversion efficiencies of around 98% were obtained. Although use of mercaptoethanol improved enzyme stability, the preparations had insufficient longevity for prolonged industrial use and recently, Gestrelius (1982) reported that the Novo Company had developed an alternative fungal penicillin V acylase which is not a thiol enzyme and is much more stable. Further advantages claimed were lower product inhibition and a broader pH optimum which would reduce the buffering requirement in recycle column reactors.

VI. PRODUCTION OF SEMISYNTHETIC PENICILLINS

A. Use of Penicillin G Acylase

The use of benzylpenicillin acylase in the reverse direction to produce benzyl-penicillin from 6-APA and phenylacetic acid was first reported in 1960 by Rolinson et al. and Kaufmann et al. (1960), who studied this reaction in great-er detail and found that it proceeded best at a pH of 4.0–6.0. Substitution of a series of other carboxylic acids, including aliphatic acid, for phenyl-acetic acid led to the synthesis of the corresponding penicillins, such as ampi-cillin and n-pentylpenicillin. Use of energy-rich derivatives led to faster rates of synthesis. The authors concluded that the enzyme was acting as an acyl transferase that was able to synthesize penicillins by transferring acyl residues from acylamino acids to 6-APA. Further studies by Cole (1969c) of the substrate specificity showed p-hydroxyphenylacetic acid to be the best substrate; ampicillin was only synthesized with energy-rich derivatives of D(−)-phenylglycine. Optimization of this reaction (Cole, 1969d) using E. coli whole cells as the enzyme source established the most suitable side-chain derivatives to be N-acylglycine and methyl ester. Increasing the molar ratio with respect to 6-APA from 1:1 to 4:1 increased the reaction rate. The opti-mum pH was 7.0, but at this value there was increased hydrolysis of the syn-thesized ampicillin; lowering to pH 6.0 during the latter part of the reaction was beneficial. The optimum temperature was 35°C. Under these conditions the conversion efficiency using a 4 M excess of side-chain methyl ester hydro-chloride from 6-APA (1%) was 60%, equivalent to 10.1 mg/ml of ampicillin.

Using fiber-entrapped benzylpenicillin acylase extracted from E. coli ATCC 9637, Marconi et al. (1975) achieved only a 45% conversion of 6-APA to ampicillin, using a threefold excess of side-chain methyl ester at pH 7.0 and 25°C. The synthesis of amoxycillin was also investigated, the highest conversion efficiency being 30.5% from a very dilute 6-APA solution (7.6 μmol/ml); increasing 6-APA concentration increased the concentration of amoxycillin in the reaction mixture, but at the expense of conversion efficiency.

Based on the studies with benzylpenicillin acylase, there seemed little prospect of developing a commercially viable process for enzymatic synthesis of semisynthetic penicillins: Conversion efficiencies were too low, efficient separation of the various components of the reaction mixture was likely to be difficult, and process economics would be adversely affected by the need for a large excess of side chain.

B. Use of Ampicillin Acylase

The need for more efficient conversion than could be achieved with benzyl-penicillin acylase led to the search for alternatives.

Among the first organisms reported to produce the so-called "ampicillin acylase" was K. citrophila KY 3641, which was selected by Nara et al. (1971) from the screening of 519 strains of bacteria, yeasts, and fungi for 6-APA production from either penicillin G or V (Sec. II.C.1). Organisms possessing acylase activity were then screened for synthesis of ampicillin from a reac-tion mixture containing cells, 6-APA, and D(−)-phenylglycine methyl ester, buffered to pH 5.5 or 6.5 and incubated at 35°C for 4 hr; ampicillin was de-tected by bioassay or thin-layer chromatography. The maximum reported conversion efficiency to ampicillin was 62.6% after 20 hr incubation (35°C, pH 6.5) of 6-APA (10 mg/ml), side-chain methyl ester (25 mg/ml), and K. citrophila cells (about 30 mg/ml).

A later publication from the same laboratory (Okachi et al., 1973) reported the selection of an improved organism *Pseudomonas melanogenum* KY 2987 (ATCC 17808), after screening strains from 11 genera directly for ampicillin production using a similar procedure to that described above for *K. citrophila*. The selected organism had a very high substrate specificity for ampicillin and neither penicillin G or V was a substrate for the enzyme. Although reported conversion yields were no higher than with *Kluyvera*, advantages for *P. melanogenum* were stated to be greater specific enzyme activity, so that only 5–8 mg/ml of cells were required, compared to 30 mg/ml for *K. citrophila*, and a lower optimum pH of 5.5–6.0 (compared to pH 6.5), for which ampicillin is more stable. A lower pH should also favor synthesis by shifting the equilibrium toward synthesis of the amide bond. Because the enzyme cannot synthesize penicillin G from 6-APA and phenylacetic acid, the authors demonstrated the feasibility of using a crude penicillin G hydrolysate (obtained in this instance by treatment with *K. citrophila*) of 6-APA and phenylacetic acid, to which was added D(−)-phenylglycine methyl ester and *P. melanogenum* cells, and the pH adjusted to 6.0 with phosphate buffer. However, there are likely to be considerable problems in efficiently separating the components of such a complex mixture. The original isolate of *P. melanogenum* also possessed some β-lactamase activity, but deficient mutants were obtained.

A simple fermentation process for this organism has been described (Kyowa Hakko Kogyo Co. Ltd., 1971). A medium (3 liters) consisting of peptone (1%), yeast extract (1%), meat extract (0.5%), sodium glutamate (0.5%), and NaCl (0.25%) was adjusted to pH 7.3 before sterilization and incubated for 24 hr at an agitation speed of 300 rpm and an airflow of 1 vol/vol per minute, with the pH maintained at 7.5 by addition of a solution of 10% glutamic and 6% sulphuric acids. After fermentation, the cells were centrifuged off and suspended in phosphate buffer (pH 6.0).

Similar enzymes have isolated from other members of the *Pseudomonadaceae* (Takeda, 1975a) by screening for ampicillin production, in a test system similar to that described above, organisms capable of growing on media containing as sole carbon source α-amino acid derivatives such as phenylglycine methyl ester, phenylglycine thioglycol ester, or N-(phenylglycyl)-glycine at a concentration of 0.1%.

The enzyme produced by *Acetobacter turbidans* ATCC 9325 has been studied (Takahashi et al., 1974), but most reports have been on the organism ultimately producing the highest yield of enzyme, *Xanthomonas citri* IFO3835 and K24. These enzymes have been described as α-amino acid ester hydrolases.

C. α-Amino Acid Ester Hydrolase from *Xanthomonas citri*

1. *Enzyme Production*

Fermentation and extraction of enzyme have been described (Takeda, 1975b; Kato et al., 1980b). A 200-liter fermenter containing 150 liters of a medium consisting of yeast extract (0.2%), peptone (0.5%), glutamate (0.2%), sucrose (2%), K_2HPO_4 (0.2%), $MgCl_2 \cdot 6H_2O$ (0.1%), and $FeSO_4 \cdot 7H_2O$ (0.01%) at pH 7.2 was inoculated from a shaken-flask seed culture grown in the same medium and incubated at 28°C, stirred at 170 rpm, and aerated with 75 liters/min air for 24 hr. The activity of the intracellular enzyme increased in parallel with cell growth. Enzyme was assayed by the method described in Sec. II.B.2. After fermentation the cells were collected by centrifugation and stored as a paste (5 kg) at −20°C until required.

In a study of the shaken-flask fermentation of this organism, Rhee et al. (1980) found that a high aeration level (e.g., 25-ml medium/500-ml flask) stimulated biosynthesis of the constitutive enzyme.

Kato et al. (1980b) extracted the enzyme by suspending the cell paste in tris buffer (pH 8.0) and lysing the cells by stirring for 2 days with egg lysozyme plus EDTA. The lysate was treated with magnesium sulfate and deoxyribonuclease for a further 24 hr to reduce viscosity, and the debris removed by centrifugation and then treated with calcium phosphate gel in order to clarify the solution to prevent interference in subsequent purification stages. After dialysis, purification was achieved by successive column chromatography on cellulose and Sephadex. Results are shown in Table 10. Less time-consuming methods for releasing the enzyme from the cells by sonication (50% release) or by surfactants (e.g., Triton X-100, 80% release) have been reported (Choi et al., 1981).

2. Properties

The isoelectric point was found to be 7.8, and the molecular weight, by both sedimentation and gel filtration, 270,000, the enzyme consisting of four subunits of molecular weight 72,000 (Kato et al., 1980b).

Extensive studies on the purified enzyme obtained from a penicillinase-deficient mutant of *X. citri* K24 have been carried out by Kato et al. (1980c). The enzyme catalyzed hydrolysis of α-amino acid esters (acyl donors) and transfer of acyl groups to amine nucleophiles (acyl acceptors) such as 7-aminocephalosporanic acid (7-ACA), 7-amino-3-deacetoxy-cephalosporanic (7-ADCA), and 6-APA. The enzyme required a free amino group on the α-carbon of the donor ester, but did not exhibit absolute stereospecificity. The enzyme also catalyzed hydrolysis of the amide bond of some β-lactam antibiotics, for example, cephalexin (Sec. VII.B.1), ampicillin, and amoxycillin (slightly), but not penicillins G or V. The acyl donor substrate specificity was extensively studied: Various esters of D(−)-phenylglycine served as substrates, the best being the methyl and n-butyl esters; n-acetyl derivatives were not hydrolyzed. D(−)-Phenylglycine methyl ester was hydrolyzed at 10 times the rate of the L(+) isomer, but on the other hand the L(+) isomer of tyrosine methyl ester was split at 20 times the rate of the D(−) isomer. The specificity of the acyl acceptors was explained on the basis of the low pK_a values of the amino groups of 7-ACA, 7-ADCA, and 6-APA. Unlike benzylpenicillin acylase, hydroxylamine and glycine did not serve as acyl acceptors. Optimum pH and temperature values were 6.4 and 35°C, respectively. Based on these catalytic properties, Kato and co-workers have named this enzyme "α-amino acid ester hydrolase" in accordance with enzyme nomenclature, which rules that enzymes that can both hydrolyze and transfer are classified as hydrolases and not as transferases.

The kinetics of acyl transfer have been studied by Kato (1980). With 7-ADCA present as an acyl acceptor, stoichiometry was observed between the reacted amount of D(−)-phenylglycine and cephalexin, which indicated that hydrolysis and transfer of the acyl group occur simultaneously. The acyl acceptor was found to inhibit the hydrolysis of the ester, depending on its concentration. Thus with 7-ADCA, at a concentration above 5mM the rate of transfer became higher than that of hydrolysis, reaching a plateau at 20 mM (Fig. 5). On the other hand, the rate at which the ester disappeared was constant, being almost equal to the sum of the rates of

Table 10 Purification of α-Amino Acid Ester Hydrolase

Step	Fraction	Volume (liter)	Total protein (g)	Total activity (×10³ units)	Specific activity (unit/mg)	Purification (fold)	Recovery (%)
1	Crude extract	21.6	255	1670	6.6	1	100
2	Calcium phosphate gel supernatant	22.1	187	1340	7.2	1.1	80
3	First CM-cellulose eluate	2.00	13.8	1030	75	11	62
4	CM-Sephadex C-50 eluate	1.47	3.20	570	178	27	34
5	Second CM-cellulose eluate	0.40	0.22	283	1310	198	17
6	Sephadex G-200 eluate	0.11	0.08	164	2050	311	10

Source: Data from Kato et al. (1980b).

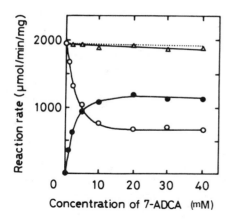

Figure 5 Effect of 7-ADCA concentration on initial rate of hydrolysis of D(−)-hydroxyphenylglycine methyl ester by α-amino acid ester hydrolase [○——○, initial rate of hydrolysis of D(−)-hydroxyphenylglycine methyl ester (formation of D(−)- p-hydroxyphenylglycine); ●——●, initial rate of transfer (formation of cephalexin); △——△, initial rate of disappearance of D(−)-p-hydroxyphenylglycine methyl ester; ···, sum of the initial rates of hydrolysis and transfer]. Data from Kato (1980).

hydrolysis and transfer, indicating that the acyl group is transferred to either 7-ADCA or water through a common acyl enzyme intermediate. This basic information has been of use in defining optimum conditions for synthesizing β-lactam antibiotics, particularly amoxycillin and cephalexin.

3. Production of Amoxycillin Using Cells

The following aspects were found by Kato et al. (1980a) to influence the efficiency of washed whole cells of *X. citri* in converting 6-APA to amoxycillin:

1. Ionic strength. When phosphate buffer was used to control reaction pH at 7.0, only 23% of the 6-APA was converted. When buffer was omitted and the pH controlled with NaOH, both reaction rate and yield increased (to about 60% conversion); plots of the degree of inhibition versus the log ionic strength of phosphate or NaCl were linear.
2. Effect of alcohols. Some alcohols were found to increase conversion efficiency, butan-2-ol being the most effective. Thus butan-2-ol (5%) led to 93% conversion of 6-APA into amoxycillin. This stimulation occurred as a result of repressing the rate of hydrolysis of the methyl ester, as shown in Table 11.
3. Optimum pH and temperature. The optimum pH was found to be 6–7; the presence of butan-2-ol lowered the optimum temperature from 35 to 25°C, owing possibly to some inactivation of the enzyme by the butanol.
4. Effect of substrate concentration. Competition between 6-APA and water as acyl acceptors explained the increase in conversion efficiency from 90 to 96% when 6-APA concentration was raised from

Table 11 Effect of Butan-2-ol on the Rate of Formation of Amoxycillin and Hydroxyphenylglycine (HPG) by α-Amino Acid Ester Hydrolase[a]

	Rate (μmol/ml per hour) of formation	
Addition	Amoxycillin	HPG
None	6.4	12.2
Butan-2-ol	8.5	8.0

[a]The reaction mixture (100 ml) containing per milliliter 92 μmol of p-hydroxyphenylglycine methyl ester, 46 μmol of 6-APA, and 35 mg of washed cells of *X. citri* K24 was incubated at pH 6.8 and 20°C with stirring. Butan-2-ol was added to 5% (vol/vol).
Source: Data from Kato et al. (1980a).

5 to 20 mg/ml (the molar ratio of side-chain ester to 6-APA was maintained at 2:1); at 20 mg/ml the conversion rate was markedly suppressed.

The time course of amoxycillin synthesis via this optimized process is shown in Figure 6. The conversion efficiency was 96%. Extraction of the liquors gave an overall yield from 6-APA to amoxycillin trihydrate of about 65%.

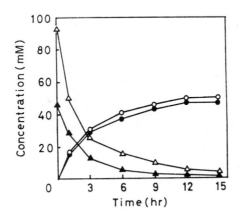

Figure 6 Time course of amoxycillin synthesis by *X. citri* K24. Reaction mixture [100 liter containing per liter D(−)-p-hydroxyphenylglycine methyl ester (92 mmol), 6-APA (46 mmol), butan-2-ol (50 ml), and washed cells (35 g) incubated with stirring at pH 6.8 and 20°C]: ●——●, amoxycillin; ○——○, D(−)-p-hydroxyphenylglycine; ▲——▲, 6-APA; △——△, D(−)-p-hydroxyphenylglycine methyl ester. Data from Kato et al. (1980a).

3. *Production of Ampicillin and Amoxycillin Using Immobilized Enzyme*

Enzyme from the primary CM-cellulose eluate described in Sec. VI.C.1 Takeda, 1975b) was attached to CM-Sephadex via the azide coupling procedure of Mitz and Summaria (1961). Optimum (e.g., 55%) enzyme fixation occurred in the presence of 20% methanol. Alternatively, crude enzyme was coupled to Sephadex G-200 or PS-1, (a polysaccharide obtained from *Alcaligenes faecalis*) via the cyanogen bromide method; coupling efficiencies were around 60% (Takeda, 1976). These preparations have been used for preparing ampicillin, amoxycillin, and cephalexin (Sec. VII.B.1), although it is evident that the optimal conditions developed for the cell system were not used in these earlier patent examples.

Ampicillin. Enzyme coupled to the PS-1 carrier was stirred for 3 hr at 25°C with 6-APA (0.5%) and D(−)-phenylglycine methyl ester (1.0%), methanol (12%), and phosphate buffer (pH 7.0). The conversion to ampicillin was 84%, and after the enzyme was filtered off, the ampicillin was extracted by adsorption onto Amberlite XAD-2, elution with 50% aqueous methanol, followed by precipitation to give ampicillin trihydrate in an overall yield of 71%. Presumably the yield would have been higher had the phosphate been omitted.

Amoxycillin. The reaction mixture was passed through a column of immobilized enzyme. Presumably because of the low solubility of amoxycillin at the reaction pH of 6.5, the concentration of 6-APA was reduced to 0.5%, the conversion efficiency was 71%; extraction by simply concentrating and adjusting the pH to crystallize the product gave amoxycillin in an overall yield of 49%.

VII. CONVERSIONS OF CEPHALOSPORINS

A. Production of Intermediates

1. *7-Aminocephalosporanic acid*

Cephalosporin C, the major product of the fermentation of *Cephalosporium acremonium*, is deacylated to produce 7-aminocephalosporanic acid (Fig. 7a), a key intermediate for semisynthetic cephalosporins. An early report that this reaction was catalyzed by enzymes from species of *Achromobacter, Brevibacterium,* and *Flavobacterium* (Walton, 1964) was not confirmed (probably owing to associated esterase activity which hydrolyzed the acetoxy group), and on the industrial scale this reaction is carried out chemically at high efficiency, using, for example, nitrosyl chloride.

A rather complicated enzymatic route from cephalosporin C via keto-adipyl and glutaryl derivatives and then hydrolysis to 7-ACA has been reported (Banyu, 1977) using D-amino acid oxidase from *Gliocladium deliquescens.* Recently, however, a strain of *Pseudomonas putida* has been reported to carry out this reaction (Meiji Seika Co. Ltd., 1978). Washed cells produced by fermentation in a conventional medium were suspended in phosphate buffer at pH 7.0, to which cephalosporin C (0.5%) was added, and the reaction continued for 7 hr at 37°C, after which time 37% of the cephalosporin C was converted to 7-ACA. This cephalosporinacylase could perhaps provide the basis for developing a commercial process for producing 7-ACA under mild conditions in a manner analogous to 6-APA production from penicillin G or V. A further report (Shibuya et al., 1981) on the isolation and properties of acylase enzyme from *P. putida* indicates that

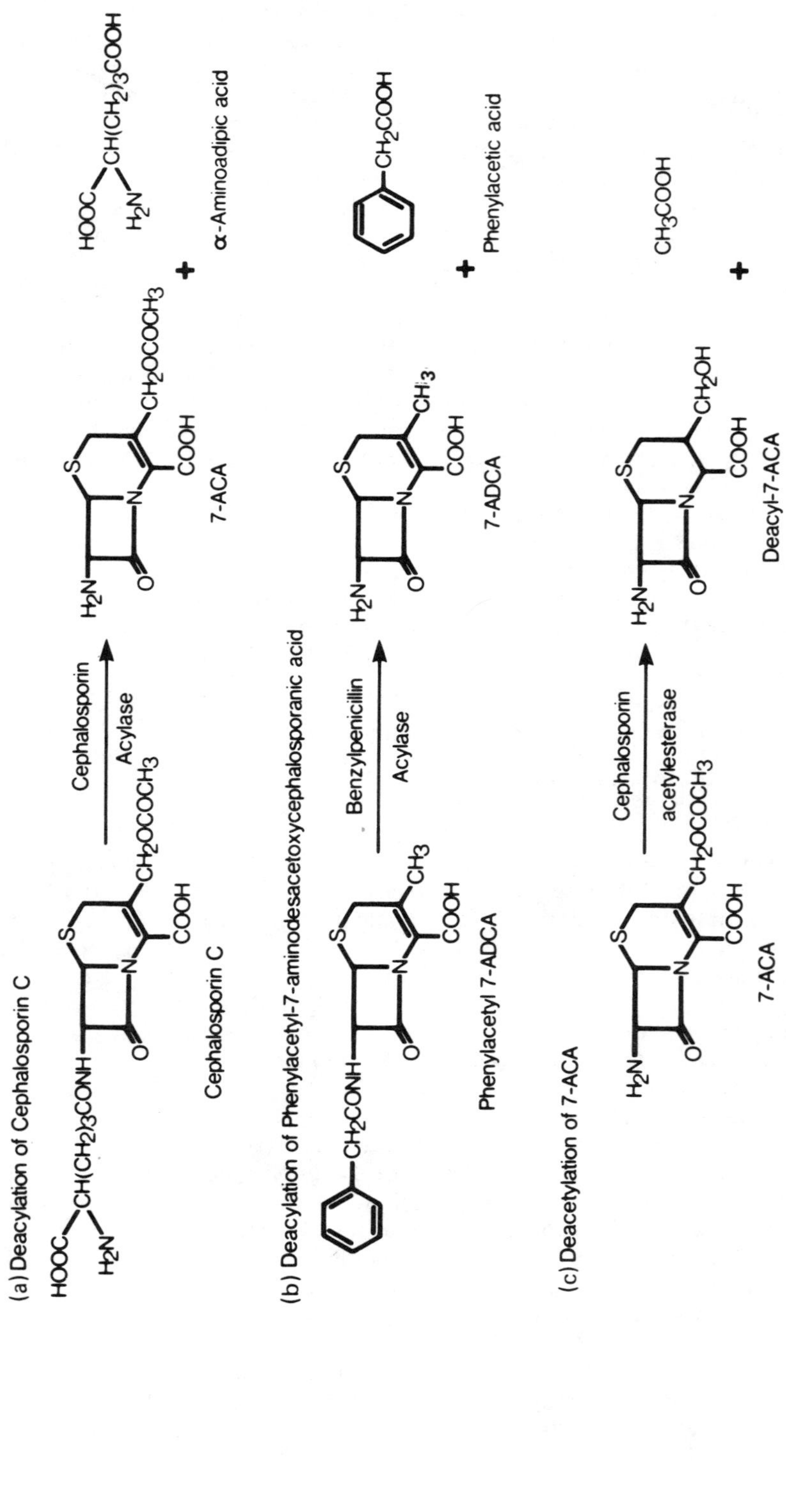

Figure 7 Enzymatic production of cephalosporin intermediates.

cephalosporin C is not a substrate and that the reaction is a two-step one involving a D-amino acid oxidase to form glutaryl-7-ACA which is rapidly hydrolyzed by the acylase.

2. *7-Aminodesacetoxycephalosporanic Acid*

Cephalosporins can be produced chemically from penicillins by expanding the five-membered thiazolidine ring of penicillin to the six membered dihydrothiazine ring of cephalosporins (Cooper and Spry, 1973). The resulting cephalosporin bears the side chain of the penicillin from which it was derived, and this side chain may be removed enzymatically to produce another key intermediate for making semisynthetic cephalosporins namely, 7-amino-3-desacetoxycephalosporanic acid, as shown in Figure 7b. These reactions are catalyzed by either benzylpenicillin or phenoxymethylpenicillin acylases, and it seems likely that any of the processes described for 6-APA production could be used for this reaction. In one process (Bayer A. G., 1975, 1977), immobilized benzylpenicillin acylase (covalent attachment to cyanogen bromide-activated dextran) was used to deacylate 7-phenyl-acetyl-ADCA (6%) at 37°C and pH 7.8 (controlled with ammonia). The reaction liquors were passed through a column containing the ion exchange resin Lewatit MP500A in the chloride form and the eluate adjusted to pH 3.7; after standing, the precipitated product was filtered off and washed successively with water and acetone and then dried to give a 94% yield of 7-ADCA. In the absence of the ion exchange treatment, the product was colored and difficult to filter.

A similar process has been developed by Fujii et al. (1976) using enzyme obtained by fermentation of *B. megaterium* in a medium containing meat extract (1%), peptone (1%), and NaCl (0.5%). After incubation for 30–41 hr at 30°C, the extracellular enzyme was adsorbed onto Celite, which was packed in a 10-liter column and used for the continuous production of 7-ADCA by flowing the substrate (0.5%, 37°C) at a rate of 5 liter/hr for 4 days. The 7-ADCA, 90% pure, was recovered (by addition of acetone and adjusting to pH 4.0) in an overall yield of 85%. The addition of toluene to minimize microbial contamination of the immobilized enzyme prolonged its useful life.

A phenoxymethylpenicillin acylase isolated from *E. aroideae* ATCC 25206 was used to deacylate 7-phenoxyacetyl-ADCA (Fleming et al., 1977). The low pH optimum of the enzyme from this organism was claimed to be advantageous (in contrast to the situation for deacylation of penicillin V, Sec. V) by minimizing base-catalyzed degradation of the β-lactam ring. To avoid excessive loss of enzyme activity during the reaction, it was found necessary to stabilize the enzyme with ferrous sulfate; the addition of a thiol, for example, β-mercaptoethanol, also improved stability. The intracellular enzyme was produced by aerobic fermentation in a conventional medium, and a 10% suspension of harvested cells in cold imidazole buffer containing ferrous ammonium sulfate and β-mercaptoethanol was disrupted on a Manton Gaulin homogenizer and the enzyme purified by fractional precipitation with isopropanol. This enzyme solution was entrapped in cellulose triacetate fibers via the procedure described in Sec. III.D, and the enzyme fibers used to hydrolyze the substrate by the same procedure as that described for penicillin deacylation.

3. Deacyl-7-Aminocephalosporanic Acid

In order to modify the substituent at the C-3 position of cephalosporins, it is first necessary to remove the acetate group. This can be achieved chemically, but yields are low owing to β-lactam degradation and lactonization between the newly formed hydroxymethyl and adjacent carboxyl group. Enzymatic hydrolysis may therefore be preferable (Fig. 7c).

Acetylesterases have been isolated from several sources, for example, plants, mammalian tissue and bacteria. Particularly useful sources of the enzyme are the *Bacillus* species (Eli Lilly Co., 1976). The enzyme from *B. subtilis* NRRL-B-558 has been extensively studied by Abbot and Fukuda (1975a,b) and Abbot et al. (1976). The organism was cultured in flasks containing tryptone soya medium for up to 22 hr at 30°C. Although most of the activity was intracellular, there was a sufficient amount of the more readily recovered extracellular enzyme which was purified by fractional precipitation with ammonium sulfate and ultrafiltration. The enzyme was assayed titrimetrically using cephalothin (Fig. 8b) as substrate. This stable enzyme had a molecular weight of 190,000, a temperature optimum between 40 and 50°C, and a pH optimum of 7.0. Both reaction products were weakly inhibitory, but the reaction readily went to completion.

The purified enzyme was immobilized to bentonite by adsorption at pH 6.3 and high ionic strength, calcium chloride (4 mg/ml) being added for this purpose. Enzyme immobilized in this way proved unstable, and hence the effect of cross-linking agents was studied. Glutaraldehyde strongly inhibited the esterase, but slight stabilization was achieved with aluminium hydroxide. Retention of the soluble enzyme within an ultrafiltration device fitted with a membrane (molecular weight cutoff, 10,000) proved to be a more successful way of reusing the enzyme. Thus reaction mixture containing esterase (5.3 mg/ml) and cephalothin (24 mg/ml) with temperature and pH controlled at 25°C and 7.0, respectively, was completely hydrolyzed in 3 hr. The enzyme was reused 20 times during an 11-day period. The

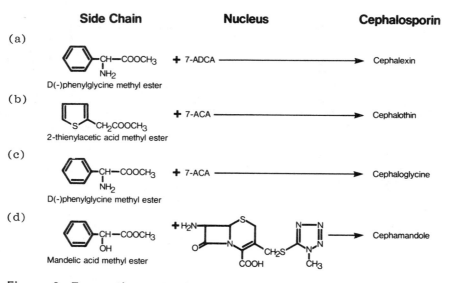

Figure 8 Enzymatic production of semisynthetic cephalosporins.

initial reaction rate decreased gradually to 52% of the initial rate, although all 20 reactions went to over 94% conversion. It was demonstrated that the reduction in reaction rate was due to a 51% loss of protein on handling, and not to decreased specific activity.

4. Miscellaneous

The chemical modification of β-lactam antibiotics often requires the carboxyl group of the nucleus to be protected by blocking groups such as nitroben-zyl or trimethylsilyl, which may be difficult or expensive to prepare and re-move. An alternative enzymatic procedure has been reported by Berry et al. (1982) who screened 7000 microorganisms for ability to cleave a methyl ester from the C4 carboxyl of 7-ADCA and found only one, *Streptomyces capilli-spira* (sp. nov.), to possess this methyl esterase activity. It is possible that the enzyme Subtilisin would be suitable for this reaction in view of its reported use in deesterifying DL-phenylglycine methyl ester (p. 214). Chester et al. (1983) patented a process using Subtilisin to remove a benzyl-protecting group from the C3 carboxyl of a semisynthetic penicillin.

B. Production of Semisynthetic Cephalosporins

1. Cephalexin

The enzymatic production of cephalexin (Fig. 8a) may have a potential ad-vantage over chemical acylation, in that protecting groups are not required because of the mild reaction conditions. Marconi et al. (1975) used the fiber-entrapped benzylpenicillin acylase described in Sec. III.D to bring about this synthesis from 7-ADCA (50 μmol/ml). With a fourfold molar excess of side-chain ester, the conversion efficiency was 75% after 1 hr at pH 7.0 (phosphate buffer) and 25°C, but this fell to 46% with a twofold excess. Longer incubation times reduced yields owing to reverse hydrolysis of the cephalexin.

A more efficient system has been described by Fujii et al. (1976) using a species of *Achromobacter*. The cell-bound enzyme was produced by fer-mentation in a glucose—yeast extract—salts medium at 26°C for 24−40 hr, and the cells used to react 7-ADCA (10.4 kg) and D(−)-phenylglycine methyl ester (41.6 kg) at pH 6.0 and 37°C for $1\frac{1}{2}$ hr in a stirred tank reac-tor. Cephalexin (10 kg) was recovered by passing the supernatant through a carbon column and eluting the product with 50% acetone. The cells were apparently reused 10 times without loss of activity.

As with the enzymatic synthesis of amoxycillin, the most efficient process reported to date is the α-amino acid ester hydrolase enzyme from *X. citri*. The optimum pH and temperature for the synthesis of cephalexin were 6.0 and 37°C, respectively. The effect of concentration of the acyl acceptor 7-ADCA is described in Sec. VI.C and Figure 5. The optimum concentration of the acyl donor D(−)-phenylglycine methyl ester was a 2 M excess when the 7-ADCA concentration was 10 mg/ml. The cell density used was equivalent to half the cell density at the end of fermentation. Aerating the reaction had no effect on the conversion. The time course of a typical reaction in a 200-liter vessel is shown in Figure 9. The con-version efficiency from 7-ADCA was 95%.

The α-amino acid ester hydrolase is also produced by *Acetobacter tur-bidans* ATCC 9325 (Takahashi et al., 1974). This enzyme was extracted and partially purified by the calcium phosphate gel treatment described in Sec. VI.C.1, and the crude enzyme preparation immobilized to cyanogen

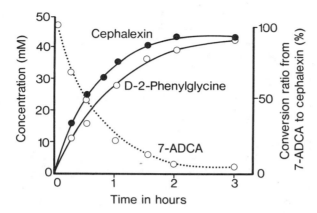

Figure 9 Time course of cephalexin synthesis by washed cells of *X. citri* IFO 3835 in a 200-liter reactor. Reaction mixture contains D(−)-phenylgly-cine methyl ester (20 mg/ml), 7-ADCA (10 mg/ml), and cells (15 mg wet wt/ml) incubated at pH 6.0 and 37°C for 3 hr. Data from Takahashi et al. (1977).

bromide-activated polysaccharide PS-1. After 20 hr reaction at pH 8 and 5°C, 93% of the enzyme activity and 45% of the total protein were immo-bilized. This immobilized enzyme preparation was used for the continuous production of cephalexin by packing it into a column through which was passed a solution containing 7-ADCA (0.5%) and D(−)-PGMe (1.5%). The reaction temperature was kept at 5°C to avoid bacterial contamination and enzyme inactivation. The initial conversion efficiency was 85%, and after 70 days continuous running the efficiency remained substantially unchanged.

Reaction kinetics for the synthesis of cephalexin using the *Xanthomonas* enzyme, both cell bound and immobilized by adsorption onto kaolin, have been reported by Rhee et al. (1980) and Choi et al. (1981).

2. Cephalothin

The *B. megaterium* enzyme used to produce 7-ADCA and cephalexin has been reported (Toyo Jozo Co. Ltd., 1973) to catalyze the reaction of 2-thienylacetic acid methyl ester and 7-ACA to produce the clinically important cephalosporin cephalothin (Fig. 8b). Thus the enzyme from 580 liters of fermentation broth was adsorbed onto Celite (6 kg), packed in a column, and used for the continuous production of cephalothin by passing a buffered solution containing side-chain ester and 7-ACA (10 mg/ml). A total of 1.0 kg of 7-ACA was converted at an efficiency of 85% to yield 1.34 kg of crystalline cephalothin.

Muneyuki et al. (1981) report that the low water solubility of side-chain esters restricts the concentrations that can be used for continuous acylation through fixed-bed reactors. They described the preparation of water-soluble organic acid esters and found them to be good enzyme sub-strates. In a typical experiment, culture supernatant from the fermentation of *Bacillus circulans* was used to provide a crude enzyme powder after frac-tional precipitation with ammonium sulfate, dialysis, and lyophilization. This enzyme preparation was further purified by DEAE cellulose chromatography and immobilized in 82% yield to granulated agar gel by standard cyanogen bromide coupling.

The gel was packed into a column through which was passed a buffered solution of an ester obtained from the reaction of 2-thienylacetic acid with a polyethyleneglycol or a polyethyleneglycol monoethyl ether at a concentration of 142 mg/ml and 7-ACA (10 mg/ml). The eluate contained cephalothin (10.9 mg/ml) equivalent to a yield of 84.4%. The column retained 95% of its initial activity after 9 days continuous operation at 37°C. The authors conclude that the use of such esters opens up new possibilities for enzymatically synthesizing a range of pharmaceutical compounds.

3. Cephaloglycine and Cephamandole

The immobilized enzyme from *B. megaterium* has also been used to synthesize cephaloglycine (Fig. 8c) by acylating 7-ACA with D(−)-phenylglycine methyl ester in a manner analogous to cephalexin production. Reported yields were, however, rather low. Similarly, the same enzyme has been used (Toyo Brewing KK, 1978) to synthesize cephamandole (Fig. 8d) in 57% yields.

VIII. OVERVIEW OF ENZYMATIC SYNTHESIS OF PENICILLINS AND CEPHALOSPORINS

It seems probable that processes for synthesizing ampicillin, amoxycillin, and cephalexin using the enzyme from *X. citri*, either as whole cells or preferably as immobilized enzyme, could be commercially viable and effectively compete with chemical processes. However, this will depend essentially on two factors: firstly, the efficiency of conversion, since reduced energy and solvent usages compared to chemical processes will not be so significant with such high-value reactants and products, and secondly, on the development of an efficient method for extracting the hydrolyzed side-chain ester for reesterification and reuse in the reaction. As an indication of this necessity, it should be noted that in the case of amoxycillin, the side-chain acid is approximately half the cost of 6-APA on a weight basis, the two together accounting for the majority of the basic manufacturing cost.

Published data for enzymatic synthesis of ampicillin, amoxycillin, or cephalexin using benzylpenicillin acylase would not indicate commercial viability, but it may be significant that all these reactions appear to have been carried out in phosphate buffer, and it would be interesting, therefore, to determine the extent to which any of the factors influencing the *Xanthomonas* enzyme reaction were applicable to the enzymes produced by *E. coli*, *K. citrophila*, or *P. melanogenum*. However, ampicillin acylases may have the advantage over the penicillin G acylase of, for example, *E. coli* in having a lower pH optimum, which will be preferable because of shifting of the reaction equilibrium away from hydrolysis. Furthermore, a lower pH would lead to less base-catalyzed hydrolysis of both the ester and the β-lactam ring.

A possible explanation for a low efficiency of conversion with the *E. coli* enzyme has been put forward by Margolin et al., (1980), who found that the hindrance to synthesis of ampicillin using D(−)-phenylglycine is due to a low reaction rate caused by a low level of the deprotonated form of the amino group, which is the form that binds to the enzyme. The thermodynamic advantages of using side-chain esters have been elucidated by Svedas et al., (1980), but they point out the limitations caused by the poor water solubility of esters. This limitation may be overcome by using the water-soluble esters of Muneyuki et al. (1981) described in Sec. VII.B.2. An alternative approach has been proposed by McDougall et al. (1982), who found that the equilibrium could be shifted in favor of acylation by changing the water activity. Thus, at a reaction temperature of 10°C and a pH of 5.2, the addition of polyethylene glycol (45% wt/vol) increased the conversion efficiency of equimolar quantities of 6-APA and phenylacetic acid from 30 to 60%. Increasing the phenylacetic acid concentration by 50% improved the conversion efficiency to 80%.

The limitations of enzymatically synthesizing peptide bonds caused by the reaction equilibrium favoring the starting amino acids have been reviewed by Semenov et al. (1981), who listed the drawbacks associated with current methods as follows:

1. Selecting products of low solubility shifts the equilibrium to the right, but limits choice of product and causes separation problems with immobilized enzymes.
2. Lowering the concentration of water by performing the reaction in a water–water miscible organic solvent mixture may increase yield, but may also reduce enzyme activity.
3. Using derivatives such as esters increases costs as well as yields.

The method proposed to overcome these drawbacks is to carry out the reaction in a water–*water-immiscible* organic solvent, thereby localizing the enzyme in the aqueous phase and thus eliminating the problem of stabilizing the enzyme against inactivation by a nonaqueous solvent. This was illustrated by the synthesis of N-acetyl-L-tryptophanyl-L-leucine amide in a biphasic system of water–ethyl acetate (from 2 to 20% vol/vol) which raised the reaction by three orders of magnitude, from 0.1% in water to 100% in the biphasic system. [Note: This approach had already been applied to resolution of side-chain acids (Bayer A. G., 1981), as described in Sec. IX.A.]

Reaction mechanisms for both *X. citri* and *E. coli* enzymes have been recently reviewed and compared by Konecny (1981).

IX. RESOLUTION OF AMINO ACIDS USED AS SIDE CHAINS IN THE PRODUCTION OF PENICILLINS AND CEPHALOSPORINS

A number of enzymatic resolution processes have been developed for producing the D(−)-phenylglycine used in the preparation of ampicillin, cephalexin, and cephaloglycine and the D(−)-hydroxyphenylglycine used in the preparation of amoxycillin. While in the case of D(−)-phenylglycine it is not certain if these enzymatic processes can compete effectively with those using chemical resolving agents, the resolution of p-hydroxyphenylglycine by chemical methods is more difficult and enzymatic processes are known to be operated on an industrial scale.

A. Processes Based on Hydrolysis of the L(+) Isomer

One of the first processes reported for resolving p-hydroxyphenylglycine (Savidge et al., 1974) took advantage of the stereospecificity of *E. coli* benzylpenicillin acylase to hydrolyze only the L(+) isomer of N-phenylacetyl-DL-p-hydroxyphenylglycine according to the reactions shown in Figure 10.

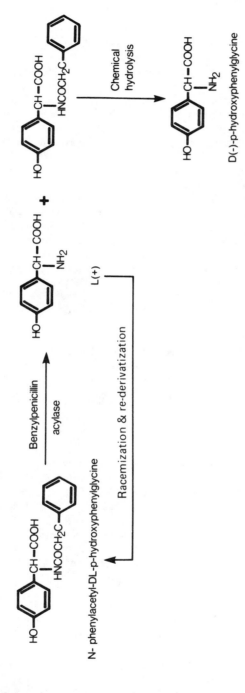

Figure 10 Enzymatic resolution of DL-p-hydroxyphenylglycine with benzylpenicillin acylase.

Whole cell or preferably immobilized enzyme was used; for example, partially purified enzyme was immobilized to the macroreticular acrylic ester resin Amberlite XAD 7 [adsorption for 16 hr in phosphate buffer at pH 5 followed by cross-linking by treatment with glutaraldehyde (5%) at pH 8 for 2 hr], and 25 g of the resulting preparation were packed into a column through which a solution of N-phenylacetyl-DL-hydroxyphenylglycine (5%) was continuously passed at a rate of 70 ml/hr. Complete conversion was sustained for 300 hr. The unhydrolyzed D(−) derivative was recovered by solvent extraction, leaving the L(+) acid in the aqueous phase. A similar type of process has been described by Neilson (1980) for the resolution of DL-phenylglycine by stereospecific hydrolysis of L(+)-phenylglycine amide by an amidase produced by *P. putida*. The free amino acid so produced had a low solubility and could therefore be separated from the reaction mixture by centrifugation for racemization and rederivatization. The remaining D(−)-phenylglycine amide was precipitated and chemically hydrolyzed to produce optically pure (99.9%) D(−)-phenylglycine.

A third variant on this type of process has been reported (Bayer A. G., 1979) which has the claimed advantage of permitting a higher concentration of starting DL mixture to be used. In this process N-acetyl-DL-phenylglycine methyl ester (10%) was stereospecifically hydrolyzed with the protease Subtilisin to give N-acetyl-L(+)-phenylglycine. The enzyme was immobilized by covalent attachment to a copolymer of methacrylate, methacrylic acid, and maleic anhydride. The reaction could also be carried out in a two-phase solvent system, for example, water and methyl isobutyl ketone (Bayer A. G., 1981), the solvent protecting the N-acylphenylglycine ester from nonspecific hydrolysis.

A disadvantage of all these processes lies in the fact that the enzymes are specific for the L(+) isomer, which necessitates chemically hydrolyzing the unchanged D(−) derivative while the L(+) isomer must be racemized and then rederivatized before reuse in the process. It would be preferable if the D(−) isomer were the substrate for the enzyme. Two such developments have been reported.

B. Processes Based on Hydrolysis of the D(−) Isomer

A D(−)-aminoacylase has been isolated from *Streptomyces olivaceous* (Sugie and Suzuki, 1978). Subsequently Sugie and Suzuki (1980) obtained higher yields of this enzyme from *Streptomyces tuirus*. The intracellular enzyme was induced by a variety of D(−)-amino acids or their N-acetyl derivatives; with DL-amino acids, twice the quantity was required to effect the same degree of induction, since L(+) isomers had no induction activity. DL-valine was selected as the preferred inducer because of its high solubility in water, the optimal concentration being 1.5%. The optimal medium composition was defined as soluble starch (2%), maltose (1%), glycerol (1%), peptone (0.5%), yeast extract (0.5%), meat extract (1.5%), corn-steep liquor (1%), NaCl (0.5%), and DL-valine (1.5%) at an initial pH of 7.0. Cultivation was for 3 days at 30°C; D-aminoacylase activity was determined by ninhydrin assay of the free acid released after 15 min incubation at 30°C, pH 7.0, with a buffered solution of N-acetyl-D(−)-phenylglycine. Crude enzyme solution was prepared by sonicating the harvested cells and removing cell debris by centrifugation. Although this solution had no racemose activity, there was some L(+)-amino acylase activity, but this was effectively eliminated by DEAE Sephadex chromatography. When the crude enzyme was

used to resolve N-acetyl-DL-phenylglycine, the optical purity of the resulting D(−) acid was 98%, but when the purified enzyme was incubated with
N-acetyl-DL-phenylglycine (2%) for 6 hr at 30°C and pH 7, the D(−) isomer
was completely hydrolyzed and optically pure (99.9%) D(−) acid recovered
in 74% yield by simply concentrating under vacuum, precipitating the crude
product which was purified by recrystallization from hot water.

In the above process the unchanged L(+) derivative has to be racemized before being used again in the process. An elegant solution to this
problem is to be found in the process developed by Snam Progetti S.P.A.
(1976), who found that a dihydroxypyrimidinase isolated from calf liver
could be used to stereospecifically hydrolyze DL-phenylhydantoin to a D(−)-
carbamoyl derivative according to the reaction shown in Figure 11. The
advantage of the method lay in the fact that the remaining L(+)-phenyl-
hydantoin underwent spontaneous racemization, so that the DL mixture was
totally converted to the D(−)-carbamoyl compound. The calf liver enzyme
was later replaced by enzyme of microbial origin following the selection of
strains which were capable of using hydantoin or hydantoin derivatives as
their sole nitrogen source, for example, *Pseudomonas* strains (Snam progetti
S.P.A., 1977). Similar activities have since been found in a wide range
of microorganisms when grown on media containing around 0.1% of a DL-
hydantoin derivative such as DL-5-(2-methylthioethyl)hydantoin (Kanegafuchi, 1978, 1980; Yamada et al., 1978). In these processes D(−)-carba-
moyl derivative was chemically hydrolyzed to give the free acid.

Subsequent developments in this field (Ajinomoto, K.K. 1978; Snam Progetti S.P.A., 1978) have led to the isolation of strains possessing both D(−)-
hydantoinase and D(−)-carbamoylase activity, so that the conversion from
DL phenylhydantoin to D(−)-phenylglycine can be achieved in a single step,
thereby avoiding the need for the rather difficult chemical hydrolysis of
the N-carbamoyl derivative. A detailed description of such a process has
been given by Olivieri et al. (1979). The organism *Agrobacterium radio-
bacter* NRRL B11291 was isolated by enrichment culture (30°C) of soil samples
on a mineral medium containing glucose (0.5%) and N-carbamoyl-D(−)-phe-
nylglycine (0.2%) as sole nitrogen source. When the strain was grown on
a simple salts and glucose medium, only D(−)-carbamoylase activity was
detected. Addition of uracil (0.2%) as sole nitrogen source led to the co-
production of the hydantoinase. After incubation in shaken flasks (220
rpm) at 30°C for 24 hr, the cells were harvested and the equivalent of 1.5
g dry weight used to convert DL-phenylhydantoin (4%) in 0.1 M phosphate
buffer at pH 7.8. Complete conversion of the substrate to D(−)-phenyl-
glycine was achieved after 20 hr at 40°C; the pH was not controlled during
the reaction. Incubation for a further 20 hr caused no racemization or
degradation of the amino acid. The amino acid was recovered (Snam
Progetti S.P.A., 1978) by precipitation of proteinaceous material with tri-
chloroacetic acid, centrifugation, and adjustment of the supernatant to the
isoelectric point (pH 5.2). From this reference it appears that the enzyme
can also be obtained by fermentation in a simple corn-steep liquor medium
(5%) adjusted to pH 7.8 with NaOH. In this example the cells were used
to hydrolyze a 10% solution of the hydantoin during a 160-hr incubation
period. Use of an unidentified thermophilic organism, CBS 303.80 (BASF,
1982) enabled hydrolysis of the hydantoin to the D(−)-carbamoyl derivative
to be carried out at 60°C, with the result that a 10% (wt/vol) solution was
hydrolyzed in 6 hr.

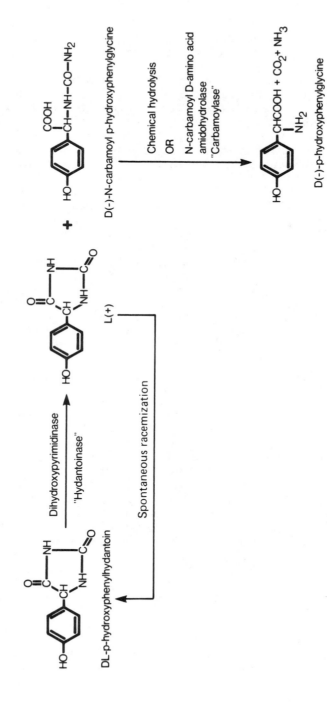

Figure 11 Enzymatic resolution of DL-p-hydroxyphenylglycine with dihydroxypyrimidinase.

X. MISCELLANEOUS REACTIONS

A. Stereospecific Synthesis of Penicillins and Cephalosporins

Recent evidence indicates the possibility of stereospecifically synthesizing cephalexin and amoxycillin by reaction of DL-phenylhydantoin with the appropriate nucleus. Reported yields are very low. Thus amoxycillin (54.6 µg/ml) was produced from 6-APA (0.5%) and DL-p-hydroxyphenylhydantoin (1%) by reacting (3 hr) with cells of *Flavobacterium hydantoinophilum* grown on a medium containing 5-methylmercaptoethylhydantoin (0.2%) as inducer (Ajinomoto, K.K. 1980).

B. Deacylation of Nocardicin C

Nocardicins are antibiotics having a monocyclic β-lactam nucleus with an N-acyl side chain. Deacylation to produce 3-aminonocardicinic acid (3-ANA) might provide the nucleus necessary for preparing semisynthetic nocardicins. Komori et al (1978) screened 100 strains of bacteria, 1000 actinomycetes, 20 yeasts, and 100 fungi, but found none which catalyzed the deacylation of nocardicin A. Nocardicin C is a minor component of nocardicin fermentations and can be prepared by chemical hydrogenation of nocardicin A. Organisms deacylating this compound were obtained fairly readily, the best being *Pseudomonas schuylkilliensis*. The reactions are shown in Figure 12. The enzyme has been extracted, purified, and characterized. It did not catalyze the reverse, synthetic reaction.

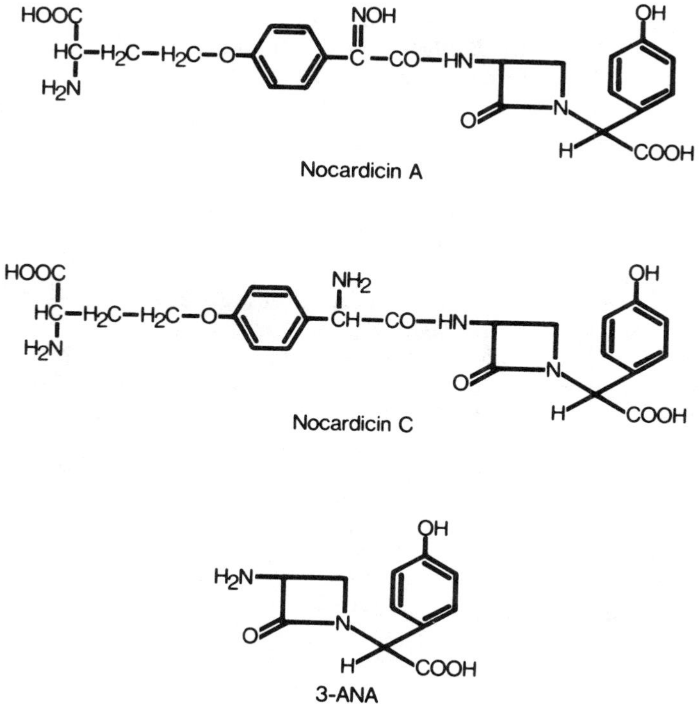

Figure 12 Structures of nocardicin A and C and 3-aminonocardinic acid (3-ANA).

C. Oxidation to Produce Cephalosporin Sulfoxides

Torri et al. (1980) reported that certain fungi, especially *Coriolus hirsutus*, were capable of oxidizing the cephalosporins, 7-phenylacetyl-3-deacetoxy-cephalosporanic acid, the 7-phenoxy compound, and cephalexin to the corresponding (R)- and (S)-sulfoxides. Unlike chemical oxidation, which produces little (R)-sulfoxide, the microbial method gave equal quantities of (R) and (S) compounds, the former possessing higher antibacterial activity.

REFERENCES

Abbot, B. J. (1976). Preparation of pharmaceutical compounds by immobilized enzymes and cells. *Adv. Appl. Microbiol. 20*:203–257.

Abbot, B. J. (1978). Immobilized cells. In *Annual Reports on Fermentation Processes*, Vol. 2. D. Perlman (Ed.). Academic, New York, pp. 91–123.

Abbot, B. J., and Fukuda, D. (1975a). Physical properties and kinetic behaviour of a cephalosporin acetylesterase produced by *Bacillus subtilis. Appl. Microbiol. 30*:413–419.

Abbot, B. J., and Fukuda, D. (1975b). Preparation and properties of a cephalosporin acetylesterase adsorbed onto bentonite. *Antimicrob. Agents Chemother. 8*:282–288.

Abbot, B. J., Cerimele, B., and Fukuda, D. (1976). Immobilization of a cephalosporin acetylesterase by containment within an ultrafiltration device. *Biotechnol. Bioeng. 18*:1033–1042.

Adelberg, E. A., Mandel, M., and Chen, G. C. C. (1965). Optimal conditions for mutagenesis by N-methyl-N'-nitro-N-nitrosoguanidine in *Escherichia coli* K12. *Biochem. Biophys. Res. Commun. 18*:788–795.

Ajinomoto, K. K. (1978). German Patent No. 2825245.

Ajinomoto, K. K. (1980). Japan Patent No. J5-5135-597.

Axen, R., Porath, J., and Ernback, S. (1967). Chemical coupling of peptides and proteins to polysaccharides by means of cyanogen halides. *Nature 214*:1302.

Balasingham, K., Warburton, D., Dunnill, P., and Lilly, M. D. (1972). The isolation and kinetics of penicillin amidase from *Escherichia coli. Biochim. Biophys. Acta 276*:250–256.

Banyu Pharmaceutical Co. Ltd. (1977). Japan Patent No. J5-2038-092.

BASF.A.G. (1982). European Patent No. 0046186.

Batchelor, F. R., Doyle, F. P., Nayler, J. H. C., and Rolinson, G. N. (1959). Synthesis of penicillin: 6-Aminopenicillanic acid in penicillin fermentations. *Nature 183*:257–258.

Batchelor, F. R., Chain, E. B., Hardy, T. L., Mansford, K. R., and Rolinson, G. N. (1961). 6-Aminopenicillanic acid. VI. Isolation and purification. *Proc. R. Soc. London B154*:498–508.

Bauer, K., Kaufmann, W., and Ludwig, S. A. (1971). Vereinfachte bestimmung der penicillin-amidase aus *Escherichia coli. Hoppe Seyler's Z. Physiol. Chem. 352*:1723–1724.

Bayer A. G. (1973a). German Patent No. 2157970.

Bayer A. G. (1973b). German Patent No. 2157972.

Bayer A. G. (1975). German Patent No. 409569.

Bayer A. G. (1977). German Patent No. 2528622.

Bayer A. G. (1979). German Patent No. 2807286.

Bayer A. G. (1982). German Patent No. 2927535.

Beecham Group Ltd. (1975). British Patent No. 1,400,468.

Beecham Group Ltd. (1976). British Patent No. 1,449,808.

Berry, D. R., Fukuda, D. S., and Abbott, B. J. (1982). Enzymatic removal of a cephalosporin methyl ester blocking group. *Enzyme Microb. Technol.* 4:80–84.

Bomstein, J., and Evans, W. G. (1965). Automated colorimetric determination of 6-aminopenicillanic acid in fermentation media. *Anal. Chem.* 37:576–578.

Brandl, E., and Knauseder, F. (1975). German Patent No. 2503584.

Carrington, T. R., Savidge, T. A., and Walmsley, M. F. (1966). British Patent No. 1,015,554.

Chester, I. R., Powell, L. W., and Roberts, D. G. (1983). European Patent No. 0051481.

Chiang, D., and Bennett, R. E. (1967). Purification and properties of penicillin amidase from *Bacillus megaterium*. *J. Bacteriol.* 93:302–308.

Choi, W. G., Lee, S. B., and Ryu, D. D. Y. (1981). Cephalexin synthesis by partially purified and immobilised enzymes. *Biotechnol. Bioeng.* 23:361–371.

Claridge, C. A., Gourevitch, A., and Lein, J. (1960). Bacterial penicillin amidase. *Nature* 187:237–238.

Cole, M. (1969a). Hydrolysis of penicillins and related compounds by the cell-bound penicillin acylase of *Escherichia coli*. *Biochem. J.* 115:733–739.

Cole, M. (1969b). Deacylation of acylamino compounds other than penicillins by the cell-bound penicillin acylase of *Escherichia coli*. *Biochem. J.* 115:741–745.

Cole, M. (1969c). Penicillins and other acylamino compounds synthesized by the cell-bound penicillin acylase of *Escherichia coli*. *Biochem. J.* 115:747–756.

Cole, M. (1969d). Factors affecting the synthesis of ampicillin and hydroxypenicillins by the cell-bound penicillin acylase of *Escherichia coli*. *Biochem. J.* 115:757–764.

Cole, M., and Sutherland, R. S. (1966). The role of penicillin acylase in the resistance of gram-negative bacteria to penicillins. *J. Gen. Microbiol.* 42:345–356.

Cole, M., Savidge, T. A., and Vanderhaeghe, H. (1975). Penicillin acylase (assay). In *Methods in Enzymology*, Vol. 43. J. H. Hash (Ed.). Academic, New York, pp. 698–705.

Cooper, R. D. G., and Spry, D. O. (1973). Rearrangements of cephalosporins and penicillins. In *Cephalosporin and Penicillins. Chemistry and Biology*. E. H. Flynn (Ed.). Academic, New York, pp. 183–254.

Delin, P. S., Ekström, B. A., Sjöberg, B. O., Thelin, K. H., and Nathorst-Westfeld, L. S. (1972). British Patent No. 1,261,711.

Diers, I. V., and Emborg, C., (1979). British Patent No. 2,021,119.

Dinelli, D. (1972). Fibre-entrapped enzymes. *Proc. Biochem.* 7:9–12.

Dunnill, P. (1980). Immobilised cell and enzyme technology. *Philos, Trans. R. Soc. London* B290:409–420.

Ekström, B., Lagerlöf, E., Nathorst-Westfeld, L., and Sjöberg, B. (1974). Ny framställningsteknik för 6-APA med immobiliserat enzym. *Sven. Farm. Tidskr.* 78:531–535.

Eli Lilly Co. (1976). U.S. Patent No. 3972-774.

Fleming, I. D., Turner, M. K., and Napier, E. J. (1977). British Patent No. 1,473,100.

Fujii, T., Matsumoto, K., and Watanabe, T. (1976). Enzymatic synthesis of cephalexin. *Proc. Biochem.* 11:21–24.

Gang, D. M., and Shaikh, K. (1976). Regulation of penicillin acylase in *Escherichia coli* by cyclic AMP. *Biochim. Biophys. Acta 425*:110–114.

Gestrelius, S. (1982). Immobilised Penicillin V acylase—Development of an industrial catalyst. *Appl. Biochem. Biotech. 7*:19–21.

Gestrelius, S. M. (1979). British Patent No. 2,019, 410.

Golub, E. I., and Bel'kind, A. M. (1977). Effect of glucose on transport of phenylacetic acid and synthesis of penicillin amidohydrolase by *Escherichia coli. Mikrobiologiya 46*:363–365.

Hetherington, P. J., Follows, M., Dunnill, P., and Lilly, M. D. (1971). Release of protein from baker's yeast (*Saccharomyces cerevisiae*) by disruption in an industrial homogeniser. *Trans. Inst. Chem. Eng. 49*:142–148.

Ho, L. Y., and Humphrey, A. E. (1970). Optimal control of an enzyme reaction subject to enzyme deactivation. I. Batch process. *Biotechnol. Bioeng. 12*:291–311.

Holt, R. J., and Stewart, G. T. (1964). Production of amidase and β-lactamase. *J. Gen. Microbiol. 36*:203–213.

Huang, H. T., English, A. R., Seto, T. A., Shull, G. M., and Sobin, B. A. (1960). Enzymatic hydrolysis of the side chain of penicillins. *J. Am. Chem. Soc. 82*:3790.

Huang, H. T., Seto, T. A., and Shull, G. M. (1963). Distribution and substrate specificity of benzylpenicillin acylase. *Appl. Microbiol. 11*:1–6.

Huber, F. M., Chauvette, R. R., and Jackson, B. G. (1972). Preparative methods for 7-aminocephalosporanic acid and 6-aminopenicillanic acid. In *Cephalosporins and Penicillins. Chemistry and Biology.* E. H. Flyn (Ed.). Academic, New York, pp. 27–73.

Kameda, Y., Kimura, Y., Toyoura, E., and Omori, T. (1961). A method for isolating bacteria capable of producing 6-aminopenicillanic acid from benzylpenicillin. *Nature 191*:1122–1123.

Kanegafuchi Chemical KK. (1978). Japan Patent No. J5-3044-690.

Kanegafuchi Chemical KK. (1980). Belgium Patent No. 881–547.

Karube, I., Suzuki, S., and Vandamme, E.J. (1984). This volume, pp. 761-780.

Kato, K. (1980). Kinetics of acyl transfer by α-amino acid ester hydrolase from *Xanthomonas citri. Agric. Biol. Chem. 44*:1083–1088.

Kato, K., Kawahara, K., Takahashi, T., and Igarasi, S. (1980a). Enzymatic synthesis of amoxicillin by the cell-bound α-amino acid ester hydrolase of *Xanthomonas citri. Agric. Biol. Chem. 44*:821–825.

Kato, K., Kawahara, K., Takahashi, T., and Kakinuma, A. (1980b). Purification of α-amino acid ester hydrolase from *Xanthomonas citri. Agric. Biol. Chem. 44*:1069–1074.

Kato, K., Kawahara, K., Takahashi, T., and Kakinuma, A. (1980c). Substrate specificity of α-amino acid ester hydrolase from *Xanthomonas citri. Agric. Biol. Chem. 44*:1075–1081.

Kaufmann, W., and Bauer, K. (1960). Enzymatische spaltung und resynthese von penicillin. *Naturwissenshaften 47*:474–475.

Kaufmann, W., Bauer, K., and Offe, H. A. (1960). Enzymatic cleavage and resynthesis of penicillins. In *Antimicrobial Agents Annual.* P. Gray, B. Tarbenkin and S. G. Bradley (Eds.). Plenum, New York, pp. 1–5.

Klein, J., and Eng, H. (1979). Immobilization of microbial cells in epoxy carrier systems. *Biotechnol. Lett. 1*:171–176.

Klein, J., and Kluge, M. (1981). Immobilization of microbial cells in polyurethane matrices. *Biotechnol. Lett. 2*:65–70.

Klein, J., and Wagner, F. (1978a). Immobilized whole cells. In *Proceedings of the 1st European Congress on Biotechnology*. DECHEMA, Frankfurt, pp. 142–164.

Klein, J., and Wagner, F. (1978b). Immobilized whole cells. *DECHEMA Monogr. 82*:142.

Klein, J., and Wagner, F. (1980). Immobilization of whole microbial cells for the production of 6-amino-penicillanic acid. *Enzyme Eng. 5*:335–345.

Klein, J., Hockel, V., Schara, P., and Eng, H. (1979). Polymer networks for entrapment of micro-organisms. *Angew. Makromol. Chem. 76/77*: 329–350.

Konecny, J. (1981). Penicillin acylases as amidohydrolases and acyl transfer catalysts. *Biotechnol. Lett. 3*:112–117.

Komori, T., Kunugita, K., Nakahara, K., Aoki, H., and Imanaka, H. (1978). Production of 3-amino-nocardicinic acid from nocardicin C by microbial enzymes. *Agric. Biol. Chem. 42*:1439–1440.

Kutzbach, C., and Rauenbusch, E. (1974). Preparation and general properties of crystalline penicillin acylase from *Escherichia coli* ATCC 11 105. *Hoppe Seyler's Z. Physiol. Chem. 35445*–53.

Kyowa Hakko Kogyo Co. Ltd. (1971). German Patent No. 2050983.

Lagerlöf, E., Nathorst-Westfeld, L., Ekström, B., and Sjöberg, B. (1976). Production of 6-aminopenicillanic acid with immobilised *Escherichia coli* acylase. In *Methods in Enzymology*, Vol. 44. K. Mosbach (Ed.). Academic, New York, pp. 759–768.

Levitov, M. M., Klapovskiya, K. I., and Kleiner, G. I. (1967). The induced acylase biosynthesis of *Escherichia coli. Mikrobiologiya 36*:912–917.

Lowe, D. A., Romancik, G., and Elander, R. P. (1981). Penicillin acylases: A review of existing enzymes and the isolation of a new bacterial penicillin V acylase. *Dev. Ind. Microbiol. 22*:163–180.

McDougall, B., Dunnill, P., and Lilly, M. D. (1982). Enzymic acylation of 6-aminopenicillanic acid. *Enzyme Microb. Technol. 4*:114–115.

Marconi, W., Cecere, F., Morisi, F., Della Penna, G., and Rappouli, B. (1973). The hydrolysis of penicillin G to 6-amino penicillanic acid by entrapped penicillin acylase. *J. Antibiot. 26*:228–232.

Marconi, W., Bartoli, F., Cecere, F., Galli, G., and Morisi, F. (1975). Synthesis of penicillins and cephalosporins by penicillin acylase entrapped in fibres. *Agric. Biol. Chem. 39*:277–279.

Margolin, A. L., Svedas, V. K., and Berezin, I. V. (1980). Substrate specificity of penicillin amidase from *E. coli. Biochim. Biophys. Acta 616*:283–289.

Mayer, H., Collins, J., and Wagner, F. (1979). Cloning of the penicillin G acylase gene of *Escherichia coli* ATCC 11105 on multicopy plasmids. In *Plasmids of Medical, Environmental and Commercial Importance*. K. N. Timmis and A. Puhler (Eds.). Elsevier/North Holland Biomedical Press, Amsterdam, pp. 459–470.

Mayer, H., Collins, J., and Wagner, F. (1980). Cloning of the penicillin G acylase gene of *Escherichia coli* ATCC 11105 on multicopy plasmids. *Enzyme Eng. 5*:61–69.

Meiji Seika Co. Ltd. (1978). Japan Patent No. J5-3094-093.

Mitz, M. A., and Summaria, L. J. (1961). Synthesis of biologically active cellulose derivatives of enzymes. *Nature 189*:576.

Morikawa, Y., Karube, I., and Suzuki, S. (1980). Enhancement of penicillin acylase activity by cultivating immobilized *Kluyvera citrophila*. *Eur. J. Appl. Microbiol. 10*:23–30.

Muneyuki, R., Mitsugi, T., and Kondo, E. (1981). Water-soluble esters: Useful enzyme substrates for the synthesis of β-lactam antibiotics. *Chem. Ind. 5*:159–161.

Nara, T., Misawa, M., Okachi, R., and Yamamoto, M. (1971). Enzymatic synthesis of D(−)-α-aminobenzylpenicillin. Part I. Selection of penicillin acylase-producing bacteria. *Agric. Biol. Chem. 35*:1676–1682.

Neilson, M. H. (1980). Enzyme technology and enzyme production. In *13th International TNO Conference*. A. Verbraeck (Ed.). Netherlands Central Organisation for Applied Scientific Research, The Hague, pp. 41–58.

Niedermeyer, A. O. (1964). Determination of phenylacetic acid in penicillin fermentation media by means of gas chromatography. *Anal. Chem. 36*: 938–939.

Okachi, R., Misawa, M., Deguchi, T., and Nara, T. (1972). Production of D(−)-α-aminobenzylpenicillin by *Kluyvera citrophila* KY 3641. *Agric. Biol. Chem. 36*:1193–1198.

Okachi, R., Kato, F., Miyamura, Y., and Nara, T. (1973). Selection of *Pseudomonas melanogenum* KY 3987 as a new ampicillin-producing bacteria. *Agric. Biol. Chem. 37*:1953–1973.

Olivieri, R., Fascetti, E., Angelini, L., and Degen, L. (1979). Enzymatic conversion of N-Carbamoyl-D-amino acids to D-amino acids. *Enzyme Microb. Technol. 1*:201–204.

Otsuka Seiyaku Co. Ltd. (1972a). Japan Patent No. 7228187.

Otsuka Seiyaku Co. Ltd. (1972b). Japan Patent No. 7228183.

Rhee, D. K., Lee, S. B., Rhee, J. S., Ryu, D. D. Y., and Hospodka, J. (1980). Enzymatic biosynthesis of cephalexin. *Biotechnol. Bioeng. 22*:1237–1247.

Rolinson, G. N., Batchelor, F. R., Butterworth, D., Cameron-Wood, J., Cole, M., Eustace, C. G., Hart, M. V., Richards, M., and Chain, E. B. (1960). Formation of 6-aminopenicillanic acid from penicillin by enzymic hydrolysis. *Nature 187*:236–237.

Rozansky, R., Biano, S., Clejan, L., Frenkel, N., Bogokovsky, B., and Altmann, G. (1969). Penicillin β-lactamase and penicillin acylase formation by gram-negative bacteria. *Israel J. Med. Sci. 5*:297–305.

Ryu, D. Y., Bruno, C. F., Lee, B. K., and Venkatasubramanian, K. (1972). Microbial penicillin amidohydrolase and the performance of a continuous enzyme reactor system. In *Fermentation Technology Today*. G. Terui (Ed.). Society of Fermentation Technology, Tokyo, p. 307.

Sato, T., Tosa, T., and Chibata, I. (1976). Continuous production of 6-aminopenicillanic acid from penicillin by immobilized microbial cells. *Eur. J. Appl. Microbiol. 2*:153–160.

Savidge, T. A., and Cole, M. (1975). Penicillin acylase (bacterial). In *Methods in Enzymology*, Vol. 43. J. Hash (Ed.). Academic, New York, pp. 705–721.

Savidge, T. A., Powell, L. W., and Lilly, M. D. (1974). British Patent No. 1,357,317.

Schneider, W. J., and Roehr, M. (1976). Purification and properties of penicillin acylase of *Bovista plumbea*. *Biochim. Biophys. Acta 452*:177–185.

Semenov, A. N., Berezin, I. V., and Martinek, K. (1981). Peptide synthesis enzymatically catalyzed in a biphasic system: Water–*water immiscible* organic solvent. *Biotechnol. Bioeng.* 23:355–360.

Shaikh, K., Talati, P. G., and Gang, D. M. (1973). Spectrophotometric method for the estimation of 6-aminopenicillanic acid. *Antimicrob. Agents Chemother.* 3:194–197.

Shibuya, Y., Matsumoto, K., and Fujii, T. (1981). Isolation and properties of 7β-(4-carboxybutanamido)cephalosporanic acid acylase-producing bacteria. *Agric. Biol. Chem.* 45:1561–1567.

Sikyta, B., and Slezak, J. (1964). Continuous cultivation of *Escherichia coli* possessing high penicillin-acylase activity. *Biotechnol. Bioeng.* 6:309–319.

Smith, R. A. G. (1976). Amphipathic enzyme–polymer conjugates. *Nature* 262:519–520.

Snam Proghetti S.P.A. (1974). British Patent No. 1,348,359.

Snam Progetti S.P.A. (1976). German Patent No. 2621076.

Snam Progetti S.P.A. (1977). German Patent No. 2631048.

Snam Progetti S.P.A. (1978). U.K. Patent No. 2,022,581.

Son, H., Mheen, T., Seong, B., and Han, M. H. (1982). Studies on microbial penicillin amidase (iv) the production of penicillin amidase from a partially constitutive mutant of *Bacillus megaterium*. *J. Gen. Appl. Microbiol.* 28:281–291.

Squibb, E. R. & Sons Inc. (1969). U.S. Patent No. 3,446,705.

Stoppok, E., Schömer, U., Segner, A., Mayer, H., and Wagner, F. (1980). Production of 6-aminopenicillanic acid from penicillin V and G by *Bovista plumbea* NRRL 3824 and *Escherichia coli* 5K (pHM 12). Poster Presentation at the 6th International Fermentation Symposium, London, Ontario, Canada.

Stoppok, E., Wagner, F., and Zadrazil, F. (1981). Identification of a penicillin V acylase processing fungus. *Eur. J. Appl. Microbiol. Biotechnol.* 13:60–61.

Sugie, M., and Suzuki, H. (1978). Purification and properties of D-aminoacylase of *Streptomyces olivaceus*. *Agric. Biol. Chem.* 42:107–113.

Sugie, M., and Suzuki, H. (1980). Optical resolution of DL-amino acids with D-aminoacylase of Streptomyces. *Agric. Biol. Chem.* 44:1089–1095.

Svedas, V. K., Margolin, A. L., and Berezin, I. V. (1980). Enzymatic synthesis of β-lactam antibiotics: A thermodynamic background. *Enzyme Microb. Technol.* 2:138–144.

Takahashi, T., Yamasaki, Y., and Kato, K. (1974). Substrate specificity of an α-amino acid ester hydrolase produced by *Acetobacter turbidans* ATCC 9325. *Biochem. J.* 137:497–503.

Takahashi, T., Kato, K., Yamasaki, Y., and Isono, M. (1977). Synthesis of cephalosporins and penicillins by enzymatic acylation. *Jpn. J. Antibiot. Suppl.* 30:S-230–S-238.

Takeda Chemical Co. Ltd. (1975a). British Patent No. 1,382,255.

Takeda Chemical Co. Ltd. (1975b). Japan Patent No. 50-11682.

Takeda Chemical Co. Ltd. (1976). Japan Patent No. 51-61686.

Torii, H., Asano, T., Matsumoto, N., Kato, K., Tsushima, S., and Kakinuma, A. (1980). Microbial oxidation of cephalosporins to cephalosporin sulfoxides. *Agric. Biol. Chem.* 44:1431–1433.

Toyo Brewing KK. (1978). Japan Patent No. J5-3118-591,

Toyo Jozo Co. Ltd. (1973). Belgium Patent No. 803-832.

Vandamme, E.J. (1983). Peptide antibiotic production through immobilized biocatalyst technology. *Enz. Microbial Technology, 5:* 403-416.

Vandamme, E. J. (1981a). Penicillin acylases and β-lactamases. In *Economic Microbiology*, Vol. 5. A. H. Rose (Ed.). Academic, New York, pp. 467–522.

Vandamme, E. J. (1981b). Use of microbial enzyme and cell preparations to synthesise oligopeptide antibiotics. *J. Chem. Technol. Biotechnol.* 31:637–659.

Vandamme, E. J., and Voets, J. P. (1973). Some aspects of the penicillin V acylase produced by *Rhodotorula glutinis var glutinis*. *Z. Allg. Mikrobiol.* 13:701.

Vandamme, E. J., and Voets, J. P. (1974). Microbial penicillin acylases. *Adv. Appl. Microbiol.* 17:311–369.

Vandamme, E. J., and Voets, J. P. (1975). Properties of the purified penicillin V acylase of *Erwinia aroideae*. *Experientia 31*:140–143.

Vanderhaeghe, H. (1975). Penicillin acylase (fungal). In *Methods in Enzymology*, Vol. 43. J. H. Hash (Ed.). Academic, New York, pp. 721–728. 721–728.

Vojtisek, V., and Slezak, J. (1975a). Penicillinamidohydrolase in *Escherichia coli*. II. Synthesis of the enzyme, kinetics and specificity of its induction and the effect of O_2. *Folia Microbiol.* 20:289–297.

Vojtisek, V., and Slezak, J. (1975b). Penicillinamidohydrolase in *Escherichia coli*. III. Catabolite repression, diauxie, effect of cAMP and nature of the enzyme induction. *Folia Microbiol.* 20:298–306.

Vojtisek, V., Slezak, J., and Culik, K. (1973). Czech Patent No. 1622-74.

Vorlop, K. D. (1978). Diplom. Thesis, Tech. Univ. Braunschweig, Germany.

Walton, R. B. (1964). Search for microorganisms producing cephalosporin C amidase. *Dev. Ind. Microbiol.* 5:349–353.

Yamada, H., Takahashi, S., Kil, Y., and Kumagai, H. (1978). Distribution of hydantoin hydrolyzing activity in microorganisms. *J. Ferment. Technol.* 56:484–491.

6

CLAVULANIC ACID: PROPERTIES, BIOSYNTHESIS, AND FERMENTATION

DENIS BUTTERWORTH *Beecham Pharmaceuticals Research Division, Chemotherapeutic Research Centre, Betchworth, Surrey, England*

I. INTRODUCTION

Clavulanic acid is a naturally occurring substance first detected in *Strepto-myces clavuligerus* (Brown et al., 1976) and shown (Howarth et al., 1976) to be a fused bicyclic β-lactam structurally different from the penicillins and cephalosporins in having an oxygen atom in place of sulphur [1].

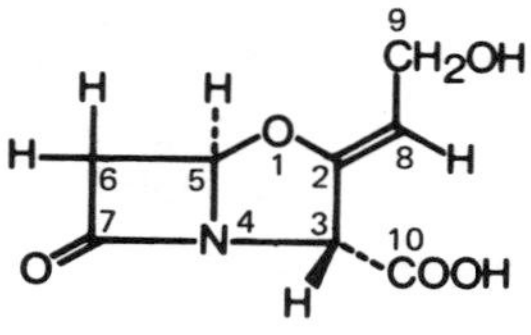

[1]

This structure is also unusual for a naturally occurring β-lactam because it has a β-hydroxyethylidene substituent at C-2 and no acylamino group at C-6. Clavulanic acid is a potent β-lactamase inhibitor and also possesses weak broad-spectrum antibacterial activity (Reading and Cole, 1977).

II. HISTORY AND DISCOVERY

The discovery of clavulanic acid, first reported in 1976 by workers at the Beecham Laboratories in England (Brown et al., 1976), arose from a microbial cultures screening program for β-lactamase inhibitors. This study was prompted by the possibility of using an inhibitor of β-lactamase in conjunction with a penicillin or cephalosporin to counter the problem of resistance to β-lactam antibiotics, since many important pathogens owe their resistance to production of β-lactamase which destroys the antibiotic.

In the agar plate test used for screening microorganisms, samples of culture filtrates were applied to agar seeded with a strain of *Klebsiella aerogenes* and containing 10 µg/ml of benzylpenicillin. The *K. aerogenes* strain used was resistant to benzylpenicillin by virtue of its ability to produce β-lactamase. Samples containing a diffusible β-lactamase inhibitor could protect the penicillin in the agar, resulting in a zone of inhibition of growth of the *Klebsiella* around the sample; in the absence of an inhibitor the *Klebsiella* grew normally, the benzylpenicillin being destroyed. *Streptomyces clavuligerus* ATCC 27064 was one of the microbial cultures giving a positive response in this test. The active component in the culture filtrate was shown (Reading and Cole, 1977) to be different from the cephalosporins previously reported to be produced by *S. clavuligerus* (Nagarajan et al., 1971; Higgens et al., 1974), and to be a novel β-lactam with the structure shown [1].

III. CLAVULANIC ACID-PRODUCING ORGANISMS

In 1971 Higgens and Kastner of the Lilly Research Laboratories described a streptomycete that had been isolated from a South American soil sample (Higgens and Kastner, 1971) and which had been reported to produce two new cephalosporin antibiotics. They concluded that the organism was quite distinct from previously described species and therefore regarded it as belonging to a new species for which they proposed the name *Streptomyces clavuligerus*. The single strain isolated by Higgens and Kastner was deposited by them in the U.S. Northern Regional Research Laboratory Collection as NRRL 3585, and in the American Type Culture Collection as ATCC 27064.

Since the first description of clavulanic acid and its production by *S. clavuligerus* there have been very few reports of its production by other organisms. Another streptomycete, *Streptomyces jumonjinensis*, was recorded as a producer of clavulanic acid in a Beecham patent (British patent No. 1,563,103). This culture had been isolated from a Japanese soil sample by a group of scientists of the laboratories of Sankyo Co. Ltd., who described it as a new species and a producer of a cephalosporin antibiotic (West German Patent No. 2,344,020). The culture was deposited by the Sankyo group in the Japanese Fermentation Research Institute (Agency of Industrial Science and Technology) as FERM No. 1545, and in the U.S. Northern Regional Research Laboratory Collection as NRRL 5741. Details of another clavulanic acid producer have been disclosed in a Japanese patent application by Takeda Chemical Industries Ltd. (Japanese Patent No. 53-104796). The strain of *Streptomyces* was isolated from a soil sample and again the organism was judged to be a new hitherto undescribed species, which was named *Streptomyces katsurahamanus*. It was deposited in the Japanese Fermentation Research Institute Collection as FERM No. 3944.

In 1980 details of a further *Streptomyces* strain which produces clavulanic acid were published in a patent application from the Japanese company Sanraku-Ocean Co. Ltd. (Japanese Patent No. 55-162993). This culture had previously been shown to produce cephalosporin C (Japanese Patent No. 51-110097), was designated *Streptomyces* sp. P6621, and is deposited in the Japanese Fermentation Research Institute Collection as FERM No. 2804; it is said to be a new species.

IV. BIOSYNTHESIS MECHANISM

The chemical structure of clavulanic acid strongly resembles those of the penicillin and cephalosporin β-lactams also synthesized by *S. clavuligerus*. This might suggest the origin of all these antibiotics from a common intermediate, but the "tripeptide theory" (Fawcett et al., 1976) of biosynthesis of penicillins and cephalosporins cannot readily be applied to clavulanic acid because of the absence of a 6-amino group and aminoadipyl side chain and the presence of an oxazolidine rather than a thiazolidine or dihydrothiazine ring (Elson and Oliver, 1978).

Studies on the incorporation of labeled precursors into clavulanic acid have been carried out by Elson and co-workers (Elson and Oliver, 1978; Stirling and Elson, 1979; Elson, 1981; Elson et al., 1981). They concluded that label from [^{13}C]acetate enters the clavulanic acid molecule via the tricarboxylic acid (TCA) cycle and 2-oxoglutarate, which provides, probably via glutamate, a five-carbon skeleton for part of the clavulanic acid molecule (carbons 9, 8, 2, 3, and 10). This conclusion was supported by evidence suggesting direct incorporation of 3,4-[^{13}C$_2$]glutamate. It was postulated that if reduction of the γ-carboxyl group of glutamic acid were to take place before closure of the oxazolidine ring, then α-amino-δ-hydroxyvalerate (AHV) might be a clavulanic acid precursor, and indeed the labeling pattern resulting from feeding 3,4-[^{13}C$_2$] AHV was consistent with this view.

Further studies showed that glycerol can provide the carbon skeleton of the β-lactam ring (carbons 5, 6, and 7) without any intermediate rearrangement of the three carbons, but that under nutrient conditions promoting carbohydrate synthesis the β-lactam carbons can also be generated from the TCA cycle by decarboxylation of malate or oxaloacetate to yield three-carbon compound precursors in the gluconeogenesis pathway. It was concluded that

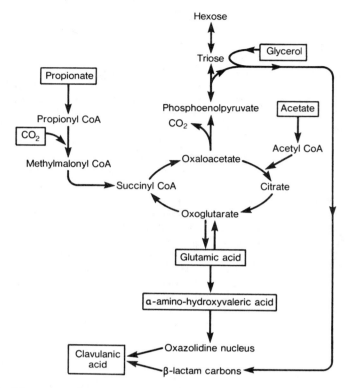

Figure 1 Summary of probable routes of incorporation of labeled precursors in clavulanic acid biosynthesis.

the glycerol carbons would have to undergo changes in oxidation levels before incorporation into the β-lactam, but propionic acid, which theoretically would be a suitable derivative, was contraindicated as a specific precursor in feeding studies using 3-[^{13}C]propionate. The observed labeling pattern from propionate was considered to result from entry into the TCA cycle via carboxylation to methylmalonate and conversion to succinate; the labeling of clavulanic acid carbons from [^{13}C]bicarbonate provided evidence in support of this supposition.

A scheme of biosynthesis of clavulanic acid suggested by the studies of Elson is indicated in Figure 1.

There are as yet no published reports as to the identity of the final precursor of clavulanic acid prior to cyclization.

Production of *S. clavuligerus* of a number of other compounds based on the clavulanic acid ring structure has been recorded. Derivatives of the clavam nucleus lacking the C-3 carboxyl group but having a substituent at C-2 different from that of clavulanic acid were discovered by Brown, et al. (1979):

[2] R=COOH
[3] R=CH$_2$OH
[4] R=CH$_2$OCHO
[5] R=CH$_2$CH(NH$_2$)COOH
[6] R=CH$_2$CH$_2$OH

where R is a carboxyl, hydroxymethyl, or formyloxymethyl group
[2–4]. A derivative in this series where R is alanyl [5] has also been described (Evans et al., 1980). More recently Wanning et al. (1981) isolated
the hydroxyethyl derivative [6] from cultures of *S. antibioticus*, and concluded that in this compound the stereochemistry at C-5 is S, opposite to that
in clavulanic acid (clavulanic acid 3R, 5R; 2 hydroxyethyl-clavam 2S, 5S).

The 3-hydroxypropionyl derivative of clavulanic acid acid [7] has been

[7]

reported in a Beecham patent (British Patent No. 1,547,222) as a further clavulanic acid derivative obtained from fermentations of *S. clavuligerus*.

V. FERMENTATION PROCESS

No manufacturing process information has yet been disclosed and therefore
the following sections are based mainly on pilot plant scale procedures described in the Beecham Group Ltd. patent specification (British Patent No.
1,508,977) which provides examples of fairly large (1500-liter) pilot-plant
operation.

A. Inoculum Preparation

It is preferable to select the highest-yielding isolate from a number of single
colonies obtained by plating out the culture collection organism (NRRL 3585
or ATCC 27064), since it is likely that some variation in yield from single-
colony isolates will initially exist. For this purpose the isolates may be tested
for clavulanic acid yield in shake-flask fermentations using 100-ml quantities
of production medium in 500-ml conical flasks which are incubated for 3–5
days at 26°C on a rotary shaker (2-in. throw, 240 rpm). The concentration
of clavulanic acid in the fermentation may be estimated using a microbiological
assay method (Brown et al., 1976), the automated enzyme inhibition assay
described by Elson and Oliver (1978) or by employing high-performance liquid
chromatography (HPLC) as described by Foulstone and Reading (1982).

The preferred isolate is preserved as a master stock either in lyophilized
form in ampoules or as soil stocks in tubes of desiccated soil. Working slant
cultures on agar are prepared from the master stock to give a supply of spore
inoculum for fermentations. Examples of solid media which are suitable for
slant cultures are as follows:

Medium A: 1% glucose, 0.1% yeast extract, 0.1% beef extract, 0.2% N-Z
Amine A, 1.5% agar, pH 7.3 (Bennett's agar)

Medium B: 1% soluble starch, 0.1% K_2HPO_4, 0.1% $MgSO_4 \cdot 7H_2O$, 0.1%
NaCl, 0.2% $(NH_4)_2SO_4$, 0.2% $CaCO_3$, 1 ml trace salts solution (trace
salts solution = 0.1% $FeSO_4 \cdot 7H_2O$, 0.1% $MnCl_2 \cdot 4H_2O$, 0.1% $ZnSO_4$
$\cdot 7H_2O$); pH 7.0–7.4

For inoculum preparation for a large-scale fermentation the medium (A or B) is set in a Roux bottle to give approximately 200 cm^2 of agar surface. This is then uniformly inoculated with a suspension of spores in sterile water and incubated for 10 days at 26°C. The sporulating aerial mycelium from a Roux bottle is used to inoculate a stirred tank seed stage which provides vegetative growth as inoculum for the final fermentation. The medium used for the seed stage contains 1.0% soybean flour, 2.0% dextrin, and 0.03% (vol/vol) antifoam consisting of 10% Pluronic L81, a block polymer of ethylene oxide and propylene oxide, dispersed in soybean oil; 75 liters of medium are prepared and steam sterilized in a 100-liter stainless steel fermenter. Agitation is by means of a flat-bladed turbine impeller, the overall impeller diameter being 19 cm.

The spores and aerial mycelium from one Roux bottle are scraped off the agar surface into sterile water and the resulting suspension is used to inoculate the seed-stage medium. The seed stage is then incubated for 72 hr at 26°C with stirring at 140 rpm and airflow of 1 vol/vol per minute.

B. Fermentation Nutrients

It was found that the most important nutrient for clavulanic acid biosynthesis was soybean protein. A suitable production fermentation medium described in the Beecham patent consisted of 1.5% soybean flour, 1.0% glycerol, 0.1% KH$_2$PO$_4$, and 0.2% (vol/vol) of 10% Pluronic L81 antifoam in soybean oil. An improved version of this medium was obtained by substituting a lipid for glycerol at the same concentration (1.0%) and omission of the antifoam: Natural oils such as lard oil, maize oil, peanut oil, and soybean oil all gave better results than glycerol, but the best clavulanic acid yield resulted from the use of Prichem P224, which is the triglyceride of a fatty acid mixture containing 65% oleic acid.

In a patent granted to Glaxo Laboratories Ltd. (British Patent No. 1,543,563) an alternative medium for pilot-plant *S. clavuligerus* fermentation is described, containing 3.0% soybean meal, 4.7% soluble starch, 0.01% FeSO$_4$ · 7H$_2$O, 0.01% K$_2$HPO$_4$, and 0.05% (vol/vol) silicone antifoam emulsion. This gave results similar to those described above.

Information available for *Streptomyces* species other than *S. clavuligerus* is limited to small-scale or laboratory fermentations. The fermentation media quoted are shown in Table 1.

C. Production Stage Procedure

In the Beecham procedure the production stage was carried out in a 2000-liter fermenter containing 1500 liters of medium. The medium was steam sterilized in the fermenter at 121°C for 30 min. The contents of a 75-liter seed stage grown for 72 hr were transferred as inoculum to the production medium. The production fermenter was a conventional stainless steel reactor fitted with baffles and mechanical agitation. The agitator was fitted with two flat-bladed turbine impellers, each with six blades and of overall diameter 48 cm, the impeller diameter—tank diameter ratio being 0.43.

After inoculation, the fermentation was incubated with constant agitation (106 rpm) at 26°C and with an airflow of 0.75 vol/vol per minute. Under these conditions and with the production medium containing lipid as carbon source (soybean flour, Prichem P224, and inorganic phosphate; pH adjusted to 7.0 before sterilization), the clavulanic acid concentration in the brew usually

Table 1 Clavulanic Acid Fermentation Media for Other Species

Streptomyces species	Medium	Reference
S. jumonjinensis	1% soybean flour, 2% glucose monohydrate, 0.2% $CaCO_3$, 0.001% $CoCl_2 \cdot 6H_2O$, 0.5% Na_2SO_4, 0.5% antifoam	British Patent No. 1,563,103
S. katsurahamanus	1.5% soybean flour, 1.5% cottonseed flour, 3% glucose, 3% cornstarch; pH 6.5	Japanese Patent No. 53-104796
Streptomyces sp. P6621	1.0% soybean flour, 2.0% soluble starch, 0.5% glycerol, 0.1% corn-steep liquor, 0.01% $FeSO_4 \cdot 7H_2O$; pH 7.0	Japanese Patent No. 55-162993

reached a maximum after approximately 90 hr incubation: Typically the maximum titer was in the region of 500 µg/ml. Evidence has been presented (British Patent No. 1,571,888) to suggest that in stirred S. clavuligerus fermentations, with a medium containing 0.52% distillers solubles, 0.52% casein hydrolysate, 2.1% soybean meal, 4.7% soluble starch, 0.78% glucose, 0.01% $FeSO_4$ $\cdot 7H_2O$, and antifoam, an improvement in the amount of clavulanic acid produced

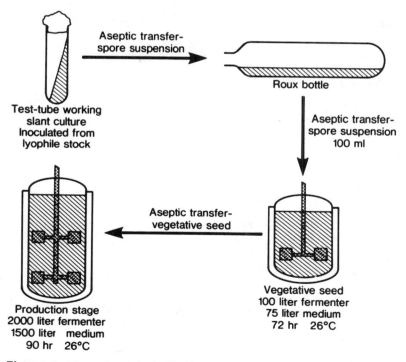

Figure 2 Flow chart to indicate clavulanic acid fermentation process for a 2000-liter fermenter.

is obtained if the pH of the fermentation is controlled at about 6.5 and within the range 6.3–6.7 throughout the fermentation.

The procedure for a 1500 liter fermentation is summarized in Figure 2.

VI. PRODUCT RECOVERY AND PURIFICATION

At the end of the fermentation the broth is first clarified by filtration or centrifugation and the *S. clavuligerus* mycelium is discarded. Primary extraction from the clarified broth may be effected by a variety of fractionation methods; two procedures (Fig. 3) suitable for larger-scale processing are (1) extraction from acidified fermentation broth into a water-immiscible organic solvent and (2) adsorption onto a strongly basic anion exchange resin followed by elution from the resin with an aqueous salt solution. Purification of the primary extract is achieved by further chromatographic procedures, particularly the use of ion exchange chromatography (Fig. 3). The final high-purity product is obtained by freeze drying or crystallization from aqueous solution.

In the procedure outlined by Reading and Cole (1977), for primary butanol extraction and subsequent purification of the sodium salt by column chromatography (Fig. 3), the culture filtrate at 2–5°C was acidified to pH 2.0 and extracted with n-butanol by means of a continuous flow multistage countercurrent extractor; the ratio of solvent to aqueous phase was 0.75. An aqueous back-extract was then obtained by efficiently mixing the butanol ex-

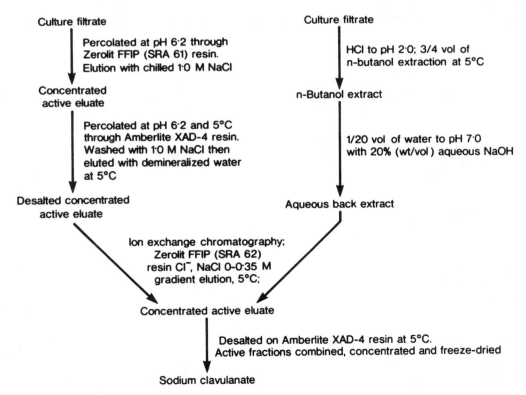

Figure 3 Scheme for product recovery and purification with two alternative methods for primary extraction.

tract with 1/20 volume of water while maintaining the pH at 7.0 using 20% sodium hydroxide solution. The squeous extract–butanol mixture was then separated in a liquid–liquid centrifugal separator. Recovery of clavulanic acid from clarified broth to squeous back-extract was approximately 40%. While the temperature of the squeous back-extract was maintained at 5°C the back-extract was percolated through a column of Permutit Zerolit FFIP (SRA62) anion exchange resin in chloride form. The column was then eluted with a sodium chloride gradient (0–0.35 M). The active fractions were combined and concentrated before desalting on a column of Bio-Rad Bio-Gel P2. Alternatively, desalting may be achieved by using a polystyrene–divinylbenzene copolymer such as Amberlite XAD-4, which adsorbs clavulanate but not inorganic salts; this operation also results in additional purification of the clavulanate (British Patent No. 1,563,103). After elution with 1.0% n-butanol solution (or demineralized water in the case of XAD-4), combined salt-free fractions containing clavulanate are vacuum concentrated to a suitable volume before freeze drying to give the solid sodium salt.

Primary extraction of clarified broth by means of a strongly basic anion exchange resin column can be achieved by a process employing Zerolit FFIP SRA61 resin (British Patent No. 1,563,103). After percolation with broth adjusted to pH 6.2, the column was washed with chilled water and then eluted with chilled 1.0 M aqueous NaCl. Combined fractions containing clavulanate were combined and adjusted to pH 6.2 before desalting on a column of Amberlite XAD4 resin. The combined active fractions were then concentrated by reverse osmosis prior to purification, for example, by the method employing SRA62 resin described above for squeous back-extract from primary n-butanol extraction (Fig. 3).

Secondary purification has also been achieved by conversion of the partially purified clavulanate to the benzyl ester, which was then dissolved in ethyl acetate and subjected to two chromatographic steps using Sephadex LH20 and silica gel; the purified benzyl ester was then hydrogenated over 10% Pd/C in the presence of sodium bicarbonate to yield sodium clavulanate tetrahydrate (British Patent No. 1,563,103).

VII. MODE OF ACTION

The biological properties and mode of action have recently been extensively reviewed by Cole (1980, 1981). Clavulanic acid by itself has only weak antibacterial activity against most bacteria, with minimum inhibitory concentration values of 25–125 µg/ml, but it is a potent inhibitor of a wide range of β-lactamase and is able to potentiate the antibacterial activity of penicillins and cephalosporins against many β-lactamase-producing resistant bacteria. The potentiation is ascribed to the preservation of adequate concentrations of antibiotic by inhibition of the bacterial β-lactamase.

The mechanism of action of the enzyme is thought to differ somewhat according to the type of β-lactamase. Initially there is a phase of competitive inhibition without chemical reaction between inhibitor and enzyme. This is followed by the progressive type of inhibition, with time-dependent inactivation of the enzyme. In some cases the inactivation is irreversible. Cole (1981) observed,

Clavulanic acid can be described as an active site-directed inhibitor. Unlike normal substrate molecules which are continuously converted to product and readily leave the enzyme surface, clavulanic acid is converted to products, some of which stay temporarily or permanently attached to the enzyme. The double bond in the side chain of the clavulanic acid molecule is essential for these reactions. The β-lactam ring of clavulanic acid mimics the β-lactam ring of the substrate and probably initially acylates the enzyme, as occurs with a substrate. This gives clavulanic acid its very high specificity for β-lactamases.

In medical practice clavulanic acid is therefore administered with a β-lactam antibiotic in order to achieve efficacy of the antibiotic against a wider range of organisms. Clavulanic acid is well absorbed in humans when administered via the oral route, and is poorly bound to serum, the presence of human serum having only a slight effect on its activity. Extensive pharmacological and clinical studies have been carried out on a formulation of clavulanic acid with amoxycillin in the ratio 1:2. The pharmacokinetics of the two components are closely matched.

VIII. MARKETING APPLICATIONS

A formulation is now marketed in the United Kingdom as an antibiotic agent for oral treatment of infections under the trademark Augmentin (Beecham Research Laboratories, 1981) in the form of tablets, each containing 125 mg clavulanic acid as the potassium salt and 250 mg amoxycillin as the trihydrate. Augmentin is bactericidal to a wide range of gram-positive and gram-negative organisms, both aerobes and anaerobes. It is indicated for common bacterial infections such as upper and lower-respiratory tract, gastrourinary tract, and other infections, for example, of skin and soft tissue, intraabdominal sepsis, and osteomyelitis.

Parenteral formulations are being developed for infections which cannot be treated by the oral route (Knudsen, 1980).

REFERENCES

Beecham Research Laboratories, Brentford, Middlesex, United Kingdom (1981). *Augmentin Data Sheet.*

Brown, A. G., Butterworth, D., Cole, M., Hanscomb, G., Hood, J. D., Reading, C., and Rolinson, G. N. (1976). Naturally occurring β-lactamase-inhibitors with antibacterial activity. *J. Antibiot.* 29:668–669.

Brown, D., Evans, J. R., and Fletton, R. A. (1979). Structures of three novel β-lactams isolated from *Streptomyces clavuligerus. J. Chem. Soc. Chem. Commun.* 282–283.

Cole, M., (1980). β-Lactams as β-lactamase inhibitors. *Philos. Trans. R. Soc. Lond. B289*:207–223.

Cole, M. (1981). Inhibitors of bacterial β-lactamases. *Drugs Future* 6:697–727.

Elson, S. W. (1981). The biosynthesis of clavulanic acid and related metabolism of *Streptomyces clavuligerus.* In *Proceedings of the Second International Symposium "Recent Advances in the Chemistry of β-Lactam Antibiotics", Royal Society of Chemistry, London, England, Special Publications No. 38,* pp. 142–150.

Elson, S. W., and Oliver, R. S. (1978). Studies on the biosynthesis of clavulanic acid I. Incorporation of ^{13}C-labelled precursors. *J. Antibiot.* 31:586–592.

Elson, S. W., Oliver, R. S., Bycroft, B. W., and Faruk, E. A. (1982). Studies on the biosynthesis of clavulanic acid III—Incorporation of DL-3,4^{13}C$_2$-glutamic acid. *J. Antibiot.* *35*:81–86.

Evans, R. H., Ax, H., Jacoby, A., Williams, T. H., Jenkins, E., and Scannell, J. P. (1980). Ro22-5417, A new clavam antibiotic from *Streptomyces clavuligerus*: II. Isolation and structure. *Abstract No. 163, 20th Interscience Conference on Antimicrobial Agents and Chemotherapy.*

Fawcett, P. A., Usher, J. J., and Abrahams, E. P. (1976). Aspects of cephalosporin and penicillin biosynthesis. *Proceedings of the Second International Symposium on the Genetics of Industrial Microorganisms* (K. D. Macdonald, ed.) Academic Press, London:129–138.

Foulstone, M., and Reading, C. (1982). Assay of amoxycillin and clavulanic acid, the components of Augmentin, in biological fluids with high performance liquid chromatography. *Antimicrob. Agents Chemother.* *22*: 753–762.

Higgens, C. E., and Kastner, R. E. (1971). *Streptomyces clavuligerus* sp. nov. a β-lactam antibiotic producer. *Int. J. Syst. Bacteriol.* *21*:326–331.

Higgens, C. E., Hamill, R. L., Sands, T. H., Hoehn, M. M., Davis, N. E., Nagarajan, R., and Boeck, L. D. (1974). The occurrence of deacetoxycephalosporin C in fungi and streptomycetes. *J. Antibiot.* *27*:298–300.

Hoffman-LaRoche Inc. (1980). *U.S. Patent No. 4,202,819.*

Howarth, T. T., Brown, A. G., and King, T. J. (1976). Clavulanic acid, a novel β-lactam isolated from *Streptomyces clavuligerus*; X-ray crystal structure analysis. *J. Chem. Soc. Chem. Commun.* 226–267.

Knudsen, E. T. (1980). In *Augmentin: Proceedings of the First Symposium, 3 and 4 July 1980.* G. N. Rolinson and A. Watson (Eds.). Excerpta Medica, Amsterdam, pp. 306–308.

Nagarajan, R., Boeck, L. D., Gorman, M., Hamill, R. L., Higgens, C. E., Hoehn, M. M., Stark, W. M., and Whitney, J. G. (1971). β-Lactam antibiotics from *Streptomyces*. *J. Amer. Chem. Soc.* *93*:2308–2310.

Reading, C., and Cole, M. (1977). Clavulanic acid: A beta-lactamase-inhibiting beta lactam from *Streptomyces clavuligerus*. *Antimicrob. Agents Chemother.* *11*:852–857.

Stirling, I., and Elson, S. W. (1979). Studies on the biosynthesis of clavulanic acid II. Chemical degradations of ^{14}C-labelled clavulanic acid. *J. Antibiot.* *32*:1125–1129.

Wanning, M., Zähner, H., Kroner, B., and Zeeck, A. (1981). Ein neues antifungisches β-Lactam-Antibioticum der Clavam-Reihe. *Tetrahed. Lett.* *22*:2539–2540.

7

CARBAPENEM COMPOUNDS; PS-SERIES ANTIBIOTICS AND THIENAMYCINS: PROPERTIES, BIOSYNTHESIS, AND FERMENTATION

YASUO FUKAGAWA AND TOMOYUKI ISHIKURA *Sanraku-Ocean Company, Ltd., Fujisawa, Japan*

I. INTRODUCTION

A long history of satisfactory use of penicillins and cephalosporins in the clinical treatment of bacterial infections has clearly shown that β-lactam antibiotics are a group of ideal chemotherapeutics which attack only pathogenic bacteria through selective action on cell wall synthesis. Numerous antibiotics have been isolated from fermentations of streptomycetes, but penicillin N had been the sole β-lactam compound produced by a streptomycete (Miller et al., 1962) until the isolation of 7-methoxycephalosporins by *Streptomyces* species was reported by the Lilly research group (Higgens and Kastner, 1971) and Merck scientists (Stapley et al., 1972). The production of 7-methoxycephalosporins by streptomycetes, together with the occurrence of an increasing number of β-lactam-resistant pathogens, accelerated screening works for other novel types of β-lactam compounds by means of new assay procedures, resulting in the discovery of nocardicins (Aoki et al., 1976) and clavulanates (Brown et al., 1976). In the more sophisticated lines of approach, results of fundamental studies on β-lactamases and the cell wall biosyntheses were effectively incorporated to devise various assay systems for possibly more sensitive and fruitful screening of new types of β-lactam antibiotics. It seems probable that the combined use of β-lactam-supersensitive detector organisms and β-lactamase inhibition tests played a significant role in the successful discovery of a novel family of β-lactam antibiotics, collectively called carbapenems, such as thienamycins, epithienamycins, olivanates, and PS-series compounds. The family of carbapenem compounds is unique in the chemical structure of desthia-1-carbapen-2-em (or 7-oxo-1-azabicyclo[3.2.0]hept-2-ene-2-carboxylate, according to the IUPAC's Recommendations on the Nomenclature of Organic Chemistry), compared with traditional penicillins and cephalosporins (Fig. 1).

Although the natural analogs of this family are chemically and biologically more labile than penicillins and cephalosporins, their highly desirable antibacterial activity against gram-positive and gram-negative bacteria and their β-lactamase-inhibiting property will effectively be exploited in the future development of new types of carbapenem drugs by chemical and biological modifications, as is the case with cephalosporin C. This chapter deals mostly with PS-series compounds and thienamycins and epithienamycins.

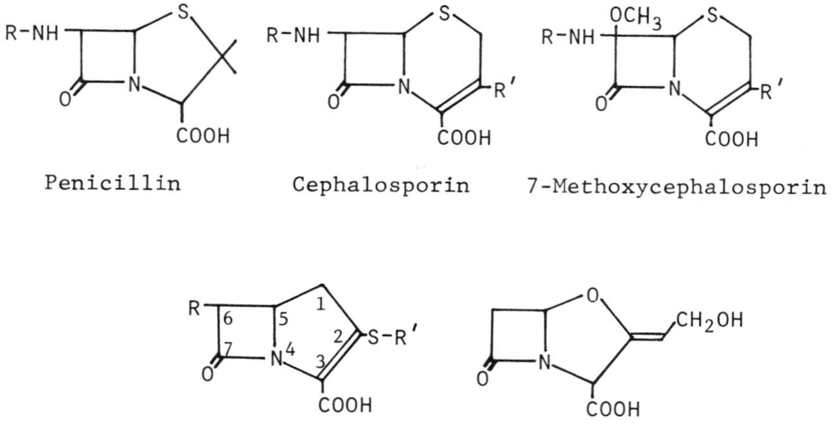

Penicillin   Cephalosporin   7-Methoxycephalosporin

Carbapenem   Clavulanate

Figure 1 Structures of β-lactam compounds.

II. SCREENING SYSTEMS

The Merck research group (Kahan et al., 1979) has not yet reported the detailed screening system for thienamycin and epithienamycin analogs, describing only that thienamycin was discovered in the course of screening soil isolated for inhibitors of peptidoglycan synthesis. It is likely, however, that
the routine use of Staphylococcus aureus ATCC 6539P as the assay organism
preferentially facilitated the isolation of thienamycin among other analogs, because thienamycin with the free amino group is antimicrobially far more active
on the assay bacterium than the other thienamycin and epithienamycin analogs
with the acetamido group.

Beecham investigators (Butterworth et al., 1979) employed an agar plate
test method for β-lactamase inhibitors which was a modification of one of the
assay systems described by Umezawa et al. (1973).

Sanraku-Ocean scientists (Okamura et al., 1978, 1979) employed a two-
step assay system for the screening of new β-lactam antibiotics. In the primary screening step, fermentation broths of soil isolates were disk assayed
on agar plates of Comamonas terrigena B-996 (a highly β-lactam-sensitive
mutant obtained from C. terrigena IFO 12685) in the presence and absence
of various types of β-lactamases. In addition, C. terrigena B-996R (a β-
lactamase-producing mutant induced from C. terrigena B-996) was supplemen-
tarily used for selective detection of 7-methoxycephalosporins (Okamura et al.,
1979). On an assay plate of the highly β-lactam-sensitive detector containing
no β-lactamase (control assay plate), a very low level of antimicrobial activity
could be detected. When such antimicrobial activity was not due to β-lactam,
halo sizes on assay plates containing various β-lactamases (β-lactamase assay
plates) were the same as that on the control assay plate. On the other hand,
if a culture contained β-lactam, inhibition zones on the β-lactamase assay
plates were rarely the same as, and often smaller than, the control halo in
size. More particularly, the extents of halo size reduction depended somewhat
on the types of β-lactamase and β-lactam products. For example, PS-5 gave
the same inhibition zone sizes on the control assay plate and the assay plate
containing Bacillus licheniformis 749/C β-lactamase (penicillinase), as the carbapenem compound is completely resistant to the enzyme. In the presence
of type II β-lactamase (cephalosporinase) of Bacillus cereus 569, in contrast,
no halo of PS-5 was seen under 100 µg/ml, because the cephalosporinase quickly decomposes PS-5 during incubation. Other β-lactamases such as type I
β-lactamase (penicillinase) of B. cereus 569 and cephalosporinases of Citro-
bacter freundii GN346 and Proteus vulgaris GN76 reduced the halo sizes of
PS-5 to a more or less significant extent, but not completely.

Broth candidates passing the primary screening step were then subjected to the secondary assay step consisting of more qualitative and quantitative tests. For bioautographic comparison with known natural β-lactam antibiotics, a more sensitive β-lactamase inhibition test system was established.
Aliquot volumes of broth filtrate were spotted on three sheets of chromatographic filter paper and developed in a suitable solvent system such as n-
propanol:water (7:3, descending). After the solvent evaporated, the chromatograms were contacted with a control assay plate of C. terrigena B-996 (assay
plate 1) and two β-lactamase assay plates (assay plates 2 and 3), respectively,
until antibiotic activities were transferred into the assay agar. After the
chromatograms were removed from the agar surfaces, only assay plate 3 was
further treated by placing a filter paper strip of a known β-lactam compound
(e.g., benzylpenicillin and cephaloridine) on the assay agar along the sup-

posedly developed channel of the said chromatogram, so that latent β-lacta-mase-inhibiting activities might be exaggeratedly revealed. The three assay plates were incubated overnight at 30°C. Plate 1 (control assay plate) gave R_f values of antibiotic substances which served for chromatographic comparison with known natural β-lactam compounds. On β-lactamase assay plate 2, the halos of the active substances were smaller than those on the control assay plate, their sizes depending on the substrate specificity of the β-lactamase added. When an active component had a substantial β-lactamase-inhibiting property, the corresponding halo on assay plate 3 was found to be apparently enlarged, as the β-lactam compound transferred from the paper strip was not decomposed in the surrounding area, resulting in a larger inhibition zone. Under controlled assay conditions, the relative enlargement of the inhibition zone seemed to be semiquantitative with respect to the concentration of an inhibitor.

Quantitative assay of β-lactamase inhibition was carried out by ultraviolet spectrophotometry using penicillins and cephalosporins as assay substrates (e.g., benzylpenicillin at 240 nm, ε 910, Δε 636; cephaloridine at 280 nm, ε 5730, Δε 5290) (Fukagawa et al., 1980b). Assay mixtures containing a fixed concentration of a substrate and various concentrations of an inhibitor were thermally equilibrated for 5 min at 30°C and a fixed amount of β-lacta-mase was added to start the assay reactions. Time courses of hydrolysis of the substrate were followed at an appropriate wavelength, preferably to complete hydrolysis. If the modes of inhibition were classical (competitive, noncompetitive, or uncompetitive), it would be easy to calculate the K_i value of the inhibitor by processing several lines of time course data (time–absorbance data) according to the integrated Michaelis–Menten equation (Fukagawa, 1980). As the mode of inhibition of carbapenem compounds such as PS-5 and PS-7 was confirmed to be nonclassical, arbitrary assay conditions were necessarily established for routine assays. First the preincubation effect of the inhibitor on the hydrolytic activity of β-lactamase was examined by varying the concentration of the inhibitor and the period of preincubation. If such an effect was observed, a suitable period of preincubation was chosen. In practice it would be more convenient to fix the period of preincubation (for example, 5–10 min) for objective comparison of β-lactamase-inhibitory activities of various inhibitor compounds. The 50% inhibition dose of the inhibitor (ID_{50}) was determined under such arbitrarily chosen assay conditions. With or without preincubation, the time courses of hydrolysis of the substrate were traced in the presence and absence of varied concentrations of the inhibitor, preferably to completion. Then an arbitrary assay time point was chosen in the seemingly linear (neither accelerating nor decelerating) range of hydrolysis of the control assay (without inhibitor). At this assay time point, amounts of product were calculated as changes in ultraviolet absorbance for all assay mixtures and plotted against the concentration of the inhibitor on a sheet of section paper. The ID_{50} value was read from the graph as the concentration of the inhibitor where the hydrolysis of substrate was inhibited by 50%. For quantitative assays, it is very important to maintain the arbitrarily chosen assay conditions such as the period of preincubation, the assay time point, temperature, pH, and the substrate concentration, as long as the types of substrate and the β-lactamase are not altered with respect to the inhibitor compound.

Utilizing similar screening strategies with some modifications and supplementary assay procedures, other research groups have recently isolated some members of the carbapenem family (Nakayama et al., 1980; Imada et al., 1980).

Table 1 Carbapenem-Producing Microorganisms

Microorganism	Products	References
Streptomyces fulvoviridis ATCC 21954	MC696-SY2-A, MC696-SY2-B	Umezawa et al. (1973)
Streptomyces cattleya NRRL 8057	Thienamycin, N-acetylthienamycin, N-acetyldehydrothienamycin	Kahan et al. (1977, 1979)
Streptomyces cattleya S-WRI-M5301	NS-5	Rosi et al. (1981)
Streptomyces flavogriseus NRRL 8139 and 8140	Epithienamycins A, B, C, and D, 890 A9, 890 A10	Cassidy et al. (1977, 1979a, b)
Streptomyces olivaceus ATCC 21379–21382	MM 4550, MM 13902	Butterworth et al. (1979),
Streptomyces olivaceus ATCC 31365	MM 17880, MM 22380	Box et al. (1979, 1981)
Streptomyces olivaceus NCIB 8238 and 8509	MM 22381, MM 22382	
Streptomyces gedanensis ATCC 4880	MM 22383	
Streptomyces argenteolus ATCC 11009		
Streptomyces flavovirens ATCC 3320		
Streptomyces flavus ATCC 3369		
Streptomyces fulvoviridis ATCC 15863		
Streptomyces sioyaensis ATCC 13989		
Streptomyces cremeus subsp. *auratilis* ATCC 31158	PS-3, PS-4, PS-5	Okamura et al. (1979a, c),
Streptomyces fulvoviridis A933	PS-6, PS-7, PS-8	Shibamoto et al. (1980),
Streptomyces olivaceus ATCC 31126		Shibamoto and Ishikura
Streptomyces flavogriseus NRRL 8139		(1981)
Streptomyces sp. SF-2050 FERM P 4358	SF-2050, SF-2050B	Nojiri et al. (1979), Ohba et al. (1979)
Streptomyces fulvoviridis FERM P 3935–3937	17927 A1	Arai et al. (1978a, b, 1980)
Streptomyces fulvoviridis ATCC 15863	17927 A2	
Streptomyces olivaceus ATCC 21379 and 31126	17927 D	
Streptomyces flavogriseus NRRL 8139 and 8140		
Streptomyces argenteolus ATCC 11009		
Streptomyces sp. KC-6643 FERM P 4467	Carpetimycins A and B	Nakayama et al. (1980)
Streptomyces griseus subsp. *cryophilus* IFO 13886	C-19393 S2 and H2	Imada et al. (1980)
Streptomyces tokunonensis FERM P 4843	PA-31088-IV	Shohji et al. (1980)

III. MICROORGANISMS

Carbapenem-producing microorganisms reported to date in the scientific references and patent specifications belong all to *Streptomyces*. Table 1 lists representative carbapenem producers together with fermentation products.

IV. STRUCTURES OF NATURAL CARBAPENEM COMPOUNDS

To date 19 compounds have been reported as the natural members of the carbapenem family, and more analogs are expected to join the family in future. Based on their chemical and biosynthetic properties, they can be classified as follows:

> Group 1. Carbapenem compounds produced by *Streptomyces cattleya* (Fig. 2)
> Group 2. Carbapenem compounds produced by other streptomycetes
> Subgroup 2.1. Carbapenem compounds with the two-carbon side chain at C6 (Fig. 3)
> Subgroup 2.2. Carbapenem compounds with the three-carbon side chain at C6 (Fig. 4)

For convenience, a synonym table for natural carbapenem antibiotics is attached (Table 2).

V. BIOSYNTHESIS AND MICROBIAL MODIFICATIONS

A. Biosynthesis

At a very early stage of research, the Merck research group revealed that glutamic acid might be a key building block for thienamycin (Albers-Schonberg et al., 1976), but no detailed information is yet available.

Using a blocked mutant of *Streptomyces olivaceus* which predominantly produced MM 22381 and MM 22383 (epithienamycins C and D), Box et al. (1979) observed the conversions of MM 22380 (epithienamycin A) to MM 17880, of MM 22382 (epithienamycin B) to MM 13902, and of MM 22382 to MM 4550.

As the available information concerning the biosynthesis of carbapenem compounds is sparse and indirect, it might be too early to discuss their biosynthetic relations in detail. However, Rose et al. (1981) reported a mutant of *S. cattleya* producing NS-5, suggesting the possibility that the initial biosynthetic route to NS-5 or PS-5 from fundamental precursors such as amino acids and sugars might be common in *S. cattleya* (group 1 producer) and other streptomycetes (group 2 producers). Thus the critical difference between the two groups of carbapenem producers would be found in the stereospecificity of 8-hydroxylase. There is no evidence supporting the epimerization of the 8-hydroxyl group, that is, the interconversion of thienamycin analogs and epithienamycin congeners.

B. Microbial Modifications

Penicillin acylase, acetyl esterase, and D-amino acid oxidase have been economically very important enzymes for the preparation of so-called semisynthetic penicillins and cephalosporins with therapeutic utility. With carbapenem compounds, several lines of enzymatic and biological modification will be possible.

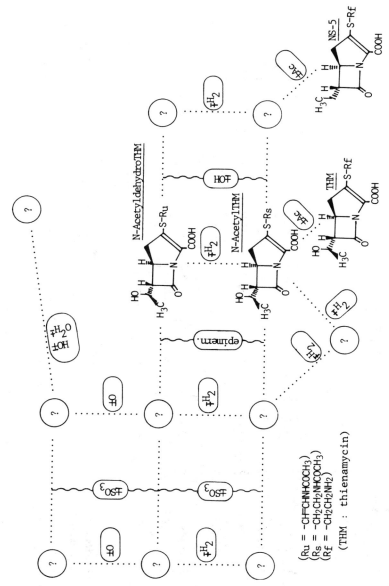

Figure 2 Carbapenem compounds: group 1.

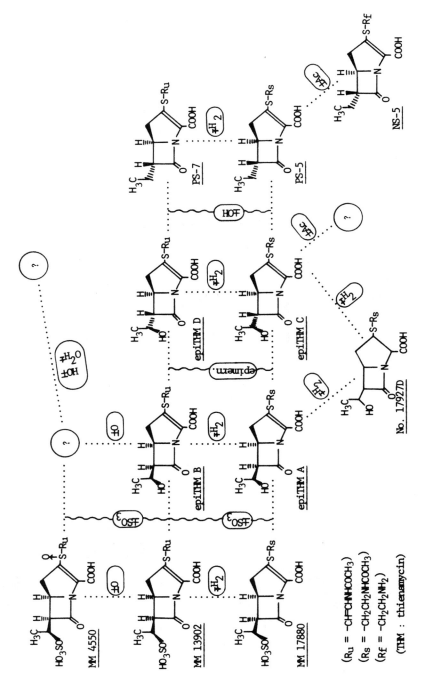

Figure 3 Carbapenem compounds: group 2; subgroup 2.1.

$(R_u = -CH=CH-NHCOCH_3)$

$(R_s = -CH_2CH_2NHCOCH_3)$

$(R_f = -CH_2CH_2NH_2)$

(THM : thienamycin)

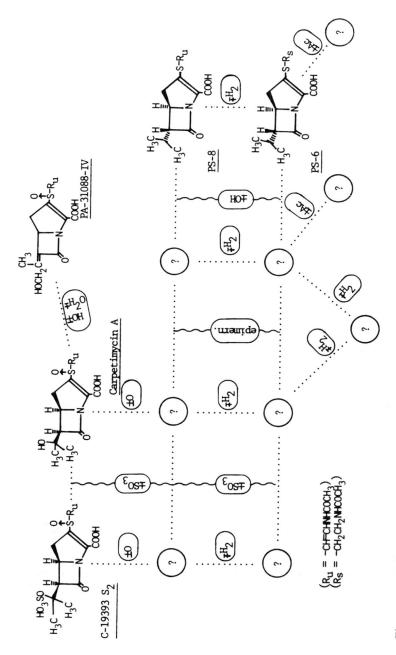

Figure 4 Carbapenem compounds: group 2; subgroup 2.2.

Table 2 Synonyms for Natural Carbapenem Antibiotics

Original compound	Synonyms
C-19393 S_2	Carpetimycin B
Carpetimycin A	C-19393 H_2
Epithienamycin A	890 A_1, MM 22380
Epithienamycin B	890 A_2, MM 22382
Epithienamycin C	890 A_3, MM 22381
Epithienamycin D	890 A_5, MM 22383, PS-4
MM 4550	MC696-SY2-A
MM 13902	890 A_9, epithienamycin E
MM 17880	890 A_{10}, epithienamycin F
N-Acetyldehydrothienamycin	
N-Acetylthienamycin	924 A_1
No. 17927D	
NS-5	
PA-31088-IV	
PS-5	
PS-6	
PS-7	
PS-8	
Thienamycin	
Analog mixture compounds	
PS-3	Epithienamycins A and C
Unidentified compounds	
MC696-SY2-B	
No. 17917 A_1	Epithienamycin A(?)
No. 17917 A_2	Epithienamycin B(?)
SF-2050	MM 17880 (?)
SF-2050B	MM 13902 (?)

1. Deacetylation

Since acylation and deacylation are ubiquitous in biological systems and are likely to be involved in the interconversion of carbapenem antibiotics, the first object of modification study was to deacetylate the acetamidoethylthio side chain. Based on the structural similarity of N-acetylethanolamine and the acetamidoethylthio side chain of N-acetylthienamycin and epithienamycins A and C, Merck scientists (Kahan and Kahan, 1977) screened soil microorganisms for N-acetylethanolamine amidohydrolase using N-acetylethanolamine as the assay substrate, and finally found that an enzyme from *Protaminobacter ruber* deacetylated not only N-acetylethanolamine, but also N-acetylthienamycin and epithienamycins A and C. In the meanwhile Fukagawa et al. (1980) discovered that L-amino acid acylase (acylase I) from hog kidney (but not from *Aspergillus* sp.) and D-amino acid acylase from *S. olivaceus* converted PS-5 to NS-5 by deacetylation. Using L- and D-acylamino acids as assay substrates, they screened methanol-utilizing bacteria, including *Protaminobacter*, for L- and D-amino acid acylases. Most of the microbes tested were found to form NS-5 from PS-5 and could be classified into the L-amino acid acylase producers and the L-amino acid acylase + D-amino acid acylase producers, depending on the amino acid substrate specificity. From cells of *Pseudomonas* sp. 1158 the two activities were isolated and purified to apparent homogeneity (Kubo et al., 1980). They showed the reasonable substrate profile on acyl amino acids and deacetylated PS-5 to NS-5, respectively. Interestingly the two enzyme activities also attacked dipeptides and tripeptides without apparent stereospecificity. In clear contrast to the N-acetylethanolamine amidohydrolase from *P. ruber* (Kahan and Kahan, 1977), neither the L-amino acid acylase nor the D-amino acid acylase from *Pseudomonas* sp. 1158 could act on N-acetylethanolamine.

2. Oxidation of 17927 D to Epithienamycins

Arai et al. (1980) described that streptomycetes of group 2 had the ability to transform 17927 D to 17927 A_1 and 17927 A_2, which were assumed to be epithienamycins A and B(?), respectively. Thus 17927 D, which displays a very weak antimicrobial activity, might be a biosynthetic precursor for epithienamycins A, B, C, and D, although its stereochemistry is not yet elucidated.

VI. FERMENTATION

A. Fermentation Studies

As the initial fermentation titers of carbapenem-forming soil isolates were very low, fermentation studies to improve productivity to a practically acceptable level were essential for chemical and biological characterization of carbapenem antibiotics. For all the carbapenem producers listed above, glucose, glycerol, and starch seem to be favorable carbon sources, while good nitrogen sources are soybean meal, yeast extract, and cottonseed meal (Kahan et al., 1979; Butterworth et al., 1979; Okamura et al., 1979a; Nakayama et al., 1980; Imada et al., 1980). As a chemically defined medium that supports a substantial level of carbapenem production is not known to date, it is very difficult to examine which compound is a key precursor for preferential biosynthesis of subgroup 2.2 carbapenems over subgroup 2.1 compounds. However, Shibamoto et al. (1980) observed that the supplementation of valine, leucine, or isoleucine to complex fermentation media led to the relative increase of PS-6 (with the three-carbon side chain) among PS-5, MM-13902, MM-17880, and epithienamycins A, B, C, and D (all with the two-carbon side chain). Although not clearly noted, Co^{2+} seems to be necessary for thienamycin fermentation by *S. cattleya* (Kahan et al., 1979). With several species of *Streptomyces* producing carbapenem compounds, Okamura and Ishikura (1979) found that the better production of carbapenem occurred in a limited concentration range of Co^{2+} (0.032–0.32 µg/ml $CoCl_2 \cdot 6H_2O$) and preferably of vitamin B_{12} (0.3–100 µg/ml), indicating that the stimulating effect of Co^{2+} was probably expressed by its bioconversion to vitamin B_{12} (cyanocobalamin). Beecham investigators also observed the stimulation of olivanate fermentation by Co^{2+} and, less significantly, Mn^{2+} (Butterworth et al., 1979). The presence of inorganic or organic sulfur compounds in fermentation media accelerates the biosynthesis of sulfated carbapenem compounds such as MM 4550, MM 13902, MM 17880, and carpetimycin B (Box et al., 1979; Nakayama et al., 1980). Together with the medium composition, the intensity of aeration was found to be influential on the type of fermentation products of group 2. PS-series compounds have been confirmed to form under less aerated conditions than epithienamycins and MM-series antibiotics. The upward shift of aeration during the course of fermentation of *Streptomyces cremeus* subsp. *auratilis* resulted in the favored production of epithienamycins A, B, C, and D in the absence of a sulfur source, while olivanates were found to be predominant under a sufficient supply of an inorganic sulfur compound such as thiosulfate.

B. Fermentation of PS-5

1. Agar Slant Culture

Streptomyces cremeus subsp. auratilis was grown on an ISP-2 agar slant (malt extract, 1.0%; glucose, 0.4%; yeast extract, 0.4%; agar 2.0%; pH 7.0) at 28°C for 2 weeks. Spores were collected from one slant culture with 10 ml of physiological saline.

2. Flask Culture

One milliliter of the spore suspension (about 10^8 cells/ml) was inoculated into a 500-ml Erlenmeyer flask containing 100 ml of seed medium SE-4 (beef extract, 0.3%; Bactotryptone, 0.5%; defatted soybean meal, 0.5%; glucose, 0.1%; soluble starch, 2.4%; yeast extract, 0.5%, $CaCO_3$, 0.4% in tap water; pH 7.5 prior to autoclaving) and shake cultured at 28°C for 48 hr on a rotary shaker (throw, 70 mm; 220 rpm) (Okamura et al., 1979a).

3. Jar Fermenter Culture

Fifteen liters of seed medium SE-4 was introduced into a 30-liter stainless steel jar fermenter and autoclaved at 120°C for 15 min. The total amount of the flask culture was transferred into the jar fermenter and cultivated at 28°C for 24 hr at 200 rpm (aeration, 0.5 vol/vol per minute).

4. Tank Culture

One liter of the jar fermenter culture was inoculated into a 200-liter stainless steel tank fermenter containing 100 liters of production medium ML-19M2 (glycerol, 4.0%; peptone, 0.5%; glucose, 0.2%; potato starch, 0.2%; defatted soybean meal, 2.5%; dry yeast, 0.5%; sodium chloride, 0.5%; calcium carbonate, 0.2%; pH 6.4 prior to sterilization) and cultivated at 28°C for 72 hr at 100 rpm under forced aeration of 0.5 vol/vol per minute. Under the specified fermentation conditions, PS-5 was found to be a substantially sole product by paper and thin-layer chromatography. A typical time course of PS-5 fermentation by S. cremeus subsp. auratilis is presented in Figure 5. As the comparative antibacterial activities of carbapenem compounds vary 10- to 100-fold against C. terrigena B-996 (highly β-lactam-sensitive assay organism), and as pure preparations of PS-5 were not available in those days, cephaloridine sodium salt was tentatively employed as the antibiotic standard. PS-5 sodium salt was later determined to be 21-fold more active than cephaloridine on a weight basis (1 µg of PS-5 sodium salt = 21 cephaloridine C. terrigena B-996 units).

C. Isolation and Purification of PS-5

In general, in comparison to traditional penicillins and cephalosporins, carbapenem compounds were very laborious to obtain in purified forms because of their unusual chemical instability, low lipophilicity, and low fermentation productivity. In practice, all the isolation and purification procedures were carried out at a low temperature below 5°C in a narrow pH range close to neu-

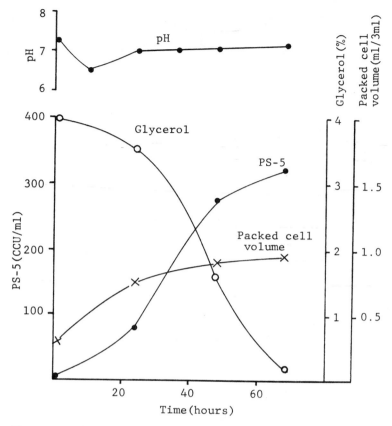

Figure 5 Time course of PS-5 fermentation.

trality. Even under such mild conditions, the overall recovery of carbapenem compounds was as low as several percent, 1.64% for thienamycin (Kahan et al., 1979) and 4.6% for PS-5 (Okamura et al., 1979).

Figure 6 summarizes an example of isolation and purification procedures for PS-5 (Okamura et al., 1979a).

VII. BIOLOGICAL PROPERTIES

Significant differences in the biological properties of the carbapenem family from penicillins, cephalosporins, and 7-methoxycephalosporins are recognized in the unusually broad spectrum of antimicrobial activity; the β-lactamase-inhibiting property; and the susceptibility to renal dipeptidases, particularly to dehydropeptidase-I. The first two characteristics are therapeutically very desirable, but the last one seems to be a serious problem to be solved in the future development of carbapenem derivatives.

A. Antimicrobial Activity In Vitro

In general, gram-negative bacteria, including *Pseudomonas*, are highly resistant to conventional penicillins and cephalosporins, and thus new β-lactam

Broth filtrate(235 CCU/ml; 160 liters)

├─ passed through a Diaion PA306 column

↓

Diaion HP20

├─ eluted with 75% methanol

↓

Diaion PA306S

├─ eluted with a NaCl concentration
│ gradient of 0-3%

↓

Diaion HP20

├─ eluted with an acetone concentration
│ gradient of 0-25%

↓

QAE-Sephadex A-25

├─ eluted with a NaCl concentration
│ gradient of 0-1.5%

↓

Diaion HP20AG

├─ eluted with an acetone concentration
│ gradient of 0-10%
├─ lyophilized

↓

Sephadex G-10

├─ eluted with distilled water

↓

QAE-Sephadex A-25

├─ eluted with a NaCl concentration
│ gradient of 0-1.5%

↓

Diaion HP20AG

├─ eluted with an acetone concentration
│ gradient of 0-10%
├─ lyophilized

↓

White powder of PS-5(95% pure; 80 mg)

Figure 6 Isolation and purification procedure for PS-5

derivatives active against gram-negative pathogens have earnestly been sought
by chemical modifications. It is expected that the seemingly intrinsic antimicrobial activity of the carbapenem family against gram-negative organisms
will be effectively exploited in the future development of more stable and potent carbapenem derivatives by chemical and biological modifications (Sakamoto et al., 1979; Kropp et al., 1980b; Basker et al., 1980; Nakayama et al.,
1980; Imada et al., 1980). Thienamycin derivatives (group 1) were found
to be active against clinical isolates of *Pseudomonas aeruginosa* (Kropp et al.,
1980b) and seem to be more antipsuedomonal than the carbapenem analogs
of group 2 (Basker et al., 1980). In addition, some inherently β-lactam-resistant pathogens belonging to *Bacteroides, Serratia, Enterobacter,* and indole-
positive *Proteus* were susceptible to carbapenem compounds (Sakamoto et al.,
1979; Basker et al., 1980; Kropp et al., 1980). Against gram-positive bacteria, the carbapenem family as a rule has potent activities comparable or superior to those of penicillins and cephalosporins.

Although it might be too early to discuss the structure–activity relations
in detail, some relevant data are available. Thienamycin is antimicrobially
two to four times as active as N-acetylthienamycin, indicating that the N-
acetylation leads to reduced antibiotic activity in spite of improved chemical
stability (Basker et al., 1980). This was also confirmed with PS-5 and NS-
5 (Fukagawa et al., 1981). In group 2 carbapenem compounds the hydroxylation at position C8 has an important effect on antimicrobial activity and must
be discussed in connection with the stereochemistry of the hydrogen atom
at C6. When PS-5 with trans protons (5R,6R) on the β-lactam ring is hydoxy-
lated to epithienamycin C with trans protons (5R,6S), the antibacterial activity
diminishes drastically (Sakamoto et al., 1979; Basker et al., 1980). However,
when epithienamycin C with trans protons (5R,6S) is epimerized to epithienamycin A with cis protons (5R,6R), recovery or improvement of the antimicrobial activity is observed (Basker et al., 1980). A similar change in antibiotic potency is also recognized among PS-7 and epithienamycins B and D
(Sakamoto et al., 1980; Basker et al., 1980). The sulfation of epithienamycins
A and B to MM 17880 and MM 13902, respectively, results in improved antimicrobial activity against gram-negative bacteria and in reduced activity
against gram-positive microbes (Basker et al., 1980). The sulfoxidation of
MM 13902 to MM 4550 significantly reduces antibacterial activity, which was
ascribed to chemical instabilization by Basker et al. (1980). It is interesting
to recall that sulfoxidation of carpetimycin A to C-19393 S_2 unexpectedly
causes no chemical instabilization, while the antimicrobial activity remains
comparable to that of PS-6 (Sakamoto et al., 1980; Nakayama et al., 1980;
Imada et al., 1980; Harada et al., 1980). Thus it is likely that the combination of sulfation and sulfoxidation in MM 4550 and C-19393 S_2 is the cause
for potency reduction, as both MM 13902 (sulfated but not sulfoxidated) and
carpetimycin A (sulfoxidated but not sulfated) are antimicrobially very active.
The length of the alkyl side chain at position C6 is less influential on antibacterial activity, although PS-5 with the ethyl side chain is slightly superior
to PS-6 with the isopropyl side chain (Sakamoto et al., 1980). The introduction of the double bond in the acetamidoethylthio side chain at position C3
also does not seem to give much influence on antibiotic potency, but the antibacterial activity against gram-negative organisms might be improved slightly,
as is observed between PS-5 and PS-7, epithienamycins C and D, epithienamycins A and B, and MM 17880 and MM 13902, respectively (Sakamoto et al.,
1979, 1980; Basker et al., 1980).

B. β-Lactamase-Inhibitory Activity and Synergism with Known Penicillins and Cephalosporins

Except for thienamycins and epithienamycins, the carbapenem family was originally isolated as β-lactamase inhibitors. However, as the types and varieties of β-lactamase are numerous, and as every laboratory has used different assay conditions for β-lactamase inhibition tests with respect to assay substrates, enzymes, analyses, reaction times, temperatures, preincubation times, definition of the inhibition index, and so on, it is practically impossible to objectively compare the assay results reported in the scientific literature. Furthermore, as is warned by Fukagawa et al. (1980b) and, in a broader sense, by Bardsley et al. (1980), inconsiderate use of the Michaelis–Menten equation in kinetic analyses of enzyme reactions, which has been common in β-lactamase assays, is highly likely to yield false conclusions on enzyme kinetics. This is very true in β-lactamases, particularly because they belong to one of the most efficient biocatalysts. To make matters worse, carbapenem compounds seem to be nonclassical inhibitors. In other words, the mode of inhibition of these antibiotics on β-lactamases cannot be discussed from the viewpoint of competitive, noncompetitive, or uncompetitive inhibition, as is the case with clavulanate (Fisher et al., 1978). Phenomenally speaking, the behaviors of carbapenem compounds on various β-lactamases can be summarized by the following observations that will be useful for a rough appraisal of the antibiotics as β-lactamase inhibitors.

1. A carbapenem compound is unsusceptible to a β-lactamase, but the enzyme is inactivated at least in a partially irreversible manner. Fukagawa et al. (1980b) described that PS-5 was completely resistant to the hydrolytic action of *Bacillus licheniformis* 749/C β-lactamase, while the enzyme activity was diminished by PS-5 to an extent depending on the duration of preincubation and the concentration of the inhibitor, following the first-order kinetics.

2. A carbapenem compound is a good substrate for a β-lactamase and is decomposed by the enzyme as rapidly as penicillins and cephalosporins. Cells of *B. cereus* 569 and, more particularly, the type II β-lactamase (cephalosporinase)—but not the type I— were found to easily open the β-lactam ring of PS-5 (Fukagawa et al., 1980b).

3. A carbapenem compound inactivates a β-lactamase at least in a partially irreversible manner, while a portion of the antibiotic is simultaneously hydrolyzed by the enzyme.

This type of kinetics was observed with PS-5 on β-lactamases from *P. vulgaris* P-5 (Okamura et al., 1980) and *P. vulgaris* GN76, *C. freundii* GN346, *B. cereus* (type I), and *Streptomyces* sp. E750-3 (Fukagawa et al., 1980).

For a practical evaluation of the possible application of carbapenem derivatives as β-lactamase inhibitors in the clinical treatment of bacterial infections, synergism with conventional penicillins and cephalosporins is one of the important criteria. Using ampicillin and cephaloridine, Okamura et al. (1979) showed the marked synergism of PS-5 in antibacterial activity against β-lactamase-producing or non-β-lactamase-producing clinical isolates. Similar findings were reported with olivanates (Butterworth et al., 1979; Basker et al., 1980) and C-19393 S_2 and H_2 (Imada et al., 1980). Since carbapenem antibiotics in themselves are potent antibacterial agents, it is difficult to differentiate the pure synergism from their intrinsic antibiotic activity. Furthermore, the permeability barrier of the bacterial cell wall complicates the interpretation of experimental results. However, most cases of synergism seem to be ascribed primarily to the significant inhibition of β-lactamase activity.

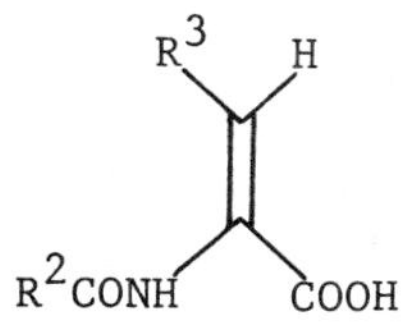

Figure 7 Specific inhibitor of renal dehydropeptidase-I.

In this connection, it is interesting to note that the combination of PS-5 with cefoxitin exhibited marked synergism in minimum inhibitory concentrations of cefoxitin (in the presence of 1 µg/ml PS-5; roughly one-fifth of the minimum inhibitory concentration of PS-5) against 19 cefoxitin-resistant clinical isolates of *Enterobacter cloacae* (all were type Ia β-lactamase producers), although cefoxitin was not attached at all by the enzyme (Yokota 1980).

C. Susceptibility to Renal Dipeptidase

Merck investigators (Kropp et al., 1980a) found that thienamycin and related compounds were extensively metabolised in vivo, predominantly in the kidney. Enzymological studies using soluble preparations from porcine and human renal cortices showed that the renal dipeptidase dehydropeptidase-I (E.C.3.4.13.11) was responsible for metabolism of carbapenem antibiotics. As a theoretical solution to raise the blood level and the urinary recovery of thienamycin and N-formimidoylthienamycin (a synthetic derivative), the researchers screened various organic compounds for specific inhibitors of renal dehydropeptidase-I and discovered that Z-2-acylamino-3-substituted propenoates (Fig. 7) were potent reversible inhibitors (Ashton et al., 1980).

When N-formimidoylthienamycin was coadministered to chimpanzees with 3-substituted-Z-2(S-2,2-dimethylcyclopropylcarboxamido)propenoate, the urinary recovery of the synthetic antibiotic was improved from 10–15% (control without inhibitor) to 65–75%, corresponding to the raised blood level.

ACKNOWLEDGMENTS

The authors are most grateful to Professor Yamada of the Tokyo College of Pharmacy for his helpful suggestions and advice during chemical studies. We thank the following colleagues for their contributions to this review: K. Okamura, N. Shibamoto, M. Sakamoto, K. Kubo, T. Takei, M. Okabe, R. Okamoto, A. Koki, K. Nakamura, H. Iguchi, T. Yoshioka, K. Yamamoto, Y. Kato, K. Isshiki, M. Nishino, and Y. Shimauchi, of Sanraku-Ocean Co., Central Research Laboratories, and J. Lein, of Panlabs, Inc.

REFERENCES

Albers-Schonberg, G., Arison, B. H., Kaczka, E., Kahan, F. M., Kahan, J. S., Lago, B., Maiese, W. M., Rhodes, R. E., and Smith, J. L. (1976). Thienamycin: Structure determination and biosynthetic data. In *16th Interscience Conference on Antimicrobial Agents and Chemotherapy, Chicago, Illinois*, Abstract No. 229. American Society for Microbiology, Washington, D.C.

Aoki, H., Sakai, H., Kohsaka, M., Konomi, T., Hosoda, J., Kubochi, Y., Iguchi, E., and Imanaka, H. (1976). Nocardicin A, a new monocyclic beta-lactam antibiotic. I. Discovery, isolation and characterization. *J. Antibiot.* 29:492-500.

Arai, M., Haneishi, T., Inukai, M., Nakashima, M., Takahashi, H., and Takiguchi, H. (1978a). Antibiotic substance No. 17927 A$_1$, its preparation method and antimicrobial drugs containing it as a principle. Japan Kokai 53-103401.

Arai, M., Haneishi, T., Inukai, M., Nakashima, M., Itoh, Y., Nakahara, M., and Takiguchi, H. (1978b). Antibiotic substance No. 17927 A$_2$. Japan Kokai 53-109997.

Arai, M., Haneishi, T., Nakashima, M., Serizawa, N., and Takiguchi, H. (1980). Substance No. 17927 D and its preparation method. Japan Kokai 55-24129.

Ashton, W. T., Barash, L., Brown, J. E., Brown, R. D., Canning, L. F., Chen, A., Graham, D. W., Kahan, F. M., Kropp, H., Sundelof, J. G., and Rogers, E. F. (1980). Z-2-Acylamino-3-substituted propenoates, inhibitors of the renal dipeptidase (dehydropeptidase-I) responsible for thienamycin metabolism. In *20th Interscience Conference on Antimicrobial Agents and Chemotherapy, New Orleans, Louisiana*, Abstract No. 271. American Society for Microbiology, Washington, D.C.

Bardsley, W. G., Leff, P., Kavanagh, J., and Waight, R. D. (1980). Deviations from Michaelis–Menten kinetics. The possibility of complicated curves for simple kinetic schemes and the computer fitting of experimental data for acetylcholin esterase, acid phosphatase, adenosine deaminase, arylsulphatase, benzylamine oxidase, chymotrypsin, fumarase, galactose dehydrogenase, beta-galactosidase, lactate dehydrogenase, peroxidase and xanthine oxidase. *Biochem. J.* 187:739-765.

Basker, M. J., Boon, R. J., and Hunter, P. A. (1980). Comparative antibacterial properties in vitro of seven olivanic acid derivatives: MM 4550, MM 13902, MM 17880, MM 22380, MM 22381, MM 22382 and MM 22383. *J. Antibiot.* 33:878-884.

Box, S. J., Hood, J. D., and Spear, S. R. (1979). Four further antibiotics related to olivanic acid produced by *Streptomyces olivaceus*: Fermentation, isolation, characterization and biosynthetic studies. *J. Antibiot.* 32:1239-1247.

Box, S. J., Hanscomb, G., and Spear, S. R. (1981). The detection of members of the olivanic acid family in *Streptomyces gedanensis*. *J. Antibiot.* 34:600-601.

Brown, A. G., Butterworth, D., Cole, M., Hanscomb, G., Hood, J. D., Reading, C., and Rolinson, G. N. (1976). Naturally-occurring beta-lactamase inhibitors with antibacterial activity. *J. Antibiot.* 29:668-669.

Butterworth, D., Cole, M., Hanscomb, G., and Rolinson, G. N. (1979). Olivanic acids, a family of beta-lactam antibiotics with beta-lactamase inhibitory properties produced by *Streptomyces* species. I. Detection, properties and fermentation studies. *J. Antibiot.* 32:287-294.

Cassidy, P. J., Stapley, E. O., Goegelman, R. T., Miller, T. W., Arison, B. H., Albers-Schonberg, G., Zimmerman, S. G., and Birnbaum, J. (1977). Epithienamycins. Isolation and identification. In *17th Interscience Conference on Antimicrobial Agents and Chemotherapy, New York*, Abstract No. 81. American Society for Microbiology, Washington, D.C.

Cassidy, P. J., Goegelman, R. T., Stapley, E. O., and Hernandez, S. (1979a). Antibiotics 890 A_2 and 890 A_5. U.S. Patent No. 4,141,986.

Cassidy, P. J., Goegelman, R. T., Stapley, E. O., and Hernandez, S. (1979b). Antibiotics 890 A_1 and 890 A_3. U.S. Patent No. 4,162,324.

Fisher, J., Charnas, R. L., and Knowles, J. R. (1978). Kinetic studies on the inactivation of *Escherichia coli* RTEM beta-lactamase by clavulanic acid. *Biochemistry* 17:2180–2184.

Fukagawa, Y. (1980). A computer-assisted spectrophotometric enzyme assay. A method for the calculation of K_m and V from a single reaction progress curve by the integrated form of the Lineweaver–Burk plot with particular reference to beta-lactamase. *Biochem. J.* 185:186–188.

Fukagawa, Y., Kubo, K., Ishikura, T., and Kouno, K. (1980a). Deacetylation of PS-5, a new beta-lactam compound. I. Microbial deacetylation of PS-5. *J. Antibiot.* 33:543–549.

Fukagawa, Y., Takei, T., and Ishikura, T. (1980b). Inhibition of beta-lactamase of *Bacillus licheniformis* 749/C by compound PS-5, a new beta-lactam antibiotic. *Biochem. J.* 185:177–185.

Fukagawa, Y., Okamura, K., Shibamoto, N., and Ishikura, T. (1981). PS-series beta-lactam antibiotics. In *Beta-Lactam Antibiotics: Drug Action and Bacterial Resistance*. S. Mitsuhashi (Ed.). Japan Scientific Societies, Tokyo, pp. 158–163.

Harada, S., Shinagawa, S., Nozaki, Y., Asai, M., and Kishi, T. (1980). C-19393 S_2 and H_2, new carbapenem antibiotics. II. Isolation and structures. *J. Antibiot.* 33:1425–1430.

Higgens, C. E., and Kastner, R. E. (1971). Description of *Streptomyces clavuligerus*, a beta-lactam-producing streptomycete. *Int. J. Syst. Bacteriol.* 21:326–331.

Imada, A., Nozaki, Y., Kintaka, K., Okonogi, K., Kitano, K., and Harada, S. (1980). C-19393 S_2 and H_2, new carbapenem antibiotics. I. Taxonomy of the producing strain, fermentation and antibacterial properties. *J. Antibiot.* 33:1417–1424.

Kahan, J. S., and Kahan, F. M. (1977). Antibiotics and their preparation method. Japan Kokai 52-65295.

Kahan, J. S., Kahan, F. M., Goegelman, R. T., Stapley, E. O., and Hernandez, S. (1977). Antibiotic 924 A_1. German Offen 2,652,681.

Kahan, J. S., Kahan, F. M., Goegelman. R., Currie, S. A., Jackson, M., Stapley, E. O., Miller, T. W., Miller, A. K., Hendlin, D., Mochales, S., Hernandez, S., Woodruff, H. B., and Birnbaum, J. (1979). Thienamycin, a new beta-lactam antibiotic. I. Discovery, taxonomy, isolation and physical properties. *J. Antibiot.* 32:1–12.

Kropp, H., Sundelof, J. G., Hajdu, R., and Kahan, F. M. (1980a). Metabolism of thienamycin and related carbapenem antibiotics by the renal dipeptidase:dehydropeptidase-I. In *20th Interscience Conference on Antimicrobial Agents and Chemotherapy, New Orleans, Louisiana,* Abstract No. 272. American Society for Microbiology, Washington, D.C.

Kropp, H., Sundelof, J. G., Kahan, J. S., Kahan, F. M., and Birnbaum, J. (1980b). MK0787(N-formimidoylthienamycin): Evaluation of in vitro and in vivo activities. *Antimicrob. Agents Chemother.* 17:993–1000.

Kubo, K., Ishikura, T., and Fukagawa, Y. (1980). Deacetylation of PS-5, a new beta-lactam compound. III. Enzymological characterization of L-amino acid acylase and D-amino acid acylase from *Pseudomonas* sp. 1158. *J. Antibiot.* 33:556–565.

Miller, I. M., Stapley, E. O., and Chaiet, L. (1962). Production of synnematin B by a member of the genus *Streptomyces*. *Bacteriol. Proc.* 1962:32.

Nakayama, M., Iwasaki, A., Kimura, S., Mizoguchi, T., Tanabe, S., Murakami, A., Watanabe, I., Okuchi, M., Itoh, H., Saino, Y., Kobayashi, F., and Mori, T. (1980). Carpetimycins A and B, new beta-lactam antibiotics. *J. Antibiot.* 33:1388–1389.

Nojiri, C., Ohba, K., Itoh, S., Ogawa, Y., Niwa, T., Totsugawa, K., Ezaki, N., Shomura, T., Inoue, S., and Yamada, Y. (1979). New antibiotic SF-2050 and its preparation method. Japan Kokai 54-109901.

Ohba, K., Nojiri, C., Ogawa, Y., Itoh, S., Totsugawa, K., Ezaki, N., Shomura, T., Niwa, T., Inoue, S., and Yamada, Y. (1979). New antibiotic SF-2050B and its preparation method. Japan Kokai 54-122203.

Okamura, K., and Ishikura, T. (1979). Method for improvement of antibiotics production. Japan Kokai 54-154598.

Okamura, K., Hirata, S., Okumura, Y., Fukagawa, Y., Shimauchi, Y., Kouno, K., Ishikura, T., and Lein, J. (1978). PS-5, a new beta-lactam antibiotic from *Streptomyces*. *J. Antibiot.* 31:480–482.

Okamura, K., Hirata, S., Koki, A., Hori, K., Shibamoto, N., Okumura, Y., Okabe, M., Okamoto, R., Kouno, K., Fukagawa, Y., Shimauchi, Y., Ishikura, T., and Lein, J. (1979a). PS-5, a new beta-lactam antibiotic. I. Taxonomy of the producing organism, isolation and physico-chemical properties. *J. Antibiot.* 32:262–271.

Okamura, K., Sakamoto, M., Fukagawa, Y., Ishikura, T., and Lein, J. (1979b). PS-5, a new beta-lactam antibiotic. III. Synergistic effects and inhibitory activity against a beta-lactamase. *J. Antibiot.* 32:280–286.

Okamura, K., Koki, A., Sakamoto, M., Kubo, K., Mutoh, Y., Fukagawa, Y., Kouno, K., Shimauchi, Y., Ishikura, T., and Lein, J. (1979c). Microorganisms producing a new beta-lactam antibiotic. *J. Ferment. Technol.* 57:265–272.

Okamura, K., Sakamoto, M., and Ishikura, T. (1980). PS-5 inhibition of a beta-lactamase from *Proteus vulgaris*. *J. Antibiot.* 33:293–302.

Rosi, D., Drozd, M. L., Kuhrt, M. R., Terminiello, L., Came, P. E., and Daum, S. J. (1981). Mutants of *Streptomyces cattleya* producing N-acetyl and deshydroxy carbapenems related to thienamycin. *J. Antibiot.* 34:341–343.

Sakamoto, M., Iguchi, H., Okamura, K., Hori, S., Fukagawa, Y., Ishikura, T., and Lein, J. (1979). PS-5, a new beta-lactam antibiotic. II. Antimicrobial activity. *J. Antibiot.* 32:272–279.

Sakamoto, M., Shibamoto, N., Iguchi, H., Okamura, K., Hori, S., Fukagawa, Y., Ishikura, T., and Lein, J. (1980). PS-6 and PS-7, new beta-lactam antibiotics: In vitro and in vivo evaluation. *J. Antibiot.* 33:1138–1145.

Shibamoto, N., and Ishikura, T. (1981). Antibiotic PS-8. Japan Kokai 56-25183.

Shibamoto, N., Koki, A., Nishino, M., Nakamura, K., Kiyoshima, K., Okamura, K., Okabe, M., Okamoto, R., Fukagawa, Y., Shimauchi, Y., Ishikura, T., and Lein, J. (1980). PS-6 and PS-7, new beta-lactam antibiotics: Isolation, physicochemical properties and structures. *J. Antibiot.* 33:1128–1137.

Shohji, J., Tanaka, K., Kawamura, Y., Hattori, M., Kondo, E., Matsumoto, K., Yoshida, T., and Tsuji, N. (1980). Novel antibiotic PA-31088-IV. Japan Kokai 55-136282.

Stapley, E. O., Jackson, M., Hernandez, S., Zimmerman, S. B., Currie, S. A., Mochales, S., Mata, J. M., Woodruff, H. B., and Hendlin, D. (1972). Cephamycins, a new family of beta-lactam antibiotics. I. Production by actinomycetes, including *Streptomyces lactamdurans* sp. n. *Antimicrob. Agents Chemother.* 2:122–131.

Umezawa, H., Mitsuhashi, S., Hamada, M., Iyobe, S., Takahashi, S., Utahara, R., Osato, Y., Yamazaki, S., Ogawara, H., and Maeda, K. (1973). Two beta-lactamase inhibitors produced by a streptomyces. *J. Antibiot.* 26:51–54.

Yokota, T. (1980). Educational lecture on beta-lactamase inhibitors. In *28th Annual Meeting of Japan Society of Chemotherapy, Tokyo*, Abstract papers pp. 38–39. Japan Society of Chemotherapy, Tokyo.

8

THE TETRACYCLINES: PROPERTIES, BIOSYNTHESIS, AND FERMENTATION

MILOSLAV PODOJIL, MARGITA BLUMAUEROVÁ, AND ZDENKO VANĚK
Institute of Microbiology, Czechoslovak Academy of Sciences, Prague, Czechoslovakia
KAREL ČULÍK *Research Institute of Antibiotics and Biotransformations, Roztoky-near-Prague, Czechoslovakia*

I. INTRODUCTION

Over the more than 30 years that have elapsed since the discovery of tetracycline antibiotics, the substances have undergone thorough study in a great
many laboratories. Biological research was focused on production organisms
(morphological and physiological studies, improvement procedures, genetic
analysis), on biosynthesis and its metabolic and genetic control, on the mechanism of antibiotic action and the origin of resistance, as well as on various
pharmacological aspects of tetracycline application. The chemistry of tetracyclines concentrated on the study of their structures, the preparation of
new derivatives, and isolation procedures and analytical methods. The results
are compiled in hundreds of publications. The limited scope of this chapter
does not permit the authors to cite all the papers, and the reader is referred
to various review articles where the references may be found.

The history of the discovery of natural tetracyclines (chlortetracycline
[1], oxytetracycline [2], tetracycline [3], and demeclocycline [4]; Table 1
and Fig. 1) has been summarized by Dürckheimer (1975). In addition to classical producers of chlortetracycline, tetracycline (*Streptomyces aureofaciens*),
and oxytetracycline (*Streptomyces rimosus*), a number of other streptomycetes
have been described to produce such substances (for a review see Di Marco
and Pennella, 1959; Hošťálek et al., 1979). In addition to these tetracyclines,
obtained through fermentation, many derivatives of this series have been prepared by chemical conversion of natural tetracyclines: methacycline [5],
doxycycline [6], minocycline [7], and rolitetracycline [8] (Reiner, 1974).
Table 1 gives the names of all the commercially important tetracyclines.

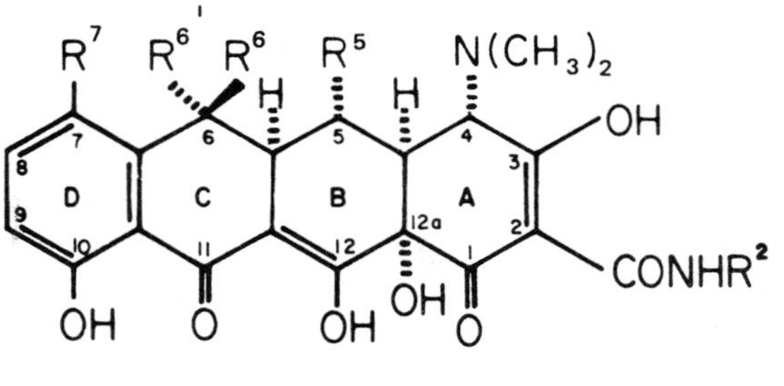

	R^2	R^5	R^6	R'	R^7
[1]	H	H	OH	CH$_3$	Cl
[2]	H	OH	OH	CH$_3$	H
[3]	H	H	OH	CH$_3$	H
[4]	H	H	OH	H$_3$CH	Cl
[5]	H	OH	=CH$_2$		H
[6]	H	OH	H	CH$_3$	H
[7]	H	H	H	H	N(CH$_3$)$_2$
[8]	-CH$_2$-N	H	OH	CH$_3$	H

Figure 1 Structures of significant tetracycline antibiotics.

II. GENETICS OF TETRACYCLINE PRODUCERS

A. Strain Improvement

Results obtained mainly with S. aureofaciens and S. rimosus in the years 1954–1973 (for references see Di Marco and Pennella, 1959; Hošťálek et al., 1974, 1979) showed that by a traditional breeding procedure (mutation and selection) one can increase the antibiotic titer by 30–500%, depending on the strain, the mutagen, and the number of selection steps. The best results were obtained with ultraviolet light either alone or in combination with ethyleneimine or x rays, whereas γ irradiation and nitrogen mustard had relatively little effect, and N-nitrosodimethylurea increased productivity only on repeated application. The efficiency of mutational improvement was often increased by controlled selection (e.g., isolation of prototrophic revertants from auxotrophs, producing revertants of nonproducing mutants, or of strains with increased resistance to tetracyclines or antibiotics with a similar mode of action). Hybridization experiments did not yield unequivocal results that would point to the expediency of this procedure as compared with mutational improvement. The recently observed possibility of fusion of S. rimosus protoplasts (Alačević et al., 1980) offers further perspectives in the field.

All the above results were obtained in basic research, using strains with antibiotic titer not exceeding 3000 µg/ml. No data on improving high-production industrial strains of S. aureofaciens and S. rimosus are available.

B. Genome Topology

The circular linkage map of S. rimosus for nutritional markers has been independently constructed in two laboratories (Alačević et al., 1973; Friend and Hopwood, 1971). Analysis of nonproduction mutants permitted a preliminary determination of the position of some loci responsible for the biosynthesis of oxytetracycline in the lower arc of the map (Alačević et al., 1976). Genes for the later steps in biosynthesis (anhydrotetracycline to oxytetracycline; see Fig. 2) were located in a small chromosome segment between the markers proA and adeA. The genes controlling the biosynthesis of cosynthetic factor 1 were also located in this region. The genes for the earlier steps of oxytetracycline formation were located in a second region, diametrically opposite to the first, between markers ribB and cysD (Rhodes et al., 1981). For S. aureofaciens the data on the genome topology are rather scant (Blumauerová et al., 1972; Hošťálek et al., 1974, 1979). No genetic analyses have been carried out with other tetracycline-producing streptomycetes.

C. Extrachromosomal DNA

Results of experiments with plasmid-curing agents indicate that extrachromosomal DNA may take part in the control of the biosynthesis of oxytetracycline (Boronin and Sadovnikova, 1972; Borisoglebskaya et al., 1979). Stepnov et al. (1978) used agarose gel electrophoresis to isolate circular plasmid DNA of molecular weight 3.7×10^7 from a production strain of S. rimosus. Plasmid DNA, both circular and open form, was recently isolated in S. aureofaciens by column chromatography on Sepharose 4B and nitrocellulose (Godány et al., 1980). The possible function of this DNA during tetracycline biosynthesis will require further study.

Genetic evidence for a plasmid controlling fertility in S. rimosus was presented by Friend et al. (1978).

Table 1 Survey of the Names of Significant Tetracycline Antibiotics

Structure[a]	Trivial name	Chemical Abstracts indexing[b]	Trade name[c]
[1]	Chlortetracycline (7-chlorotetracycline)	7-Chloro-4-(dimethylamino)-1,4,4a,5,5a,6,11,12a-octahydro-3,6,10,12,12a-pentahydroxy-6-methyl-1,11-dioxo-2-naphthacenecarboxamide	Aureomycin (Lederle), Aureomykoin (Spofa)
[2]	Oxytetracycline (5-hydroxytetracycline)	4-(Dimethylamino)-1,4,4a,5,5a,6,11,12a-octahydro-3,5,6,10,12,12a-hexahydroxy-6-methyl-1,11-dioxo-2-naphthacene-carboxamide	Terramycin (Pfizer), Terravenös (Pfizer), Oxymykoin (Spofa)
[3]	Tetracycline	4-(Dimethylamino)-1,4,4a,5,5a,6,11,12a-octahydro-3,6,10,12,12a-pentahydroxy-6-methyl-1,11-dioxo-2-naphthacenecarboxamide	Tetracyn (Pfizer), Achromycin (Lederle), Hostacycline (Hoechst), Ambramycin (Lepetit), Mediacyclin (Médial), Tetracyclin (Bayer)
[4]	Demethylchlortetracycline (7-chloro-6-demethyl-tetracycline)	7-Chloro-4-(dimethylamino)-1,4,4a,5,5a,6,11,12a-octahydro-3,6,10,12,12a-pentahydroxy-1,11-dioxo-2-naphthacene-carboxamide	Demeclocycline (Pfizer), Ledermycin (Lederle)

[5]	Methacycline (6-methylene-5-hydroxy- tetracycline)	4-(Dimethylamino)-1,4,4a,5,5a,6, 11,12a-octahydro-3,5,10,12,12a- pentahydroxy-6-methylene-1,11- dioxo-2-naphthacenecarboxamide	Rondomycin (Pfizer)
[6]	Doxycycline (α-6-deoxy-5-hydroxy- tetracycline)	4-(Dimethylamino)-1,4,4a,5,5a,6, 11,12a-octahydro-3,5,10,12,12a- pentahydroxy-6-methyl-1,11- dioxo-2-naphthacenecarboxiamide	Vibramycin (Pfizer)
[7]	Minocycline (7-dimethylamino-6- demethyl-6-deoxy- tetracycline)	4,7-Bis (dimethylamino)-1,4,4a,5, 5a,6,11,12a-octahydro-3,10,12, 12a-tetrahydroxy-1,11-dioxo-2- naphthacenecarboxamide	Minocin (Lederle)
[8]	Rolitetracycline (pyrrolidinomethyl- tetracycline)	4-(Dimethylamino)-1,4,4a,5,5a,6, 11,12a-octahydro-3,6,10,12,12a- pentahydroxy-6-methyl-1,11- dioxo-N-(1-pyrrolidinylmethyl-2- naphthacene-carboamide)	Reverin (Hoechst)

aSee Figure 1.
bFrom *Chemical Index Guide* (1977-1981).
cFrom Reiner (1974).

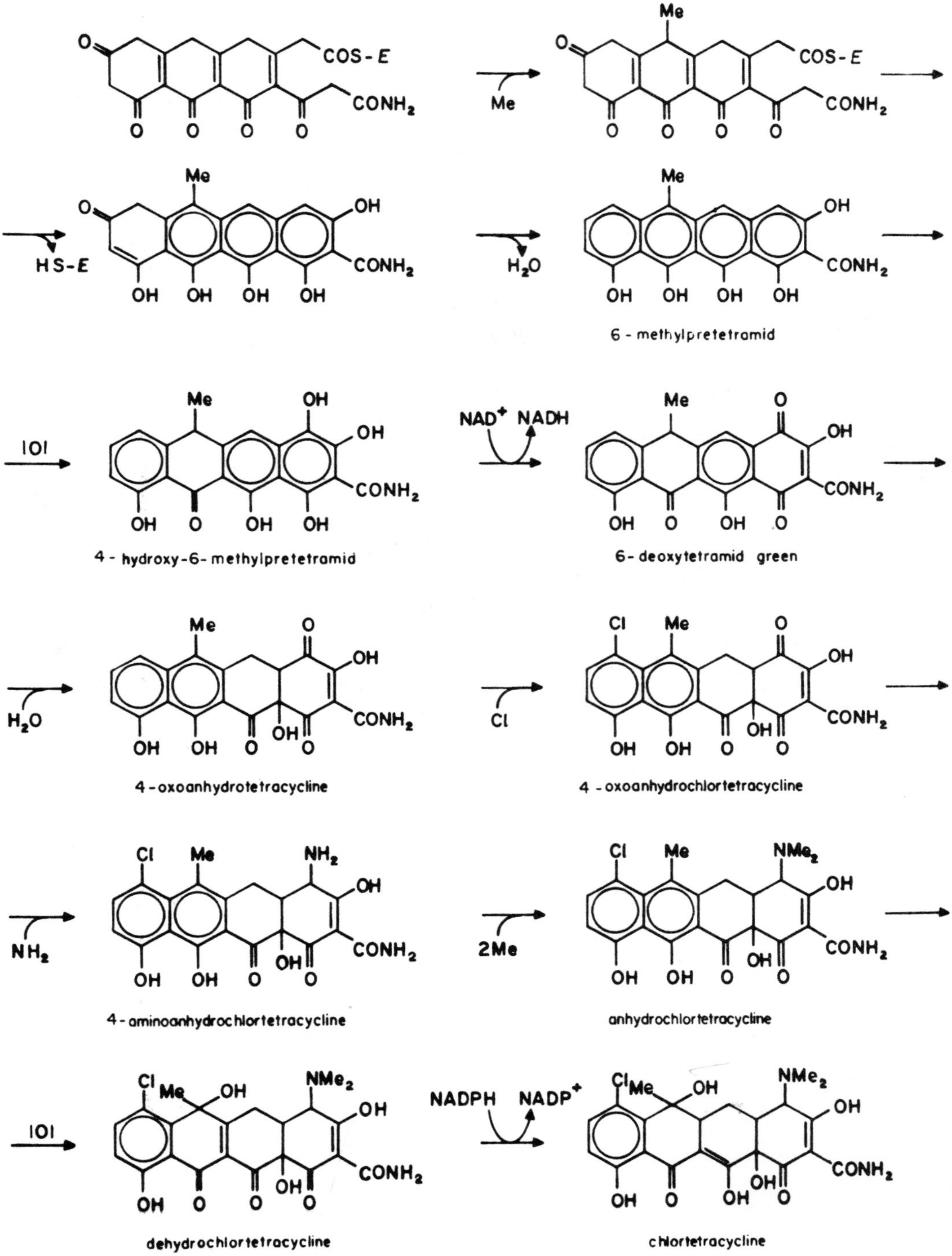

Figure 2 Scheme for chlortetracycline [1] biosynthesis in *S. aureofaciens*. The first three compounds as well as 6-deoxytetramid green are hypothetical. After Hošťálek et al. (1979).

III. BIOSYNTHESIS

A. Basic Steps

The biosynthesis of natural tetracycline antibiotics is treated in several out-standing reviews by McCormick (1965, 1967, 1969). Together with other re-view articles (Mitscher, 1968; Vaněk et al., 1971; Hlavka and Boothe, 1973; Hošťálek et al. 1974, 1979; Hošťálek and Vaněk, in press), they collate the results of dozens of papers devoted to this topic. Their analysis indicates the following:

1. The experimental results best fit the concept of synthesis of a linear oligoketidamide chain (from which the skeleton of tetracycline antibiotics is formed in further reaction steps) through head-to-tail condensation of 1 unit of malonamic acid and 8 units of malonic acid.

2. At the level of the linear chain, the methylated-series antibiotics undergo methylation at C6 (numbering as in tetracyclines), reduction at C8, dehydration of the alcohol to a C7–C8 olefin, and cyclization (based on aldo-lization in positions C4a–C12a, C5a–C11a, and C6a–C10a and on an intra-molecular acylation at C2–C3 by a thiolester) to a tetracycline derivative.

3. The C2 building units (malonyl coenzyme A) required for the forma-tion of the oligoketidamide linear chain are formed by conversion of glucose. Carboxylation of acetyl coenzyme A (CoA) (which has its source in oxidative decarboxylation of pyruvic acid) under catalysis by acetyl-CoA carboxylase is not theoretically the unique pathway for malonyl-CoA formation. Běhal et al. (1977) showed that during the biosynthesis of tetracycline, malonyl-CoA is also formed by oxidative decarboxylation of oxaloacetate.

Malonamoyl is formed through amidation of malonyl-CoA or through carba-moylation of acetyl-CoA. Other possibilities are offered by transamination of glutamic acid and transamination or deamidation of asparagine. These last two reactions were studied by Podojil et al. (1973), who showed a nonuniform imcorporation of [U-^{14}C]asparagine into the carboxamide carbon and into the C and D rings of the tetracycline molecule. The nonuniform distribution of radioactivity in tetracycline suggests that the carboxamide group is a part of a unit which differs in its origin from units forming the basic skeleton. This view is in agreement with the observation of Běhal et al. (1974), that the biosynthetic pathway for the formation of malonate units of the main skele-ton of tetracycline differs from that of the malonamoyl group.

4. The indicated sequence of reactions results in pretetramides, the first substances of the biosynthetic sequence that can be demonstrated experi-mentally. Up to the formation of pretetramide or 6-methylpretetramide, the biosynthesis of tetracyclines is of a speculative nature.

All the following biosynthetic intermediary steps leading to tetracyclines do not take place before the tetracyclic amide is formed.

5. Point-blocked mutants accumulating intermediates of the biosynthetic pathway proved to be the most valuable material for the study of secondary alterations of pretetramids. Determination of the chemical structures of inter-mediary products of the main or side pathway and the study of their precursor activity by the cosynthetic method suggested a biosynthetic scheme and the sequence of the respective intermediates in the course of formation of chlor-tetracycline and tetracycline.

After the origin of pretetramids, C4-hydroxylation, oxidation of ring A, C4a,12a-hydration, C7-chlorination, C4-transamination, double N-methyla-tion, C6-hydroxylation, and reduction of the 5a–11a double bond follow. Hydrogenation of the 5a–11a double bond required the presence of the cosynthetic factor, which was found in a number of strains of S. aureofaciens Miller et al., 1960). Low-producing cultures of S. aureofaciens supplemented with this factor considerably increased the yields of tetracyclines.

The biosynthesis of oxytetracycline apparently proceeds in the same way as that of chlortetracycline and tetracycline, but C7-chlorination is blocked and C5-hydroxylation is effected at the level of 5a,11a-dehydrotetracycline before the final reduction (Miller et al., 1965). As an illustrative example, a general scheme for chlortetracycline biosynthesis, including the individual steps, is given in Figure 2.

B. Shunt Products

Here one should mention protetrone and its methylanthrone analog, described by McCormick's groups. Their formation is effected by a block at the site of the final cyclization of the oligoketide chain. Among the overflow metabolic products of high-production strains of S. aureofaciens, one should mention ekatetrone (Podojil et al., 1978), which is chemically related to protetrone (Přikrylová et al., 1978). Another interesting metabolite is the antibiotically inactive glucoside aureovocin, which accompanies chlortetracycline during standard fermentations of S. aureofaciens strains. Its aglycone aureovocidin is apparently a product of overflow metabolism of 4-hydroxy-6-methylpreteramide, with a partial block in the oxidation of ring A (Blumauerová et al., 1972).

C. Enzymes of Metabolic Pathways

Modification of the pretetramide skeleton to tetracyclines is catalyzed in the individual reaction steps by enzymes. Three enzymes of the final stages of tetracycline biosynthesis have been described, and an in vitro reaction supported the biosynthetic pathways proposed on the basis of cosynthesis and experiments with blocked mutant strains.

1. Methylation of the Amino Group at C4

This reaction is catalyzed by S-adenosylmethionine:dedimethylaminoanhydrotetracycline N-methyltransferase, described by Miller and Hash (1975). The enzyme is present in strains of S. aureofaciens and S. rimosus, and eventually in their blocked mutants. Bioconversion of dedimethylamino-4-aminoanhydrotetracyclines to anhydrotetracyclines is followed isotopically by measuring the incorporation of radiolabeled S-adenosylmethionine into the amino group of anhydrotetracyclines.

2. C6-Hydroxylation

Běhal et al. (1979) described the enzyme anhydrotetracycline oxygenase, showed it to catalyze C6-hydroxylation of the tetracycline ring, and observed that its activity in cell-free extracts of S. aureofaciens strains is directly proportional to the kinetics of tetracycline formation. The specific activity of the enzyme is inhibited by inorganic phosphate and increased by benzyl thiocyanate (Běhal, et al., 1982). Quantitative determination of anhydrotetracycline oxygenase is based on a change of the absorption spectrum in the visible range. A decrease in absorption at 440 nm indicates a conversion of anhydrotetracycline to 5a,11a-dehydrotetracycline, the formation of which is indicated by an increase of absorbance at 400 nm (Běhal et al., 1979).

3. Reduction of the 5a—11a Double Bond

This reaction is the final one in the biosynthetic pathway leading to tetra-cyclines; reduction of the double bond of 5a,11a-dehydrotetracyclines by NADP:tetracycline 5a,11a-dehydrogenase yields tetracyclines. For an in vitro reaction the substrate used was 7-chloro-5a,11a-dehydrotetracycline. A crude enzyme preparation was obtained from *S. aureofaciens* ATCC 13192. A posi-tive reaction is indicated by an increase in the antibiotic activity of chlor-tetracycline from a biologically inactive precursor (Miller and Hash, 1975).

D. Regulation

Numerous data on the environmental factors affecting the production capacity of *aureofaciens* and *S. rimosus* cultures during submerged fermentation (sources of C and N, inorganic phosphate, metal ions, various stimulants and inhibitors, temperature, pH, aeration, and the like) were summarized by Di Marco and Pennella (1959) and Hoštálek et al. (1979). The latter group also discussed their application to the controlled biosynthesis of tetracyclines (directed toward a general increase of efficiency or optimization of the quan-titative ratio between chlortetracycline and tetracycline in the mixture pro-duced).

The inhibitory effect of inorganic phosphate, studied in a great many laboratories, indicates a close relation between carbohydrate metabolism and tetracycline biosynthesis. Data on the activity of enzymes of primary meta-bolism at different stages of development of production versus nonproduction cultures of *S. aureofaciens* (for references see Hoštálek et al., 1979; Hoštálek and Vaněk, in press) permit one to draw the following conclusions on the metabolic control of biosynthesis: The rate of formation of the naphthacene skeleton of tetracyclines, dependent on a sufficient supply of the primary building blocks (provided by the glycolytic pathway) is indirectly related to the activity of competing metabolic systems, first of all of the tricarboxylic acid cycle. The low level of enzymes of this cycle is accompanied by a low level of adenosine triphosphate and polyphosphates in the mycelium, which indicates a low level of energy metabolism in high-production strains. Changes in the activity of enzymes capable of phosphorylating hexoses, just as the alternative possibility of malonyl-CoA formation, then indicate that the forma-tion of the intermediates of the tetracycline pathway is catalyzed by other enzyme systems than is the formation of the same intermediates utilized in primary metabolism.

IV. FERMENTATION PROCESS

A. General Aspects

Compared with the abundant literature on the medical and technical applica-tions of tetracycline antibiotics, references on process technology are rather scant and most of them are included in the patent literature.

In this context one should quote Evans (1968): "Completely lacking in the reference library, however, is any work on the manufacture and produc-tion of these drugs. This is surprising as the tetracyclines are in the hundred million dollars fraternity." An explanation may be sought in the fact that the basic biosynthetic procedures and particularly the effective production strains have been protected for a longer time by solid patents of only three or four firms in the whole world. The fermentation technology of five of the principal tetracycline antibiotics was protected by 33 U.S. patents, and it was only after several years that the first patents of European and Japanese authors were applied for.

Not all the patented procedures of tetracycline biosynthesis were later used in industrial production, even if they opened a new way for preparing the antibiotic. In other cases, rapid scientific progress eliminated the less economical technologies.

Exhaustive reviews of the patent literature from the aspect of the technology of biosynthetically produced tetracycline antibiotics were published by Brunner (1963) and, from the point of view of U.S. patents, Evans (1968). The first author mentions 30 patent applications from the year 1956 to 1960 concerning the technology of the fermentative production of a single antibiotic—tetracycline. Such width of technologies is only apparent and, after the patents have expired, the world production of tetracycline antibiotics is based mainly on the most economical, least complicated technology, making use of high-production strains for the direct fermentation of a single desired product (Growich and Deduck, 1963).

Whereas the principal patents report tetracycline yields of up to hundreds or thousands of micrograms per milliliter of fermentation broth, the existing industrial yields are above 20,000 µg/ml in about 200 hr of fermentation.* A role in the increasing yields is certainly played by the general progress in fermentation technology, particularly in the design of fermenters and stirrers, in improved monitoring equipment, in the recognition of the shear effect during stirring, and in the use of fed-batch techniques, even if their application has been described earlier for other antibiotics.

B. Technological Equipment

During the production of tetracycline antibiotics one can use preinoculation, inoculation, and fermentation tanks, designed universally for the production of various antibiotics, for example, penicillin. They must provide for sufficient oxygen transfer (0.4−0.8 µmol/liter per minute) effected in fermenters of 100−150 m^3 working volume by three open turbines 1460−2100 mm in diameter with maximum speeds of 80 rpm, with a power input of 300 kW at the axel, which corresponds to up to 3 kW/m^3 of fermented broth (Heine and Heck, 1977).

Variable revolutions of the stirrer are better suited for solving the inverse relationship between oxygen transfer into the liquid by the stirrer and the shear stress of the biomass, and they can better cope with froth formation when the stirer cannot be switched off. The minimum supply with oxygen is of particular importance with producers of tetracycline antibiotics (repeatedly interrupting the oxygen supply for more than 10 min results in stopping the production of chlortetracycline or tetracycline), but the same holds for the frequency, concentration, and composition of fed batches (Matelová et al., 1955).

A special system must be used for effecting and controlling various sterile-fed batches during fermentation (Müller and Kieslich, 1966; Oldshue et al., 1977).

---

*Offer of licence (1979) of KRKA Farmaceutika Co., 68000 Novo Mesto, Yugoslavia.

C. Preparation of Inoculum

Strains for industrial fermentation are maintained for prolonged periods either freeze dried or at liquid nitrogen temperatures as spore stock. An example of the preparation of *S. aureofaciens* (producer of chlortetracycline) inoculum is shown here (analogous procedures apply to producers of other tetracyclines):

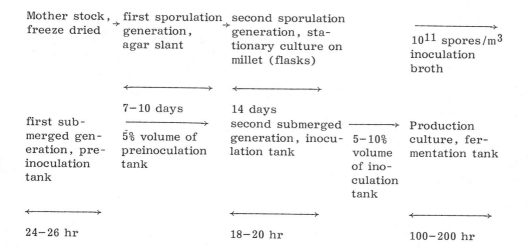

Mother stock, freeze dried → first sporulation generation, agar slant → second sporulation generation, stationary culture on millet (flasks) → 10^{11} spores/m^3 inoculation broth

7–10 days

first submerged generation, preinoculation tank → 5% volume of preinoculation tank → 14 days second submerged generation, inoculation tank → 5–10% volume of inoculation tank → Production culture, fermentation tank

24–26 hr   18–20 hr   100–200 hr

All cultivations take place at 29°C. The composition of the inoculation medium is similar to that in the production tank (to ensure a short lag phase of the main culture) and contains carbohydrates (sucrose and molasses, 2.5% wt/vol), a source of organic nitrogen (soybean meal and corn-steep liquor, 1.7%), a buffer (calcium carbonate, 0.2–0.3%), inorganic salts (NaCl and KH_2PO_4), and vegetable oil (0.2%).

The culture is monitored for pH, residual sugar, respiration (CO_2), and increase of biomass both in volume and as to morphology.

D. Nutrients

Of the technical cheap carbon sources, production strains preferentially utilize sucrose (molasses), starch, or technical glucose (Wise, 1950), starch as a polysaccharide being particularly suitable for prolonged fermentations of about 200 hr. Corn-steep, soybean meal, or peanut meal are the main technical sources of organic nitrogen. Calcium carbonate, besides its effect on pH, binds the formed antibiotic from the heterogeneous phase of complexes insoluble above 1500 µg/ml and thus decreases the inhibitory effect of their own product on the producer (Di Marco, 1956). Addition of various inorganic salts and a controlled level of phosphate are required for successful fermentation. Animal or vegetable lipids are used as antifoaming agents and even as carbon sources. One may also use synthetic antifoaming agents such as polypropylene glycol and silicones.

Chloride ions serve as precursors of chlortetracycline biosynthesis, while benzylthiocyanate is an inhibitor of undesirable metabolic pathways during lack of oxygen (Pecák et al., 1959).

The principal fermentation broth takes care of the start and lag phases of cultures; subsequent feeding is necessary for the growth and production phases (Szumski, 1964). Regulation of the growth rate is based on the selection of C and N sources in broth, their ready or slow assimilation must be tuned to the technical possibility of oxygen transfer by the apparatus, which represents the absolute limit. These relations were expressed for penicillin by Ryu and Humphrey (1972) and Queener and Schwartz (1979).

E. Parameters of Fermentation

Liquid nutrient broths are sterilized either at 120°C for 40 min or at 140°C in 1-min intervals (C and N sources separately). Stirring uses $2.5-3 \text{ kW/m}^3$ medium, revolutions can be varied up to a peripheral velocity of the propeller equal to 500 m/min, aeration ranges up to 0.8 medium vol/min, and oxygen overpressure up to 0.1 MPa. The temperature ($29 \pm 1°C$), pH, oxygen content

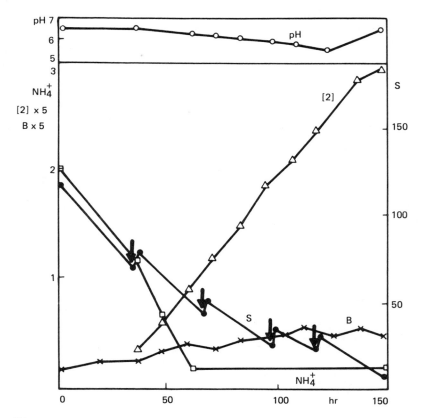

Figure 3 Pilot-plant fermentation of oxytetracycline [2]. Here NH_4^+ is the concentration of ammonium ions (mg/ml; regulation at the level 0.2 mg/ml in the period 60–150 hr); S stands for starch (mg/ml; the arrows indicate the feeding of 1% starch); [2], production (mg/ml × 5); and B, biomass (% vol/ vol × 5). Note: Ryu and Hospodka (1977) mention the experimental value of penicillin yield per gram of glucose as 0.073 g. It is of interest that in the present case the value for oxytetracycline referred to starch is 0.091 g.

of the medium (at least 20% saturation), and CO_2 in the outcoming air are monitored continuously. The medium composition, biomass content, antibiotic production, carbohydrate concentration, NH_4^+ and NH_2-N, the macro- and microscopic states of culture, and sterility are checked periodically. As an example we show here the biosynthesis of oxytetracycline in a pilot-plant scale fermenter. The medium contained is an follows: starch, 12 + 4% (additional feeding); technical amylase, 0.1%; yeast (dry wt), 1.5%; $CaCO_3$, 2%, ammonium sulfate, 1.5%, lactic acid, 0.13%, lard oil, 2%, and total inorganic salts, 0.01% (Fig. 3) (Anonymous, 1980b).

V. PRODUCT RECOVERY AND PURIFICATION

A. Natural Tetracyclines

Isolation methods must take into account the amphoteric nature of these substances and the possibilities of their polymerization or rearrangement. The isolation of natural tetracyclines of commercial importance, chlortetracycline, oxytetracycline, tetracycline, and demeclocycline, was exhaustively reviewed by Evans (1968), who compiled dozens of cases mainly from the patent literature:

1. Adsorption on diatomaceous earth or active charcoal with subsequent chromatography or selective extraction.
2. Extraction from acid or alkaline medium. The most frequently used extraction agent is 1-butanol, owing to its suitable partition coefficient and economic availability.
3. Direct mash extraction based on solubilizing the antibiotic by acidification, precipitation of Ca^{2+} with ammonium oxalate, addition of quaternary ammonium compounds as carriers, and extraction of the metabolite with an organic solvent, usually one of the methylalkyl ketone type.
4. Precipitation (dry salt) process based on precipitating the antibiotic from dilute aqueous solutions of aryl azosulfonic acid dyes. Tetracyclines are precipitated as complexes with alkaline earth metal compounds or with primary and secondary alkyl amines.
5. Solvent extraction of the antibiotic with salt, based on salting out (NaCl) the antibiotic from the aqueous to the organic phase (1-butanol). This method is also suited for refining a crude product.

To isolate oxytetracycline, Maleninský et al. (1974) worked out a procedure combining the release of oxytetracycline into the medium by acidification, precipitation of ballast compounds with $K_4Fe(CN)_6$ and $ZnSO_4$, extraction of lipid fractions with butyl acetate, and precipitation of Ca^{2+} with oxalic acid; after adjusting the filtrate with EDTA, Na_2SO_3, and citric acid, oxytetracycline was obtained as a crystalline base.

The preparations are purified by crystallization as salts (e.g., hydrochlorides) or bases. Particularly efficient is crystallization from boiling solvents, such as lower alcohols, ketones, or aliphatic ethers of ethylene glycol, which yields nonhygroscopic preparations. The specific crystal surface depends on the temperature, stirring, and pH value. The residual amounts of antibiotic in the mother liquor are increased by oxalate and chloride anions, while sulfate anions have the opposite effect. Crystallization is most efficiently performed at 2°C for 3 hr (Linkov and Zhukovskaya, 1975a,b).

Among auxiliary operations one should mention separation of mycelium from the culture filtrate before the actual isolation of antibiotic. Its success depends on the filtration properties of the mycelium, which are determined by the microorganism, the composition of the medium, and conditions of biosynthesis (Zhukovskaya and Gorskaya, 1974).

B. Semisynthetic Tetracyclines

Methacycline [5] is prepared from oxytetracycline [2] by a three-step synthesis and may be isolated as a sulfosalicylate which is converted to a hydrochloride of methacycline by crystallization from an acid mixture of acetone and methanol in a 62% yield (Blackwood et al., (1963).

During synthesis of doxycycline [6], the reaction resulting in methacycline is extended by two further steps when CH_2 in position 6 is reduced to CH_3. Doxycycline is obtained as sulfosalicylate and is further purified by conversion to a base in a 63% yield (referred to methacycline). The analytical preparation is isolated as a hydrochloride from the base of methacycline by crystallization from acid ethanol at 0°C (Stephens et al., 1963). The configuration of the substituents in the C6 position of the tetracycline nucleus is decisive of biological activity. The 6-stereoisomer has a low efficiency. The patent literature contains two examples of the biosynthetic preparation of doxycycline by fermentation of S. rimosus CR_{1244}/UV, a variant derived from S. rimosus ATCC 10970 (Pelleia da Luz, 1974) and by biotransformation of 5- hydroxyanhydrotetracycline by strains of Streptomyces alboflavus or Streptomyces lusitanus. Conversion to doxycycline attains about 50% of the starting substrate (Villax, 1974).

Minocycline [7] is the reaction product of a five-stage synthesis, the starting substance being a natural isolate of 6-demethyltetracycline. Undesirable by-products are separated by liquid—liquid partition chromatography (Church et al., 1971).

Rolitetracycline [8] is prepared by Mannich's reaction on suspending tetracycline in tert-butyl alcohol, with an addition of pyrrolidine and formaldehyde (Gottstein et al., 1959). The reaction mixture yields rolitetracycline on cooling to 30°C in a 70% amount based on tetracycline.

C. Analytical Methods

The basic analytic assays of tetracycline antibiotics in the course of their fermentation, isolation, and purification consist in dilution, diffusion, or line tests in vitro (Reiner, 1974). Two microbiological assay methods have been described, a turbidimetric method using Micrococcus pyogenes var. aureus ATCC 6538-P, and a cylinder-plate method using Bacillus cereus var. mycoides ATCC 9634 (Grove and Randall, 1955).

Chemical assays of tetracyclines based on metal complexes utilize the characteristic absorption or fluorescence of the complex formed. Using this method, one can analyze the principal tetracyclines both in biological material and in commercial preparations (Poigner and Schatter, 1976). Polarographic assay of tetracyclines is based on their reduction in a boric acid—sodium borate buffer (Olliff and Chatten, 1977).

In the analysis of tetracyclines during the fermentation and isolation process, use can be made of paper chromatography (Blinov and Khokhlov, 1970; Wagman and Weinstein, 1973; Betina, 1975), thin-layer chromatography (Aszalos and Frost, 1975), and gel chromatography (Ragazzi and Veronese, 1977).

Recent analytical methods include high-pressure liquid chromatography of tetracyclines in both regular and reverse phase (Mack and Ashworth, 1978; Nelis and DeLeenheer, 1980; Eksborg, 1981) and on ion exchangers (Butterfield et al., 1973).

VI. MODE OF ACTION

In view of their chemical relations, all tetracyclines possess a similarly wide spectrum of antibiotic efficiency (Walter and Heilmeyer, 1969; Dürckheimer, 1975). They are active against a number of gram-positive and gram-negative bacteria, some large viruses, rickettsiae, spirochetes, and mycoplasmas. They are active against some species of *Pseudomonas*, *Proteus*, and *Salmonella*. *Mycobacteria*, protozoa, and fungi are resistant.

The study of the mechanism of action of tetracycline antibiotics shows them to block various enzymes essential for basic functions of bacterial and mammalian cells (Laskin, 1967; Sasykin, 1972). These reactions require antibiotic concentrations that are higher than bacteriostatic, and they are thus not responsible for the antibiotic effect itself.

Gale and Paine (1950) observed that the antibiotic effect at bacteriostatic concentrations of tetracyclines is produced by an inhibition of protein synthesis. The key discovery at the molecular level was the inhibition of protein synthesis directly at the ribosomes (Connamacher and Mandel, 1965; Day, 1966). One could thus assume that tetracyclines effect the translational phase of protein synthesis. These conceptions have been fully proved, for example, by Carter and McCarty (1966), Suzuka et al. (1966), and Jonák and Rychlík (1970).

At higher tetracycline concentrations the peptide bond formation catalyzed by peptidyl transferase taking place at the ribosomal complex in the elongation phase of translation becomes inhibited. Tetracyclines also block the termination of the peptide chain, which may be taken as evidence that peptidyl transferase participates in splitting off the nascent peptide from peptidyl-tRNA during termination (Capecchi and Klein, 1969).

At tetracycline concentrations higher than those inhibiting protein synthesis, cell permeability is affected (Baloun and Hudák, 1979) and their transport through the cell membrane is hindered (Ball et al., 1977). This results in a suppression of DNA replication while RNA synthesis proceeds in a normal fashion (Pato, 1977).

The development of resistance to tetracyclines (see Belousova, 1980; Chopra and Howe, 1978; Shales et al., 1980) is regulated by both chromosomal and plasmid genes, their level being increased generally on amplification of these genes. With enterobacteria, two types of plasmid-determined resistance are known to occur: with type A (where sensitivity to tetracycline but not to minocycline is affected) the accumulation of the antibiotic in cells is diminished by a decrease of its transport; with type B (where sensitivity to tetracycline and minocycline is affected) not only the uptake of the antibiotic is decreased, but also the structure of 30-S ribosomal subunits is altered. Type A resistance, connected with a modification of membrane proteins and lipids, is inducible (it is achieved through an adaptive alteration of cells rather than by selection of resistant variants from a heterogeneous population). The induction process is inhibited by chloramphenicol. In clinical practice, the symptoms of this resistance may be suppressed by administration of tetracycline in combination with membranotropic agents or chloramphenicol. Another path of research would be in developing novel tetracycline derivatives that penetrate better into cells.

In connection with the mechanism of action of tetracyclines, one should also mention their effect on the producer itself. With *S. aureofaciens*, exogenous chlortetracycline depresses respiratory activity and affects protein synthesis and the level of nucleic acids. On addition to young cultures, it suppresses even the biosynthesis of tetracycline (see Hošťálek et al., 1979). Differences in the biosynthetic activity of various strains are given, among other things, by the degree of their sensitivity to the antibiotic produced. The resistance of high-production strains is connected with a modification of the membrane and translational system and with the formation of physiologically inactive ribosomal aggregates.

With *S. rimosus* resistance to oxytetracycline is inducible and regulated by extrachromosomal DNA (Boronin and Sadovnikova, 1972; Borisoglebskaya et al., 1979).

VII. APPLICATIONS AND ECONOMY

Official lists (Anonymous, 1979a,1980a) mention pharmaceutical forms with only the four principal fermentation-produced tetracyclines (chlortetracycline, oxytetracycline, tetracycline, and demeclocycline). The principal tetracyclines have been modified to more than 3000 semisynthetic derivatives (Perlman, 1974). Wide therapeutical application exists mainly with methacycline, minocycline, rolitetracycline, and especially doxycycline.

According to a British estimate, the world consumption of antibiotics (excluding the socialist countries) in 1977 was about $5 billion, with tetracyclines, at $900 million, in third place after semisynthetic cephalosporins and penicillins. Production of tetracycline antibiotics in the United States rose from 1700 tons in 1967 to 3000 tons in 1977, the rise being 10% during the last year of the decade (Anonymous, 1978a,b).

After penicillins, tetracyclines are the most frequently applied antibiotics in human medicine in the United States, a position held to the end of the 1970s (Anonymous, 1979b). In many countries tetracyclines remain in use in technical and veterinary practice. In the case of chlortetracycline, whose use for technical purposes has been criticized, worldwide consumption (as a growth promotor of domestic animals or as premedicated fodder) still amounts to more than 2500 tons. A wide application for veterinary and technical purposes, including preservation of foods, exists in the case of oxytetracycline.

With continuing technical progress the price of tetracyclines decreases in spite of worldwide inflation, the principal substance being sold for about $25/kg. The most complicated semisynthetic doxycycline, is sold for $180–190/kg. The substance undergoes a sharp price increase on being converted to a drug form. One ampoule containing 0.1 g of the active substance (the actual price of which is about 2 cents) is sold in pharmacies for about $6.50 (Anonymous, 1979b).

To obtain a more detailed view of the production costs of tetracycline antibiotics, one can quote Heine and Heck (1977), who published a cost balance sheet of the fermentation process using 10 fermenters of 120-m^3 capacity. The raw materials for the fermentation broth for oxytetracycline, the composition of which was given in the preceding section, cost about $97.5 per cubic meter of medium, this representing about 50% of the total cost. Thus 1 liter of the fermentation broth costs about 19.5 cents, which is equal to or less than the cost of the traditional fermentation product, that is, beer.

At a rate of biosynthesis equal to 110–120 µg/ml per hour,* the cost of 1 g of oxytetracycline contained in the fermentation broth is about 1 cent or, as pure isolated substance, about 2 cents, this including the losses in yield during processing and the treatment of the fermented broth.

REFERENCES

Alačević, M., Strašek-Vešligaj, M., and Sermonti, G. (1973). The circular linkage map of *Streptomyces rimosus*. *J. Gen. Microbiol.* 77:173–185.

Alačević, M., Pigac, J., and Vešligaj, M. (1976). Mapping of morphological and otc genes in *Streptomyces rimosus*. In *Abstracts of the Vth International Fermentation Symposium*. H. Dellweg (Ed.). Berlin, p. 195.

Alačević, M., Hranueli, B., and Pigac, J. (1980). Analysis of progeny obtained by protoplast fusion and conventional crosses in *Streptomycetes*. In *Abstracts of the VIth International Fermentation Symposium*. London, p. 24.

Anonymous. (1978a). Speciality organics: Output mostly higher. *C & EN* 56:49.

Anonymous. (1978b). Most successful groups of antibiotics. Scrip 324:19.

Anonymous. (1979a). Tetracycline. In *American Drug Index*. S. M. Billups (Ed.). Lippincott, Philadelphia, pp. 626–627.

Anonymous. (1979b). Antibiotika/Tetracycline. B8. *Rote Liste 1979*. Editio Cantor, Aulendorf/Württ., pp. 10161B–10207B.

Anonymous. (1980a). Tetracyklin. In *FASS 1980 Farmacevtiska specialiteter i Sverige*. Läkemedelsinformation AB, Stockholm, pp. 694–695.

Anonymous. (1980b). Research report. Research Institute of Antibiotics and Biotransformations, Roztoky near Prague, Czechoslovakia.

Aszalos, A., and Frost, D. (1975). Thin-layer chromatography of antibiotics. In *Methods in Enzymology*, Vol. 43. H. Hash (Ed.). Academic, New York, pp. 186–201 and references cited therein.

Ball, P. R., Chopra, J., and Eccles, S. J. (1977). Accumulation of tetracyclines by *Escherichia coli* K-12. *Biochem. Biophys. Res. Commun.* 77:1500–1507.

Baloun, J., and Hudák, J. (1979). Nuclear degeneration induced by chlortetracycline. *Experientia* 35:201–202.

Běhal, V., Podojil, M., Hošťálek, A., Vaněk, Z., and Lynen, F. (1974). Regulation of biosynthesis of excessive metabolites. XVI. Origin of the terminal group of tetracyclines. *Folia Microbiol.* 19:146–150.

Běhal, V., Jechová, V., Vaněk, Z., and Hošťálek, Z. (1977). Alternate pathways of malonylCoA formation in *Streptomyces aureofaciens*. *Phytochemistry* 16:347–350.

Běhal, V., Hošťálek, Z., and Vaněk, Z. (1979). Anhydrotetracycline oxygenase in *Streptomyces aureofaciens*. *Biotechnol. Lett.* 1:177–182.

Běhal, V., Grégrová-Prušáková, J., and Hošťálek, Z. (1982). Effect of inorganic phosphate and benzyl thiocyanate on the activity of anhydrotetracycline oxygenase in *Streptomyces aureofaciens*. *Folia Microbiol.* 27:102–106.

---

*Offer of licence (1979 of KRKA Farmaceutika Co., 68000 Novo Mesto, Yugoslavia.

Belousova, I. I. (1980). Tetracycline resistance of microorganisms. *Antibiotiki 25*:777–793.

Betina, V. (1975). Paper chromatography of antibiotics. In *Methods in Enzymology*, Vol. 43. J. H. Hash (Ed.). Academic, New York, pp. 100–172. and references cited therein.

Blackwood, R. K., Beereboom, J. J., Rennhard, H. H., Schach von Wittenau, M., and Stephens, C. R. (1963). 6-Methylenetetracyclines. III. Preparation and properties. *J. Am. Chem. Soc. 85*:3943–3953.

Blinov, N. O., and Khokhlov, A. S. (1970). Tetracyclines. In *Paper Chromatography of Antibiotics*. L. K. Sokolova (Ed.). Nauka, Moscow, pp. 130–137.

Blumauerová, M., Hošťálek, Z., and Vaněk, Z. (1972). Biosynthesis of tetracyclines: Problems and perspectives of genetic analysis. In *Fermentation Technology Today*. G. Terui (Ed.). Society for Fermentation Technology, Osaka, Japan, pp. 223–232.

Borisoglebskaya, A. N., Perebityuk, A. N., and Boronin, A. M. (1979). Study on resistance of *Actinomyces rimosus* to oxytetracycline. *Antibiotiki 24*:883–888.

Boronin, A. M., and Sadovnikova, I. G. (1972). Elimination by acridine dyes of oxytetracycline resistance in *Actinomyces rimosus*. *Genetika (Moscow) 8*:174–176.

Brunner, R. (1963). Tetracycline. In *Die Antibiotica*, Vol. 1. R. Brunner and G. Machek (Eds.). Verlag Hans Carl, Nürnberg, pp. 311–623.

Butterfield, A. G., Hughes, D. W., Pound, N. Y., and Wilson, W. L. (1973). Separation and detection of tetracyclines by high-speed liquid chromatography. *Antimicrob. Agents Chemother. 7*:11–15.

Capecchi, M. R., and Klein, H. A. (1969). Characterization of three proteins involved in polypeptide chain termination. *Cold Spring Harbor Symp. Quant. Biol. 34*:469–477.

Carter, W., and McCarty, K. S. (1966). Molecular mechanism of antibiotic action. *Ann. Intern. Med. 64*:1087–1113.

Chemical Abstracts Index Guide (1977–1981). *Chem. Abstr.* 10th Coll. Index 86–95:245G, 813G, 917G, 1105G, 1243G.

Chopra, I., and Howe, T. G. B. (1978). Bacterial resistance to the tetracyclines. *Microbiol. Rev. 42*:707–724.

Church, R. F. R., Schaub, R. E., and Weiss, M. J. (1971). Synthesis of 7-dimethylamino-6-deoxytetracycline (Minocycline) via 9-nitro-6-demethyl-6-deoxytetracycline. *J. Org. Chem. 36*:723–725.

Connamacher, R. H., and Mandel, H. G. (1965). Binding of tetracycline to the 30S ribosomes and to polyuridylic acid. *Biochem. Biophys. Res. Commun. 20*:98–103.

Day, L. E. (1966). Tetracycline inhibition of cell-free protein synthesis. I. Binding of tetracycline to components of the system. *J. Bacteriol. 91*: 1917–1923.

DiMarco, A. (1956). Metabolism of *Streptomyces aureofaciens* and biosynthesis of chlortetracycline. *G. Microbiol. 2*:285–300.

DiMarco, A., and Pennella, P. (1959). The fermentation of the tetracyclines. In *Progress in Industrial Microbiology*, Vol. 1. D. J. D. Hockenhull (Ed.). Heywood, London, pp. 45–92.

Dürckheimer, W. (1975). Tetracyclines: Chemistry, biochemistry, and structure–activity relations. *Angew. Chem. 14*:721–734.

Eksborg, S. (1981). Reversed-phase ion-pair chromatography of tetracyclines on a LiChrosorb NH_2 column. *J. Chromatogr. 208*:78–82.

Evans, R. C. (1968). The technology of the tetracyclines. Quadrangle, New York, pp. i, 60–136, 168–201, 324–398.

Friend, E. J., and Hopwood, D. A. (1971). The linkage map of *Streptomyces rimosus. J. Gen. Microbiol. 68*:187–197.

Friend, E. J., Warren, M., and Hopwood, D. A. (1978). Genetic evidence for a plasmid controlling fertility in an industrial strain of *Streptomyces rimosus. J. Gen. Microbiol. 106*:201–206.

Gale, E. F., and Paine, T. F. (1950). Effect of inhibitors and antibiotics on glutamic acid accumulation and on protein synthesis in *Staphylococcus aureus. Proc. Biochem. Soc. 47*:26.

Godány, A., Zelinková, E., Zelinka, J., and Štokrová, J. (1980). Plasmid DNA from *Streptomyces aureofaciens. Bull. Czech. Biochem. Commun. 8*:18.

Gottstein, W. Y., Minor, W. F., and Cheney, L. C. (1959). Carboxamido derivatives of the tetracyclines. *J. Am. Chem. Soc. 81*:1198–1201.

Growe, D. C., and Randall, W. A. (1955). Assay methods of antibiotics. A laboratory manual. In *Antibiotics Monographs*, No. 2. H. Welch and F. Ibáñez (Eds.). Medical Encyclopedia, New York, pp. 48–65.

Growich Jr., J. A., and Deduck, N. (1963). Tetracycline fermentation. U.S. Patent No. 3,092,556.

Heine, H., and Heck, G. (1977). Wirtschaftliche Betrachtungen zur Anwendung von O_2 bei der Antibiotikafermentation. In *DECHEMA Monogr. 81*:217.

Hlavka, J. J., and Boothe, J. H. (1973). Tetracyclines. *Fortschr. Arzneimittelforsch. 17*:210–240.

Hošťálek, Z., and Vaněk, Z. (in press). Biosynthesis of tetracyclines. In *Handbook of Experimental Pharmacology*. J. H. Booth and J. T. Hlavka (Eds.). Springer-Verlag.

Hošťálek, Z., Blumauerová, M., and Vaněk, Z. (1974). Genetic problems of the biosynthesis of tetracycline antibiotics. In *Advances in Biochemical Engineering*, Vol. 3. T. K. Ghose, A. Fiechter, and N. Blakebrough (Eds.). Springer-Verlag, Berlin, pp. 13–67

Hošťálek, Z., Blumauerová, M., and Vaněk, Z. (1979). Tetracycline antibiotics. In *Economic Microbiology*, Vol. 3. A. H. Rose (Ed.). Academic, London, pp. 293–354.

Jonák, J., and Rychlík, I. (1970). Role of messenger RNA in binding of peptidyl transfer RNA to 30-S and 50-S ribosomal subunits. *Biochim. Biophys. Acta 199*:421–434.

Laskin, A. J. (1967). Tetracyclines. In *Antibiotics*, Vol. 1. D. Gottlieb and P. D. Shaw (Eds.). Springer-Verlag, Berlin, pp. 331–359, 752–754.

Linkov, G. I., and Zhukovskaya, S. A. (1975a). Effect of certain factors on crystallization of tetracyclines. *Antibiotiki 20*:423–426.

Linkov, G. I., and Zhukovskaya, S. A. (1975b). Effect of the conditions for oxytetracycline crystallization on the indexes of the process. *Antibiotiki 20*:591–595.

McCormick, J. R. D. (1965). Biosynthesis of tetracyclines. In *Biogenesis of Antibiotic Substances*. Z. Vaněk and Z. Hošťálek (Eds.). Publishing House of the Czechoslovak Academy of Sciences, Prague, pp. 73–92.

McCormick, J. R. D. (1967). Tetracyclines. In *Antibiotics*, Vol. 2. D. Gottlieb and P. D. Shaw (Eds.). Springer-Verlag, Berlin, pp. 113–122.

McCormick, J. R. D. (1969). Point-blocked mutants and the biogenesis of tetracyclines. In *Genetics and Breeding of Streptomycetes*. G. Sermonti, and M. Alačević (Eds.). Yugoslav Academy of Science and Arts, Zagreb, pp. 163–176.

Mack, G. D., and Ashworth, R. B. (1978). A high performance liquid chromatographic system for the analysis of tetracycline drug standards, analogs, degradation, products and other impurities. *J. Chromatogr. Sci. 16*: 93–101.

Maleninský, S. Čulík, K., and Hilbert, O. (1974). Isolation of exytetracycline. Czechoslovak Patent No. 152,853.

Matelová, O., Musílková, M., Nečásek, J., and Šmejkal, F. (1955). The influence of the interruption of oxygen supply on the production of chlorotetracycline. *Preslia 27*:27–34.

Miller, P. A., and Hash, J. H. (1975). Tetracyclines. In *Methods in Enzymology*, Vol. 43. J. H. Hash (Ed.). Academic, New York, pp. 603–607.

Miller, P. A., Sjolander, N. O., Nalesnyk, S., Arnold, N., Johnson, S., Doerschuk, A. P., and McCormick, J. R. D. (1960). Cosynthetic factor I, a factor involved in hydrogen-transfer in *Streptomyces aureofaciens*. *J. Am. Chem. Soc. 82*:5002–5003.

Miller, P. A., Hash, J. H., Lincks, M., and Bohonos, N. (1965). Biosynthesis of 5-hydroxytetracycline. *Biochem. Biophys. Res. Commun. 18*:325–331.

Mitscher, L. A. (1968). Biosynthesis of tetracycline antibiotics. *J. Pharm. Sci. 57*:1633–1649.

Müller, R., and Kieslich. (1966). Technologie der Darstellung organischer Substanzen mit Mikroorganismen. *Chem. Ing. Technol. 38*:813–825.

Nelis, H. J. C. F., and DeLeenheer, A. P. (1980). Retention mechanisms of tetracyclines on a C_8 reversed-phase material. *J. Chromatogr. 195*:35–42.

Oldshue, J. Y., Coyle, C. K., Brügger, K., and Zemke, H. J. (1977). Scale-Up-Problematik bei begasten Rührgefässen. In *DECHEMA Monogr. 81*: 115–121.

Olliff, C. J., and Chatten, L. G. (1977). Polarography for tetracycline analysis. *J. Pharm. Sci. 66*:1564–1566.

Pato, M. L. (1977). Tetracycline inhibits propagation of deoxyribonucleic acid replication and alters membrane properties. *Antimicrob. Agents Chemother. 11*:318–323.

Pecák, V., Čížek, S., Musil, J., Čerkes, L., Herold, M., Bělík, E., and Hoffman, J. (1959). Process for enhancement of tetracycline antibiotics production. Czechoslovak Patent No. 90,962.

Pelleia da Luz, A. (1974). Verfahren Zur Herstellung von 6-deoxy-5-hydroxytetrazyklin. Japanese Patent No. 50-142.788.

Perlman, D. (1974). Evolution of the antibiotics industry 1940–1975. *ASM News 40*:910–916.

Pigac, J., Hranueli, D., Smokvina, T., and Alačević, M. (1982). Optimal cultural and physiological conditions for handling *Streptomyces rimosus* protoplasts. *Appl. Environ. Microbiol. 44*:1178–1186.

Podojil, M., Vaněk, Z., Běhal, V., and Blumauerová, M. (1973). Regulation of biosynthesis of excessive metabolites. XIV. Incorporation of (U-14C)-asparagine into the molecule of tetracycline. *Folia Microbiol. 18*:415–417.

Podojil, M., Vaněk, Z., Přikrylová, V., and Blumauerová, M. (1978). Isolation of ekatetrone, a new metabolite of producing variants of *Streptomyces aureofaciens*. *J. Antibiot. 31*:850–854.

Poigner, H., and Schatter, C. (1976). Fluorometric determination of tetracyclines in biological materials. *Analyst 101*:808–814.

Přikrylová, V., Podojil, M., Sedmera, P., Vokoun, J., Vaněk, Z., and Hassall, C. H. (1978). The structure of ekatetrone, a metabolite of strains of *Streptomyces aureofaciens*. *J. Antibiot. 31*:855–862.

Queener, S., and Schwartz, R. (1979). Penicillins: Biosynthetic and semi-synthetic. In *Economic Microbiology*, Vol. 3. A. H. Rose (Ed.). Academic, London, pp. 35–122.

Ragazzi, E., and Veronese, G. (1977). Gel chromatography of tetracycline antibiotics. *J. Chromatogr. 134*:223–229.

Reiner, R. (1974). *Antibiotica und ausgewählte Chemotherapeutica*. Thieme Verlag, Stuttgart, pp. 11–16, 150–152.

Rhodes, P. M., Winskill, N., Friend, E. J., and Warren, M. (1981). Biochemical and genetic characterization of Streptomyces rimosus mutants impaired in oxytetracycline biosynthesis. *J. Gen. Microbiol. 124*:329–338.

Ryu, D. Y., and Hospodka, J. (1980). Quantitative physiology of *Penicillium chrysogenum* in penicillin fermentation. *Biotechnol. Bioeng. 22*:289–298.

Ryu, D. Y., and Humphrey, A. E. (1972). A reassessment of oxygen-transfer rates in antibiotics fermentations. *J. Ferment. Technol. 50*:424–431.

Sasykin, Ju.O. (1972). Die Tetracycline und ihr Einfluss auf die Eiweisssynthese und auf andere biochemische Prozesse bei Mikroorganismen und in tierischen Zellen. In *Antibiotika als Inhibitoren biochemischer Prozesse*. P. Westerman (Ed.). Akademie Verlag, Berlin, pp. 196–225.

Shales, S. W., Chopra, I., and Ball, P. R. (1980). Evidence for more than one mechanism of plasmid-determined tetracycline resistance in *Escherchia coli*. *J. Gen. Microbiol. 121*:221–229.

Stephens, C. R., Beereboom, J. J., Rennhard, H. H., Gordon, P. N., Murai, K., Blackwood, R. K., and Schach von Wittenau, M. (1963). 6-Deoxytetracyclines. IV. Preparation, C-6-stereochemistry, and reactions. *J. Am. Chem. Soc. 85*:2643–2652.

Stepnov, V. P., Garaev, M. M., Fedotov, A. R., and Golub, E. I. (1978). Plasmids in *Actinomycetes* producing oxytetracycline and neomycin. *Antibiotiki 23*:892–895.

Suzuka, I., Kaji, H., and Kaji, A. (1966). Binding of specific RNA to 30S ribosomal subunits: Effect of 50S ribosomal subunits. *Proc. Nat. Acad. Sci. U.S.A. 55*:1483–1490.

Szumski, S. A. (1964). Fermentative preparation of tetracycline and 7-chlortetracycline. U.S. Patent No. 3,121,670.

Vaněk, Z., Cudlín, J., Blumauerová, M., and Hošťálek, Z. (1971). How many genes are required for the synthesis of chlortetracycline? *Folia Microbiol. 16*:225–240.

Villax, I. (1974). Procédé de transformation biologique des anhydrotétracyclines en 6-désoxytétracyclines. French Patent No. 2,187,301.

Wagman, G. H., and Weinstein, M. J. (1973). Tetracyclines. In *Chromatography of Antibiotics*. Elsevier, Amsterdam, pp. 181–190.

Walter, A. M., and Heilmeyer, L. (1969). Tetracycline. In *Antibiotika-Fibel*. Thieme Verlag, Stuttgart, pp. 260–297.

Wise, N. S. (1950). The measurement of the aeration of biological culture media. *J. Gen. Microbiol. 4*:xv.

Zhukovskaya, S. A., and Gorskaya, S. V. (1974). Effect of fermentation conditions on the filtration characteristics of fermentation broths. *Biotechnol. Bioeng. Symp. 4*:925–942.

9

THE RIFAMYCINS: PROPERTIES, BIOSYNTHESIS, AND FERMENTATION

ORESTE GHISALBA, JOHN A. L. AUDEN, THOMAS SCHUPP, AND JAKOB NÜESCH *Central Research Laboratories and Pharmaceutical Research Department, Ciba-Geigy Ltd., Basel, Switzerland*

We dedicate this chapter to Professor Piero Sensi, the true father of the rifamycins.

I. INTRODUCTION

The discovery of the rifamycins produced by *Streptomyces mediterranei,* as the organism was then called, by Sensi et al. was first announced in 1959. A comprehensive review on the production of the rifamycins was last written by Sensi and Thiemann (1967). In it, the early history of the rifamycins, their discovery, fermentation, and isolation are excellently presented. However, much has been written since then on the ansamycins in general and on the rifamycins in particular. The majority of these publications has been concerned with the discovery of new rifamycins and rifamycin biosynthesis.

Hartmann et al. (1967) were the first to discover that the mode of action of rifamycins was to inhibit specifically bacterial RNA polymerase and that this occurred in the presence of very low concentrations of the antibiotic. At about the same time, after an intensive cooperative effort between Ciba-Geigy and Lepetit, the new semisynthetic oral rifamycin rifampicin was introduced. Rifampicin, which was in fact isolated in the Lepetit laboratories, is a rifamycin SV derivative with a (4-methylpiperazinyl) iminomethyl group in the 3-position and was synthesized from rifamycin SV via 3-formylrifamycin SV (Fig. 1). The synthesis of an orally active rifamycin, whereas previously only the injectable rifamycin SV had been available, gave great impetus to the work on the improvement of the fermentation, recovery, and synthetic processes involved in its manufacture. It presented the first opportunity to treat tuberculosis with a relatively nontoxic oral antibiotic, albeit in combination with other drugs.

As is so often the case, good fortune played a role in the early development of the rifamycin B fermentation. If rifamycin B, which has only a slight antibiotic activity, had not been readily degraded to the very active derivative S, it might well have slipped through the Lepetit antibiotic screening process without being noticed. Sensi et al. (1960), working with the wild-type strain, reported that in fermentation media lacking barbiturate a number of rifamycins are produced, usually A, B, C, D, and E; but B, which is the most stable, was only present in small amounts. The very early discovery by Margalith and Pagani (1961), also working with the wild type, that the presence of sodium barbiturate in the fermentation medium resulted in the almost exclusive production of rifamycin B, paved the way for the development of the rifamycin fermentation process. One might question why barbiturate was selected for this purpose at all. A providential publication by Ferguson

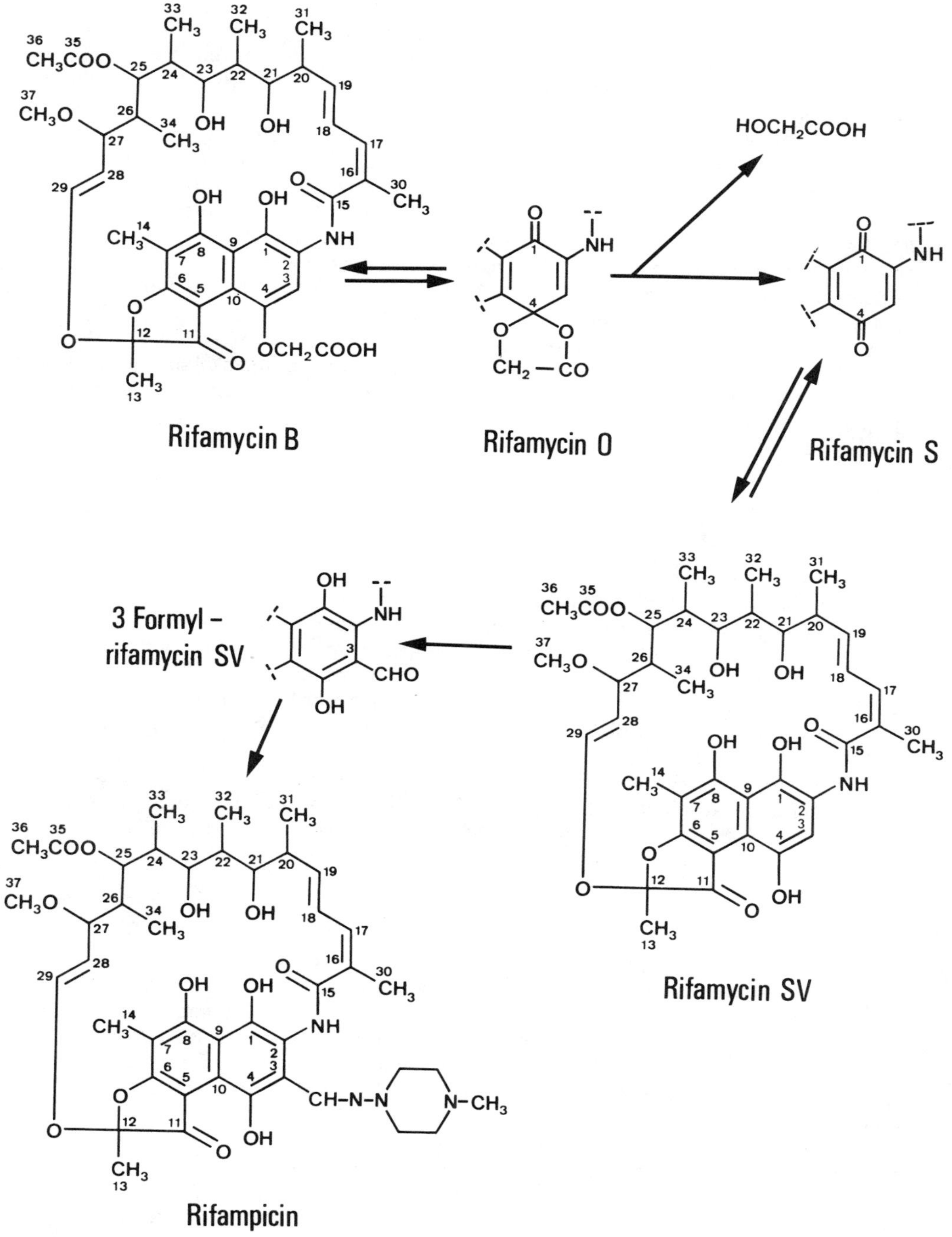

Figure 1 The structural formulas of the most important rifamycins.

284 / Ghisalba, Auden, Schupp, and Nüesch

et al. (1957), in which they showed that streptomycin production by *Streptomyces griseus* in synthetic medium is greatly stimulated in the presence of barbital, might have been crucial. Four successive phage infections, which occurred during the early stages of the strain improvement program, although they caused a considerable shock at the time, resulted in an effective yield increase of 100% in a very short period. Be all that as it may, the discoverers and those responsible for the early developmental work deserved a few strokes of good fortune and they certainly knew how to take full advantage of them.

It is impossible to say in how many firms or in which countries rifamycins are being produced at present, and for this reason it is difficult to estimate the volumes of sales worldwide. We are unfortunately obliged to present strain and fermentation process development somewhat superficially, because much of the information is confidential. In any case very little has been published in this field since 1967. The process for the manufacture of rifamycin SV from rifamycin B has already been described in detail before (Sensi and Thiemann, 1967). On the other hand, we shall present the biosynthesis of the rifamycins in greater detail.

II. STRAIN DEVELOPMENT

A. The Production Strain

Originally the production strain was known as *Streptomyces mediterranei*, after Margalith and Beretta (1960). Subsequently, as a result of comparative cell wall analysis of the strain with typical nocardias, it was proposed that it should be reclassified as a Nocardia (Thiemann et al., 1969). Very recently more exhaustive cell wall analysis indicated that the strain has still not found its final systematic niche, although a new generic name has not yet been proposed (Alderson et al., 1979). Nowadays the strain is known as *Nocardia mediterranei*.

The current production strains are maintained in the form of freeze-dried mycelium of selected high-producing colonies in sterile, sealed glass ampoules. These ampoules can be kept at 4°C almost indefinitely without strain degeneration or viability loss.

B. Strain Improvement

The strategy of strain improvement in rifamycin fermentation has tended to change according to current production necessities. The earliest yield improvement came providentially through phage infection. Subsequent yield improvements were achieved by mutagenic treatment, usually with nitrosoguanidine or ultraviolet radiation, as can be seen from the family tree (Fig. 2). A recurrent difficulty has been that higher-producing mutants have tended to produce an extremely viscous broth. The problems of adequate mixing and oxygen supply in the bioreactor could often only be solved by selection of strains with a nocardial morphology. Problems such as low productivity, irregular supplies of raw materials, and cooling difficulties may be solved by a variety of shake-flask screening programs, all of which have been used from time to time (Table 1).

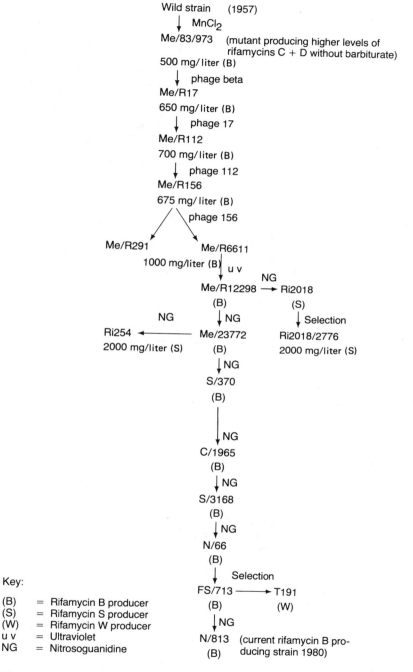

Figure 2 Family tree of *N. mediterranei* (strain improvement 1957–1980).

Table 1 Selective Mutant Screening in Shake Flasks

Problem	Effect on production	Possible screening technique
High O$_2$ demand (inherent or due to viscosity)	High power input requirement (perhaps limiting)	Lower rate of shaking or increased medium concentration Higher rate of shaking or lower medium concentration Select for less viscous growth
Long lag phase Long fermentation time	Low productivity	Select for high productivity by testing at two different times and calculating the productivity rate
Raw material costs and availability, particularly organic nitrogen sources	Production bottleneck	Select for mutant able to produce effectively with several different organic nitrogen sources
Fermentation temperature	Low temperature requirement produces high cooling costs	Select for more thermophilic strains

III. GENETICS

A. Mutation: Applications and Methods

The application of genetics plays an important role in the quantitative and qualitative aspects of rifamycin synthesis. Mutation and selection have contributed greatly to the improvement of strains for industrial rifamycin production. In addition, mutants blocked in the rifamycin biosynthesis have been successfully used for the detection of new rifamycins and elucidation of rifamycin biosynthesis.

For induction of mutants of *N. mediterranei*, two methods have been applied, namely, exposure to ultraviolet light or treatment with N-methyl-N'-nitro-N-nitrosoguanidine (NTG). Ultraviolet irradiation is carried out by exposing mycelial fragments suspended in water to a dose of 600–800 ergs/mm^2, giving a survival of 0.1–0.8% (Schupp, 1973). Treatment with NTG is performed under the conditions recommended by Delić et al. (1970), with 1 mg of NTG per milliliter in 0.05 M tris (hydroxymethyl) aminoethane hydrochloride–maleic acid buffer (pH 9.0) and a 60–120 min incubation period.

B. Recombination and Linkage Map

Genetic recombination can also be used for creating new genotypes of *N. mediterranei*. Recombination analysis together with the knowledge acquired from a linkage map offers a tool for determining the arrangement of genes involved in rifamycin biosynthesis. It has been shown that genetic recombination occurs when two marked strains of *N. mediterranei* are grown in mixed culture (Schupp, 1973). For the gene transfer, contact between viable cells is required, which indicates that in *N. mediterranei*, as in many other actinomycetes, recombination results from a process of conjugation. The crosses yield about one recombinant per 10,000 cells of parental genotypes and 40–80% of the recombinants behave as haploid genotypes. In addition, a variable number of recombinant colonies are found which show a mixture of phenotypes. These colonies very probably arise from multinucleate plating units, which is not unexpected, since the plating units are actually fragments of mycelium and not spores.

For the construction of the linkage map of *N. mediterranei* haploid recombinants were analyzed. The linkage analysis applied is based on the analysis of the segregation of nonselected markers in samples of progeny selected in different ways. This method allows efficient mapping, since all chromosomal markers show linkage. The mapping data could be combined to give a unique sequence of auxotrophic markers on a circular linkage map (see Fig. 3) (Schupp et al., 1975). In order to obtain basic information on the genetic control of rifamycin biosynthesis, crosses involving blocked mutants have been performed (Schupp and Nüesch, 1979). It has been possible to show that the final three steps in the rifamycin B biosynthesis are chromosomally determined and that these genes are closely linked to each other (Fig. 3). Preliminary linkage analysis with the two mutants A8 (transketolase⁻) and A10 (aro⁻) show that these two genes, which form a connection between primary metabolism (shikimate pathway) and rifamycin biosynthesis (see Fig. 3), are not closely linked to the three mapped genes of rifamycin biosynthesis.

C. Plasmids

Linkage analysis clearly shows that the mutations which give rise to blocks in the final steps of rifamycin synthesis are not located on a plasmid. Experiments with the curing agents acridine orange and ethidium bromide or cultivation at elevated temperature (37°C) also indicate that plasmids do not play a significant role in rifamycin synthesis. These experiments have shown

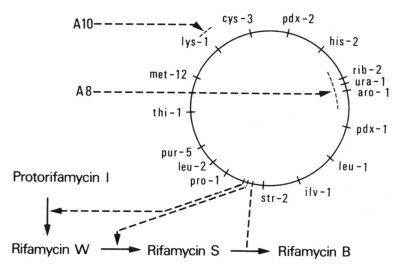

Figure 3 Genetic map of *N. mediterranei* ATCC 13685.

that the occurrence of low-producing variants is not influenced by the curing treatment. As far as we are aware, no plasmid has yet been isolated from *N. mediterranei* ATCC 13685 and its mutants. In our laboratory the two methods we used for plasmid detection gave negative results. The first was the separation of plasmid DNA by a CsCl–ethidium bromide gradient in the ultracentrifuge, and the second, selective alkaline denaturation and analysis by agarose gel electrophoresis (Birnboim and Doly, 1979).

D. Applications of Genetic Recombination

It could be shown by Schupp et al. (1981) that it is possible to obtain strains that produce new rifamycins by genetic recombination. The parental strains used were mutant strains derived from one ancestral strain, *N. mediterranei* ATCC 13685. Figure 4 shows the family tree of the strains used. One parental strain is blocked in rifamycin B biosynthesis; it is a rifamycin W-producing mutant of the high rifamycin B producer FS713. The other parental strain T104 is itself a recombinant strain which was selected for its properties as a multimarked strain with adequate rifamycin B production. The isolated recombinant strain R21 produces seven new rifamycins (Fig. 5) and the known rifamycins B, O, S, W, and P. Two possible explanations for this increased biosynthetic capacity generated by the recombination process can be postulated. Gene rearrangement may activate "silent genes," for example, by removal of efficient repression or by joining to an active promotor; or the intracellular concentration of rifamycin precursors could be significantly increased by recmbination involving mutants blocked in rifamycin B biosynthesis. The increased concentration of such a precursor can then give rise to side reactions leading to new products which are not detectable in the parental strains with a low concentration of the precursor.

A second possible application of genetic recombination is for the improvement in antibiotic yield. The principal and most frequently applied method

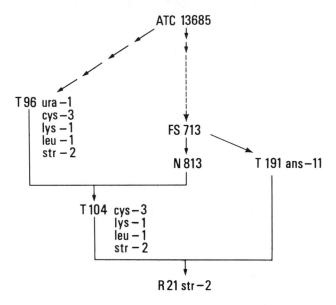

Figure 4 Family tree of the strain *N. mediterranei* R21.

for improvement of antibiotic-producing strains is still random mutation and careful selection. It seems probable that this kind of selection by mutation will continue to be the method of choice for obtaining improved strains in the shortest possible time, especially with strains that have not already been highly developed genetically. We believe that genetic recombination can also contribute to strain development, particularly when the strains have already reached a high production level. Figure 6 shows the results of a cross be-

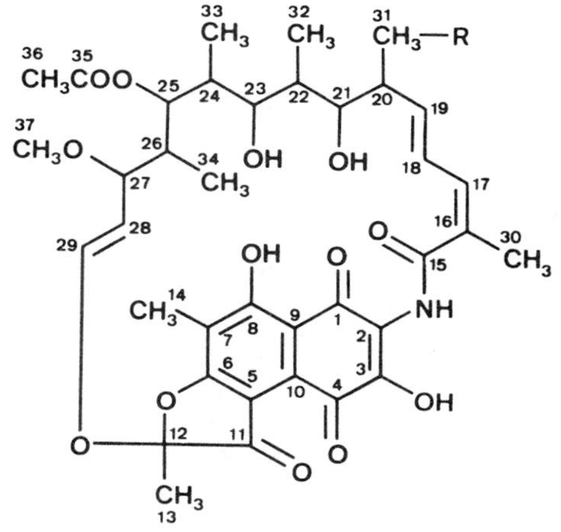

R=H: 3-Hydroxyrifamycin S
R=OH: 3,31-Dihydroxyrifamycin S

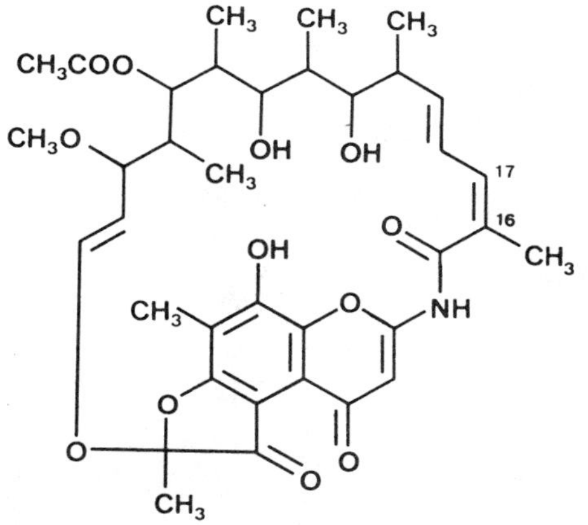

16,17-Dehydrorifamycin G

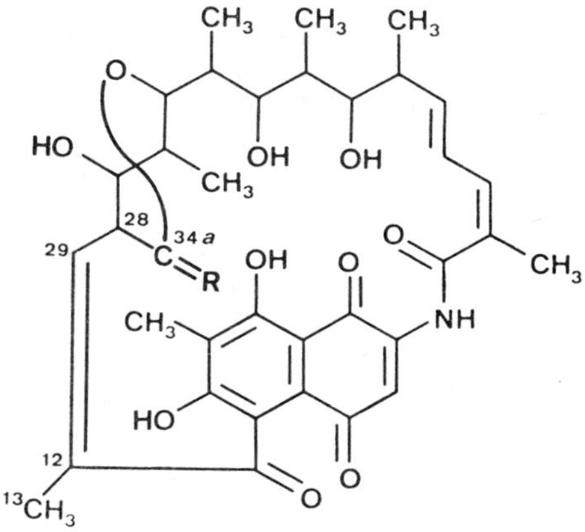

R=O: Rifamycin W-lactone
R=H,OH: Rifamycin W-hemiacetal

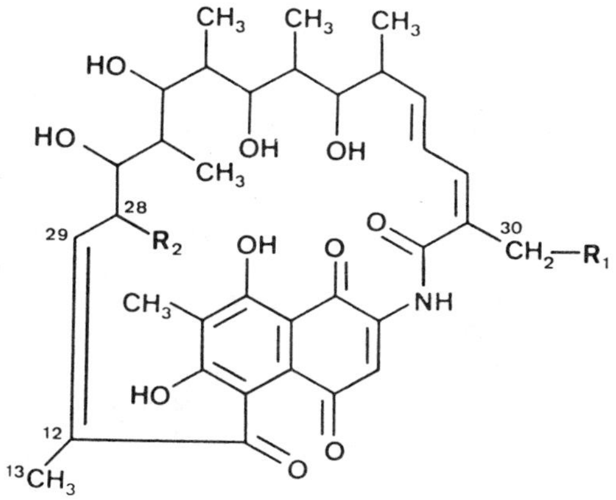

R₁=OH R₂=CH₂OH: 30-Hydroxyrifamycin W
R₁=OH R₂=OH: 28-Dehydroxymethyl-
                28,30 dihydroxyrifamycin W
R₁=H   R₂=CH₂OH: Rifamycin W

Figure 5 New rifamycins produced by the recombinant strain R21.

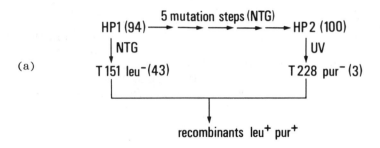

(a)

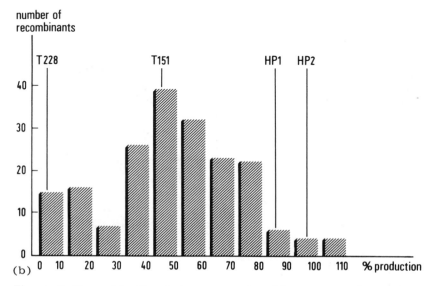

(b)

Figure 6 Recombination between auxotrophic mutants of two high-producing strains of *N. mediterranei*. (a) Relationship of the strains. Numbers in parentheses give % productivity. (b) Productivity of the isolated recombinants. Parental strains are given as references.

tween two mutants of high-producing *N. mediterranei* strains. It can be seen that this method is a possible approach for strain improvement. The very low productivity due to the auxotrophy of the parents need not be a problem, since the auxotrophic markers are selected out after the cross.

After the recombination has been made, a careful selection of the higher producers must be performed in the same way as after a routine treatment with a mutagen, and the stability of recombinant strains always has to be carefully checked.

IV. BIOSYNTHESIS AND NEW RIFAMYCINS

With the present available biosynthetic information we are able to depict a general pathway for the biosynthesis of all the known ansamycins. Biosynthetic data are available for different groups of ansamycins and related antibiotics. These data show clearly that one general pathway is operating which, in the later biosynthetic steps, splits off into several branches leading to the different ansamycin types. For this reason we will not deal with rifamycin biosynthesis as an isolated pathway, although our own investigations were restricted to the rifamycins.

A. Groups of Ansamycins

1. Rifamycins from Nocardia Mediterranei

The first rifamycin complex containing the rifamycins A, B, C, D, and E (Fig. 7) was isolated by Sensi et al. (1959). The rifamycins A, C, D, and E are very unstable and were therefore never obtained in high purity and their structures have not been elucidated (Sensi et al., 1960); however, they were reported to be biogenetically derived from rifamycin S (White et al., 1975).

Rifamycin B shows good stability and is also biogenetically derived from rifamycin S (Lancini and Sensi, 1967; Lancini et al., 1969). It can be degraded by chemical oxidation to rifamycin O and subsequent hydrolysis to rifamycin S (Fig. 1). This degradation also occurs spontaneously and is responsible for the so-called aging of rifamycin B solutions, which show increased biological activity on standing (Sensi et al., 1961).

Rifamycin SV is obtained by reduction of rifamycin S, but can also be isolated from mutant strains of N. mediterranei (Lancini and Hengeller, 1969; Birner et al., 1972).

The structures of the rifamycins B, O, S, and SV were elucidated by means of spectroscopy and chemical degradation (Prelog, 1963a,b; Oppolzer et al., 1964; Leitich et al., 1964; Oppolzer and Prelog, 1973), as well as by x-ray crystallography (Brufani et al., 1964).

Rifamycin L was only found as a transformation product in transformation experiments with rifamycin S or SV and washed mycelium of N. mediterranei (Lancini et al., 1969). It has never been isolated directly from fermentation broth. Perhaps rifamycin L should be interpreted as a transformation artifact (Lancini and White, 1976).

Rifamycin G is an inactive component of the original rifamycin complex and it was shown to be a metabolite of rifamycin S (Lancini and Sartori, 1976). C1 of the rifamycin chromophore is lost in the transformation reaction starting from rifamycin S and then replaced by oxygen. A possible intermediate in this transformation, 16,17-dehydrorifamycin G, has quite recently been isolated from a recombinant strain of N. mediterranei (Schupp et al., 1981; Traxler et al., 1981). From a mutant strain of N. mediterranei unable to transform rifamycin SV into rifamycin B, the rifamycins P, Q, R, and Verde were isolated (White et al., 1975; Lancini and Parenti, 1978; Martinelli et al., 1978; Cricchio et al., 1980) and shown to be derived from rifamycin S. The inactive rifamycin Y is an oxidation product of rifamycin B. C20 in this case bears an additional hydroxyl group and the oxygen function at C21 is oxidized to a keto group (Lancini et al., 1967). The Y-type rifamycins derived from rifamycin S, SV, and O have also been described (Leitich et al., 1967).

As is depicted in Figure 7, rifamycin S is a key intermediate for a variety of transformation reactions leading to the rifamycins finally excreted by N. mediterranei. The pathway leading to rifamycin S and some details about the transformations of rifamycin S into the later rifamycins will be discussed in the ensuing sections of this review.

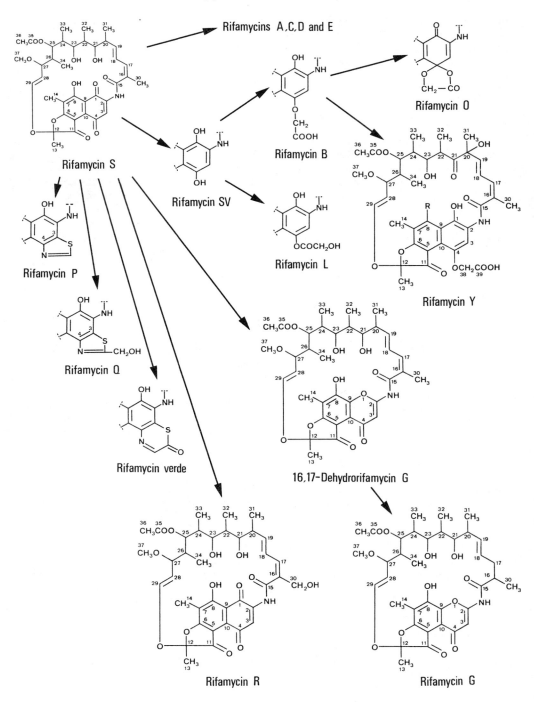

Figure 7 Rifamycins from *N. mediterranei*, with rifamycin S as a key intermediate.

2. Rifamycins and Rifamycin Derivatives from Other Actinomycetes

Rifamycins are not only produced by *N. mediterranei* but also by other micro-organisms of the order *Actinomycetales* (Fig. 8):

> *Streptomyces tolypophorus* produces tolypomycin O, rifamycin B, and rifamycin O (Kishi et al., 1972).
>
> *Micromonospora lacustris* sp. produces rifamycin S, rifamycin SV, 3-thiomethylrifamycin S, and 3-thiomethylrifamycin SV (Celmer et al., 1975).
>
> *Micromonospora halophytica* produces the halomicins A, B, C, and D (Ganguly et al., 1974, 1977).

A number of other actinomycetes have been reported to produce rifamycin S, SV, O, or B.

The thiomethylrifamycins S and SV, tolypomycin O, and the halomicins are probably derived from rifamycin S; however, no transformation studies have been reported so far.

3. Further Ansamycins with Naphthoquinone Chromophores (Fig. 9)

A streptovaricin complex containing the streptovaricins A, B, C, D, and E was isolated in 1956 from *Streptomyces spectabilis* (Siminoff et al., 1957). The correct structures of these streptovaricins have only been known since 1971 (Rinehart et al., 1971). Later the streptovaricins F, G, H, I, and J were described (Rinehart, 1972; Rinehart et al., 1974).

Naphthomycin was isolated from a species of *Streptomyces* and a preliminary structure was published by Williams (1975). Later this structure was revised (Deshmukh et al., 1976; Brufani et al., 1979).

Actamycin is also a product of a *Streptomyces* sp. and shows structural similarities with naphthomycin (Kibby et al., 1980; Allen et al., 1981; McDonald and Rickards, 1981). The antibiotics rubradirin and rubradirin B contain the ansamycin moieties rubransarol A and B (Hoeksema et al., 1978).

4. Ansamycins with Benzenoid or Benzoquinoid Chromophores (Fig. 10)

Geldanamycin is a product of *Streptomyces hygroscopicus var. geldanus* (De Boer et al., 1970; Sasaki et al., 1970). Herbimycin A and B, two ansamycins related to geldanamycin, were recently isolated from a *S. hygroscopicus* strain (Omura et al., 1979; Iwai et al., 1980).

The maytansinoids are a group of ansamycins isolated from higher plants, namely, *Maytenus ovatus*, *Maytenus buchananii*, *Maytenus serrata*, *Putterlickia verrucosa*, and *Colubrina texensis* (Kupchan et al., 1977).

The ansamitocins, a group of ansamycins very closely related to the maytansinoids, have been isolated from a species of *Nocardia* (Higashide et al., 1977). The macbecins I and II are products of a species of *Nocardia* (Muroi et al., 1980).

B. Biosynthesis of the Ansa Chain

Among the ansamycins presented in Sec. IV.A, biogenetic data on ansa chain synthesis are available for rifamycins, streptovaricins, geldanamycin, and herbimycins.

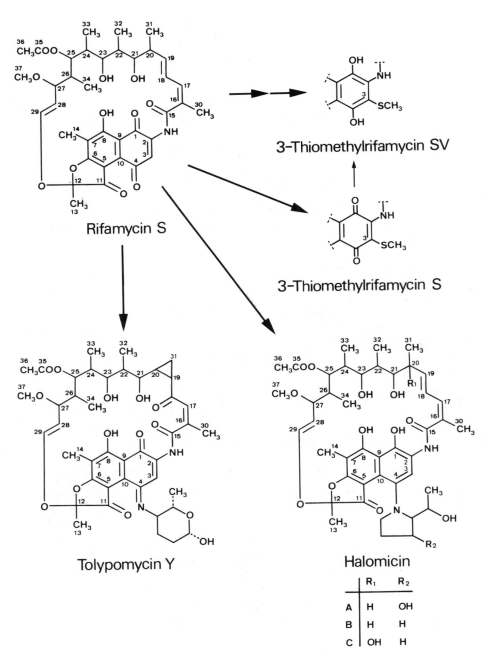

Figure 8 Rifamycins from other actinomycetes probably derived from rifamycin S.

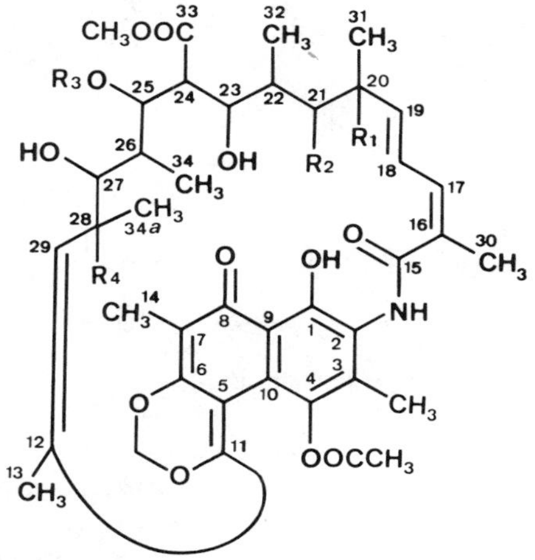

	R₁	R₂	R₃	R₄
A	OH	OH	OCCH₃	OH
B	H	OH	OCCH₃	OH
C	H	OH	H	OH
D	H	OH	H	H
E	H	=O	H	OH
G	OH	OH	H	OH
J	H	OCCH₃	H	OH
F	OH		H	OH

LACTONE BETWEEN
C 21 AND C 24

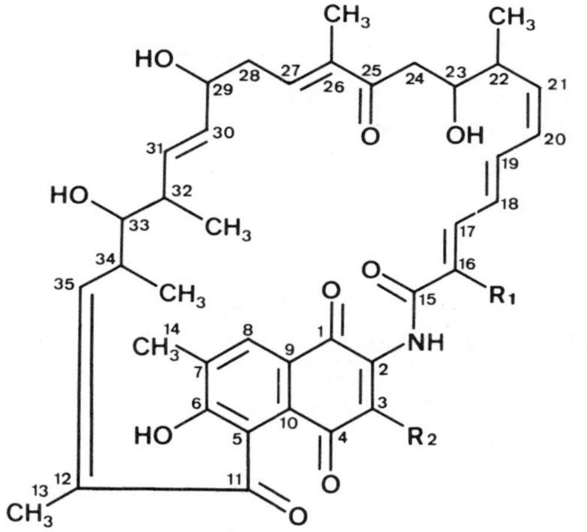

Streptovaricin

	R₁	R₂
Naphthomycin	CH₃	Cl
Actamycin	H	OH

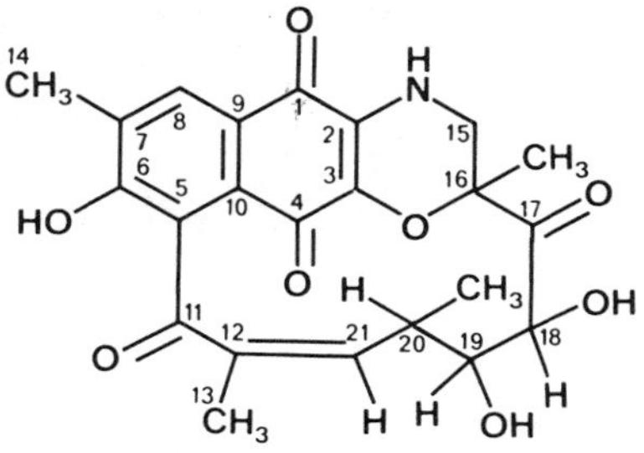

Rubransarol A

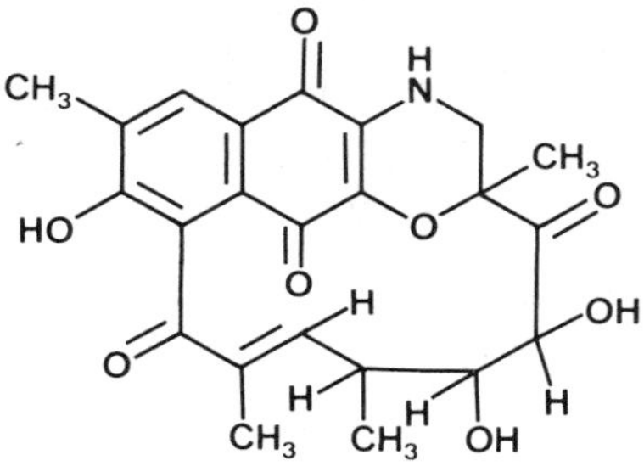

Rubransarol B

Figure 9 Streptovaricins, naphthomycin, actamycin, and rubransarols (numberings analogous to the rifamycins).

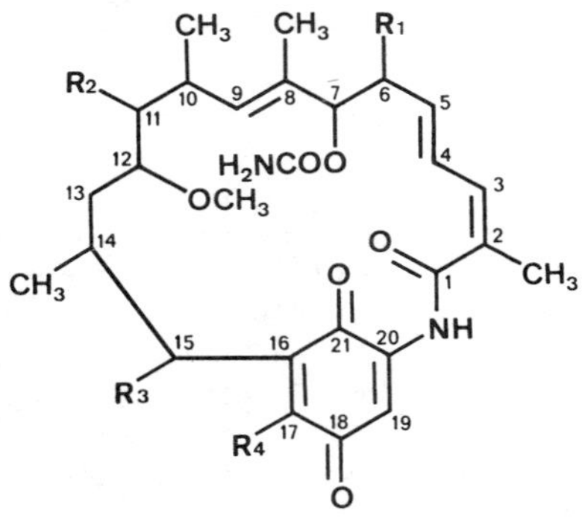

	R₁	R₂	R₃	R₄
Geldanamicin	OCH₃	OH	H	OCH₃
Herbimycin A	OCH₃	OCH₃	OCH₃	H
Herbimycin B	OCH₃	OH	H	H
Macbecin I	CH₃	OCH₃	OCH₃	H
Macbecin II = Macbecin I – Hydroquinone

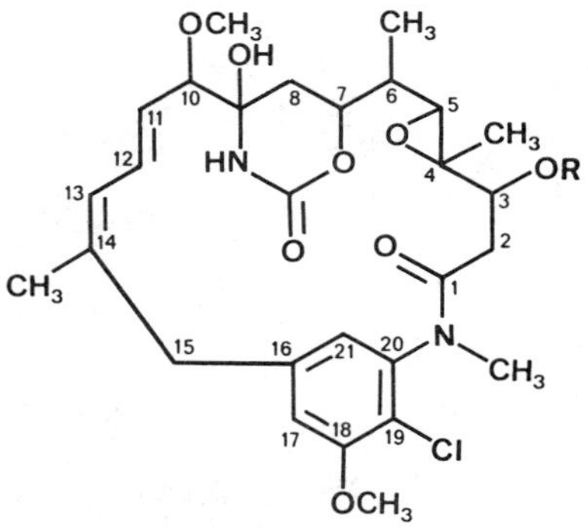

Ansamitocin

	R
P–0	H
P–1	COCH₃
P–2	COCH₂CH₃
P–3	COCH(CH₃)₂
P–3′	COCH₂CH₂CH₃
P–4	COCH₂CH(CH₃)₂

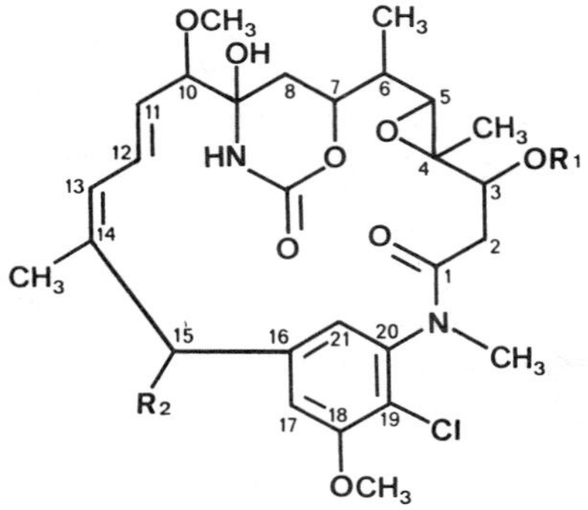

	R₁	R₂
Maytansine	COCH(CH₃)N(CH₃)COCH₃	H
Maytanprine	COCH(CH₃)N(CH₃)COCH₂CH₃	H
Maytanbutine	COCH(CH₃)N(CH₃)COCH(CH₃)₂	H
Maytanvaline	COCH(CH₃)N(CH₃)COCH₂CH(CH₃)₂	H
Maytanbutacine	COCH(CH₃)₂	OCOCH₃
Maytanacine	COCH₃	H
Maytansinol	H	H
Colubrinol	COCH(CH₃)N(CH₃)COCH(CH₃)₂	OH
Colubrinolacetate	COCH(CH₃)N(CH₃)COCH(CH₃)₂	OCOCH₃

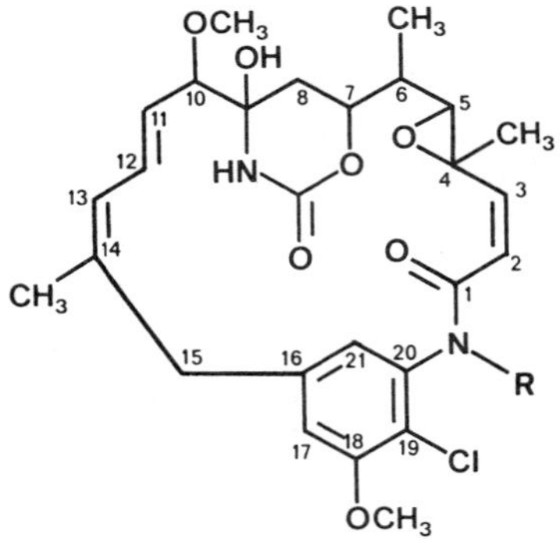

	R
Maysine	CH₃
Normaysine	H

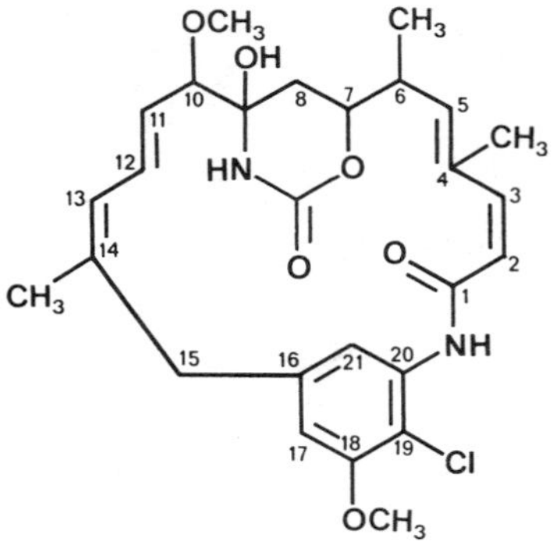

Maysenine

Figure 10 Geldanamycin, herbimycins, macbecins, maytansinoids and ansamitocins (numberings analogous to geldanamycin.

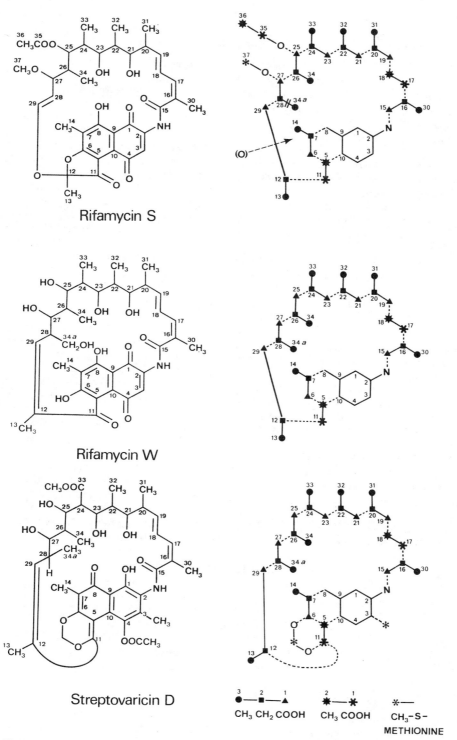

Figure 11 Incorporation of acetate, propionate, and methyl from methionine into rifamycin S, rifamycin W, and streptovaricin D.

1. The Ansa Chain of the Rifamycins and Streptovaricins

By incorporation of ^{14}C precursors followed by chemical degradation and by incorporation of ^{13}C precursors combined with nuclear magnetic resonance spectroscopy, it has been shown that the ansa chain of rifamycin S is derived from acetate, propionate, and methionine, as is depicted in Figure 11 (Brufani et al., 1973; Karlsson et al., 1974; White et al., 1973). The biogenetic origin of 30 of the 37 carbon atoms in rifamycin S was established using this incorporation pattern. The incorporation of acetate (except the O-acetyl group which originates directly from acetyl coenzyme A) and propionate units into rifamycin takes place via malonyl coenzyme A (CoA) or methylmalonyl-CoA, respectively. A seven-carbon amino unit including C1–C4 and C8–C10 of the naphthoquinone part of rifamycin S is not derived from acetate/propionate units. As C27 was labeled by [1-^{13}C]propionate and C28 by [2-^{13}C]propionate, it was concluded that [3-^{13}C]propionate would label a carbon C34a attached to C28. C34a is lost in the later steps of the rifamycin biosynthesis. The ansa chain was interpreted as a clockwise condensation of acetate and propionate units with a seven-carbon amino starter unit. C12, C13, and C29 originate from the same propionate unit which is later split off by the introduction of oxygen between C12 and C29. From a mutant strain of N. mediterranei a biogenetic precursor of rifamycin S was isolated and designated rifamycin W (Martinelli et al., 1974). In a transformation experiment it was demonstrated that radioactive labeled rifamycin W was transformed into radioactive rifamycin B via rifamycin S. The structure of rifamycin W (Fig. 11) is in accordance with the biogenetic model derived from the incorporation pattern for acetate/propionate into rifamycin S. C34a is still present in rifamycin W and the carbons C12, C13, and C29 are not yet separated by the introduction of oxygen. Rifamycin W showed an incorporation pattern for ^{13}C-labeled acetate and propionate identical to the pattern found earlier for rifamycin S (White et al., 1974).

Some incorporation data are also available for streptovaricin D (Fig. 11). [^{14}CH$_3$]methionine, [1-^{14}C]malonate, and [1-^{14}C]propionate were well incorporated into streptovaricin D, and [1-^{13}C]propionate was found to label C15, C19, C21, C23, C25, C27, C29, and C6 (numbering analogous to the rifamycins) (Milavetz et al., 1973). The [1-^{13}C]propionate incorporation pattern for streptovaricin D is thus identical to the pattern for the rifamycins. C34a is still present and C12, C13, and C29 originate from the same propionate unit as in rifamycin W. The methyl group attached to C3 and the methylenedioxy group between C6 and C11 are derived from methionine (Johnson et al., 1974). As in the case of rifamycin, we are left with a seven-carbon amino unit which is not derived from acetate/propionate units. Based on the identical incorporation patterns, a common progenitor for some of the naphthoquinoid ansamycins was proposed (White et al., 1973, 1974). In a ^{13}C-labeling experiment with actamycin and [1-^{13}C]propionate, C21, C25, C31, C33, C35, and C6 were found to be enriched. The seven-carbon amino unit C1–C4 and C8–C10 was again unlabeled (McDonald and Rickards, 1981).

2. The Ansa Chain of Geldanamycin and Herbimycin

Incorporation studies with geldanamycin and ^{14}C and ^{13}C precursors have shown that, as in the rifamycins, the ansa chain is derived from acetate and propionate units (O-methyl groups from methionine), with the exception that two of the three C$_2$ units in the ansa chain are not labeled by acetate or malo-

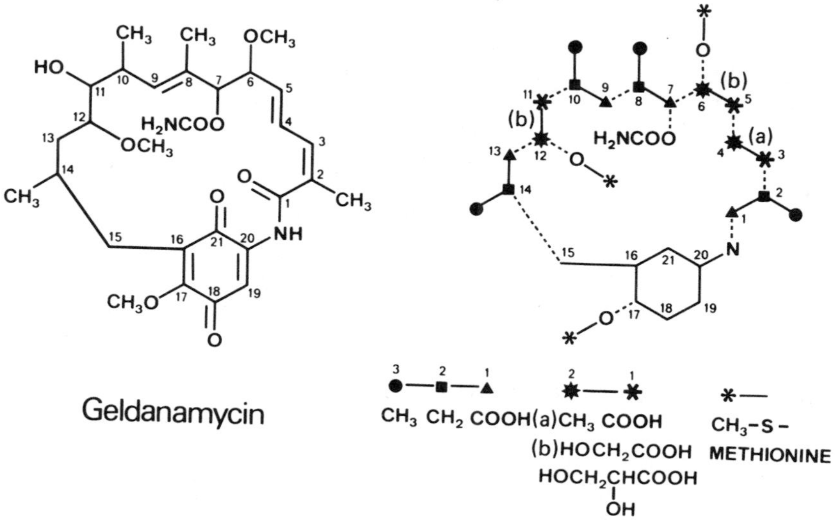

Figure 12 Incorporation of acetate, glycolate, glycerate, propionate, and methyl from methionine into geldanamycin.

nate, but only by glycerate and glycolate (Fig. 12). The direct incorporation of glycolate is proposed for these two C_2 units. The carbamoyl group in the ansa chain is labeled by the guanido group of arginine (Johnson et al., 1973, 1974; Haber et al., 1977). As in streptovaricin and rifamycin, we are left with a seven-carbon amino unit which is not derived from acetate/propionate units.

The incorporation pattern with [1-^{13}C]propionate and herbimycin A was found to be identical to the pattern of geldanamycin (Omura et al., 1979).

For the maytansinoids and ansamitocins no biosynthetic data are available, but one would expect to find a similar situation.

C. The Origin of the Seven-Carbon Amino Unit

All the ansamycins investigated so far contain a seven-carbon amino unit which is not derived from acetate/propionate units. A similar C_7N unit was found in the mitomycins. Only a few incorporation experiments have been published.

1. Incorporation Studies on Rifamycin, Geldanamycin, and Mitomycin Using Classic Precursors

Incorporation studies with rifamycin (Karlsson et al., 1974; White and Martinelli, 1974) led to the hypothesis that the seven-carbon amino unit of the ansamycins derives from an intermediate of the shikimate pathway. D[1-^{13}C]-Glucose labels C1 and C10 of rifamycin S. C1 of glucose is known to label C2 and C6 of shikimic acid, 3-dehydroshikimic acid, or 3-dehydroquinic acid (Srinivasan et al., 1956; Haslam, 1974), which would correspond to C1 and C10 of the rifamycin chromophore. D[1-^{13}C] glycerate labels C3 and C8 of rifamycin. C1 of glycerate originates from C3 or C4 of glucose, which are known to label C4 and C7 of shikimic acid corresponding to C3 and C8 of the

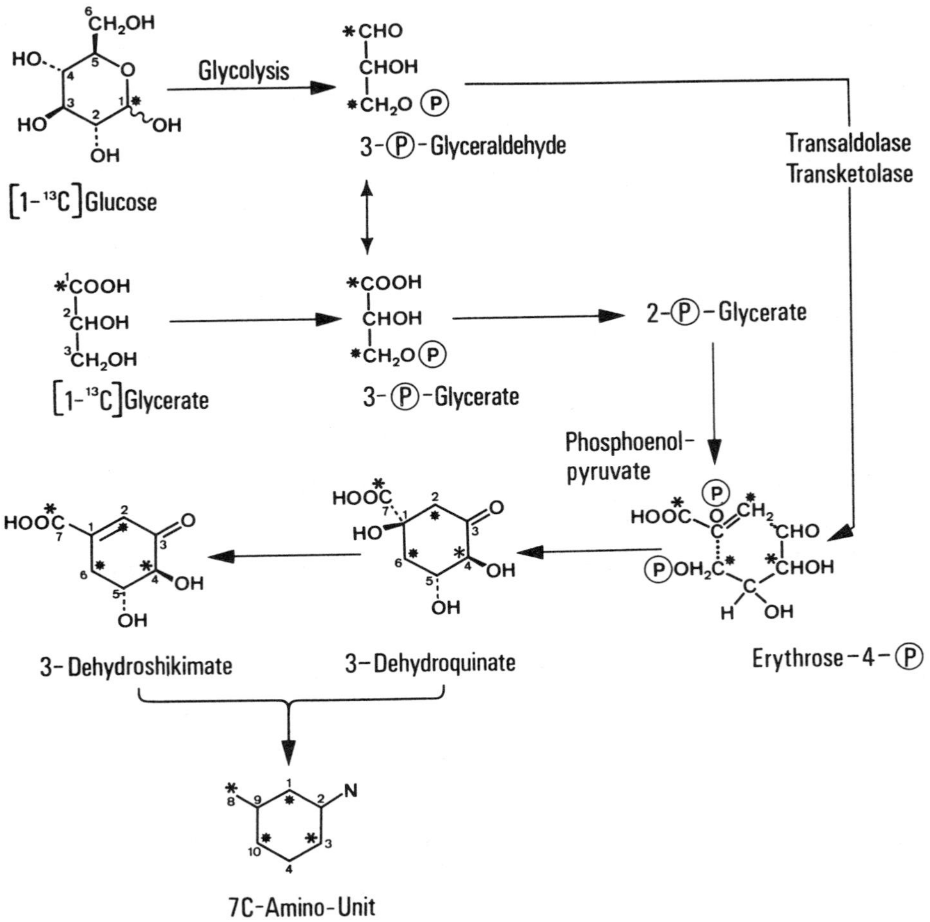

Figure 13 Incorporation of [1-13C]glucose and [1-13C]glycerate into the seven-carbon amino unit of rifamycin.

rifamycin chromophore. The incorporation pattern of D[1-13C]glucose and D[1-13C]glycerate would therefore be in accordance with a shikimate-type origin of the seven-carbon amino unit (Fig. 13). Benzoate, tryptophan, tyrosine, and phenylalanine (all 14C-labeled) are not incorporated into the rifamycin chromophore. [U-14C]Shikimic acid is not incorporated either, but this is no proof for it not being a precursor of the seven-carbon amino unit of rifamycin, as postulated by Karlson et al. (1974). Bruggisser (1975) and Ghisalba (1978) could show that shikimate is not able to penetrate into the cells of *N. mediterranei* (rifamycin producer) within a pH range of 4.0–9.0. The addition of DMSO did not improve the uptake of shikimic acid into the cells. Karlson et al. (1974) suggested that 3-dehydroquinate or 3-dehydroshikimate was the most probable precursor for the seven-carbon amino unit.

Studies with geldanamycin (Haber et al., 1977) have shown that D[6-13C]glucose labels C15, C17, and C21 in the aromatic nucleus of geldanamycin, but C17 and C21 (corresponding to C1 and C10 in rifamycin) were much more highly enriched than C15. As C6 and C1 of glucose label the same carbon atoms of shikimic acid, this incorporation pattern is in agreement with that for [1-^{13}C]glucose and rifamycin. In the incorporation experiment with [1-^{13}C]glycerate, C15 and C19 of geldanamycin (corresponding to C3 and C8 of rifamycin) are labeled (enriched), which is again in agreement with the incorporation of [1-^{13}C]glycerate into rifamycin S. [U-^{14}C]Shikimic acid was reported not to be incorporated into geldanamycin, and for this reason 3-dehydroquinic acid or 3-dehydroshikimic acid is proposed as a precursor of the C$_7$N unit. However, the uptake of shikimic acid into the cells of S. hygroscopicus was not investigated.

Incorporation studies on mitomycin (a group of antibiotics from Streptomyces caespitosus, Streptomyces ardus sp. nov., and Streptomyces verticillatus; for structure see Fig. 23) have shown that this type of antibiotic contains a C$_7$N unit similar to that of the ansamycins (Bezanson and Vining, 1971; Hornemann et al., 1974, 1980). The data are in agreement with a shikimate-type origin of this unit. [U-^{14}C]Shikimate and [7-^{14}C]3-dehydroquinate were not incorporated into the mitomycins. In the case of 3-dehydroquinate the uptake into the cells was not studied, and in the case of shikimate, uptake is reported, but not only mitomycin C but also phenylalanine and tyrosine were so poorly labeled that one could consider this uptake as an experimental artifact (perhaps the cells were not washed prior to counting).

2. A Genetic Approach to the Biosynthesis of the Seven-Carbon Amino Unit

Because of the inability of shikimic acid (and probably also of its precursors 3-dehydroquinic acid and 3-dehydroshikimic acid) to penetrate into the cells of N. mediterranei and because the selection of ^{13}C precursors available for further incorporation studies is very limited, we chose a genetic approach for further investigations.

Two aromatic amino acid-deficient mutants of N. mediterranei were isolated and characterized so far. These two mutants provide genetic evidence that intermediates of the shikimate pathway must be involved in the biosynthesis of the rifamycin chromophore. The mutant strain N. mediterranei A8 (Ghisalba and Nüesch, 1978a) is auxotrophic for the aromatic amino acids phenylalanine and tyrosine and shows greatly reduced rifamycin production compared with the parent strain N813 (high rifamycin B producer). Although it is phenotypically an aro$^-$ mutant, no block in the enzymes of the shikimate pathway was found, but only a block in the transketolase activity. A mixture of pentoses with D(−)-ribulose as the major product was found to be accumulated by mutant A8 in the culture broth. The mutant must be somewhat leaky, because growth could be observed on an incomplete supplement of aromatic amino acids containing only phenylalanine and tyrosine but no tryptophan. A very small amount of shikimate pathway precursors must therefore be synthesized which can produce tryptophan and a trace of rifamycin B. The mutation was shown by reversion to be a single-point one. The fact that the production of rifamycin B in N. mediterranei N813 is dependent on the formation of D-sedoheptulose-7-phosphate (Su7P) and the presence of transketolase activity (see pathway in Fig. 14) shows clearly that the seven-carbon amino unit of the rifamycin chromophore must be derived from an intermediate of the shikimate pathway. No other pathway for synthesizing Su7P except the transketolase reaction is known so far. Likewise, D-erythrose-4-phosphate (E4P) is only known to be synthesized from Su7P by means of the transaldolase reaction or from fructose-6-phosphate and glyceraldehyde-3-phosphate

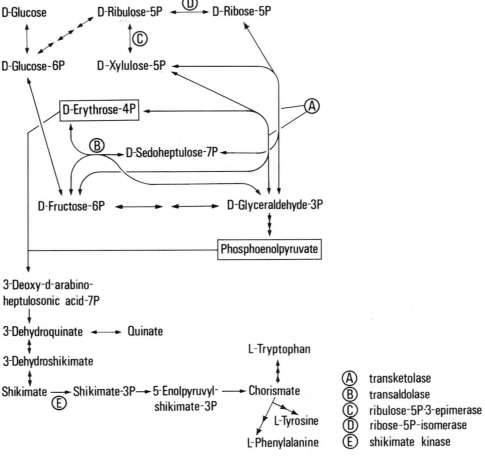

Figure 14 Biosynthetic pathways: pentose shunt, glycolysis, and shikimate pathway.

by the transketolase reaction. Thus a mutant lacking the transketolase activity is not able to synthesize both E4P and Su7P.

No other pathway except the shikimate one is known which starts from Su7P or E4P and leads to aromatic compounds. The mutant strain *N. mediterranei* A10 (Ghisalba and Nüesch, 1978b) is auxotrophic for the aromatic amino acids phenylalanine, tyrosine, and tryptophan and produces the same amount of rifamycin B as the parent strain N813, at least under suboptimal conditions. The mutant accumulates shikimic acid and 3-dehydroshikimic acid in the culture broth. The enzymatic and auxanographic studies showed that mutant A10 is blocked in one of the enzymatic steps leading from shikimic acid to chorismic acid (see Fig. 14). The missing enzyme is most likely to be the shikimate kinase. Under optimal conditions in an industrial production medium it yields less rifamycin B than the parent; however, this might be due to an inhibitory effect caused by the large amount of shikimate excreted by strain A10, or the direct precursor of the seven-carbon amino unit might be partially withdrawn from the pool owing to the excretion of shikimate. As mutant A10 is only defective in the biosynthesis of aromatic amino acids and not in the biosynthesis of rifamycin, it would appear that the seven-carbon amino unit of the rifamycin chromophore must be derived from an intermediate of the shikimate pathway not behind shikimate. If one combines the results from mutants A8 and A10, the origin of this moiety can be localized between Su7P and shikimate.

A direct cyclization of Su7P to yield a precursor of the seven-carbon amino unit can definitely be excluded because, as mentioned before, Karlsson et al. (1974) and White and Martinelli (1974) have shown that D[1-^{13}C]glucose enriches C1 and C10 of rifamycin, which would correspond to C2 and C6 of 3-dehydroquinate, 3-dehydroshikimate, or shikimate. If one assumes a direct cyclization of Su7P, its C3 and C7 would correspond to C1 and C10 of rifamycin. It is known that C1 and C7 of Su7P, but not C3, originate from C1 of glucose. Therefore Su7P is excluded as a precursor of the seven-carbon amino unit. During the biosynthesis of 3-deoxy-D-arabinoheptulosonic acid-7-phosphate (DAHP) from Su7P, an inversion of the triose fragment C1–C3 of Su7P takes place and C3 of DAHP corresponds to C1 of Su7P originating from C1 of glucose. DAHP is the first intermediate of the shikimate pathway showing the incorporation pattern needed for a precursor of the seven-carbon amino unit.

The selection of possible precursors is now reduced to only four intermediates of the shikimate pathway, namely, DAHP, 3-dehydroquinate, 3-dehydroshikimate, and shikimate. DAHP cannot be definitely excluded because it is not known if the 3-dehydroquinate synthetase reaction is the only way to achieve a cyclization of DAHP.

A number of non-rifamycin-producing ultraviolet mutants derived from N. mediterranei strains N813 and A10 were found to accumulate an identical complex of aromatic components instead of rifamycin B (Ghisalba et al., 1981). The main component of this aromatic complex, product P8/1-OG, was isolated from six of these P⁻ mutant strains and identified spectroscopically as a very early precursor in the biosynthesis of rifamycins. The compound was isolated in two forms, either as 2,6-dimethyl-3,5,7-trihydroxy-7-(3'-amino-5'-hydroxyphenyl)-2,4-heptadienoic acid or as its 1,5-lactone (see Fig. 15).

By structural comparison with rifamycin S or rifamycin W, product P8/1-OG can easily be recognized as a very early ansamycin precursor containing the seven-carbon amino starter unit and three initial acetate/propionate units of the ansa chain (Fig. 16). 3-Amino-5-hydroxybenzoyl-CoA might act as a starter molecule for the biosynthesis of product P8/1-OG and of the rifamycins. To this seven-carbon amino starter unit, first a propionate unit (via methylmalonyl-CoA), then an acetate unit (via malonyl-CoA), and finally another propionate unit are added by condensation and decarboxylation. The resulting aromatic triketide is then converted into product P8/1-OG by hydrogenation of the keto group C7 and by enolization of the keto groups C3 and C5. The CoA is then split off (possibly during the excretion of the product). If we compare this biogenetic model for P8/1-OG with the well-known incorporation patterns for [^{13}C]acetate and [^{13}C]propionate into rifamycin S or rifamycin W (Figs. 11 and 16), we can deduce that the structures must have the same biogenetic origin. The methyl groups C14 and C13 in the rifamycins correspond to the methyl groups C8 and C9 in P8/1-OG.

The isolation of product P8/1-OG provides strong evidence that the seven-carbon amino starter unit for the ansamycin biosynthesis is 3-amino-5-hydroxybenzoic acid or its CoA derivative.

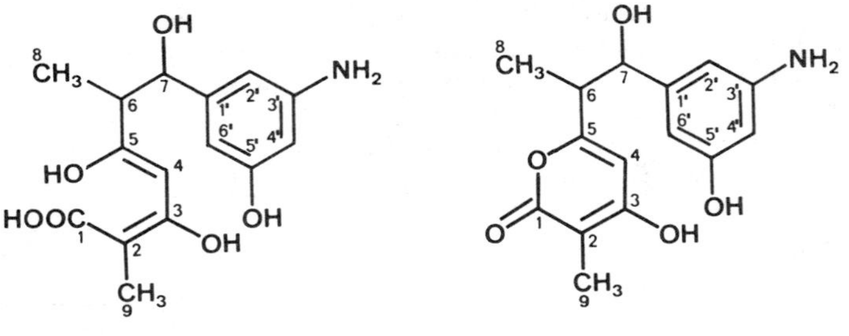

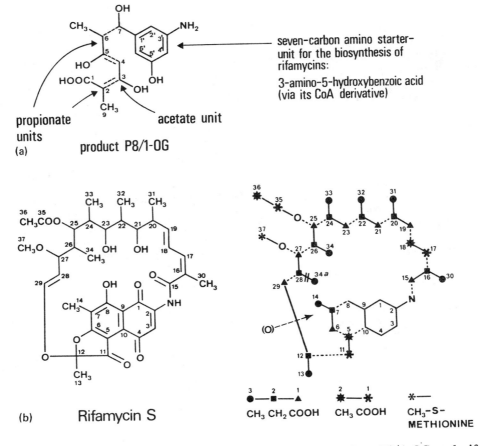

product P8/1-OG   product P8/1-OG-lactone

Figure 15 Structure of product P8/1-OG.

(a)

product P8/1-OG

propionate units

acetate unit

seven-carbon amino starter-unit for the biosynthesis of rifamycins:

3-amino-5-hydroxybenzoic acid (via its CoA derivative)

(b)   Rifamycin S

Figure 16 Biosynthetic interrelationship between product P8/1-OG and rifamycin.

3. Supplementation and Incorporation Studies with 3-Amino-5-Hydroxybenzoic Acid

3-Amino-5-hydroxybenzoic acid has been investigated for its ability to induce rifamycin biosynthesis in an appropriate mutant (strain A8) of *N. mediterranei* and identified as a direct precursor of the seven-carbon amino starter unit for the biosynthesis of ansamycins (Ghisalba and Nüesch, 1981). In a cosynthesis experiment with strains *N. mediterranei* A8 (transketolase⁻ mutant) and P14, which produces product P8/1-OG, it was demonstrated that a metabolite excreted by strain P14 is utilized by strain A8 for the biosynthesis of rifamycin. Product P8/1-OG in a supplementation experiment with strain A8 did not stimulate the production of rifamycin B and can therefore be excluded as the factor responsible for the increase of the rifamycin B production in the cosynthesis experiment. The failure of product P8/1-OG to stimulate rifamycin B production might be explained in several ways. P8/1-OG might not be able to penetrate into the cells of *N. mediterranei* A8 or it cannot be reactivated to P8/1-OG-CoA for further polyketide synthesis. Thus an earlier metabolite of strain P14 must be utilized by strain A8, for example, 3-amino-5-hydroxybenzoic acid.

Supplementation studies with 3-amino-5-hydroxybenzoic acid and *N. mediterranei* A8 showed clearly that this compound can indeed substitute for the seven-carbon amino unit. The original rifamycin production capacity of the parent strain N813 can be restored for strain A8 by supplementation with 3-amino-5-hydroxybenzoic acid. Supplementation studies with 13 other commercially available mono-, di-, or trisubstituted benzoic acids did not lead to any positive results. This indicates that the activation of 3-amino-5-hydroxybenzoic acid with coenzyme A is a highly specific enzymatic reaction. It seems to be a strict requirement for the activation step that the right substituents (one hydroxyl and one amino group) are in the correct positions at C3 and C5 of benzoic acid.

Incorporation experiments with [carboxy-^{14}C]-3-amino-5-hydroxybenzoic acid and *Streptomyces* sp. E/784 demonstrated that this precursor specifically labels the seven-carbon amino unit in actamycin (see Fig. 9) (Kibby et al., 1980).

[Carboxy-^{13}C]-3-amino-5-hydroxybenzoic acid was shown to enrich the C6 methyl group of porfiromycin in an incorporation experiment with *S. verticillatus*. With this result, 3-amino-5-hydroxybenzoic acid is established as a precursor of the methylbenzoquinone nucleus (C_7N unit) of the mitomycin antibiotics (Anderson et al., 1980). Quite recently [carboxy-^{13}C]-3-amino-5-hydroxybenzoic acid was also shown to label the C_7N unit of ansamitocin P-3 (Hatano et al., 1982).

D. Intermediates in the Biosynthesis of Rifamycin and Streptovaricin and Novel Rifamycins

Rifamycin W, the first isolated biogenetic precursor of rifamycin S, has already been mentioned in Sec. IV.B (Martinelli et al., 1974; White et al., 1974; see Fig. 11).

A mutant strain *N. mediterranei* F1/24 was found to accumulate a number of 8-deoxyansamycins. The main component of this 8-deoxyansamycin complex, protorifamycin I (8-deoxyrifamycin W), is a direct biogenetic precursor of rifamycin W (Ghisalba et al., 1978). In transformation experiments the partial conversion of protorifamycin I to rifamycin W and rifamycin B was ob-

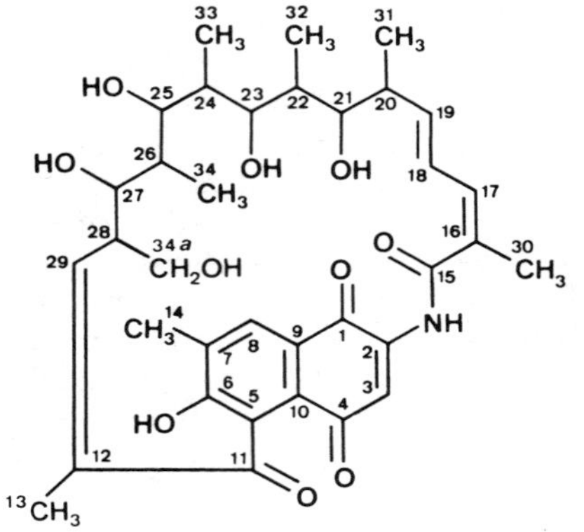

Protorifamycin I

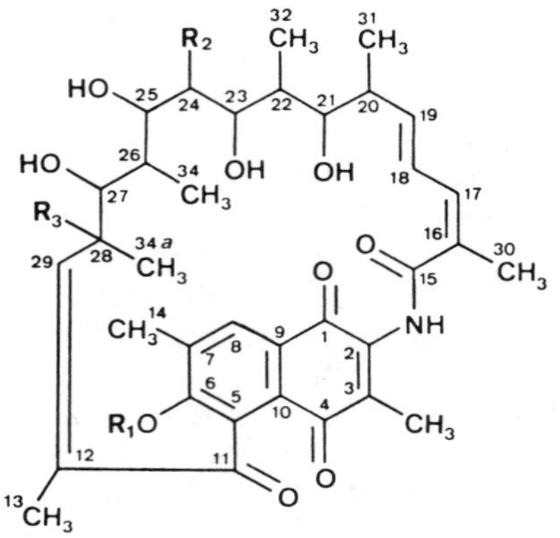

Protostreptovaricin

	R₁	R₂	R₃
I	H	CH₃	H
II	CH₃	CH₃	H
III	H	CH₃	OH
IV	CH₃	CH₃	OH
V	H	H	H

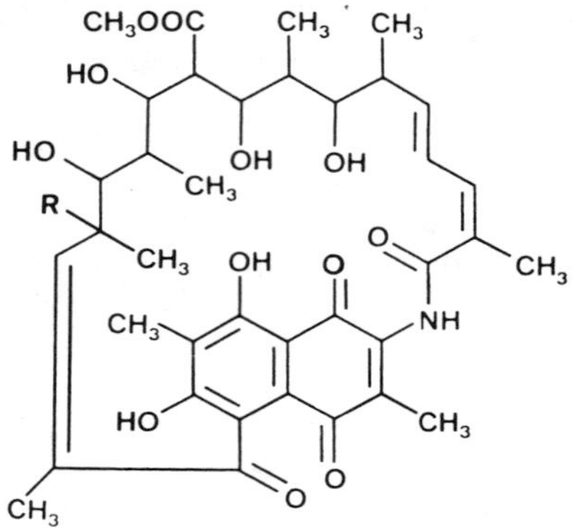

	R
C	OH
D	H

Damavaricin

Figure 17 Structures of protorifamycin I, protostreptovaricins I–V, and damavaricins C and D.

served. Protorifamycin I is structurally related not only to rifamycin W, but also to the protostreptovaricins (Fig. 17).

The protostreptovaricins I–V were isolated as minor components of the streptovaricin complex (Deshmukh et al., 1976).

Another group of apparent streptovaricin precursors, made up of the damavaricins C and D, is structurally related to rifamycin W (Rinehart et al., 1976). Protostreptovaricins are considered to be precursors for both damavaricins and streptovaricins. Protostreptovaricin I is the earliest precursor for the ansamycins of the streptovaricin group known so far.

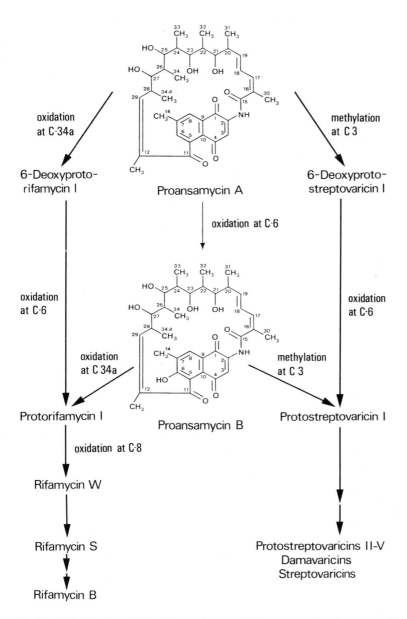

Figure 18 Possible pathways for the biosynthesis of protorifamycin I and protostreptovaricin I starting from the hypothetical common progenitors "proansamycin A" and "proansamycin B."

Comparing the structures depicted in Fig. 17, it is obvious that proto-rifamycin I is very closely related to protostreptovaricins I and III. Proto-streptovaricin I lacks the hydroxyl group at C34a and has an additional methyl group at C3 originating from methionine; protostreptovaricin III has a hydroxyl group at C28 instead of C34a and an additional methyl group at C3 (numbering analogous to the rifamycins). Protorifamycin I and protostreptovaricin I thus differ only in the substitutions at C3 and C34a and should therefore be closely related to the common progenitor for some of the naphthalenic ansamycins (rifamycins, streptovaricins, tolypomycin, halomicin) postulated by White et al. (1973, 1974).

Two hypothetical structures for this common progenitor have been proposed (Ghisalba et al., 1978), namely, "proansamycin A" and "proansamycin B" (see Fig. 18). From "proansamycin B" (34a-deoxyprotorifamycin I or 3-demethylprotostreptovaricin I), protorifamycin I and protostreptovaricin I could easily be synthesized by simple one-step transformations. From "proansamycin A" (6,34a-dideoxyprotorifamycin I or 3-demethyl-6-deoxyproto-streptovaricin I) different two-step transformations would lead to protorifamy-cin I and protostreptovaricin I. If we assume that the methylation at C3 for the ansamycins of the streptovaricin type and the oxidation at C34a for the ansamycins of the rifamycin type take place only after the ring closure between C5 and C11 and after the formation of the peptide bond between the chromo-phore and the ansa chain, the "proansamycins A and B" are the only two remaining possibilities for a common progenitor having the basic structure for the rifamycins, streptovaricins, tolypomycins, and halomicins. As all the naphthalenic ansamycins known so far (see Fig. 23) bear an oxygen function at C6 of the chromophore, one might predict that "proansamycin B" is the correct structure.

In order to discover one of the proansamycins, the minor compounds of the 8-deoxyansamycin complex from *N. mediterranei* F1/24 were investigated. Neither of these two proansamycins has been isolated yet. However, the two metabolites, proansamycin B-M1 and protorifamycin I-M1, have been identified (Ghisalba et al., 1979). Proansamycin B-M1 and protorifamycin I-M1 (see Fig. 19) were the first members of the ansamycin group with an opened ansa chain to be reported. In these two metabolites the bond between C5 and C11 is cleaved and the aliphatic chain is attached to the chromophore by the amide bond only. As is known from the ^{13}C incorporation studies (see Sec. IV.B), that C5 and C11 of rifamycin S (or W) originate from one acetate unit (C5 from C2 and C11 from C1 of acetate or malonyl-CoA, respectively), proansamycin B-M1 and protorifamycin I-M1 are more likely to be metabolites or degradation products of early ansamycin precursors than to be early precursors themselves. This was confirmed by transformation experiments. Proansamycin B-M1 is thus a degradation product of the hypothetical proansamy-cin B. The only difference between proansamycin B-M1 and proansamycin B is the cleavage of the bond between C5 and C11. The existence of proansamycin B-M1 is thus an indirect argument that proansamycin B also exists and a support for the biosynthetic hypothesis depicted in Figure 18. By analogy, protorifamycin I-M1 must be interpreted as a degradation product of proto-rifamycin I.

Shortly after the publication of the discovery of proansamycin B-M1 and protorifamycin I-M1 the isolation of streptovaricin U was reported (Knöll et al., 1980). Streptovaricin U (see Fig. 19) shows an opened ansa chain too and is a degradation product of protostreptovaricin I. The cleavage of the bond between C5 and C11 seems to be a nonspecific (enzymatic) reaction

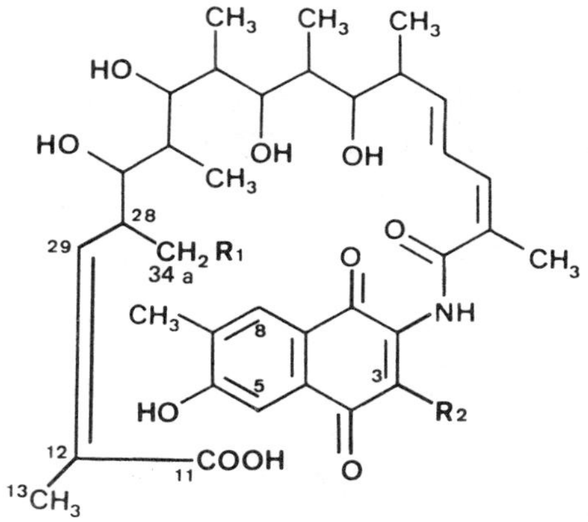

	R_1	R_2
Proansamycin B-M1	H	H
Protorifamycin I-M1	OH	H
Streptovaricin U	H	CH_3

Figure 19 Structures of proansamycin B-M1, protorifamycin I-M1, and streptovaricin U.

for ansamycins with a protorifamycin/protostreptovaricin-type chromophore. Nine further products from *N. mediterranei* F1/24, all of them 8-deoxyansamycins, were identified (Ghisalba et al., 1980). Six of these minor compounds are direct modifications of protorifamycin I, namely, protorifamycin I-lactone, 23-ketoprotorifamycin I, 30-hydroxyprotorifamycin I, 20-hydroxyprotorifamycin I, 13-hydroxyprotorifamycin I, and 23-acetoxyprotorifamycin I (see Fig. 20). Three of the identified minor compounds are defective rifamycins, namely, 8-deoxyrifamycin B, 8-deoxyrifamycin S (SV), and 8-deoxy-3-hydroxyrifamycin S (SV).

From a recombinant strain *N. mediterranei* R21 seven novel rifamycins have been isolated (Fig. 20) (Schupp et al., 1981; Traxler et al., 1981): rifamycin W-lactone, rifamycin W-hemiacetal, 30-hydroxyrifamycin W, 28-dehydroxymethyl-28, 30-dihydroxyrifamycin W, 16,17-dehydrorifamycin G (see Fig. 7), 3-hydroxyrifamycin S, and 3,31-dihydroxyrifamycin S.

From another mutant strain of *N. mediterranei*, 27-demethylrifamycin SV, 25-deacetyl-27-demethylrifamycin SV, and 25-demethylrifamycin B were isolated (Lancini et al., 1970). 25-Deacetylrifamycin SV was found together with the rifamycins P, Q, R, and Verde (White et al., 1975; Lancini and Parenti, 1978).

The structures of the rifamycins isolated from *N. mediterranei* mutant strains and recombinant strains supply important biosynthetic information (see Fig. 21).

Starting from protorifamycin I and rifamycin W, a number of modifications in the ansa chain have been observed, such as the oxidation of the C23 hydroxyl group, hydroxylation at C13, C20, or C30, and acetylation of the C23 hydroxyl group. Similar modifications are known for some end products of the rifamycin group. These data on the rifamycins together with those on the other ansamycin groups indicate that hydroxylation is possible at al-

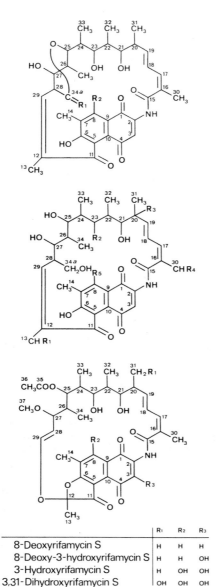

	R_1	R_2
Protorifamycin I-lactone	=O	H
Rifamycin W-lactone	=O	OH
Rifamycin W-hemiacetal	H,OH	OH

	R_1	R_2	R_3	R_4	R_5
23-Ketoprotorifamycin I	H	=O	H	H	H
30-Hydroxyprotorifamycin I	H	OH	H	OH	H
20-Hydroxyprotorifamycin I	H	OH	OH	H	H
13-Hydroxyprotorifamycin I	OH	OH	H	H	H
23-Acetoxyprotorifamycin I	H	OAc	H	H	H
30-Hydroxyrifamycin W	H	OH	H	OH	OH

	R_1	R_2	R_3
8-Deoxyrifamycin S	H	H	H
8-Deoxy-3-hydroxyrifamycin S	H	H	OH
3-Hydroxyrifamycin S	H	OH	OH
3,31-Dihydroxyrifamycin S	OH	OH	OH

8-Deoxyrifamycin B

28-Dehydroxymethyl-28,30-dihydroxyrifamycin W

Figure 20 Further rifamycins from *N. mediterranei* strains F1/24 and R21.

310

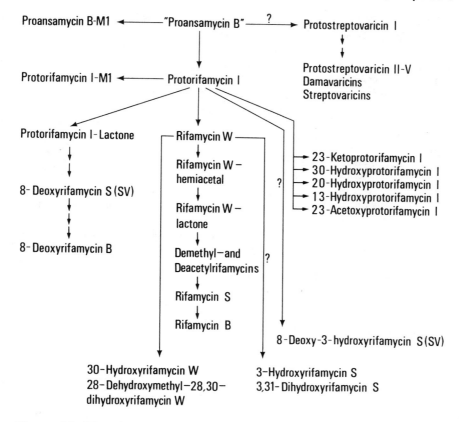

Figure 21 The biogenetic relationships among the known 8-deoxyansamycins and the rifamycins (streptovaricins).

most every methyl group of the ansa chain and even hydroxylation at trisubstituted carbons of the ansa chain can occur. It seems that the hydroxylating enzymes are not very specific, because the hydroxylations occur at many different positions and also with ansamycins having very different chromophores, such as protorifamycin I, rifamycin S, or streptovaricins.

The existence of rifamycin W-hemiacetal, protorifamycin I-lactone, and rifamycin W-lactone indicates that the elimination of C34a during the transformation rifamycin W → rifamycin S or protorifamycin I → 8-deoxyrifamycin S follows the normal route for elimination of methyl by decarboxylation:

$$28 \diagdown\!\!\!\!\!\overset{34a}{\diagup} \text{CH}_3 \rightarrow \diagdown\!\!\!\!\!\diagup \text{CH}_2\text{OH} \rightarrow \diagdown\!\!\!\!\!\diagup \text{CHO} \rightarrow \diagdown\!\!\!\!\!\diagup \text{COOH} \rightarrow \diagdown\!\!\!\!\!\diagup \text{H}$$

$$\underset{\text{hemiacetal}}{} \quad \underset{\text{lactone}}{}$$

The fact that proansamycin B-M1 possesses a C34a methyl group and rifamycin W a C34a hydroxymethyl group supports this hypothesis.

The formation of protorifamycin I-lactone together with 8-deoxyrifamycin S, 8-deoxy-3-hydroxyrifamycin S, and 8-deoxyrifamycin B by strain F1/24 proves that the enzyme systems for the elimination of C34a, the ring extention or introduction of oxygen between C12 and C29, the formation of the five-membered ring in the chromophore, and the transformation 8-deoxyrifamycin S → 8-deoxyrifamycin B or rifamycin S → rifamycin B accept both the 8-deoxy and the 8-hydroxy products as substrates. Protorifamycin I, however, is not as good a substrate as rifamycin W and the transformation protorifamycin I → 8-deoxyrifamycin B is very slow. In *N. mediterranei* F1/24 only about 10% of the synthesized protorifamycin I is transformed into protorifamycin I-lactone, 8-deoxyrifamycin S, 8-deoxy-3-hydroxyrifamycin S, and 8-deoxy-rifamycin B during the fermentation. The rest is accumulated in the culture broth or modified by hydroxylation to about 5%. This contrasts with the original strain N813 (rifamycin B producer), where the transformation rifamycin W → rifamycin B is almost complete and no significant accumulation of precursors, such as the corresponding rifamycin W, is observed. In the strepto-varicin-producing organism *S. spectabilis* all the enzyme systems mentioned above seem to be absent (or nonfunctional), because all the known ansamycins of the streptovaricin type bear an unmodified C28 methyl group C34a and have an uncleaved carbon ansa chain with no oxygen between C12 and C29, no five-membered ring in the chromophore, and no glycolic acid at C4.

For the introduction of the hydroxyl group at C3, in the case of 8-deoxy-3-hydroxyrifamycin S, 3-hydroxyrifamycin S, and 3,31-dihydroxyrifamycin S, different biogenetic origins can be postulated: C3 hydroxylation of 8-deoxyrifamycin S or rifamycin S, C3 hydroxylation of a precursor such as protorifamycin I, rifamycin W, or proansamycin B followed by transformation to the final rifamycins, or biosynthesis beginning with a seven-carbon amino starter unit already hydroxylated in the correct position. In transformation experiments (Schupp et al., 1981) it was found that 3-hydroxyrifamycin S is not a metabolite of rifamycin S. The C3 hydroxyl group must therefore be introduced at an earlier stage in rifamycin biosynthesis.

Mutasynthesis experiments with *Nocardia mediterranei* A8 (Traxler and Ghisalba, 1982) demonstrated that only 3-amino-5-hydroxybenzoic acid but not 4-substituted 3-amino-5-hydroxybenzoic acids such as 3-amino-4, 5-di-hydroxybenzoic acid, 3-amino-4-hydroxy-5-methoxybenzoic acid, or 3-amino-4-methyl-5-hydroxybenzoic acid can substitute for the seven-carbon amino starter-unit in the biosynthesis of rifamycins. This indicates that the activating enzyme system for the seven-carbon amino starter-unit must be highly specific. Not even traces of 3-hydroxyrifamycin S (SV), 3-hydroxy-4-methoxyrifamycin SV, or 3-methylrifamycin S (SV) could be detected in the mutasynthesis experiments. Thus, the most probable route for the insertion of 3-substituents into rifamycin is C3 hydroxylation of a precursor such as protorifamycin I, rifamycin W or proansamycin B.

If we assume comparable substrate specificities for the 3-amino-5-hydroxybenzoic acid-activating enzymes in all the other ansamycin-producing actinomycetes it can be concluded from the engative results with the incorporation of 3-amino-4-methyl-5-hydroxybenzoic acid that the 3-methyl group in antibiotics of the streptovaricin-type, and, most likely, also the 3-chlorine in other ansamycins (e.g., maytansins) is not present in the seven-carbon amino starter-unit but introduced at a much later biosynthetic stage.

E. Transformation Reactions Starting from Rifamycin S

As already mentioned in Sec. IV.A (see Fig. 7), rifamycin S is a key intermediate in the biosynthesis of a considerable number of rifamycins, namely, the rifamycins A, B, C, D, E, SV, B, L, O, Y, G, R, P, Q, and Verde. For the thiazorifamycins P, Q, and Verde, transformation studies have been reported (Cricchio et al., 1980). The thiazorifamycins are derived from rifamycin S and cysteine. The rifamycins P and Verde are formed simply by chemical reactions, but the biosynthesis of rifamycin Q involves obligatory enzymatic assistance. None of the three thiazorifamycins is the precursor of the other two.

The origin of the glycolic acid moiety in the rifamycins B and L was investigated by ^{14}C incorporation experiments (Lancini and Sensi, 1967; Lancini et al., 1969). It was found that only C_3 precursors such as glycerol or sugars such as glucose and ribose were used for the glycolic acid moiety of rifamycin B, but not C_2 precursors such as glycolate, glyoxylate, glycine, or ethanol. $[1-^{14}C]$glucose, $[6-^{14}C]$glucose, and $[1-^{14}C]$ribose predominantly labeled the hydroxymethyl group (80–90% of the total glycolic acid radioactivity), whereas $[2-^{14}C]$glucose, $[3,4-^{14}C]$glucose, and $[1-^{14}C]$glycerol labeled both the hydroxymethyl group (40–60%) and the carboxyl group (40–60%). Only one incorporation experiment was carried out with rifamycin L. With $[1-^{14}C]$-glucose, 90% of the radioactivity was found in the hydroxymethyl group, as in the experiment with rifamycin B. From this result it was concluded that the origin of the glycolic acid moiety is the same for both rifamycins B and L. Rifamycin L was excluded as a precursor of rifamycin B. Rifamycin O was proposed as a possible common precursor for both rifamycins B and L. A reinvestigation of the transformation of rifamycin S into the rifamycins B and L gave different results (Ghisalba et al., 1982; Roos and Ghisalba, 1980). The reduction of rifamycin S to rifamycin SV was shown to be NADH dependent, but not specific for the rifamycin biosynthetic pathway. This reduction functions also using *Escherichia coli* cells. Rifamycin O was excluded as a precursor for rifamycins B and L. A mutant strain producing rifamycins SV and L but not B was found to convert rifamycin O into rifamycin B, and not into rifamycin L, as one would expect based on the hypothesis by Lancini et al. (1969). Even with spirolactone-labeled rifamycin O, no traces of radioactive rifamycin L were found in the transformation assay with strain N813, only labeled rifamycin B. The transformation of rifamycin S into B and L is completely inhibited by the thiamine antagonist oxythiamine chloride hydrochloride, indicating the participation of thiamine or thiamine pyrophosphate in both transformation reactions. A thiamine-dependent strain of *N. mediterranei* was found to grow and to transform rifamycin S into B and L only when thiamine was added to the medium. Therefore a thiamine pyrophosphate-dependent decarboxylating enzyme system is postulated for both transformation reactions. Based on the hypothesis by Lancini et al. (1969), one would expect to find identical levels of radioactivity incorporated for both rifamycin B and L in transformation experiments with labeled C_3 precursors. However, this was found not to be the case either with $[U-^{14}C]$-glycerol or with different labeled pyruvates. Glycerol predominantly labels the glycolic acid moiety of rifamycin B, and pyruvate the glycolic acid moiety of rifamycin L. This indicates that two completely different biosynthetic pathways using different C_3 precursors for the biosynthesis of the glycolate ether rifamycin B and for the glycolate ester rifamycin L must be operating (see Fig. 22).

In contrast to rifamycin S, 3-hydroxyrifamycin S and 8-deoxy-3-hydroxyrifamycin S are not transformed into the corresponding 3-hydroxyrifamycins of the B or L type by *N. mediterranei*.

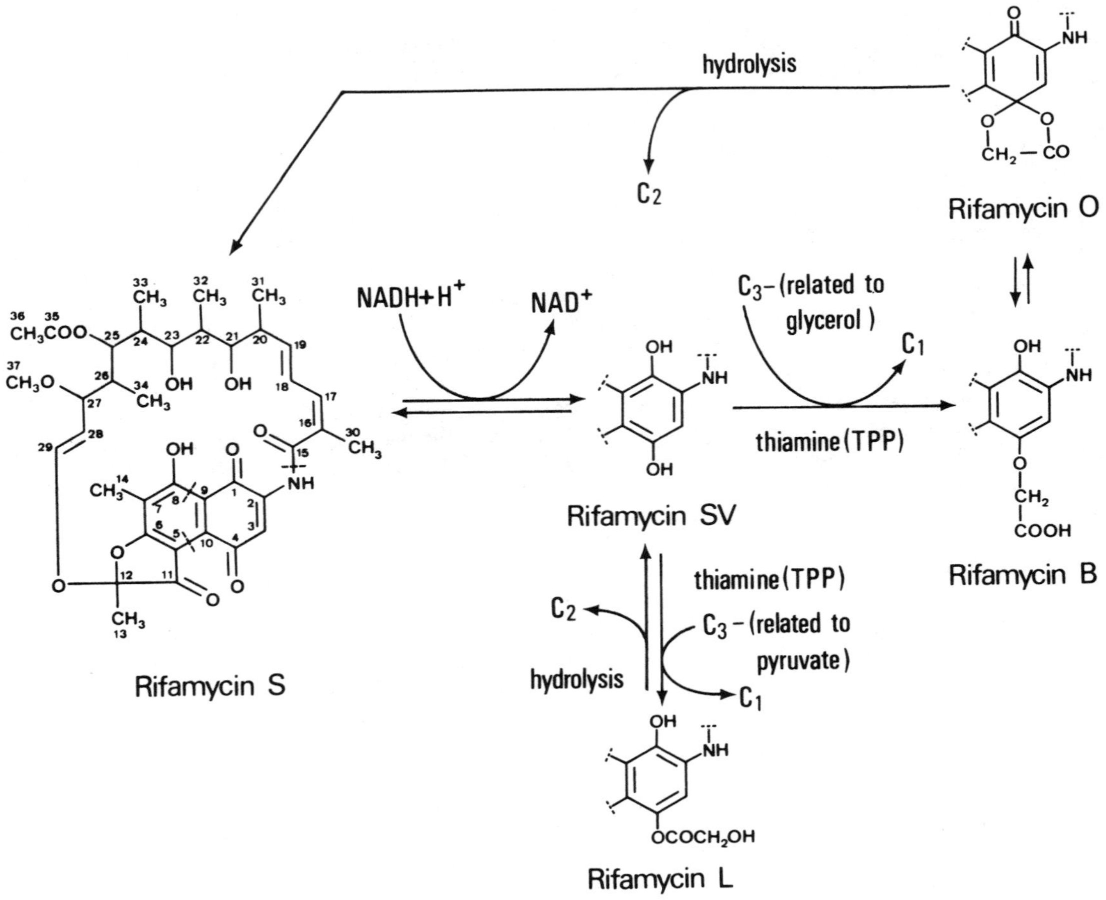

Figure 22 Revised biosynthetic model for the transformation of rifamycin S into the rifamycins B and L.

F. A General Pathway for the Biosynthesis of Ansamycins

With all the available biosynthetic data and by analyzing structural analogies, we can depict a biogenetic model for all the known ansamycins (Fig. 23). The seven-carbon amino starter unit for the ansamycin biosynthesis is derived from an intermediate of the shikimate pathway between 3-deoxy-D-arabino-heptulosonic acid 7-phosphate and shikimate. The exact branch point remains to be identified. Three possible pathways for the biosynthesis of 3-amino-5-hydroxybenzoic acid are shown in our model, starting from 3-deoxy-D-ara-binoheptulosonic acid 7-phosphate (as proposed by Hornemann et al., 1980, for the mitomycins), 3-dehydroquinic acid, or 3-dehydroshikimic acid. 3-Amino-5-hydroxybenzoic acid is the direct precursor for the seven-carbon amino starter unit which can now be identified as 3-amino-5-hydroxybenzoyl-CoA.

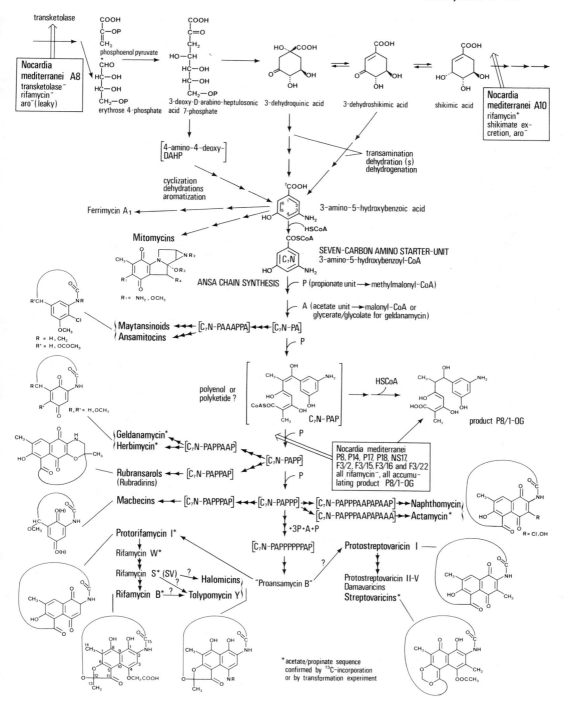

Figure 23 Possible pathways for the biosynthesis of ansamycins starting from shikimate pathway intermediates via 3-amino-5-hydroxybenzoic acid as the direct precursor of the seven-carbon amino starter unit.

The antibiotics of the mitomycin type containing a seven-carbon amino unit similar to that of the ansamycins are derived directly from 3-amino-5-hydroxybenzoic acid (the carboxyl group is reduced to a methyl group in one of the following biosynthetic steps).

In the case of the antibiotic ferrimycin A_1, 3-amino-5-hydroxybenzoic acid itself is a structural element of the compound (Bickel et al., 1966). Starting with 3-amino-5-hydroxybenzoyl-CoA, a polyketide chain (ansa chain) is built up by subsequent condensation with propionate and acetate units, via methylmalonyl-CoA and malonyl-CoA, respectively. Different biosynthetic branch points in the polyketide synthesis can be located by analyzing structural analogies of the ansamycin types (ansa methyl groups indicate a propionate unit!) or by the known incorporation patterns for [^{13}C]acetate and [^{13}C]propionate. A first ansamycin branch derived from the sequence C_7N-propionate–acetate (C_7N-PA) leads to the maytansinoids and ansamitocins. The sequence C_7N-PAP is excreted as its deactivated form (product P8/1-OG) by nine independent production$^-$ mutants of *N. mediterranei*. The following ansamycin types are derived from polyketides behind the sequence C_7N-PAP: geldanamycin and herbimycin, rubradirins (containing the ansamycin moieties rubransarol A or B), macbecins, streptovaricins via proto-streptovaricins and damavaricins, rifamycins via proansamycin B and proto-rifamycin I, halomicins and tolypomycin Y, naphthomycin, and actamycin.

The structural features of 3-amino-5-hydroxybenzoic acid are clearly visible in all these ansamycins. In all cases the nitrogen function and the oxygen function are both present in the meta position to the carbon atom originating from the carboxy group of 3-amino-5-hydroxybenzoic acid (C8 in the case of rifamycin). For tolypomycin Y and the halomicins the oxygen function at C4 is replaced by nitrogen function. The oxygen function originating from C7 of 3-amino-5-hydroxybenzoic acid is lost in ansamitocins, maytansinoids (except colubrinol and colubrinol acetate), geldanamycins, rubransarols, naphthomycin, actamycin, protorifamycin I, and protostreptovaricins, but is still present in herbimycin and macbecins. For the naphthalenic ansamycins, this oxygen function must be eliminated during the formation of the left aromatic ring. In later biosynthetic steps leading to the formation of rifamycins, halomicins, tolypomycin Y, damavaricins, and streptovaricins, the oxygen at C8 (rifamycin numbering) is again introduced (protorifamycin I → rifamycin W, protostreptovaricins → damavaricins and streptovaricins). The oxygen function at C1 of the chromophore (rifamycin numbering; originating from C2 of 3-amino-5-hydroxybenzoic acid) which is present in most of the ansamycins listed above is still absent in maytansinoids and ansamitocins, thus indicating that this oxygen must be absent in the seven-carbon amino starter unit. In some of the ansamycins additional substituents such as chlorine, hydroxyl, methoxy, or methyl are introduced into the chromophore, probably in later biosynthetic steps (in positions corresponding to C4 or C6 of 3-amino-5-hydroxybenzoic acid).

V. FERMENTATION PROCESS

A. The Vegetative Growth Cycle

One of the characteristics of the vegetative growth cycle of our strains is its considerable duration. A total of 6 days in the laboratory and a further 5 days in the plant before the production fermenter is inoculated is customary.

Table 2 Composition of Laboratory and Plant Vegetative Media (g/liter)

Laboratory medium		Plant medium	
Meat extract	5.0	Soya meal,	
Peptone	5.0	Peanut meal, fish meal	
Yeast extract	5.0	(singly or in combination)	$\sim$40 (total)
Enzyme hydrolyzed		Glucose	20-40
Casein	2.5	$(NH_4)_2SO_4$	2-4
Glucose	20.0	$CaCO_3$	6
		Trace elements Fe, Zn, Mn,	
		Co, Cu, Mo	

With the possible advent of new, faster-growing strains the situation could change. The exact composition of the vegetative medium in the plant varies (Table 2), but the laboratory medium is still virtually the same as that quoted by Thiemann and Sensi in their 1967 review. This medium is expensive so that it is not suitable for use in the plant, but it does have the advantage of possessing no solids and it is translucent. The vegetative medium in the plant must of necessity be cheaper and its composition changes from strain to strain as its optimization is an important stage in the process development. Basically the plant vegetative medium contains a meal suspension or a mixture of meals, glucose as the carbon source, ammonium sulfate with a suitable quantity of calcium carbonate to buffer the effect of its metabolism, and a supplement of trace elements in solution (Table 2).

B. The Production Process

1. General Observations

If some of the observations reported by Sensi and Thiemann (1967) no longer apply today, this is not in the least surprising. Anyone who has been associated with a strain and fermentation development process program which continues over a period of years will know that old problems disappear just as new problems appear to take their place. Since the appearance of phage 156 in 1960, no further phage infections have been experienced, so that a calcium-sequestering agent is no longer considered to be necessary in the vegetative media. This is a considerable advantage, because there are now no restrictions in the choice of the vegetative medium on the grounds of its calcium content. Moreover, the present production strains are far more stable than their ancestors were, so that maintenance on lyo ampoules and the use of agar slants, which can be kept for several months at 4°C without severe loss of viability or culture rundown, is standard.

In the earlier production strains, the unwanted rifamycin Y, which is chemically similar to rifamycin B, was produced in quite high amounts, especially in the presence of added phosphate. The present production strains only produce rifamycin Y in very small amounts, regardless of the level of phosphate in the fermentation medium. No special strain-screening program was necessary to achieve a reduction of rifamycin Y production. It happened gradually in the course of the routine strain improvement program.

2. The Reactor and Its Instrumentation

The fermentation has been developed and is now generally carried out in conventional bioreactors equipped with turbine stirrers, baffles, and ring spargers. The ratio of the height of the medium to the diameter of the reactor can vary between 3:1 and 2:1. During the fed-batch fermentation, this ratio will change considerably. Some very limited studies with an air-lift system did not result in yield improvement or show any economic advantages (J. Gruner, Ciba-Geigy, personal communication). In view of the fact that our production plants are equipped with reactors of conventional design, development work has been largely confined to process and strain improvement, where the best prospects for obtaining an increase in productivity lie. Very little attention has been given to the effects of the design of the bioreactor.

The power input requirement of the fermentation varies very much with the strain. In the course of the strain improvement program the nature of the growth has varied very considerably, from relatively fluid to extremely viscous. The reactors are equipped with the necessary oxygen and pH electrodes to ensure the successful running of a fed-batch fermentation.

3. Rifamycin B Fermentation

Batch fermentations with the present production strains still have a very similar pattern to that shown by Sensi and Thiemann (1967) (Fig. 24). In the course of the development of the fermentation process a more sophisticated fed-batch fermentation has been evolved which can last up to 300 hr. The pH level is maintained between relatively narrow limits (6.0–6.5) and sources of sugar and organic and inorganic nitrogen are fed batchwise or continuously. Relatively large amounts of added nitrogen are necessary to stimulate production, whereas the addition of sugar alone does not have any stimulatory effect. The fermentation is considered to be monophasic, in that growth is essential for production to take place. Unlike many other secondary metabolite fermentations, rifamycin synthesis is not directly regulated to any marked degree by phosphate. The production strains are versatile in their ability to produce well with different organic nitrogen sources. The best nitrogen sources are fish meal, soya meal, peanut meal, and cottonseed meal. According to the organic nitrogen source used, the amount of barbiturate in the medium has to be varied. The exact regulation or the growth achieved in this way is very critical for optimal production.

The effect of barbiturate on rifamycin production has been investigated quite thoroughly. Kluepfel et al. (1965) showed that it did not act as a precursor. We have been able to produce the same amount of rifamycin SV with a carefully balanced medium with or without barbiturate (J. Gruner, Ciba-Geigy, unpublished data). The most recent work by Polish investigators (Ruczaj et al., 1972) attributes relatively complicated regulatory effects to barbiturate in the production medium. They could measure changes in the pentose concentration in the cells of two mutants, in one of which antibiotic production was stimulated and in the other depressed. They do not report whether their strains were producing rifamycin B or SV.

4. Rifamycin SV Fermentation

The chemical conversion of rifamycin B to rifamycin SV, which is a necessary step in the synthesis of rifampicin, can only be carried out in relatively low yields. Thus if strains could be isolated which synthesize only rifamycin SV, then a potentially more efficient process for rifampicin production might be realized. Lancini and Hengeller (1969) described a strain which produced 75% rifamycin S apart from other rifamycins. Mutants producing almost exclusively rifamycin S were isolated in the Ciba-Geigy research laboratories in 1970 (unpublished data).

Küenzi et al. (1977) describe the fermentation of one of these strains which was characterized by some interesting features not observed in the rifamycin B fermentation, although the optimized fermentation conditions were very similar to those for the rifamycin B production. Growth and production occurred in two phases. Both were initially relatively rapid and in the second phase they slowed down appreciably. It was demonstrated that rifamycin S itself was responsible for growth inhibition as well as inhibition of rifamycin

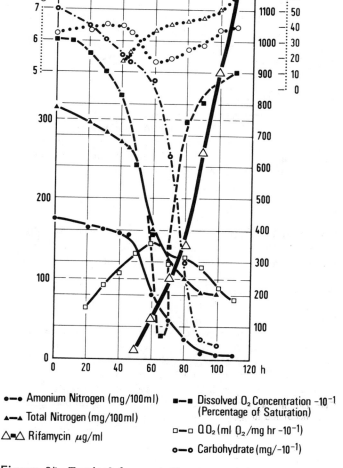

● ─● Amonium Nitrogen (mg/100ml) ■ ─■ Dissolved O₂ Concentration -10⁻¹
                                          (Percentage of Saturation)
▲ ─▲ Total Nitrogen (mg/100ml)

△ ─△ Rifamycin μg/ml   □ ─□ Q O₂ (ml O₂/mg hr -10⁻¹)

                                    ○ ─○ Carbohydrate (mg/-10⁻¹)

Figure 24 Typical fermentation curves for *N. mediterranei* in complex medium.

synthesis by growing cells. Conversely, rifamycin B does not inhibit growth or rifamycin synthesis. Attempts to improve rifamycin S yields in the Ciba-Geigy laboratories were unsuccessful, almost certainly because it was not possible to isolate rifamycin S-resistant strains. The inhibitory effect of rifamycins was found to be nonspecific and was not confined to the inhibition of the RNA polymerase (T. Schupp, unpublished laboratory data).

Chinese investigators (Jiao Rui-shen et al., 1979) reported on the role of nitrate in the rifamycin SV fermentation. They came to the conclusion that nitrate stimulated production because of its regulatory effect on the lipid and rifamycin biosynthetic pathways. Their strain did not utilize nitrate as a nitrogen source. We do not observe this effect with our own rifamycin-producing strains. It is probable that the nitrate effect is strain specific. From this Chinese publication it can be deduced that rifamycin is probably being fermented worldwide as SV as well as B.

C. Conclusion

With the development of the rifamycin fermentation process the characteristics of our production strains have changed in many ways, and not only in their ability to produce more antibiotic. According to where the process has been developed, these characteristics will vary and thus the publications which have appeared and will appear in the future will often seem to be contradictory, particularly when the source of the strain being used is not clearly specified.

Improvement of the industrial production process is only one facet of the activities involved in the fermentation of the rifamycins. Biosynthetic studies supported by sophisticated genetic techniques, including genetic engineering, could lead to the discovery of other ansamycins with important, different antibacterial spectra. However, the main purpose of our studies on rifamycin biosynthesis was to improve our understanding of the fermentation and by this means to make logical improvements in the fermentation process.

V. RECOVERY AND PURIFICATION

The recovery and purification of rifamycin SV from rifamycin B is described by Sensi and Thiemann (1967). The procedure for the transformation of rifamycin SV to rifampicin was originally described by Maggi et al. (1966) (see Figure 1).

Rifamycin B is oxidized to rifamycin O with persulfate and can then be easily extracted in a suitable solvent. The glycolic acid residue is hydrolyzed by sulfuric acid in a solvent mixture. The resulting rifamycin S is the most important starting material for the preparation of semisynthetic derivatives such as rifampicin. This product was first prepared via 3-formyl rifamycin SV, which was obtained by treatment of rifamycin S with a secondary amine and formaldehyde and by oxidation of the Mannich base so formed with weak oxidizing agents, such as alkyl nitrites or lead tetraacetate, in a solvent. A second method for preparing 3-formyl rifamycin SV consists in treating rifamycin S with formaldehyde and a primary lower alkyl amine in the presence of an oxidizing agent such as manganese dioxide. The 3-formyl rifamycin SV can be converted to rifampicin by reacting it with 1-amino-4-methyl-piperazine. Various other methods of making rifampicin have been recently described.

VII. (BIO)ASSAY METHODS

Various assay methods have been developed for the monitoring of the extraction and purification process and for the testing of the final product, rifampicin. The basic methods were described some time ago by Gallo et al., 1960, 1962). A summary of all the assay procedures is given by Sensi and Thiemann (1967). The methods have subsequently been adapted to the different types of analytical equipment available.

Rifamycins B, O, S, and SV are usually estimated spectrophotometrically using a continuous autoanalyzer system.

The methods make use of the dramatic shifts in the electron spectrum, which occur when the hydroquinone forms are converted to the quinone form or vice versa by means of oxidation–reduction reactions.

The bioassay of rifampicin is carried out either turbidimetrically or on plates using well-known standard techniques. The various impurities which may be present in the final product or in clinical samples are often biologically active so that it is important to select the test organism most specific for rifampicin when assaying the final product. The strain most frequently used is *E. coli* ATCC 10536. Other strains are employed when testing for the presence of specific biologically active rifamycin degradation products in the blood or in urine (Fürész and Scotti, 1961).

VIII. MODE OF ACTION AND APPLICATION

The mode of action of the rifamycins has been described in detail by Wehrli (1977) and Wehrli and Staehelin (1971).

The rifamycins specifically inhibit the synthesis of RNA in bacteria by inactivating RNA polymerase. This occurs at very low concentrations (0.01 ng/ml, 10^{-8} M). Wehrli (1977) points out that as a considerable number of rifamycin derivatives have been synthesized and tested, some knowledge of the structural requirements of the molecule for effective RNA polymerase inhibition has been acquired. He lists three of the main ones, namely, free hydroxyl or keto groups at C1 and C8, an unbroken ansa bridge, and free hydroxyl groups at C21 and C23.

The structure of the derivative, as is to be expected, strongly influences its penetration in bacteria and its pharmacokinetics. Rifamycin B, which, as has already been pointed out, is a very weak antibiotic, is nevertheless a very strong inhibitor of bacterial RNA polymerase. Rifamycin SV has no bacterial activity when administered orally, whereas rifampicin is a powerful oral antibiotic.

The rifamycins show antiviral and antitumor activity, but only at very high concentrations. The mode of action is nonspecific. However, ansamitocins, new maytansinoid antitumor antibiotics, have been produced recently by fermentation (Tanida et al., 1980).

Medoff et al. (1972) showed that rifampicin in combination with amphotericin B inhibited growth of *Saccharomyces cerevisiae* and *Candida albicans*. Alone, neither of the antibiotics were active against these organisms when tested at the same concentrations. In this case the mode of action is not clearly understood, because it is known that rifampicin has no effect on eukaryotic RNA polymerases.

The clinical application of the most important semisynthetic rifamycin, rifampicin, has been principally in the treatment of tuberculosis in combination with other drugs (Hobby and Lenert, 1970).

Owing to the fact that spontaneous resistance to the rifamycins occurs in vitro (Kradolfer and Schnell, 1970), the use of rifampicin alone in the treatment of bacterial infections has been limited, although numerous reports of successful treatment of a variety of other diseases are to be found in the literature, a survey of which has been published by Nessi and Fowst (1979).

REFERENCES

Alderson, G., Goodfellow, M., Wellington, E. M., Williams, S. T., Minikin, S. M., and Minikin, D. E. (1979). Chemical and numerical taxonomy of *Nocardia mediterranei*. Paper presented at 8th FEMS Symposium on Actinomycete Biology, Berlin.

Allen, M. S., McDonald, I. A., and Rickards, R. W. (1981). The ansamycin antibiotic actamycin. I. Definition of structural features by deuterium labelling. *Tetrahedron Lett. 22*:1145–1148.

Anderson, M. G., Kibby, J. J., Rickands, R. W., and Rothschild, J. M. (1980). *J. Chem. Soc. Chem. Commun. 1980*:1277–1278.

Bezanson, G. S., and Vining, L. C. (1971). Studies on the biosynthesis of mitomycin C by *Streptomyces verticillatus*. *Can. J. Biochem. 49*:911–918.

Bickel, H., Mertens, P., Prelog, V., Seibl, J., and Walser, A. (1966). Stoffwechselprodukte von Mikroorganismen-53 Ueber die Konstitution von Ferrimycin A$_1$. *Tetrahedron Suppl. 8*:171–179.

Birnboim, H. C., and Doly, J. (1979). A rapid alkaline extraction procedure for screening recombinant plasmid DNA. *Nucleic Acids Res. 7*:1513–1523.

Birner, J., Hodgson, P. R., Lane, W. R., and Baxter, E. H. (1972). An Australian isolate of *Nocardia mediterranei* producing rifamycin SV. *J. Antibiot. 25*:356–359.

Brufani, M., Fedeli, W., Giacomello, G., and Vaciago, A. (1964). The x-ray analysis of the structure of rifamycin B. *Experientia 20*:339–342.

Brufani, M., Kluepfel, D., Lancini, G. C., Leitich, J., Mesentsev, A. S., Prelog, V., Schmook, F. P., and Sensi, P. (1973). Ueber die Biogenese des Rifamycins S. *Helv. Chim. Acta 56*:2315–2323.

Brufani, M., Cellai, L., and Keller-Schierlein, W. (1979). Degradation studies of naphthomycin. *J. Antibiot. 32*:167–168.

Bruggisser, S. (1975). Zur Biosynthese des Rifamycin-Chromophors. Dissertation No. 5435, Federal School of Technology, Zürich, Switzerland.

Celmer, W. D., Sciavolini, F. C., Routien, J. B. and Cullen, W. P. (1975). Verfahren zur Herstellung von Antibiotika. German Patent No. 25 00 898.

Cricchio, R., Antonini, P., and Sartori, G. (1980). Thiazorifamicins. III. Biosynthesis of rifamycins P, Q and Verde, novel metabolites from a mutant of *Nocardia mediterranei*. *J. Antibiot. 33*:842–846.

De Boer, C., Meulman, P. A., Wnuk, R. J., and Peterson, D. H. (1970). Geldanamycin, a new antibiotic. *J. Antibiot. 23*:442–447.

Delić, V., Hopwood, D. A., and Friend, E. J. (1970). Mutagenesis by N-methyl-N'-nitro-N-nitrosoguanidine (NTG) in *Streptomyces coelicolor*. *Mutat. Res. 9*:167–182.

Deshmukh, P. V., Kakinuma, K., Ameel, J. J., Rinehart, K. L., Jr., Wiley, P. F., and Li, L. H. (1976). Protostreptovaricins I–V. *J. Am. Chem. Soc. 98*:870–872.

Ferguson, J. H., Huang, H. T., and Davidson, J. W. (1957). Stimulation of streptomycin production by a series of synthetic organic compounds. *Appl. Microbiol. 5*:339–344.

Fürész, S., and Scotti, R. (1961). Further studies on rifomycin SV in vitro activity, adsorption and elimination in man. *Farmaco Ed. Sci.* 16:262–271.

Gallo, G. G., Sensi, P., and Redaelli, P. (1960). VII. Analisi spectrofotometrica della rifamicina B. *Farmaco Ed. Prat.* 15:283–291.

Gallo, G. G., Chiesa, L., and Sensi, P. (1962). Amperometric titrations of rifamycin B, rifamycin O, rifamycin S and rifamycin SV. *Farmaco Ed. Sci.* 17:668–678.

Ganguly, A. K., Szmulewicz, S., Sarre, O. Z., Greeves, D., Morton, J., and McGlotten, J. (1974). Structure of halomicin B. *J. Chem. Soc. Chem. Commun.* 1976:395–396.

Ganguly, A. K., Liu, Y. T., Sarre, O. Z., and Szmulewicz, S. (1977). Structures of halomicins A and C. *J. Antibiot.* 30:625–627.

Ghisalba, O. (1978). Untersuchungen über den Zusammenhang der Rifamycin-Chromophor-Biosynthese mit der Aromaten-Biosynthese über den Shikimisäureweg. Dissertation, University of Basel, Basel, Switzerland.

Ghisalba, O., and Nüesch, J. (1978a). A genetic approach to the biosynthesis of the rifamycin-chromophore in *Nocardia mediterranei*. I. Isolation and characterization of a pentose-excreting auxotrophic mutant of *Nocardia mediterranei* with drastically reduced rifamycin-production. *J. Antibiot.* 31:202–214.

Ghisalba, O., and Nüesch, J. (1978b). A genetic approach to the biosynthesis of the rifamycin-chromophore in *Nocardia mediterranei*. II. Isolation and characterization of a shikimate excreting auxotrophic mutant of *Nocardia mediterranei* with normal rifamycin-production. *J. Antibiot.* 31:215–225.

Ghisalba, O., and Nüesch, J. (1981). A genetic approach to the biosynthesis of the rifamycin-chromophore in *Nocardia mediterranei*. IV. Identification of 3-amino-5-hydroxybenzoic acid as a direct precursor of the seven-carbon amino starter-unit. *J. Antibiot.* 34:64–71.

Ghisalba, O., Traxler, P., and Nüesch, J. (1978). Early intermediates in the biosynthesis of ansamycins. I. Isolation and identification of protorifamycin I. *J. Antibiot.* 31:1124–1131.

Ghisalba, O., Traxler, P., Fuhrer, H., and Richter, W. J. (1979). Early intermediates in the biosynthesis of ansamycins. II. Isolation and identification of proansamycin B-M1 and protorifamycin I-M1. *J. Antibiot.* 32:1267–1272.

Ghisalba, O., Traxler, P., Fuhrer, H., and Richter, W. J. (1980). Early intermediates in the biosynthesis of ansamycins. III. Isolation and identification of further 8-deoxyansamycins of the rifamycin-type. *J. Antibiot.* 33:847–856.

Ghisalba, O., Fuhrer, H., Richter, W. J., and Moss, S. (1981). A genetic approach to the biosynthesis of the rifamycin-chromophore in *Nocardia mediterranei*. III. Isolation and identification of an early aromatic ansamycin-precursor containing the seven-carbon amino starter-unit and three initial acetate/propionate-units of the ansa chain. *J. Antibiot.* 34:58–63.

Ghisalba, O., Roos, R., Schupp, T., and Nüesch, J. (1982). Transformation of rifamycin S into the rifamycins B and L. A revision of the current biosynthetic hypothesis. *J. Antibiot.* 35:74–800.

Haber, A., Johnson, R. D., and Rinehart, K. L., Jr. (1977). Biosynthetic origin of the C_2 units of geldanamycin and distribution of label from D[6-13C]glucose. *J. Am. Chem. Soc.* 99:3541–3544.

Hartmann, G., Honikel, K. O., Knüsel, F., and Nüesch, J. (1967). The specific inhibition of the DNA directed RNA synthesis by rifamycin. *Biochim. Biophys. Acta* 145:843–844.

Hatano, K., Akiyama, S. I., Asai, M., and Rickards, R. W. (1982). Biosynthetic origin of aminobenzenoid nucleus (C_7N-Unit) of ansamitocin, a group of novel maytansinoid antibiotics. *J. Antibiotics* 35:1415–1417.

Haslam, E. (1974). *The Shikimate Pathway*. Butterworths, London, pp. 1–11.

Higashide, E., Asai, M., Ootsu, K., Tanida, S., Kozi, Y., Hasegawa, T., Kishi, T., Sugino, Y., and Yoneda, M. (1977). Ansamitocin, a group of novel maytansinoid antibiotics with antitumour properties from *Nocardia*. *Nature* 270:721–722.

Hobby, G. L., and Lenert, T. F. (1970). The action of rifampicin alone and in combination with other antituberculosis drugs. *Am. Rev. Resp. Dis.* 10Z:462–465.

Hoeksema, H., Chidester, C., Mizsak, S. A., and Baczynski, L. (1978). The chemistry of the rubradirins. I. The structures of rubransarols A and B. *J. Antibiot.* 31:1067–1069.

Hornemann, U., Kehrer, J. P., and Eggert, J. H. (1974). Pyruvic acid and D-glucose as precursors in mitomycin biosynthesis by *Streptomyces verticillatus*. *J. Chem. Soc. Chem. Commun.* 19741045–1046.

Hornemann, U., Eggert, J. H., and Honor, D. P. (1980). Role of D-[4-^{14}C]-erythrose and [3-^{14}C] pyruvate in the biosyntheses of the meta $C–C_6$–N unit of the mitomycin antibiotics in *Streptomyces verticillatus*. *J. Chem. Soc. Chem. Commun.* 1980:11–13.

Iwai, Y., Nakagawa, A., Sadakane, N., Omura, S., Oiwa, H., Matsumoto, S., Takahashi, M., Ikai, T., and Ochiai, Y. (1980). Herbimycin B, a new benzoiquinoid ansamycin with anti-TMV and herbicidal activities. *J. Antibiot.* 33:1114–1119.

Jiao Rui-shen, Chen Yu-mei, Wu Meng-gan, and Gu Weiling. (1979). Studies on the metabolic regulation of biosynthesis of rifamycin by *Nocardia mediterranei*. *Acta Phytophysiol. Sinica* 5:395–401.

Johnson, R. D., Haber, A., and Rinehart, K. L., Jr. (1973). The biosynthesis of geldanamycin. *Abstr. Pap. Am. Chem. Soc.* 166:124.

Johnson, R. D., Haber, A., and Rinehart, K. L., Jr. (1974). Geldanamycin biosynthesis and carbon magnetic resonance. *J. Am. Chem. Soc.* 96:3316–3317.

Karlsson, A., Sartori, G., and White, R. J. (1974). Rifamycin biosynthesis: further studies on origin of the ansa chain and chromophore. *Eur. J. Biochem.* 47:251–256.

Kibby, J. J., McDonald, I. A., and Rickards, R. W. (1980). 3-Amino-5-hydroxybenzoic acid as a key intermediate in ansamycin and maytansinoid biosynthesis. *J. Chem. Soc. Chem. Commun.* 1980:768–769.

Kishi, T., Yamana, H., Muroi, M., Harada, S., Asai, M., Hasegawa, T., and Mizuno, K. (1972). Tolypomycin, a new antibiotic. III. Isolation and characterization of tolypomycin Y. *J. Antibiot.* 25:11–15.

Kluepfel, D., Lancini, G. C., and Sarton, G. (1965). Metabolism of Barbital by *Streptomyces mediterranei*. *Appl. Microbiol.* 13:600–604.

Knöll, W. M. J., Rinehart, K. L., Wiley, P. F., and Li, L. H. (1980). Streptovaricin U, an acyclic ansamycin. *J. Antibiot.* 33:249–251.

Kradolfer, F., and Schnell, R. (1970). Incidence of resistant pulminary tuberculosis in relation to initial bacterial load. *Chemotherapy* 15:242–249.

Küenzi, M. T., Gruner, J., Fiechter, A., and Nüesch, J. (1977). Product inhibition during the rifamycin S fermentation. *Experientia* 33:141.

Kupchan, S. M., Komoda, Y., Branfman, A. R., Sneden, A. T., Court, W. A., Thomas, G. J., Hintz, H. P. J., Smith R. M., Karim, A., Howie, G. A., Verma, A. K., Nagao, Y., Dailey, R. G., Jr., Zimmerly, V. A., and Sumner, W. C., Jr. (1977). The maytansinoids. Isolation, structural elucidation, and chemical interrelation of novel ansa macrolides. *J. Org. Chem. 42*:2349–2357.

Lancini, G., Gallo, G. G., Sartori, G., and Sensi, P. (1969). Isolation and structure of rifamycin L and its biogenetic relationship with other rifamycins. *J. Antibiot. 22*:369–377.

Lancini, G., and Hengeller, C. (1969). Isolation of rifamycin SV from a mutant *Streptomyces mediterranei* strain. *J. Antibiot. 22*:637–638.

Lancini, G., and Sartori, G. (1976). Rifamycin G, a further product of *Nocardia mediterranei* metabolism. *J. Antibiot. 29*:466–468.

Lancini, G. C., and Sensi, P. (1967). Studies on the final steps in Rifamycin biosynthesis. *Proceedings of the 5th International Congress on Chemotherapy*, Vol. 1. K. H. Spitey (Ed.). Verlag Wiener Medizin Akademie, Vienna, pp. 41–47.

Lancini, G., and White, R. J. (1976). Rifamycin biosynthesis. In *Second International Symposium on Genetics of Industrial Microorganisms*. K. D. McDonald (Ed.). Academic, New York, pp. 139–153.

Lancini, G., Thiemann, J. E., Sartori, G., and Sensi, P. (1967). Biogenesis of rifamycins. The conversion of rifamycin B into rifamycin Y. *Experienta 23*:899–900.

Lancini, G., Hengeller, C., and Sensi, P. (1970). New naturally occurring rifamycins. In *Proceedings of the International Congress on Chemotherapy 6th*, 1969, Vol. 2. University of Tokyo Press, Tokyo, pp. 1166–1173.

Lancini, G. C., and Parenti, F. (1978). Rifamycin biogenesis. In *FEMS Symposium V: Antibiotics and Other Secondary Metabolites; Biosynthesis and Production*. R. Hütter, T. Leisinger, J. Nüesch, and W. Wehrli (Eds.). Academic, London, pp. 129–139.

Leitich, J., Oppolzer, W., and Prelog, V. (1964). Ueber die Konfiguration des Rifamycins B und verwandter Rifamycine. *Experientia 20*:343–344.

Leitich, J., Prelog, V., and Sensi, P. (1967). Rifamycin Y und seine Umwandlungsprodukte. *Experientia 23*:505–507.

McDonald, I. A., and Rickards, R. W. (1981). The ansamycin antibiotic actamycin. II. Determination of the structure using carbon-13 biosynthetic labelling. *Tetrahedron Lett. 22*:1149–1152.

Maggi, N., Pasqualucci, C. R., Ballotta, R., and Sensi, P. (1966). Rifampicin, a new orally active rifamycin. *Chemotherapia 11*:285–297.

Margalith, P., and Beretta, G. (1960). Rifomycin XI. Taxonomic studies on *Streptomyces mediterranei*. nov. sp. *Mycopathol. Mycol. Appl. 13*:4.

Margalith, P., and Pagani, H. (1961). Rifomycin XIV. Production of rifomycin B. *Appl. Microbiol. 9*:325–333.

Martinelli, E., Gallo, G. G., Antonini, P., and White, R. J. (1974). Structure of rifamycin W. A novel ansamycin from a mutant of *Nocardia mediterranei*. *Tetrahedron 30*:3087–3091.

Martinelli, E., Antonini, P., Cricchio, R., Lancini, G., and White, R. J. (1978). Rifamycin R, a novel metabolite from a mutant of *Nocardia mediterranei*. *J. Antibiot. 31*:949–951.

Medoff, G., Kobayashi, G. S., Kwan, C. N., Schlessinger, D., and Venkov, P. (1972). Potentiation of rifampicin and 5-fluorocystosine as antifungal antibiotics by amphotericin B. *Proc. Nat. Acad. Sci. U.S.A. 69*:196–199.

Milavetz, B., Kakinuma, K., Rinehart, K. L., Jr., Rolls, J. P., and Haak, W. J. (1973). Carbon-13 magnetic resonance spectroscopy and the biosynthesis of streptovaricin. *J. Am. Chem. Soc. 95*:5793–5795.

Muroi, M., Haibara, K., Asai, M. and Kishi, T. (1980). The structures of macbecin I and II, new antitumour antibiotics. *Tetrahedron Lett. 21*: 309–312.

Nessi, R., Fowst, G. (1979). Clinical use of rifampicin in combination for nonmycobacterial infections. A survey of published evidence. *J. Int. Med. Res. 7*:179–185.

Omura, S., Nakagawa, A., and Sadakane, N. (1979). Structure of herbimycin, a new ansamycin antibiotic. *Tetrahedron Lett. 20*:4323–4326.

Oppolzer, W., and Prelog, V. (1973). Ueber die Konstitution und die Konfiguration der Rifamycine B, O, S und SV. *Helv. Chim. Acta 56*:2287–2314.

Oppolzer, W., Prelog, V., and Sensi, P. (1964). Konstitution des Rifamycins B und verwandter Rifamycine. *Experientia 20*:336–339.

Prelog, V. (1963a). Constitutions of rifamycins. *Pure Appl. Chem. 7*:551–564.

Prelog, V. (1963b). Ueber die Konstitution der Rifamycine. *Chemotherapia 7*:133–136.

Rinehart, K. L., Jr. (1972). Antibiotics with ansa rings. *Acc. Chem. Res. 5*:57–64.

Rinehart, K. L., Jr., Maheshwari, M. L., Antosz, F. J., Mathur, H. H., Sasaki, K., and Schacht, R. J. (1971). Chemistry of the streptovaricins. VIII. Structures of streptovaricins A, B, C, D, E, F, and G. *J. Am. Chem. Soc. 93*:6273–6274.

Rinehart, K. L., Jr., Antosz, F. J., Sasaki, K., Martin, P. K., Naheshwari, M. L., Reusser, F., Li, L. H., Moran, D., and Wiley, P. F. (1974). Relative biological activities of individual streptovaricins and streptovaricin acetates. *Biochemistry 13*:861–867.

Rinehart, K. L., Jr., Antosz, F. J., Deshmukh, P. V., Kakinuma, K., Martin, P. K., Milavetz, B. I., Sasaki, K., Witty, T. R., Li, L. H., and Reusser, F. (1976). Identification and preparation of damavaricins, biologically active precursors of streptovaricins. *J. Antibiot. 29*:201–203.

Roos, R., and Ghisalba, O. (1980). Transformation of rifamycin S into the rifamycins B and L. A revision of the current biosynthetic hypothesis. *Experientia 36*:486.

Ruczaj, Z., Ostrowska-Krysiak, B., Sawnor-Korszynska, D., and Raczynska-Brjanowska, K. (1972). On the effect of barbital on *Streptomyces mediterranei. Acta Microbiol. Pol. B 4*:201–209.

Sasaki, K., Rinehart, K. L., Jr., Slomp, G., Grostic, M. F., and Olson, E. C. (1970). Geldanamycin. I. Structure assignment. *J. Am. Chem. Soc. 92*:7591–7593.

Schupp, T. (1973). Genetik von *Nocardia mediterranei*. Dissertation No. 5153, Federal School of Technology, Zürich, Switzerland.

Schupp, T., and Nüesch, J. (1979). Chromosomal mutation in the final step of rifamycin B biosynthesis. *FEMS Microbiol. Lett. 6*:23–27.

Schupp, T., Hütter, R., and Hopwood, D. A. (1975). Genetic recombination in *Nocardia mediterranei. J. Bacteriol. 121*:128–136.

Schupp, T., Traxler, P., and Auden, J. A. L. (1981). New rifamycins produced by a recombinant strain of *Nocardia mediterranei. J. Antibiot. 34*:965–970.

Sensi, P., and Thiemann, J. E. (1967). Production of rifamycins. *Prog. Ind. Microbiol.* 6:21–59.

Sensi, P., Margalith, P., and Timbal, M. T. (1959). Rifomycin, a new antibiotic.—Preliminary report. *Farmaco Ed. Sci.* 14:146–147.

Sensi, P., Greco, A. M., and Ballotta, R. (1960). Rifomycin. I. Isolation and properties of rifomycin B and rifomycin complex. *Antibiot. Annu. 1959–1960*:262–270.

Sensi, P., Ballotta, R., Greco, A. M., and Gallo, G. G. (1961). Rifomycin. XV—Activation of rifomycin B and rifomycin O.—Production and properties of rifomycin S and rifomycin SV. *Farmaco Ed. Sci.* 16:165–180.

Siminoff, P., Smith, R. M., Sokolski, W. T., and Savage, G. M. (1957). Streptovaricin. I. Discovery and biologic activity. *Am. Rev. Tuberc. Pulm. Dis.* 75:576–583.

Srinivasan, P. R., Shigeura, H. T., Sprecher, M., Sprinson, D. B., and Davis, B. D. (1956). The biosynthesis of shikimic acid from D-glucose. *J. Biol. Chem.* 220:477–497.

Tanida, S., Hasegawa, T., Hatano, K., Higashide, E., and Yoneda, M. (1980). Ansamitocins, maytansinoid antitumor antibiotics. Producing organism, fermentation and antimicrobial activities. *J. Antibiot.* 33:192–198.

Thiemann, J. E., Zucco, G., and Pelizza, G. (1969). A proposal for the transfer of *Streptomyces mediterranei* (Margalith and Beretta) to the genus *Nocardia mediterreanea*. *Arch. Mikrobiol.* 67:147–151.

Traxler, P., and Ghisalba, O. (1982). A genetic approach to the biosynthesis of the rifamycin-chromophore in *Nocardia mediterranei*. V. Studies on the biogenetic origin of 3-substituents. *J. Antibiotics* 35:1361–1366.

Traxler, P., Schupp, T., Fuhrer, H., and Richter, W. J. (1981). 3-Hydroxy-rifamycin S and further novel ansamycins from a recombinant strain R-21 of *Nocardia mediterranei*. *J. Antibiot.* 34:971–979.

Wehrli, W. (1977). Ansamycins. Chemistry, biosynthesis and biological activity. *Top. Curr. Chem.* 72:21–49.

Wehrli, W., and Staehelin, M. (1971). Actions of the rifamycins. *Bacteriol. Rev.* 35:290–309.

White, R. J., and Martinelli, E. (1974). Ansamycin biogenesis: Incorporation of [1-^{13}C]glucose and [1-^{13}C]glycerate into the chromophore of rifamycin S. *FEBS Lett.* 49:233–236.

White, R. J., Martinelli, E., Gallo, G. G., Lancini, G., and Beynon, P. (1973). Rifamycin biosynthesis studied with ^{13}C enriched precursors and carbon magnetic resonance. *Nature* 243:273–277.

White, R. J., Martinelli, E., and Lancini, G. (1974). Ansamycin biogenesis: Studies on a novel rifamycin isolated from a mutant strain of *Nocardia mediterranei*. *Proc. Nat. Acad. Sci. U.S.A.* 71:3260–3264.

White, R. J., Lancini, G., and Sensi, P. (1975). New natural rifamycins. In *Proceedings of the First Intersectional Congress of IAMS*, Vol. 3. T. Hasagawa (Ed.). Science Council of Japan, Tokyo, pp. 483–492.

Williams, T. H. (1975). Naphthomycin, a novel ansamacrocyclic antimetabolite. Proton NMR spectra and structure elucidation using lanthanide shift reagent. *J. Antibiot.* 28:85–86.

10

THE AMINOGLYCOSIDES: PROPERTIES, BIOSYNTHESIS, AND FERMENTATION

RYO OKACHI *Pharmaceuticals Research Laboratory, Kyowa Hakko Kogyo Company Ltd., Shizuokaken, Japan*
TAKASHI NARA *Pharmaceuticals Division, Kyowa Hakko Kogyo Company Ltd., Tokyo, Japan*

I. INTRODUCTION

Aminoglycoside antibiotics are a representative group of antibiotics discovered during the early age of antibiotic research; they became important, clinically useful therapeutic agents for the maintenance of human, animal, and plant health. This group of antibiotics, which comprises over 100 compounds produced by diverse genera of microorganisms, have typical chemical and antimicrobial features in common: They are all mixtures of closely related water-soluble basic pseudosaccharides that display a broad antibacterial spectrum.

The first aminoglycoside antibiotic discovered, streptomycin, could successfully exterminate tuberculosis from human life, and it stimulated generally antibiotic research. As second-generation aminoglycosides, kanamycin and related antibiotics proved to be very useful for the treatment of serious infections with various gram-negative pathogenic bacteria. They were followed by gentamicins, which expanded the antibacterial spectrum of aminoglycoside antibiotics to the Pseudomonadaceae, hardly affected by other antibiotics. The medical use of aminoglycoside antibiotics is limited to parenteral doses, because of its poor absorption from the gastrointestinal tract. Oto- or nephro-toxicity common to most of the aminoglycoside antibiotics is still an unsolved problem; however, the remarkable bactericidal activity and stability rarely found in natural products has stimulated the search and developmental studies for these antibiotics. Besides the naturally isolated compounds, semisynthetic derivatives of aminoglycoside antibiotics were developed to conquer resistant pathogens carrying aminoglycoside-inactivating enzymes.

Some other aminoglycoside antibiotics, such as, kasugamycin, have been applied for agricultural use.

Recently, fermentation (Nara, 1977, 1978; Claridge, 1979), biosynthesis (Rinehart, 1980; Cleophax et al., 1980; Pearce and Rinehart, 1981), chemical synthesis (Reden and Dürckheimer, 1979; Umezawa, 1980; Suami, 1980), and pharmacological studies (Barza and Scheife, 1977) of clinically important aminoglycoside antibiotics have been reviewed in detail. During the study of aminoglycoside antibiotics, numerous successes were achieved in adjacent research fields such as mutasynthesis, derived from biosynthetic studies, elucidation of resistance mechanisms of pathogens, and also chemical modification of aminoglycoside antibiotics. This chapter mainly deals with the current status of the fermentation process and production of aminoglycoside antibiotics.

II. HISTORY AND DEVELOPMENTS

The dawn of aminoglycoside antibiotics was announced by the discovery of streptomycin in 1944 by Selman Waksman (Schatz et al., 1944) in the New Jersey Agricultural Experiment Station of Rutgers University. The story of that epoch-making antibiotic was described by Waksman (1949) and others (Weinstein and Ehrenkranz, 1958). Streptomycin was the first antibiotic produced by a *Streptomyces* (*Streptomyces griseus*), one of the genera of the Actinomycetales. This presented the driving force to stimulate further investigation of the genus *Streptomyces* in the search for antibiotic-producing microorganisms. Streptomycin analogs were later reported and are summarized in Figure 1.

In 1949, a second aminoglycoside, neomycin, was discovered and found to consist as a complex of three components, neomycins A, B, and C, each having different antimicrobial activities; it was produced by *Streptomyces fradiae*. The properties, fermentation production, and practical applications of neomycin have been reviewed by Waksman (1953). Neomycin B is the major and most active component in the mixture, and commercial preparations consist almost entirely of neomycin B. In medical practice, it is used for external application, owing to its serious nephrotoxicity. The neomycin group of antibiotics includes many related substances which have a 4,5-substituted deoxy-streptamine moiety in common, as is shown in Figure 2. Rinehart (1964) reviewed the neomycins and related antibiotics, and later he and his colleagues (Shier et al., 1969) first reported on the mutasynthesis of hybrimycins using deoxystreptamine-requiring mutants of *S. fradiae* for neomycin production. Biologically converted neomycins (hybrimycins) were produced depending on the type of cyclitol added instead of deoxystreptamine.

Kanamycin was the next successful chemotherapeutic agent reported by Umezawa et al. in 1957 (Fig. 3). Both kanamycins A and B are useful antibiotics and each of them is important as starting material for semisynthetic derivatives with improved properties.

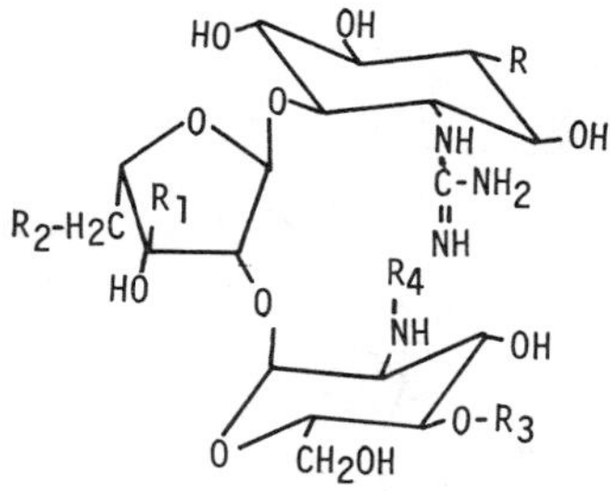

	R	R_1	R_2	R_3	R_4
[1]	NHC(=NH)NH$_2$	CHO	H	H	CH$_3$
[2]	NHC(=NH)NH$_2$	CH$_2$OH	H	H	CH$_3$
[3]	NHC(=NH)NH$_2$	CHO	H	✢	CH$_3$
[4]	NHC(=NH)NH$_2$	CHO	H	H	H
[5]	NHC(=NH)NH$_2$	CHO	OH	H	CH$_3$
[6]	NHC(=NH)NH$_2$	CHO	OH	✢	CH$_3$
[7]	OCONH$_2$	CH$_2$OH	H	H	CH$_3$

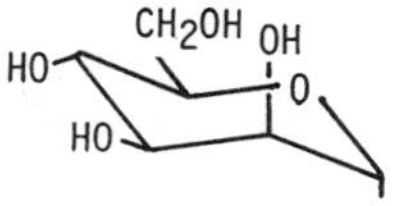

Figure 1 The streptomycin group. Streptomycin [1] *Streptomyces griseus* (Schatz et al., 1944), *S. olivaceus, S. rameus, S. bikiniensis, S. galbus, S. mashuensis, S. erythrochromogenes* var. *narutoensis.* Dihydrostreptomycin [2] *S. humidus* (Tatsuoka et al., 1957). Mannosidostreptomycin (=Streptomycin B) [3] *S. griseus* (Fried and Titus, 1947). N-Demethylstreptomycin [4] *S. griseus* (Heding, 1968). Hydroxystreptomycin [5] *S. griseocarneus* (Benedict et al., 1950), *S. rubrireticuli.* Mannosidohydroxystreptomycin [6] *S. sp.* (Arcamone et al., 1968). Bluensomycin [7] *S. bluensis* var. *bluensis* (Mason et al., 1962) = Glebomycin; *S. hygroscopicus* var. *glebosus* (Okanishi et al., 1962).

Nebramycins were described in 1967 (Fig. 3) as tenebrimycins consisting of three major components, factors 2, 4, and 5', and some minor components. Factor 6, tobramycin, chemically derived form factor 5', is used in clinical medicine. Factor 2, apramycin, has been introduced in veterinary medicine.

In 1963 a new antibiotic mixture, gentamicin, produced by a new species of *Micromonospora*, was identified as the first aminoglycoside antibiotic from this genus (Fig. 4). The gentamicin C complex (C_1, C_{1a}, C_2) was demonstrated to be very useful for anti-*Pseudomonas* therapy. It has stimulated the research for other useful antibiotics from *Micromonospora* sp. It was soon followed by sisomicin and related substances (Fig. 5), and by sagamicin with numerous minor components. Sisomicin and sagamicin are already currently used in clinical practice.

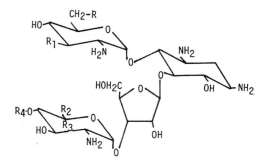

	R	R₁	R₂	R₃	R₄
[8]	NH₂	OH	H	CH₂NH₂	H
[9]	NH₂	OH	CH₂NH₂	H	H
[10]	OH	OH	H	CH₂NH₂	H
[11]	OH	OH	CH₂NH₂	H	H
[12]	OH	H	H	CH₂NH₂	a
[13]	OH	H	H	CH₂NH₂	H
[14]	OH	OH	H	CH₂NH₂	a

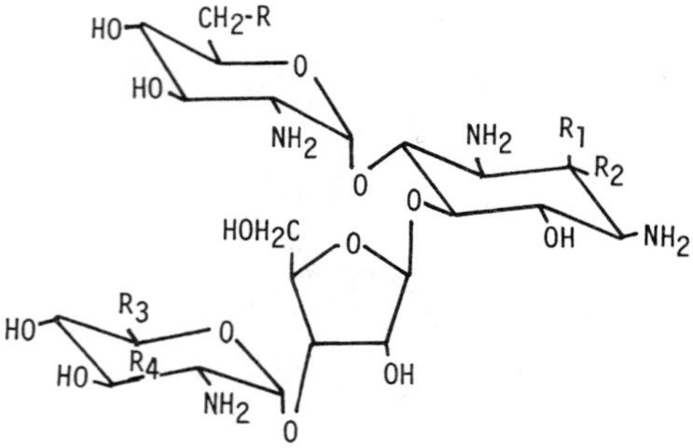

	R	R₁	R₂	R₃	R₄
[19]	NH₂	H	OH	H	CH₂NH₂
[20]	NH₂	H	OH	CH₂NH₂	H
[21]	NH₂	OH	H	H	CH₂NH₂
[22]	NH₂	OH	H	CH₂NH₂	H
[23]	OH	H	OH	H	CH₂NH₂
[24]	OH	H	OH	CH₂NH₂	H

a
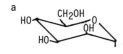

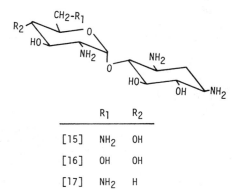

	R₁	R₂
[15]	NH₂	OH
[16]	OH	OH
[17]	NH₂	H

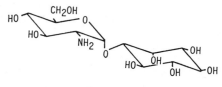

[18]

In 1970, another aminoglycoside antibiotic, ribostamycin, was added to the series (Fig. 6). Although it displays a lower specific activity and is less active against *Pseudomonas*, it hardly shows ototoxicity, which is a common and serious side effect of aminoglycoside antibiotic therapy. It is clinically utilized in Japan as a therapeutic agent and is safer than kanamycin.

Ribostamycin was followed by the discovery of butirosins, produced by *Bacillus* sp. Butirosins possess a unique and useful side chain (2-hydroxy-aminobutyric acid) on the 1-N position of deoxystreptamine (Fig. 6). Clinically isolated pathogens harboring aminoglycoside phosphotransferase were resistant to ribostamycin, but those clinical isolates were sensitive to the naturally modified ribostamycin, butirosin. It was later revealed that the 2-hydroxyaminobutyric acid (HABA) moiety attached on the 1-N position of deoxystreptamine was effective in promoting resistance to such inactivating enzymes. Although the butirosins themselves were not used in clinical medicine, they stood model for 1-N-HABA modifications, by chemical means of most of the deoxystreptamine-containing antibiotics. This procedure yielded amikacin (1-N-HABA kanamycin A) as a second important semisynthetic aminoglycoside antibiotic, next to the successful development of dibekacin. Dibekacin, also resistant to aminoglycoside phosphotransferase, was developed by removing the hydroxyl moieties at the 3'- and 4'-positions of 2,6-diamino-2,6-dideoxy-glucose of kanamycin B. Incredibly enough, a 4'-deoxy sugar-containing antibiotic complex, designated as seldomycin, was subsequently isolated in 1977 from *Streptomyces hofunensis* (Fig. 7).

All of the useful antibiotics mentioned above are composed as trisaccharides and some neomycins and paromomycins as tetrasaccharides. In 1977, fortimicin was reported as a pseudodisaccharide aminoglycoside antibiotic (Fig. 8) produced by *Micromonospora* sp. Its peculiar structure and excellent antibacterial spectrum aroused much interest. Fortimicin was not inactivated by aminoglycoside-modifying enzymes, except for aminoglycoside acetyltransferase (3)-I. Fortimicin A, among many minor components (McAlpine et al., 1980; Shirahata et al., 1980), is currently under development for clinical use; it is also of interest that the compound exhibits very low ototoxicity and renal toxicity. Its ototoxicity, initially investigated with newborn guinea pigs,

---

Figure 2 The neomycin group. Neomycin B (=Framycetin, Streptothricin B$_{II}$) [8] *Streptomyces fradiae* Waksman and Lechevalier, 1949), *S. albogriseolus*. Neomycin C (=Streptothricin B$_I$ [9] *S. fradiae*. Neomycin component G, K; *S. fradiae* (Claes et al., 1974). Paromycin I (=Aminosidine I, Catenulin) [10] *S. rimosus* forma *paromomycinus* (Haskell et al., 1959) *S. chrestomyceticus aurantiodeus*. Paromomycin II (=Aminosidine II) [11]. Lividomycin A (=Quintomycin B) [12] *S. lividus* (Oda et al., 1971). Lividomycin B (=Quintomycin D) [13]. Mannosylparomomycin (=Quintomycin A, Antibiotic 2330-C) [14]. Neamine (=Neomycin A) [15] *S. fradiae* (Waksman and Lechevalier, 1949). Paromamine (=Neomycin D) [16] *S. fradiae* (Hessler et al., 1970), *S. kanamyceticus* (Kojima et al., 1968). Seldomycin factor 2 (=XK-88-2) [17] *S. hofunensis* (Nara et al., 1977a). X-14847 [18] *Micromonospora echinospora* (Maehr et al., 1980). Hybrimycin A$_1$ [19], A$_2$ [20], B$_1$ [21], B$_2$ [22], C$_1$ [23], C$_2$ [24] *S. fradiae* (Shier et al., 1969).

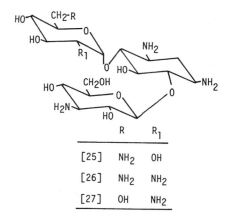

	R	R$_1$
[25]	NH$_2$	OH
[26]	NH$_2$	NH$_2$
[27]	OH	NH$_2$

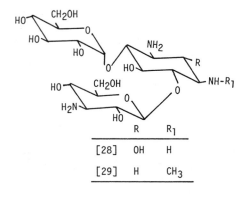

	R	R$_1$
[28]	OH	H
[29]	H	CH$_3$

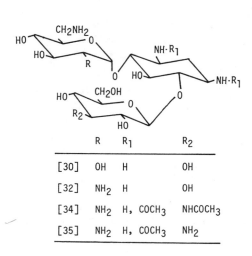

	R	R$_1$	R$_2$
[30]	OH	H	OH
[32]	NH$_2$	H	OH
[34]	NH$_2$	H, COCH$_3$	NHCOCH$_3$
[35]	NH$_2$	H, COCH$_3$	NH$_2$

	R	R$_1$
[31]	NH$_2$	OH
[33]	OH	NH$_2$

	R
[36]	H
[37]	OH

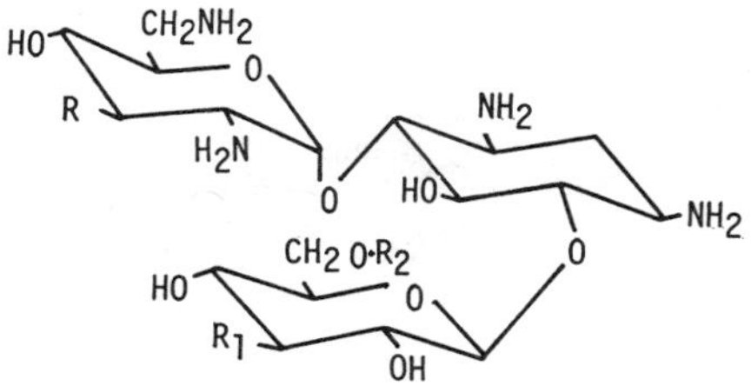

	R	R1	R2
[38]	OH	OH	H
[39]	OH	NH2	CONH2
[40]	H	NH2	CONH2
[41]	H	NH2	H
[45]	H	OH	H

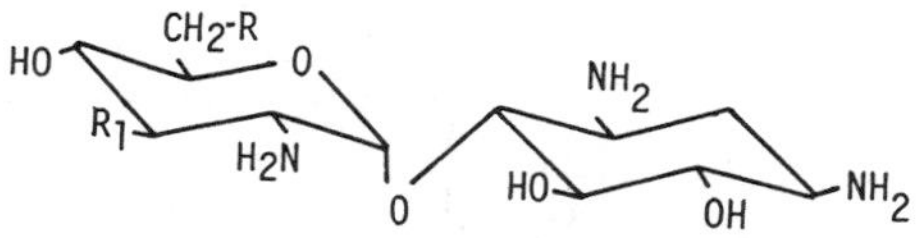

	R	R1
[42]	NH2	H
[43]	OH	H

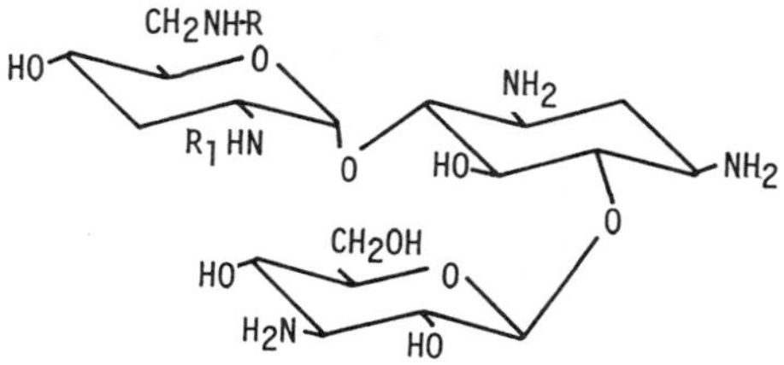

	R	R1
[44]	H	CONH2
[46]	CONH2	H

Figure 3 Kanamycin and the nebramycin group. Kanamycin A [25] *Streptomyces kanamyceticus* (Umezawa et al., 1957); Kanamycin B [26] (Schmitz et al., 1958); Kanamycin C [27] (Maeda et al., 1957); 2-Hydroxy-6'-deaminokanamycin [28] (Yasuda et al., 1975); 1-N-Methyl-6'-deaminokanamycin [29]; NK-1001 [30], NK-1003 [31], NK-1012-1 [32], NK-1012-2 [33], NK-1012-3; NK-1013-1 [34], NK-1013-2 [35] (Murase et al., 1970). Nebramycin complex; *S. tenebrarius* (Stark et al., 1967); *Streptoverticillium hindustanus* (Kawaguchi et al., 1977); factor 2 (=Apramycin) [36] *S. tenebrarius*, *Saccharopolyspora hirsuta* (Kamiya et al., 1980); factor 3 [38], factor 4 [39], factor 5' [40], factor 5 [26]; factor 6 (=Tobramycin) [41], factor 7 (=Oxyapramycin) [37] (Dorman et al., 1976); factor 8 (=Nebramine) [42], factor 9 (=Lividamine) [43]; factor 10 (=Neamine) [15], factor 11 [44], factor 12 [45]; factor 13 [46] (Koch et al., 1978).

is extremely low and its renal toxicity is lower than that of amikacin or gentamicin (Saito et al., 1980). Subsequently much energy was put in the search for similar compounds, resulting in the discovery of many analogous antibiotics from various actinomycetes, for example, sporaricins from *Saccharopolyspora*, sannamycins and istamycins from *Streptomyces*, and dactimicins from *Dactylosporangium*.

Besides these large groups of aminoglycoside antibiotics, miscellaneous related aminoglycoside antibiotics have been reported (Fig. 9); some of them find medical or agricultural application.

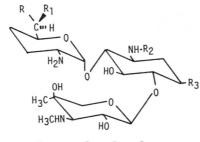

	R	R_1	R_2	R_3
[47]	$NHCH_3$	CH_3	H	NH_2
[48]	NH_2	CH_3	H	NH_2
[49]	NH_2	H	H	NH_2
[50]	NH_2	*	H	NH_2
[51]	$NHCH_3$	H	H	NH_2
[52]	NH_2	H	CH_3	NH_2
[53]	$N(CH_3)_2$	H	H	NH_2
[54]	$NHCH_3$	H	H	OH

* isomer of C-6'

	R	R_1
[55]	CH_3	NH_2
[56]	H	NH_2
[57]	CH_3	OH
[58]	H	OH

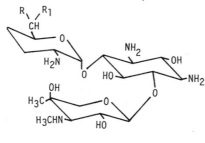

	R	R_1
[70]	$NHCH_3$	CH_3
[71]	NH_2	CH_3

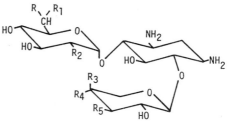

	R	R_1	R_2	R_3	R_4	R_5
[59]	H	OH	NH_2	H	OH	$NHCH_3$
[60]	H	OH	NH_2	OH	H	$NHCH_3$
[61]	H	OH	NH_2	H	OH	OH
[62]	H	NH_2	OH	OH	H	$NHCH_3$
[63]	H	OH	NH_2	H	OH	$N(CHO)CH_3$
[64]	H	NH_2	OH	OH	CH_3	$NHCH_3$
[65]	CH_3	NH_2	NH_2	OH	CH_3	$NHCH_3$
[66]	H	OH	NH_2	OH	CH_3	$NHCH_3$
[67]	H	NH_2	NH_2	OH	CH_3	$NHCH_3$
[68]	CH_3	NH_2	NH_2	OH	CH_3	$NHCH_3$
[69]	CH_3	OH	NH_2	OH	CH_3	$NHCH_3$

III. BIOSYNTHESIS

Aminoglycoside antibiotics can be classified in distinct groups (as shown in Table 1) according to structural similarities. For each group of antibiotics, the related compounds are biosynthesized from common precursors passing through common intermediates, and finally diverge to the individual end products.

Antibiotics are generally considered as secondary metabolites (idiolites), nonessential by-products for the producing microorganisms. For some aminoglycoside antibiotics, the biosynthetic route seems to be related to that of cell wall peptidoglycan biosynthesis (Barabas et al., 1978, 1980; Nimi et al., 1981). Up to now, all aminoglycoside antibiotics are known to be produced only by microorganisms with a bacteria-like cell wall peptidoglycan such as Actinomycetales (*Streptomyces, Micromonospora*, etc.) and Eubacteriales (*Bacillus, Pseudomonas*).

Common approaches utilized in the study of the biosynthetic mechanism of aminoglycoside antibiotics are

1. Microbiological conversion studies, based on the addition of postulated labeled or nonlabeled precursors to intact cells or cell-free enzymic systems of producing organisms
2. Mutational studies aimed at the isolation of blocked mutants, especially in individual steps of the secondary metabolite pathway

Based on these approaches, tremendous progress has been made toward the elucidation of the biosynthesis of neomycins, and these aspects have been reviewed together with those of some of the other aminoglycoside antibiotics (Rinehart and Stroshane, 1976; Rinehart, 1980; Sepulchre et al., 1980).

The biosynthesis mechanism of streptomycin presents an old problem, still important at present. Several reviews have appeared in recent years on this subject (Rinehart and Stroshane, 1976; Grisebach, 1978; Claridge, 1979) including one of the A factor (Khokhlov et al., 1976), which stimulates streptomycin accumulation while influencing the ultrastructure of the producing microorganism (Zaslavsky et al., 1979). Walker (1980) has surveyed the recent status of streptomycin biosynthesis.

Figure 4 The gentamicin group. Gentamicin C complex; *Micromonospora purpurea, M. echinospora* (Weinstein et al., 1963); *Dactylosporangium thailandense* (Fujii et al., 1980); C_1 [47], C_2 [48], C_{1a} [49], C_{2a} [50] (Daniels, 1975). Sagamicin (=XK-62-2) [51] *M. sagamiensis* (Nara et al., 1975) =Gentamicin C_{2b} (Daniels et al., 1975); XK-62-3 [52], XK-62-4 [53], SUM-3 [54] (Okachi and Nara, 1980). Antibiotic IB_1 (=Combimicin B_1) [55] *Micromonospora* sp.; IB_2 (=Combimicin B_2) [56] *M. echinospora* (Oka et al., 1979, 1981); IA_1 [57], IA_2 (=Combimicin A_2) [58]. Gentamicin A [59] (Maehr and Schaffner, 1967), A_1 [60] (Nagabhushan et al., 1975), A_2 [61], A_3 [62], A_4 [63], B [64], B_1 [65], X_2 [66] (Wagman et al., 1972); JI-20A [67], JI-20B [68] (Testa and Tilley, 1975). Antibiotic G-418 [69] *M. rhodorangea* (Wagman et al., 1974). 2-Hydroxygentamicin C_1 [70], C_2 [71] *M. purpurea* (Rosi et al., 1977).

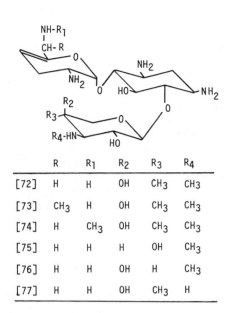

	R	R_1	R_2	R_3	R_4
[72]	H	H	OH	CH_3	CH_3
[73]	CH_3	H	OH	CH_3	CH_3
[74]	H	CH_3	OH	CH_3	CH_3
[75]	H	H	H	OH	CH_3
[76]	H	H	OH	H	CH_3
[77]	H	H	OH	CH_3	H

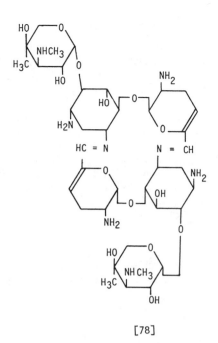

[78]

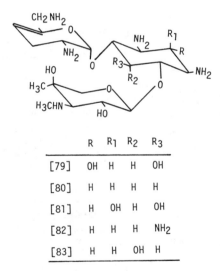

	R	R_1	R_2	R_3
[79]	OH	H	H	OH
[80]	H	H	H	H
[81]	H	OH	H	OH
[82]	H	H	H	NH_2
[83]	H	H	OH	H

Figure 5 The sisomicin group. Sisomicin (=Antibiotic 66-40) [72] *Micromonospora inyoensis* (Weinstein et al., 1970) *M. cyaneogranulata* (Kotani et al., 1979a); *Dactylosporangium thailandense* (Fujii et al., 1980). Verdamicin (-Sch 15666) [73] *M. rhodorangea*, *M. grisea* (Weinstein et al., 1975b), *M. cyaneogranulata* (Kotani et al., 1979b). G-52 (=Sch 17726) [74] *M. zionensis* (Marquez et al., 1976). 66-40C [78] *M. inyoensis* (Davies et al., 1977); 66-40B [75], 66-40D [76] (Davies et al., 1975), 66-40G [77] (Kugelman et al., 1978). Mutamicin 1 [79] *M. inyoensis* (Testa et al., 1974); 2 [80], 4 [81], 5 [82], 6 [83] (Weinstein et al., 1975a).

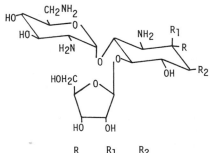

	R	R_1	R_2	R_3	R_4
[84]	NH_2	OH	H	H	OH
[85]	NH_2	OH	H	OH	H
[86]	NH_2	OH	⚹	OH	H
[87]	NH_2	OH	⚹	H	OH
[88]	OH	OH	⚹	OH	H
[89]	OH	OH	⚹	H	OH
[90]	NH_2	H	⚹	OH	H
[91]	NH_2	H	⚹	H	OH
[92]	OH	OH	H	H	OH

⚹ $COCHOHCH_2CH_2NH_2$

[95]

	R	R_1	R_2
[96]	H	H	$NHCH_3$
[97]	OH	H	NH_2
[98]	H	OH	NH_2

	R	R_1	R_2	R_3	R_4
[93]	CH_3	H	H	H	OH
[94]	CH_3	H	H	OH	H

Figure 6 Ribostamycin and the butirosin group. Ribostamycin [84] *Streptomyces ribosidificus* (Shomura et al., 1970); *Bacillus circulans* (Fujiwara et al., 1979); Xylostasin [85] *B. vitellinus* (Horii et al., 1974). Butirosin A [86] *B. circulans* (Howells et al., 1972); Butirosin B [87]; Bu-1709E$_1$ [88], Bu-1709E$_2$ [89] (Tsukiura et al., 1973); Bu-1975C$_1$ [90], Bu-1975C$_2$ [91] (Kawaguchi et al., 1974); LL-BM408$_\alpha$ [92] *S. canus* (Kirby et al., 1977a); 3',4'-dideoxy-6'-N-methylbutirosin A [93] *B. circulans* (Takeda et al., 1978a); 3',4'-Dideoxy-6'-N-methylbutirosin B [94]. 3',4'-Dideoxy-3'-eno-ribostamycin [95] *S. ribosidificus* (Yasuda et al., 1975); 1-N-Methylribostamycin [96], 2-Hydroxy-scyllo-ribostamycin [97]; 2-Hydroxy-myo-ribostamycin [98].

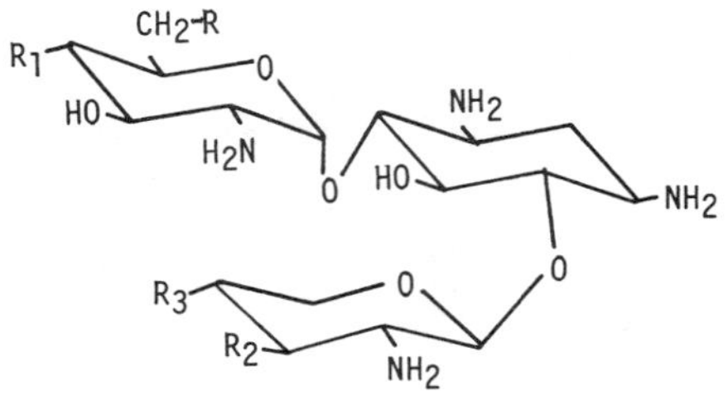

	R	R$_1$	R$_2$	R$_3$
[99]	OH	OH	OH	OH
[100]	NH$_2$	OH	OH	OH
[101]	NH$_2$	H	NH$_2$	OCH$_3$

Figure 7 The seldomycin group. Seldomycin factor 1 (=XK-88-1) [99] *Streptomyces hofunensis* (Nara et al., 1977a) factor 2 [17], factor 3 [100], factor 5 [101].

Mutual interactions of the multiple components involved in the biosynthesis of gentamicin C mixtures have been elucidated and the role of the cobalt ion in each of the methylation steps has been discussed (Testa and Tilley, 1976). The role of cobalt ions and vitamin B$_{12}$ on C- and N-methylation steps during antibiotic synthesis is of special interest with relation to the biosynthesis of fortimicin (Yamamoto et al., 1977) and seldomycin (Shimizu et al., 1978).

Some species of aminoglycoside antibiotic-producing microorganisms exhibit identical antibiotic-inactivating systems such as acetylation or phosphorylation reactions as clinically isolated aminoglycoside-resistant gram-negative pathogens. The resistance mechanisms of such clinical isolates was revealed to be controlled by R plasmids, which might originate from the antibiotic-producing organisms (Courvalin et al., 1977). The role of such antibiotic-inactivating systems in the producing microorganisms seems to be one of a

Figure 8 The fortimicin group. Fortimicin A (=XK-70-1) [102] *Micromonospora olivasterospora* (Nara et al., 1977b); Fortimicin B [103]; Fortimicin C [104], D [105], KE [106] (Sugimoto et al., 1979). Sporaricin A [107] *Saccharopolyspora hirsuta* (Deushi et al., 1979a); Sporaricin B [108]; Sporaricin C [109]; D [110], (Mori et al., 1978). Sannamycin A [111] *Streptomyces sannanensis* (Deushi et al., 1979b); Sannamycin B [112]; Sannamycin C [113] (Deushi et al., 1980). Istamycin A (=Sannamycin A) [111] *S. tenjimariensis* (Okami et al., 1979); Istamycin B [114]; Istamycin A$_0$ (=Sannamycin B) [112]; Istamycin B$_0$ [115] (Umezawa et al., 1981). Dactimicin (=SF-2052) [116] *Dactylosporangium matsuzakiense*)Inouye et al., 1979). SF-1854 (=SF-2052B) [117] *Micromonospora* sp. (Inouye et al., 1980).

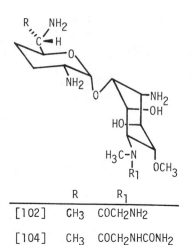

	R	R$_1$
[102]	CH$_3$	COCH$_2$NH$_2$
[104]	CH$_3$	COCH$_2$NHCONH$_2$
[105]	H	COCH$_2$NH$_2$

	R	R$_1$
[103]	CH$_3$	H
[106]	H	H

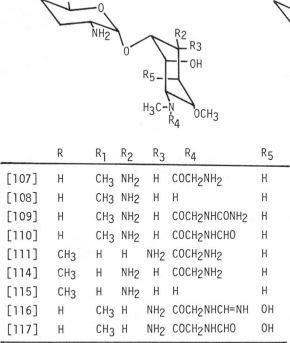

	R	R$_1$	R$_2$	R$_3$	R$_4$	R$_5$
[107]	H	CH$_3$	NH$_2$	H	COCH$_2$NH$_2$	H
[108]	H	CH$_3$	NH$_2$	H	H	H
[109]	H	CH$_3$	NH$_2$	H	COCH$_2$NHCONH$_2$	H
[110]	H	CH$_3$	NH$_2$	H	COCH$_2$NHCHO	H
[111]	CH$_3$	H	H	NH$_2$	COCH$_2$NH$_2$	H
[114]	CH$_3$	H	NH$_2$	H	COCH$_2$NH$_2$	H
[115]	CH$_3$	H	NH$_2$	H	H	H
[116]	H	CH$_3$	H	NH$_2$	COCH$_2$NHCH=NH	OH
[117]	H	CH$_3$	H	NH$_2$	COCH$_2$NHCHO	OH

	R	R$_1$
[112]	H	OCH$_3$
[113]	OCH$_3$	H

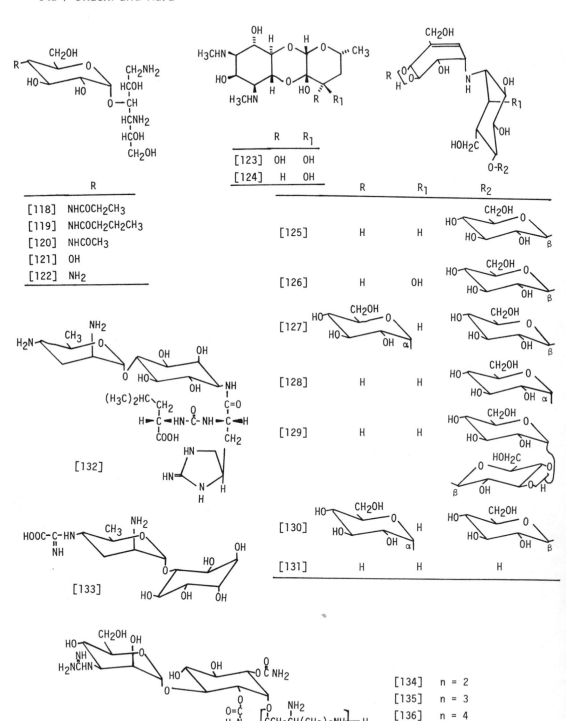

	R	R_1
[123]	OH	OH
[124]	H	OH

	R
[118]	$NHCOCH_2CH_3$
[119]	$NHCOCH_2CH_2CH_3$
[120]	$NHCOCH_3$
[121]	OH
[122]	NH_2

	R	R_1	R_2
[125]	H	H	
[126]	H	OH	
[127]		H	
[128]	H	H	
[129]	H	H	
[130]		H	
[131]	H	H	H

[132]

[133]

[134]	n = 2
[135]	n = 3
[136]	n = 4
[137]	n = 5

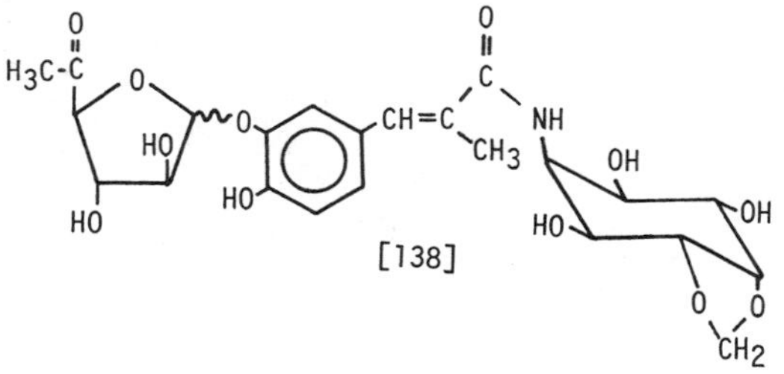

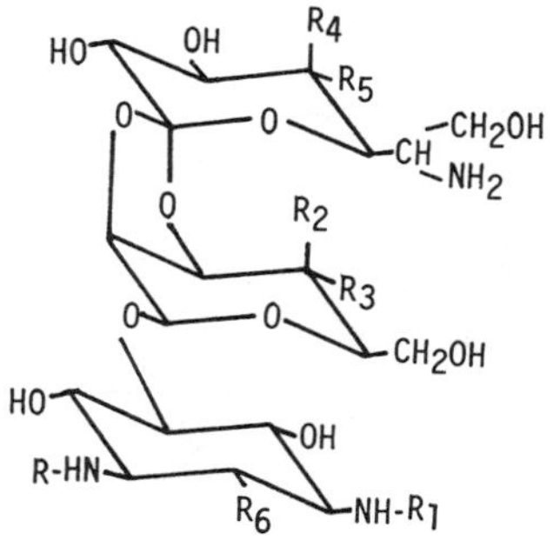

	R	R_1	R_2	R_3	R_4	R_5	R_6
[139]	CH_3	H	OH	H	OH	H	H
[140]	CH_3	CH_3	H	OH	H	OH	H
[141]	CH_3	CH_3	OH	H	OH	H	H
[142]	H	CH_3	OH	H	OH	H	H
[143]	H	H	OH	H	OH	H	OH
[144]	H	H	OH	H	OH	H	H

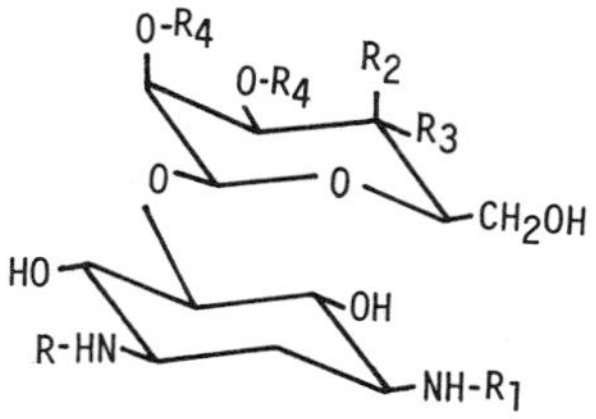

	R	R_1	R_2	R_3	R_4
[145]	CH_3	CH_3	H	OH	✶
[146]	H	H	H	OH	H
[147]	H	H	OH	H	H

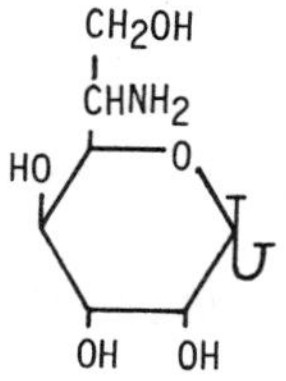

Figure 9 Other aminoglycosides. Sorbistin A_1 [118] *Pseudomonas sorbicinii* (Tsukiura et al., 1976); =LL-AM31β; *Streptoverticillium netropsis* (Kirby et al., 1977b); Sorbistin A_2 [119], B (=LL-AM31γ) [120], C [121], D (=LL-AM31α_ [122]. Spectinomycin [123] *Streptomyces spectabilis* (Mason et al., 1961); *S. flavopersicus, S. hygroscopicus* (Yamamoto et al., 1974); Dihydrospectinomycin [124] (Knight and Hoeksema, 1975). Validamycin A [125] *S. hygroscopicus* var. *limoneus* (Iwasa et al., 1970); B [126], C [127], D [128], E [129], F [130], Validoxylamine A [131]. Minosaminomycin [132] *S. aureomonopodiales* (Hamada et al., 1974). Kasugamycin [133] *S. kasugaensis* (Umezawa et al., 1965). Myomycin [134] *Nocardia* sp. (French et al., 1973); LL-BM782α₂ [135], LL-BM782α₁ [136], LL-BM782α₁ₐ [137] (McGahren et al., 1981). Hygromycin A [138] *S. hygroscopicus* (Pittenger et al., 1953); *S. noboritoensis, Corynebacterium equi* (Wakisaka et al., 1980). Destomycin A [139] *S. rimofaciens* (Kondo et al., 1965); B [140], C [141] (Shimura et al., 1975) =AB-74; *S. aquacanus* (Tamura et al., 1976); Hygromycin B [142] *S. hygroscopicus* (Mann and Bromer, 1958); SS-56C [143] *S. eurocidicus* (Inouye et al., 1973); SS-56D [144] =A-396-1 (Shoji et al., 1970), A-16316C [145] (Tamura et al., 1975); SS-56A [146], SS-56B [147]. 4-Amino-4-deoxy-α,α-trehalose [148] *S. cirratus, S.* sp. (Naganawa et al., 1974). Mannosylglucosaminide [149] *S. virginiae* (Uramoto et al., 1967). Trehalosamine [150] *Streptomyces* sp. (Arcamone and Bizioli, 1957). 84B-3 [151] *S. rameus* (Harada et al., 1971). 3-Amino-3-deoxyglucose [152] *S. lansus* (Dolak et al., 1980); *Bacillus aminoglucosidicus* (Umezawa et al., 1967). SF-1993 (=N-Carbamoyl-D-glucosamine) [153] *S. halstedii* (Shomura et al., 1979).

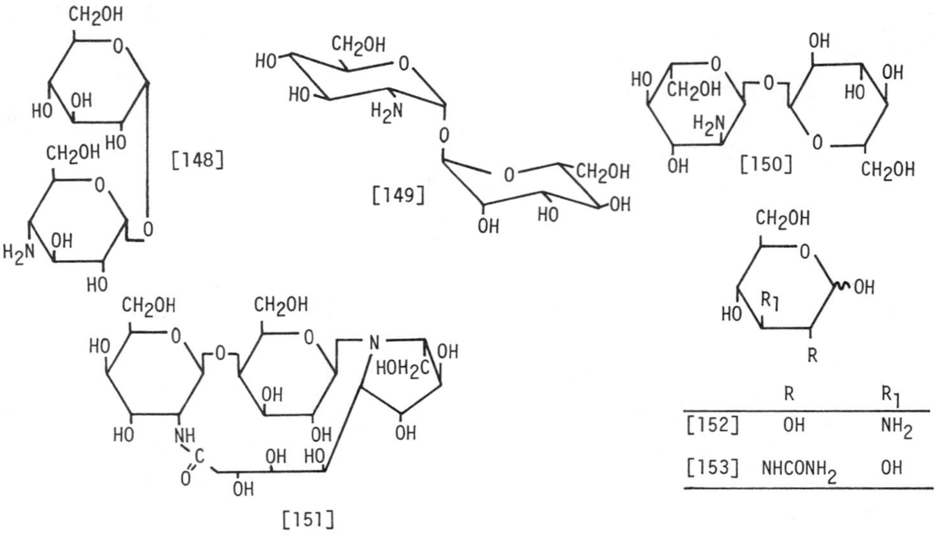

	R	R_1
[152]	OH	NH_2
[153]	$NHCONH_2$	OH

Figure 9 (Continued)

detoxification system for their own intermediates or end products to prevent suicide. The kanamycin 6'-N-acetyltransferase and 6'-N-acetylkanamycin amidohydrolase were reported in *Streptomyces kanamyceticus*, and aminoglycoside 3'-phosphotransferase was shown to be present in the producer microorganisms of 4,5-disubstituted deoxystreptamine antibiotics.

Table 1 Aminoglycoside Antibiotics Classified According to Biosynthetic Similarities

Deoxystreptamine-containing antibiotics
- 4,5-Disubstituted
  - Neomycin, paromomycin, lividomycin, ribostamycin, xylostasin, butirosin, and related substances
- 4,6-Disubstituted
  - Kanamycin, nebramycin, seldomycin, gentamicin, and related substances
- Monosubstituted
  - Destomycin, hygromycin B, and related substances

Streptidine (bluensidine) -containing antibiotics
- Streptomycin and related substances

Actinamine-containing antibiotics
- Spectinomycin

Monoaminocyclitol-containing antibiotics
- Validamycin, hygromycin A, minosaminomycin, kasugamycin, fortimicin, and related substances

As to the butirosins produced by *Bacillus circulans*, which have a unique structure (1-N-4-amino-2-hydroxybutyryl deoxystreptamine), the biosynthetic routes have been established by means of blocked mutants and mutational biosynthesis techniques (Takeda et al., 1979).

Table 2 Mutasynthesis of Aminoglycoside Antibiotics

Antibiotics	Producer	Main products	References
Neomycins	*Streptomyces fradiae* (DOS⁻)[a]	Hybrimycins	Shier et al. (1969,1974)
Ribostamycin	*Streptomyces ribosidificus*	2-Hydroxy ribostamycin 3',4'-Dideoxy ribostamycin	Kojima and Satoh (1973)
Kanamycin	*Streptomyces kanamyceticus*	6'-Hydroxy-6'-deamino-2-epihydroxy kanamycin 6'-Hydroxy-6'-deamino-1-N-methyl kanamycin	Kojima and Satoh (1973)
Paromomycins	*Streptomyces rimosus*	6'-Deoxy paromomycin	Shier et al. (1974) Cleophax et al. (1976)
Butirosin	*Bacillus circulans*	2-Hydroxy butirosin 3',4'-Dideoxy butirosin	Taylor and Schmitz (1976) Takeda et al. (1978a)
Sisomicin	*Micromonospora inyoensis*	Mutamicins	Testa et al. (1974)
Gentamicins	*Micromonospora purpurea*	2-Hydroxy gentamicin	Daum et al. (1977) Rosi et al. (1977)
Streptomycin	*Streptomyces griseus* (streptamine⁻)	Streptomutin A	Nagaoka and Demain (1975)
Spectinomycin	*Streptomyces spectabilis* (actinamine⁻)	Not isolated	Slechta and Coats (1974)

[a]DOS⁻, Deoxystreptamine-requiring mutant.

Mutational biosynthesis (mutasynthesis) was first introduced in amino-glycoside antibiotic research and yielded fruitful results. Since the first such experiments with *S. fradiae* in 1969, there have been numerous examples (as shown in Table 2) of successful mutasynthesis of aminoglycoside antibiotics. The most promising antibiotic produced by this technique is 2-hydroxygenta-micin, which proved to be less toxic than its parent antibiotic, without decreasing in vivo activity. Problems related to the application of this technique for industrial production involve the high cost of precursors and the overall yield of the final product compared with analogous chemical modification processes of the parent antibiotics.

IV. FERMENTATION PROCESS

All the practical aspects involved in the commercially important aminoglycoside antibiotics production are behind the curtain of industrial secrecy; therefore it is only possible to describe general procedures applied for the fermentation of aminoglycoside antibiotics. The producer microorganisms have widely broadened from *Streptomyces*, the original producer strains of aminoglycoside antibiotic, to other genera of actinomycetes and even to bacteria. As carbon sources, these microorganisms utilize starch, dextrin, glucose, glycerol, and other economically available materials, solely or as a mixture. Natural agricultural by-products, soybean meal, corn-steep liquor, cottonseed flour, casein hydrolysates, or yeast and its extracts, are supplied as nitrogen sources in fermentation media. Sometimes inorganic nitrogen-containing salts, such as ammonium sulfate and ammonium nitrate, are used additionally. Besides the above ingredients, certain mineral salts and vegetable or animal oils are often fed to the fermenter.

The fermentation process is carried out in submerged culture in most cases. An example of the sagamicin fermentation process is shown in Figure 10. Common assay methods for aminoglycoside antibiotics during fermentation and isolation processes are based on microbioassays using sensitive bacterial strains. Recently gas—liquid chromatography or high-performance liquid chromatography with ion exchange resin, silicic acid, or a reversed-phase column has been applied to detect the potency of fermentation broths. The accumulation of the antibiotic usually occurs during the late phase of logarithmic growth or at the initial stationary phase of fermentation (Fig. 11).

Though the producing organism is normally rather resistant to its own antibiotic, too high a level of antibiotic in the fermentation broth exhibits adverse effects on the increase in potency. Aminoglycoside antibiotics generally present a higher killing activity on a weight basis to the own producer organisms than other antibiotics normally do. In these respects, increasing the total potency of the industrial fermentation of aminoglycoside antibiotics by means of improving cultural conditions or manipulating the producer organism presents a serious problem

The aminoglycoside antibiotics are generally produced in a complex of numerous active factors consisting of structurally similar components. Qualitative fermentation improvement is directed toward the preferential synthesis of the most effective or useful component of the complex. In this respect, manipulation of fermentation conditions and strain improvement through mutation have been effective in certain cases. An extensive review on these interesting aspects of the nebramycin fermentation has been published (Stark et al., 1980). Nebramycin is produced by parent strains as a complex con-

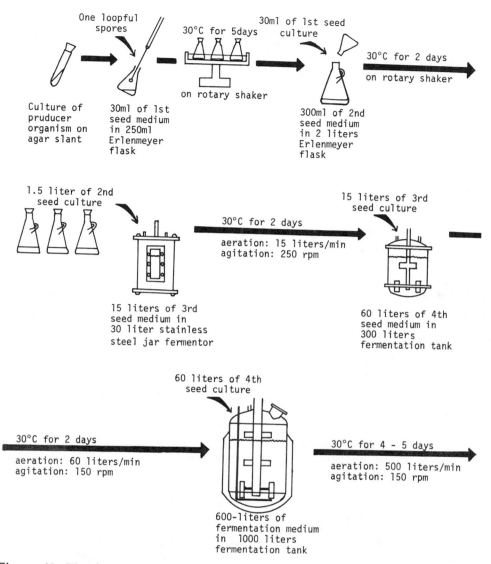

Figure 10 The fermentation process of sagamicin.

sisting of three major factors (factors 2, 4, and 5') and some minor factors. Firstly, fermentation conditions were selected to increase the global titer of all antibiotics in the fermentation broth, without markedly varying the ratio of different antibiotic factors (Stark et al., 1976). When tobramycin, which could be derived very easily from factor 5', was recognized to be effective in human medicine, efforts were concentrated to specifically increase the amount of factor 5' in the fermentation broth. The successful isolation of a mutant strain that produces factor 5' as a sole major component was even followed by the isolation of another mutant which accumulates factor 2, known as apramycin, which is applied in veterinary practice.

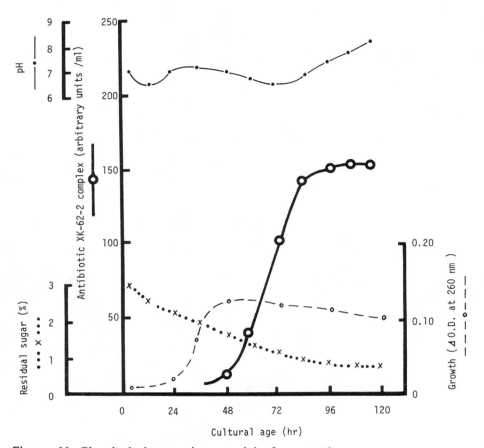

Figure 11 Chemical changes in sagamicin fermentation.

Because of the commercial importance of kanamycin, some studies have been reported concerning its fermentation conditions, strain improvement, biosynthesis process, including mutasynthesis, and a role of plasmids in the fermentation production (Hotta et al., 1977). Recently the influence of intercalating agents on kanamycin production by industrial strains was reported (Chang et al., 1980).

The streptomycin fermentation is still intensively studied from various viewpoints, but especially from those related to cell wall formation and enzymatic release of antibiotic-containing peptidoglycan fragments from streptomycin-producing *S. griseus* strains (Barabas et al., 1980). Investigations on enzyme systems involved in streptomycin production, such as glycosyltransferase (Kniep and Grisebach, 1980a), and on enzymatic formation of streptomycin-6-phosphate (Kniep and Grisebach, 1980b) have recently been reported. Energy balances related to anabolic–catabolic glucose utilization and streptomycin formation have also been discussed (Bormann and Christner, 1980). Such efforts might lead to an increase in streptomycin titer to much higher levels than those now reported (maximum 8500 μg/ml) (Singh et al., 1976).

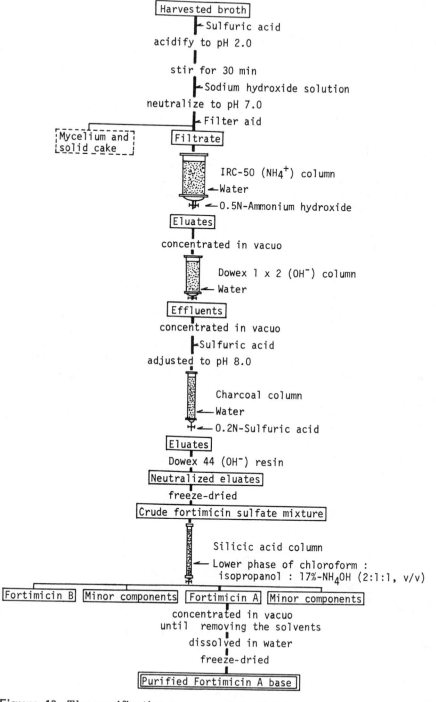

Figure 12 The purification process of fortimicin A.

V. RECOVERY AND PURIFICATION

Most of the aminoglycoside antibiotics are produced as mixtures of structurally related compounds. Normally it is necessary to isolate from the fermentation broth only the useful component and to purify it to allow medical application. The one exception among clinically important antibiotics is gentamicin; it is a mixture of three chemically analogous major components and several minor factors. Sometimes aminoglycoside antibiotics are coproduced with quite different types of antibiotic; for instance, caerulomycin is coproduced with nebramycin, and aureothricin with kasugamycin. Aminoglycoside antibiotics are easy to separate from those by-products, based on their typical physicochemical properties.

The isolation and purification of certain deoxystreptoamine- (Marquez and Kershner, 1978) or streptamine- (Perlman and Ogawa, 1978) containing aminoglycoside antibiotics have already been reviewed comprehensively.

One example of the purification processes of fortimicin A is given in Figure 12.

Aminoglycoside antibiotics are, without exception, all water soluble and exhibit basic properties. They are generally stable toward pH and temperature changes, usually encountered through isolation and purification processes. Cation exchange resins (as shown in Table 3) are conventionally used for the first recovery step from the fermentation broth of most aminoglycoside antibiotics. Later in the purification process, identical resins (small mesh) are applied for chromatographic separation. Recently, a trial of foam separation process adding certain surfactant into the gentamicin solution was proposed for the recovery of gentamicin (Lee and Ryu, 1979). Certain aminoglycoside antibiotics are absorbed on the surface of the mycelium of the producing organism (Reiblein et al., 1973); acidic or alkaline treatment of the whole broth prior to isolation is reported to be effective in releasing these antibiotics into the fermentation broth.

The procedure for the isolation of a certain component from its analogs is based on column-chromatographic separation with alumina, silica gel, cellulose, or carbon. Recently ion exchange cellulose (e.g., carboxymethylcellulose, phosphocellulose, and sulfoethyl cellulose) or ion exchange gels (e.g., CM Sephadex, SP Sephadex, and CM Sepharose) have been applied for a stricter re-

Table 3 Ion Exchange Resin Used for the Isolation of Aminoglycoside Antibiotics

Ion exchanger	Manufacturer
Amberlite IR-120 Amberlite IRC-50 Amberlite CG-50	Rohm & Haas Co.
Diaion SK-1B Diaion WK-10 Diaion PK-204	Mitsubishi Kasei Ind., Co. (Japan)
Dowex 50W X2	The Dow Chemical Co.
Duolite C-3 Duolite ES-26 Duolite CC-3	Chemical Process Co.

fining. In certain case, such as for the gentamicins, the Craig distribution method was applied to separate each component (Byrne et al., 1977).

Most of the aminoglycoside antibiotics display no characteristic ultraviolet absorption; however, refractive index monitoring is convenient for tracing the antibiotic in eluates of chromatographic columns.

Aminoglycoside antibiotics—in free base form—are difficult to crystallize; organic or inorganic acid salt derivatives, such as Reineckate, helianthate, picrate, and p-hydroxyazobenzenesulfonate, are prepared to carry out crystallization. Finally they are transformed into hydrochlorides or sulfates as end products.

VI. APPLICATION AND ECONOMY

The first aminoglycoside antibiotic discovered, streptomycin, has been used as an effective tuberculostatic drug all over the world. It succeeded in conquering the most tremendous disease of mankind, and has contributed to the extension of the human life-span. This antibiotic is still important in developing countries for the treatment of tuberculosis. Another important role of aminoglycoside antibiotics in medical use is the control of infectious diseases due to gram-negative Enterobacteriaceae.

Though several semisynthetic β-lactam antibiotics, including the newer cephalosporins with broader antibacterial spectra, have been approved for clinical use, the typical bactericidal activity of aminoglycoside antibiotics against aerobic Enterobacteriaceae is still not surpassed by other antibiotics.

Aminoglycoside antibiotics, including semisynthetic ones, applied for clinical use are listed in Table 4. Among the gram-negative rods which cause

Table 4 Clinically Important Aminoglycoside Antibiotics for Medical Use

Antibiotic	Manufacturers
Streptomycin	Merck & Co., Inc.
Kanamycin Ribostamycin Dibekacin[a]	Meiji Seika Kaisha, Ltd. (Japan)
Neomycin	Pfizer, Inc.
Paromomycin	Parke-Davis & Co.
Lividomycin	Rhône-Poulenc S. A. (France)
Gentamicin Sisomicin Netilmicin[a]	Schering Corp.
Spectinomycin	The Upjohn Co., Abbott Labs.
Tobramycin	Eli Lilly & Co.
Amikacin[a]	Bristol Myers & Co.
Sagamicin	Kyowa Hakko Kogyo Co. Ltd. (Japan)

[a]Semisynthetic aminoglycoside.

Table 5 Aminoglycoside-Inactivating Enzymes Found in Clinical Isolates

Inactivating enzymes*		Streptomycin	Ribostamycin	Butirosin	Neomycin	Paromomycin	Lividomycin	Spectinomycin	Kanamycin A	Kanamycin B	Kanamycin C	Tobramycin	Dibekacin	Amikacin	Gentamicin C1a	Gentamicin C2	Gentamicin C1	Sagamicin	Sisomicin	Netilmicin	Fortimicin	Clinical isolates which produce each enzyme**
Phosphotransferase [APH]	APH(3')-I		○	○	○	○	⑤		○	○	○											E.coli, P.aeruginosa, S.aureus
	APH(3')-II		○	○	○	○			○	○	○			○								E.coli
	APH(3')-III		○	○	○	○	⑤		○	○	○			○								P.aeruginosa
	APH(2")								○	○	○	○	○		○	○	○	○	○			S.aureus
	APH(3")	○																				E.coli, P.aeruginosa
	APH(5")		○				○															P.aeruginosa
	APH(6)	○																				P.aeruginosa
Acetyltransferase [AAC]	AAC(6')-I				○				○	○		○	○	○	○	○		○	○	○		P.aeruginosa
	AAC(6')-II				○				○	○		○	○						○	○		Moraxella sp.
	AAC(6')-III				○				○	○		○	○						○			P.aeruginosa
	AAC(6')-IV								○	○		○	○						○		①	P.aeruginosa
	AAC(3)-I														○	○	○	○	○			P.aeruginosa
	AAC(3)-II											○			○	○	○	○	○			E.coli
	AAC(3)-III								○						○	○	○	○	○			P.aeruginosa
	AAC(3)-IV								○	○		○			○	○	○	○	○	○		E.coli, K.pneumoniae, Arizona sp.
	AAC(2')				○					○		○	○		○	○	○	○	○	○		Providencia sp.
Adenylyltransferase [AAD]	AAD(2")								○	○		○	○		○	○	○	○	○			E.coli, K.pneumoniae
	AAD(3")	○						⑨														E.coli
	AAD(4')		○	○	○	○			○	○		○	○	○								S.epidermidis, B.brevis
	AAD(6)	○											④									S.aureus

○ : Substrates inactivated by the enzymes. Numbers in a circle means position of inactivation.

* [APH] : Aminoglycoside phosphotransferase
  [AAC] : Aminoglycoside acetyltransferase
  [AAD] : Aminoglycoside adenylyl (or nucleotidyl) transferase

** E : Escherichia
   P : Pseudomonas
   S : Staphylococcus
   K : Klebsiella
   B : Bacillus

infectious diseases, *Pseudomonas aeruginosa* has been resistant to most of the currently known antibiotics. Some of the semisynthetic penicillins like carbenicillin or surbenicillin, certain peptides, polymyxin (colistin), and, recently, certain semisynthetic cephalosporins like cefsulodin or ceftazidime (GR 20263) are clinically being applied as antipseudomonal chemotherapeutic agents. But, in practice, aminoglycoside antibiotics like gentamicin, sagamicin, and dibekacin remain the first choice of antipseudomonal drugs owing to their marked reliable bactericidal activity.

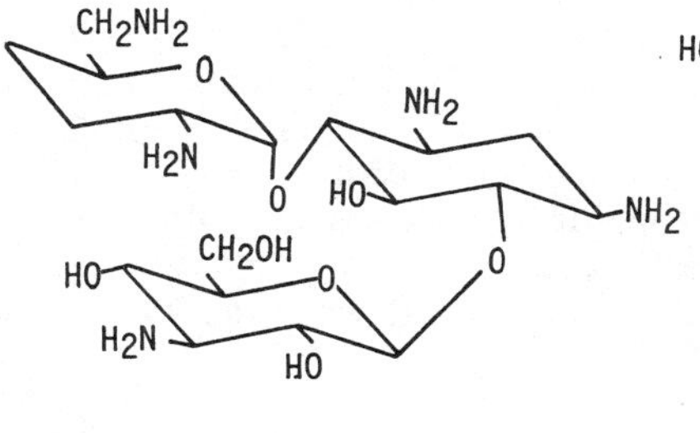

Dibekacin

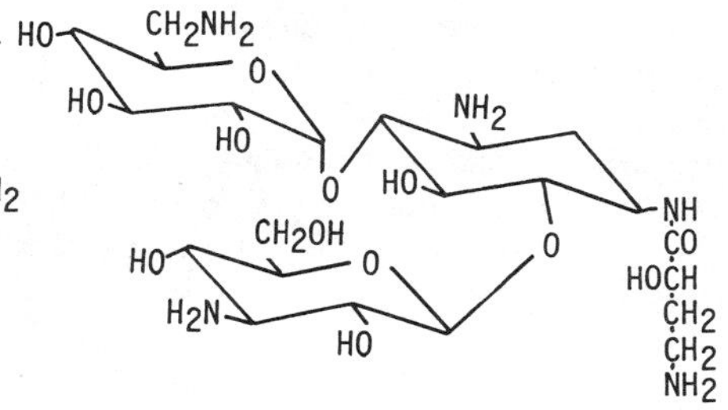

Amikacin

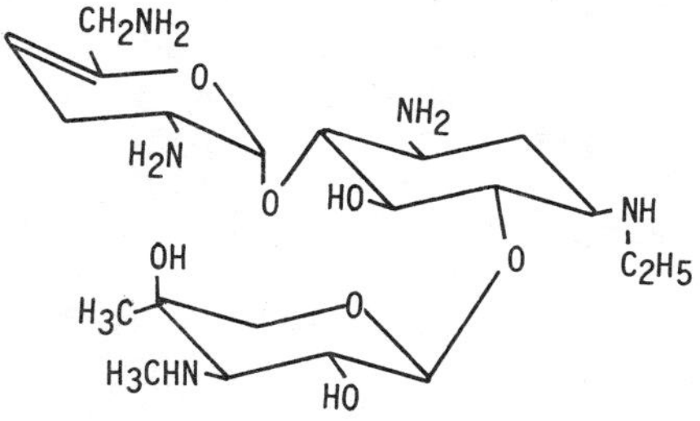

Netilmicin

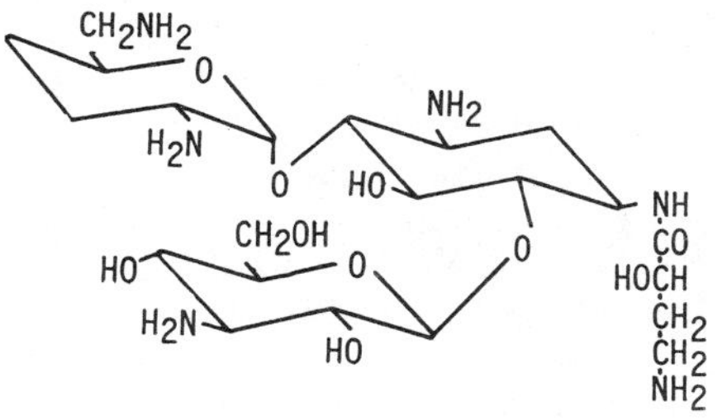

Habekacin

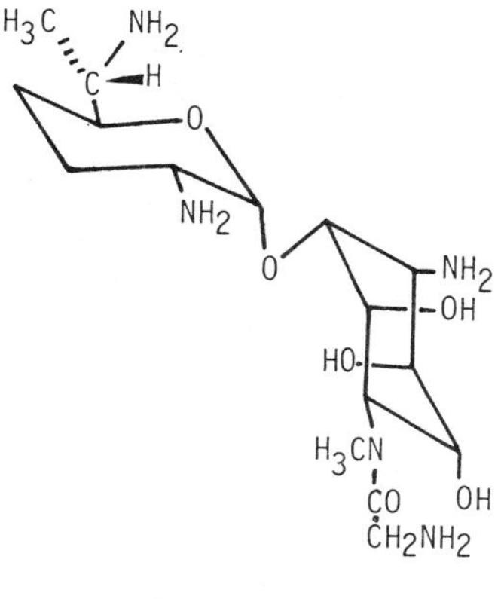

3-O-Demethylfortimicin A

Figure 13 Semisynthetic aminoglycosides.

Since kanamycin has been applied in clinical practice in Japan, some resistant pathogens have appeared, especially among gram-negative rods which enzymatically inactivate functional groups of the antibiotic molecule, yielding a modified antibiotic devoid of antibacterial activity. The relations between aminoglycoside antibiotics and inactivating enzymes produced by clinical isolates are summarized in Table 5.

Numerous studies on aminoglycoside antibiotic-inactivating enzymes led to an idea of how to protect the parent antibiotic against these pathogens by modifying it. This resulted in the synthesis of semisynthetic antibiotics, which are not inactivated by these enzymes. The first successful example was 3',4'-didehydroxylation of kanamycin B. Kanamycin was inactivated by phosphorylation of 3'- and/or 4'-hydroxyl groups. Umezawa and his colleagues succeeded in removing these hydroxyl groups without decreasing the antibacterial activity of the parent antibiotic. The resulting dibekacin, obtained by knowledge of the resistance mechanisms of resistant pathogens, opened the way to develop a new field in aminoglycoside chemistry (Umezawa, 1980; Suami, 1980).

Amikacin, 1-N-L(−)-4-amino-2-hydroxybutyryl kanamycin A, synthesized from kanamycin A, displayed improved antibacterial activities against various aminoglycoside resistant organisms by interfering with the binding of inactivating enzymes to the antibiotic. The aminoglycoside group of antibiotics present an example of natural products which are still very valuable as unmodified clinical drugs, but recently successes with dibekacin and amikacin have proved the possibilities of making up new useful drugs with chemical modifications on natural products. Success in obtaining antibiotics with improved characteristics also largely depends on sugar chemistry (Price et al., 1977). Several semisynthetic aminoglycoside antibiotics such as 1-N-ethyl-sisomicin (netilmicin), derived from sisomicin, 1-N-L(−)-4-amino-2-hydroxy-butyryl dibekacin (habekacin), and 3-O-demethylfortimicin or propikacin are currently being applied in clinical practice (Fig. 13).

Aminoglycoside antibiotics application for medical use is limited only to parenteral administration, because all of them are not well absorbed through oral routes. Some of the aminoglycosides are used for sterilization of intestines, and paromomycin and a few others are used as anthelmintic agents

Table 6 Anthelmintic Aminoglycoside Antibiotics

Antibiotic	Reference
Hygromycin B	Mann and McGuire (1962)
Destomycin A	Kondo et al. (1965)
Paromomycin	Waitz et al. (1966) Cavier and Leger (1970)
Gentamicin X	Panitz (1975)
G-418	Loebenberg et al. (1975)
Homomycin	Sumiki et al. (1955)
Kanamycin Neomycin	Garin et al. (1971)

Table 7 Aminoglycoside antibiotics for agricultural and veterinary use

	Plant and vegetables			Poultry and cattle		
Antibiotics	Use	Manufacturer		Antibiotics	Use	Manufacturer
Kasugamycin	Rice blast disease (*Piricularia oryzae*)	HokkoKagaku Kogyo Co. (Japan)		Hygromycin B	Anthelmintic, feed additive	Eli Lilly & Co.
Validamycin	Sheath blight disease (*Pellicularia sasakii*)	Takeda Chemical Ind. (Japan)		Destomycin A	Anthelmintic, feed additive	Meiji Seika Kaisha, Ltd. (Japan)
Streptomycin	Bacterial disease of vegetables (*Xanthomonas*, etc.)	Merck & Co., Inc.		Streptomycin	Bacterial infectious diseases	Merck & Co., Inc.
				Dihydro-streptomycin	Bacterial infectious diseases	Merck & Co., Inc.
				Kanamycin	Bacterial infectious diseases	Meiji Seika Kaisha, Ltd. (Japan)
				Neomycin	Bacterial infectious diseases	Pfizer, Inc.
				Apramycin	Bacterial infectious diseases	Eli Lilly & Co.

against intestinal parasites, including the tapeworm or ascaris, in humans and animals (Table 6).

Another large-application field of aminoglycoside antibiotics lies in agricultural practice (Table 7). Hygromycin B and destomycin A are currently used as feed additives to control helminths and to increase body weight in cattle. Streptomycin and some other aminoglycoside antibiotics are used alone or in combination with other antibiotics like penicillins or tetracyclines for protection against and curing of animal diseases. Kasugamycin and validamycin are used in Japan to control rice blast disease and sheath blight disease of the rice plant, respectively.

The uses in the class of aminoglycoside antibiotics are still numerous and interest in the field will increase.

REFERENCES

Arcamone, F., and Bizioli, F. (1957). Isolation and constitution of trehalosamine, a new aminosugar from streptomycete. *Gazz. Chim. Ital. 87*:896–902.

Arcamone, F., Cassinelli, G., D'Amico, G., and Orezzi, P. (1968). Mannosidohydroxystreptomycin from *Streptomyces* sp. *Experientia 24*:441–442.

Barabas, G., Szabo, I., and Szabo, G. (1978). Enzymatically released cell wall fragments with antibiotic activity from aminoglucoside-producing streptomycetes. In *Current Chemotherapy, Proceedings of the 10th International Congress of Chemotherapy*, Vol. 1, W. Siegenthaler and R. Lüthy (Eds.). American Society of Microbiology, Washington, D.C., pp. 507–509.

Barabas, G., Szabo, I., Ottenberger, A., Z.-Nagy, V., and Szabo, G. (1980). Enzymatic release of antibiotic-containing peptidoglycan fragments from streptomycin-producing *Streptomyces griseus*. *Can. J. Microbiol. 26*:141–145.

Barza, M., and Scheife, R. T. (1977). Antimicrobial spectrum, pharmacology, and therapeutic use of antibiotics. IV. Aminoglycosides. *N. Engl. Med. Center Hosp. Forum 6*:17.

Benedict, R. G., Stodola, F. H., Shotwell, O. L., Borud, A. M., and Lindenfelser, L. A. (1950). A new streptomycin. *Science 112*:77–78.

Bormann, E. J., and Christner, A. (1980). Anabolic-catabolic glucose utilization and product formation of *Streptomyces griseus*. *Z. Allg. Mikrobiol. 20*:367–374.

Byrne, K. M., Kershner, A. S., Maehr, H., Marquez, J. A., and Schaffner, C. P. (1977). Separation of gentamicin C-complex into five components by Craig distribution. *J. Chromatogr. 131*:191–203.

Cavier, R., and Leger, N. (1970). Antihelmintic properties of paromomycin sulfate. *Ann. Pharmacol. Fr. 28*:729–732.

Chang, L. T., Behr, D. A., and Elander, R. P. (1980). The effects of intercalating agents on kanamycin production and other phenotypic characteristics in *Streptomyces kanamyceticus*. In *Develop. Indust. Microbiol., Proc. 36th General Meeting Soc. Ind. Microbiol.*, Vol. 21. L. A. Underkofler and M. L. Wulf (Eds.). Soc. Ind. Microbiol., Virginia, pp. 233–243.

Claes, P. J., Compernolle, F., and Vanderhaeghe, H. (1974). Chromatographic analysis of neomycin. Isolation and identification of minor components. *J. Antibiot. 27*:931–942.

Claridge, C. A. (1979). Aminoglycoside antibiotics. In *Economic Microbiology*, Vol. 3. A. H. Rose (Ed.). Academic, New York, pp. 151–238.

Cleophax, J., Gero, S. D., Leboul, J., Akhtar, M., Barnett, J. E. G., and Pearce, C. J. (1976). A chiral synthesis of D-(+)-2,6-dideoxystreptamine and its microbial incorporation into novel antibiotics. *J. Am. Chem. Soc.* 98:7110–7112.

Cleophax, J., Roland, A., Colas, C., Castellanos, L., Gero, S. D., Sepulchre, A. M., and Quiclet, B. (1980). Synthesis and mutasynthesis of pseudosaccharides related to aminocyclitol-glycoside antibiotics. *ACS Symp. Ser.* 125:393–411.

Courvalin, P., Weisblum, B., and Davies, J. (1977). Aminoglycoside-modifying enzyme of an antibiotic-producing bacterium acts as a determinant of antibiotic resistance in *Escherichia coli*. *Proc. Nat. Acad. Sci. U.S.A.* 74:999–1003.

Daniels, P. J. L., Luce, C., Nagabhushan, T. L., Jaret, R. S., Schumacher, D., Reimann, H., and Ilavsky, J. (1975). The gentamicin antibiotics. 6. Gentamicin C_{2b}, an aminoglycoside antibiotic produced by *Micromonospora purpurea* mutant JI-33. *J. Antibiot.* 28:35–41.

Daum, S. J., Rosi, D., and Goss, W. A. (1977). Production of antibiotics by biotransformation of 2,4,6/3,5-pentahydroxycyclohexanone and 2,4/3,5-tetrahydroxycyclohexanone by a deoxystreptamine-negative mutant of *Micromonospora purpurea*. *J. Am. Chem. Soc.* 99:283–284.

Davies, D. H., Greeves, D., Mallams, A. K., Morton, J. B., and Tkach, R. W. (1975). Structures of the aminoglycoside antibiotics 66-40B and 66-40D produced by *Micromonospora inyoensis*. *J. Chem. Soc. Perkin I*: 814–818.

Davies, D. H., Mallams, A. K., McGlotten, J., Morton, J. B., and Tkach, R. W. (1977). Structure of aminoglycoside 66-40C, a novel unsaturated imine produced by *Micromonospora inyoensis*. *J. Chem. Soc. Perkin I*: 1407–1411.

Deushi, T., Iwasaki, A., Kamiya, K., Kunieda, T., Mizoguchi, T., Nakayama, M., Itoh, H., Mori, T., and Oda, T. (1979a). A new broad-spectrum aminoglycoside antibiotic complex, sporaricin I. Fermentation, isolation and characterization. *J. Antibiot.* 32:173–179.

Deushi, T., Iwasaki, A., Kamiya, K., Mizoguchi, T., Nakayama, M., Itoh, H., and Mori, T. (1979b). New aminoglycoside antibiotics, sannamycins. *J. Antibiot.* 32:1061–1065.

Deushi, T., Yamaguchi, T., Kamiya, K., Iwasaki, A., Mizoguchi, T., Nakayama, M., Watanabe, I., Itoh, H., and Mori, T. (1980). A new aminoglycoside antibiotic, sannamycin C and its 4-N-glycyl derivative. *J. Antibiot.* 33:1274–1280.

Dolak, L. A., Castle, T. M., Dietz, A., and Laborde, A. L. (1980). 3-Amino-3-deoxyglucose produced by a *Streptomyces* sp. *J. Antibiot.* 33:900–901.

Dorman, D. E., Paschal, J. W., and Merkel, K. E. (1976). [15]N nuclear magnetic resonance spectroscopy. The nebramycin aminoglycosides. *J. Am. Chem. Soc.* 98:6885–6888.

French, J. C., Bartz, Q. R., and Dion, H. W. (1973). Myomycin, a new antibiotic. *J. Antibiot.* 26:272–283.

Fried, J., and Titus, E. (1947). Streptomycin B, an antibiotically active constituent of streptomycin concentrates. *J. Biol. Chem.* 168:391–392.

Fujii, T., Satoi, S., Muto, N., Kodama, A., and Kotani, M. (1980). Methods for the production of antibiotic gentamicin C_{1a}. Japanese Kokai No. 55-156,592.

Fujiwara, T., Matsumoto, K., and Kondo, E. (1979). Methods for production of ribostamycin. Japanese Kokai No. 54-84,095.

Garin, J. P., Despeignes, J., and Vincent, G. (1971). Effect of four antibiotics of the oligosaccharide group on *Taenia saginata*. *Lyon Med.* *225*: 123-129.

Grisebach, H. (1978). Biosynthesis of sugar components of antibiotic substances. In *Advances in Carbohydrate Chemistry and Biochemistry*, Vol. 35. R. S. Tipson and D. Horton (Eds.). Academic, New York, pp. 81-126.

Hamada, M., Kondo, S., Yokoyama, T., Miura, K., Iinuma, K., Yamamoto, M., Maeda, K., Takeuchi, T., and Umezawa, H. (1974). Minosaminomycin, a new antibiotic containing *myo*-inosamine. *J. Antibiot.* *27*:81-83.

Harada, Y., Kumabe, K., Sato, Y., Miyamura, Y., Kato, F., and Tanimoto, M. (1971). Methods for production of antiviral substance 84-B-3. Japanese Kokai No. 46-19,593.

Haskell, T. H., French, J. C., and Bartz, Q. R. (1959). Paromomycin. I. Paromamine, a glycoside of D-glucosamine. *J. Am. Chem. Soc.* *81*:3480-3481.

Heding, H. (1968). N-Demethylstreptomycin. I. Microbiological formation and isolation. *Acta Chim. Scand.* *22*:1649-1654.

Hessler, E. J., Jahnke, H. K., Robertson, J. M., Tsuji, K., Rinehart, Jr., K. L., and Shier, W. T. (1970). Neomycins D, E and F: Identity with paromamine paromomycin I and paromomycin II. *J. Antibiot.* *23*:464-466.

Horii, S., Nogami, I., Mizokami, N., Arai, Y., and Yoneda, M. (1974). New antibiotic produced by bacteria, 5-β-D-xylofuranosylneamine. *Antimicrob. Agents Chemother.* *5*:578-581.

Hotta, K., Okami, Y., and Umezawa, H. (1977). Elimination of the ability of a kanamycin-producing strain to biosynthesize deoxystreptamine moiety by acriflavine. *J. Antibiot.* *30*:1146-1149.

Howells, J. D., Anderson, L. E., Coffey, G. L., Senos, G. D., Underhill, M. A., Vogler, D. L., and Ehrlich, J. (1972). Butirosin, a new aminoglycoside antibiotic complex: Bacterial origin and some microbiological studies. *Antimicrob. Agents Chemother.* *2*:79-83.

Inouye, S., Shomura, T., Watanabe, H., Totsugawa, K., and Niida, T. (1973). Isolation and gross structure of a new antibiotic SS-56C and related compounds. *J. Antibiot.* *26*:374-385.

Inouye, S., Ohba, K., Shomura, T., Kojima, M., Tsuruoka, T., Yoshida, J., Kato, N., Ito, M., Amano, S., Omoto, S., Ezaki, N., Ito, T., Niida, T., and Watanabe, K. (1979). A novel aminoglycoside antibiotic, substance SF-2052. *J. Antibiot.* *32*:1354-1356.

Inouye, S., Shomura, T., Ohba, K., Watanabe, H., Omoto, S., Tsuruoka, T., Kojima, M., and Niida, T. (1980). Isolation and identification of N-formylfortimicin A. *J. Antibiot.* *33*:510-513.

Iwasa, T., Yamamoto, H., and Shibata, M. (1970). Studies on validamycins, new antibiotic. I. *Streptomyces hygroscopicus* var. *limoneus* nov. var., validamycin-producing organism. *J. Antibiot.* *23*:595-602.

Kamiya, K., Iwasaki, A., Deushi, T., Watanabe, I., and Mori, T. (1980). Methods for the production of apramycin. Japanese Kokai No. 55-1-2,397. 102,397.

Kawaguchi, H., Tomita, K., Hoshiya, T., Miyaki, T., Fujisawa, K., Kimeda, M., Numata, K., Konishi, M., Tsukiura, H., Hatori, M., and Koshiyama, H. (1974). Aminoglycoside antibiotics. V. The 4'-deoxybutirosins (Bu-1975C$_1$ and C$_2$), new aminoglycoside antibiotics of bacterial origin. *J. Antibiot.* 27:460-470.

Kawaguchi, H., Tsukiura, H., Tomita, K., Konishi, M., Saito, K., Kobaru, S., Numata, K., Fujisawa, K., Miyaki, T., Hatori, M., and Koshiyama, H. (1977). Tallysomycin, a new antitumor antibiotic complex related to bleomycin I. Production, isolation and properties. *J. Antibiot.* 30:779-788.

Khokhlov, A. S., Tovarova, I. I., and Anisova, L. N. (1976). Regulators of streptomycin biosynthesis and development of *Actinomyces streptomycini. Nova Acta Leopold. Suppl.* 7:289-298.

Kirby, J. P. (1980). Microbiological, chemical and clinical findings of aminoglycoside antibiotic research. *Proc. Biochem.* 15:14-23.

Kirby, J. P., Borders, D. B., and VanLear, G. E. (1977a). Structure of LL-BM408, an aminocyclitol antibiotic. *J. Antibiot.* 30:175-177.

Kirby, J. P., VanLear, G. E., Morton, G. O., Gore, W. E., Curran, W. V., and Borders, D. B. (1977b). LL-AM31 antibiotic complex: Aminoalditol antibiotics from a *Streptoverticillium. J. Antibiot.* 30:344-347.

Kniep, B., and Grisebach, H. (1980a). Biosynthesis of streptomycin. Purification and properties of a DTDP-L-dihydrostreptose-streptidine-6-phosphate dihydrostreptosyltransferase from *Streptomyces griseus. Eur. J. Biochem.* 105:139-144.

Kniep, B., and Grisebach, H. (1980b). Biosynthesis of streptomycin. Enzymatic formation of dihydrostreptomycin 6-phosphate from dihydrostreptosyl streptodine 6-phosphate. *J. Antibiot.* 33:416-419.

Knight, J. C., and Hoeksema, H. (1975). Reduction products of spectinomycin. *J. Antibiot.* 28:136-142.

Koch, K. F., Merkel, K. E., O'Connor, S. C., Occolowitz, J. L., Paschal, J. W., and Dorman, D. E. (1978). Structures of some of the minor aminoglycoside factors of the nebramycin fermentation. *J. Org. Chem.* 43:1430-1434.

Kojima, M., and Satoh, A. (1973). Microbial semi-synthesis of aminoglycosidic antibiotics by mutants of *S. ribosidificus* and *S. kanamyceticus. J. Antibiot.* 26:784-786.

Kojima, M., Yamada, Y., and Umezawa, H. (1968). Studies on the biosynthesis of kanamycins. Part I. Incorporation of ^{14}C-glucose or ^{14}C-glucosamine into kanamycins and kanamycin-related compounds. *Agric. Biol. Chem.* 32:467-473.

Kondo, S., Sezaki, M., Koike, M., Shimura, M., Akita, E., Satoh, K., and Hara, T. (1965). Destomycins A and B, two new antibiotics produced by a *Streptomyces. J. Antibiot.* A18:38-42.

Kotani, M., Muto, N., and Satoi, S. (1979a). Methods for the production of sisomicin. Japanese Kokai No. 54-62,392.

Kotani, M., Muto, N., and Satoi, S. (1979b). Methods for the production of verdamicin. Japanese Kokai No. 54-62,393.

Kugelman, M., Jaret, R. S., Mittelman, S., and Gau, W. (1978). The structure of aminoglycoside antibiotic 66-40G produced by *Micromonospora inyoensis. J. Antibiot.* 31:643-645.

Lee, S. B., and Ryu, D. D. Y. (1979). Separation of gentamicin by foaming. *Biotechnol. Bioeng.* 21:2045-2059.

Loebenberg, D., Counelis, M., and Waitz, J. A. (1975). Antibiotic G 418, a new *Micromonospora*-produced aminoglycoside with activity against protozoa and helminths: Antiparasitic activity. *Antimicrob. Agents Chemother.* 7:811–815.

McAlpine, J. B., Egan, R. S., Stanaszek, R. S., Cirovic, M., Mueller, S. L., Carney, R. E., Collumn, P., Fager, E. E., Goldstein, A. W., Grampovnik, D. J., Kurath, P., Martin, J. R., Post, G. G., Seely, J. H., and Tadanier, J. (1980). The structures of minor components of the fortimicin complex. In *ACS Symp. Ser.* 125:295–308.

McGahren, W. J., Hardy, B. A., Morton, G. O., Lovell, F. M., Perkinson, N. A., Hargreaves, R. T., Borders, D. B., and Ellestad, G. A. (1981). (β-Lysyloxy)myoinositol guanidino glycoside antibiotics. *J. Org. Chem.* 46:792–799.

Maeda, K., Ueda, M., Yagishita, K., Kawaji, S., Kondo, S., Murase, M., Takeuchi, T., Okami, Y., and Umezawa, H. (1957). Studies on kanamycin. *J. Antibiot.* A 10:228–232.

Maehr, H., and Schaffner, C. P. (1967). The chemistry of gentamicins. I. Characterization and gross structure of gentamicin A. *J. Am. Chem. Soc.* 89:6787–6788.

Maehr, H., Liu, C. -M., Hermann, T., Prosser, B. L. T., Smallheer, J. M., and Palleroni, N. J. (1980). Microbial products. IV. X-14847, a new aminoglycoside from *Micromonospora echinospora*. *J. Antibiot.* 33:1431–1436.

Mann, R. L., and Bromer, W. W. (1958). The isolation of a second antibiotic from *Streptomyces hygroscopicus*. *J. Am. Chem. Soc.* 80:2714–2716.

Mann, R. L., and McGuire, J. M. (1962). Hygromycin B, its production and treatment of intestinal parasites. U.S. Patent 3,018,220.

Marquez, J. A., and Kershner, A. (1978). 2-Deoxystreptamine-containing antibiotics. In *J. Chromatogr. Library*, Vol. 15. M. J. Weinstein and G. H. Wagman (Eds.). Elsevier, New York, pp. 159–214.

Marquez, J. A., Wagman, G. H., Testa, R. T., Waitz, J. A., and Weinstein, M. J. (1976). A new broad spectrum aminoglycoside antibiotic, G-52, produced by *Micromonospora zionensis*. *J. Antibiot.* 29:483–487.

Mason, D. J., Dietz, A., and Smith, R. M. (1961). Actinospectacin, a new antibiotic. I. Discovery and biological properties. *Antibiot. Chemother.* 11:118–122.

Mason, D. J., Dietz, A., and Hanka, L. J. (1962). U-12898, a new antibiotic. I. Discovery, biological properties, and assay. *Antimicrob. Agents Chemother.* 1962:607–613.

Mori, T., Deushi, T., Iwasaki, A., Kunieda, T., Mizoguchi, T., Kamiya, K., Nakayama, M., Itoh, H., and Oda, T. (1978). Antibiotic KA-6606. German Offen. No. 2,813,021.

Murase, M., Ito, T., Fukatsu, S., and Umezawa, H. (1970). Studies on kanamycin related compounds produced during fermentation by mutants of *Streptomyces kanamyceticus*. Isolation and properties. In *Progress in Antimicrobial and Anticancer Chemotherapy*, Vol. 2. University of Tokyo Press, Tokyo, pp. 1098–1110.

Nagabhushan, T. L., Turner, W. N., Daniels, P. J. L., and Morton, J. B. (1975). The gentamicin antibiotics. 7. Structures of the gentamicin antibiotics A_1, A_3, and A_4. *J. Org. Chem.* 40:2830–2834.

Naganawa, H., Usui, N., Takita, T., Hamada, M., Maeda, K., and Umezawa, H. (1974). 4-Amino-4-deoxy-α,α-trehalose, a new metabolite of a streptomyces. *J. Antibiot.* 27:145–146.

Nagaoka, K., and Demain, A. L. (1975). Mutational biosynthesis of a new antibiotic, streptomutin A, by an idiotroph of *Streptomyces griseus*. *J. Antibiot. 28*:627–635.

Nara, T. (1977). Aminoglycoside antibiotics. In *Annual Reports on Fermentation Processes*, Vol. 1. D. Perlman and G. T. Tsao (Eds.). Academic, New York, pp. 299–326.

Nara, T. (1978). Aminoglycoside antibiotics. In *Annual Reports on Fermentation Processes*, Vol. 2. D. Perlman and G. T. Tsao (Eds.). Academic, New York, pp. 223–266.

Nara, T., Kawamoto, I., Okachi, R., Takasawa, S., Yamamoto, M., Sato, S., Sato, T., and Morikawa, A. (1975). New antibiotic XK-62-2 (sagamicin). II. Taxonomy of the producing organism, fermentative production and characterization of sagamicin. *J. Antibiot. 28*:21–28.

Nara, T., Yamamoto, M., Takasawa, S., Sato, S., Sato, T., Kawamoto, I., Okachi, R., Takahashi, I., and Morikawa, A. (1977a). A new aminoglycoside antibiotic complex—The seldomycins. I. Taxonomy, fermentation and antibacterial properties. *J. Antibiot. 30*:17–24.

Nara, T., Yamamoto, M., Kawamoto, I., Takayama, K., Okachi, R., Takasawa, S., Sato, S., and Sato, T. (1977b). Fortimicins A and B, new aminoglycoside antibiotics. I. Producing organism, fermentation and biological properties of fortimicins. *J. Antibiot. 30*:533–540.

Nimi, O., Kawashima, H., Ikeda, A., Sugiyama, M., and Nomi, R. (1981). Biosynthetic correlation between streptomycin and mucopeptide in utilization of D-glucosamine as a common precursor. *J. Ferment. Technol. 59*:91–96.

Oda, T., Mori, T., Ito, H., Kunieda, T., and Munakata, K. (1971). Studies on new antibiotic lividomycins I. Taxonomic studies on the lividomycin-producing strain *Streptomyces lividus* nov. sp. *J. Antibiot. 24*:333–338.

Oka, Y., Ishida, H., Morioka, M., Numasaki, Y., Yamafuji, Y., Osono, Y., and Umezawa, H. (1979). Methods for the production of new antibiotics. Japanese Kokai No. 54-98,741.

Oka, Y., Ishida, H., Morioka, M., Numasaki, Y., Yamafuji, Y., Osono, Y., and Umezawa, H. (1981). Combimicins, new kanamycin derivative bioconverted by some *Micromonosporas*. *J. Antibiot. 34*:777–781.

Okachi, R., and Nara, T. (1980). Current trends in antibiotic fermentation research in Japan. *Biotechnol. Bioeng. Suppl. 1*:65–81.

Okachi, R., Takasawa, S., Sato, T., Sato, S., Yamamoto, M., Kawamoto, I., and Nara, T. (1977). Fortimicins A and B, new aminoglycoside antibiotics II. Isolation, physico-chemical and chromatographic properties. *J. Antibiot. 30*:541–551.

Okami, Y., Hotta, K., Yoshida, M., Ikeda, D., Kondo, S., and Umezawa, H. (1979). New aminoglycoside antibiotics, istamycins A and B. *J. Antibiot. 32*:964–966.

Okanishi, M., Koshiyama, H., Ohmori, T., Matsuzaki, M., Ohashi, S., and Kawaguchi, H. (1962). Glebomycin, a new member of the streptomycin class. I. Biological studies. *J. Antibiot. A15*:7–14.

Panitz, E. (1975). Anthelmintic effect of the gentamicin complex and coproduced antibiotics against tapeworms of lambs and cats. *J. Parasit. 61*:157–158.

Pearce, C. J. and Rinehart, K. L., Jr. (1981). Biosynthesis of aminocyclitol antibiotics. In *Antibiotics*, Vol. IV. J. W. Corcoran (Ed.). Springer-Verlag, Berlin, pp. 74–100.

Perlman, D., and Ogawa, Y. (1978). Streptamine-containing antibiotics. In *J. Chromatogr. Library*, Vol. 15. M. J. Weinstein and G. H. Wagman (Eds.). Elsevier, New York, pp. 587–616.

Pittenger, R. C., Wolfe, R. N., Hoehn, M. M., Marks, P. N., Daily, W. A., and McGuire, J. M. (1953). Hygromycin I. Preliminary studies on the production and biologic activity of a new antibiotic. *Antibiot. Ann. 1953/54*:157–166.

Price, K. E., Godfrey, J. C., and Kawaguchi, H. (1977). Effect of structural modification on the biological properties of aminoglycoside antibiotics containing 2-deoxystreptamine. In *Structure–Activity Relationships Among the Semisynthetic Antibiotics*. D. Perlman (Ed.). Academic, New York, pp. 239–355 and 357–395.

Reden, J., and Dürckheimer, W. (1979). Aminoglycoside antibiotics. Chemistry, biochemistry, structure–activity relationships. In *Top. Curr. Chem. 83*:105–170.

Reiblein, W. J., Watkins, P. D., and Wagman, G. H. (1973). Binding of gentamicin and other aminoglycoside antibiotics to mycelium of various actinomycetes. *Antimicrob. Agents Chemother. 4*:602–606.

Rinehart, K. L., Jr. (1964). *The Neomycins and Related Antibiotics*. Wiley, New York,

Rinehart, K. L., Jr. (1980). Biosynthesis and mutasynthesis of aminocyclitol antibiotics. In *ACS Symp. Ser. 125*:335–370.

Rinehart, K. L., Jr., and Stroshane, R. M. (1976). Biosynthesis of aminocyclitol antibiotics. *J. Antibiot. 29*:319–353.

Rosi, D., Goss, W. A., and Daum, S. J. (1977). Mutational biosynthesis by idiotrophs of *Micromonospora purpurea* I. Conversion of aminocyclitols to new aminoglycoside antibiotics. *J. Antibiot. 30*:88–97.

Saito, A., Ueda, Y., and Akiyoshi, M. (1980). Experimental studies on the ototoxicity and nephrotoxicity of fortimicin A. In *Current Chemotherapy and Infectious Disease, Proc. 11th Intern. Cong. Chemoth. & 19th Intersci. Conf. Antimicr. Ag. & Chemother.*, Vol. 1. J. D. Nelson and C. Grassi (Eds.). American Society for Microbiology, Washington, D.C., pp. 401–403.

Schatz, A., Bugie, E., and Waksman, S. A. (1944). Streptomycin, a substance exhibiting antibiotic activity against gram-positive and gram-negative bacteria. *Proc. Soc. Exp. Biol. Med. 55*:66–69.

Schmitz, H., Fardig, O. B., O'Herron, F. A., Rousche, M. A., and Hooper, I. R. (1958). Kanamycin. III. Kanamycin B. *J. Am. Chem. Soc. 80*: 2911–2912.

Sepulchre, A. -M., Quiclet, B., and Gero, S. D. (1980). Bioconversion dans le domaine des antibiotiques aminocyclitolglycosidiques. *Bull. Soc. Chim. Fr. 1-2*:II-56–II-65.

Shier, W. T., Rinehart, K. L., Jr., and Gottlieb, D. (1969). Preparation of four new antibiotics from a mutant of *Streptomyces fradiae*. *Proc. Nat. Acad. Sci. U.S.A. 63*:198–204.

Shier, W. T., Schaefer, P. C., Gottlieb, D., and Rinehart, K. L., Jr. (1974). Use of mutants in the study of aminocyclitol antibiotic biosynthesis and the preparation of the hybrimycin C complex. *Biochemistry 13*:5073–5078.

Shimizu, M., Takahashi, I., and Nara, T. (1978). Some problems involved in seldomycin fermentation. *Agric. Biol. Chem. 42*:653–658.

Shimura, M., Sekizawa, Y., Iinuma, K., Naganawa, H., and Kondo, S. (1975). Destomycin C, a new member of destomycin family antibiotics. *J. Antibiot.* 28:83–84.

Shirahata, K., Shimura, G., Takasawa, S., Iida, T., and Takahashi, K. (1980). The structures of new fortimicins having double bonds in their purpurosamine moieties. In *ACS Symp. Ser.* 125:309–320.

Shoji, J., Kozuki, S., Mayama, M., Kawamura, Y., and Matsumoto, K. (1970). Isolation of a new water-soluble basic antibiotic A-396-I. *J. Antibiot.* 23:291–294.

Shomura, T., Ezaki, N., Tsuruoka, T., Niwa, T., Akita, E., and Niida, T. (1970). Studies on antibiotic SF-733, a new antibiotic. I. Taxonomy, isolation and characterization. *J. Antibiot.* 23:155–161.

Shomura, T., Yoshida, J., Amano, S., Kojima, M., Inouye, S., and Niida, T. (1979). Studies on *Actinomycetales* producing antibiotics only on agar culture I. Screening, taxonomy and morphology–productivity relationship of *Streptomyces halstedii*, strain SF-1993. *J. Antibiot.* 32:427–435.

Singh, A., Bruzelius, E., and Heding, H. (1976). Streptomycin. A fermentation study. *Eur. J. Appl. Microbiol.* 3:97–101.

Slechta, L., and Coats, J. H. (1974). Studies of the biosynthesis of spectinomycin. In *Abst. Papers, 14th Intersci. Conf. Antimicr. Ag. & Chemoth.*, San Francisco, No. 294.

Stark, W. M., Hoehn, M. M., and Knox, N. G. (1967). Nebramycin, a new broad-spectrum antibiotic complex I. Detection and biosynthesis. *Antimicrob. Agents Chemother.* 1967:314–323.

Stark, W. M., Knox, N. G., Wilgus, R., and DuBus, R. (1976). The nebramycin fermentation: Culture and fermentation development. In *Develop. Indust. Microbiol.*, Vol. 17. L. A. Underkofler (Ed.). American Institute of Biological Science, Washington, D.C., pp. 61–77.

Stark, W. M., Wilgus, R. M., and DuBus, R. (1980). Fermentation studies with aminoglycoside-producing microorganisms. In *Develop. Indust. Microbiol.*, Vol. 21. L. A. Underkofler and M. L. Wulf (Eds.). Soc. Indust. Microbiol., Virginia, pp. 77–89.

Suami, T. (1980). Modification of aminocyclitol antibiotics. In *ACS Symp. Ser.* 125:43–73.

Sugimoto, M., Ishii, S., Okachi, R., and Nara, T. (1979). Fortimicin C, D and KE, new aminoglycoside antibiotics. *J. Antibiot.* 32:868–873.

Sumiki, Y., Nakamura, G., Kawasaki, M., Yamashita, S., Anzai, K., Isono, K., Serizawa, Y., Tomiyama, Y., and Suzuki, S. (1955). A new antibiotic, homomycin. *J. Antibiot.* A 8:170.

Takeda, K., Okuno, S., Ohashi, Y., and Furumai, T. (1978a). Mutational biosynthesis of butirosin analogs. I. Conversion of neamine analogs into butirosin analogs by mutants of *Bacillus circulans*. *J. Antibiot.* 31:1023–1030.

Takeda, K., Kinumaki, A., Hayasaka, H., Yamaguchi, Y., and Ito, Y. (1978b). Mutational biosynthesis of butirosin analogs. II. 3',4'-Dideoxy-6'-N-methylbutirosins, new semisynthetic aminoglycosides. *J. Antibiot.* 31:1031–1038.

Takeda, K., Aihara, K., Furumai, T., and Ito, Y. (1979). Biosynthesis of butirosins. I. Biosynthetic pathways of butirosins and related antibiotics. *J. Antibiot.* 32:18–28.

Tamura, A., Furuta, R., and Kotani, H. (1975). Antibiotic A-16316-C, a new water-soluble basic antibiotic. *J. Antibiot.* 28:260–265.

Tamura, A., Furuta, R., and Naruto, S. (1976). Isolation of an antibiotic AB-74, related to destomycin C. *J. Antibiot.* 29:590-591.

Tatsuoka, S., Kusaka, T., Miyake, A., Inoue, M., Hitomi, H., Shiraishi, Y., Iwasaki, H., and Imanishi, M. (1957). Studies on antibiotics. XVI. Isolation and identification of dihydrostreptomycin produced by a new Streptomyces, *Streptomyces humidus* nov. sp. *Chem. Pharm. Bull. (Tokyo)* 5:343-349.

Taylor, H. D., and Schmitz, H. (1976). Antibiotics derived from a mutant of *Bacillus circulans*. *J. Antibiot.* 29:532-535.

Testa, R. T., and Tilley, B. C. (1975). Biotransformation, a new approach to aminoglycoside biosynthesis: I. Sisomicin. *J. Antibiot.* 28:573-579.

Testa, R. T., and Tilley, B. C. (1976). Biotransformation, a new approach to aminoglycoside biosynthesis: II. Gentamicin. *J. Antibiot.* 29:140-146.

Testa, R. T., Wagman, G. H., Danields, P. J. L., and Weinstein, M. J. (1974). Mutamicins; biosynthetically created new sisomicin analogues. *J. Antibiot.* 27:917-921.

Tsukiura, H., Saito, K., Kobaru, S., Konishi, M., and Kawaguchi, H. (1973). Aminoglycoside antibiotics. IV. Bu-1709E_1 and E_2, new aminoglycoside antibiotics related to the butirosins. *J. Antibiot.* 26:386-388.

Tsukiura, H., Hanada, M., Saito, K., Fujisawa, K., Miyaki, T., Koshiyama, H., and Kawaguchi, H. (1976). Sorbistin, a new aminoglycoside antibiotic complex of bacterial origin. I. Production, isolation and properties. *J. Antibiot.* 29:1137-1146.

Umezawa, H., Ueda, M., Maeda, K., Yagishita, K., Kondo, S., Okami, Y., Utahara, R., Osato, Y., Nitta, K., and Takeuchi, T. (1957). Production and isolation of a new antibiotic, kanamycin. *J. Antibiot.* A10:181-188.

Umezawa, H., Okami, Y., Hashimoto, T., Suhara, Y., Hamada, M., and Takeuchi, T. (1965). A new antibiotic, kasugamycin. *J. Antibiot.* A18:101-103.

Umezawa, H., Okami, Y., and Kondo, S. (1981). Methods for production of istamycin A_0 and/or istamycin B_0. Japanese Kokai No. 56-43,295.

Umezawa, S. (1980). Synthesis of aminocyclitol antibiotics. In *ACS Symp. Ser.* 125:15-41.

Umezawa, S., Umino, K., Shibahara, S., Hamada, M., and Omoto, S. (1967). Fermentation of 3-amino-3-deoxy-D-glucose. *J. Antibiot.* A20:355-360.

Uramoto, M., Otake, N., and Yonehara, H. (1967). Mannosyl glucosaminide, a new antibiotic. *J. Antibiot.* A20:236-237.

Wagman, G. H., Marquez, J. A., Bailey, J. V., Cooper, D., Weinstein, J., Tkach, R., and Danields, P. (1972). Chromatographic separation of some minor components of the gentamicin complex. *J. Chromatogr.* 70:171-173.

Wagman, G. H., Testa, R. T., Marquez, J. A., and Weinstein, M. J. (1974). Antibiotic G-418, a new *Micromonospora*-produced aminoglycoside with activity against protozoa and helminths: Fermentation, isolation, and preliminary characterization. *Antimicrob. Agents Chemother.* 6:144-149.

Waitz, J. A., McClay, P., and Thompson, P. E. (1966). Effects of paromomycin on tapeworms of mice, rats and cats. *J. Parasitol.* 52:830-831.

Wakisaka, Y., Koizumi, K., Nishimoto, Y., Kobayashi, M., and Tsuji, N. (1980). Hygromycin and epihygromycin from a bacterium, *Corynebacterium equi* No. 2841. *J. Antibiot.* 33:695-704.

Waksman, S. A. (1949). *Streptomycin, Nature and Practical Applications.* Williams & Wilkins, Baltimore.

Waksman, S. A. (1953). *Neomycin.* Rutgers University Press, New Jersey.

Waksman, S. A., and Lechevalier, H. A. (1949). Neomycin, a new antibiotic active against streptomycin-resistant bacteria, including tuberculosis organisms. *Science 109*:305–307.

Walker, J. B. (1980). Biosynthesis of aminoglycoside antibiotics. In *Develop. Indust. Microbiol.*, Vol. 21. L. A. Underkofler and M. L. Wulf (Eds.). Soc. Indust. Microbiol., Virginia, pp. 105–113.

Weinstein, L., and Ehrenkranz, N. J. (1958). In *Antibiotics Monographs No. 10, Streptomycin and Dihydrostreptomycin.* Medical Encyclopedia, New York.

Weinstein, M. J., Leudemann, G. M., Oden, E. M., and Wagman, G. H. (1963). Gentamicin, a new broad-spectrum antibiotic complex. *Antimicrob. Agents Chemother. 1963*:1–7.

Weinstein, M. J., Marquez, J. A., Testa, R. T., Wagman, G. H., Oden, E. M., and Waitz, J. A. (1970). Antibiotic 6640, a new *Micromonospora*-produced aminoglycoside antibiotic. *J. Antibiot. 23*:551–554.

Weinstein, M. J., Daniels, P. J. L., Wagman, G. H., and Testa, R. T. (1975a). Methods for the preparation of semi-synthetic aminocyclitol aminoglycoside antibiotics. German Offen. No. 2,437,159.

Weinstein, M. J., Wagman, G. H., Marquez, J. A., Testa, R. T., and Waitz, J. A. (1975b). Verdamicin, a new broad spectrum aminoglycoside antibiotic. *Antimicrob. Agents Chemother. 7*:246–249.

Yamamoto, M., Okachi, R., Takasawa, S., Kawamoto, I., Kumakawa, H., Sato, S., and Nara, T. (1974). Production of spectinomycin by a new subspecies of *Streptomyces hygroscopicus. J. Antibiot. 27*:78–80.

Yamamoto, M., Okachi, R., Kawamoto, I., and Nara, T. (1977). Fortimicin A production by *Micromonospora olivoasterospora* in a chemically defined medium. *J. Antibiot. 30*:1064–1072.

Yasuda, S., Suami, T., Ishikawa, T., Umezawa, S., and Umezawa, H. (1975). Methods for the production of aminocyclitol derivatives. Japanese Kokai No. 50-25,793.

Zaslavsky, P. L., Zhukov, V. G., Kornitsky, E. Y., Tovarova, I. I., and Khokhlov, A. S. (1979). Influence of A-factor on the ultrastructure of the A-factor deficient mutant of *Streptomyces griseus. Microbios 25*: 145–153

11

THE STREPTOTHRICINS: PROPERTIES, BIOSYNTHESIS AND FERMENTATION

HEINZ THRUM *Central Institute of Microbiology and Experimental Therapy, Academy of Sciences of the German Democratic Republic, Jena, German Democratic Republic*

I. INTRODUCTION

The streptothricin antibiotics as a group are among the most abundant of the antibiotic substances produced by the actinomycetes. Following the initial discovery of streptothricin by Waksman and Woodruff in 1942, numerous examples of streptothricin substances were found by investigators throughout the world during their search for microbial products with antibiotic activity. Difficulties were encountered in characterizing the streptothricin preparations described in the literature until it was shown by chromatographic methods that most of the isolated antibiotic preparations of this group are mixtures of several closely related single streptothricins.

Streptothricins exhibit strong inhibitory action on gram-positive, gram-negative, and acid-fast bacteria, as well as on some fungi. Several streptothricin preparations also display antiviral activity. Furthermore, other physiological activities such as taeniacidal, fish toxicity, and inhibitory effects on plant growth have been described.

Though streptothricin antibiotics have a broad spectrum of antibiotic and physiologic activity, they have not been brought into medical use, mainly because of their strong delayed toxicity.

Their use in agriculture seems somewhat more promising, both for the suppression of certain phytopathogenic bacteria and for addition into animal feed.

II. HISTORY AND DISCOVERY

A. Preparations and Producing Strains

Streptothricin was first isolated from *Streptomyces lavendulae* by Waksman and Woodruff (1942) as a basic water-soluble substance with activity against gram-positive, gram-negative, and acid-fast bacteria. Furthermore, it moderately inhibits some fungi. Subsequently, other *Streptomyces* species produced several streptothricin complexes which differed to varying degrees from each other in terms of physical properties and component composition. Paper-chromatographic studies (Horowitz and Schaffner, 1958; Betina, 1975) revealed that most of these preparations were mixtures of closely related substances and that some of the individual components were common to several streptothricin preparations.

Khokhlov and co-workers (Khokhlov and Reshetov, 1964; Reshetov and Khokhlov, 1964; Reshetov and Khokhlov, 1965; Reshetov et al., 1965; Voronina et al., 1969) succeeded in isolating and characterizing at least seven single components of several natural streptothricin preparations which were designated streptothricins F, E, D, C, B, A, and X. All seven compounds were found to contain one residue of peculiar heterocyclic β-amino acid streptolidine (roseonine, geamine) and one residue of D-gulosamine but they differ in β-lysine content, which can vary from 1 (streptothricin F) to 7 units (streptothricin X) per molecule. On the basis of earlier structural investigations by several groups (Nakanishi et al., 1954; Brockmann and Musso, 1955; Van Tamelen et al., 1961; Johnson and Westley, 1962), Khokhlov and Shutova (1972) elucidated the following constitutions [1] for streptothricin components A—X (Fig. 1)

Structure [1] is now a generally accepted one, but the location of a carbamoyl moiety at position 6 of the D-gulosamine residue has not yet been proven. It was suggested by Borders (1975) that streptothricin F, as well as a number of streptothricin-like antibiotics, studied later, contains this carbamoyl group at the third or fourth carbon atom of the amino sugar moiety. This suggestion corresponds better with the structure of streptothricins [2].

The findings allowed an exact comparison of the natural streptothricin complexes in terms of their contents of the streptothricin components F—X, and it was found that most of them are mixtures varying in the number or relative amounts of components, as summarized in Table 1.

A review of the isolation, purification, and structural elucidation of the single streptothricins A—X, with a detailed description, as well as of naturally formed streptothricin mixtures and streptothricin-like antibiotics has been recently published (Khokhlov, 1978).

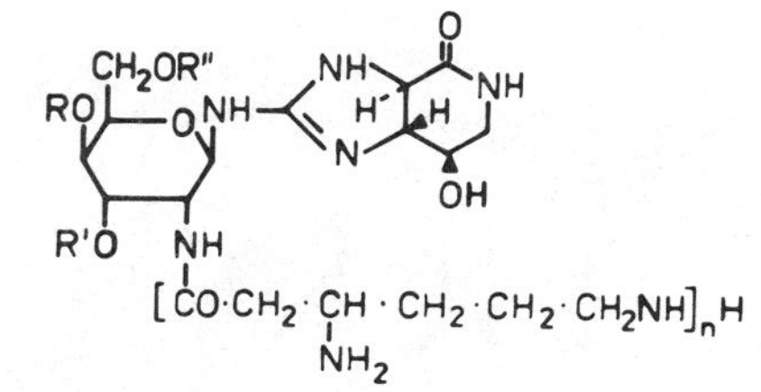

Streptothricin F; n = 1
              E; n = 2
              D; n = 3
              C; n = 4
              B; n = 5
              A; n = 6
              X; n = 7

[1] R, R' = H
     R" = CONH$_2$

[2] R, R' = H, CONH$_2$
     R" = H

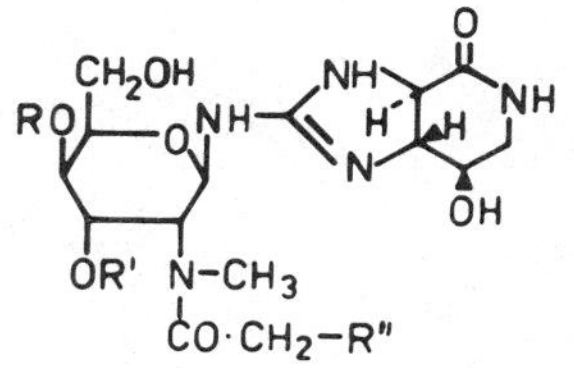

[3] R, R' = H, CONH$_2$
     R" = NH–CH
              ‖
              NH

[4] R, R' = H, CONH$_2$
     R" = NH$_2$

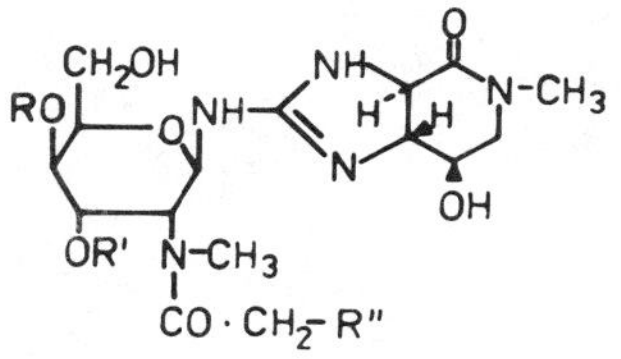

[5] R, R' = H, CONH$_2$
     R" = NH –CH
              ‖
              NH

[6] R, R' = H, CONH$_2$
     R" = NH$_2$

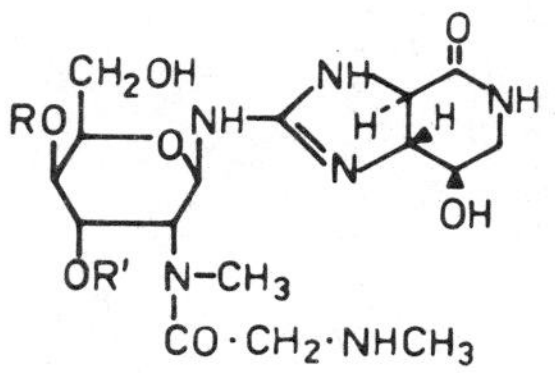

[7] R, R' = H, CONH$_2$

Figure 1 Chemical constitution of streptothricins and related antibiotics: Streptothricins F–X [1,2] according to Khokhlov and Shutova (1972) and Borders (1975), respectively. Antibiotic LL-AC 541 [3], deforminino LL-AC 541 [4], antibiotic LL-AB 664 [5], deformimino LL-AB 664 [6], and antibiotic SF-701 [7]. (From Khokhlov, 1978.)

Table 1 Streptothricins and Streptothricin Mixtures Produced by *Streptomyces* Strains

Antibiotic	Microorganism	Components	Reference
Streptothricin	S. lavendulae	F	Waksman and Woodruff (1942)
Streptolin	S. lavendulae 11	F,D,(C,B)	Rivett and Peterson (1947)
A-136	S. lavendulae 136B	F,D,C	Bohonos et al. (1947)
Streptothricin IV	S. lavendulae 3516	F,D	Hutchinson et al. (1949)
Roseothricin	S. roseochromogenes	F,D	Hosoya et al. (1949)
Pleocidin	S. lavendulae 272	F,E,D,(C,B)	Charney et al. (1952)
Geomycin	S. xanthophaeus	(D),C,B,(A)	Brockmann and Musso (1954)
Grassieromycin	S. lavendulae	F	Ueda et al. (1955)
Grisin (grisemin)	Actinomyces griseus	F,D	Krassilnikov et al. (1957)
Racemomycin	S. racemochromogenes	F,E,D,C,(B)	Sugai (1956)
Mycothricin	S. lavendulae 3716 and 3717	F,E,D,C	Rangaswami et al. (1956)
Phytobacteriomycin	Actinomyces 696	E,D,C	Semenova et al. (1960)
Polymycin	Actinomyces polymycini	C,B,A,X	Solovieva et al. (1960)
Nourseothricin	S. noursei JA 3890b	F,E,D,(C)	Bradler and Thrum (1963)
Virothricin	S. lavendulae var. virothricinus	F,E,D,C (streptomycin in addition)	Thrum and Bradler (1966)
Yazumycins	S. lavendulae IN-183,T	F,E	Akasaki et al. (1968)

Boseimycins	Streptomyces sp. AC$_6$-569		Sinha and Nandi (1968)
Akimycin	S. lavendulae E	19% F, 83% D, (E)	Arai et al. (1960)
A-8265	Streptomyces A 8265	F	Lumb et al. (1962)
A 3698	Unidentified Actinomyces	15% F, 15% E, 60% D, 5% C, 5% B	Lumb et al. (1962)
A 4788	Streptomyces sp. A 4788	F,E,D,C,B	Lumb et al. (1962)
A 3885	Streptomyces sp. A 3885	40% F, 10% E, 40% D, 5% C, 4% B, 1% A	Lumb et al. (1962)
A 3967	Streptomyces sp. 3967	5% F, 5% E, 35% D, 25% C, 20% B, 10% A	Lumb et al. (1962
A 4714-12	S. lavendulae	23% F, 3% E, 40% D, 10% C, 20% B, 4% A	Khokhlov (1978)
A 11-8	S. flaveolus	19% F, 17% E, 48% D, 16% C	Khokhlov (1978)
A 4562-3	Unidentified Streptomyces	D,C,B,A	Khokhlov (1978)
A 4786-14	S. lavendulae	F,E,D,C	Khokhlov (1978)
A 4850-16	S. lavendulae	D.C	Khokhlov (1978)
A 5438-10	S. griseus	F,E,D,C	Khokhlov (1978)
A 5491-17	Unidentified Streptomyces	D,C,B	Khokhlov (1978)
A S 15-1	S. purpeofuscus	D,C,B	Khokhlov (1978)
		60% F, 30% E, (D)	Kawamura et al. (1976b)

*( ) = minor compound.

In addition to those streptothricin antibiotics listed in Table 1, the following streptothricin-like antibiotics have been reported.

Antibiotic LL-AC 541 [3] from *Streptomyces hygroscopicus*, has been reported by Borders et al. (1967, 1970) and Zbinovsky et al. (1968). Three further antibiotics were described: BY-81 from *Streptomyces olivoreticuli* nov. var. MCRL-0358 (Ito et al., 1968), E-749-c from *S. hygroscopicus* (Shoji et al., 1968), and 1483-A (citromycin) from *Streptomyces* IN 1483 (Taniyama and Sawada, 1971). These were revealed to be identical with LL-AC 541.

Deformimino-LL-AC 544 [4] is produced along with LL-AC 541 by *S. hygroscopicus*. It was also formed by letting LL-AC 541 stand in an aqueous solution of pH 8 for 1 hr at room temperature, or in methanolic solution for about 1 week (Zbinovsky et al., 1968).

Antibiotic LL-AB 664 [5] from *Streptomyces candidus* has been reported by Sax et al. (1968).

Antibiotic BD-12, from *Streptomyces luteocolor* (Ito et al., 1968) has proved to be identical with LL-AB 664.

Deformimino-LL-AB 664 [6], from *S. candidus* has been reported by Sax et al. (1968).

Glycinothricin, from *Streptomyces griseus* proved to be identical with deformimino-LL-AB 664 (Sawada et al., 1977c).

Antibiotic SF-701, from *Streptomyces griseochromogenes* (Tsuruoka et al., 1968), is identical to antibiotic LL-BL 136 [7] from *Streptomyces* BL 136 (Borders et al., 1972).

The structure of sclerothricin, from *Streptomyces sclerogranulatus* (Kono et al., 1969), is unknown.

The structure of racemomycin O, from *Streptomyces racemochromogenes* (Takemura, 1960a–c), has been proposed to be the same as structure [8]. It probably contains glucosamine instead of gulosamine.

The structure of fucothricin, from *Streptomyces fradiae* strain MYC-19, is unknown (Thirumalachar et al., 1971). It probably contains fucosamine instead of gulosamine.

Antibiotic R4H, from *Streptomyces lavendulae* strain R4 (Sawada et al., 1974), has been proposed to be a natural derivative of streptothricin F.

B. Biological Properties

Streptothricin attracted widespread attention because of its strong inhibitory action on gram-positive, gram-negative, and acid-fast bacteria, as well as some fungi (Ryabova et al., 1965; Taniyama et al., 1971). Furthermore, streptothricins, especially those with elongated β-lysyl peptide chains, show in addition to the above-mentioned antibacterial actions definite antiviral activities (Germanova and Goncharovskaya, 1965; Taniyama et al., 1971).

These findings were supported by studies with the virothricin complex, which revealed in vitro inhibition of several viruses. Plaque formation by vaccinia, Newcastle disease (ND), and pseudorabies viruses was inhibited. In chorioallantoic membrane cultures, the multiplication of influenza APR8 and ND viruses proved to be affected when measured by hemagglutination and infectivity titrations. In embryonated eggs infected with APR8 or ND virus, the titers were reduced. Virothricin proved to be nonvirucidal (Küchler and Küchler, 1966). The streptothricin antibiotic S 15-1 (SQ 21,704), which consists mainly of components F and E, also showed activity against ND virus (Arima and Kawamura, 1972; Arima et al., 1972) and taeniacidal effect (Brown et al., 1977; Liu et al., 1981).

Antitumor action against human cancer cells was found in streptothricin preparations with a high content of components A and B (e.g., polymycin) (Navashin et al., 1961).

However, inherent to all streptothricins are delayed toxicity and nephrotoxicity, which has so far prevented their use in medicine (Germanova and Goncharovskaya, 1965; Taniyama et al., 1971; Inamori et al., 1978; Kato et al., 1981).

Their use in agriculture seems somewhat more promising, both for combating some phytopathogenic bacteria and fungi and as additions into animal feed. Phytobacteriomycin has been applied to combat infections of plants and vegetables (Solovieva et al., 1961; Chaban and Poberezkina, 1971). Grisin (grisemin) was effective in bacterial diseases of plants (Krassilnikov et al., 1957).

Kormogrisin (grisin) was tested in three trials with store pigs and in two trials with fattening pigs. With the exception of trial 1 with store pigs, doses of 1, 2, 4, 10, 20, and 40 mg of kormogrisin per kilogram of feed had a positive effect on daily live weight gain and feed input per kilogram of live weight (Lüdke and Knobloch, 1978/1979). The antibiotic was also effective in chickens (Maslovskii et al., 1976; Hennig et al., 1979).

Recent studies with racemomycin D showed further physiological activities:

Insecticidal effects on *Blattella germanica*, *Musca domestica*, *Nilaparvata lugens*, *Laodelphax striatella*, and *Plutella xylostella*

Growth inhibition of the plants *Brassica campestris* L. subsp. *Napus Hook fil et Anders* var. *nippo-okifera*, *Arctium Lappa* L., *Petrocelium sativum*, and *Raphanus sativus* L. var. *acanthiformis* with 500 ppm (with an inhibitory ratio of 0.1, relative to 1.0 for the control group)

Fish toxicity to the *Misgurnus anguillicaudatus*, *Carassius auratus*, and *Oriziae latipes* (Takemoto et al., 1980).

III. STRAIN IMPROVEMENT

Strain S 15-1, identified as *Streptomyces fuscus* Yamaguchi and Saburi, produces the antibiotics S 15-1-A and S 15-1-B (Brown et al., 1977), which were identified with racemomycins A and C (Liu et al., 1981).

For antibiotic production soluble starch was found to be superior to dextrin and glucose as a carbon source. Meat extract, soybean meal, and cornsteep liquor were found to be superior to other organic nitrogen sources tested. The addition of 2% of each nitrogen source was the most suitable concentration for production of the antibiotic mixture S 15-1. The latter was not produced in media where organic nitrogen sources were omitted. The time course of antibiotic formation was investigated in a basal medium containing 3% soluble starch, 0.4% $NaNO_3$, 0.2% KH_2PO_4, 0.1% KCl, 0.1% $MgSO_4 \cdot 7H_2O$, and 0.002% $FeSO_4 \cdot 7H_2O$. Two percent meat extract, soybean meal, or corn-steep liquor was added. Fermentation conditions were 26.5°C, 115 rpm, 100 ml of medium (pH 7) in 500-ml flasks, and the inoculum size was 5% of the seed broth. Maximum production of antibiotic S 15-1 occurred after 72–96 hr incubation. The most suitable organic nitrogen source was meat extract (maximum activity; 260 μg/ml).

Mutant strains which possessed the capacity for higher productivity of antibiotic S 15-1 were induced by ultraviolet irradiation of the original strain. Almost all (98%) of the mutants obtained by ultraviolet irradiation produced antibiotic S 15-1 at the same level as the original strain.

Table 2 Production of Antibiotic S 15-1 by High-Yield Mutant Substrains[a]

Mutant substrains	Fermentation time			
	6 days		7 days	
	pH	S 15-1 (μg/ml)	pH	S 15-1 (μg/ml)
A-60-126	7.7	1840	7.6	1780
A-60-128	7.6	1210	7.7	1510
A-60-169	7.8	1820	7.5	1260
A-60-181	7.6	1540	7.8	1000
A-60-207	7.8	1650	8.0	900
A-60-225	7.6	1650	7.9	1350
A-60-227	7.8	1650	7.6	1780
A-60-251	8.1	2100	7.8	980
A-60-262	7.3	1540	7.6	1960
A-60-716	7.6	2030	7.8	2480
A-60-13	7.8	1740	7.8	1100
Control (Parent)	7.6	250	7.6	200

[a]Medium: starch, 3.0%; soybean meal, 2.0%; $NaNO_3$, 0.4%; K_2HPO_4, 0.2%; KCl, 0.1%; $MgSO_4 \cdot 7H_2O$, 0.1%; and $FeSO_4 \cdot 7H_2O$, 0.002% (pH 7.0). Fermentation conditions: 26.5°C, 6 days, reciprocal shaker 115 strokes per minute; 100 ml of medium in 500-ml flask; inoculum size, 5.0% of seed broth; bioassay using *Bacillus subtilis* strain PCl 219.
Source: Kawamura et al. (1976b).

A total of 11 mutant substrains which possessed the capacity for high productivity were tested in the same medium mentioned above. Mutant strain A-60-716 (see Table 2) produced 2480 μg/ml of the antibiotic S 15-1, while the control yielded only about 250 μg/ml (Kawamura et al., 1976a). In kormogrisin production wheat barley flour in the medium ensures the intense biosynthesis of grisin up to 60,000 units per milliliter (Solntsev et al., 1970).

IV. BIOSYNTHESIS: MECHANISMS AND REGULATION

Several efforts were made to obtain information about the biosynthetic pathway of streptothricin antibiotics. According to Voronina and Khokhlov (1972), Voronina et al. (1973), Carter et al., (1974), and Sawada et al. (1976a), the β-lysine moieties are formed in streptomycetes via isomerization of α-lysine. On the other hand, it is likely that the gulosamine part is synthesized from glucose (Johnson and Westley, 1962; Sawada et al., 1977a) or D-glucosamine (Sawada et al., 1977b) as a precursor. Streptolidine as the third constituent of streptothricins was found to be produced biosynthetically from acetate in a strain of *S. lavendulae* ISP-5069 (Sawada et al., 1976a). This was also supported by studies with carboxyl and methyl carbon-13-labeled acetates (Sawada et al., 1977b). Other incorporation experiments with the nourseothricin-

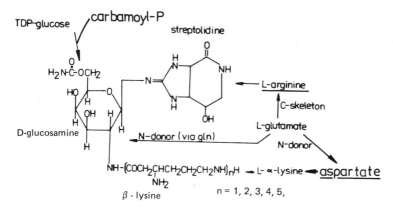

Figure 2 Biogenesis of streptothricin antibiotics. (Courtesy of U. Gräfe, Central Institute of Microbiology and Experimental Therapy, Jena.)

producing strain *Streptomyces noursei* JA 3890b (Bradler and Thrum, 1963; Thrum and Bradler, 1966) using ^{14}C-labeled amino acids showed, however, that the streptolidine moiety of nourseothricin is formed via the dehydroarginine pathway (Fig. 2) (Gräfe et al., 1977).

Further investigations with *S. lavendulae* strain OP-2 producing racemomycin C also showed preferential incorporation of [[14]D]arginine into streptolidine in racemomycin C (Sawada et al., 1977d). This led to the conclusion that streptolidine biosynthesis by *S. lavendulae* strains is due to at least two pathways (Sawada et al., 1978).

Catabolite repression was reported in streptothricin-producing strains of *S. lavendulae* ISP-5096 (Sawada et al., 1976b) and streptomycete S-2, which produced a streptothricin with a higher content of β-lysin (Inamori et al., 1976). Antibiotic formation was found to be repressed in the presence of glucose in the medium, although glucose was the best carbon source for growth of *S. lavendulae* ISP-5069. The repression by glucose was gradually restored by the addition of amino acid mixtures.

The formation of nourseothricin, a mixture of streptothricins F, E, D, and C (Bradler and Thrum, 1963; Thrum and Bradler, 1966) is strongly stimulated by o-aminobenzoic acid (OABA) (Thrum and Bocker, 1965; Bocker and Thrum, 1966). The regulatory influence of this aromatic acid has been extensively studied in a series of papers. It was shown that OABA interferes with amino acid metabolism and influences the efficiency of the precursor-supplying system in the nourseothricin biosynthesis. Thus mycelium grown in the presence of OABA exhibits markedly increased activities of NADP-specific glutamate dehydrogenase. Simultaneously, this effector prevents the induction of alanine dehydrogenase in productive cultures, suggesting that the enhanced formation of alanine interferes with antibiotic biosynthesis (Gräfe et al., 1974). In continuing these efforts, the participation of the glutamine synthetase/glutamate synthase pathway in the regeneration of glutamic acid was established. Glutamate synthase of this organism requires 2-ketoglutarate, glutamine, and NADH for glutamate production (Gräfe et al., 1977). OABA was also found to strongly increase the NADH/NAD[+] ratio in growing mycelium of *S. noursei* 3890b, indicating that this effector is capable of interfering with the function of the respiratory chain. In parallel, a complex shift of metabolism was shown to be induced by simultaneous alteration of the mycelial activities of alanine dehydrogenase, glutamine synthetase, and glutamate dehydrogenase. These changes may be responsible for the observed delay of amino acid catabolism and may improve the precursor supply to the secondary metabolism (Gräfe et al., 1978).

Further papers deal with the instability of the metabolism and mode of action of OABA (Gräfe et al., 1979a), as well as with the effect of OABA on cytochrome levels and amino acid transport (Gräfe et al., 1980). An additional paper reported on an alcohol-induced switch of the metabolic flux in S. noursei 3890b from preferential oxidative deamination of alanine toward reinforced acquisition of NH_4^+. These changes were related to the decrease of the ratio of saturated to olefinic fatty acids in the mycelium, suggesting that alcohol and other polar lipophilic compounds might interfere with the biosynthesis and formation of the cytoplasmatic membrane of streptomycetes (Gräfe et al., 1979c). From these foregoing studies arose an investigation on the regulation of glutamine synthetase formation in several species of streptomycetes. The results support the formerly described findings with S. noursei 3890b (Gräfe et al., 1974). Thus the repressive influence of NH_4^+ on glutamine synthetase formation may be interpreted as if the enzyme takes part in the fixation of small amounts of NH_4^+ on the glutamine synthetase/glutamate synthase pathway in streptomycetes (Gräfe et al., 1979b).

V. FERMENTATION PROCESS

Owing to a lack of large-scale production experiments with these antibiotics, some examples of fermentation of special streptothricins under laboratory and pilot-plant conditions are given.

Example 1. Streptolin Production in 30-Liter Fermentations (Rivett and Peterson, 1947). The medium consists of 1% soybean meal, 0.25% Curbay BG, 2% glucose, 1% NaCl, 0.1% $CaCO_3$. Fermentation was for 65 hr. In an effort to increase the yield of streptolin in this medium in the fermenters, a series of runs were made in which agitation and aeration were varied. The results (Table 3) show that there is an optimum aeration rate at about 0.25 vol of air per minute per volume of medium with an agitation rate of 195 rpm. Both increasing and decreasing the aeration or agitation from this optimum resulted in a decrease in yield.

Table 3 Effect of Agitation and Aeration on Streptolin Production in 30-Liter Fermenters

Agitation (rpm)	Aeration (vol/vol per min)	Maximum yield (units/ml)	Age (hr)
70	0.25	500	58
195	0.66	23,900	60
195	0.25	27,300	65
195	0.25	26,400	60
375	0.25	23,400	83
375	0.66	15,300	83

Source: Rivett and Peterson (1947).

The main production of streptolin occurred after most of the glucose had been used and during the period of most rapid non-ammonium-nitrogen utilization and growth. A rise in pH, probably due to ammonia liberation, also occurred at the time of streptolin formation. It was found that soybean meal and corn-steep liquor were the best nitrogen sources, while glucose and glycerol were the best carbon sources. Sodium chloride was essential for high yields of streptolin on media containing soybean meal.

Example 2. Racemomycin A Production by *S. lavendulae* ISP-5069 (Sawada et al., 1977c). One milliliter of the precultivation medium (containing 1% glucose, 1% yeast extract, 1% polypeptone, 0.2% NaCl, 0.01% $(NH_4)_2SO_4$, 0.01% KH_2PO_4, 0.01% K_2HPO_4, 0.01% $MgSO_4$; pH 7.2) of *S. lavendulae* was inoculated into 100 ml of fermentation medium (consisting of 2% maltose, 0.5% polypeptone, 0.5% meat extract, 0.3% yeast extract, 0.3% NaCl, 0.1% $MgSO_4$, 1 ml of trace salts solution per 100 ml; pH 7.4) in a 500-ml flask. This was run at 160 rpm at 27°C for 40 hr. This strain produced about 200 µg/ml of racemomycin A when it was shake cultured under these conditions.

Example 3. Nourseothricin Production by *S. noursei* JA 3890b (Gräfe et al., 1974). Three milliliters of the precultivation medium (containing 2% glucose, 2% soybean meal, 0.5% NaCl, 0.1% KH_2PO_4, 0.3% $CaCO_3$) of *S. noursei* JA 3890b were inoculated into 80 ml of fermentation medium (consisting of 2.9% glucose, 1.3% soybean meal, 0.5% NaCl, 0.3% $CaCO_3$; pH 6.5) in a 500-ml flask, and this flask was shaken at 240 rpm for 96 hr at 25°C. The yield was 80 µg/ml. Addition of 7.5 mM o-aminobenzoic acid enhanced the nourseothricin formation up to 10-fold.

VI. PRODUCT RECOVERY AND PURIFICATION

Streptothricins and streptothricin-like antibiotics are extremely water-soluble basic compounds which can be recovered from fermentation broth filtrate either by precipitation or adsorption procedures. The preferred method is absorption on cation exchange resins and elution with diluted mineral acids or weak bases. Further steps of purification consist in desalting and decoloring the crude products (Khokhlov, 1978).

To isolate single streptothricin components, Khokhlov and Reshetov successfully applied ion exchange chromatography on carboxy methyl cellulose with elution in NaCl gradient. This method allows a rather quick separation and good yields (Khokhlov and Reshetov, 1964).

Some selected examples of isolation and purification of streptothricin preparations are given below.

Example 1. Production and Isolation of Streptothricin F (St-F) (Haupt et al., 1978). *Streptomyces lavendulae* JA 2254, which produces ST-F free of other streptothricin components, was cultivated under submerged conditions in a medium containing 1% soybean meal, 2% glucose, 0.5% sodium chloride, 0.3% calcium carbonate, and 0.5% corn-steep liquor at 28°C for 96 hr.

The harvested mash (70 liters) with 70 µg/ml was acidified to pH 3 with dilute hydrochloric acid and the mycelium removed by centrifugation. The resulting filtrate was passed through a column of the weakly acidic cation exchange resin Wofatit CP-300 (Na^+ form) after adjustment to pH 7 with dilute sodium hydroxide. The resin was thoroughly washed with water followed by 0.01 N acetic acid. The antibiotic was eluted with 0.05 N hydrochloric acid and the acidic effluent neutralized by addition of the anion exchange resin Wofatit L-150 (OH^- form). The active fractions of the eluate were combined and concentrated in vacuo to 10% of the volume. The concentrate was decolored by treatment with activated carbon (EPN-Kohle), filtered, and evaporated to dryness. The resulting residue was reprecipitated twice with methanol acetone. The yield was 2.1 g.

Example 2. Production and Isolation of Streptothricin Antibiotic R4H (Sawada et al., 1974). *Streptomyces lavendulae* strain R4 was cultivated on a medium (20 liters, lab fermenter) containing 0.1% glucose, 1% peptone, 0.5% yeast extract, 0.3% NaCl, 0.1% KH_2PO_4, and 0.05% $MgSO_4$ (pH 6.0), at 28°C for 20 hr, with an aeration of 20 liters per minute and agitation at 200 rpm. Antimicrobial activity reached a maximum after 18 hr.

Culture filtrate was passed through a column of Amberlite IRC-50 (Na^+ form) after adjustment of the pH to 7.0, and the column was then washed with water and eluted with 1 N acetic acid. The concentrated solution of active fractions was filtered to remove the inactive precipitate, and acetone was added to the filtrate to precipitate a brownish crude powder which, on reprecipitating thrice with methanol acetone, gave 8.7 g of a powder containing the antibiotic acetate.

Crude complex (2 g) was purified on a column (25 mm in diameter; bed volume, 400 ml) of SE-Sephadex C-25 (Pharmacia Fine Chemicals, Uppsala) by stepwise elution with buffers (0.1, 0.5, 1.0, and 3.0 M pyridine acetic acid; pH 5.0). Fractions containing R4H were pooled and evaporated in vacuo to a small volume and precipitated with acetone to yield 128 mg of a white powder.

Example 3. Fermentation and Isolation of Glycinothricin (Sawada et al., 1977e). Glycinothricin was produced by *S. griseus* 979 in a 600-liter fermenter containing 300 liters of medium consisting of 3% starch, 2% corn-steep liquor, 1.5% Pharmamedia, 1% meat extract, and 0.01% Defoam CB-442 (pH 7.0 before sterilization). The potency obtained after 72 hr of cultivation was 1100 μg/ml. The culture broth (320 liters) was adjusted to pH 4.0 and filtered with the aid of infusorial earth. The filtrate (270 liters) was adjusted to pH 8.5 with NaOH and adsorbed on activated carbon (4.35 kg). The carbon cake was washed with 50 liters of 50% aqueous acetone (pH 8.5 with NaOH). The adsorbate was eluted twice with a total of 120 liters of 50% acidic aqueous acetone (pH 4.0 with HCl after agitation). The eluate was concentrated in vacuo to give 4 liters of the solution containing 83.5 g glycinothricin in 29.0% recovery from the culture filtrate.

A 400-ml aliquot of the concentrate was adjusted to pH 7.0 with dilute NaOH and treated with 10 parts of acetone. The precipitate formed was dissolved in water and further purified. The antibiotic was adsorbed on a column with CM-Sephadex C-25 equilibrated with 0.1 M ammonium formiate and eluted with 0.4 and 0.6 M ammonium formiate, successively. The active fractions were concentrated and passed through a column of activated carbon followed by elution with 50% aqueous acetone of pH 2.0. The active fractions were then chromatographed on a column with Sephadex LH-20, with water as an eluant. The glycinothricin-containing fractions were concentrated and, after addition of 0.3 N HCl, treated with 10 parts of acetone.

A final purification was performed by column chromatography on Sephadex LH-20 with water as an eluant. The active solution was lyophilized to yield 1.5 g of glycinothricin powder.

VII. MODE OF ACTION

Several efforts were made to study the mode of action of streptothricin antibiotics. As early as 1967, the interaction of streptothricin F and a natural preparation (containing mainly components D and C) with DNA was investigated with spectrophotometric melting, titration, and sedimentation measurements. It was found that with increasing antibiotic concentrations, precipitation of insoluble DNA–antibiotic complexes was observed. Soluble DNA–antibiotic complexes are formed at low ionic strength from stabilization of the DNA secondary structure against thermal denaturation. In addition, spectrophotometric acid titration profiles shift to lower pH values, and sedimentation coefficients increase with increasing antibiotic concentration. These results suggest that at low ionic strength these antibiotics are loosely bound to the negatively charged groups of the DNA by electrostatic interaction. This bonding can be reversed by increasing the ionic strength to moderate and high salt concentrations (Zimmer et al., 1967).

Boseimycin, a natural mixture of streptothricins, was reported to act on the translation level of the protein-synthesizing system of bacteria. There is only a transient inhibition of protein synthesis rather than any irreparable damage caused to the cellular components (Misra and Sinha, 1971). This was confirmed in studies on the inhibition of protein biosynthesis by different streptothricins in Nirenberg cell-free system containing *Escherichia coli* ribosomes (Telesnina et al., 1973). The inactivation of protein biosynthesis depends strongly on the β-lysine content in streptothricins: The activity of streptothricin B is commensurate with that of chloramphenicol. In addition, it was found that antibacterially inactive products obtained by mild acid hydrolysis of streptothricin have an effect on protein biosynthesis similar to that of the parent antibiotics. Probably, the difference in the actions of these products on intact microbial cells and on ribosomal systems, prepared from these cells, is a result of their way of transport through bacterial cell walls and membranes.

Furthermore, the effect of streptothricin F (Str-F) on macromolecular syntheses in intact cells and on the cell-free protein synthesis of *E. coli* was studied (Haupt et al., 1978). The results indicate that protein synthesis is the primary site of inhibition by Str-F in growing cells of *E. coli*. Cell-free polypeptide synthesis in *E. coli* directed by poly(U) was inhibited, while poly(A)- and poly(C)-directed polypeptide syntheses were both stimulated by this drug. Furthermore, Str-F caused a misreading of the translation of poly(U)-, poly(A)-, and poly(C)-directed protein syntheses of *E. coli* systems. The extent of miscoding increased with increasing drug concentrations. However, in liver extract protein synthesis, directed by poly(U) or endogenous mRNA, was not inhibited.

In continuing these efforts on the mode of action of Str-F, we used the factor-free and factor-dependent translation systems (Haupt et al., 1980). Str-F inhibited polyphenylalanine synthesis in both systems. Thus it was concluded that the antibiotic acts via interaction with the ribosome and not by direct interaction with the elongation factors. On the other hand, Str-F significantly inhibits translocation and to a somewhat lesser degree binding of AA-tRNA while it does not inhibit peptide bond formation.

With respect to inhibition of translation and induction of misreading, Str-F resembles aminoglycoside antibiotics, such as neomycin, paromomycin, kanamycin, gentamicin, and hygromycin, though it differs completely as to its chemical structure from this group of antibiotics.

VIII. APPLICATIONS

Kormogrisin is used as a food additive in animal nutrition, especially for chickens, piglets, and calves. It is manufactured in the Soviet Union on a large scale with a grisin- (grisemin) forming strain *Actinomyces (Streptomyces) griseus*. The applied mixtures contain 5 and 10 mg grisin/kg food, respectively.

REFERENCES

Akasaki, H., Abe, H., Seino, A., and Shirato, S. (1968). Yazumycin, a new antibiotic produced by Streptomyces lavendulae. *J. Antibiot. Ser. A (Tokyo)* 21:98–105.

Arai, T., Koyama, Y., Honda, H., and Hayashi, M. (1960). Simultaneous production of two antibiotics by S. lavendulae E 20-27. *Ann. Rep. Inst. Food Microbiol.* 13:39–44.

Arima, K., and Kawamura, T. (1972). The analysis of a new antiviral substance S15-1, streptothricin group antibiotic. *J. Antibiot. Ser. A (Tokyo)* 25:471-472.

Arima, K., Kawamura, T., and Beppu, T. (1972). A new antiviral substance S15-1, streptothricin group antibiotic. *J. Antibiot. Ser. A (Tokyo)* 25:387–392.

Betina, M. (1975). Paper chromatography of antibiotics. In *Methods in Enzymology*, Vol. 43. J. H. Hash (Ed.). Academic, New York, pp. 123–172.

Bocker, H., and Thrum, H. (1966). Stimulation of nourseothricin production by aminobenzoic acids. In *Antibiotics—Advances in Research, Production and Clinical Use*. M. Herold and Z. Gabriel (Eds.). Butterworth, London, pp. 584–587.

Bohonos, N., Emerson, R. L., Whiffen, A. J., Nash, M. P., and DeBoer, C. (1947). A new antibiotic produced by a strain of *Streptomyces lavendulae*. *Arch. Biochem.* 15:215–225.

Borders, D. B. (1975). Ion-exchange chromatography of streptothricin-like antibiotics. In *Methods in Enzymology*, Vol. 43. J. H. Hash (Ed.). Academic, New York, pp. 256–263.

Borders, D. B., Hausmann, W. K., Wetzel, E. R., and Patterson, E. L. (1967). Partial structure of antibiotic LL-AC 541. *Tetrahedron Lett.* 1967:4187–4192.

Borders, D. B., Sax, K. J., Lancaster, J. E., Hausmann, W. K., Mitscher, L. A., Wetzel, E. R., and Patterson, E. L. (1970). Structure of LL-AC 541 and LL-AB 664 new streptothricin-type antibiotics. *Tetrahedron 26*:3123–3133.

Borders, D. B., Kirby, J. P., Wetzel, E. R., Davies, M. C., and Hausmann, W. K. (1972). Analytical method for streptothricin-type antibiotics: Structure of antibiotic LL-BL 136. *Antimicrob. Agents Chemother. 1*:403–407.

Bradler, G., and Thrum, H. (1963). Nourseothricin A and B, zwei neue antibakterielle Antibiotika einer *Streptomyces-noursei*-Variante. *Z. Allg. Mikrobiol.* 3:105–112.

Brockmann, H., and Musso, H. (1954). Geomycin, ein neues, gegen gramnegative Bakterien wirksames Antibiotikum. *Naturwissenschaften 41*:451–452.

Brockmann, H., and Musso, H. (1955). Hydrolytischer Abbau der Geomycine; Geomycin, III. Mitteil. Antibiotika aus Actinomyceten. XXX. Mitteil. *Chem. Ber.* 88:648–661.

Brown, W. E., Szanto, J., Meyers, E., Kawamura, T., and Arima, K. (1977). Taeniacidal activity of streptothricin antibiotic complex S15-1 (SQ 21704). *J. Antibiot. Ser. A (Tokyo)* 30:886–889.

Chaban, L. M., and Poberezkina, D. Ya. (1971). Experimental data on stability of streptothricin antibiotics: Phytobacteriomycin and polymycin. *Antibiotiki 16*:741–743.

Charney, J., Roberts, W. S., and Fisher, W. P. (1952). Pleocidin, a new antibiotic related to streptothricin. *Antibiot. Chemother. 2*:307–310.

Carter, J. H., DuBus, R. H., Dyer, J. R., Floyd, J. C., Rice, K. C., and Shaw, P. D. (1974). Biosynthesis of viomycin. II. Origin of β-lysine and viomycidine. *Biochemistry 13*:1227–1233.

Germanova, K. I., and Goncharovskaya, T. Ya. (1965). Antiviral properties of streptothricins A, B, C, D, F. *Antibiotiki 14*:48–51.

Germanova, K. I., Goncharovskaya, T. Ya., Delova, I. D., Iliinskaya, S. A., Melnikova, A. A., Oreshnikova, T. P., Reshetov, P. D., Rudaya, S. D., Sinitsyna, Z. T., Solovieva, N. K., and Khokhlov, A. S. (1965). Composition and antiviral properties of some streptothricins. *Antibiotiki 10*: 117–122.

Gräfe, U., Bocker, H., Reinhardt, G., and Thrum, H. (1974). Regulative Beeinflussung der Nourseothricinbiosynthese durch o-Aminobenzoesäure in Kulturen des Streptomyces noursei JA 3890b. *Z. Allg. Mikrobiol. 14*: 659–673.

Gräfe, U., Bocker, H., and Thrum, H. (1977). Regulative influence of o-aminobenzoic acid on the biosynthesis of nourseothricin in cultures of *Streptomyces noursei* JA 3890b. II. Regulation of glutamine synthetase and the role of the glutamine synthetase/glutamate synthase pathway. *Z. Allg. Mikrobiol. 17*:201–209.

Gräfe, U., Reinhardt, G., Bocker, H., and Thrum, H. (1977). Biosynthesis of streptolidine moiety of streptothricins by *Streptomyces noursei* JA 3890b. *J. Antibiot. Ser. A (Tokyo) 30*:106–110.

Gräfe, U., Bocker, H., Reinhardt, G., and Thrum, H. (1978). Regulative influence of o-aminobenzoic acid on the biosynthesis of nourseothricin in cultures of *Streptomyces noursei* JA 3890b. III. Change of redox state of nicotinamide-adenine-dinucleotides in the presence of aminobenzoic acids. *Z. Allg. Mikrobiol. 18*:479–486.

Gräfe, U., Bocker, H., and Thrum, H. (1979a). Regulative influence of o-aminobenzoic acid on the biosynthesis of nourseothricin in cultures of *Streptomyces noursei* JA 3890b. IV. Bistability of metabolism and the mechanism of action of aminobenzoic acids. *Z. Allg. Mikrobiol. 19*:235–246.

Gräfe, U., Bocker, H., and Thrum, H. (1979b). Regulation der Glutamin-Synthetasebildung in Streptomyceten. *Z. Allg. Mikrobiol. 19*:663–666.

Gräfe, U., Schade, W., Bocker, H., and Thrum, H. (1979c). Alcohol-induced switching over of metabolic flux in *Streptomyces noursei* JA 3890b. *Z. Allg. Mikrobiol. 19*:721–726.

Gräfe, U., Steudel, A., Bocker, H., and Thrum, H. (1980). Regulative influence of o-aminobenzoic acid (OABA) on the biosynthesis of nourseothricin in cultures of *Streptomyces noursei* JA 3890b. V. Effect of OABA on cytochrome levels and amino acid transport. *Z. Allg. Mikrobiol. 20*:185–194.

Haupt, I., Hübener, R., and Thrum, H. (1978). Streptothricin F, an inhibitor of protein synthesis with miscoding activity. *J. Antibiot. Ser. A (Tokyo) 31*:1137–1142

Haupt, I., Jonak, J., Rychlik, I., and Thrum, H. (1980). Action of streptothricin F on ribosomal functions. *J. Antibiot. Ser. A (Tokyo) 33*:636–641.

Hennig, A., Jeroch, H., and Flachowski, G. (1979). Zum gegenwärtigen Stand des Einsatzes der Ergotropika, insbesondere der Antibiotika, als Futterzusatz. *Mh. Vet. Med.* *34*:343–351.

Horowitz, M. I., and Schaffner, C. P. (1958). Paper chromatography of streptothricin antibiotics. Differentiation and fractionation studies. *Anal. Chem.* *30*:1616–1620.

Hosoya, S., Soeda, M., Komatsu, N., Inamura, S., Iwasaki, M., Sonoda, Y., and Okada, K. (1950). The antibiotic substances produced by Streptomyces. *Jpn. J. Exp. Med.* *20*:121–133.

Hutchinson, D., Swart, E. A., and Waksman, S. A. (1949). Production, isolation and antimicrobial, notably antituberculosis, properties of streptothricin IV. *Arch. Biochem.* *22*:16–30.

Inamori, Y., Sunagawa, S., Sawada, Y., and Taniyama, H. (1976). Inhibitory effect of glucose on streptothricin antibiotic production. *J. Ferment. Technol.* *54*:795–800.

Inamori, Y., Sunagawa, S., Tsuruga, M., Sawada, Y., Taniyama, H., Saito, G., and Daigo, K. (1978). Toxicological approaches to streptothricin antibiotics. I. Implications of delayed toxicity in mice. *Chem. Pharm. Bull.* *26*:1147–1152.

Ito, Y., Ohashi, Y., Sakurai, Y., Sukurazawa, M., Yoshida, H., Awataguchi, S., and Okuda, T. (1968). New basic water soluble antibiotics BD-12 and BY-81. II. Isolation, purification and properties. *J. Antibiot. Ser. A (Tokyo)* *21*:307–312.

Johnson, A. W., and Westley, J. W. (1962). The streptothricin group of antibiotics. Part I. The general structural pattern. *J. Chem. Soc. 1962*: 1642–1655.

Kato, Y., Kubo, M., Morimoto, K., Saito, G., Morisaka, K., Inamori, Y., Hama, I., Maekawa, T., Mazaki, H., Ishimasa, T., Sawada, Y., and Taniyama, H. (1981). Toxicological approaches to streptothricin antibiotics. IV. Toxicity of streptothricin antibiotics to the blood. *Chem. Pharm. Bull.* *29*:580–584.

Kawamura, T., Tago, K., Beppu, T., and Arima, K. (1976a). Antiviral antibiotic S15-1. Taxonomy of the producing strain and study of conditions for production of the antibiotic. *J. Antibiot. Ser. A (Tokyo)* *29*:242–247.

Kawamura, T., Kimura, T., Tago, K., Beppu, T., and Arima, T. (1976b). The identity of S15-1-A and B with racemomycins A and C. *J. Antibiot. Ser. A (Tokyo)* *29*:844–846.

Khokhlov, A. S. (1978). Streptothricins and related antibiotics. In *Journal of Chromatography Library*, Vol. 15. M. J. Weinstein and G. H. Wagman (Eds.). Elsevier, Amsterdam, pp. 617–713.

Khokhlov, A. S., and Reshetov, P. D. (1964). Chromatography of streptothricins on carboxymethylcellulose. *J. Chromatogr.* *14*:495–496.

Khokhlov, A. S., and Shutova, K. I. (1972). Chemical structure of streptothricins. *J. Antibiot. Ser. A (Tokyo)* *25*:501–508.

Kono, Y., Makino, S., Takeuchi, S., and Yonehara, H. (1969). Sclerothricin, a new basic antibiotic. *J. Antibiot. Ser. A (Tokyo)* *22*:583–589.

Krassilnikov, N. A., Belosersky, A. N., Rautenstein, J. I., Koréniako, A. I., Nikitina, N. I., Sokolova, A. I., and Urisson, S. O. (1957). The antibiotic grizin (grizemin) and its producers. *Mikrobiologiya* *26*:418–415.

Küchler, C., and Küchler, W. (1966). Studies on antiviral activity of virothricin. *Acta Virol.* *10*:195–199.

Liu, W. C., Astle, G. L., Wells, J. S., Cruthers, L. R., Platt, T. B., Brown, W. E., and Linkenheimer, W. H. (1981). Separation and biological activities of individual components of S15-1, a streptothricin class antibiotic. *J. Antibiot. Ser. A (Tokyo)* 34:292–297.

Lüdke, H., and Knobloch, F. (1978/79). Untersuchungen zum Einsatz des Antibiotikums Kormogrisin in der Fütterung der Läufer und Mastschweine. *Tierernährung und Fütterung—Erfahrungen, Ergebnisse, Entwicklungen* 11:109–114.

Lumb, M., Chamberlain, N., Gross, T., Macey, P. E., Spyvee, J., Uprichard, J. M., and Wright, R. D. (1962). An antibiotic complex (A 4788) containing streptothricin and having activity against powdery mildews. *J. Sci. Food Agric.* 13:343.

Maslovskii, K. S., Chulkov, P. A., Bolotin, P. F., Digaltseva, T. N., Vinnikova, M. F., Biryukova, M. I., and Voskoboinikov, V. G. (1976). Evaluation of the effectiveness of feed grisin and its combination with trace elements in raising chicks. *Doklady TSKhA*, 215:73–76.

Misra, T. K., and Sinha, R. K. (1971). Mechanism of action of boseimycin. *Experientia (Basel)* 27:642–644.

Nakanishi, K., Ito, T., and Hirata, Y. (1954). Structure of a new amino acid obtained from roseothricin. *J. Am. Chem. Soc.* 76:2845–2846.

Navashin, S. M., Fomina, I. P., and Koroleva, V. G. (1961). A study of the antitumor action of actinoxanthine and polymycin in cultures of human cancer cells. *Antibiotiki* 6:912–918.

Rangaswami, G., Schaffner, C. P., and Waksman, S. A. (1956). Isolation and characterization of two mycothricin complexes. *Antibiot. Chemother.* 6:675–683.

Reshetov, P. D., and Khokhlov, A. S. (1964). Research of streptothricins by ion-exchange chromatography. *Antibiotiki* 9:197–201.

Reshetov, P. D., and Khokhlov, A. S. (1965). Research of streptothricins. VI. Isolation and properties of individual streptothricins. *Khim. Prir. Soedin.* 1:42–52.

Reshetov, P. D., Egorov, C. A., and Khokhlov, A. S. (1965). Studies of streptothricins. VII. Determination of ninhydrin-positive parts of streptothricin antibiotics. *Khim. Prir. Soedin.* 2:117–122.

Rivett, R. W., and Peterson, W. H. (1947). Streptolin, a new antibiotic from a species of *Streptomyces*. *J. Am. Chem. Soc.* 69:3006–3009.

Ryabova, I. D., Reshetov, P. D., Zhdanov, G. L., and Khokhlov, A. S. (1965). Antimicrobial activity of streptothricins. *Antibiotiki* 10:1066–1068.

Sawada, Y., Taniyama, H., Hanyuda, N., Hayashi, H., and Ishida, T. (1974). A new streptothricin antibiotic, R4H. *J. Antibiot. Ser. A (Tokyo)* 27:535–543.

Sawada, Y., Kubo, T., and Taniyama, H. (1976a). Biosynthesis of streptothricin antibiotics. I. Incorporation of ^{14}C-labeled compound into racemomycin-A and distribution of radioactivity. *Chem. Pharm. Bull.* 24:2163–2167.

Sawada, Y., Sakamoto, H., Kubo, T., and Taniyama, H. (1976b). Biosynthesis of streptothricin antibiotics. II. Catabolite inhibition of glucose on racemomycin-A production. *Chem. Pharm. Bull.* 24:2480–2485.

Sawada, Y., Nakashima, S., and Taniyama, H. (1977a). Biosynthesis of streptothricin antibiotics. VI. Mechanism of β-lysine and its peptide formation. *Chem. Pharm. Bull.* 25:3210–3217.

Sawada, Y., Kawakami, S., Taniyama, H., and Inamori, Y. (1977b). Incorporation of carboxyl and methyl carbon-13 labeled acetates into racemomycin A by *Streptomyces lavendulae* ISP 5069. *J. Antibiot. Ser. A (Tokyo) 30*:630–632.

Sawada, Y., Nakashima, S., Taniyama, H., and Inamori, Y. (1977c). Biosynthesis of streptothricin antibiotics. III. Incorporation of D-glucosamine into D-gulosamine moiety of racemomycin-A. *Chem. Pharm. Bull. 25*: 1478–1481.

Sawada, Y., Nakashima, S., Taniyama, H., Inamori, Y., Sunagawa, S., and Tsuruga, M. (1977d). Biosynthesis of streptothricin antibiotics. IV. On the incorporation of L-arginine into streptolidine moiety by *Streptomyces lavendulae* OP-2. *Chem. Pharm. Bull. 25*:1161–1163.

Sawada, Y., Kawakami, S., Taniyama, H., Hamano, K., Enokita, R., Iwado, S., and Arai, M. (1977e). Glycinothricin, a new streptothricin-class antibiotic from *Streptomyces griseus. J. Antibiot. Ser. A (Tokyo) 30*: 460–467.

Sawada, Y., Nakashima, S., Taniyama, H., Inamori, Y., Sunagawa, S., and Tsuruga, M. (1978). Biosynthesis of streptothricin antibiotics. VII. Origin of streptolidine moiety in antibiotics from *Streptomyces* species. *Chem. Pharm. Bull. 26*:885–892.

Sax, K. J., Monnikendam, P., Borders, D. B., Shu, P., Mitscher, L. K., Hausmann, W. K., and Patterson, E. L. (1968). LL-AB 664, a new streptothricin-like antibiotic. *Antimicrob. Agents and Chemother. 1968*:442–448.

Semenova, V. A., Solovieva, N. K., Buyanovskaya, I. S., Dmitrieva, V. S., Trakhtenberg, D. M., Rodinovskaya, E. I., Cherenkova, L. V., Khokhlov, A. S., Bychkova, M. M., and Ginzburg, G. N. (1960). Phytobacteriomycin, an antibiotic effective against plant bacteriosis. *Tr. Vses. Nauchno Issled. Inst. Skh. Mikrobiol. 17*:131–139; cited in *Chem. Abstr. 57*:6442 (1962).

Shoji, J., Kozuki, S., Ebata, M., and Otsuka, H. (1968). A water-soluble basic antibiotic E-749-C identical with LL-AC 541. *J. Antibiot. Ser. A (Tokyo) 21*:509–511.

Sinha, R. K., and Nandi, P. N. (1968). A new antibacterial agent produced by *Streptomyces* sp. Ac$_6$ 569. *Experientia (Basel) 24*:795–796.

Solntsev, K. M., Chumachenko, V. E., Smolskaya, R. G., Piotukh, Z. A., and Selskova, E. N. (1970). Improving production technology of kormogrizin. *Nauchn. Osn. Razv. Zhivotnovod. BSSR 1970*:210–213; cited in *Chem. Abstr. 77*:7, 46666 (1972).

Solovieva, N. K., Delova, I. D., Germanova, K. I., Savelieva, A. M., Khokhlov, A. S., Mamiofe, S. M., Sinitsyna, Z. T., Petrova, M. A., Koroleva, V. A., Navashin, S. M., Fomina, I. P., Buyanovskaya, I. S., Vasilenko, O. S., Efremova, S. A., Berezina, E. K., Weiss, R. A., Dmitrieva, V. S., Semenov, S. M., and Shneerson, A. N. (1960). Polymycin as a new antibiotic from the group of streptothricins. *Antibiotiki 5*:(12) 5–9.

Solovieva, N. K., Semenova, V. E., Buyanoskaya, I. S., Bychkova, M. M., and Ginzburg, T. H. (1961). Phytobacteriomycin, an antibiotic of the streptothricin group effective in combating bacteriosis of cotton and leguminous plants. *Primen. Antibiot. Rastenievodstve (Erevan: Akad. Nauk Arm. SSR) Sb. 1961*:103–106; cited in *Chem. Abstr. 59*:6, 6743d (1963).

Sugai, T. (1956). New antibiotics 229 and 229 B of colorless, water-soluble and basic nature. *J. Antibiot. Ser. A (Tokyo) 9*:170–179.

Takemoto, T., Inamori, Y., Kato, Y., Kubo, M., Morimoto, K., Morisaki, K., Sakai, M., Sawada, Y., and Taniyama, H. (1980). Physiological activity of streptothricin antibiotics. *Chem. Pharm. Bull.* 28:2884–2891.

Takemura, S. (1960a). Racemomycin. 5. Degradation products of racemomycin O. *Chem. Pharm. Bull.* 8:154–156.

Takemura, S. (1960b). Antibiotics produced by actinomycetes. IX. Racemomycin. 6. Structure of new degradation product of racemomycin O. *Chem. Pharm. Bull.* 8:574–577.

Takemura, S. (1960c). Antibiotics produced by actinomycetes. X. Racemomycin. 7. Structure of racemomycin O. *Chem. Pharm. Bull.* 8:578–582.

Taniyama, H., and Sawada, Y. (1971). The identity of citromycin with LL-AC 541, E-749-C and BY-81. *J. Antibiot. Ser. A. (Tokyo)* 24:708–710.

Taniyama, H., Sawada, Y., and Kitagawa, T. (1971). Characterization of racemomycins. *Chem. Pharm. Bull.* 19:1627–1634.

Telesnina, G. N., Ryabova, I. D., Shutova, K. I., and Khokhlov, A. S. (1973). Mechanism of action of streptothricins. *Dokl. Akad. Nauk. SSSR* 213:743–745.

Thirumalachar, M. J., Deshmukh, P. V., Sukapure, R. S., and Rahalkar, P. W. (1971). Fucothricin, a new streptothricin-like antibiotic. *Hind. Antibiot. Bull.* 14:4–10.

Thrum, H., and Bocker, H. (1965). Amino- and nitrobenzoic acid metabolism in actinomycetes and its influence on synthesis of antibiotics. In *Biogenesis of Antibiotic Substances, Materials from the Panel Discussion "Basic Research and Practical Aspects of Antibiotic Production," Congress on Antibiotics, Prague, June 19, 1964.* Z. Vanek and Z. Hostalek (Eds.). Czechoslovak Academy of Sciences, Prague, pp. 233–239.

Thrum, H., and Bradler, G. (1966). Studies on virothricin. In *Antibiotics— Advances in Research, Production, and Clinical Use.* M. Herold and Z. Gabriel (Eds.). Butterworth, London, pp. 410–413.

Tsuroka, T., Shoumura, T., Ezaki, N., Niwa, T., and Niida, T. (1968). SF-701, a new streptothricin-like antibiotic. *J. Antibiot. Ser. A (Tokyo)* 21:237–238.

Ueda, K., Okimoto, Y., Sakai, H., Arima, K., Yonehara, H., and Sakagami, Y. (1955). An antibiotic against silkworm jaundice virus, grasseriomycin, produced by *Streptomyces* species. *J. Antibiot. Ser. A (Tokyo)* 8:91–95.

Van Tamelen, E. E., Dyer, J. R., Whaley, H. A., Carter, H. E., and Whitfield, Jr. G. B. (1961). Constitution of the streptolin-streptothricin group of *Streptomyces* antibiotics. *J. Am. Chem. Soc.* 83:4295–4296.

Voronina, O. I., and Khokhlov, A. S. (1972). Pathways of biosynthesis of peptide moieties of streptothricin antibiotics. *Postepy Hig. Med. Dosw.* 26:541–548.

Voronina, O. I., Shutova, K. I., Tovarova, I. I., and Khokhlov, A. S. (1969). Strepthricin X—A new antibiotic of streptothricin group produced under specific conditions of biosynthesis. *Antibiotiki* 14:1063–1069.

Voronina, O. I., Tovarova, I. I., and Khokhlov, A. S. (1973). Transformation of α-lysine into β-lysine during the process of polymycin biosynthesis. *Izv. Akad. Nauk. SSSR (Moscow)* 2:228–232.

Waksman, S. A., and Woodruff, H. B. (1942). Streptothricin, a new selective bacteriostatic and bactericidal agent particularly against gram-negative bacteria. *Proc. Soc. Exp. Biol. Med.* 49:207–209.

Zbinovsky, V., Hausmann, W. K., Wetzel, E. R., Borders, D. B., and Patterson, E. L. (1968). Isolation and characterization of antibiotic LL-AC 541. *Appl. Microbiol. 16*:614–616.

Zimmer, Ch., Triebel, H., and Thrum, H. (1967). Interaction of streptothricin and related antibiotics with nucleic acids. *Biochim. Biophys. Acta 145*:742–751.

12

CHLORAMPHENICOL: PROPERTIES, BIOSYNTHESIS, AND FERMENTATION

LEO C. VINING *Dalhousie University, Halifax, Nova Scotia, Canada*
DONALD W. S. WESTLAKE *University of Alberta, Edmonton, Alberta, Canada*

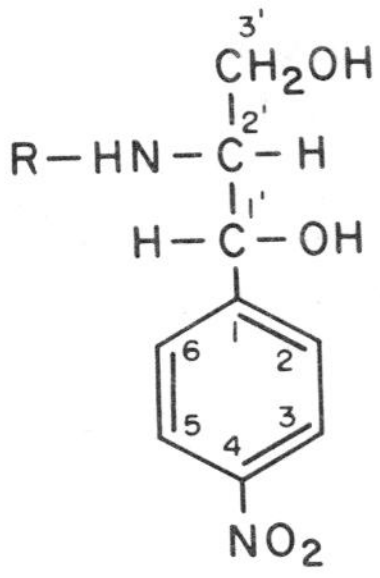

$$
\begin{array}{c}
\overset{3'}{CH_2}OH \\
| \\
R-HN-\overset{2'}{C}-H \\
| \\
H-\overset{1'}{C}-OH
\end{array}
$$

Figure 1 Chemical structures of p-nitrophenylserinol (R = H–), chloramphenicol (R = Cl₂CHCO–), and corynecins I (R = CH₃CO–), II (R = CH₃CH₂CO–), and III [R = (CH₃)₂CHCO–]. (From Rebstock et al., 1949; Suzuki et al., 1972).

I. INTRODUCTION

Chloramphenicol was discovered independently by a group at Parke Davis and Company (Ehrlich et al., 1947), another at the University of Illinois (Gottlieb et al., 1948), and one in Japan (Umezawa et al., 1948). It proved to be a relatively simple compound (Rebstock et al., 1949), with the structure shown in Figure 1. Chemical syntheses were soon devised (Controulis et al., 1949; Long and Troutman, 1949; Shemyakin et al., 1952) and furnished the product more economically than did the microbiological route. As a result, chloramphenicol became the only widely used antibiotic to be made commercially by a wholly chemical process. Because of this early switch from microbiological production to chemical synthesis, the chloramphenicol fermentation did not undergo the intensive development usually associated with large-scale production and marketing of a new antibiotic. There are few reports on the systematic search for optimum culture conditions, on selection of higher-yielding strains or on pilot-plant production. The best yields obtained by fermentation are in the range of 200–300 mg/liter.

Chloramphenicol has been the subject of several reviews. Brock (1961), Smith and Hinman (1963), and Malik (1972) have given broad coverage; its mode of action was dealt with by Hahn (1967), Gale and colleagues (1972), and recently again by Pongs (1979). The present position of chloramphenicol in antimicrobial therapy was described by Bartlett and Tally (1981), and summaries of work on its biosynthesis are available (Vining et al., 1968; Westlake and Vining, 1969, 1981).

II. PRODUCING STRAINS

The organisms from which chloramphenicol was first isolated in the United States resembled one another and were designated as a new species, *Streptomyces venezuelae* (Ehrlich et al., 1948). Okami (1948) suggested that the chlorampenicol-producing strain discovered in Japan be named *Streptomyces phaeochromogenes* var. *chloromyceticus*, but, in direct comparisons, the strain was found to be indistinguishable from *S. venezuelae* (Umezawa et al., 1949a). Other actinomycetes subsequently reported to make the antibiotic are *Streptomyces omiyaensis* (Umezawa et al., 1949b) and *Streptosporangium viridogriseum* (Tamura et al., 1971). Of the unnamed producers (Smith, 1958; Ogata et al.,

1951), strain 13s from *Streptomyces* species 3022a has been identified as *S. venezuelae* (Chatterjee and Vining, 1981). *Corynebacterium hydrocarboclastus* produces a group of compounds (Fig. 1) differing from chloramphenicol only in the N-acyl group (Suzuki et al., 1972).

III. BIOSYNTHESIS

The phenylpropanoid structure of chloramphenicol suggests a biosynthetic relationship to the aromatic amino acids formed by the shikimate pathway. Since numerous plant alkaloids and other secondary metabolites containing a phenylpropanoid skeleton are derived via phenylalanine or tyrosine, early biosynthetic studies explored whether this was also true for chloramphenicol. The idea received impetus from a stimulation of yield observed when cultures grown in a defined medium with nitrate as the nitrogen source were supplemented with phenylalanine or tyrosine (Gottlieb et al., 1954). However, structurally unrelated amino acids, such as leucine and norleucine, also increased production, and subsequent work with radioactive supplements showed that none of these compounds served as biosynthetic intermediates (Gottlieb et al., 1962; Vining and Westlake, 1964). Even p-nitrophenylserinol, which had also appeared to stimulate antibiotic production (Gottlieb et al., 1954), was not converted to chloramphenicol but, instead, was N-acetylated (Gottlieb et al., 1956).

First attempts to determine whether chloramphenicol was a product of the usual pathway for the biosynthesis of aromatic compounds were frustrated by the low uptake of [U-^{14}C]shikimate (Gottlieb et al., 1962). However, with a modified feeding regimen, Vining and Westlake (1964) observed incorporation of radioactivity into chloramphenicol and the phenylpropanoid amino acids of cellular protein. The molecules were labeled with similar efficiencies and stepwise chemical degradations located all of the radioactivity in the C6–C1 portions. Cultures supplemented with ^{14}C-labeled glycerol, glucose, or glycine also gave chloramphenicol and phenylalanine with the isotopic distributions expected if the compounds had been incorporated via the shikimate pathway. Additional evidence that the antibiotic is formed by this route was obtained by detailed analysis through chemical degradations of the labeling patterns obtained when cultures were supplemented with glucose, acetate, or pyruvate bearing ^{14}C at specific locations (O'Neill et al., 1973). [6-^{13}C]-Glucose has also been supplied as a precursor and the isotopic distribution shown by ^{13}C nuclear magnetic resonance spectroscopy to be consistent with biosynthesis by the shikimate pathway (Munro et al., 1975).

When [U-^{14}C]shikimic acids labeled specifically with tritium at C6 were fed to *S. venezuelae*, the ^{14}C:^{3}H ratio in chloramphenicol isolated from the cultures indicated that only the pro-6S hydrogen was retained (Emes et al., 1974). Thus the stereochemical course of the conversion is the same as in the enzymic formation of chlorismate from this substrate. The chloramphenicol was chemically degraded to 1-amino-2,4,6-tribromobenzene with retention of ^{3}H. Since the chemical reactions did not alter the ^{14}C-specific activity in the compounds, the carboxyl group of shikimic acid had been eliminated during biosynthesis of the antibiotic. These results established the relationship of the aromatic ring substituents in chloramphenicol to those in the carbocyclic ring of shikimic acid (Fig. 2).

Evidence of the independent origin of phenylalanine (or tyrosine) and chloramphenicol from shikimate suggested that the aromatic pathway in *S. venezuelae* contains an additional branching reaction. Information about the

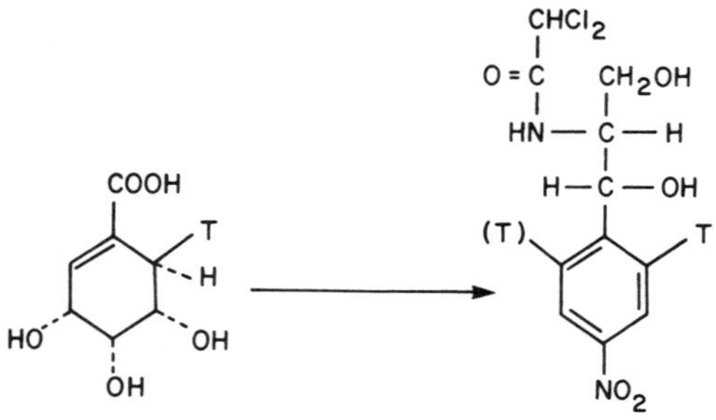

Figure 2 Incorporation of [6S-^{3}H]shikimic acid into chloramphenicol. (From Emes et al., 1974.)

nature of this reaction and subsequent steps in the unique pathway leading to the antibiotic was obtained from precursor feeding experiments with ^{14}C-labeled potential intermediates (Table 1). p-Aminophenyl[α-^{14}C]alanine was incorporated efficiently into the antibiotic, but not into other cell constituents (Siddiqueullah et al., 1967). A small pool of the L-amino acid was shown, by an isotope-trapping experiment, to be present within the mycelium of cultures engaged in chloramphenicol biosynthesis (McGrath et al., 1967). L-p-Aminophenyl[α-^{14}C]alanine labeled the antibiotic more efficiently than did the D-isomer and all of the radioactivity was localized at C2' (McGrath et al., 1968). Only the L form of p-aminophenyl [α-^{14}C; α-^{15}N]alanine was incorporated with retention of both labels. Although ^{15}N was diluted 43% more than ^{14}C during incorporation of the L form, all of the ^{15}N was lost during incorporation of ^{14}C supplied as the D-amino acid. These results implicated p-aminophenylpyruvate as an intermediate.

Feeding experiments with ^{14}C-labeled chorismate and prephenate to pinpoint the chloramphenicol branch point were unsuccessful because of the impermeability and instability of these compounds (Emes et al., 1974). The branching reaction was eventually identified through studies with cell-free systems. Extracts from antibiotic-producing mycelium catalyzed the conversion of chorismate to p-aminophenylalanine (Jones and Westlake, 1974; Jones and Vining, 1976). This arylamine synthetase was absent from nonproducing

Table 1 Incorporation of ^{14}C-Labeled Compounds into Chloramphenicol by *Streptomyces venezuelae.*

Compound	Specific incorporation (%)
L-p-Aminophenyl[2-^{14}C]alanine	53
D-p-Aminophenyl[2-^{14}C]alanine	7.0
DL-p-Nitrophenyl[2-^{14}C]alanine	0.13
DL-threo-p-Aminophenyl[1-^{14}C]serine	11
DL-threo-p-Nitrophenyl[1-^{14}C]serine	0.27
D-threo-p-Aminophenyl[3-^{14}C]serinol	0.03
N-Dichloroacetyl D-threo-p-aminophenyl[3-^{14}C]serinol	36

Source: McGrath et al. (1968) and Siddiqueullah et al. (1967).

strains of *S. venezuelae*. It resembled anthranilate and p-aminobenzoate synthetases in requiring glutamine as a co-substrate, but differed from them
in showing increased activity when NAD^+ was added. Crude enzyme preparations contained an α-ketoglutarate-dependent aminotransferase with broad
specificity for phenylalanine, p-aminophenylalanine, and tyrosine. When partially purified, this enzyme (molecular weight, 90,000) exhibited a requirement
for pyridoxal phosphate. Jones and co-workers (1978) proposed that it is
responsible for the formation of p-aminophenylalanine from p-aminophenylpyruvate; the latter may be formed from chorismate in *S. venezuelae* by a
sequence of reactions similar to those shown in Figure 3.

Arylamine synthetase and aminotransferase activity can be readily separated by chromatography on DEAE cellulose, a result causing apparent loss
of the former if activity is measured by formation of p-aminophenylalanine
(Francis and Westlake, 1979). The arylamine synthetase fraction is not homogeneous and has been separated into three components by gel filtration on
Sephadex G-100. Fraction a, with a molecular weight of 100,000, converted
chorismate to a polar aromatic amine that could not be extracted into organic
solvents. Fractions b, of 80,000 daltons, and c, of 68,000 daltons, converted
chorismate to extractable nonpolar aromatic amines. The cells from producing
cultures also contained smaller molecules (molecular weights, 27,000 and 12,500)
that stimulated and modified the activity of the larger components. The
12,500-dalton molecule also modified the activity of glutaminase-free anthranilate synthetase so that it formed a new arylamine product. The overall characteristics suggest that a core protein interacts with several smaller molecules
to generate activities that furnish a wide range of aromatic amines in *S. venezuelae*.

Because threo-p-aminophenyl DL-[carboxyl-^{14}C]serine labeled chloramphenicol efficiently and specifically at C1', whereas p-nitrophenyl L-[α-
^{14}C]alanine did not (Table 1), the reaction following synthesis of p-aminophenylalanine is believed to be a C1' hydroxylation. The enzyme may be in-

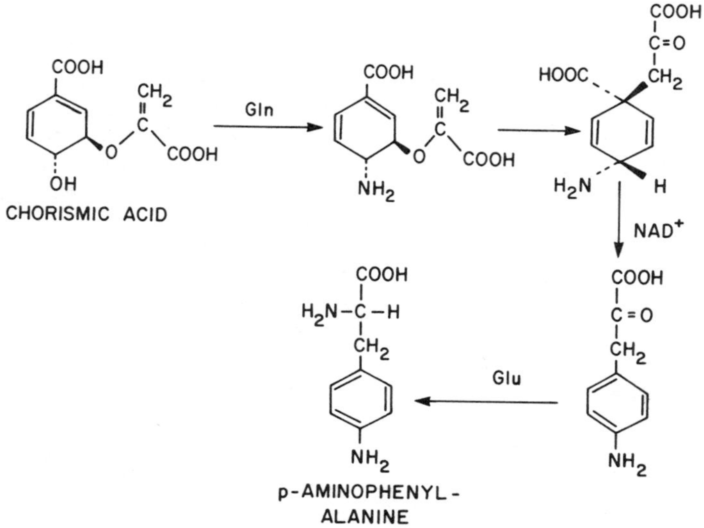

Figure 3 Possible reaction sequence catalyzed by arylamine synthetase complex.

hibited by p-nitrophenylserinol, since cultures supplemented with small amounts of this compound produced less chloramphenicol and accumulated an abnormal metabolite lacking a C1' hydroxyl group (Wat et al., 1971).

Neither D-threo-p-nitrophenyl[3-^{14}C]serine nor D-threo-p-aminophenyl-[3-^{14}C]serinol served as precursors of chloramphenicol when added to producing cultures and the intermediate following p-aminophenylserine in the pathway is believed to be an N-acyl derivative. Although the evidence remains inconclusive, the dichloroacetyl group does not appear to be introduced directly—for example, by transfer from the coenzyme A ester. An early report (Wang et al., 1959) that unlabeled dichloroacetic acid diluted the incorporation of [^{36}Cl]chloride ions into chloramphenicol has not been supported by precursor feeding experiments with the ^{14}C-, ^{2}H-, or ^{3}H-labeled acid (Gottlieb et al., 1956; Simonsen et al., 1978). Trapping experiments with ^{36}Cl^{-} enriched cultures also failed to detect free dichloroacetic acid in the mycelium of S. venezuelae engaged in chloramphenicol synthesis (Lemieux and Vining, unpublished results.) A plausible conclusion is that chlorination takes place after acylation of p-aminophenylserine. Small amounts of N-acetyl, N-propionyl, and N-butyryl congeners of chloramphenicol accumulate in normal cultures (Stratton and Rebstock, 1963) and they are formed exclusively in cultures deprived of halogens (Smith, 1958). A similar spectrum of nonhalogenated products is produced by C. hydrocarboclastus (Suzuki et al., 1972). Thus the acylating enzyme appears to have low specificity for the acyl donor. Studies with C. hydrocarboclastus have implicated amino acids as the source of the acyl groups. [U-^{14}C]Threonine labeled corynecin II exclusively in the propionyl group, suggesting that the acyl substituent was derived via α-ketobutyric acid and propionyl coenzyme A (Nakano et al., 1976a).

The biogenetic origin of the dichloroacetyl substituent has been explored with isotopically labeled precursors. [1-^{14}C]Acetate and [2-^{14}C]acetate introduced radioactivity into the corresponding carbons of the dichloroacetyl group (Gottlieb et al., 1962). Extensive randomization of the label from [2-^{14}C]acetate could be accounted for by Krebs cycle activity. [6-^{13}C]Glucose labeled predominantly the dichloromethine carbon and, consistent with the precursor having been incorporated via acetyl coenzyme A, also enriched the carbonyl carbon to some extent (Munro et al., 1975). Analysis by ^{13}C nuclear magnetic resonance spectroscopy of the labeling pattern generated by [1,2-^{13}C]acetate showed that most of the isotope entered the dichloroacetyl component only after scission of the original ^{13}C–^{13}C bond (Simonsen et al., 1978). Overall incorporation was low and the 4:1 imbalance in isotopic distribution in favor of the dichloromethine carbon suggested that the carbonyl group had exchanged with carbon dioxide before being linked to the phenylpropanoid intermediate. This is supported by the observation (Gottlieb et al., 1962) that [^{14}C] formate and [^{14}C] carbon dioxide label the dichloroacetyl carbonyl group exclusively. Among mechanisms for equilibrating this group with carbon dioxide, those that involve Krebs cycle intermediates are excluded, since [2,3-^{13}C]succinate enriched only the carbonyl carbon of dichloroacetic acid (Simonsen et al., 1978).

If one assumes that the chlorine atoms are introduced into N-acyl p-aminophenylserine by the action of a chloroperoxidase (Shaw and Hager, 1959), the acyl group forming a substrate for this enzyme should contain a strongly electrophilic carbon at the chlorination site. N-Malonyl p-aminophenylserine is a plausible candidate. To account for the isotopic labeling results, the malonyl coenzyme A formed by carboxylation of acetate must equilibrate with

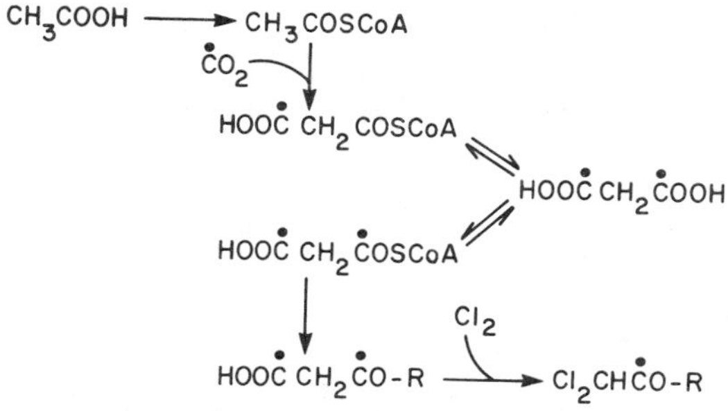

Figure 4 Possible reaction sequence introducing the dichloracetyl group into chloramphenicol (R = p-aminophenylserine). (From Simonsen et al., 1978.)

free malonic acid (Fig. 4). Whether this occurs in *S. venezuelae* has not yet been determined.

After acylation, the two steps needed to complete synthesis of the antibiotic are (1) reduction of the carboxyl group and (2) oxidation of the p-amino group. Since small amounts of N-dichloroacetyl-p-aminophenylserinol have been found to accumulate in chloramphenicol-producing cultures (Stratton and Rebstock, 1963) and a ^{14}C-labeled specimen was efficiently converted to chloramphenicol without randomization (McGrath et al., 1968), oxidation is thought to be the terminal step, giving the overall pathway shown in Figure 5.

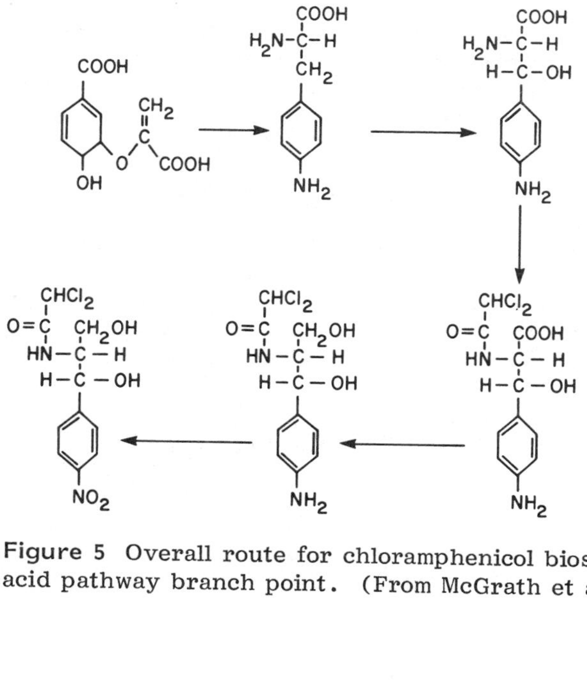

Figure 5 Overall route for chloramphenicol biosynthesis from the shikimic acid pathway branch point. (From McGrath et al., 1968).

IV. REGULATION OF SHIKIMATE PATHWAY

In suitable media, the amount of chloramphenicol synthesized by *S. venezuelae* is at least as great as the amount of phenylalanine or tyrosine made for incorporation into proteins; consequently, antibiotic production represents an appreciable drain on the output of the shikimate pathway. Lowe and Westlake (1971, 1972) have examined the mechanisms by which the flow of intermediates through this pathway is controlled. The enzymes DAHP synthetase, chorismate mutase, and prephenate dehydratase were shown to be present, but no prephenate dehydrogenase activity was detected. It is known from radiotracer studies that tyrosine cannot be formed in this organism by hydroxylation of phenylalanine (Chandra and Vining, 1968). Lack of prephenate dehydrogenase may therefore mean that tyrosine is synthesized by the pretyrosine sequence as in cyanobacteria and other microorganisms (Jensen and Stenmark, 1975).

The enzyme DAHP synthetase has been isolated from *S. venezuelae* cells as a single protein species of 88,000 daltons (Lowe and Westlake, 1971). No evidence of isozymic forms was found and its synthesis was not repressed by protein—amino acid end products or by intermediates of the pathway. The enzyme also failed to show feedback inhibition of activity by pathway end products tested in vitro. On this evidence, *S. venezuelae* joins other *Streptomyces* species in the group of organisms classified by Jensen and Rebello (1970) as having an unregulated entry point to the shikimate pathway.

Chorismate mutase in *S. venezuelae* is a single protein species of 75,000 daltons (Lowe and Westlake, 1972). It is neither inhibited nor repressed by pathway intermediates and end products. By contrast, prephenate dehydratase and anthranilate synthetase, the two branch-point enzymes detected, were inhibited in vitro by their specific pathway end products. Prephenate dehydratase was characterized as a single protein species of 225,000 daltons. Its synthesis was not repressed by end products added to the culture medium. Anthranilate synthetase was separated into two nonidentical proteins, each contributing to the overall reaction (Francis et al., 1978). Both its synthesis and its activity were negatively modulated by anthranilic acid; tryptophan and histidine also repressed enzyme synthesis. Although flow of precursors into the shikimate pathway is not prevented at the entry point, excessive accumulation of at least two of the end-product amino acids is controlled by feedback effects at individual branching reactions.

V. REGULATION OF CHLORAMPHENICOL BIOSYNTHESIS

Like other shikimate pathway end products, chloramphenicol exerts no feedback effects on DAHP synthetase (Lowe and Westlake, 1971). Unlike the amino acids, it does not inhibit the activity of its specific branch-point enzyme (Jones and Westlake, 1974). This may not be significant, since the antibiotic does not accumulate within the cells of producing cultures (Legator and Gottlieb, 1953). Arylamine synthetase activity is inhibited 50—60% by p-aminophenylalanine at 1 mM concentration and the existence of such a high degree of product inhibition suggests that control of this key enzyme is exerted through sequential backing up of cytoplasmic pathway intermediates when there is a restraint on excretion of the end product.

Addition of p-aminophenylalanine, threo-p-aminophenylserine, or chloramphenicol to growing cultures of *S. venezuelae* caused a sharp decrease in arylamine synthetase activity within the mycelium (Jones and Westlake, 1974).

The effect was specific in that branching enzymes for the other pathway end products were unaffected. Adding chloramphenicol at a range of concentrations to cultures beginning production of the antibiotic does not alter appreciably the maximum titer eventually reached (Legator and Gottlieb, 1953). While the existence of an upper limit to the amount of chloramphenicol accumulated suggested that the biosynthetic machinery was negatively influenced by the level of product in the environment, the results could also be explained by the presence of a drug-inactivating mechanism. Firm evidence for a feedback control system was subsequently obtained by analysis of cultures grown with [6-^{3}H]glucose as a carbon source and supplemented with [3'-^{14}C]chloramphenicol (Malik and Vining, 1972a). The ^{14}C:^{3}H ratios in reisolated chloramphenicol showed that the supplements had caused a marked concentration-dependent suppression of de novo synthesis. The effectiveness of this feedback control could be seen equally well with the p-methylthio analog (Fig. 6).

Evidence of feed-back inhibition is also contained in the marked yield increase observed when halides are omitted from the culture medium. Under these conditions, a mixture of nonhalogenated N-acyl-p-nitrophenylserinols is formed. Smith (1958) suggested that these products have less severe feed-back effects than chloramphenicol. At a limiting low chloride concentration, incremental addition of bromide ions to the medium at first caused a drop in

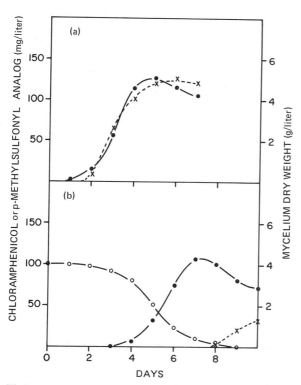

Figure 6 Growth of *S. venezuelae* and chloramphenicol production (a) in GSL medium and (b) in GSL medium supplemented with the p-methylthio analog of chloramphenicol at 100 mg/liter:mycelium dry weight (●), chloramphenicol (×), p-methylthioanalog (○). (From Malik and Vining, 1972a.)

the overall yield of acylated p-nitrophenylserinols, followed by an increase
at higher bromide levels. Smith ascribed this to the initial formation of the
monobromochloroacetyl derivative, which resembles chloramphenicol in feed-
back activity, and subsequent accumulation of the dibromoacetyl derivative
with much weaker feed-back action.

Since intracellular concentrations of chloramphenicol are negligible, its
feed-back effect must be exerted indirectly. A plausible route would be
through the terminal enzyme that appears to be associated with vectorial transport of the antibiotic through the cell membrane. It is noteworthy that *S.
venezuelae* in halide-free media and *C. hydrocarboclastus* in normal conditions
produced exceptionally high yields of p-nitrophenylserinol derivatives (Smith,
1958; Suzuki et al., 1974), suggesting that nonchlorinated analogs might have
less affinity for the feed-back sensor and thus a weaker inhibitory effect than
chloramphenicol. Higher-producing strains of *C. hydrocarboclastus* have
been obtained by selecting colonies for resistance to the inhibitory effects
of chloramphenicol (Tomita et al., 1974). The biochemical basis for the increased productivity has not been determined, but reduced sensitivity to feed-
back control may well be responsible. This possibility is strengthened by
the observation (Nakano et al., 1974a) that a mutant yielding a higher proportion of the N-acetyl congener (corynecin I), which is substantially less toxic
to the producer than longer-chain homologs, produced twice as much of the
antibiotic complex as did the parent strain. Nakano and co-workers (1974b)
reported that p-aminophenylethanol, unlike p-aminophenylalanine, inhibits
corynecin biosynthesis without suppressing growth. When added to cultures
producing corynecins, it caused the accumulation of several aromatic amines,
including p-aminophenylalanine and N-acyl derivatives of p-aminophenylserinol
(Nakano et al., 1976b). p-Aminophenylethanol is thought to function as the
analog of a natural pathway intermediate in inhibiting the terminal biosynthetic
reaction. Taken together, the results of these studies suggest that the cytoplasmic signal restricting chloramphenicol production is a buildup of N-dichloroacetyl-p-aminophenylserinol.

VI. NUTRITIONAL REQUIREMENTS

In one of the first full accounts of the production and properties of chloramphenicol, Smith and co-workers (1948) included a study of media promoting
high antibiotic yields. With a complex medium containing a proteinaceous nitrogen source and yeast fermentation solubles to supply mineral nutrients,
they found glycerol to be superior to glucose, maltose, or lactose as a source
of carbon. Although the best nitrogen source was tryptone, it was replaced
by a less expensive hog stomach residue preparation on the basis of cost.
With this change, molasses was found to give higher titers than yeast fermentation solubles, and optimum yields were achieved after 3 instead of 4 days.
However, growth was accompanied by a rapid increase in the acidity of the
culture which, at the yield maximum, reached inhibitory values near pH 5.
Other studies of the nutritional requirements for optimum chloramphenicol production in complex media have been reported by the Japanese group (Umezawa
et al., 1948; Umezawa and Maeda, 1949). With a medium containing 0.5% beef
extract as well as 0.5% peptone and 0.5% sodium chloride, glycerol at 1% was
superior to maltose or dextrin as a carbon source. Glucose and lactose were
unsuitable. Yields were also sensitive to the sodium chloride concentration
and were best at 2%, the highest level tested.

Table 2 Composition of Gottlieb and
Diamond's Medium

Ingredient	Amount (g/liter)
Glycerol	10.0
DL-Serine	5.0
Sodium lactate	11.0
$K_2HPO_4 \cdot 3H_2O$	2.39
KH_2PO_4	1.39
$MgSO_4 \cdot 7H_2O$	1.0
NaCl	3.0

Source: Gottlieb and Diamond (1951).

To meet the need for a simple and reproducible medium for biosynthetic studies, Gottlieb and Diamond (1951) selected the nutrient mixture shown in Table 2. Its composition was based on a study in which the ingredients of a complex glycerol–tryptone–brewer's yeast–distiller's solubles–sodium chloride medium were systematically replaced by simple sources of carbon, nitrogen, and mineral nutrients. The yield of antibiotic was insensitive to changes in the concentration of potassium phosphate, but markedly dependent on the high level of lactate in the medium. Replacing lactate with other organic acids caused a sharp loss of production. However, neither glycerol nor DL-serine was uniquely necessary, since the former could be replaced by glucose and the latter by alanine with no sacrifice in yield.

Matsuoka and co-workers (1953) reported that soil isolates of S. vene-zuelae generally required organic nitrogen sources for chloramphenicol production. To examine the influence of carbon sources, they isolated a mutant strain that synthesized chloramphenicol in a medium containing 0.2% sodium nitrate, 0.1% dispotassium hydrogen phosphate, 0.05% magnesium sulfate, 0.05% potassium chloride, 1% sodium chloride, and a carbon source at 3%. Glycerol again gave a higher titer than glucose, maltose, or starch, and the peak yield was reached in 3 days. For physiological studies, Chatterjee et al. (1983) have also devised a simple medium containing only single sources of carbon and nitrogen. Because of the insensitivity of chloramphenicol production to phosphate, the nutrient solution could be heavily buffered with phosphate salts, allowing growth on inorganic ammonium salts without inhibitory pH changes. With ammonium as the nitrogen source, a comparison of carbon sources showed galactose or cellobiose to be the best. Glycerol gave relatively poor yields.

Chloramphenicol production in a glycerol–sodium nitrate medium could be increased appreciably by adding various nitrogenous supplements (Yagishita and Umezawa, 1951). Aromatic amino acids were particularly effective, but the response was not concentration dependent. DL-Methionine was unique in giving an increment when added with other supplements. Similar yield increases were obtained with certain amino acids in a glycerol-sodium nitrate medium containing sodium lactate (Gottlieb et al., 1954). Phenylalanine and DL-norleucine were the most effective. A later study in which a variety of nitrogen sources replaced DL-serine in GSL medium, which was based on the one devised by Gottlieb and Diamond (1951), indicated that amino acids supporting slow growth gave the highest yields of antibiotic (Westlake et al., 1968). D-Serine and L-serine were metabolized at different rates and provided a steady supply of nitrogen over a reasonable fermentation period of 7 days.

The influence of nitrogen sources in determining the level of chloramphenicol production in a simple phosphate-buffered medium with glucose as a nonlimiting source of carbon is similar to that observed in GSL medium (Chatterjee et al., 1983). Ammonia, asparagine, and glutamine, which allowed fast growth, gave poor yields; nitrate proline, threonine, and alanine supported growth at a modest rate and yields equalled those in GSL medium in a 4-5 day fermentation; phenylalanine, leucine, and isoleucine, which allowed slow growth, gave high yields of antibiotic, but incubation times of up to 15 days were necessary for optimum yields. Substitution of a less rapidly metabolized carbon source, such as galactose, in this medium enhanced the yield, with nitrogen sources supporting rapid growth. With those supporting an intermediate growth rate, the effect was small, while with very poor nitrogen sources it was negative.

Zinc and iron are the only trace elements required in chloramphenicol fermentations. Using a calcium carbonate-treated medium containing glycerol, nitrate, lactate, and salts, Gallicchio and Gottlieb (1958) found that omitting both of these cations almost eliminated antibiotic production. By contrast, omitting copper from the medium allowed an increase in yield. The optimum concentrations of iron and zinc were 10^{-3} and 10^{-4} M, respectively. With all other elements optimal, the concentration of magnesium sulfate promoting maximum chloramphenicol production in this medium was surprisingly low at 0.1 mM.

Using *Streptomyces* sp. 3022, a soil isolate that produces chloramphenicol as the sole antibiotic, and a medium containing glycerol, nitrate, lactate, norleucine, and salts from which all halide ions were removed, Smith (1958) found that a mixture of N-acetyl-, N-propionyl-, N-butyryl-, and a less polar, unidentified N-acyl-p-nitrophenyserinol were produced. The overall yield, at 400 μg/ml, was much higher than the normal yield of chloramphenicol. With the gradual addition of chloride ions to the medium, increasing amounts of chloramphenicol were produced but the overall yield declined. At 1.7 mM chloride, where chloramphenicol became the dominant product, the yield was only 150 μg/ml. The effect of adding bromide ions depended upon the amount of chloride already present. Ratios above 25:1 were toxic. At lower values the response was similar to that obtained by gradually removing chloride ions. Beside the decrease in chloramphenicol yield and disproportionate increase in N-acyl analogs, some monobromochloro- and dibromoacetyl derivatives were produced. The interaction between bromide and chloride ions appeared to be competitive.

VII. RELATIONSHIP TO GROWTH

The fermentation pattern most commonly reported is one where growth and chloramphenicol production proceed simultaneously (Oyaas et al., 1950; Malik and Vining, 1972a). Under some conditions, maximum antibiotic titers are not reached until the culture is in the stationary phase (Umezawa et al., 1948; Legator and Gottlieb, 1953). Measurement of growth as increased nucleic acid or protein content of cultures rather than as increased biomass has shown that substantial amounts of chloramphenicol are often made after true growth has ended (Lowe and Westlake, 1972). In media where S. venezuelae grows very rapidly, chloramphenicol is produced mainly in the stationary phase

(Malik and Vining, 1970). In poorly buffered media (e.g., the 1% glycerol–0.5% tryptone–0.5% distiller's solubles medium used by Legator and Gottlieb) early termination of growth can be attributed to a drop in pH to inhibitory values.

The growth-linked pattern of antibiotic synthesis has been obtained in both complex and defined media (Fig. 7). Oyaas and colleagues (1950) used a medium similar in composition to that of Legator and Gottlieb, but with the addition of 0.5% beef extract. The pH of the culture remained close to 7.5 and cell mass peaked along with the chloramphenicol titer at 3 days. These maxima corresponded closely with exhaustion of glycerol from the medium; only 40% of the total nitrogen content of the medium was used. The frequently used GSL medium, which supports growth at a fairly slow rate, also gives a growth-linked fermentation pattern. Depending upon the size and history of the inoculum and the conditions used to sterilize the medium, *S. venezuelae* may show an initial growth lag of up to 48 hr. This can be eliminated with a small supplement of yeast extract, allowing maximum antibiotic titers to be attained in 5 days (Malik and Vining, 1972a). Chloramphenicol is synthesized very rapidly in the early growth period; the rate decreases before growth peaks, possibly owing to feed-back inhibition, and maximum titers are reached in early stationary phase.

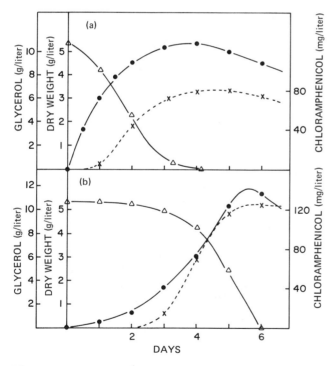

Figure 7 Growth of *S. venezuelae*, production of chloramphenicol, and utilization of glycerol in (a) a medium containing (per liter) 10 ml glycerol, 5 g peptone, 5 g corn-steep liquor, 5 g sodium chloride, 1 g potassium monohydrogen phosphate, adjusted to pH 7.5 with sodium hydroxide before autoclaving, and in (b) the GSL medium described in Table 2: growth as dry weight of cells (●), chloramphenicol titer in culture filtrate (×), glycerol in culture filtrate (△). (From Malik and Vining, 1972a; Oyaas et al., 1950.)

Streptomyces venezuelae has been grown in continuous culture using either GSL medium or a complex glycerol–tryptone–distiller's solubles medium (Sala and Westlake, 1966). Higher titers were achieved in the defined medium, but these cultures were more subject to selection and titers eventually declined to very low values. A medium replacement time of 32 hr imposed severe selection for faster utilization of nutrients to achieve a higher growth rate and antibiotic production was lost rapidly. At a replacement time of 54 hr, the chloramphenicol titer remained steady during 18 replacements. Under both conditions the cultures appeared to be carbon limited, since glycerol concentrations were near zero and there was evidence that serine was being used as a carbon source. Lactic acid, on the other hand, was not used in appreciable amounts. In the complex medium, chloramphenicol titers slowly declined after 40 medium replacements in cultures grown with a 27-hr replacement time.. Surprisingly, production was well maintained with a faster replacement time of 13 hr. This was attributed to heavy wall growth which was not displaced in the cyclone fermenter owing to foaming and which provided a continuous inoculum from a population not subject to selection pressure for fast growth.

VIII. METABOLIC CONTROL

Antibiotics and other secondary metabolites are not usually formed under conditions where the producing organism has free access to readily used sources of carbon and nitrogen as well as an abundant supply of mineral nutrients; not uncommonly, the onset of production is linked to exhaustion of a specific nutrient from the culture medium (Demain, 1972; Martin and Demain, 1980). One possible explanation is that systems regulating nutrient utilization have a dual role, controlling not only pathways that supply essential starting material for cell synthesis, but also pathways to secondary metabolites that protect the species against competition when growth is restricted. Control of nutrient utilization is generally exercised in bacteria by systems for nutrient transport, enzyme induction, and catabolite repression. Indications of a possible linkage with secondary metabolism should be revealed by the response of such control systems during derepression of coexisting catabolic and secondary metabolic pathways.

Cultures of *S. venezuelae* grown in a basal salts medium with a variety of single carbon sources have been used for such studies (Chatterjee and Vining, 1981, 1982a,b). Maltose induced high levels of α-glucosidase which were not suppressed by glucose. Cultures grown on a glucose–maltose mixture used both carbon sources simultaneously and also produced chloramphenicol rapidly while glucose was being metabolized. Thus glucose did not catabolite repress either maltose utilization or chloramphenicol biosynthesis. β-Glucosidase activity was exhibited by two enzymes, one with a substrate preference for cellobiose and the other more active on aryl β-glucosides. Each was preferentially induced by its substrate. Cellobiose induction was strongly suppressed by glucose. Cyclic adenosine 3',5'-monophosphate added to the culture did not reverse this effect, but the significance of the result is uncertain since nucleotides may not be transported intact into the cells of *Streptomyces* (Martin and Demain, 1977). Cultures of *S. venezuelae* grown on a mixture of glucose and cellobiose used glucose preferentially and showed a diauxic pause at the changeover. However, chloramphenicol was produced during consumption of both substrates. Similar results were obtained with a mixture of glucose and lactose. Although β-galactosidase induction was suppressed by glucose and the two substrates were used sequentially, chloramphenicol was produced throughout the entire growth period. Intracellular cyclic adenosine 5'-monophosphate concentrations showed no significant change at the diauxie step from the average value of 5.1 ± 0.8 pmol/mg protein. These results indicate that although utilization of some carbon sources is controlled by substrate preference mechanisms, the induction and repression systems operating on catabolic pathways do not regulate chloramphenicol biosynthesis. Moreover, it is unlikely that cyclic adenosine 5'-monophosphate is involved.

The observation that slowly metabolized nitrogen sources favor antibiotic production prompted a study of possible relationships between control of nitrogen utilization and chloramphenicol production (Shapiro and Vining, 1983). In *S. venezuelae*, the usual pathways for introducing inorganic nitrogen into organic molecules are present. At low concentrations, ammonia is incorporated by the action of glutamine synthetase and glutamate synthase. With changes in the availability of ammonia, many organisms exhibit large differences in the cellular level of glutamine synthetase which reflect the extent to which nitrogen-scavenging catabolic enzymes are ammonia repressed (Magasanik, 1976). In *S. venezuelae*, the specific activity of glutamine synthetase was relatively constant with changing availability of ammonia, and no correlation was observed with the rate of chloramphenicol synthesis. At high concentrations, ammonia is assimilated by glutamate dehydrogenase. Intracellular levels of this enzyme were directly related to the amount of ammonia in the medium and inversely related to the rate at which chloramphenicol was accumulated. Thus there may be a regulatory link between nitrogen assimilation through this pathway and antibiotic production.

IX. RECOVERY AND PURIFICATION

Procedures reported for isolating chloramphenicol from fermentation beers are limited to those used in laboratory-scale research experiments (e.g., Umezawa and Maeda, 1949; Vining and Westlake, 1964). The antibiotic is readily extracted from clarified culture fluid at slightly alkaline pH with a solvent such as ethyl acetate. Acidification of the broth before removal of solids avoids troublesome emulsions but yields a less pure product. The residue left after removing the solvent is freed from lipid by leaching with petroleum ether. If it is strongly colored, the pigmented impurity may be removed by passage of a solution in diethyl ether through a short column of alumina. The product can be crystallized from water, ethylene dichloride, or an ether-petroleum ether mixture.

X. ASSAY

A considerable choice of methods is available for measuring chloramphenicol concentrations. A typical microbiological assay uses the agar diffusion method with *Sarcina lutea* as test organism (Grove and Randall, 1955). Colorimetric procedures based on reduction of the nitro group to an amine which can be diazotized and coupled to the Bratton-Marshall reagent (Bessman and Stevens, 1950; Levine and Fischbach, 1951) are reliable but do not distinguish between the active antibiotic and some of its metabolites. The presence of certain sulfonamide drugs may also interfere while the synthetic analog, thiamphenicol does not react. Alternative procedures measure the yellow color formed by reaction with isonicotinic acid hydrazide in alkaline solution (Shah et al., 1968), the blue color produced by heating with α-naphthol in 6N sodium hydroxide (Masterson, 1968) or the violet color obtained with ferric chloride and hydroxylamine (Karawya and Ghourab, 1970). Chloramphenicol may also be assayed spectrophotometrically by measuring the absorbance of solutions at its 278-nm maximum (Grove and Randall, 1955), but preliminary fractionation is usually necessary to avoid interference.

More selective procedures have been developed using gas–liquid chromatography, high-pressure liquid chromatography, or the enzyme chloramphenicol acetyltransferase. Gas-chromatographic procedures require preliminary extraction of the antibiotic from aqueous solutions and its conversion to the trimethylsilyl derivative (Resnick et al., 1966; Margosis, 1970). By using an electron capture detector, gas chromatography can achieve very high sensitivity as well as specificity. High-pressure liquid chromatography is advantageous in that it does not require the derivatization step and has been developed into a fast and convenient method for routine analysis of serum (Crechiolo and Hill, 1979; Sample et al., 1979), as well as other preparations (Vigh and Inczedy, 1976; White et al., 1977). Radiochemical assays based on the enzymic acetylation of chloramphenicol are also specific and convenient for routine analyses (Robison et al., 1978; Weber, 1981).

XI. MODE OF ACTION

1-(p-Nitrophenyl)-2-(2,2-dichloroacetamido)-1,3-propanediol contains two centers of asymmetry and so can exist in four stereoisomeric forms. The molecule with the D-threo-configuration is chloramphenicol and it alone is a useful antibacterial agent. The L-threo isomer is less than 0.5% as active, whereas the erythro isomers exhibit no activity. A large number of structural analogs have been prepared and the results show that modification of the aromatic portion is much less critical for antibiotic action than altering the propanoid moiety (Hahn, 1968; Shemyakin et al., 1956). Thiamphenicol, the analog in which the nitro group is replaced by a methylsulfonyl group, has an antibiotic spectrum similar to that of the natural compound, but significant differences in therapeutic response have led to considerable use in Europe. The risk of aplastic anemia is reported to be much lower than in treatments with chloramphenicol (Keiser et al., 1972).

In the propanoid moiety the presence of hydroxyl groups on C1' and C3' is important for activity, but other polar functions such as halogen or carbonyl groups may be substituted. Masking of the hydroxyl groups, for example, by acylation, causes inactivation and is a common biological mechanism of acquiring resistance. Extension or branching of the propanoid chain also causes loss of activity. The characteristics of the N-acyl substituent are important in allowing the antibiotic to penetrate bacterial cell envelopes. Replacement of the dichloroacetyl group with substituents of increased size or different polarity reduce the drug's effectiveness; however, these changes have much less influence on the inhibition of in vitro protein biosynthesis (Hahn and Gund, 1975).

Chloramphenicol is a bacteriostatic agent causing inhibition of protein synthesis (Hahn and Wisseman, 1951). It binds reversibly to the 50-S ribosomal subunit of prokaryotes and to the equivalent subunit in organelles, but not to components of the 80-S eukaryotic ribosome (Vasquez, 1964; Freeman, 1970). Attachment of the antibiotic to the 50-S subunit interferes with protein biosynthesis by blocking the formation of peptide bonds (Pestka, 1977). The site of inhibition has been identified as the peptidyltransferase reaction (Cundliffe and McQuillen, 1967), with the receptor to which the antibiotic binds inconclusively located by competition and inhibition studies at either the peptidyl acceptor or donor site (Pongs, 1979). Reconstitution of ribosomal subunits omitting proteins singly and affinity-labeling experiments with monobromo- and monoiodoamphenicol indicate that protein L16 is closely connected with antibiotic binding (Nierhaus and Nierhaus, 1973; Pongs et al., 1973). This protein has been shown to also bind aminoacyl-tRNA (Pellegrini et al., 1974), thus pinpointing a receptor for chloramphenicol at the peptidyl acceptor site. However, there is evidence that the antibiotic can bind at more than one site and that a second location, associated with proteins L2, L24, and L27, involves the peptidyl donor site. A plausible explanation (Pongs, 1979) is that the chloramphenicol receptor is a domain of interacting proteins surrounding the peptidyltransferase center.

The inhibitory effect of the drug on polypeptide synthesis in vitro using well-defined systems is lower than the effect in vivo and has been found to depend on the base composition of the template present (Kucan and Lipmann, 1964). The response is particularly weak with poly(U) and improves markedly with poly(UC) as the proportion of cytidylic acid is increased. However, the most sensitive in vitro systems are those constituted with natural mRNA (Hahn and Gund, 1975; Pongs, 1977). At least part of the differences in response is due to differences in the solubility of the polypeptides formed. This leads to variation in the recovery of reaction products (Pestka, 1969). There are, nevertheless, significant nonartifactual differences in the susceptibility of protein-synthesizing systems to chloramphenicol and these extend to in vivo situations as well. Proteins involved in DNA replication, for instance, continue to be synthesized with 30 µg/ml chloramphenicol present (Lark and Lark, 1966). Association of the ribosomal machinery with the cell membrane may be involved (Pongs, 1979).

XII. PRODUCER TOLERANCE

As in other prokaryotes, chloramphenicol binds to ribosomes in S. venezuelae and inhibits protein biosynthesis. In vitro systems are susceptible, whether they are prepared from cells engaged in chloramphenicol biosynthesis or from nonproducing mycelium. However, differences have been found in the sensitivity in vivo (Malik and Vining, 1972b). Cells from cultures producing the antibiotic were not affected by concentrations that inhibited protein biosynthesis and prevented growth in nonproducing cultures. Growth inhibition in the latter was temporary and its duration was related to the amount of antibiotic added. When growth recommenced, the culture had acquired tolerance to antibiotic concentrations up to but not exceeding the exposure level. This could be maintained by subculture in media containing chloramphenicol, but was lost during a single passage in unsupplemented media.

While the culture is adapting, the antibiotic titer decreases, but resumption of growth does not depend upon detoxication of the medium. The cells of S. venezuelae contain a hydrolytic enzyme that deactivates chloramphenicol by removing the dichloroacetyl substituent (Malik and Vining, 1971). The enzyme is constitutive. Since its activity does not change in response to the presence of antibiotic in the surroundings, chloramphenicol entering the cell is degraded at a steady rate. Development of tolerance is believed to be the result of reduced uptake of the antibiotic. Impairment of the transport mechanism appears to be a graded response to increasing concentrations of chloramphenicol in the environment and insulates the cytoplasmic ribosomal system from their lethal effects. Chloramphenicol hydrolase, located in the cytoplasm, scavenges antibiotic that has entered the cells before the permeability change or that enters more slowly afterward. Since the action of chloramphenicol is bacteriostatic, growth resumes when the intracellular concentration of the antibiotic has been reduced enough to allow renewed protein biosynthesis. That the cells remain leaky is indicated by the steady decrease in antibiotic titer after the culture has resumed growth, as well as by the accumulation in such cultures of N-acetyl p-nitrophenylserinol and p-nitrobenzylalcohol, the expected products of chloramphenicol degradation (Malik and Vining, 1970).

Cultures of *S. venezuelae* in some media produce inhibitory concentrations of chloramphenicol early in the growth phase. Nevertheless, growth and further antibiotic syntheses continue at an accelerating pace (Malik and Vining, 1972a). After the onset of production, chloramphenicol may be added to the culture without inhibiting growth or protein biosynthesis. Since the amount of antibiotic in the mycelium during production is extremely small (Legator and Gottlieb, 1953), the metabolite is either irreversibly transported through the cell membrane as fast as it is formed or is synthesized by a membrane protein and excreted at the outer surface. Evidence that p-aminobenzoic acid is oxidized to p-nitrobenzoic acid in *S. venezuelae* while covalently bound to a cell macromolecule favors the latter possibility (Siddiqueullah et al., 1968).

That chloramphenicol excreted by producing cultures is unable to reenter the cells may be deduced from the absence of N-acetyl p-nitrophenylserinol or degradation products in producing cultures. Experiments with radioactive chloramphenicol have shown that antibiotic added before production begins

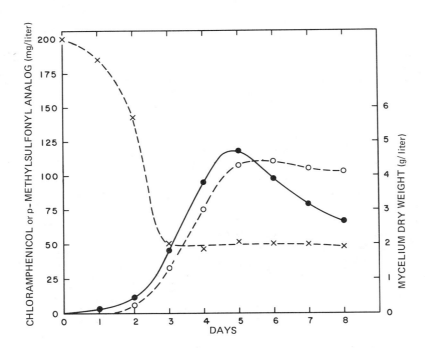

Figure 8 Growth of *S. venezuelae* in GSL medium supplemented with the p-methylsulfonyl analog of chloramphenicol at 200 mg/liter: mycelium dry weight (●), chloramphenicol (○), p-methylsulfonyl analog (×). (From Malik and Vining, 1972a.)

is degraded, whereas additions after this point remain intact. In contrast to tolerant nonproducing cells, mycelium producing the antibiotic does not allow the product outside the cells to come in contact with the degrading enzyme in the cytoplasm. The apparent loss of permeability at the onset of antibiotic synthesis is also observed with the p-methylsulfonyl analog, which can be conveniently measured by gas-liquid chromatography in the presence of biosynthetic chloramphenicol (Fig. 8). Since the p-methylsulfonyl analog does not suppress chloramphenicol production, the mechanisms for exclusion and feed-back inhibition cannot be identical; feed-back is the more specific phenomenon. A plausible explanation for the prevention of net uptake during biosynthesis is that vigorous functioning of an efflux mechanism associated with the terminal reaction effectively purges the cells of antibiotic leaking in from the outside.

XIII. CLINICAL APPLICATIONS

Chloramphenicol is a broad-spectrum antibacterial agent that inhibits most species at concentrations of 10 μg/ml or less. Gram-negative organisms are generally more sensitive than the gram-positive ones; *Salmonella* and *Haemophilus* species are particularly susceptible. The drug is active against most anaerobes, rickettsiae, chlamydia, and mycoplasma. It is not effective against eukaryotic organisms, but does interfere with mitochondrial and chloroplast functions. Although it is a very useful clinical agent and is the preferred antibiotic for treating typhoid fever and other systemic *Salmonella* infections, its use has been curtailed because of the incidence of bone marrow suppression during prolonged therapy (Yunis and Bloomberg, 1964). Hematological toxicity is of two types, a dose-dependent form that is reversible and one that is unrelated to dose and irreversible. The latter is usually lethal and occurs with a risk of not more than 1 out of 20,000 cases; its onset is often delayed and its cause is unknown.

Because of the hazard of aplastic anemia, clinical use of chloramphenicol is restricted to serious infections by organisms resistant to or not readily treated by other agents. Besides bacteremias and enteric fevers due to *Salmonella*, infections of the central nervous system, sepsis caused by anaerobes such as *Bacteroides fragilis*, and rickettsial diseases such as typhus, Rocky Mountain spotted fever, and Q fever are frequently treated with chloramphenicol (Bartlett and Tally, 1980). One of its characteristics leading to rapid and effective therapy is its high rate of absorption after oral administration, wide distribution in body tissues, and outstanding ability to penetrate to the brain and cross other lipid barriers. It is also readily cleared from the bloodstream by conjugation with glucuronic acid in the liver and excretion in the urine. Its half-life is 2-3 hr.

REFERENCES

Barlett, J. G., and Tally, F. P. (1980). Chloramphenicol. In *Antimicrobial Chemotherapy*, 3rd ed. B. M. Kagan (Ed.). Saunders, Philadelphia, pp. 127-136.

Bessman, S. P., and Stevens, S. (1950). A colorimetric method for the determination of chloromycetin in serum or plasma. *J. Lab. Clin. Med. 35:* 129-135.

Brock, T. D. (1961). Chloramphenicol. *Bacteriol. Rev. 25:*32-48.

Chandra, P., and Vining, L. C. (1968). Conversion of phenylalanine to tyrosine by microorganisms. *Can. J. Microbiol. 14*:573–578.

Chatterjee, S., and Vining, L. C. (1981). Nutrient utilization in actinomycetes. Induction of α-glucosidase in *Streptomyces venezuelae*. *Can. J. Microbiol. 27*:639–645.

Chatterjee, S., and Vining, L. C. (1982a). Catabolite repression in *Streptomyces venezuelae*. Induction of β-galactosidase, chloramphenicol production, and intracellular cyclic adenosine 3',5'-monophosphate concentrations. *Can. J. Microbiol. 28*:311–317,

Chatterjee, S., and Vining, L. C. (1982b). Glucose suppression of β-glucosidase activity in a chloramphenicol producing strain of *Streptomyces venezuelae*. *Can. J. Microbiol. 28*:593–599.

Chatterjee, S., Vining, L. C., and Westlake, D. W. S. (1983). Nutritional requirements for chloramphenicol biosynthesis in *Streptomyces venezuelae*. *Can. J. Microbiol. 29*:247-253.

Controulis, J., Rebstock, M. C., and Crooks, H. M., Jr. (1949). Chloramphenicol (chloromycetin) V. Synthesis. *J. Am. Chem. Soc. 71*:2463–2468.

Crechiolo, J., and Hill, R. E. (1979). Determination of serum chloramphenicol by high-performance liquid chromatography. *J. Chromatog 162*:480–484.

Cundliffe, E., and McQuillen, K. (1967). Bacterial protein synthesis. The effects of antibiotics. *J. Mol. Biol. 30*:137–146.

Demain, A. L. (1972). Cellular and environmental factors affecting the synthesis and excretion of metabolites. *J. Appl. Chem. Biotechnol. 22*:345–362.

Ehrlich, J., Bartz, Q. R., Smith, R. M., and Joslyn, D. A. (1947). Chloromycetin, a new antibiotic from a soil actinomycete. *Science 106*:417.

Ehrlich, J., Gottlieb, D., Burkholder, P. R., Anderson, L. E., and Pridham, T. G. (1948). *Streptomyces venezuelae* n. sp., the source of chloromycetin. *J. Bacteriol. 56*:467–477.

Emes, A., Floss, H. G., Lowe, D. A., Westlake, D. W. S., and Vining, L. C. (1974). Biosynthesis of chloramphenicol in *Streptomyces* species 3022a. Isotope incorporation experiments with [G-^{14}C]chorismic, [G-^{14}C] prephenic and [G-^{14}C, 6-^{3}H]shikimic acids. *Can. J. Microbiol. 20*:347–352.

Francis, M. M., and Westlake, D. W. S. (1979). Biosynthesis of chloramphenicol in *Streptomyces* species 3022a: The nature of the arylamine synthetase system. *Can. J. Microbiol. 25*:1408–1415.

Francis, M. M., Vining, L. C., and Westlake, D. W. S. (1978). Characterization and regulation of anthranilate synthetase from a chloramphenicol-producing streptomycete. *J. Bacteriol. 134*:10–16.

Freeman, K. B. (1970). Inhibition of mitochondrial and bacterial protein synthesis by chloramphenicol. *Can. J. Biochem. 48*:479–485.

Gale, E. F., Cundliffe, E., Reynolds, P. E., Richmond, M. H., and Waring, M. J. (1972). *The Molecular Basis of Antibiotic Action*. Wiley, London, pp. 325–332.

Gallicchio, V., and Gottlieb, D. (1958). The biosynthesis of chloramphenicol III. Effects of micronutrients on synthesis. *Mycologia 50*:490–496.

Gottlieb, D., and Diamond, L. (1951). A synthetic medium for chloromycetin. *Bull. Torrey Bot. Club 78*:56–60.

Gottlieb, D., Bhattacharyya, P. K., Anderson, H. W., and Carter, H. E. (1948). Some properties of an antibiotic obtained from a species of *Streptomyces*. *J. Bacteriol. 55*:409–417.

Gottlieb, D., Carter, H. E., Legator, M., and Gallicchio, V. (1954). The biosynthesis of chloramphenicol 1. Precursors stimulating the synthesis. *J. Bacteriol.* *68*:243–251.

Gottlieb, D., Carter, H. E., Robbins, P. W., and Burg, R. W. (1962). Biosynthesis of chloramphenicol IV. Incorporation of carbon 14-labeled precursors. *J. Bacteriol.* *84*:888–895.

Gottlieb, D., Robbins, P. W., and Carter, H. E. (1956). The biosynthesis of chloramphenicol. II. Acetylation of p-nitrophenylserinol. *J. Bacteriol.* *72*:153–156.

Grove, D. C., and Randall, W. A. (1955). *Assay Methods of Antibiotics: A Laboratory Manual.* Medical Encyclopedia, New York, pp. 66–75.

Hahn, F. E. (1967). Chloramphenicol. In *Antibiotics*, Vol. 1. D. Gottlieb and P. D. Shaw (Eds.). Springer-Verlag, New York, pp. 308–330.

Hahn, F. E. (1968). Relationship between the structure of chloramphenicol and its action upon peptide synthetase. *Experientia 24*:856–864.

Hahn, F. E., and Gund, P. (1975). A structural model of the chloramphenicol receptor site. In *Drug Receptor Interactions in Antimicrobial Therapy*, Vol. 1. J. Drews and F. E. Hahn (Eds.). Springer-Verlag, New York, pp. 245–266.

Hahn, F. E., and Wisseman, C. L. (1951). Inhibition of adaptive enzyme formation by antimicrobial agents. *Proc. Soc. Exp. Biol. Med. 76*:533–535.

Jensen, R. A., and Rebello, J. L. (1970). Comparative allostery of microbial enzymes at metabolic branch-points: Evolutionary implications. *Dev. Ind. Microbiol. 11*:105–121.

Jensen, R. A., and Stenmark, S. L. (1975). Ancient origin of a second microbial pathway for L-tyrosine biosynthesis in prokaryotes. *J. Mol. Evol. 4*:249–259.

Jones, A., and Vining, L. C. (1976). Biosynthesis of chloramphenicol in *Streptomyces* sp 3022a. Identification of p-amino-L-phenylalanine as a product from the action of arylamine synthetase on chorismic acid. *Can. J. Microbiol. 22*:237–244.

Jones, A., and Westlake, D. W. S. (1974). Regulation of chloramphenicol synthesis in *Streptomyces* sp. 3022a. Properties of arylamine synthetase, an enzyme involved in antibiotic biosynthesis. *Can. J. Microbiol. 20*: 1599–1561.

Jones, A., Francis, M. M., Vining, L. C., and Westlake, D. W. S. (1978). Biosynthesis of chloramphenicol in *Streptomyces* sp. 3022a. Properties of an amino transferase accepting p-aminophenylalanine as a substrate. *Can. J. Microbiol. 22*:237–244.

Karawya, M. S., and Ghourab, M. G. (1970). Assay of chloramphenicol and its esters in formulations. *J. Pharm. Sci. 59*:1331–1333.

Keiser, G., Bolli, P., and Buchegger, U. (1972). Hematologische Nebenwirkungen von Chloramphenicol and Thiamphenicol. *Schweiz. Med. Wochenschr. 102*:1595.

Kucan, Z., and Lipmann, F. (1964). Differences in chloramphenicol sensitivity of cell-free amino acid polymerization systems. *J. Biol. Chem. 239*:516–530.

Lark, K. G., and Lark, C. (1966). Protein required for initiation of DNA replication. *J. Mol. Biol. 20*:9–19.

Legator, M., and Gottlieb, D. (1953). The dynamics of chloramphenicol synthesis. *Antibiot. Chemother. 3*:809–817.

Levine, J., and Fischbach, H. (1951). The chemical determination of chloramphenicol in biological materials. *Antibiotic. Chemother. 1*:59-62.

Long, L. M., and Troutman, H. D. (1949). Chloramphenicol (chloromycetin) VII. Synthesis through p-nitroacetophenone. *J. Am. Chem. Soc. 71*: 2473–2475.

Lowe, D. A., and Westlake, D. W. S. (1971). Regulation of chloramphenicol synthesis in *Streptomyces* sp. 3022a. 3-Deoxy-D-*arabino*-heptulosonate-7-phosphate synthetase. *Can. J. Biochem. 49*:448–455.

Lowe, D. A., and Westlake, D. W. S. (1972). Regulation of chloramphenicol synthesis in *Streptomyces* sp. 3022a. Branchpoint enzymes of the shikimic acid pathway. *Can. J. Biochem. 50*:1064–1073.

McGrath, R., Siddiqueullah, M., Vining, L. C., Sala, F., and Westlake, D. W. S. (1967). *para*-Aminophenylalanine and *threo*-p-aminophenylserine: Specific precursors of chloramphenicol. *Biochem. Biophys. Res. Commun. 29*:576–581.

McGrath, R., Vining, L. C., Sala, F., and Westlake, D. W. S. (1968). Biosynthesis of chloramphenicol III. Phenylpropanoid intermediates. *Can. J. Biochem. 46*:587–594.

Magasanik, B. (1976). Classical and post-classical modes of regulation of the synthesis of degradative bacterial enzymes. In *Progress in Nucleic Acid Research*, Vol. 18. W. E. Cohen (Ed.). Academic, New York, pp. 99–115.

Malik, V. S. (1972). Chloramphenicol. *Adv. Appl. Microbiol. 15*:297–336.

Malik, V. S., and Vining, L. C. (1970). Metabolism of chloramphenicol by the producing organism. *Can. J. Microbiol. 16*:173–180.

Malik, V. S., and Vining, L. C. (1971). Metabolism of chloramphenicol by the producing organism. Some properties of chloramphenicol hydrolase. *Can. J. Microbiol. 17*:1287–1290.

Malik, V. S., and Vining, L. C. (1972a). Effect of chloramphenicol on its biosynthesis by *Streptomyces* species 3022a. *Can. J. Microbiol. 18*:137–143.

Malik, V. S., and Vining, L. C. (1972b). Chloramphenicol resistance in a chloramphenicol-producing *Streptomyces*. *Can. J. Microbiol. 18*:583–590.

Margosis, M. (1970). Analysis of antibiotics by gas chromatography II. Chloramphenicol. *J. Chromatogr. 47*:341–347.

Martin, J. F., and Demain, A. L. (1977). Effect of exogenous nucleotides on the candicidin fermentation. *Can. J. Microbiol. 23*:1334–1339.

Martin, J. F., and Demain, A. L. (1980). Control of antibiotic biosynthesis. *Microbiol. Rev. 44*:230–251.

Masterson, D. S., Jr. (1968). Colorimetric assay for chloramphenicol using 1-naphthol. *J. Pharm. Sci. 57*:305–308.

Matsuoka, M., Yagashita, K., and Umezawa, H. (1953). Studies on the intermediate metabolism of chloramphenicol production. II. On the carbohydrate metabolism of *Streptomyces venezuelae*. *Jpn. J. Med. Sci. Biol. 6*:161–169.

Munro, M. H. G., Taniguchi, M., Rinehart, K. L., Jr., and Gottlieb, D. (1975). A cmr study of the biosynthesis of chloramphenicol. *Tetrahedron Lett.* 2659–2662.

Nakano, H., Tomita, F., and Suzuki, T. (1974a). Production of corynecins by mutants defective in glycolipid synthesis and increased in corynecin I. *Agric. Biol. Chem. 38*:2471–2475.

Nakano, H., Tomita, F., and Suzuki, T. (1974b). Incorporation of shikimic acid into corynecins and its regulation. *Agric. Biol. Chem. 38*:2505–2509.

Nakano, H., Tomita, F., and Suzuki, T. (1976a). Biosynthesis of corynecins by *Corynebacterium hydrocarboclastus*: On the origin of the N-acyl group. *Agric. Biol. Chem. 40*:331–336.

Nakano, H., Tomita, F., and Suzuki, T. (1976b). Role of p-aminophenylalanine in biosynthesis of corynecins and aromatic amino acids by *Corynebacterium hydrocarboclastus*. *Agric. Biol. Chem. 40*:207–212.

Nierhaus, D., and Nierhaus, K. H. (1973). Identification of the chlorampheni-
col-binding protein in *Escherichia coli* ribosomes by partial reconstruc-
tion. *Proc. Nat. Acad. Sci. U.S.A. 70*:2224–2228.

Ogata, K., Shibata, M., Ueno, T., and Nakazawa, K. (1951). Studies on
Actinomyces and its antibiotics. VII. Production of chloramphenicol 2.
Tank culture of a strain producing high potency. *J. Antibiot. 4A*:44–47.

Okami, Y. (1948). Studies on the characters of a chloromycetin-producing
strain (O-163). *Jpn. Med. J. 1*:499–503.

O'Neill, W. P., Nystrom, R. F., Rinehart, K. L., Jr., and Gottlieb, D.
(1973). Biosynthesis of chloramphenicol. Origin and degradation of the
aromatic ring. *Biochemistry 12*:4775–4784.

Oyaas, J. E., Ehrlich, J., and Smith, R. M. (1950). Chemical changes during
chloramphenicol (chloromycetin) fermentation. *Ind. Eng. Chem. 42*:
1775–1776.

Pellegrini, M., Oen, H., Eilat, D., and Cantor, C. R. (1974). The mechanism of covalent reaction of bromacetylphenylalanyl-transfer RNA with
the peptidyl-transfer RNA binding site of the *Escherichia coli* ribosome.
J. Mol. Biol. 88:809–829.

Pestka, S. (1969). Studies on the formation of transfer ribonucleic acid–
ribosome complexes. *J. Biol. Chem. 244*:1533–1539.

Pestka, S. (1977). Inhibitors of protein synthesis. In *Molecular Mechanisms
of Protein Biosynthesis*. H. Weissbach and S. Pestka (Eds.). Academic,
New York, pp. 467–553.

Pongs, O. (1977). The receptor site for chloramphenicol in vitro and in vivo.
In *Drug Action at the Molecular Level*. G. C. K. Roberts (Ed.). Mac-
Millan, London, pp. 190–200.

Pongs, O. (1979). Chloramphenicol. In *Antibiotics*, Vol. 5. F. E. Hahn
(Ed.). Springer-Verlag, New York, pp. 26–42.

Pongs, O., Bald, R., and Erdmann, V. A. (1973). Identification of chlor-
amphenicol-binding protein in *Escherichia coli* ribosomes by affinity labeling. *Proc. Nat. Acad. Sci. U.S.A. 70*:2229–2233.

Rebstock, M. C., Crooks, H. M., Controulis, J. and Bartz, Q. R. (1949).
Chloramphenicol (chloromycetin) IV. Chemical studies. *J. Am. Chem.
Soc. 71*:2458–2462.

Resnick, G. L., Corbin, D., and Sandberg, D. H. (1966). Determination of
serum chloramphenicol utilizing gas-liquid chromatography and electron
capture spectrometry. *Anal. Chem. 38*:582–585.

Robison, L. R., Seligsohn, R., and Lerner, S. A. (1978). Simplified radio-
enzymatic assay for chloramphenicol. *Antimicrob. Agents Chemother.
13*:25–29.

Sala, F., and Westlake, D. W. S. (1966). Strain degeneration during continuous culture of a chloramphenicol-producing *Streptomyces venezuelae*.
Can. J. Microbiol. 12:817–829.

Sample, R. H. B., Glick, M. R., Kleiman, M. B., Smith, J. W., and Oei,
T. O. (1979). High-pressure liquid chromatographic assay of chloramphenicol in biological fluids. *Antimicrob. Agents Chemother. 15*:491–
493.

Shah, R. C., Raman, P. V., and Sheth, P. V. (1968). Determination of chloramphenicol in blood. *Indian J. Pharm. 30*:68–69.

Shapiro, S., and Vining, L. C. (1983). Nitrogen metabolism and chloramphenicol production in *Streptomyces venezuelae*. Can. J. Microbiol. (in press).

Shaw, P. D., and Hager, L. P. (1959). Biological chlorination. IV. Peroxidative nature of enzymic chlorination. *J. Am. Chem. Soc. 81*:6527–6528.

Shemyakin, M. M., Bamdas, E. M., Vinogradova, E. I., Karapetyan, M. G., Kolosov, M. M., Khokhlov, A. S., Shvetsov, Y. B., and Shchukina, L. A. (1952). Paths of synthesis of optically active analogues of D-threo-1-(p-nitrophenyl)-2-(dichloroacetamido)-1,3-propanediol. *Dokl. Akad. Nauk, S.S.S.R. 86*:565–568.

Shemyakin, M. M., Kolosov, M. N., Levitov, M. M., Germanova, K. I., Karapetyan, M. G., Shvetsov, I., and Bamdas, E. M. (1956). Researches into the chemistry of chloromycetin (levomycetin) VIII. Dependency of antimicrobial activity of chloromycetin on its structure and the mechanism of effect of chloromycetin. *Zh. Obshch. Khim. 26*:773–785.

Siddiqueullah, M., McGrath, R., Vining, L. C., Sala, F., and Westlake, D. W. S. (1967). Biosynthesis of chloramphenicol II. p-Amino-phenylalanine as a precursor of the p-nitrophenylserinol moiety. *Can. J. Biochem. 45*:1881–1889.

Siddiqueullah, M., McGrath, R., Vining, L. C., Sala, F., and Westlake, D. W. S. (1968). Metabolism of p-aminobenzoic acid by a chloramphenicol-producing *Streptomyces* sp. *Can. J. Biochem. 46*:9–14.

Simonsen, J. N., Paramasigamani, K., Vining, L. C., McInnes, A. G., Walter, J. A., and Wright, J. L. C. (1978). Biosynthesis of chloramphenicol. Studies on the origin of the dichloroacetyl moiety. *Can. J. Microbiol. 24*:136–142.

Smith, C. G. (1958). Effect of halogens on the chloramphenicol fermentation. *J. Bacteriol. 75*:577–583.

Smith, C. G., and Hinman, J. W. (1963). Chloramphenicol. *Prog. Ind. Microbiol. 4*:137–163.

Smith, R. M., Joslyn, D. A., Gruhzit, O. M., McLean, I. W., Jr., Penner, M. A., and Ehrlich, J. (1948). Chloromycetin; biological studies. *J. Bacteriol. 55*:425–428.

Stratton, C. D., and Rebstock, M. C. (1963). A new metabolite of *Streptomyces venezuelae*: D-threo-1-p-Aminophenyl-2-dichloroacetamido-1,3-propanediol. *Arch. Biochem. Biophys. 103*:159–163.

Suzuki, T., Honda, H., and Katsumata, R. (1972). Production of antibacterial compounds analogous to chloramphenicol by a n-paraffin-grown bacterium. *Agric. Biol. Chem. 36*:2223–2228.

Suzuki, T., Tomita, F., and Nakano, H. (1974). Fermentative production of chloramphenicol analogs (corynecins) from sucrose. *Agric. Biol. Chem. 38*:2477–2481.

Tamura, A., Takeda, I., Naruto, S., and Yoshimura, Y. (1971). Chloramphenicol from *Streptosporangium viridogriseum* var. *kofuense*. *J. Antibiot. 24*:270.

Tomita, F., Nakano, H., Honda, H., and Suzuki, T. (1974). Production of corynecins by chloramphenicol-resistant mutants of *Corynebacterium hydrocarboclastus*. *Agric. Biol. Chem. 38*:2183–2188.

Umezawa, H., and Maeda, K. (1949). Studies on the fermentation and extraction of chloromycetin. *J. Antibiot. 3*:49–52.

Umezawa, H., Tazaki, T., Kanari, H., Okami, Y., and Fukuyama, S. (1948). Isolation of crystalline antibiotic substance from a strain of *Streptomyces* and its identity with chloromycetin. *Jpn. Med. J. 1*:358–363.

Umezawa, H., Tazaki, T., and Fukuyama, S. (1949a). Resistances of antibiotic strains of *Streptomyces* to chloromycetin and a rapid isolation method of chloromycetin-producing strains. *Jpn. Med. J.* 2:73–78.

Umezawa, H., Tazaki, T., Okami, Y., and Fukuyama, S. (1949b). On a new source of chloromycetin, *Streptomyces omiyaensis*. *Jpn. Med. J.* 2:207–211.

Vasquez, D. (1964). The binding of chloramphenicol by ribosomes from *Bacillus megaterium*. *Biochem. Biophys. Res. Commun.* 15:464–468.

Vigh, G., and Inczedy, J. (1976). Separation of some chloramphenicol intermediates by high-pressure liquid chromatography. *J. Chromatogr.* 116:472–474.

Vining, L. C., and Westlake, D. W. S. (1964). Biosynthesis of the phenylpropanoid moiety of chloramphenicol. *Can. J. Microbiol.* 10:705–716.

Vining, L. C., Malik, V. S., and Westlake, D. W. S. (1968). Biosynthesis of chloramphenicol. *Lloydia* 31:355–363.

Wang, E. L., Izawa, M., Miura, T., and Umezawa, H. (1959). Studies on metabolism of a chloramphenicol-producing *Streptomyces* with chlorine-36. *J. Antibiot.* 3A:81–85.

Wat, C. K., Malik, V. S., and Vining, L. C. (1971). Isolation of 2(S)-dichloroacetamido-3-(p-acetamidophenyl)propan-1-ol from a chloramphenicol-producing *Streptomyces* species. *Can. J. Chem.* 49:3653–3656.

Weber, A. F., Opheim, K. E., Koup, J. R., and Smith, A. L. (1981). Comparison of enzymatic and liquid chromatographic chloramphenicol assays. *Antimicrob. Agents Chemother.* 19:323–325.

Westlake, D. W. S., and Vining, L. C. (1969). Biosynthesis of chloramphenicol. *Biotechnol. Bioeng.* 11:1125–1134.

Westlake, D. W. S., and Vining, L. C. (1981). Enzymology of chloramphenicol biosynthesis. In *Advances in Biotechnogy, Vol. III. Fermentation Products*. M. Moo-Young, C. Vezina, and K. Singh (Eds.). Pergamon, Toronto, pp. 89-94.

Westlake, D. W. S., Sala, F., McGrath, R., and Vining, L. C. (1968). Influence of nitrogen source on formation of chloramphenicol in cultures of *Streptomyces* sp. 3022a. *Can. J. Microbiol.* 14:587–593.

White, E. R., Carroll, M. A., and Zarembo, J. E. (1977). Reverse-phase high-speed liquid chromatography of antibiotics II. Use of high-efficiency small-particle columns. *J. Antibiot.* 30:811–818.

Yagishita, K., and Umezawa, H. (1950). Studies on the intermediate metabolism of chloramphenicol production 1. Changes of amino acids during fermentation and utilization of amino acids for chloramphenicol production. *Jpn. Med. J.* 3:289–297.

Yunis, A. A., and Bloomberg, G. R. (1964). Chloramphenicol toxicity, clinical features and pathogenesis. *Progr. Hematol.* 4:138–159.

13

THE COUMERMYCINS: PROPERTIES, BIOSYNTHESIS, AND FERMENTATION

CHARLES A. CLARIDGE, RICHARD P. ELANDER, AND KENNETH E. PRICE
Bristol-Myers Company, Syracuse, New York

I. INTRODUCTION

Coumermycin (also called Sugordomycin and Notomycin) was discovered independently in the early 1960s by workers at Hoffmann-La Roche Laboratories (Berger et al., 1966) and at the Bristol-Banyu Research Institute in Japan (Kawaguchi et al., 1965b). The Hoffmann-La Roche group isolated an organism from a soil sample collected in Gaspé, Canada, which produced an antibiotic complex they called Sugordomycin. They designated the producing organism *Streptomyces hazeliensis* var. *hazeliensis* (Whitaker, 1968). The Japanese group had named their producing strain *Streptomyces rishiriensis* after Rishieri Island, Hokkaido, Japan, where the organism had been isolated. A direct comparison of the two cultures has shown them to be identical, with the latter name being preferred because of prior description (Godfrey and Price, 1972). Both groups reported a multiplicity of related antibiotics produced by the fermentations, the structures of which are shown in Figure 1 (Berger and Batcho, 1978). However, it was soon realized that one component, designated coumermycin A_1, was more potent than the others. The closely related fraction coumermycin A_2, altered only in the absence of the

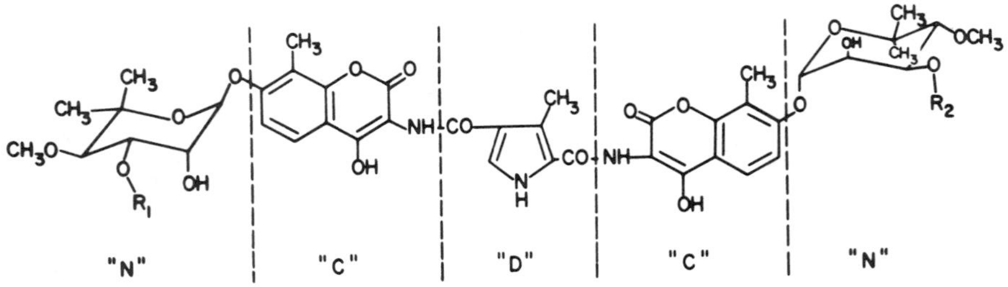

| | | Nomenclature | | Relative Antibiotic |
R_1	R_2	Bristol-Banyu	Hoffman-LaRoche	Activity[a]
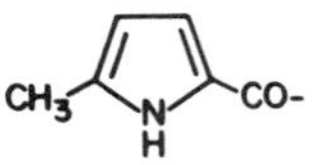 CH₃-pyrrole-CO-	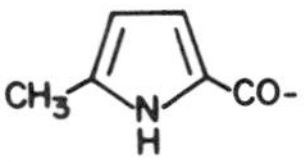 CH₃-pyrrole-CO-	A_1	D-1a	100
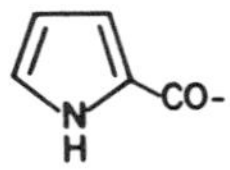 pyrrole-CO-	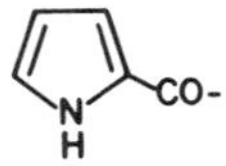 pyrrole-CO-	A_2	D-1d	3.6
CH₃-pyrrole-CO-	pyrrole-CO-		D-1b	25
pyrrole-CO-	CH₃-pyrrole-CO-		D-1c	25
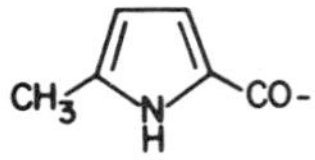 CH₃-pyrrole-CO-	H	B	D-2	20
H	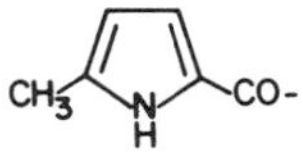 CH₃-pyrrole-CO-		Iso-D-2	20
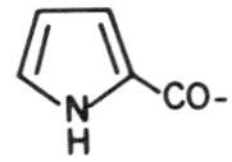 pyrrole-CO-	H	C	D-3	1.6
H	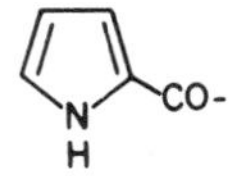 pyrrole-CO-		Iso-D-3	1.6
H	H	D	D-4	0.01

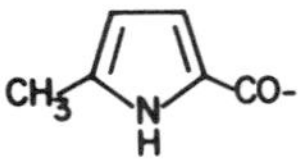

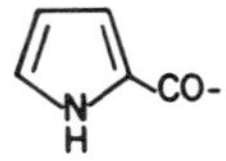

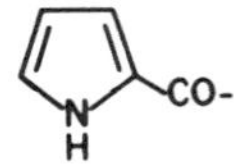

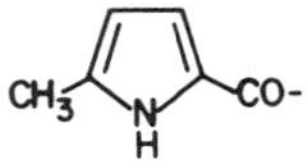

Figure 1 Structures of the coumermycins.
[a]Cup Plate Assay with *S. aureus* ATCC 6538P.

Table 1 Coumermycin Antimicrobial Spectra

Organism	Coumermycin A_1	Coumermycin A_2
Staphylococcus aureus Smith	0.004[a]	0.25
Staphylococcus aureus Smith + serum	1.6	50
Streptococcus pyogenes	0.062	0.5
Bacillus subtilis	6.3	12.5
Escherichia coli	12.5	>100
Klebsiella pneumoniae	6.3	12.5
Proteus morganii	6.2	25
Proteus mirabilis	12.5	>100
Proteus vulgaris	3.1	>100
Psudeomonas aeruginosa 8206A	12.5	>100
Pseudomonas aeruginosa Yale	25	>100
Salmonella enteritidis	12.5	50
Salmonella typhi	12.5	>100

[a]All values give as the minimal inhibitory concentration (μg/ml) in heart infusion broth.
Source: Cron et al. (1970).

two 5-methyl groups in the terminal 2-pyrrole carboxylic ester moieties, accompanied the main fermentation component and showed remarkably different microbial properties (Table 1).

II. FERMENTATION

The coumermycin complex is readily produced by *S. rishiriensis* in laboratory shake-flask fermentations or in submerged fermentations in larger pilot-scale tanks (see Sec. III). Because the desired component is coumermycin A_1, fermentation studies have been directed to maximizing the production of this antibiotic. The levels of the various components can readily be monitored by a number of chromatographic procedures (Claridge et al., 1966; Scannel, 1968). The Hoffmann-La Roche workers have reported that media ingredients such as bean meals, cottonseed flour, peptones, and distiller's dried solubles supported good growth and produced potencies up to 580 units/ml of whole broth (Berger et al., 1966). With a yellow pea meal as a nitrogen source, these same workers reported that dextrin, glucose, and mannitol were almost as satisfactory as starch as a carbon source. But in a similar medium, sucrose, sorbitol, sorbose, glycerol, and xylose were quite inferior. Furthermore, no antibiotic was produced when 0.25% levels of citric, fumaric, malic, acetic, lactic, gluconic, succinic, or glutamic acid were added as a sodium salt to the yellow pea medium.

The Japanese workers have reported that a typical fermentation medium contained hydrolyzed corn starch, pharmamedia, glycerin, yeast extract, $CaCO_3$, KH_2PO_4, and KCl, and that maximum yields were produced in 6–8 days with adequate aeration and agitation at 26–30°C (Kawaguchi et al., 1965b). Although no production levels were reported in the original Japanese work, a subsequent publication (Claridge et al., 1966) listed potencies as high as 548 units/ml in a medium composed of starch, lard oil, cottonseed meal, yeast, KH_2PO_4, and $CaCO_3$.

A. Influence of Cobalt on Coumermycin Fermentation

Because the desired coumermycin A_1 differed from coumermycin A_2 by being methylated, Claridge et al. (1966) reported the results of their attempts to enhance methylation in the fermentation by the addition of methionine, betaine, and lecithin at levels ranging from 0.05 to 0.5%. These compounds had no effect on coumermycin production. At the time of these studies, experimental fermentations in the pilot plant at Bristol Laboratories revealed that solids obtained from fermentations in stainless steel tanks were always more potent than those obtained from fermentations carried out in carbon steel tanks. The methylation of the pyrrole-2-carboxylic acid to produce 5-methyl-pyrrole-2-carboxylic acid found in coumermycin A_1 had suggested the possible role of vitamin B_{12} either at this stage or at a precursor level during the biosynthesis. This vitamin contains cobalt in its molecular structure and the stainless steel used in the fabrication of the fermentation vessel was known to contain traces of cobalt. It was hypothesized that some mineral, presumably cobalt, was leaching out during fermentation at a level sufficient to lead to the exclusive formation of coumermycin A_1. This proved to be the case. In a medium without added cobalt, coumermycin A_1 yields ranged from 40 to 75% of the total coumermycin isolated. However, the addition of as little as 0.01 μg/ml of cobalt resulted in the production of over 93% coumermycin A_1, the limit of sensitivity of the assay. Various other metallic ions, Fe^{2+}, Fe^{3+}, Mn^{2+}, Al^{3+}, Cr^{3+}, Cd^{3+}, Cu^{2+}, Mg^{2+}, and Pb^{2+}, at levels varying from 0.04 to 80 μg/ml, were without effect.

Table 2 Effect of Co^{2+}, Ni^{2+}, and Vitamin B_{12} on the 5-Methylpyrrole-2-Carboxylic Acid (MP) Content of Coumermycin A

Addition	Amount (μg/ml)	Medium N1-72[a] Coumermycin A (units/ml)	MP (%)	Medium N6-113[a] Coumermycin A (units/ml)	MP (%)
None		212	76	460	60
Co^{2+}	0.001	196	75	436	90
	0.005	216	90	436	91
	0.01	204	>93	440	>93
Ni^{2+}	1	176	89	476	89
	5	160	92	444	89
	10	160	89	528	91
	20	304	>93	548	>93
Vitamin B_{12}	0.4	212	67	288	82
	2	264	84	468	70
	4	356	>93	420	88

[a]Medium N1-72 contains corn syrup, cottonseed meal, soybean meal, yeast, ammonium phosphate, and calcium carbonate; medium N6-113 contains starch, lard oil, cottonseed meal, yeast, potassium phosphate, and calcium carbonate. *Source*Claridge et al. (1966).

Of the metallic ions tested, the only other ion producing an effect similar to cobalt was nickel, which, however, had to be present in the medium at a concentration of 20 µg/ml. Since it was subsequently found that the nickel chloride used contained 0.4% cobalt, the observed effect could be explained by the presence of this contaminant (Table 2).

Vitamin B_{12} was also added to the fermentation medium to determine possible stimulation of coumermycin A_1 formation. The high level of this vitamin required to produce the same effect as cobalt could again be explained by the cobalt content (4.34%). It is probable that the producing organism, however, has a requirement for vitamin B_{12} in order to carry out the methylation of the precursor to the pyrrole-2-carboxylic acid moiety, and that the trace of cobalt ensures the presence of enough vitamin B_{12} to complete that sequence (Godfrey and Price. 1972).

B. Assays of Coumermycin

Total coumermycin can be measured microbiologically by a standard plate assay employing *Staphylococcus aureus* ATCC 6438P. A more rapid spectrophotometric assay in which the whole fermentation broth is first extracted with methyl ethyl ketone and then diluted in ethyl alcohol containing 0.5% HCl for measurement of the absorption at 340 nm has been described by Claridge et al. (1966).

These same workers described a differential assay that allowed the direct measurement of the 5-methyl-pyrrole-2-carboxylic acid and the pyrrole-2-carboxylic acid contents of the methyl ethyl ketone extracts of the fermentation broths. Because it was recognized that coumermycin A_1 was the desired component of the coumermycin A fraction, the use of this assay enabled the Bristol Laboratories workers to develop media leading to the exclusive production of the A_1 component (Claridge et al., 1966).

An even more satisfactory assay has been developed by Scannel (1968). This assay permits one to determine quantitatively the ratio of pyrrole to methyl pyrrole liberated from crude extracts of fermentation broths and, semiquantitatively, the total concentration of coumermycin in the extracts. The procedure used is based upon a low-temperature pyrolysis followed by a gaschromatographic analysis to separate the two components.

C. Strain Improvement

The original report on the isolation and fermentation of coumermycin A_1 by *S. rishiriensis* did not reveal levels of production (Kawaguchi et al., 1965b); however, a search of the patent literature resulting from the Bristol-Banyu discovery disclosed that potencies as high as 1000 µg/ml were obtained (Kawaguchi et al., 1965a).

A media improvement and culture mutation program instituted at Bristol Laboratories resulted in an increase in potency up to 4000 µg/ml in laboratory shake flasks after 6–8 days incubation at 250 rpm at 27°C (C. A. Claridge and J. A. Bush, unpublished experiments). For the mutation work, spores of the *S. rishiriensis* culture were treated with nitrous acid followed by ultraviolet light, or with N-methyl-N[1]-nitro-N-nitrosoguanidine. The resultant survivors were plated out and isolates screened in selected fermentation media. The fermentations were monitored by the spectrophotometric assay and the coumermycin A_1 content determined by the $Ba(OH)_2$ hydrolysis procedure previously described (Claridge et al., 1966).

III. LARGE-SCALE FERMENTATIONS

A. Slant Culture Propagation

The organism used for large-scale (44–4000 liter) fermentations is a strain of *S. rishiriensis* designated BU-620. The strain is propagated on agar slants incubated at 28°C, and abundant conidiation was observed following 7 days incubation. The agar slant medium has the following composition: yeast extract, 1.0 g/liter; beef extract, 1.0 g/liter; tryptose, 2.0 g/liter; $FeSO_4 \cdot 7H_2O$ 0.03 g/liter; glucose, 10 g/liter; agar, 15 g/liter; distilled water, 1.0 liter; pH adjusted to 7.2.

B. Seed Development

The composition of the seed medium used for the various seed propagation stages is shown in Table 3. The typical seed propagation stages for successfull 44- and 4000-liter fermentations are, respectively,

$$\text{Slant} \xrightarrow{3 \text{ ml}} \text{flask (125 ml; vol:25 ml)} \xrightarrow{2.5 \text{ ml}} \text{flask (500 ml; vol:100 ml)}$$

$$\xrightarrow{100 \text{ ml}} \text{vessel (24 liter)} \xrightarrow{200 \text{ ml}} \text{vessel (44 liter)}$$

$$\text{Slant} \xrightarrow{3 \text{ ml}} \text{Erlenmeyer flask (500 ml)} \xrightarrow{300 \text{ ml}} \text{vessel (4000 liter)}$$

$$\xrightarrow{150 \text{ liter}} \text{vessel (4000 liter)}$$

C. Seed Tank Conditions

	44-liter vessel	4000-liter vessel
Volume (liters)	37	2960
Temperature (°C)	27	27
Airflow rate	40 liters/min	70 cfm
Pressure [psi (gauge)]	3	3
Agitation rate (rpm)	none	none
Age at inoculation (hr)	48	35

Table 3 Fermentation Media Used for Coumermycin Production in 44- and 4000-Liter Vessels—Seed Medium Composition

Constituent	Percentage
Cerelose	2
Cottonseed protein	1
Corn-steep liquor (50% solids)	1
$(NH_4)_2SO_4$	0.3
$ZnSO_4 \cdot 7H_2O$	0.003
$CaCO_3$	0.4

Table 4 Fermentation Media Used for Coumermycin Production in 44- and 4000-Liter Vessels—Fermentation Media Composition

| Medium no. 1 (44 liters) | | Medium no. 2 (4000 liters) | |
Constituent	Percentage	Constituent	Percentage
Soluble starch	7.0	Soluble starch	4.0
Cottonseed protein	2.0	Cottonseed protein	2.5
Debittered whole yeast	1.0	Debittered whole yeast	2.5
Lard oil	2.0	Lard oil	4.0
$(NH_4)_2SO_4$	0.4	$(NH_4)_2HPO_4$	0.1
$CaCO_3$	0.5	$CaCO_3$	0.5
K_2HPO_4	0.25	K_2HPO_4	0.1
$CoCl_2 \cdot 6H_2O$	0.0008	$CoCl_2 \cdot 6H_2O$	0.0008
Antifoam (PL-61)	0.05		

D. Fermentations

The fermentation media composition used for the 44- and 4000-liter fermentors are shown in Table 4. The physical conditions used for the respective fermentations can be summarized as follows:

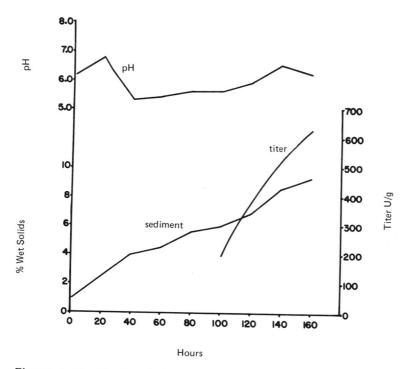

Figure 2 Profile for the coumermycin fermentation in a 44-liter fermenter.

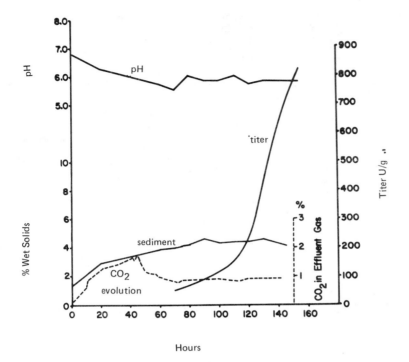

Hours

Figure 3 Profile for the coumermycin fermentation in a 4000-liter fermenter.

	44-liter vessel	4000-liter vessel
Volume (liters)	37	2960
Temperature (°C)	27	27
Airflow rate	120 liters/min	75 cfm
Pressure [psi (gauge)]	15	15
Agitation rate (rpm)	370	155
Cycle time (hr)	160	144

Typical fermentation profiles for the 44- and 4000-liter fermentation vessels are shown in Figures 2 and 3. Fermentation titers at harvest range from 600 to 800 units/g of whole broth based on turbidometric assay (H. J. Palocz, personal communication).

IV. BIOSYNTHESIS

The chemical similarity of portions of coumermycin A_1 to that of novobiocin has been used to make analogies regarding common biosynthetic pathways. Scannel and Kong (1970) proposed that glucose is the precursor of the noviose moieties as reported by Birch et al. (1962); that tyrosine is responsible for coumarin moieties (Bunton et al., 1963); and finally that methionine is a source

Table 5 Incorporation of Labeled Precursors into Coumermycin A_1

| | Percentage of incorporation | | | |
| | Washed cells | | Growing cells | |
Precursor	Whole molecule	Pyrroles	Whole molecule	Pyrroles
L-Methionine	14.9	12.2	10.0	7.9
L-Proline	7.0	6.9[a]	7.6	8.0[a]
δ-Aminolevulinic acid	0.5	1.0	0.9	0.4
D-Glucose	0.01	0	1.0	0

[a]Corrected for pyrolytic loss of carboxyl label, assuming uniform labeling of the molecule.
Source: Scannel and Kong (1970).

of the coumarin C-methyl and noviose O-methyl groups, as well as one methyl of the gem dimethyl group (Birch et al., 1960).

Scannel and Kong (1970) showed that methyl-labeled [^{14}C]methionine was readily incorporated into coumermycin A_1 by a strain of *S. hazeliensis* and that [^{14}C]L-proline was incorporated to the same extent in the pyrrole portions of the molecule (Table 5). In the case of porphyrin biosynthesis, Lascelles (1964) showed that δ-aminolevulinic acid (ALA) is the preferred precursor of the pyrrole group. When ALA was tested as a precursor with *S. hazeliensis*, it was incorporated only at 1/10 of the level of proline. Moreover, in a cold precursor competition experiment, ALA had no effect on proline incorporation, whereas proline lowered ALA incorporation (Scannel and Kong, 1970).

Although Table 5 shows that the incorporation of radioactive carbon from glucose was low compared to that of the other precursors, this may merely be a reflection of the rapid utilization of glucose for other metabolic purposes.

V. PREPARATION OF ANALOGS

It was recognized quite early that coumermycin A_1 had an irritation liability that was too great for it to be used parenterally, that it was remarkably susceptible to serum binding and possible inactivation, and, additionally, that it was poorly absorbed following oral administration. Because of these drawbacks, a program of chemical modification was initiated at Bristol Laboratories. The results of this program have been extensively reported by Godfrey and Price (1972). The most promising of the chemically modified coumermycins was BL-C43 (Fig. 4), which, when tested in humans, was found to be well absorbed. Studies utilizing the mono- and disodium salts of coumermycin A_1 consistently gave blood levels of less than 0.1 μg/ml after an oral dose of 500 mg, whereas with an equivalent dose of BL-C43 peak blood levels of over 15 μg/ml were reached in 3 hr and they were still high after 10 hr. However, phase 1 clinical studies revealed that there were undesirable side effects after 10 days administration at doses that were only 10–20% of the highest dose used in dog toxicology studies. Further investigations were terminated (Godfrey and Price, 1972).

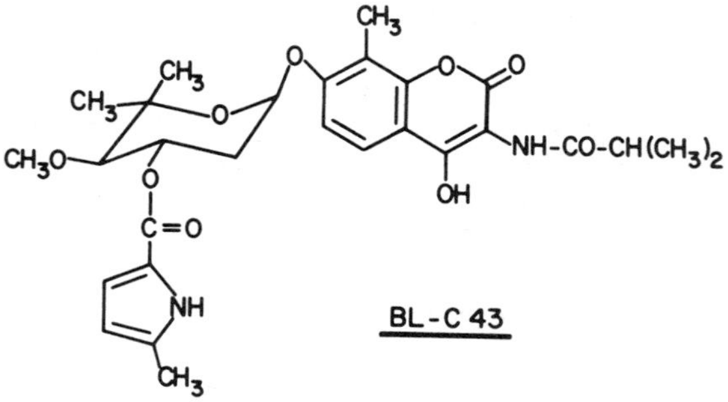

Figure 4 Structure of BL-C43.

Early in the chemical synthesis studies when new methods for the formation of coumermycin analogs were being sought, a program was initiated at
Bristol Laboratories to search for mutants of *S. rishiriensis* that were blocked
in antibiotic synthesis (C. A. Claridge and V. Z. Rossomano, unpublished experiments). A number of mutants were isolated, but one in particular, labeled
27-505, was shown to incorporate several coumermycin fragments as well as
some coumarin amine derivatives to yield a bioactive product. Some of the pre-

Table 6 Incorporation of Coumarin Fragments by *S. rishiriensis* Blocked
Mutant 27-505

Fragment added[a]		Antibiotic produced	
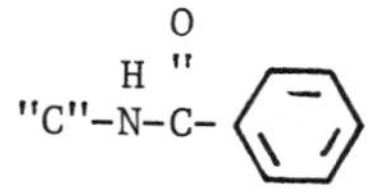	BL-C3	Compound 1[b]	
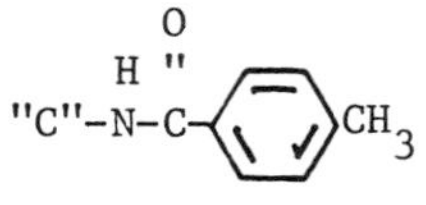	BL-C4	Compound 4[b]	
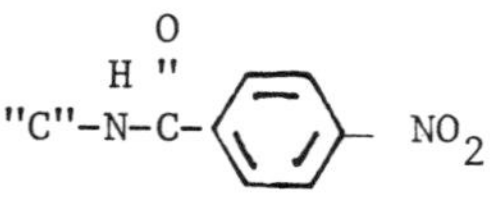	BL-C14	Compound 14[b]	
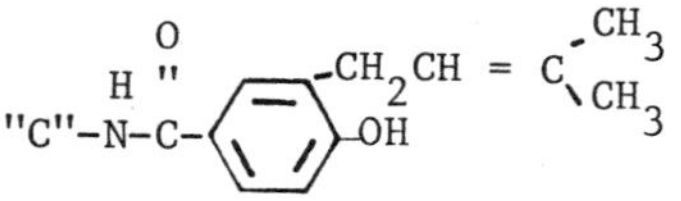	BL-C44	Compound 4[c]	
"CDC"	Coumermycin A		
"NCDCN"	Coumermycin A		

[a]From Figure 1: "C" = coumarin portion of coumermycin, "D" = 3,5-dicarboxyl-
pyrrole portion of coumermycin, "N" - noviose portion of coumermycin.
[b]From Keil et al. (1968).
[c]From Schmitz et al. (1968).

cursors that were found active and the products produced are presented in Table 6. These results show that the mutant was able to incorporate certain fragments and was able also to add either the pyrrole or the pyrrole-noviose portion to produce a bioactive molecule. The antibiotics that were produced were identified by direct chromatographic comparison to the authentic compounds that had been previously synthesized chemically (Keil et al., 1968; Schmitz et al., 1968).

VI. MODE OF ACTION

Although a relatively large number of antibiotics inhibit chromosome replication by interacting with the DNA template or by inhibiting the synthesis of purine and pyrimidine base precursors (Corcoran and Hahn, 1975; Kersten and Kersten, 1974; Gale et al., 1972), relatively few antibiotics interfere with actual DNA polymerization reactions. Coumermycin A_1, an antibiotic related to novobiocin, inhibits nucleic acid synthesis in intact *Escherichia coli*, with replication being more sensitive to coumermycin A_1 than transcription. Ultra-violet repair synthesis is only partially inhibited under conditions where DNA replication was eliminated by coumermycin A_1 (Ryan, 1976). Coumermycin A_1 similarly affects a DNA-dependent RNA polymerase of *E. coli* bacteriophage N 4 (Falco et al., 1978). The antibiotic inhibits supercoiling of *Micrococcus luteus* DNA via a specific gyrase inhibition (Liu and Wang, 1978). Similar effects have also been reported for coumermycin on genetic recombination and DNA gyrase inhibition in *E. coli* phage λ (Hays and Boehmer, 1978).

On the molecular level, coumermycin A_1 and the related antibiotic novo-biocin inhibit the supercoiling of circular double-stranded DNA (Gellert et al., 1976). The enzyme has been designated as DNA gyrase, an enzyme in cells of *E. coli* which introduce superhelical turns into DNA. The enzyme reaction required adenosine triphosphate and Mg^{2+} and is stimulated by spermidine. The *E. coli* enzyme also acts equally well on relaxed closed-circular colicin E_1 DNA, phage DNA, and simian virus 40 (SV40) DNA. The superhelical density of the DNA following catalysis by the DNA gyrase is usually considerably greater than that found in intercellular supercoiled DNA. Coumermycin A_1 also inhibits the replication of phage T4 and T7 DNA via a specific gyrase (Itoh and Tomizawa, 1977; McCarthy, 1979).

DNA gyrase also functions in DNA replication, specifically in the replication of double-stranded circular DNA. The enzyme either acts as a swivel to counteract the positive superhelical turns introduced by replication, or pretensions DNA to produce a state of negative supercoiling strain in DNA. In the latter role, DNA gyrase activity leads to a persistent torsional stress, thereby resulting in an unwinding in local regions of the DNA double helix, probably at the replication forks (Castora and Simpson, 1979).

Folded chromosomes isolated from *E. coli* following treatment with coumermycin A_1 were found to have reduced superhelical densities, with the loss of supercoiling closely paralleling inhibition of DNA synthesis (Orlica and Snyder, 1978). Coumermycin also produces a loss of supercoiling in non-replicating chromosomes that have been synchronized by amino acid starvation. No effect of coumermycin was observed on supercoiling of *E. coli* chromosomes isolated from a mutant strain which had coumermycin-resistant gyrase activity (McCarthy, 1979). Thus the correlation between coumermycin inhibition of cell growth, DNA synthesis, and in vitro gyrase activity extends to loss of chromosomal DNA supercoiling, indicating that the loss of supercoiling following coumermycin treatment arises from the action of DNA-relaxing activity.

REFERENCES

Berger, J., and Batcho, A. D. (1978). Coumarin-glycoside antibiotics. In *Antibiotics, Isolation, Separation, and Purification*. M. J. Weinstein and G. H. Wagman (Eds.). Elsevier, New York, pp. 101–144.

Berger, J., Schocher, A. J., Batcho, A. D., Pecherer, B., Keller, O., Maricq, J., Karr, A. E., Vaterlaus, B. P., Furlenmeier, A., and Speigelberg, H. (1966). Production, isolation, and synthesis of the Coumermycins (Sugordomycins), a new streptomycete antibiotic complex. *Antimicrob. Agents Chemother.* 1965:778–785.

Birch, A. J., Cameron, D. E., Holloway, P. W., and Rickards, R. W. (1960). Further examples of biological C-methylation: Novobiocin and actinomycin. *Tetrahedron Lett.* 25:26–31.

Birch, A. J., Holloway, P. W., and Rickards, R. W. (1962). The biosynthesis of Noviose, a branched chain monosaccharide. *Biochim. Biophys. Acta* 57:143–145.

Bunton, C. A., Kenner, G. W., Robinson, M. J. T., and Webster, B. R. (1963). Experiments related to the biosynthesis of novobiocin and other coumarins. *Tetrahedron* 19:1001–1010.

Castora, F. J., and Simpson, M. V. (1979). Evidence for a DNA gyrase in rat liver mitochondria. *Fed. Proc.* 38:779.

Claridge, C. A., Rossomano, V. Z., Buono, N. S., Gourevitch, A., and Lein, J. (1966). Influence of cobalt on fermentative methylation. *App. Microbiol.* 14:280–283.

Corcoran, J. W., and Hahn, F. E. (Eds.). (1975). Mechanism of action of antimicrobial and antitumor agents. In *Antibiotics*, Vol. 3. Springer-Verlag, New York.

Cron, M. J., Godfrey, J. C., Hooper, I. R., Keil, J. G., Nettleton, D. E., Price, K. E., and Schmitz, H. (1970). Studies on semisynthetic antibiotics derived from coumermycin. *Prog. Antimicrob. Anticancer Chemother. Proc. Int. Congr. Chemother.* 2:1069–1082.

Falco, S. C., Zivin, R., and Rothman-Denes, L. B. (1978). Novel template requirements of N4 virion RNA polymerase. *Proc. Nat. Acad. Sci. U.S.A.* 75:3220–3224.

Gale, E. F., Cundliffe, E., Reynolds, P. E., Richmond, M. H., and Waring, M. J. (Eds.). (1972). *The Molecular Basis of Antibiotic Action*. Wiley, New York.

Gellert, M., Mizuuchi, K., O'Dea, M. H., and Nash, H. A. (1976). DNA gyrase: An enzyme that introduces superhelical turns into DNA. *Proc. Nat. Acad. Sci. U.S.A.* 73:3872–3876.

Godfrey, J. C., and Price, K. E. (1972). Structure–activity relationships in coumermycins. *Adv. Appl. Microbiol.* 15:231–296.

Hays, J. B., and Boehmer, S. (1978). Antagonists of DNA gyrase inhibit repair and recombination of UV-irradiated phase λ. *Proc. Nat. Acad. Sci. U.S.A.* 75:4125–4129.

Itoh, T., and Tomizawa, J. (1977). Involvement of DNA gyrase in bacteriophage T7 DNA replication. *Nature (London)* 270:78–80.

Kawaguchi, H., Okanishi, M., and Miyaki, T. (1965a). Coumermycin and salts thereof. U.S. Patent No. 3,201,386, August 17, 1965.

Kawaguchi, H., Tsukiura, H., Okanishi, M., Miyaki, T., Ohmori, T., Fujisawa, K., and Koshiyama, H. (1965b). Studies on coumermycin, a new antibiotic. I. *J. Antibiot.* 17:1–10.

Keil, J. G., Hooper, I. R., Cron, M. J., Fardig, O. B., Nettleton, D. E., O'Herron, F. A., Ragan, E. A., Rousche, M. A., Schmitz, H. A.,

Schreiber, R. H., and Godfrey, J. C. (1968). Semisynthetic coumermycins. I. Preparation of 3-acylamido-4-hydroxy-8-methyl-7-[3-0-(5-methyl-2-pyrroylcarbonyl) noviosyloxy] coumarins. *J. Antibiot.* 21: 551–566.

Kersten, H., and Kersten, W. (1974). *Inhibitors of Nucleic Acid Synthesis.* Springer-Verlag, New York.

Lascelles, J. (1964). *Tetrapyrrole Biosynthesis and Its Regulation.* Benjamin, New York.

Liu, L. F., and Wang, J. C. (1978). *Micrococcus luteus* DNA gyrase: Active components and a model for its supercoiling of DNA. *Proc. Nat. Acad. Sci. U.S.A.* 75:2098–2102.

McCarthy, D. (1979). Gyrase-dependent initiation of bacteriophage T4 DNA replication: Interactions of *Escherichia coli* gyrase with novobiocin, . coumermycin and phase DNA-delay gene products. *J. Mol. Biol.* 127: 265–283.

Orlica, K., and Snyder, M. (1978). Superhelical *Escherichia coli* DNA: Relaxation by coumermycin. *J. Mol. Biol.* 120:145–154.

Ryan, M. J. (1976). Coumermycin A$_1$: A preferential inhibitor of replicative DNA synthesis in *Escherichia coli*. I. In vivo characterization. *Biochemistry* 18:3769–3777.

Scannel, J. (1968). Characterization and analysis of the antibiotic coumermycin by means of pyrolysis-vapor-phase chromatography and ultraviolet spectra. *Antimicrob. Agents Chemother. 1967:*470–474.

Scannel, J., and Kong, Y. L. (1970). Biosynthesis of coumermycin A$_1$: Incorporation of L-proline into the pyrrole groups. *Antimicrob. Agents Chemother. 1969:*139–143.

Schmitz, H., Devault, R. L., McDonnell, C. D., and Godfrey, J. C. (1968). Semisynthetic coumermycins. II. Preparation and properties of 3-(substituted benzamido)-4-hydroxyl-8-methyl-7-[3-0(5-methyl-2-pyrrolyl-carbonyl)noviosyloxy] coumarins. *J. Antibiot.* 21:603–610.

Whitaker, W. D. (1968). Antibiotic substances, sugordomycins, and a process for their manufacture, and an analog sugordomycin. British Patent No. 1,111,511, May 1, 1968.

14

FUSIDIC ACID: PROPERTIES, BIOSYNTHESIS, AND FERMENTATION

WELF VON DAEHNE, SVERRE JAHNSEN, INGER KIRK, ROBERT LARSEN, AND HENNING LÖRCK *Leo Pharmaceutical Products, Ballerup, Denmark*

I. INTRODUCTION

Fusidic acid [1], an antibiotic particularly useful in the treatment of staphylococcal infections, is produced by fermentation of a strain of the fungus *Fusidium coccineum*. Chemically, it belongs to a group of tetracyclic triterpenoic acids based on the structure of the hypothetic hydrocarbon fusidane [2].

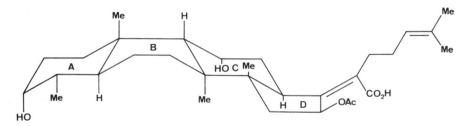

This hydrocarbon contains structural features also present in many natural steroids and triterpenes, that is, a cyclopentanoperhydrophenanthrene ring system connected at C17 with a side chain consisting of eight carbon atoms. However, the stereochemistry of the fusidane ring system deviates fundamentally from that of the steroids by the trans–syn–trans arrangement of rings A, B, and C, which forces ring B into a boat conformation, as evident from the three-dimensional representation of fusidic acid in Figure 1. The unusual stereochemistry of the ring system in fusidane-type antibiotics, including the previously discovered fungal metabolites helvolic acid and cephalosporin P_1, complicated their structural elucidation, since a correlation with triterpenes of known structure was not possible.

II. HISTORY AND DISCOVERY

In the 1950s, the increased occurrence of penicillinase-producing staphylococci created such severe clinical problems that the need for improved peni-

Figure 1 Conformation of fusidic acid.

cillins became urgent. However, a feasible way for preparing new semisynthetic penicillins was not indicated until the isolation of 6-aminopenicillanic acid (6-APA) from precursor-starved *Penicillium chrysogenum* fermentations was achieved in 1959.

At about that time, programs were initiated at the Leo Research Laboratories to screen fungi for (1) the production of 6-APA from phenoxymethylpenicillin and (2) the dehydrogenation of Δ^4-3-keto steroids to the corresponding $\Delta^{1,4}$ analogs (Lorck and Bremer, 1962) by means of enzymes produced by these microorganisms. In an additional program, the culture fluids from these experiments were tested for antibiotic activity. In the course of these studies, Lorck noticed in June 1960 that under aerobic fermentation conditions a strain of the imperfect fungus *Fusidium coccineum* produced a principle which strongly inhibited the growth of various bacteria, including penicillin-sensitive and penicillinase-producing staphylococci (Godtfredsen et al., 1963).

The active compound was easily extracted from the culture filtrate with organic solvents and subsequently obtained in a pure crystalline form. It turned out to be a weak carboxylic acid, not previously encountered, and was named fusidic acid. The acid was slightly soluble in water, but could readily be converted into crystalline water-soluble salts, for example, the sodium or diethanolamine salt. Its molecular formula was found to be $C_{31}H_{48}O_6$ from elemental analysis, and evidence of basic structural features was provided by chemical studies (Godtfredsen et al., 1962a).

Fusidic acid was found to have high in vitro activity against a variety of gram-positive bacteria, in particular staphylococci, corynebacteria, and clostridia, whereas gram-negative organisms, with the exception of the genus *Neisseria*, were resistant, and this was also true of fungi (Godtfredsen et al., 1962a,b). Complete cross-resistance was demonstrated among fusidic acid and the related antibiotics helvolic acid, isolated in 1943 from the culture fluid of a strain of *Aspergillus fumigatus* (Chain et al., 1943), and cephalosporin P_1, isolated in 1951 from the culture broth of the same strain of *Cephalosporium acremonium* which also produced penicillin N and cephalosporin C (Burton and Abraham, 1951). The high activity of fusidic acid against *Staphylococcus aureus*, including the penicillinase-producing strains, and its lack of cross-resistance with any other antibiotic in clinical use were considered to be of special clinical interest. Acute and chronic toxicity studies in animals revealed that fusidic acid was practically nontoxic (Godtfredsen et al., 1962a,b). Studies on the pharmacokinetic properties of fusidic acid indicated that the drug, after oral administration of its sodium salt to human volunteers, was (1) efficiently absorbed from the gastrointestinal tract, providing high and sustained serum levels, (2) mainly excreted with the bile, and (3) well tolerated when dosed repeatedly for several days.

On the basis of encouraging results obtained from extensive bacteriological, pharmacological, and toxicological studies, clinical trials were initiated in 1961, and in 1962 a series of reports on the successful use of fusidic acid in the treatment of staphylococcal infections were published. Since then the antibiotic has been used clinically as an important part of the antistaphylococcal armory.

Based on available chemical evidence, a first approach to the structure of fusidic acid was presented in 1962 (Godtfredsen and Vangedal, 1962). As a result of further chemical and spectroscopic studies, the full structure [1] of the antibiotic, including its relative and absolute stereochemistry, was reported in 1964 (Arigoni et al., 1964; Godtfredsen et al., 1965) and the correctness of formula [1] was subsequently confirmed by an x-ray crystallographic analysis of the 3-p-bromobenzoate of methyl fusidate (Cooper, 1966; Cooper and Hodgkin, 1968).

Chemical modification of fusidic acid provided a series of new derivatives whose structure—activity relationships were examined (Godtfredsen et al., 1966a,b). The metabolism of fusidic acid in man was studied, and several metabolites with reduced antibacterial activity were isolated from bile (Godtfredsen and Vangedal, 1966). In 1966–1967, the correct structures of cephalosporin P_1 (Halsall et al., 1966; Oxley, 1966) and helvolic acid (Okuda et al., 1967; Iwasaki et al., 1970) were finally established, indicating their close structural relation to fusidic acid. The structural correlation between fusidic acid and helvolic acid was subsequently established by their conversion to a common transformation product by a combination of chemical and microbiological methods (von Daehne et al., 1968). The isolation and identification of a number of cometabolites of fusidic acid, produced during the industrial fermentation of *F. coccineum*, was recently reported (Godtfredsen et al., 1979).

The extensive experimental and clinical studies with fusidic acid have been the subject of several review articles (Garrod et al., 1981; Tanaka, 1975) and a monograph (Godtfredsen, 1967) and also structure—activity relationships among fusidic acid-type antibiotics have been reviewed comprehensively (von Daehne et al., 1979).

III. STRAINS AND PROGRAMS

In early fermentations, relatively low levels of fusidic acid (50–250 µg/ml) were produced by the original strain of *F. coccineum* (Godtfredsen et al., 1963), and extensive efforts were therefore directed toward the development of optimal conditions for the economical manufacture of the antibiotic. These efforts included the search for other fusidic acid-producing microorganisms, investigations on the physical and nutritional requirements of the antibiotic-producing strain, and the development of high-yielding variants of the wild strain by mutation and selection.

A. Occurrence of Fusidic Acid

1. Fusidium Coccineum: Taxonomy

The original fusidic acid-producing fungus was isolated in 1953 by the Japanese mycologist Tubaki from a sample of monkey dung collected in Minomo, Settu, and identified to be a strain of *F. coccineum* belonging to the imperfect fungi family of *Moniliaceae* (Tubaki, 1954). Tubaki, who did apparently not observe its antibiotic-producing ability, deposited the fungus at the Centraal Bureau voor Schimmelcultures in Baarn, The Netherlands (CBS 197.55).

Morphological characteristics of the fungus include the formation of one-celled, fusiform conidia, $1.6–2.0 \times 4.5–6.3$ µm in size, originating from terminal chains on the conidiophores (Fig. 2). According to Rabenhorst (1907), *F. coccineum* is synonymous with *Ramularia coccinea*, a fungus pathogenic to the plant *Veronica officinalis*. However, morphological studies revealed that the spores of *R. coccinea* are frequently two celled and differ considerably in size ($1.9–2.7 \times 18.1–21.2$ µm) from those of Tubaki's fungus (Buchwald, 1963, unpublished observations). These findings indicated that Tubaki's designation of the species is erroneous. The generic name *Fusidium* has been criticized by Rabenhorst (1907) and, more recently, by Nicot (1968)

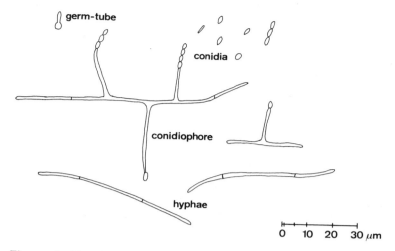

Figure 2 Morphological characteristics of *Fusidium coccineum* Tubaki (×400).

and Gams (1972), who referred to Tubaki's fungus as *Paecilomyces fusidioides* and *Acremonium fusidioides*, respectively.

If the generic name *Fusidium* is maintained for practical reasons, Tubaki's fungus should be designated either *Fusidium coprophilium*, with reference to its origin, or *Fusidium antibioticum*, according to its ability to produce fusidic acid.

2. Genera, Species, Strains

In addition to *F. coccineum*, a number of other organisms were demonstrated to produce fusidic acid. The formation of the antibiotic by certain strains of *Cephalosporium*, for example, *Cephalosporium lamellaecola*, was reported in a patent (Godtfredsen et al., 1964). Isolation of an antibiotic mixture called ramycin from the culture broth of a phycomycete, *Mucor ramannianus*, was described in 1958 (van Dicjk and de Somer, 1958), but it was not until 1965 that the identity of one of its components with fusidic acid was established (Vanderhaeghe et al., 1965). In the late 1960s, the production of low levels of fusidic acid (10-20 µg/ml) by various dermatophytes was reported. These included strains of *Epidermophyton floccosum* (Wallerström, 1970; Haller and Loeffler, 1969), *Calcarisporium antibioticum* (Haller and Loeffler, 1969), *Keratinomyces longifusus* (Haller and Loeffler, 1969), and *Microsporum gypseum* (Elander et al., 1969; Haller and Loeffler, 1969). The presence of an antibiotic resembling fusidic acid in the culture fluid of an ascomycete, *Isaria kogane*, was demonstrated in 1972 (Hikino et al., 1972).

Soviet scientists reported on the production of fusidic acid by a strain of *F. coccineum* originating from their own collection (Kuznetsov et al., 1967) and described later some cultural characteristics of the original strain, as well as improved mutants (Bartoshevich et al., 1973; Penzikova et al., 1977; Telesnina et al., 1978).

Various species of the genus *Fusidium* available from culture collections were screened in our laboratories for the synthesis of fusidic acid. Only a strain of *F. coccineum* (CBS 191.60) was found to produce small amounts of the antibiotic. Species of *Fusidium* incapable of producing fusidic acid included strains of *Fusidium aeruginosum*, *Fusidium candidum*, *Fusidium parasiticum*, *Fusidium terricola*, and *Fusidium viride*.

In our search for fusidic acid-producing microorganisms, a program concerned with the screening of fungi and actinomycetes was initiated at an early stage. From the screening of 6000 strains, isolated from soil samples collected in different parts of the world, 18 organisms were found to be fusidic acid producers. These included species of *Cephalosporium* or *Fusidium* (13 strains), *Microsporum* (2 strains), *Mucor* (2 strains), and *Chrysosporium* (1 strain). With the exception of *Cephalosporium spinosum* (320 µg/ml), the strains produced only low levels of the antibiotic. In addition, several species of imperfect fungi were demonstrated to produce active principles which exhibited cross-resistance with fusidic acid. However, the slight antibiotic titers present in the broths hampered any further identification of the active compounds.

B. Microbiological Transformations

Modification of the structure and antibacterial properties of fusidic acid was attempted by microbiological processes. During a program, a total of 2254 isolates, mainly fungi, streptomycetes, and bacteria, was incubated with fusidic acid (25-50 µg/ml) in shake-flask cultures and the broths subsequently screened for antibiotic activity against selected gram-positive and coliform bacteria. None of the cultures exhibited a modified spectrum of activity, but 65 (2.9%) of the organisms were found to inactivate fusidic acid.

In a similar program, a variety of microorganisms capable of transforming steroid substrates by means of bio-oxidative and bioreductive processes were incubated with fusidic acid. Conversion of the antibiotic was observed with the majority of strains tested, and transformation products were eventually isolated and identified. The results, summarized in Table 1, indicate the

Table 1 Microbiological Bioconversion of Fusidic Acid

Organism	Source	Product structure
Eubacterials		
Bacillus sp.	Leo B 74	[5]
Corynebacterium simplex	ATCC 13260	[3]
Pseudomonads		
Pseudomonas cruciviae	ATCC 13262	[3]
Pseudomonas testosteroni	ATCC 11996	[3,4]
Actinomycetales		
Micromonospora chalcea	Leo HL 214	[5]
Streptomyces cauvorensis	Leo MZ 1158	[5]
Nocardia corallina	ATCC 13259	[3]
Nocardia restrictus	ATCC 14887	[3]
Nocardia sp.	Leo HL 3 IV	[3]
Eumycetes		
Aspergillus sp.	Leo AK 1452	[3–5]
Colletotrichum atrament.	CBS 164.49	[3,4]

frequent conversion of fusidic acid into its 3-keto derivative [3], also reported elsewhere (Dvonch et al., 1966). Further transformation products were 3-epifusidic acid [4], presumably formed by reduction of structure [3], and 16-deacetylfusidic acid [5], produced by enzymatic hydrolysis of the parent compound.

[3]	R^1=O; R^2=H, α-OH	
[4]	R^1=H, β-OH; R^2=H, α-OH	[5]
[10]	R^1=H, α-OH; R^2=O	

In connection with the structural correlation between fusidic acid and helvolic acid, introduction of oxygen functions into positions 6 and 7 of fusidic acid by a strain of the fungus *Acrocylindrium oryzae* was demonstrated (von Daehne et al., 1968).

Compared with fusidic acid, all transformation products showed markedly reduced antibacterial activity.

C. Strain Requirements

Optimal physical and nutritional conditions during the fermentation process are of particular importance for proper vegetative growth and high antibiotic productivity of *F. coccineum*. Therefore the influence of various physical parameters on the fusidic acid fermentation was studied, and nutritional requirements of the original strain of the fungus were identified.

1. Physical Parameters

The effect of variations in temperature, pH, aeration, and agitation on the growth characteristics and antibiotic production of the wild strain of *F. coccineum* was studied in a series of laboratory fermentations. The fungus was cultured on a complex medium based on soybean meal and sucrose, and concentrations of biomass, carbohydrate, and fusidic acid were measured at suitable intervals during the process. Fermentations carried out at 22–27°C and pH 6.5–7.5 with efficient agitation (470 rpm) for 5–6 days produced maximum levels of fusidic acid, almost unaffected by the various rates of aeration (0.1–0.5 liter of air per liter of fluid per minute) used in these experiments. No changes in growth characteristics were observed in the range 22–32°C, but antibiotic production decreased significantly at 32°C. At a reduced rate of agitation (184 rpm), the fungus showed decreased vegetative development, the consumption of sucrose was minimal, and the synthesis of fusidic acid ceased.

2. Nutritional Requirements

Auxanographic techniques (Pontecorvo, 1949) were used to identify growth factor requirements of *F. coccineum*. Incubation of the wild strain with a variety of amino acids, vitamins, and nucleic acid constituents on a nutritionally deficient agar medium revealed that biotin was an essential growth factor, while the presence of specific amino acids was not required. Cultivation of the fungus in laboratory fermenters on a minimal medium containing increasing levels of biotin (0.01–200 µg/liter) indicated further that the amount of this compound necessary for optimal growth and fusidic acid production is approximately 1 µg/liter. Complex culture media containing corn-steep liquor, soybean meal, or similar organic nitrogen sources generally contain a sufficient amount of the growth factor.

The stimulating effect of organic nitrogen sources on the vegetative development and antibiotic productivity of *F. coccineum* Tubaki was studied in laboratory fermenters. Increasing amounts of these materials were added to glucose media, and concentrations of dry cell weight, carbohydrate, and fusidic acid were monitored during the fermentations. With respect to fusidic acid yields, it could be demonstrated that corn-steep liquor, soybean meal, and milk powder were superior to other materials such as meat–bone meal, pease meal, or fish meal. However, antibiotic yields were markedly affected by the amount of added nitrogen source; for example, soybean meal provided optimal fusidic acid levels at 5–10 g/liter.

Table 2 Utilization of Carbon Sources by *F. coccineum* Tubaki

Carbon source	Fusidic acid (µg/ml)
Sodium acetate	2
Sodium citrate	3
Glycerol	123
D-Arabinose	22
D-Xylose	69
D-Ribose	0
D-Glucose	80
D-Galactose	79
D-Fructose	102
D-Mannose	85
Sorbose	5
Sucrose	123
Lactose	4
Maltose	76
Raffinose	51
Starch	3
Dulcitol	9
Mannitol	123
Palmitic acid	45
Lard oil	85

Several synthetic fermentation media were developed containing chemically defined nitrogen sources such as sodium or ammonium nitrate, ammonium sulfate, and urea, but none of these were found to be superior to the complex organic media.

Obviously, the yield of fusidic acid is the crucial point in selecting the most favorable nitrogen source, but other factors, including foaming, rheology, product recovery, and prices, must also be considered.

Carbohydrate utilization properties of *F. coccineum* were studied in shake-flask cultures on a medium containing soybean meal as the nitrogen source. Fusidic acid levels were determined after incubation at 25°C for 4 days. As evident from the results summarized in Table 2, sucrose, glycerol, and mannitol produced the highest antibiotic yields.

In view of the terpenoid nature of fusidic acid, it was attempted to improve fermentation yields by adding compounds known to be intermediates in the biosynthesis of steroids and polycyclic triterpenes, for example, acetate, mevalonate, isopentenol, geranol, farnesol, and squalene (see Sec. IV). Although some of these compounds were efficiently incorporated into fusidic acid, none of them stimulated the production of the antibiotic.

D. Strain Improvement

Based on the low productivity of the original strain of *F. coccineum*, a program concerned with the induction and selection of superior fusidic acid-producing variants was immediately initiated (Kirk, 1965).

Cytological investigations of the fungus disclosed that it forms uninucleate conidia. The spores were exposed to mutagenic treatment, and generated mutants were subsequently examined for increased productivity in shake-flask fermentations. Mutations were performed by treatment with various chemical mutagens, as well as irradiation with ultraviolet light (253.7 nm) and ^{60}Co γ rays. With respect to increased fusidic acid productivity, ultraviolet irradiation was found to be the best mutagen. Throughout the program

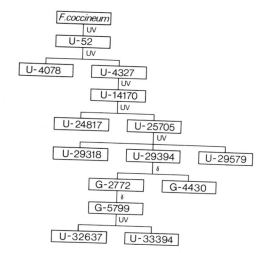

Figure 3 Genealogy of a series of improved mutants of *Fusidium coccineum*. Mutations were brought about by ultraviolet irradiation (wavelength, 253.7 nm) and by irradiation with ^{60}Co γ rays.

it was noted that lower doses were more effective in yielding a high rate of improved mutants than doses which would effect the highest rate of kill.

Intensive selection work resulted in a gradual development of improved mutants which produced increasing amounts of fusidic acid and concomitantly reduced the formation of by-products and colored impurities. The genealogy of a series of superior fusidic acid-producing mutants developed from the original strain of *F. coccineum* is presented in Figure 3.

As previously noticed in connection with strain improvement of other microorganisms, for example, the penicillin producer *Penicillium chrysogenum* (Backus and Stauffer, 1955) and the cephalosporin C producer *C. acremonium* (Elander, 1979), growth characteristics of high-yield mutants differed considerably from those of the original strain. Improved variants of *F. coccineum* showed a progressive reduction in colony diameter and sporulation vigor in surface culture and also decreased vegetative development in submerged culture. On the other hand, it was possible to extend advantageously the duration of fermentation. In Figure 4, fusidic acid synthesis by a typical series

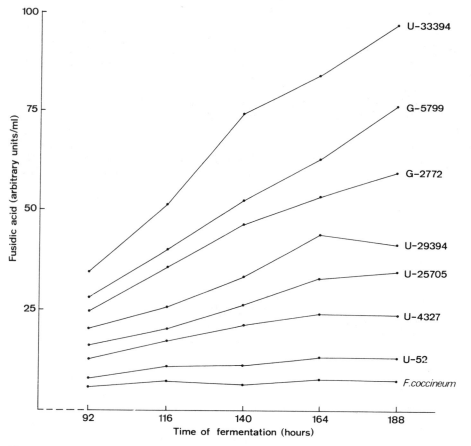

Figure 4 Synthesis of fusidic acid by the original strain of *Fusidium coccineum* and a series of higher-yield mutants in shake-flask cultures agitated on a rotating shaker board at 27°C.

of improved mutants of *F. coccineum* is compared with that of the original
strain in shake-flask fermentations under standardized conditions. It is evident from the curves that the original strain had reached maximum productivity after incubation for 116 hr, whereas, for example, strains G-5799 and
U-33394 still showed increasing antibiotic titers after 188 hr of fermentation.

IV. BIOSYNTHESIS

In view of the triterpenoid structure, the biosynthesis of fusidic acid was
expected to follow the general pathway for the biogenesis of sterols and polycyclic triterpenes, that is, acetate → mevalonate → isopentenyl pyrophosphate
→ geranyl pyrophosphate → farnesyl pyrophosphate → squalene. Experimental
evidence for the correctness of this expectation was provided by the efficient
incorporation of [1-^{14}C]acetate as well as [2-^{14}C]mevalolactone into fusidic
acid by the fungus *F. coccineum* (Godtfredsen, 1967), later also demonstrated
with [1-^{13}C]acetate (Riisom et al., 1974). Degradation of ^{14}C-labeled fusidic
acid, biosynthesized from [2-^{14}C]mevalolactone, showed that the distribution
of radioactivity, as evident from Figure 5, conformed with the assumption
of a squalene precursor (Arigoni, 1964; Visconti, 1968).

The discovery of 2,3-oxidosqualene as an intermediate in the biosynthesis
of cholesterol (Corey et al., 1966; van Tamelen et al., 1966) made it of interest to investigate whether this compound also served as a precursor of fusidic
acid. That this, in fact, was the case could be demonstrated by incubation
of *F. coccineum* with ^{3}H-labeled 2,3-oxidosqualene yielding efficient incorporation of the precursor into fusidic acid (Godtfredsen et al., 1968). Subsequent degradation of the labeled antibiotic indicated that the distribution
of radioactivity conformed with the expected pattern.

In the biosynthesis of cholesterol from 2,3-oxidosqualene the first stable
intermediate is lanosterol (Fig. 6). According to theoretical considerations
(Eschenmoser et al., 1955), the squalene precursor is folded in a potential chair—
boat—chair—boat conformation by the cyclizing enzyme. The cyclization is
thought to be initiated by enzyme-bound H$^+$ and to proceed through a sequence
of synchronous antiplanar additions to the double bonds leading to the carbenium ion [6]. A concerted rearrangement then takes place with stereospecific
1,2-migrations of two hydride ions and two methyl groups, whereupon the
formation of lanosterol is completed by elimination of the C9 proton to generate the C8 double bond.

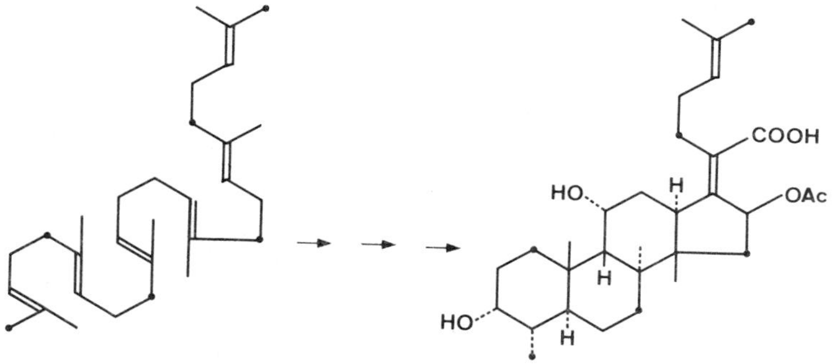

Figure 5 Incorporation of ^{14}C into squalene and fusidic acid biosynthesized
from [2-^{14}C]mevalolactone.

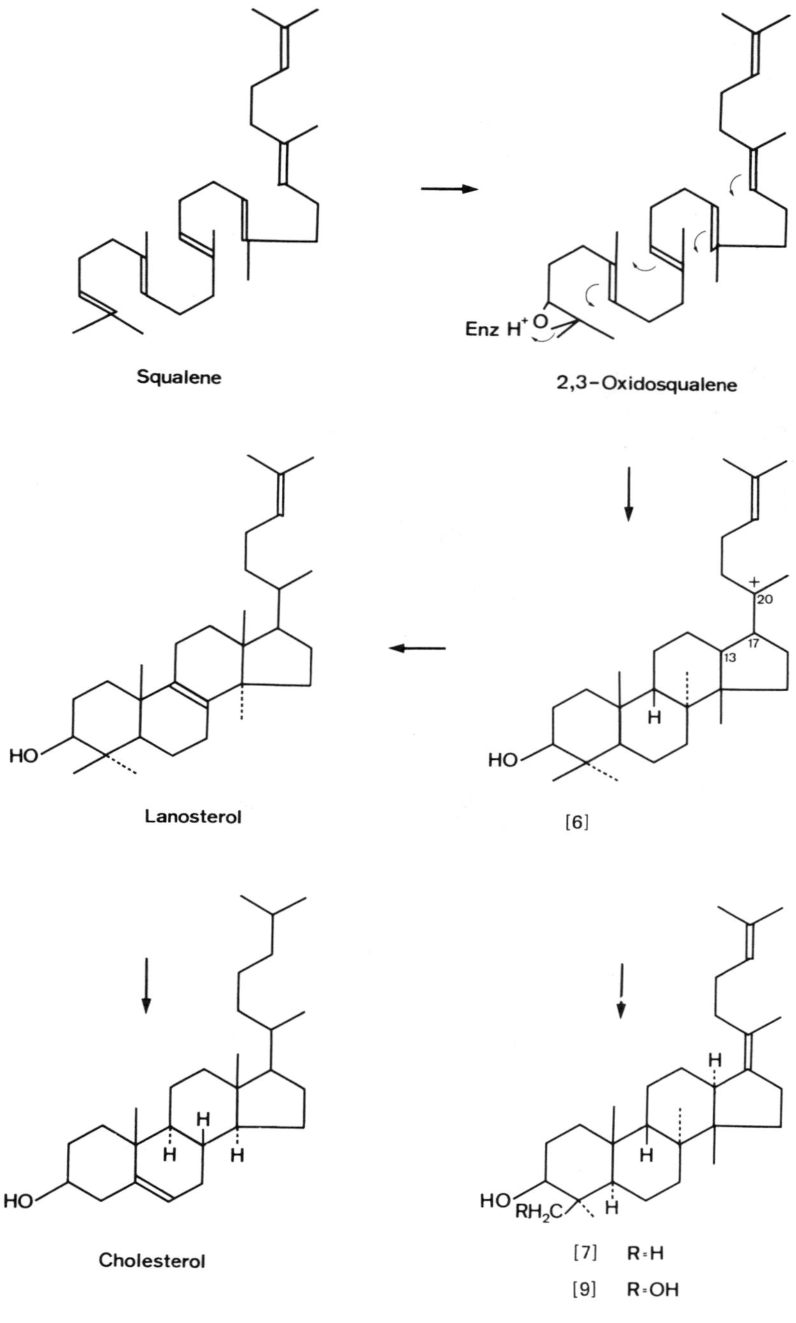

Figure 6 Conversion of squalene into sterols and protosterols.

In the biosynthesis of fusidic acid the intermediacy of the same carbenium ion [6] was suggested, and its direct stabilization by extrusion of the C17 proton to form the precursor [7] was postulated (Godtfredsen et al., 1965). In the search for such intermediates or others, the mycelium of *F. coccineum* was extracted with acetone, and from the nonsaponifiable lipid fraction of the residue a number of metabolites were isolated and identified (Visconti, 1968). These included the protosterol [8] and minor amounts of a triterpene diol whose structure [9] was proposed on the basis of spectroscopic evidence and biogenetic considerations. It could further be shown that the dihydro-acetate derived from structure [8] rearranged under the influence of mineral acid to dihydrolanosterol acetate.

[8]

A similar examination of the mycelium of *Cephalosporium caerulens* for intermediates in the biosynthesis of the related antibiotic helvolic acid resulted in the isolation and structural elucidation of three triterpene alcohols (Okuda et al., 1968; Hattori et al., 1969). Two of these metabolites were found to be identical in every respect with the protosterols [8] and [9] previously isolated from the mycelium of *F. coccineum*. Chemical and spectral data obtained for the remaining metabolite were consistent with structure [7], proposed earlier for the precursor of fusidic acid on the basis of theoretical considerations (see Fig. 6). The correctness of structure [7] was subsequently confirmed by the chemical correlation between metabolite [7] and the diol [9]. Incorporation of ^{3}H-labeled diol [9] into helvolic acid by *C. caerulens* was demonstrated (Okuda et al., 1968), and it could further be shown that cell-free extracts of the helvolic acid-producing fungus *Emericellopsis salmosynnemata* were able to incorporate [2-^{14}C]mevalonic acid into the protosterol [7] as well as lanosterol (Kawagushi and Okuda, 1970).

A careful reinvestigation of the nonsaponfiable lipid fraction of the mycelium of *F. coccineum* led to the isolation of an additional metabolite which proved to be completely identical with the triterpene alcohol [7], previously isolated by Okuda's group as a precursor of helvolic acid. Subsequently, efficient incorporation of the ^{3}H-labeled protosterol [7] into fusidic acid by incubation with *F. coccineum* was demonstrated (Godtfredsen, 1969).

The formation of common intermediates in the biosyntheses of fusidic acid and helvolic acid indicated that the principal biogenetic processes leading to these antibiotics are identical.

The mechanism of oxidative cyclization of squalene concerning the formation of intermediates in the biosynthesis of fusidic acid by *F. coccineum* was studied in detail (Mulheirn and Caspi, 1971; Caspi et al., 1973; Ebersole et al., 1973, 1974). Degradation of labeled fusidic acid, prepared biosynthetically from [^{14}C,^{3}H]mevalonic acid, demonstrated the retention of the C4 and C2 protons of mevalonic acid incorporated into fusidic acid at C13 and C22, respectively, as well as the absence of isotopic hydrogen at C3 of the antibiotic. These findings excluded the intermediacy of $\Delta^{13(17)}$ and $\Delta^{20(22)}$ precursors and indicated that the removal of the 4β-methyl group probably proceeds through a 3-keto derivative.

V. FERMENTATION PROCESS

Large-scale production of fusidic acid is carried out by deep-culture fermentation of high-yield mutants of *F. coccineum* under optimal physical and nutritional conditions. The process follows the traditional route used for industrial manufacturing of antibiotics by batch fermentation, the several steps being kept strictly sterile. The specific characteristics of the mutant strains, media, and fermentation conditions used in this process as well as the yields of the product cannot be described in detail since they are considered proprietary. However, general principles of inoculum preparation, media, fermentation equipment, and fermentation conditions associated with the production of fusidic acid are reviewed here.

A. Inoculum Preparation

The size of production fermenters requires that the inoculum be developed from the original spore suspension in several stages of increasing volume. Lyophilized spores of a high-yield mutant strain of *F. coccineum* are seeded on solid medium and incubated at 27°C for 3 weeks. The spores generated from this culture are suspended in sterile water, and a specific amount is inoculated into 400 ml of liquid seed medium in a 3-liter Erlenmeyer flask which is subsequently incubated at 25°C for 72 hr on a reciprocating shaker. The content of the flask is then transferred into 1200 liters of seed medium in a 2000 liter seed tank and cultured at 28°C for 96–120 hr with aeration (0.6 m^3/min) and agitation (1.0 hp/m^3). The broth of one or two seed tanks, used as inoculum for 25,000 liters of fermentation medium, is a sufficient amount to provide a proper rate of growth and antibiotic production in the final fermenter.

B. Medium Preparation and Substrates

The media for the various vegetative stages of *F. coccineum* are designed for the requirements of the high-yield mutants and were developed empirically. The several seed and fermentation media have a similar composition, based on glucose or sucrose as the source of carbohydrates and corn-steep liquor or soybean meal as the nitrogen source. Further ingredients include potassium dihydrogen phosphate, magnesium sulfate, and the antifoam agents lard oil and silicone oil. The substrates of the seed media are mixed with water, adjusted to pH 6.6–6.8, and sterilized at 120°C for 45 min in the respective culture vessels. In contrast, the substrates of the fermentation media are sterilized in two portions; that is, corn-steep liquor or soybean meal with an admixture of the inorganic salts and the carbohydrate solution are sterilized separately at 120°C for 45 min in a special sterilization tank. After aseptic transfer of the two portions into a previously sterilized fermenter, the mixture is adjusted to pH 6.6–6.8 and inoculated with the content of one or two seed tanks, providing to the productive stage an inoculum of 5–10%.

C. Fermentation Equipment

The fermenters for fusidic acid production used in the plant of Leo Pharmaceutical Products, Ltd., have a total volume of 35,000 liters (H/D ratio 2.6), providing a working capacity of about 28,000 liters, and are fitted with an agitator and a ring-shaped sparger for air distribution. Each unit is further equipped with facilities for adding sterile nutrients and antifoam agents and provided with control systems to measure and monitor the conditions of temperature, airflow, pH, pO_2, and foam during the process.

A constant working temperature in the fermenter is maintained by automatic control, regulating the circulation of cooling water through a system of coils. The considerable aeration is supplied by compressors and the air is sterilized by filtration before passing into the fermenter. Foaming during the fermentation process is controlled by addition of sterilized antifoam agents from a special container by means of an automatic arrangement.

D. Fermentation Procedure

Batch fermentation of fusidic acid is carried out at 27–28°C for 180–200 hr with efficient aeration (0.5 m^3 of air per cubic meter of fluid per minute) and vigorous agitation (2.0 hp/m^3), brought about by five impellers. In Figure 7, the time course of a typical fusidic acid fermentation is presented graphically. After growth for about 48 hr, the culture assumes the typical characteristics of a non-Newtonian fluid, a well-known property of highly filamentous organisms, and simultaneously a decrease in the concentration of dissolved oxygen is observed. In the growth phase, which lasts about 60 hr, the major amount

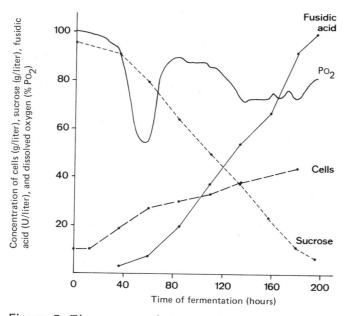

Figure 7 Time course of changes in concentration of sucrose, dry cell weight (including insoluble substrate), fusidic acid (arbitrary units/liter), and dissolved oxygen (% pO_2) during a typical fusidic acid fermentation.

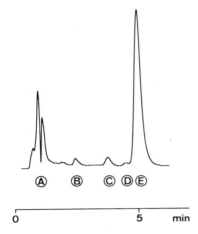

Ⓐ   Ⓑ   ⒸⒹⒺ

0   5   min

Figure 8 High-pressure liquid chromatogram of *Fusidium coccineum* fermentation broth: Column, 12.5 cm × 4.6 mm inner diameter, Lichrosorb 18, 10 μm; eluent, acetonitrile–methanol–1% aqueous acetic acid, 60:10:30; flow rate, 2 ml/min; detection, ultraviolet at 235 nm (A, impurities from broth; B, 24,25-dihydroxyfusidic acid 21,24-lactone [12]; c, 3- and 11-letofusidic acids [3, 10]; D, 16-epideacetylfusidic acid [11]; E, fusidic acid [1]).

of the cell mass necessary for high yields of fusidic acid is formed. During the following production phase, lasting about 140 hr, vegetative growth is markedly reduced. In the course of the process, the initial pH value of the fermentation mixture rises slowly and at the end of the cycle the broth is at pH 7.5–8.0.

Normally, none of the parameters are adjusted during the process, and no precursors are used. The concentration of fusidic acid in broth is monitored by high-pressure liquid chromatography, indicating the presence of approximately 10% of related compounds (Fig. 8). A number of these compounds were isolated in small amounts from the culture filtrate, and nine cometabolites of fusidic acid have recently been identified (Godtfredsen et al., 1979), including 3-ketofusidic acid [3] and 11-ketofusidic acid [10]. In contrast, two further compounds isolated from the broth [11,12] were considered to be artifacts derived from fusidic acid under the environmental conditions of fermentation.

[11]   [12]

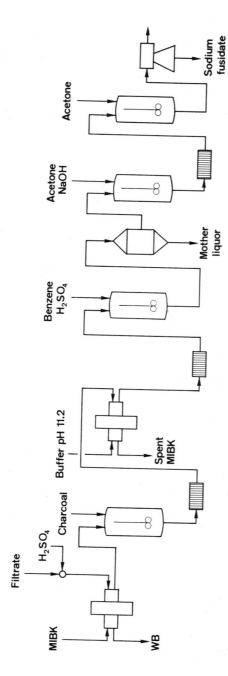

Figure 9 Flow sheet of the extraction process used in fusidic acid production at the plant of Leo Pharmaceutical Products, Ltd. (WB, waste broth; MIBK, methyl isobutyl ketone).

VI. PRODUCT RECOVERY AND PURIFICATION

Based on the solubility and ionic characteristics of the antibiotic, solvent extraction techniques were found to be most suitable for the recovery and purification of fusidic acid. The various stages of the process are schematically depicted in Figure 9.

At the completion of the fermentation, the contents of the tank are transferred to a rotary vacuum filter to remove the mycelium and other solids. The filtered broth is adjusted to pH 6.8 with sulfuric acid and extracted with about 10% volume of methyl isobutyl ketone using a countercurrent centrifugal extractor. The fusidic acid-containing organic extract is treated with charcoal to remove colored impurities, and the antibiotic is then back-extracted into an aqueous phase with about 20% volume of buffer solution at pH 11.2. To the clarified aqueous solution is added 20% volume of benzene, and the squeous phase is adjusted to pH 6.3 with sulfuric acid to precipitate fusidic acid in the form of its benzene solvate. The crystalline product is filtered off, washed with benzene followed by water, and dried at 60°C to afford crude fusidic acid containing about 3% of related compounds. A solution of the crude material in acetone is adjusted to pH 10.0 with methanolic sodium hydroxide and subsequently sterilized by filtration. Crystalline sodium fusidate is precipitated from the filtrate by further addition of sterile acetone until turbidity appears. The crystals are collected by centrifugation, washed with acetone, and dried at 40°C in vacuo to provide colorless sodium fusidate with a purity of at least 98% and containing less than 1.5% of related compounds.

A modified procedure for the precipitation of fusidic acid which avoids the use of benzene has recently been developed. The crude antibiotic is separated by addition of sodium chloride to the final aqueous extract and subsequently isolated as a crystalline methanol solvate.

Other forms of the antibiotic used in pharmaceutical preparations are prepared from pure fusidic acid methanol solvate.

VII. MODE OF ACTION

Fusidic acid inhibits in vitro and in vivo protein synthesis in both prokaryotes and eukaryotes, and it is now well established that the step which is inhibited during protein synthesis is the translocation process, that is, the movement of the ribosome relative to the messenger RNA from the 5'- to the 3'-end (Tanaka et al., 1968; Pestka, 1968; Haenni and Lucas-Lenard, 1968).

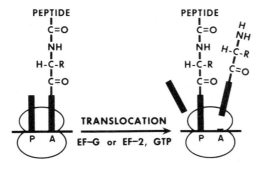

Figure 10 Schematic representation of translocation. The solid bars represent the tRNA molecule. (From Brot, 1977.)

The process, depicted schematically in Figure 10, shows a ribosome with a peptidyl-tRNA on the acceptor site (A site) and a deacylated tRNA on the peptidyl site (P site). During translocation, the deacylated tRNA is ejected from the P site, the peptidyl-tRNA moves from the A to the P site, and the ribosome moves precisely one codon closer to the 3'-end of the messenger RNA. This results in the appearance of a new codon in the A site, allowing a new aminoacyl-tRNA to bind to the vacant A site. In both prokaryotes and eukaryotes, translocation involves hydrolysis of guanosine triphosphate (GTP) to guanosine diphosphate (GDP) and is catalyzed by supernatant proteins called elongation factor G (EF-G) and elongation factor 2 (EF-2), respectively.

In prokaryotes, EF-G and GTP bind to the ribosome, forming a labile complex (I), which with liberation of inorganic phosphate (P_i) is hydrolyzed to a ribosome·EF-G·GDP ternary complex (II). The latter subsequently dissociates to ribosome, EF-G, and GDP, whereupon a new cycle can take place:

$$\text{Ribosome} + \text{EF-G} + \text{GTP} \rightleftharpoons \text{Ribosome·EF-G·GTP}$$
(complex I)

Fusidic acid

$$\text{Ribosome} + \text{EF-G} + \text{GDP} \rightleftharpoons \text{Ribosome·EF-G·GDP} + P_i$$
(complex II)

Fusidic acid stabilizes complex II and thus inhibits catalytic GTP hydrolysis by preventing the reutilization of EF-G and GDP (Bodley et al., 1979a; Brot et al., 1971). However, the antibiotic allows one round of GTP hydrolysis along with one round of translocation. Intact cells, when treated with fusidic acid, accumulated peptidyl-tRNA in the P site of the ribosomes, indicating that protein synthesis stopped after translocation had occurred (Cundliffe, 1972; Celma et al., 1972). It was concluded from these studies that fusidic acid is not a direct inhibitor of translocation, but inhibits protein synthesis by preventing the functional binding of aminoacyl-tRNA to the ribosome.

In eukaryotes, the analogous ribosome·EF-2·GDP complex is more stable than the prokaryote complex, and therefore the effect of fusidic acid is less pronounced (Bodley et al., 1970b; Chuang and Weissbach, 1972).

The effect of a large number of structurally modified fusidic acid derivatives on protein synthesis and stabilization of complex II has been studied in a sensitive *Escherichia coli* system. It could be concluded from this study that the most essential structural requirements for optimal activity are the α,β-unsaturated carboxylic acid and the 16β-oriented acetoxyl group, whereas the nature and stereochemistry of other functional groups are contributory but less critical (Bodley and Godtfredsen, 1972).

The numerous papers dealing with the mode of action of fusidic acid have been covered in recent reviews (Tanaka, 1975; Brot, 1977).

VIII. APPLICATION

Fusidic acid is primarily indicated for the treatment of staphylococcal infections. The high activity of the antibiotic against staphylococci, including multiple resistant strains, and its special ability for tissue penetration are valuable properties in the treatment of fulminating staphylococcal infections, where it is difficult to achieve bactericidal concentrations at the site of infection.

Systemic application of fusidic acid, alone or in combination with other antibiotics, has been used successfully in the treatment of septicemia, endocarditis, staphylococcal pneumonia, acute and chronic osteomyelitis, cystic fibrosis, and wound infections. Topically applied fusidic acid has proved outstandingly effective in the treatment of a wide range of staphylococcal and streptococcal skin infections, as well as traumatic and surgical wounds, burns, and ulcers.

For clinical use, fusidic acid is available in various forms of pharmaceutical presentation, all known as Fucidin. Orally, the antibiotic is administered either in the form of tablets containing sodium fusidate or as an aqueous suspension of fusidic acid, whereas it may be used parenterally by slow intravenous infusion of diethanolamine fusidate in sterile buffer solution. Topical Fucidin is applied in the form of an ointment containing 2% sodium fusidate or as intertulle, a sterile gauze impregnated with the ointment.

REFERENCES

Arigoni, D. (1964). In *Biogenesi delle Sostanze Naturali*, Accademia Nazionale dei Lincei, Rome, p. 1 ff.

Arigoni, D., Daehne, W. von, Godtfredsen, W. O., Melera, A., and Vangedal, S. (1964). The stereochemistry of fusidic acid. *Experientia 20*344–347.

Backus, M. P., and Stauffer, J. F. (1955). The production and selection of a family of strains in *Penicillium chrysogenum*. *Mycologia 47*:429–463.

Bartoshevich, Yu. E., Petrukhina, T. Yu., Dmitrieva, S. V., Novikova, N. D., and Goldshtein, V. L. (1973). Degradation of cultural features in *Fusidium coccineum* during selection. *Antibiotiki* Moscow *18*:981–986.

Bodley, J. W., and Godtfredsen, W. O. (1972). Studies on translocation. XI. Structure–function relationships of fusidane-type antibiotics. *Biochem. Biophys. Res. Commun. 46*:871–877.

Bodley, J. W., Zieve, F. J., and Lin, L. (1970a). Studies on translocation IV. The hydrolysis of a single round of guanosine triphosphate in the presence of fusidic acid. *J. Biol. Chem. 245*:5662–5669.

Bodley, J. W., Lin, L., Salas, M. L., and Tao, M. (1970b). Studies on translocation. V. Fusidic acid stabilization of a eukaryotic ribosome–translocation factor–GDP complex. *FEBS Lett. 11*:153–156.

Brot, N. (1977). Translocation. In *Molecular Mechanisms of Protein Synthesis*. H. Weissbach and S. Pestka (Eds.). Academic, New York, pp. 375–411.

Brot, N., Spears, C., and Weissbach, H. (1971). Interaction of transfer factor G, ribosomes, and guanosine nucleotides in the presence of fusidic acid. *Arch. Biochem. Biophys. 143*:286–296.

Burton, H. S., and Abraham, E. P. (1951). Isolation of antibiotics from a species of *Cephalosporium*. Cephalosporin P_1, P_2, P_3, P_4, and P_5. *Biochem. J. 50*:168–174.

Caspi, E., Ebersole, R. C., Mulheirn, L. J., Godtfredsen, W. O., and Daehne, W. von. (1973). Mechanism of squalene cyclization: The chiral origin of the C7 and C15 hydrogen atoms of fusidic acid. *J. Steroid Biochem. 4*:433–437.

Celma, M. L., Vazquez, D., and Modolell, J. (1972). Failure of fusidic acid and siomycin to block ribosomes in the pretranslocated state. *Biochem. Biophys. Res. Commun. 48*:1240–1246.

Chain, E., Florey, H. W., Jennings, M. A., and Williams, T. I. (1943).

Helvolic acid, an antibiotic produced by *Aspergillus fumigatus*, mut. *helvola* Yuill. *Br. J. Exp. Pathol.* 24:108–119.

Chuang, D. M., and Weissbach, H. (1972). Studies on elongation factor II from calf brain. *Arch. Biochem. Biophys.* 152:114–124.

Cooper, A. (1966). Crystal structure of fusidic acid methyl ester 3-p-bromo-benzoate. *Tetrahedron* 22:1379–1381.

Cooper, A., and Hodgkin, D. C. (1968). The crystal structure and absolute configuration of fusidic acid methyl ester 3-p-bromobenzoate. *Tetrahedron* 24:909–922.

Corey, E. J., Russey, W. E., and Ortiz de Montellano, P. R. (1966). 2,3-Oxidosqualene, an intermediate in the biological synthesis of sterols from squalene. *J. Am. Chem. Soc.* 88:4750–4751.

Cundliffe, E. (1972). The mode of action of fusidic acid. *Biochem. Biophys. Res. Commun.* 46:1794–1801.

Daehne, W. von, Godtfredsen, W. O., and Lorck, H. (1968). Microbiological transformations of fusidane-type antibiotics. A correlation between fusidic acid and helvolic acid. *Tetrahedron Lett.* 1968:4843–4846.

Daehne, W. von, Godtfredsen, W. O., and Rasmussen, P. R. (1979). Structure–activity relationships in fusidic acid-type antibiotics. *Adv. Appl. Microbiol.* 25:95–146.

Dijck, P. J. van, and Somer, P. de. (1958). Ramycin. A new antibiotic. *J. Gen. Microbiol.* 18:377–381.

Dvonch, W., Greenspan, G., and Alburn, H. E. (1966). Microbiological oxidation of fusidic acid. *Experientia* 22:517.

Ebersole, R. C., Godtfredsen, W. O., Vangedal, S., and Caspi, E. (1973). Mechanism of oxidative cyclization of squalene. Evidence for cyclization of squalene from either end of the squalene molecule in the in vivo biosynthesis of fusidic acid by *Fusidium coccineum*. *J. Am. Chem. Soc.* 95:8133–8140.

Ebersole, R. C., Godtfredsen, W. O., Vangedal, S., and Caspi, E. (1974). Mechanism of oxidative cyclization of squalene. Concerning the mode of formation of the 17(20) double bond in the biosynthesis of fusidic acid by *Fusidium coccineum*. *J. Am. Chem. Soc.* 96:6499–6507.

Elander, R. P. (1979). Mutations affecting antibiotic synthesis in fungi producing β-lactam antibiotics. In *Genetics of Industrial Microorganisms*. O. K. Sebek and A. I. Laskin (Eds.). American Society for Microbiology, Washington, D. C., pp. 21–35.

Elander, R. P., Gordee, R. S., Wilgus, R. M., and Gale, R. M. (1969). Synthesis of an antibiotic closely resembling fusidic acid by imperfect and perfect dermatophytes. *J. Antibiot.* 22:176–178.

Eschenmoser, A., Ruzicka, L., Jeger, O., and Arigoni, D. (1955). Zur Kenntnis der Triterpene. Eine stereochemische Interpretation der biogenetischen Isoprenregel bei den Triterpenen. *Helv. Chim. Acta 38:* 1890–1904.

Gams, W. (1972). In *Centraalbureau voor Schimmelcultures—List of Cultures*, Baarn, The Netherlands.

Garrod, L. P., Lambert, H. P., and O'Grady, F. (1981). Fusidanes. In *Antibiotics and Chemotherapy*, 5th ed. Churchill Livingstone, Edinburgh, pp. 220–225.

Godtfredsen, W. O. (1967). *Fusidic Acid and Some Related Antibiotics*, Copenhagen.

Godtfredsen, W. O. (1969). Steric aspects of the biosynthesis of fusidic acid. *5. EUCHEM Conference on Stereochemistry*, Bürgenstock (Lucerne).

Godtfredsen, W. O., and Vangedal, S. (1962). The structure of fusidic acid. *Tetrahedron* 18:1029–1048.

Godtfredsen, W. O., and Vangedal, S. (1966). On the metabolism of fusidic acid in man. *Acta Chem. Scand.* 20:1599–1607.

Godtfredsen, W. O., Jahnsen, S., Lorck, H., Roholt, K., and Tybring, L. (1962a). Fusidic acid. A new antibiotic. *Nature (London)* 193:987.

Godtfredsen, W. O., Roholt, K., and Tybring, L. (1962b). Fucidin. A new orally active antibiotic. *Lancet* 1:928–931.

Godtfredsen, W. O., Lorck, H., and Jahnsen, S. (1963). British Patent No. 930,786.

Godtfredsen, W. O., Lorck, H., and Jahnsen, S. (1964). Canadian Patent No. 687,898.

Godtfredsen, W. O., Daehne, W. von, Vangedal, S., Arigoni, D., Marquet, A., and Melera, A. (1965). The stereochemistry of fusidic acid. *Tetrahedron* 21:3505–3530.

Godtfredsen, W. O., Daehne, W. von, Tybring, L., and Vangedal, S. (1966a). Fusidic acid derivatives. I. Relationship between structure and antibacterial activity. *J. Med. Chem.* 9:15–22.

Godtfredsen, W. O., Albrethsen, C., Daehne, W. von, Tybring, L., and Vangedal, S. (1966b). Transformations of fusidic acid and the relationship between structure and antibacterial activity. *Antimicrob. Agents Chemother.* 1965:132–137.

Godtfredsen, W. O., Lorck, H., Tamelen, E. E. van, Willett, J. D., and Clayton, R. B. (1968). Biosynthesis of fusidic acid from squalene 2,3-oxide. *J. Am. Chem. Soc.* 90:208–209.

Godtfredsen, W. O., Rastrup-Andersen, N., Vangedal, S., and Ollis, W. D. (1979). Metabolites of *Fusidium coccineum*. *Tetrahedron* 35:2419–2431.

Haenni, A. L., and Lucas-Lenard, J. (1968). Stepwise synthesis of a tripeptide. *Proc. Nat. Acad. Sci. U.S.A.* 61:1363–1369.

Haller, B., and Loeffler, W. (1969). Stoffwechselprodukte von Mikroorganismen. 71. Mitteilung. Fusidinsäure aus Dermatophyten und anderen Pilzen. *Arch. Mikrobiol.* 65:181–194.

Halsall, T. G., Jones, E. R. H., Lowe, G., and Newall, C. E. (1966). Cephalosporin P_1. *Chem. Commun.* 1966:685–687.

Hattori, T., Igarashi, H., Iwasaki, S., and Okuda, S. (1969). Isolation of 3β-hydroxy-4β-methylfusida-17(20)[16,21-cis],24-diene (3β-hydroxyprotosta-17(20)[16,21-cis],24-diene) and a related triterpene alcohol. *Tetrahedron Lett.* 1969:1023–1026.

Hikino, H., Asada, Y., Arihara, S., and Takemoto, T. (1972). Fusidic acid, a steroidal antibiotic from *Isaria kogane*. *Chem. Pharm. Bull.* 20:1067–1069.

Iwasaki, S., Sair, M. I., Igarashi, H., and Okuda, S. (1970). Revised structure of helvolic acid. *Chem. Commun.* 1970:1119-1120.

Kawagushi, A., and Okuda, S. (1970). Incorporation of [2-^{14}C]-mevalonic acid into the prototype sterol 3β-hydroxyprotosta-17(20),24-diene with cell-free extracts of *Emericellopsis* sp. *Chem. Commun.* 1970:1012–1013.

Kirk, I. (1965). British Patent No. 999,794.

Kuznetsov, V. D., Suprun, T. P., Smirnova, A. D., Pivovarova, E. V., and Melekhova, N. P. (1967). Selection of natural variants, biological properties, inoculum preparation of fusidin-producing organism and procedure for antibiotic activity determination. *Antibiotiki* 12:13–17.

Lorck, H., and Bremer, O. (1962). Microbial 1-dehydrogenation of steroid hormones. *VIII. Intern. Congr. Microbiol.*, Montreal, Abstr. C 23.5, p. 72.

Mulheirn, L. J., and Caspi, E. (1971). Mechanism of squalene cyclization. The biosynthesis of fusidic acid. *J. Biol. Chem.* 246:2494–2501.

Nicot, J. (1968). Sur l'identité de l'organisme producteur de l'acide fusidique, antibiotique antistaphylococcique. *C. R. Acad. Sci.* 267:290–292.

Okuda, S., Iwasaki, S., Sair, M. I., Machida, Y., Inoue, A., Tsuda, K., and Nakayama, Y. (1967). Stereochemistry of helvolic acid. *Tetrahedron Lett.* 1967:2295–2302.

Okuda, S., Sato, T., Hattori, T., Igarashi, H., Tsuchiya, T., and Wasada, N. (1968). Isolation of 3β-hydroxy-4β-hydroxymethylfusida-17(20)[16, 21-cis],24-diene. *Tetrahedron Lett.* 1968:4769–4772.

Oxley, P. (1966). Cephalosporin P_1 and helvolic acid. *Chem. Commun. 1966:* 729–730.

Penzikova, G. A., Stepanova, N. E., and Levitov, M. M. (1977). Study of the peculiar properties of fusidin biosynthesis by *Fusidium coccineum,* strain 257 A. *Antibiotiki* 22:29–32.

Pestka, S. (1968). Studies on the formation of transfer ribonucleic acid–ribosome complexes. V. On the function of a soluble transfer factor in protein synthesis. *Proc. Nat. Acad. Sci. U.S.A.* 61:726–733.

Pontecorvo, G. (1949). Auxanographic techniques in biochemical genetics. *J. Gen. Microbiol.* 3:122–126.

Rabenhorst, L. (1907). *Kryptogamen-Flora von Deutschland, Oesterreich und der Schweiz,* 2nd ed. E. Kummer-Verlag, Leipzig, pp. 65 and 495.

Riisom, T., Jacobsen, H. J., Rastrup-Andersen, N., and Lorck, H. (1974). Assignment of the ^{13}C NMR spectra of fusidic acid derivatives. Biosynthetic incorporation of sodium [1-^{13}C]-acetate into fusidic acid. *Tetrahedron Lett.* 1974:2247–2250.

Tamelen, E. E. van, Willett, J. D., Clayton, R. B., and Lord, K. E. (1966). Enzymic conversion of squalene 2,3-oxide to lanosterol and cholesterol. *J. Am. Chem. Soc.* 88:4752–4754.

Tanaka, N. (1975). Fusidic acid. In *Antibiotics,* Vol. 3. J. W. Corcoran and F. E. Hahn (Eds.). Springer-Verlag, Berlin, pp. 436–447.

Tanaka, N., Kinoshita, T., and Masukawa, H. (1968). Mechanism of protein synthesis inhibition by fusidic acid and related antibiotics. *Biochem. Biophys. Res. Commun.* 30:278–283.

Telesnina, G. N., Bartoshevich, Yu.E., Mironov, V. A., Sazykin, Yu.O., and Zvyagilskaya, R. A. (1978). Studies on energy metabolism in *Fusidium coccineum* strains with different levels of antibiotic production. *Antibiotiki* 23:109–114.

Tubaki, K. (1954). Studies on Japanese hyphomycetes. I. Coprophilous group. *Nagaoa Mycol. J. Nagao Inst.* 4:7–8.

Vanderhaeghe, H., Dijck, P. van, and de Somer, P. (1965). Identity of ramycin with fusidic acid. *Nature (London)* 205:710–711.

Visconti, G. (1968). Chemische und biogenetische Untersuchungen einiger Metaboliten von *Fusidium coccineum.* Ph.D. thesis No. 4156, Eidgen. Technische Hochschule, Zurich.

Wallerström, A. (1970). Production of antibiotics by *Epidermophyton floccosum.* 5. Relation between the epidermophyton factor (EPF) and some steroid antibiotics, particularly fucidin. *Acta Pathol. Microbiol. Scand. Sect. B* 78:395–400.

15

THE MACROLIDES: PROPERTIES, BIOSYNTHESIS, AND FERMENTATION

EIJI HIGASHIDE *Takeda Chemical Industries, Ltd., Osaka, Japan*

I. INTRODUCTION

The term *macrolide* was originally applied to a group of lipophilic basic antibiotics with a macrocyclic lactone ring in their chemical structure (Woodward,
1957). These antibiotics are highly active against gram-positive bacteria and
mycoplasmas and hence are effective, when administered orally, against infectious diseases caused by these microorganisms. When administered to mammals,
these drugs are characteristically distributed to the liver, lung, and spleen
and effectively halt microbial respiratory or biliary tract infection. To cure
an adult, 1 g of erythromycin, oleandomycin, spiramycin, or kitasamycin (leucomycin) is administered orally on a daily basis (Walter and Heilmeyer, 1975).
They are therapeutically and economically important drugs.

Among the numerous macrolide antibiotics discovered during the past
30 years, erythromycin, oleandomycin, leucomycin, spiramycin, josamycin,
midecamycin, and tylosin have been produced industrially. Many reviews
on the chemistry (Inouye, 1973, 1974, 1976a,b; Ōmura and Nakagawa, 1975),
biosynthesis (Grisebach, 1978; Ōmura and Takeshima, 1975a–c; Ōmura and
Kitao, 1979), biotransformation (Ōmura and Nakagawa, 1975; Shibata and
Uyeda, 1978; Ōmura and Kitao, 1979), and mode of action (Oleinick, 1975;
Vazquez, 1975) or macrolide antibiotics have been described. The present
review will focus mainly on their discovery, strain improvement, fermentation,
and the more recent work on their biosynthesis and mode of action.

II. HISTORY AND DISCOVERY

Pikromycin (Brockmann and Henkel, 1950) produced by *Streptomyces* sp. strain
326 was the first macrolide antibiotic reported (Brockmann and Henkel, 1951).
Strain 326 was identified with *Streptomyces felleus* (Lindenbein, 1952). Erythromycin, which is more active but less toxic than pikromycin, was described
by McGuire et al. (1952). The same year, carbomycin, the first 16-membered
macrolide, produced by *Streptomyces halstedii*, was reported (Tanner et al.,
1952; Hochstein and Murai, 1954), and the following year leucomycin, produced
by *Streptomyces kitasatoensis*, was detected (Hata et al., 1953; Hata and Sano,
1955); the structure of the latter compound was elucidated by Ōmura and Nakagawa (1975). Most of the more than 50 macrolides discovered during the in-
tervening 30 years are basic compounds. The first neutral macrolide, lankamycin, produced by *Streptomyces violaceoniger*, was reported by Gäumann
et al. (1960); its 14-membered structure was elucidated by Muntwyler and
Keller-Schierlein (1972). The first macrolide antibiotic produced by a member
of the genus *Micromonospora*, *Micromonospora megalomicea*, megalomicin (a
14-membered macrolide), was reported in 1969 (Weinstein et al., 1969; Marquez
et al., 1969; Mallams et al., 1969). The megalomicin antibiotics are very active against gram-positive bacteria. Recently, B-41 (milbemycin) and C-076
(avermectin) produced by *Streptomyces hygroscopicus* subsp. *aureolacrimosus*
(Aoki et al., 1974; Takiguchi et al., 1980) and *Streptomyces avermitilis* (Albers-Schonberg, 1977; Burg et al., 1979; Miller et al., 1979), respectively,
were discovered as new types of 16-membered macrolides with insecticidal
and anthelminthic activity. There are indeed many types of macrocyclic compounds known to be produced by fungi as well as by bacteria (Inouye, 1973,
1974, 1976a,b).

In this chapter, the basic and neutral macrolide antibiotics which exhibit strong activity against gram-positive bacteria and mycoplasmas are described. Although the macrolides have been classified by their characteristic absorption spectra (Hütter et al., 1961), a classification based on chemical structure is preferable (Table 1).

A. 12-Membered Macrolides

In this group, methymycin, produced by *Streptomyces* sp. M-2140, was reported first by Donin et al. (1953) and its structure [1] was elucidated by Djerassi and Zderic (1956). Neomethymycin [2] and YC-17 [3] were reported by Djerassi and Halpern (1957), Djerassi et al. (1958), and by Kinumaki and Suzuki (1972), respectively.

		R_1	R_2	R_3
Methymycin	[1]	S	OH	H
Neomethymycin	[2]	S	H	OH
YC – 17	[3]	S	H	H

B. 14-Membered Macrolides

Pikromycin [13], the first macrolide discovered, belongs to this group. Erythromycin [6], produced by *Streptomyces erythreus* (McGuire et al., 1952; Harris et al., 1965), and oleandomycin [12], produced by *Streptomyces antibioticus* (Sobin et al., 1955; Celmer, 1965a,b; Celmer et al., 1958), have been produced commercially. Erythromycin and oleandomycin were also produced by *S. griseoplanus* (Thompson and Strong, 1971), *Micromonospora* sp. 1225 (Wagman and Weinstein, 1980), and *Streptomyces olivochromogenes* (Higashide et al., 1965), respectively. Erythromycins B (Pettinga et al., 1954), C (Wiley et al., 1957), and D (Majer et al., 1977) are produced as minor components together with erythromycin A. Erythromycin E [7] is derived from erythromycin A, B, or C when fed to the fermentation broth of a blocked mutant (Abbott 2NU 153) of *S. erythreus* (Martin et al., 1975). 3-O-Oleandorosyl-5-O-desosaminylerythronolide A oxime, which is very stable in acidic solution compared with erythromycin A, was obtained from a culture broth of an oleandomycin producer incubated with erythronolide A oxime for 60 hr (LeMahieu et al., 1976). O-Demethyloleandomycin was detected in the fermentation broth of *Streptomyces oleandros* (Celmer, 1971).

Megalomicin [10] produced by the genus *Micromonospora* also belongs to this group. The antibiotic was produced concomitantly with XK-41B$_2$ (4"-O-propionyl megalomicin A) and erythronolide B in the culture broth of *Micromonospora inositola* ATCC 21733, FERM 1317 (Kawamoto et al., 1974; Nara et al., 1973; Okachi et al., 1974).

The same year oleandomycin was discovered, narbomycin [11] produced by *Streptomyces narbonensis* was found (Corbaz et al., 1955). This antibiotic

Table 1 Macrolide Antibiotics

Antibiotics (synonym)	Producer strains	Structure	References
12-membered macrolides			
Methymycin	Streptomyces sp. M-2140	[1]	Djerassi and Zderic (1956), Donin et al. (1953)
Neomethymycin	Streptomyces sp. M-2140	[2]	Djerassi et al. (1958), Djerassi and Halpern (1957)
YC-17 (10-deoxy-methymycin)	S. venezuelae MCRL-0376	[3]	Kinumaki and Suzuki (1972)
14-membered macrolides			
Albocycline	S. bruneogriseus, S. roseo-cinereus, S. roseochromo-genes var. abocyclini, S. maigenus	[4]	Furumai et al (1967), Nagahama et al. (1967), Nagahama et al. (1971)
Cineromycin B (7-O-demethylalbocycline)	S. cinerochromogenes	[5]	Miyairi et al. (1966)
Erythromycins A,B,C,D	S. erythreus, S. griseoplanus Micromonospora sp. 1225	[6]	Harris et al. (1965), Majer et al. (1977), Wagmen (1980), McGuire et al. (1952), Pettinga et al. (1954), Thompson and Strong (1971), Wiley et al. (1957)
Erythromycin E	S. erythreus Abbott 2NU 153	[7]	
Kujimycin (desacetyl-lankamycin)	S. spirochromogenes var. kujimyceticus	[8]	Martin et al. (1975) Namiki et al. (1969)
Lankamycin	S. violaceoniger	[9]	Gäumann et al. (1960), Muntwyler and Keller-Schierlein (1972)
Megalomicins A,B,C₁,C₂	M. megalomicea NRRL 3724, M. megalomicea var. nigra NRRL 3725	[10]	Mallams et al. (1969), Marquez et al. (1969), Weinstein et al. (1969)
Narbomycin (12-desoxypikromycin)	S. narbonensis, S. fradiae B-62169	[11]	Corbaz et al. (1955), Yamamoto et al. (1969), Suzuki and Aota (1969)

Compound		Species	References
Oleandomycin	[12]	S. antibioticus	Els et al. (1958), Higashide et al. (1965), Sobin et al. (1955)
O-Demethyloleandomycin		S. olivochromogenes	Celmer (1971)
Pikromycin (amaromycin, 12-hydroxynarbomycin)	[13]	S. oleandros	Brockmann and Henkel (1950), Brockmann and Henkel (1951), Lindenbein (1952), Muxfeldt et al. (1968)
		S. felleus, S. venezuelae, S. flavochromogenes	
23672 RP	[14]	S. chryseus DS12370 NRRL 3892	Arnox et al. (1980)
T-2636 B (acetyldemethyl-yl-lankamycin)		S. rochei var. volubilis	Harada et al. (1969), Higashide et al. (1971)
XK-41 B₂ (4''-O-propionylmegalomicin A)		M. inositola	Nara et al. (1973), Okachi et al. (1974)
16-membered macrolides			
Angolamycin (shincomycin A)	[15]	S. eurythermus, S. antibioticus, S. phaeochromogenes, S. flavochromogenes	Kinumaki and Suzuki (1972), Nishimura et al. (1965)
Aldgamycins E,F	[16]	S. lavendulae	Achenbach and Karl (1975a), Achenbach and Karl (1975b), Kunstmann et al. (1964)
Carbomycin A	[17]	S. halstedii	Tanner et al. (1952)
Carbomycin B	[18]	S. macrosporeus Stv. albireticuli	Hochstein and Murai (1954) Hütter et al. (1961)
Chalcomycin (aldgamycin D)	[19]	S. tendae, S. thermotolerans S. bikiniensis	Nakazawa (1955) Forhardt et al. (1962), Woo et al. (1964)
Cirramycins A,B	[20]	S. cirratus	Koshiyama et al. (1963)
Deltamycins A₁,A₂,A₃	[21]	S. halstedii subsp. deltae FERM P-2504	Shimauchi et al. (1978a,b)
Espinomycin	[22]	S. fungicidicus var. espino- myceticus FERM 0351	Umezawa et al. (1969)
Juvenimicins A₁,A₂,A₃, B₁,B₂,B₃	[23]	M. chalcea subsp. izumensis ATCC 21561	Hatano et al. (1976), Kishi et al. (1976)
Josamycin (leucomycin A₃)	[24]	S. narbonensis var. josamyceti- cus ATCC 17835, NIHJ 440	Ōmura et al. (1970), Osono et al. (1967)

Table 1 (continued)

Antibiotics (synonym)	Producer strains	Structure	References
Leucomycins A1,A3,A4, A5,A6,A7,A8,A9,AU,AV	Stv. kitasatoensis NRRL 2486, 2487	[24]	Hata et al. (1953), Hata and Sano (1953), Ōmura and Nakagawa (1975)
Maridomycins I,II,III, IV,V,VI,VII	S. hygroscopicus IFO 12995, ATCC 21582	[25]	Higashide et al. (1969), Muroi et al. (1973), Ono et al. (1973), Uchida et al. (1975a)
Midecamycins (SF 837) A1,A2,A3,A4	S. mycarofaciens ATCC 21454	[26]	Inouye et al. (1971), Niida et al. (1971), Tsuruoka et al. (1969
Mycinamicins I,II,III, IV,V	M. griseorubida A 11725 NRRL 11452	[27]	Satoi et al. (1980)
Neutramycin (4-desmethylchalcomycin)	S. rimosus	[28]	Lefemine et al. (1963), Mitscher and Kunstmann (1969)
Niddamycin (desacetylcarbomycin B)	S. djakartensis	[29]	Huber et al. (1962)
Platenomycins A,B,C,W (YL-704)	S. platensis var. malvinus MCRL 0388, FERM 289, MCRL 0373, FERM 998	[30]	Furumai et al. (1974), Kinumaki et al. (1974a,b), Okuda and Awata (1969)
Relomycin	S. hygroscopicus NRRL 3017, S. griseospiralis	[31]	Whaley et al. (1963)

Compound	Organism	Ref	Citations
Rosamicin	M. rosaria NRRL 3178	[32]	Reiman and Jaret (1972), Wagman et al. (1972)
Spiramycins I,II,III	S. ambofaciens	[33]	Kenhne and Benson (1965), Pinnert-Sindico (1954), Pinnert-Sindico et al. (1955)
Staphcoccomycin	Streptomyces sp. ASNG-16	[34]	Shimi et al. (1979)
Turimycin (macrolide JA 6599)	S. hygroscopicus JA 6599	[35]	Forberg et al. (1979), Fricke et al. (1979), Veb. Jenapharm (1969)
Tylosins A,B,C,D	S. fradiae M48-E-2724, S. hygroscopicus	[36]	Hamill et al. (1961), McGuire et al. (1961), Morin et al. (1970)
A-6888 C,X	S. flocculus NRRL 11459	[37]	Nasch et al. (1980)
B-58941 (acumycin, diacetyl cirramycin B$_1$)	S. fradiae var. acinicolor	[38]	Kusaka et al. (1970), Suzuki (1970)
M-4365 A,G	M. capillata MCRL-0940, FERMP-2598	[39]	Furumai et al. (1977), Kinumaki et al. (1977)
SCH 23831	M. rosaria	[40]	Puar et al. (1979)
17-membered macrolides			
Bundlin B (T-2636A)	S. griseofuscus, Streptomyces sp. 6642-GC$_1$	[41]	Sakamoto et al. (1962), Uramoto et al. (1969)
Lankacidin (T-2636C)	S. violaceoniger	[42]	Gäumann et al. (1960)
T-2636 D,E,F	S. rochei var. volubilis	[42]	Harada et al. (1969)
	S. rochei var. volubilis	[42]	Harada et al. (1969), Higashide et al. (1971)

was also produced by *Streptomyces fradiae* B-62169 (Yamamoto et al., 1969) and *Streptomyces venezuelae* MCRL 0376 (Hori et al., 1971). The accumulation of narbonolide was reported in a culture of the latter strain fed organic acids (Maezawa et al., 1974a) and arsenite (Maezawa et al., 1974b). The structure of albocycline [4] (Nagahama et al., 1967), produced by *Streptomyces bruneogriseus, Streptomyces roseocinereus,* and *Streptomyces roseochromogenes* var. *albocyclini,* was elucidated as a macrolide aglycon (Nagahama et al., 1971). Cineromycin B [5] produced by *Streptomyces cinerochromogenes* (Miyairi et al., 1966) was identified with 7-O-demethylalbocycline (Nagahama et al., 1971).

The structure of the first neutral macrolide, lankamycin [9] (Gäumann et al., 1960), was described by Muntwyler et al. (1972); the structures of antibiotic T-2636 B (Harada et al., 1969), produced by *Streptomyces rochei* var. *volubilis* (Higashide et al., 1971), and of kujimycin [8], produced by *Streptomyces spirochromogenes* var. *kujimyceticus* (Namiki et al., 1969), are related to lankamycin. Recently, a new antibiotic, 23672RP [14], belonging to this group was reported by Arnox et al. (1980). The drug displays interesting characteristics because it is active not only against gram-positive bacteria but also against mycobacteria and it has already proved effective against both types of microorganisms in vitro and in vivo.

14 Membered Macrolide

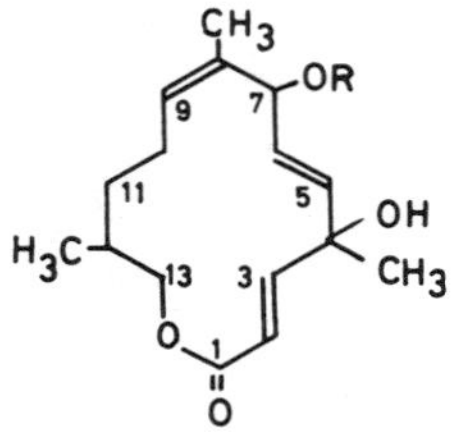

	R
Albocycline [4]	CH₃
Cineromycin B [5]	H

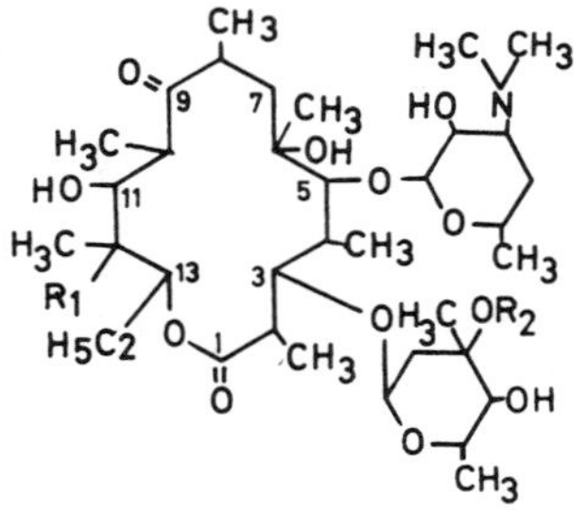

Erythromycin [6]

	R₁	R₂
A	OH	CH₃
B	H	CH₃
C	OH	H
D	H	H

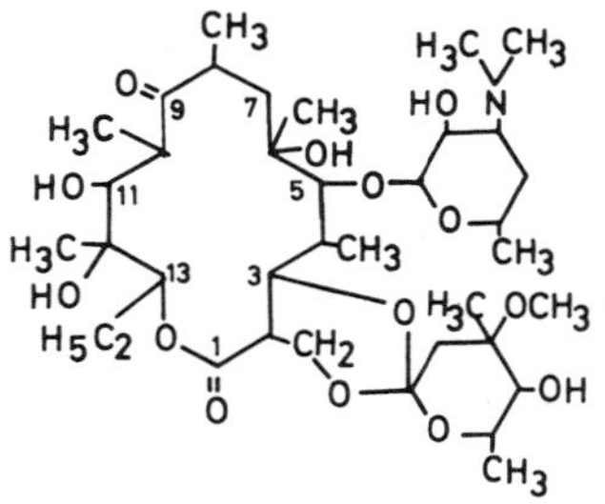

Erythromycin E [7]

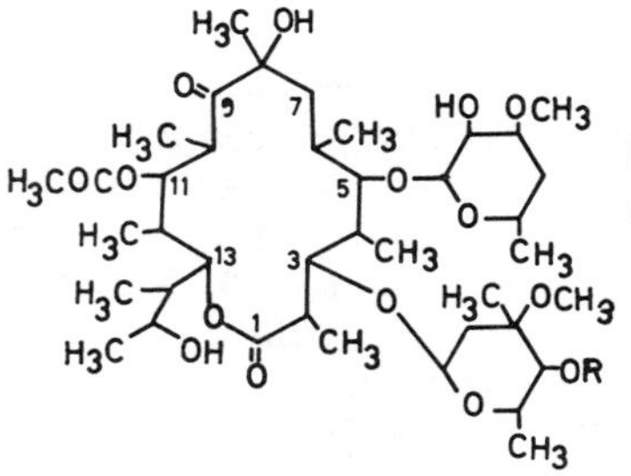

Kujimycin [8]

Lankamycin [9]

	R
Kujimycin [8]	H
Lankamycin [9]	COCH3

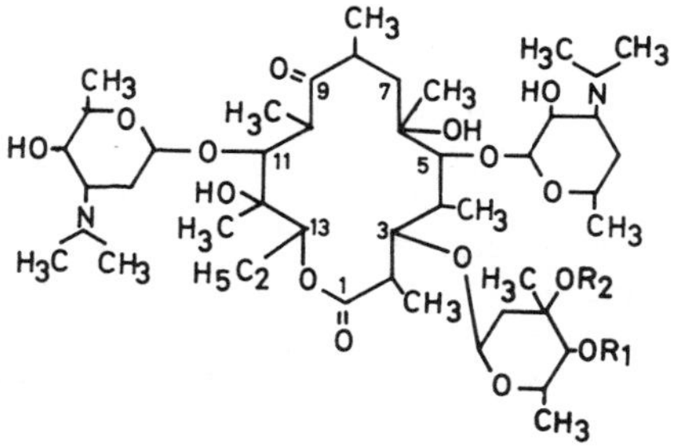

Megalomicin [10]

	R1	R2
A	H	H
B	COCH3	H
C1	COCH3	COCH3
C2	COC2H5	COCH3

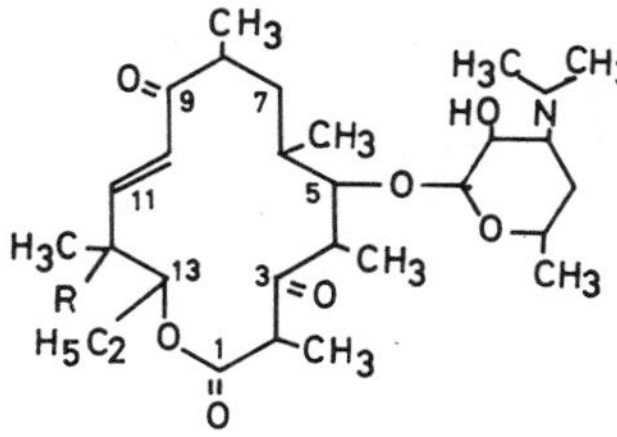

Narbomycin [11]

Pikromycin [13]

	R
Narbomycin [11]	H
Pikromycin [13]	OH

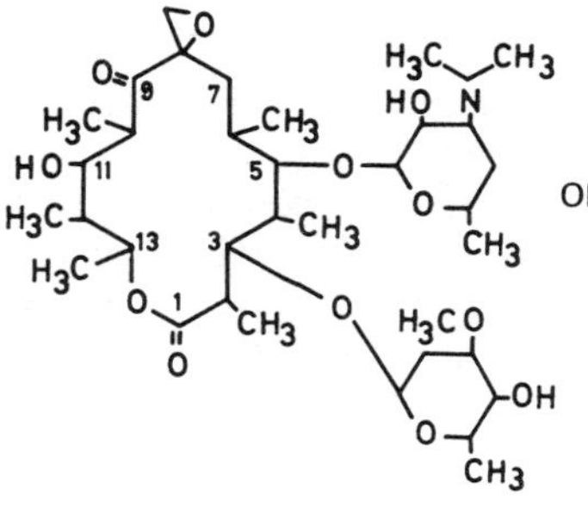

Oleandomycin [12]

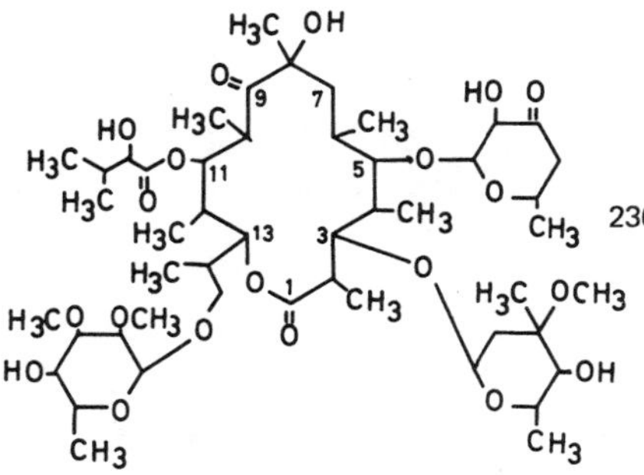

23672 RP [14]

C. 16-Membered Macrolides

Carbomycin A [17], produced by *S. halstedii*, was the first known 16-membered macrolide; carbomycin B [18] was detected in the culture broth of *S. halstedii*. Both antibiotics have since been found in cultures of various other microorganisms. Leucomycin [24] (Hata and Sano, 1953) and spiramycin [33] (Pinnert-Sindico et al., 1955) were discovered in cultures of *S. kitasatoensis* NRRL 2486 and *Streptomyces ambofaciens* (Pinnert-Sindico, 1954), respectively. These antibiotics have been produced industrially and their structures were elucidated by Ōmura and Nakagawa (1975) and Kenhne and Benson (1965), respectively. Tylosin [36] (McGuire et al., 1961; Hamill et al., 1961), first reported in 1959 (Hamill et al., 1959), is very active against mycoplasmas and is produced industrially for animal use. Its structure was elucidated by Morin et al. (1970).

The discovery of tylosin was followed by that of niddamycin [29] (Huber et al., 1962), cirramycin [20] (Koshiyama et al., 1963; Tsukiura et al., 1969), relomycin [31] (Whaley et al., 1963), and angolamycin (shinocomycin) [15] (Nishimura et al., 1965; Kinumaki and Suzuki, 1972).

Garrod (1957) reported on the drug resistance of staphylococci toward erythromycin; the mechanism of the resistance to macrolide was elucidated by mitsuhashi in 1967. The development of a 16-membered macrolide was required and in 1967 josamycin [24] (Osono et al., 1967) was found and subsequently produced industrially [it is identical to leucomycin A$_3$ (Ōmura et al., 1970)]. In 1969 the following new macrolides were discovered: SF 837 (Midecamycin) [26] (Tsuruoka et al., 1971), macrolide complex (Turimycin) [35] (VEB Jenapharm, 1970), YL-704 (Platenomycin) [28] (Okuda and Awata, 1971), K-231 F-1 (Kanda and Abe, 1972), espinomycin [22] (Umezawa et al., 1971), and B-5050 (Maridomycin) [25] (Higashide et al., 1972). Subsequent research on producer strains, structure, and so on, was carried out on compound SF-837 by Niida et al. (1971), Tsuruoka et al. (1971), and Inouye et al. (1971), on macrolide complexes by Forberg et al. (1979) and Fricke et al. (1979), on YL-704 by Furumai et al. (1974) and Kinumaki et al. (1974a,b), and on B-5050 by Muroi et al. (1973) and Ono et al. (1973). Subsequently, B-58941 [38] (Kusaka et al., 1970; Suzuki, 1970), deltamycin [21] (Shimauchi et al., 1978a,b, 1979), staphcoccomycin [34] (Shimi et al., 1979), and A-6888 [37] (Nash et al., 1980) were reported as basic macrolide antibiotics.

The following 16-membered macrolides are neutral: chalcomycin [19] (Forhardt et al., 1962; Woo et al., 1964), neutramycin [28] (Lefemine et al., 1963; Mitscher and Kunstmann, 1969), and aldgamycins E and F [16] (Kunstmann et al., 1964; Achenbach and Karl, 1975a,b).

Rosamicin [32] (Wagman et al., 1972; Reiman and Jaret, 1972) and juvenimicin [23], coproduced with everinomicin (Hatano et al., 1976; Kishi et al., 1976) by the genus *Micromonospora*, are active against both gram-positive and gram-negative bacteria. The discoveries of M-4365 [35] and mycinamicin [27] (A-11725) were reported by Furumai et al. (1977) and Satoi et al. (1980), respectively. A characteristic 16-membered macrolide, Sch23831 [40], was isolated as a by-product from an early-phase culture of a rosamicin producer (Puar et al., 1979).

16 Membered Macrolide

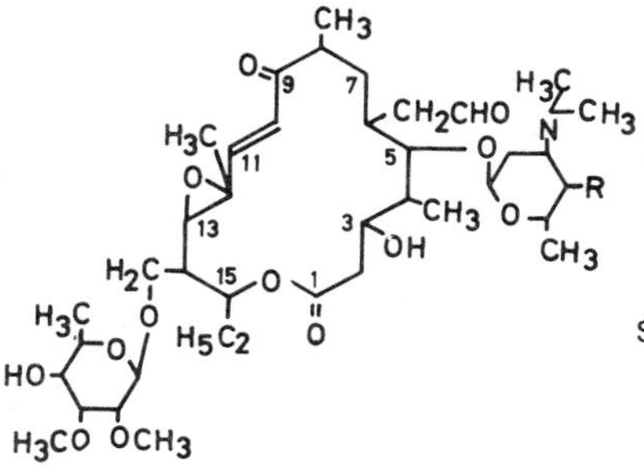

Angolamycin
(Shincomycin)
[15]

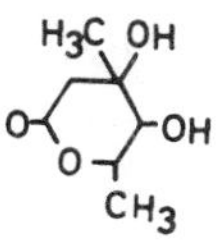

Staphcoccomycin
[34]

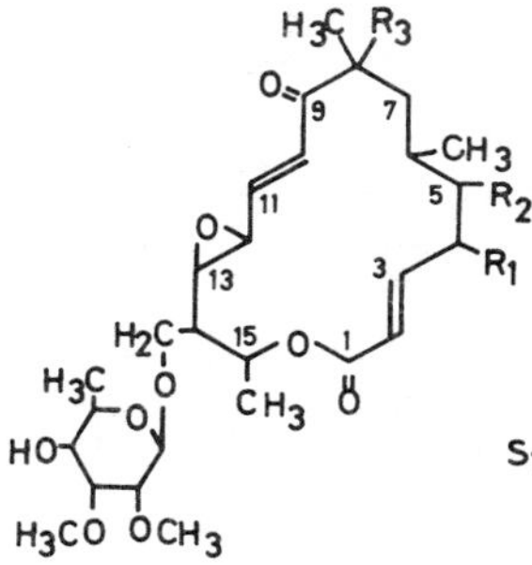

	R_1	R_2	R_3
Aldgamycin E	CH_3	S_1	OH
F [16]	CH_3	S_1	H
Chalcomycin [19]	CH_3	S_2	OH
Neutramycin [28]	H	S_2	H

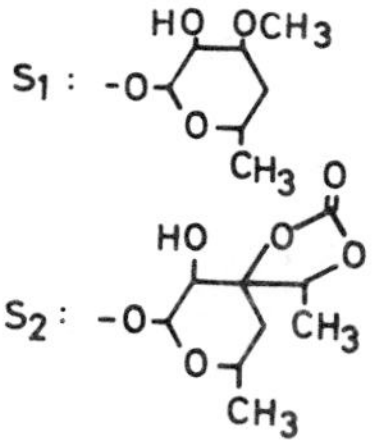

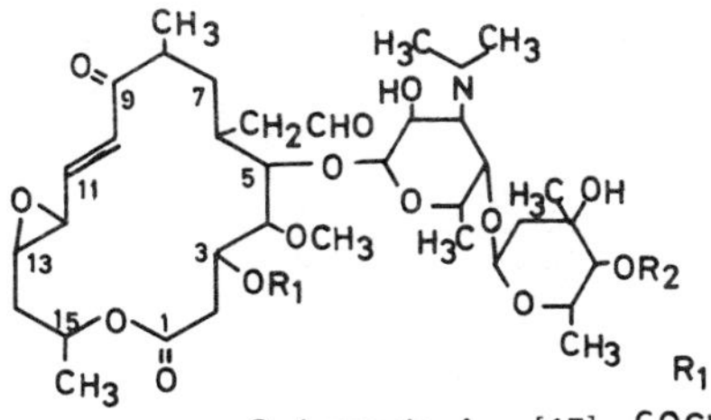

S_1 :

S_2 :

	R_1	R_2
Carbomycin A [17]	COCH_3	COCH_2CH(CH_3)_2
Deltamycin A_1	COCH_3	COCH_3
[21] A_2	COCH_3	COCH_2CH_3
A_3	COCH_3	COCH_2CH_2CH_3

		R₁	R₂
Carbomycin B	[18]	$COCH_3$	$COCH_2CH(CH_3)_2$
Midecamycin A₃ (SF 837)	[26]	COC_2H_5	$COCH_2CH_3$
A₄		COC_2H_5	$COCH_2CH_2CH_3$
Niddamycin	[29]	H	$COCH_2CH(CH_3)_2$
Platenomycin W₁ (YL 704)	[30]	COC_2H_5	$COCH_2CH(CH_3)_2$
W₂		COC_3H_7	$COCH_2CH(CH_3)_2$

S_1 :

S_2 :

		R₁	R₂	R₃
Cirramycin A₁	[20]	CHO	CH_3	OH
Juvenimicin A₂	[23]	H	CH_3	H
A₄		CH_2OH	CH_3	H
Rosamicin	[32]	CHO	CH_3	H
A 6888 C	[37]	CHO	H	S_1
X		CH_2OH	H	S_1
B 58941	[38]	CHO	CH_3	S_2
M 4365 A₁	[39]	CH_3	CH_3	H

		R₁	R₂
Juvenimicin B₁	[23]	CH_2OH	H
B₃		CH_2OH	OH
M 4365 G₁	[39]	CH_3	H
G₂		CHO	H

		R$_1$	R$_2$
Espinomycin A$_2$	[22]	COC$_2$H$_5$	COCH(CH$_3$)$_2$
Leucomycin A$_1$		H	COCH$_2$CH(CH$_3$)$_2$
(Josamycin) A$_3$		COCH$_3$	COCH$_2$CH(CH$_3$)$_2$
A$_4$		COCH$_3$	COC$_3$H$_7$
A$_5$		H	COC$_3$H$_7$
A$_6$	[24]	COCH$_3$	COC$_2$H$_5$
A$_7$		H	COC$_2$H$_5$
A$_8$		COCH$_3$	COCH$_3$
A$_9$		H	COCH$_3$
U		COCH$_3$	H
V		H	H
Midecamycin A$_1$	[26]	COC$_2$H$_5$	COC$_2$H$_5$
(SF-837) A$_2$		COC$_2$H$_5$	COC$_3$H$_7$
Platenomycin A$_1$		COC$_2$H$_5$	COCH$_2$CH(CH$_3$)$_2$
(YL-704) C$_2$	[30]	COC$_2$H$_5$	COCH$_3$
Turimycin Type I	[35]	COCH$_3$ or COC$_2$H$_5$	COCH$_3$,COC$_2$H$_5$,COCH(CH$_3$)$_2$ or COCH$_2$CH(CH$_3$)$_2$

		R$_1$	R$_2$
Maridomycin	I	COC$_2$H$_5$	COCH$_2$CH(CH$_3$)$_2$
[25]	II	COCH$_3$	COCH$_2$CH(CH$_3$)$_2$
	III	COC$_2$H$_5$	COC$_2$H$_5$
	IV	COCH$_3$	COC$_2$H$_5$
	V	COC$_2$H$_5$	COCH$_3$
	VI	COCH$_3$	COCH$_3$
	VII	COC$_2$H$_5$	COC$_3$H$_7$

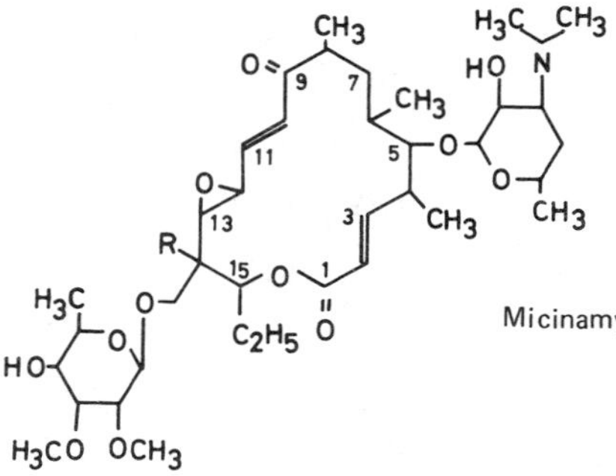

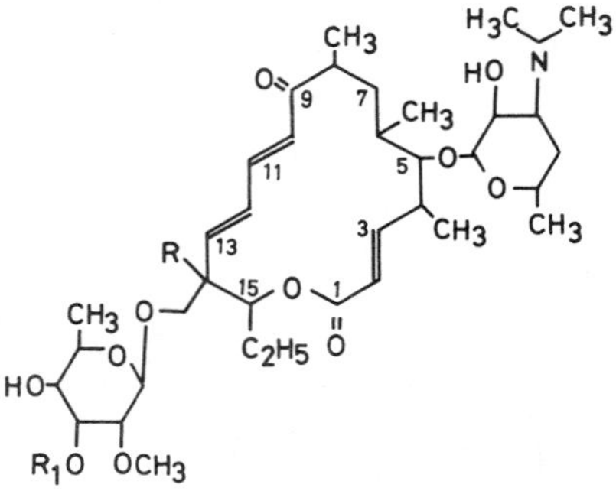

Micinamycin [27]

	R	R$_1$
I	H	
II	OH	
III	H	H
IV	H	CH$_3$
V	OH	CH$_3$

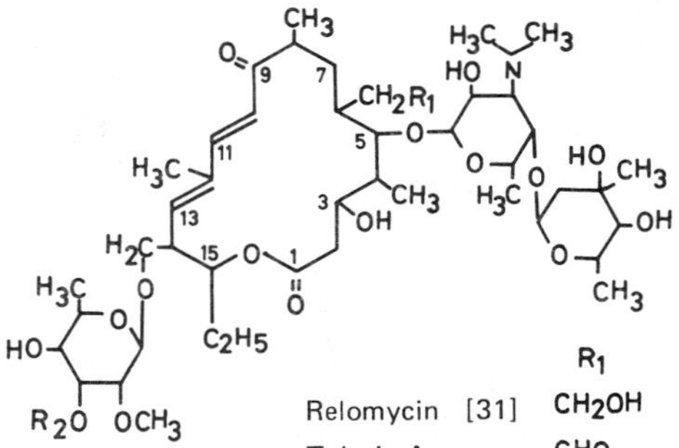

	R$_1$	R$_2$
Relomycin [31]	CH$_2$OH	CH$_3$
Tylosin A [36]	CHO	CH$_3$
C	CHO	H

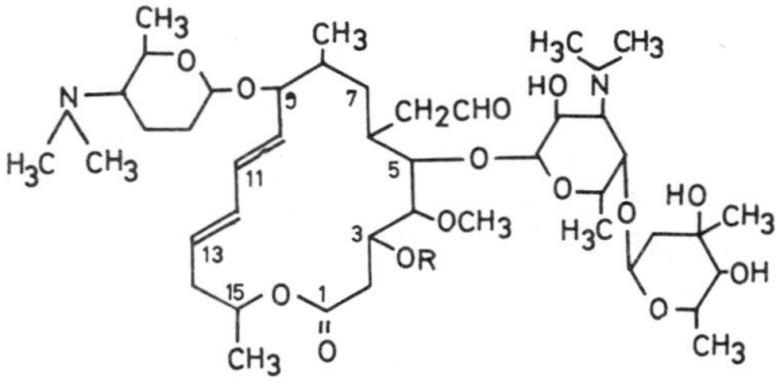

Spiramycin

	R
I	H
II [33]	COCH$_3$
III	COCH$_2$CH$_3$

Sch 23831 [40]

D. 17-Membered Macrolides

Lankacidin [42], produced by *S. violaceoniger*, was the first antibiotic (Gäu-mann et al., 1960) of this group to be described. Following this, bundlin B, produced by *Streptomyces griseofuscus* together with coproducts lankaci-din and moldcidins A and B (pentaene antibiotic), was discovered by Sakamoto et al. (1962) and the chemical structure of bundlin B* [41] from *Streptomyces* sp. 6642-GC$_1$ was elucidated by Uramoto et al. (1969). T-2636 D, E, and F [43], produced by *S. rochei* var. *volubilis*, were formed together with lanka-cidin, T-2636 A (Bundlin B*) and T-2636 B (a new neutral macrolide), and a pentaene antibiotic (Harada et al., 1969; Higashide et al., 1971). *Strepto-myces rochei* var. *volubilis* also produces a characteristic esterase which re-sults in specific acylation or deacylation at the C14 position of T-2636 C (lan-kacidin) (Fugono et al., 1970; Higashide et al., 1971). Although these 17-membered macrolides are not glycosides, they are active against gram-positive bacteria and mycoplasmas. In addition, the primary mode of action of T-2636 C is the inhibition of bacterial protein synthesis (Tsuchiya et al., 1971) and hence resembles a typical macrolide. On the other hand, T-2636 C displays interesting antitumor activity; Ootsu and Matsumoto (1973) reported that T-2636 C (lankacidin), T-2636 C8-propionate, and C14-propionate showed an inhibitory effect on the growth of leukemia L-1210, lymphosarcoma 6C3HED/OG, and melanoma B-16 in mice.

		R$_1$	R$_2$
Bundlin B	[41]	COCH$_3$	O
Lankacidin	[42]	H	O
T-2636 D	[43]	COCH$_3$	H, OH
T-2636 F		H	H, OH

III. SCREENING PROGRAMS AND STRAINS

A. Screening Programs

If a screening program is set up to yield a new macrolide, it must usually first select a new producing microorganism. Antibiotic-producing microorganisms can be obtained from institutions such as the American Type Culture Collection, Rockville; the Northern Regional Research Laboratories, Peoria; the Centraal Bureau voor Schimmelcultures, Baarn, The Netherlands; and the Institute for Fermentation, Osaka, Japan. One can also isolate strains from a natural source such as soil, plants, animals, water, and so on.

Microorganisms from culture collections usually produce low levels of antibiotic, and since they are well known, a really novel antibiotic is unlikely to be found. Generally, stock cultures are an unfavorable source of new antibiotic producers. On the other hand, with the enrichment culture or crowded plate technique (Stokes and Woodward, 1942), the isolation of microorganisms having antibiotic activity from a natural source has been very successful. More than 4000 antibiotics have been discovered and microbiology, medicine, and agriculture as well as other sciences have benefited from it.

There are two methods of isolating novel producer strains. First, a microorganism with peculiar morphology or metabolism is isolated and then its products are further investigated. To achieve his goal, the investigator searches for unique habitats and employs atypical substrates in the medium with the hope of isolating a strange microorganism. For example, Higashide et al. (1977) isolated a new *Nocardia* species, the producer of a new antitumor antibiotic ansamitocin, from a plant, and Okami et al. (1976) isolated a new antibiotic (aplasmomycin) producer from a soil sample from the sea bottom. Kitano et al. (1976a) isolated a new antibiotic producer by using n-paraffin, ethanol, or acetate as the sole carbon source. Applying unfavorable conditions (i.e., pH, temperature) for the growth of the microorganism may prove very useful. In this respect, Imada et al. (1981) were able to isolate a new β-lactam antibiotic (sulfazecin) producer on an acidified (pH 4.5) agar plate.

The second method involves the isolation of microorganisms with desired physiological characteristics, based on the fact that an antibiotic producer is generally resistant to its own antibiotic type. Using this approach, Hatano et al., (1976) successfully isolated the juvenimicin [22] producer. Recently Bibikova et al. (1980) reported on the drug resistance of microorganisms isolated from soil samples. They investigated the sensitivity of 36 strains of *Micromonospora* sp. to various antibiotics. All strains showed high sensitivity to penicillin, ristomycin, tetracycline, rifampicin, streptomycin, and olivomycin, and two strains were resistant to erythromycin and lincomycin. The products of two strains showed cross-resistance to erythromycin and lincomycin and, according to chromatographic behavior, resembled macrolide antibiotics. As shown in Table 2, the erythromycin producer displayed a higher "indicator" enzyme activity than the tetracycline and the polyene producer (Raczynska-Bojanowska et al., 1973). These results suggest that it might be possible to select a macrolide producer by looking for "indicator" enzyme activities.

When a product is investigated, it is necessary to be aware of the presence of coproducts. *Streptomyces olivochromogenes* 69895 produced oleandomycin and chromomycin simultaneously (Higashide et al., 1965), and *Micromonospora chalcea* var. *izumensis* formed juvenimicin together with everinomicin (Hatano et al., 1976).

A test organism sensitive to the antibiotic can be used for screening with the "overlayer technique." Kitano et al. (1976b) reported that screening

Table 2 Propionate and Acetate Kinase Activity and Their Ratios in Antibiotic-Producing *Streptomyces*

| Species | Activity | | Propionate–acetate activity ratio |
	Propionate	Acetate	
S. erythreus (erythromycin)	0.558	0.041	13:1
S. aureofaciens (tetracycline)	0.375	0.468	0.8:1
S. noursei var. *polifungin* (polyene)	0.381	0.328	1.1:1

Source: Raczynska-Bojanowska et al. (1973).

for a new β-lactam antibiotic was successful because of the use of a hypersensitive mutant of *Pseudomonas aeruginosa*. It should be possible to use a strain sensitive to a macrolide to obtain macrolide producers. Frequently an active strain screened on agar plates turns inactive in liquid medium. Shomura et al. (1979) showed that this caused by fragmentation of the microorganism in liquid medium, and a diluted medium or a nonfragmented mutant leads to successful antibiotic production by the filamentous mycelia.

B. Strain Maintenance

Preservation of producer microorganisms is of course important for screening programs and antibiotics production. Generally, it is well known that the

Table 3 Productivity of a Leucomycin Producer after Long-Term Slant Culture Preservation

Strain	After 1 month (A)	Productivity after 3 months (µg/ml) (B)	Ratio B/A × 100
A	1390	690	49.6
B	1000	940	94.0
C	1325	740	55.8
D	1400	570	40.7
E	1215	955	78.6
F	1435	100	7.0
G	1350	265	19.6
H	1295	120	9.3
I	1345	0	0
J	1435	100	7.0
Average	1319	492.8	37.4

Source: Aizawa et al. (1975a).

Table 4 Life-Span of Lyophilized Ampule Cultures of a Leucomycin Producer

Strain	Preservation time (months)	Control	Productivity preservation (µg/ml)	Percentage
B	20	940	1200	127.7
C	22	920	1100	119.6
D	23	920	1150	125
E	24	850	550	64.7
F	25	780	770	98.7
G	25	840	880	104.8
H	26	1020	1150	112.7
I	27	1010	1000	99.0
Average	24	910	975	107.1

Source: Aizawa et al. (1975a).

productivity of an antibiotic producer decreases by repeated culture transfer. Maintaining an antibiotic producer on an agar slant for a long period may cause a reduction in productivity. Aizawa et al. (1975a) reported that when slant cultures of a leucomycin producer were preserved for 3 months, the productivity fell to an average of 37% (range, 0-94%) of the original (Table 3). Recently the preservation of actinomycetes was reviewed in detail by Seino (1977). Usually the strain is preserved by deep freezing in a freezer or in liquid nitrogen, by lyophilization, and L-drying (drying from liquid state) (Iijima and Sakane, 1973). Stock cultures prepared by these methods remain viable, depending on the composition of the sporulation medium used, the temperature and incubation time of the culture, spore concentration, and addition or protective agents. In this respect, Aizawa et al. (1975a) reported on the preservation of strains of lyophilized leucomycin producers. As shown in Table 4, microorganisms preserved for 27 months showed excellent productivity and the average of eight strains was somewhat higher than that of the control. Such improvement of productivity has also been observed with other antibiotic producers.

IV. STRAIN IMPROVEMENT

One of the most important requirements in industrial microbiology is the need for high-production microorganisms. Recently Suzuki and Nakao (1979) and Kikuchi (1980) published reviews on strain improvement for antibiotic production. Strain improvement of macrolide antibiotic producers will be discussed here in more detail.

Generally, selection in the presence of an antibiotic after random mutation with various mutagens may be effective in obtaining high-titer producing strains; a high-titer strain is generally more resistant than a low-titer one (Aizawa, 1975b) (Table 5). In the early stages of strain improvement research for leucomycin production, such combined mutagen antibiotic treatment was applied with success. A high-titer strain, M227 (250 µg/ml), was obtained after five generations of selection (Sugawara and Takeda, 1955). The sporulation ability of the bicyclomycin (Miyoshi et al., 1980) and spiramycin producers (Ōmura et al., 1979b) might be related to antibiotic productivity; plasmids might be responsible for this ability. An interrelation between sporula-

Table 5   Interrelation Between Productivity and Resistance[a]

Strain	Productivity (μg/ml)	Resistance to leucomycin (μg/ml)
1-163	70	1,500
68-61-4	2,540	6,000
68-67-1	5,100	6,000
68-73-1	7,500	8,500
70-9-3	11,000	>12,000

[a]Incubation was for 5 days at 30°C on Waksman agar.
Source: Aizawa et al. (1975b).

tion and antibiotic production is suggested in the report by Aizawa et al. (1975b) on leucomycin production. As shown in Figure 1, no improvement in production was obtained in strains with poor aerial mycelium after 50 generations, whereas a high leucomycin production strain [71-15-1] (14.2 mg/ml) was obtained from a rich sporulation strain in 35 generations.

A biosynthetic "indicator" enzyme might also be used for strain improvement. With respect to erythromycin production, a higher-titer producing organism displayed higher propionate kinase activity and a lower K_m value for propionate than did a lower-titer organism (Tables 2 and 6). Revertants from a nonproducing mutant, nutritional mutants (auxotrophs) (Dulaney, 1955), phage-resistant mutants (Sensi and Thieman, 1965), protoplast fusion (Hopwood et al., 1977), transformation (Bibb et al., 1978), and transduction (Stuttard, 1979) have also been used for strain improvement.

Maridomycin [25] produced by *S. hygroscopicus* B-5050 contains seven components that have different acyl groups at the C3 and C4" positions (Ono et al., 1973; Muroi et al., 1973; Uchida et al., 1975a). Uchida et al. (1975b) were successful in obtaining a strain improved toward selective production of maridomycin III. The effect of the addition of various organic acids, fatty acids, alcohols, and amino acids to the medium on maridomycin III production was investigated. The amino acids (homoserine, threonine, methionine, and isoleucine) and fatty acids (α-aminobutyric and α-ketobutyric acids) from which propionyl coenzyme A (CoA) can be synthesized increased the amount of maridomycin III predominantly without affecting the final production level. Maridomycin III reached a level twice that of the control (Table 7). Based on tracer experiments with 4" depropionyl maridomycin III and leucomycin A_7, it was confirmed that those amino acids and fatty acids were the precurs-

Table 6   Kinase Activity with Propionate in *S. erythreus* Mutants

Strain	Erythromycin yield (%)	Activity	Propionate–acetate activity ratio	K_m for propionate
I	100	0.558	13:1	5×10^{-4}
II	30	0.231	32:1	—
III	8	0.190	34:1	5×10^{-3}

Source: Raczynska-Bojanowska et al. (1973).

ors of the propionyl group connected with the C3 and C4" position of maridomycin (Miyagawa et al., 1979a). As a consequence, strain improvement for
the predominant production of maridomycin III was tried. The effect of amino
acids on the growth of *S. hygroscopicus* B-5050 showed that this strain was
very sensitive to valine and to a lesser degree to serine, homoserine, α-aminobutyric acid, and leucine. Growth suppression by valine was reversed by
the addition of isoleucine, α-ketobutyric acid, or α-aminobutyric acid. These
results suggest that it might be possible to search for a valine-resistant strain
that would accumulate propionyl-CoA and produce predominantly maridomycin
III; meanwhile, valine-resistant strains (V and AV) that displayed this ability
were found (Table 8). As shown in Figure 2, the production of components
other than maridomycin III was markedly suppressed in strain AV; as a result,
a simpler purification procedure for the isolation of maridomycin III was possible and the yield was increased.

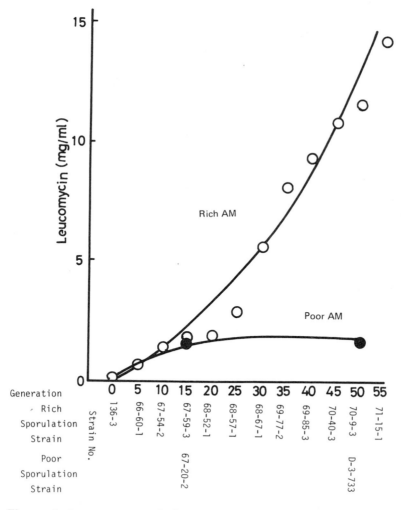

Figure 1 Improvement of rich and poor sporulation strains. (From Aizawa
et al., 1975b)

Leucomycin [24] is composed of 14 components; when the fermentation of *S. kitasatoensis* B-896 is carried out in complex medium, a titer of 3000 µg/ml is reached and the antibiotic mixture contained 10% of A_1 and A_3, 60% of A_4 and A_5, and 30% of the other components. An improved strain which produced predominantly leucomycins A_1 and A_3 has been selected. Strain B-896 was sensitive to 4-azaleucine and resistant to α-aminobutyric acid and valine. Growth inhibition by 4-azaleucine could be reversed by L-leucine, but not by valine, isoleucine, threonine, or methionine. Resistant mutants (resistant to 3 mg/ml of 4-azaleucine) were further improved (Vezina et al., 1979). Selection of a trifluoroleucine-resistant mutant resulted in the predominant accumulation of maridomycin I (B-5050 A) (Suzuki et al., 1975).

These few results demonstrate how the control mechanism of primary metabolism (amino acid metabolism) can affect secondary metabolism (antibiotic production).

Table 7 Effect of Various Compounds on Maridomycin (MDM) III Fermentation[a]

Compound added	Total maridomycin (µg/ml)	Maridomycin III (µg/ml)	MDM III/MDM (%)
Acetate	4300	1630	38
Propionate	4250	1830	43
n-Butyrate	4250	1910	45
n-Valerate	4000	2120	53
n-Caproate	3900	1560	40
Ethyl alcohol	4350	1740	40
n-propyl alcohol	4250	1620	38
n-Butyl alcohol	4150	1580	38
n-Amyl alcohol	4000	1600	40
Propionamide	4350	1960	45
Succinate	4600	2070	45
α-Ketoglutarate	4280	1630	38
Citrate	4300	1720	40
Glutamate	4450	1780	40
Glycine	4380	1660	38
Arginine	4450	1780	40
Aspartate	4300	1940	45
Lysine	4300	1720	40
Homoserine	4000	2800	70
Threonine	4300	3010	70
Methionine	4600	3220	70
Isoleucine	4450	3340	75
α-Amino-n-butyrate	3750	3000	80
α-Ketobutyrate	3680	2940	80
Leucine	4400	1320[b]	30
None	4500	1800	40

[a]Each compound was added at a concentration of 0.1% after 24 hr of fermentation, except for amino acids, which were added at a concentration of 0.2% upon inoculation.
[b]An increase of MDMs I and II was observed.
Source: Miyagawa et al. (1979a).

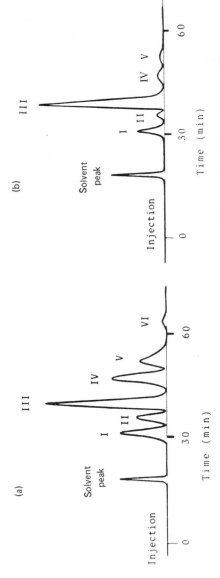

Figure 2 Liquid chromatograms of maridomycin produced by strains HA and AV. (From Miyagawa et al., 1979b).

Table 8 Maridomycin (MDM) and Maridomycin III (MDM III) Productivity of Valine-Resistant Mutants

Strain	Maridomycin (μg/ml)	Maridomycin III (μg/ml)	MDM III/MDM (%)
V-9	4200	3000	71
V-18	4400	3050	69
V-125	4550	3000	66
V-178	4550	3100	68
V-201	4500	3000	68
V-226	4700	3350	71
V-265	4600	3200	70
V-355	4500	3000	67
V-410	4400	3050	69
AV	4700	3525	75
HA	4500	1800	40

Source: Miyagawa et al. (1979b).

V. BIOSYNTHESIS

Many reports (Grisebach, 1978; Inouye, 1973; Ōmura and Takeshima, 1975a–c) on the biosynthesis of macrolide antibiotics have appeared concomitant with the progress in analytical methods using labeled compounds (^{3}H, ^{14}C), 13C-nuclear magnetic resonance (NMR), blocked mutants, and inhibitory agents. Recently molecular microbiology has been applied to the elucidation of macrolide pathways.

As a result of progress in this area, Grisebach (1978) classified the biogenetic origins of macrolides into three groups: (1) macrolides biosynthesized from propionate units (erythromycin), (2) macrolides derived from acetate and propionate units (methymycin), and (3) macrolides containing, in addition to acetate and propionate units, butyrate or its biological equivalent (carbomycin). As mentioned above, many reviews have been published; hence the focus here will be on recent work.

A. 12-Membered Macrolides

Birch et al. (1964) investigated the biosynthesis of methymycin [1] by adding [CH$_3$-^{14}C]methionine, [1-^{14}C] and [2-^{14}C]propionic acid, [2-^{14}C]pyruvic acid, [1-^{14}C]formic acid, [1-^{14}C]acetic acid, diethyl[2-^{14}C]malonate, and [1-^{14}C]formic acid to a growing culture of S. venezuelae. As shown in Table 9, most of [CH$_3$-^{14}C]methionine was incorporated into the molecule of desosamine and the N-methyl group was derived from methionine while the C-methyl group was not. [1-^{14}C]Propionic acid and [2-^{14}C]propionic acid were incorporated exclusively in the aglycone. Incorporation of 21.77% pyruvic acid into the aglycone suggested that pyruvic acid was transformed to propionyl or methylmalonyl-CoA. From these results it appears that the aglycone is synthesized by the polyketide route from five propionic acid units and one acetic acid unit (Fig. 3).

Table 9 Percentage Distribution of Radioactivity in Methymycin Preparations

Compound assayed	Percentage of the relative molar activity present according to tracer used				
	[Me-^{14}C] Methionine	[1-^{14}C] Propionic acid	[2-^{14}C] Propionic acid	[2-^{14}C] Pyruvic acid	[1-^{14}C] Formic acid
Methymycin	100	100	100	100	100
Desosamine hydrochloride	87.6	—	0.15	0	1.94
Lactonic acic	1.66	62.0	58.7	—	—
Propionaldehyde 2,4-dinitrophenylhydrazone	0.15	21.7	19.9	17.9	—
Kuhn-Roth acetic acid		a	b	b	b
Methyl-carbon	—	0.48	2.17	7.72	0.8
Carboxyl-carbon	—	0.18	20.5	8.25	1.4

[a]Acetic acid obtained by oxidation of lactonic acid.
[b]Acetic acid obtained by oxidation of methynolide.
Source: Birch et al. (1964).

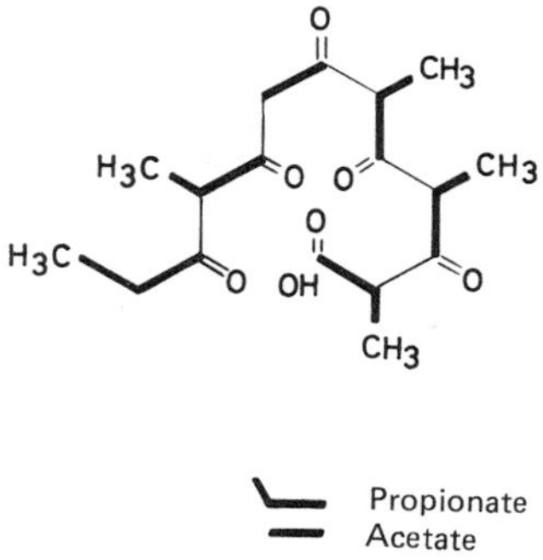

Figure 3 Biosynthesis of methymycin.

B. 14-Membered Macrolides

In the erythromycin fermentation, erythromycin A is usually produced as the main product, with erythromycins B, C, and D as minor components. Degradation of the labeled macrolide showed that the lactone ring of erythromycin A is derived from seven propionate units (Kaneda et al., 1962). Friedman et al. (1964) proved that only one propionate unit was incorporated directly into the ring; the other six propionate units were incorporated via methyl malonate. Biosynthesis of erythromycin was also studied with blocked mutants (Queener et al., 1978). Blocked mutant 2NU 153 (blocked between propionate and 6-deoxyerythronolide B) was able to convert exogenously fed 6-deoxy-erythronolide B, as well as erythronolide B, 3-α-L-mycalosylerythronolide B, erythromycin C, and erythomycin D, to erythromycin A. With these compounds added to the culture, complete disappearance of erythronolide B occurred after 24 hr, and that of 3-α-L-erythonolide B after 96 hr (Martin and Rosenbrook, 1968). Majer et al. (1977) isolated erythromycin D [6] from mother liquors after crystallization of erythromycin A produced by S. ery-threus. A S-adenosyl-L-methionine-dependent transmethylase of S. ery-threus CA 340 (Abbott) converted erythromycin D to erythromycins A and B (Cocran and Majer, 1975; Majer et al., 1977). The transmethylase showed a very high degree of substrate specificity, and 3-O-methyloxyerythrono-lide, mycarose, anomeric methyl-L-mycarosides, tylosin, and nidamycin were not able to serve as substrates (Corcoran, 1975). Based on these results, the biosynthetic pathway of erythromycin shown in Figure 4 was proposed (Martin and Goldstein, 1970; Martin et al., 1975). Erythronolide B was isolated from a culture of M. inositola XK-41 that produced megalomicins A, B, C_1, and C_2 and XK-41 B_2 (Nara et al., 1973), and this suggested that the megalomicin group of antibiotics might be synthesized via erythronolide B. O-Demethyloleandomycin was found in a culture of S. oleandros by Celmer (1971); in the biosynthetic pathway of oleandomycin, the step O-de-methyloleandomycin → oleandomycin may be similar to the erythromycin C → erythromycin A step in erythromycin biosynthesis. Erythronolide A oxime incubated with S. antibioticus ATCC 11891 was converted to a new metabolite, 3-O-oleandrosyl-5-desosaminyl erythronolide A oxime (Fig. 5) (LeMahieu et al., 1976). The addition of sodium arsenite or sodium acetate to a culture of S. venezuelae MCRL-0376 caused inhibition of antibiotic production and simultaneous accumulation of narbonolide (Maezawa et al., 1974a,b). Hori et al. (1970, 1971) reported that when [^{2}H]narbonolide was added to washed

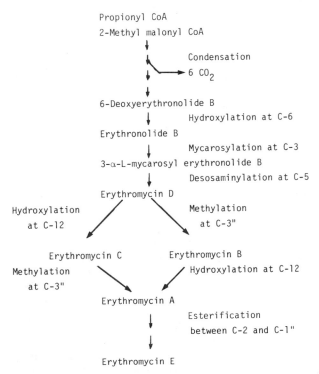

Figure 4 Biosynthetic pathway of erythromycin. (From Martin et al., 1975).

mycelium of *S. narbonensis* ISP 5016, it was converted to [^{2}H]narbomycin. The latter compound was converted to [^{2}H]picromycin when added to washed mycelium of a picromycin-producing strain *Streptomyces* sp. MCRL 0405. When *Streptomyces flavochromogenes*, a picromycin-producing strain, was incubated with [1-^{13}C]acetate and [1-^{13}C]propionate for 26–36 hr, the ^{13}C-NMR spectrum of the picromycin isolated from the culture showed that the incorporation pattern of these building units into the aglycone of picromycin was confirmed, as shown in Figure 6 (Ōmura et al., 1976). From these results, a biosynthetic pathway for narbomycin and picromycin was proposed by Ōmura et al. (1976) and Hori et al. (1971) (Fig. 7).

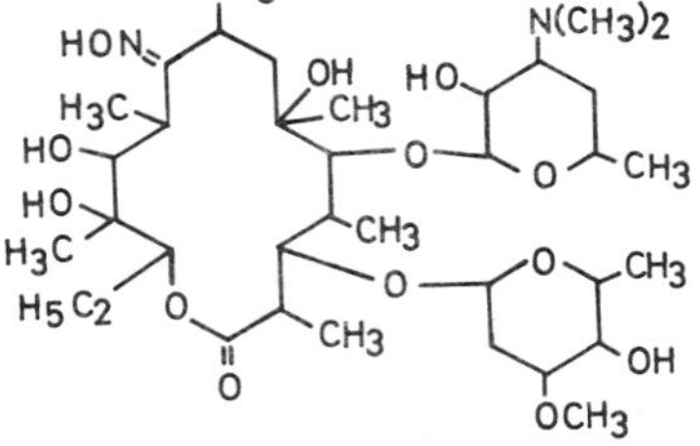

Figure 5 Structure of 3-O-oleandrosyl-5-O-desosaminyl erythronolide A oxime. (From LeMahieu et al., 1976).

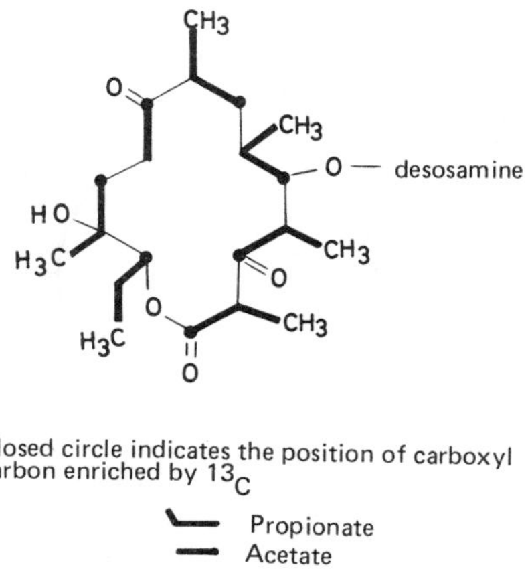

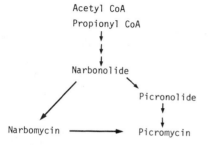

Closed circle indicates the position of carboxyl carbon enriched by ^{13}C

⌐ Propionate
— Acetate

Figure 6 Biosynthesis of picromycin. (From Ōmura et al., 1976.)

C. 16-Membered Macrolides

According to Srinivasan and Srinivasan (1967), the aglycone carbons of mag-namycin B are derived from nine acetate, one propionate, and one methionine units (Fig. 8a). The biosynthesis of leucomycin A$_3$, tylosin, and rosamicin was investigated in a recent systematic ^{13}C-NMR study. *Streptomyces kita-satoensis* 66-14-3, a leucomycin producer, was cultured and various kinds of ^{13}C precursors were added after a 7-hr incubation. Cultivation was con-tinued at 27°C for 48 hr. ^{13}C-NMR spectra suggested that leucomycin A$_3$ was derived from five acetate, one propionate, one butyrate, one methionine, and an unknown two-carbon compound units (Ōmura et al., 1975a) (Fig. 8b). *Streptomyces fradiae* C-373, the tylosin producer, was incubated with ^{13}C-labeled precursors, and the ^{13}C-NMR spectrum of tylosin obtained from the culture showed that the aglycone of tylosin was derived from two acetate, five propionate, and one butyrate units (Ōmura et al., 1975b) (Fig. 8c). *Micro-monospora rosaria*)NRRL 3718) was incubated with various labeled precursors (^{13}C or ^{14}C). The aglycone ring of rosamicin is biosynthesized from one buty-rate, two acetate, and five propionate units. The methyl groups of the di-methylamino group of desosamine are derived from L-methionine (Ganguly et al., 1976) (Fig. 8d). Incorporation of ^{14}C compounds into maridomycin showed that the aglycone might be derived from propionate, acetate, and butyrate;

```
        Acetyl CoA
        Propionyl CoA
              ↓
              ↓
              ↓
        Narbonolide
           ╱      ╲
          ╱        Picronolide
         ╱            ↓
        ╱             ↓
Narbomycin ────────→ Picromycin
```

Figure 7 Biosynthetic pathway of narbomycin and pikromycin. (From Hori et al., 1971.)

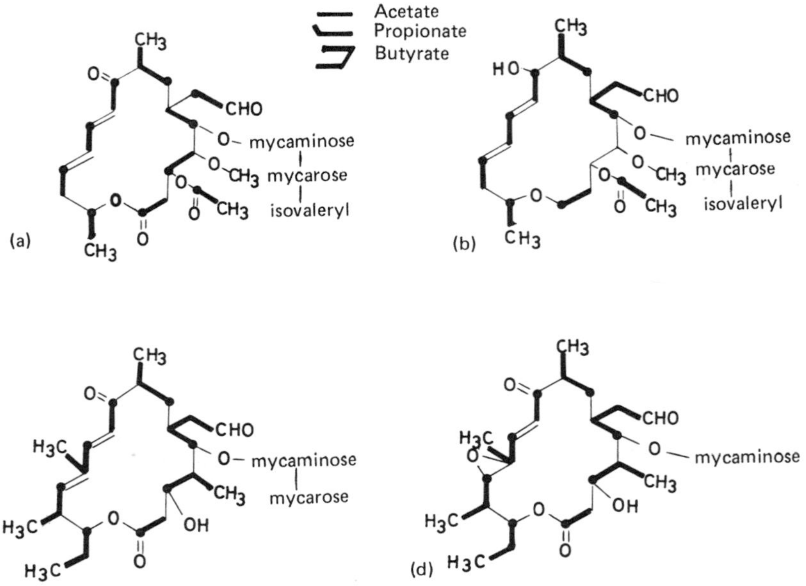

Figure 8 Biosynthesis of (a) magnamycin B (from Srinivasan and Srinivasan, 1967), (b) leucomycin A$_3$ (from Ōmura et al., 1975a), (c) tylosin (from Ōmura et al., 1975b), and (d) rosamicin (from Ganguly et al., 1976).

degradation studies revealed that one methyl group in the aglycone, two methyl groups in mycaminose, and one methyl group in mycarose were derived from L-methionine (Ono et al., 1974). From these results it is reasonable to suggest that carbon atoms at the 5-, 6-, 15-, and 20-positions in magnamycin B may also be derived from butyrate. In the biosynthesis of turimycin L-[CH$_3$-^{14}C]methionine and n-[1-^{14}C]butyrate were exclusively incorporated (Gersch et al., 1977). Recently, Puar et al. (1981) reported that AR-5 antibiotics (mycinamicins) were biosynthesized from three acetate and five propionate units, unlike the biosynthesis of rosamicin or tylosin.

Streptomyces fradiae produced tylosin together with macrocin, desmycosin, lactenocin, and relomycin (Hamill et al., 1961; Hamill and Stark, 1964; Jensen et al., 1964) (Fig. 9). When S. fradiae was incubated with D-[U-^{14}C]glucose, L-[CH$_3$-^{14}C]methionine, and [2-^{14}C] Na-propionate, labeled tylosin, macrocin, and relomycin were isolated from the culture. Labeled lactenocin and desmycosin were prepared chemically from labeled macrocin and tylosin, respectively. The data obtained revealed that macrocin and desmycosin were direct precursors of tylosin, whereas lactenocin was an immediate precursor of both macrocin and desmycosin. A proposed scheme of tylosin biosynthesis is presented in Figure 10 (Seno et al., 1977; Ōmura et al., 1975a; Pape and Brillinger, 1973). The biosynthesis of mycarose was revealed by using a cell-free extract of Streptomyces rimosus, a tylosin producer (Pape and Brillinger, 1973). The pathway of macrocin to tylosin seems to resemble the pathway of erythromycin C to erythromycin A with respect to the O-methylation of sugar. The synthesis of the aglycone of tylosin was inhibited by adding cerulenin to the culture of the tylosin producer S. fradiae C-373. Mycaaminosyl tylonolide, mycaminosyl relonolide, desmycosin (demycarosyl tylosin),

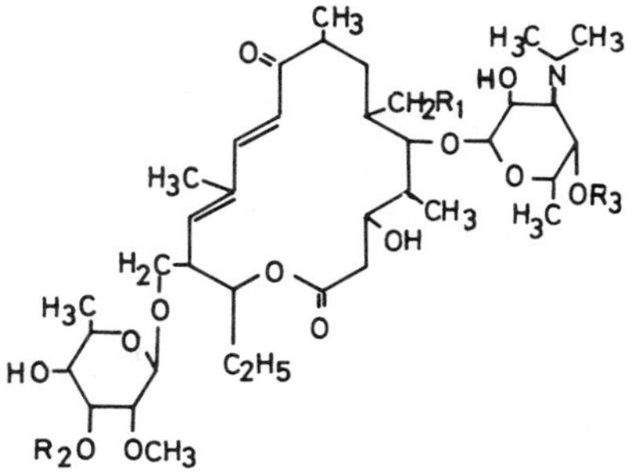

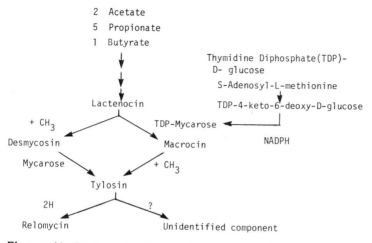

Figure 9 Chemical structure of the tylosin fermentation components.

and demycalosyl relomycin were transformed predominantly to tylosin in the early phase of the culture, and predominantly to relomycin in the later phase. Interconversion between tylosin and relomycin rarely occurred (Ōmura et al., 1978) (Fig. 11).

Furumai and Suzuki (1975a) and Furumai et al. (1973) obtained 24 mutants of a stable platenomycin nonproducer strain from *Streptomyces platensis* subsp. *malvinus* MCRL-0388 (NRRL 3761) and classified them into eight groups (A–G and indefinable one) according to their cosynthesis behavior Delic and Pigac (1969). Furumai and Suzuki (1975b) have isolated four glycosidic compounds, three major and one minor, from a culture of the blocked mutant (N-90). These were identi- fied as 3-O-propionyl-5-O-mycaminosyl platenolide I and II (PPL-I-MC, PPL- II-MC), 9-dehydrodemycarosyl platenomycin (DDM-PLM), and demycarosyl platenomycin (DM-PLM). Two new macrocyclic lactones, platenolide I (PL- I) and II (PL-II) were isolated as the major products from a culture of a blocked mutant, U-92 belonging to group B (Furumai and Suzuki, 1975c). Based on these compounds a scheme for the biosynthesis of platenomycin was proposed (Fig. 12). The 4"-isovaleryl unit of platenomycin A_1 and the 4"-propionyl unit of platenomycin B_1 were derived from L-leucine and L-isoleucine, re- spectively (Furumai and Suzuki, 1975c).

Figure 10 Proposed scheme of tylosin biosynthesis. (From Ōmura et al., 1975a, Pape and Brillinger, 1973, and Seno et al., 1977.)

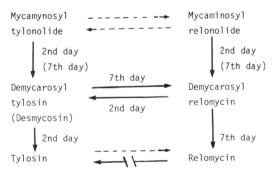

Figure 11 Later stages in the biosynthetic pathway of tylosin and relomycin. (From Ōmura et al., 1976.)

Cerulenin, an inhibitor of fatty acid and polyketide biosynthesis, inhibited biosynthesis of leucomycin without inhibiting growth, biosynthesis of protein and RNA, and incorporation of acetate into the mycelium (Takeshima et al., 1977). The acylation of leucomycin A₁ was tested with a suspension of resting cells in which leucomycin production was inhibited by the addition of cerulenin. The enzyme for OH acylation at the C3 position of leucomycin A₁ was induced by glucose and inhibited by butyrate (Ōmura et al., 1979a; Kitao et al., 1979a). Acylation at the C3 and C4" positions of leucomycin was investigated by feeding leucomycins V, U, A₁, and A₃ to a culture in which leucomycin formation was inhibited by cerulenin. From these results, a pathway of leucomycin analogs was proposed, and it was suggested that acylation occurred predominantly at the C4" rather than at the C3 position (Ōmura et al., 1979a).

Streptomyces hygroscopicus B-5050 produced maridomycins together with leucomycins and carbomycins A and B (Ono et al., 1973). Suzuki et al.

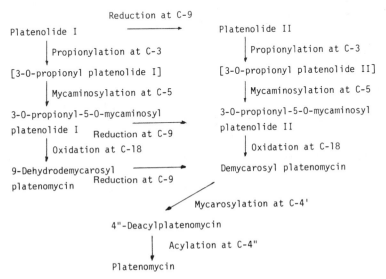

Figure 12 Scheme for platenomycin biosynthetic pathway. (From Furumai and Suzuki, 1975c.)

(1979) derived various blocked mutants from the maridomycin producer and studied the interconversion among carbomycins A and B, leucomycin A_3, and maridomycin II, using intact cells of these mutants. These experiments led to the conclusion that maridomycin II was synthesized as follows:

$\rightarrow$ carbomycin B $\rightleftarrows$ carbomycin A / leucomycin A_3 $\rightarrow$ maridomycin II

On the basis of these results and those of other reports (Ōmura et al., 1975a; Furumai and Suzuki, 1975c), the biosynthetic pathway of the 16-membered macrolides was proposed (Fig. 13). The acetyl group at the C3 position and the propionyl group at the C4" position were confirmed to be derived from acetyl-CoA and propionyl-CoA, respectively (Miyagawa et al., 1979a).

The biosynthesis of spiramycin was investigated using *S. ambofaciens* KA-1028, the spiramycin producer, by feeding spiramycin analogs after addition of cerulenin. A scheme of the biosynthetic pathway of spiramycin is shown in Figure 14 (Ōmura et al., 1979a). The formation of the enzyme for the acy-

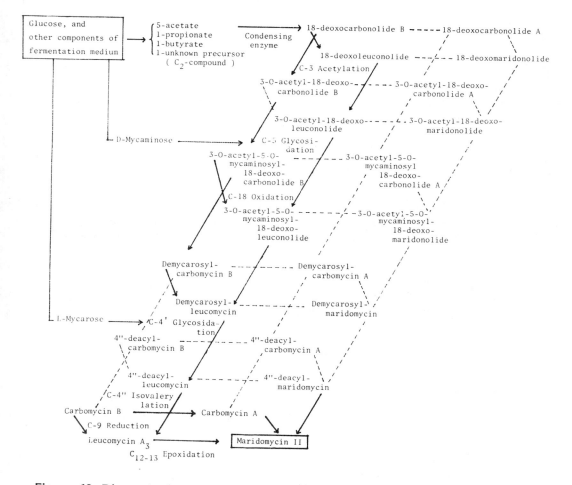

Figure 13 Biosynthetic pathway of 16-membered macrolides. (From Suzuki et al., 1979.)

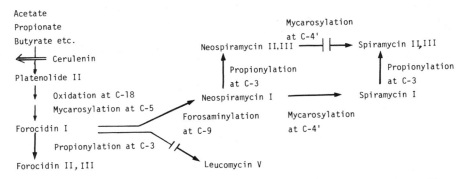

Figure 14 Scheme for biosynthetic pathway of spiramycin. (From Ōmura et al., 1979a.)

lation of the hydroxyl group at the C3 position was induced by glucose and cerulenin and inhibited by butyrate (Kitao et al., 1979a).

D. 17-Membered Macrolides

The lankacidin group (lankacidin [42], bundlin B [41], and T-2636 D, F [43]) comprises 17-membered macrocyclic lacton antibiotics. Hatano et al. (1975) investigated the biosynthesis of this group by incorporating various ^{14}C-labeled compounds into the growing cells of *S. rochei* var. *volubilis*. Significant incorporation (22.6–25.4%) was observed with [^{14}C-CH$_3$]-L-methionine; acetate, propionate, glycine, and pyruvate were also incorporated. It was assumed that two allylic methyl moieties (C5 and C11) of lankacidin were derived from propionate, and that the methyl group at C2 or C7 was derived from methionine.

The ^{13}C-labeled antibiotics were prepared by feeding experiments (Uramoto et al., 1978). It was confirmed that the formation of a linear polyketide chain was initiated by glycine and was followed by the incorporation of eight acetic acid units, with methionine responsible for the four branching methyl groups. Enrichment by propionate was not observed and propionate might not participate in the formation of lankacidin (Fig. 15).

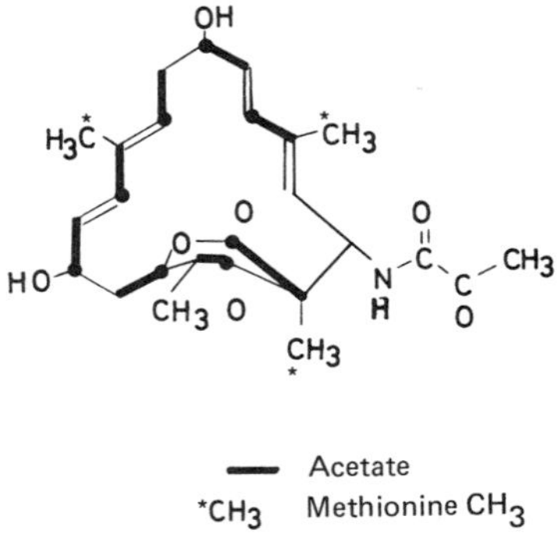

Figure 15 Biosynthesis of lankacidin.

Hopwood (1978) reviewed the plasmid-mediated actinomycete antibiotics in detail. Only two recent reports will be mentioned here. Hayakawa et al. (1979) detected a linear plasmidlike DNA in a lankacidin producer, *Streptomyces* sp. 7434-AN$_4$. Recently Ōmura et al. (1981) established a procedure for the rapid extraction of streptomycete plasmid DNA. They analyzed 21 strains of streptomycetes which produced various 12-, 14-, and 16-membered macrolide antibiotics. Plasmid DNAs were found in the following five strains: *S. ambofaciens* KA-1028 (spiramycin), *Streptomyces bikiniensis* KA-421 (chalcomycin), *Streptomyces cirratus* JTB-3 (cirramycin A$_1$), *S. hygroscopicus* KCC-80439 (tylosin and relomycin), and *Streptomyces* sp. AM-4939 (angolamycin). The molecular weights of covalentry closed circular (CCC) DNAs as calculated by electron microscopic and gel-electrophoretic analysis were 19 $\times$ 10^6–115 $\times$ 10^6.

On the basis of these fascinating findings, it would appear that in the near future genetics and molecular biology will contribute much to the progress of industrial microbiology.

VI. FERMENTATION PROCESS

There are only a few reports on antibiotic fermentation processes which have reached industrial-scale production, contrary to the many reports of low-titer antibiotic production in publications on the discovery and biosynthesis of antibiotics. In this section, fermentation processes that have reached industrial level or which yield a titer of more than 1000 µg/ml of antibiotic will be described.

A. Inoculum Preparation

In industrial fermentation, the inoculum should be prepared very carefully. Furthermore, it yield a high titer and stable fermentation process. Plasmid involvement in actinomycetes antibiotic synthesis has been suggested by Hopwood (1978), and aerial mycelium and pigment formation were found to be associated with spiramycin and bicyclomycin production, respectively (Ōmura et al., 1979a; Miyoshi et al., 1980). Such properties are affected by the composition of the medium and the temperature and duration of incubation. Generally, antibiotic productivity of actinomycetes tends to decrease when a metal ion-deficient medium is used and when the incubation of the inoculum lasts a long time and is carried out at a high temperature. To obtain a stable inoculum, at least 10^8 viable cells per milliliter of suspension are necessary. For sporulation of actinomycetes the following media are generally used: glucose asparagine agar (D-glucose, 10 g; L-asparagine, 0.5 g; K$_2$PHO$_4$, 0.5 g; pH 6.6–7.0; agar, 15–20 g; distilled water, 1000 ml), glycerol asparagine agar (glycerol, 10 g; L-asparagine, 1 g; K$_2$PHO$_4$, 1 g; trace metal solution, 1 ml—composed of FeSO$_4$·7H$_2$O, 0.1 g; MnCl$_2$·4H$_2$O, 0.1 g; ZnSO$_4$·7H$_2$O, 0.1 g; distilled water, 100 ml—pH 7.0–7.4; agar, 15–10 g; distilled water, 1000 ml), yeast extract—malt extract agar (yeast extract, 4 g; malt extract, 10 g; D-glucose, 4 g; pH 7.3; agar, 15–20 g; distilled water, 1000 ml), oatmeal agar (oatmeal, 10 g; distilled water, 1000 ml; pH 7.2), and sporulation agar ATCC No. 5 (D-glucose, 10 g; meat extract, 1 g; yeast extract, 1 g; tryptose, 2 g; FeSO$_4$·7H$_2$O, traces; pH 7.2; distilled water, 1000 ml; agar, 20 g).

Just a few examples of inoculum preparation are given here. Sporulation of *S. erythreus* (Lilly No. C-233) for erythromycin fermentation was carried out on tryptone agar slants and the cells were harvested in sterile water

(Smith et al., 1962). *Streptomyces fradiae* (NRRL 2702), the tylosin producer, was inoculated on agar slants (1% cerelose, 1% phytone, 0.0001% biotin, 0.1% sodium thiosulfate, 2.5% Mear agar), incubated at 28°C for 10 days, and subsequently preserved at 4°C (Seno et al., 1977). *Streptomyces kitasatoensis*, the leucomycin producer, was inoculated on a modified oatmeal agar (6% oatmeal, 1.5% agar, pH 6.8–7.0) and incubated at 30°C for 8–9 days. After abundant sporulation had occurred, the cells were harvested by scraping with a platinum needle into 0.1% Tween 60 aqueous solution (Aizawa et al., 1975a). *Streptomyces hygroscopicus* B-5050 (strain AV), the maridomycin producer, was inoculated on a sporulation medium (1% soluble starch, 0.04% peptone, 0.02% meat extract, 0.02% yeast extract, 0.02%N-Z amine [type A], 2% agar, pH 7.0) and incubated at 28°C for 7 days. The cells were harvested from agar plates and the resultant suspension (5×10^6 viable cells/ml) in sterile water was preserved at 4°C (Miyagawa et al., 1979a).

B. Medium Preparation

The seed culture media should be optimal for rapid growth and display high respiratory and enzyme activity. The best seed culture media for erythromycin, leucomycin, and maridomycin are 0.5% corn-steep solid, 1.5% soybean oil meal, 0.5% dry yeast, 0.5% NaCl, 0.3% $CaCO_3$ (Smith et al., 1962); 2% soluble starch, 0.5% Pharmamedia (from cottonseed meal), 0.5% NaCl, 0.5% meat extract, 0.3% dry yeast, 0.3% $CaCO_3$ (precipitated), at pH 7.0–7.2 (Aizawa et al., 1975a); and 3% glucose, 3% corn-steep liquor, 0.05% $MgSO_4 \cdot 7H_2O$, 0.3% $CaCO_3$ (precipitated), at pH 6.6 (Miyagawa et al., 1979a), respectively.

Attainment of high antibiotic production requires proper composition and concentration of the components of the fermentation medium and optimum pH and viscosity. Inorganic phosphate, nitrogen, and metal ions, as well as organic nitrogen and carbon, and substrate precursors are important for high antibiotic fermentation. Fermentation media for macrolide production are presented in Table 10.

C. Regulation of the Fermentation Process

The following principles are important for regulation of industrial fermentataions: (1) increase of the genetic ability for metabolite production, (2) detection of the primary physiological parameter for the production, (3) detection of the scale-up parameter, (4) design of improved fermenters, and (5) simplification or automation of the operation for each fermentation and purification step. This chapter will mainly describe the second principle.

Antibiotic production occurs in general after the growth reaches the stationary phase or has passed an optimal level. The growth and production phases are called trophophase and idiophase, respectively (Bu'Lock et al., 1975). It is important to adapt growth of both phases to antibiotic production, because production depends primarily on growth. In the trophophase, growth should be controlled to attain a suboptimal level as soon as possible; the mycelium should have high respiratory and enzyme activity (e.g., phosphatase, protease, amylase). In the idiophase, a high level of antibiotic productivity depends on the maintenance of suboptimal growth, the concentration and consumption rate of inorganic phosphate, ammonium nitrogen, and metal ions, as well as of carbon and nitrogen sources, and a high concentration of dissolved oxygen. Optimum pH, temperature, and viscosity should be controlled to attain at a high titer of production.

Table 10 Fermentation Media for Macrolide Fermentation[a]

Erythromycin[b]		Tylosin[c]		Leucomycin[d]		Maridomycin[e]	
Ingredient	Percentage	Ingredient	Percentage	Ingredient	Percentage	Ingredient	Percentage
Sucrose	5	NaCl	0.2	Soluble starch	4.3	Glucose	12
Corn-steep liquor	0.5	$MgSO_4 \cdot 7H_2O$	0.05	Shoyu oil	2.5	Proflo	2
Soybean oil meal	1.5	$CoCl_2 \cdot 5H_2O$	0.0001	Defatted germ	1.5	Soybean oil	0.5
Yeast	1.0	$ZnSO_4 \cdot 7H_2O$	0.001	Corn-steep liquor	0.9	NH_4Cl	0.5
NaCl	0.5	Ferric ammonium citrate	0.3	Glucose	0.9	KH_2PO_4	0.12
$CaCO_3$ precipitate	0.3	Betain HCl	0.5	Molasses	0.46	$MnSO_4 \cdot 4-6H_2O$	0.05
		L-Sodium glutanate	2.0	NH_4Cl	0.4	$MgSO_4 \cdot 7H_2O$	0.02
		Glucose	3.5	Dry yeast	0.29	$FeSO_4 \cdot 7H_2O$	0.1
		$CaCO_3$	0.3	$CaCO_3$ precipitate	0.13	$ZnSO_4 \cdot 7H_2O$	0.05
		Methyl oleate	0.3	KH_2PO_4	0.09	$CaCO_3$	1.2
		K_2HPO_4	2.3	$MgSO_4 \cdot 7H_2O$	0.044		
				$MnSO_4 \cdot 5H_2O$	0.0026		

[a] pH 7.0–7.2.
[b] Smith et al. (1962); 2000 µg/ml.
[c] Vu-Thron et al. (1980); (also, this volume, Chap. 27); 2200 µg/ml.
[d] Aizawa et al. (1975b); 14,000 µg/ml.
[e] Miyagawa et al. (1979a); 5000 µg/ml.

The concentrations of soluble and insoluble inorganic phosphate in the medium regulates growth during trophophase and idiophase, respectively (Martin, 1977). Many antibiotic-producing organisms are able to grow adequately in the presence of 0.3–500 mM of inorganic phosphate; however, the phosphate range for antibiotic production is generally quite narrow (Martin, 1977). In the maridomycin fermentation, the concentration of inorganic phosphate required for growth of S. hygroscopicus B-5050 was 5–15 mM, while for maridomycin production the concentration required was limited to the narrow range of 5–7 mM.

In oleandomycin biosynthesis by S. antibioticus, orthophosphate inhibited the hexose monophosphate pathway, decreasing the rates of hexose utilization and pyruvate formation (Torbochikina et al., 1964). Since the major pathway of carbohydrate metabolism in the producer appears to be the hexose monophosphate pathway, its inhibition by phosphate could account for the inhibition of oleandomycin synthesis (Torbochikina and Dormidoshina, 1964).

The intracellular level of cyclic adenosine 5'-monophosphate (cAMP) in a high-titer producing mutant of S. hygroscopicus showed two peaks, one at the trophophase and the other at the idiophase, a fact which did not occur in the wild-type strain. The addition of 10 mM phosphate inhibited turimycin production, and the second peak of cAMP disappeared. The addition of 5 mM cAMP to the culture restored turimycin production and caused the reappearance of a protein which may be responsible for turimycin biosynthesis (Gersch et al., 1979; Gersch, 1980).

In the biosynthesis of tylosin by S. rimosus, Thymidine diphospho (TDP)-D-glucose oxidoreductase and transmethylase are required in the pathway from TDP-glucose to TDP-L-mycarose. These two enzymes are only formed during the idiophase, as shown in Figure 16 (Matern et al., 1973). As antibiotic production is controlled by these enzymes, it is suppressed in the trophophase and appears only in the idiophase. Excess inorganic phosphate during the

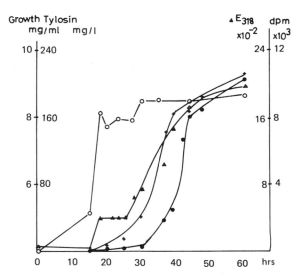

Figure 16 Mycelial growth, concentration of tylosin, and specific activities of TDP–glucose–oxidoreductase and "transmethylase" in *Streptomyces rimosus*. (From Matern et al., 1973.)

idiophase might inhibit this enzyme's formation and subsequent antibiotic production.

Glucose is usually an excellent carbon source for antibiotic production and, as previously described, is used for macrolide production, though it interferes with the biosynthesis of many antibiotics (Martin and Demain, 1980). During studies on fermentation medium improvement, polysaccharides or oligosaccharides such as starch, dextrin, lactose, and fructose are often found to be better than glucose as carbon sources for antibiotic production. These carbon sources are used more slowly than glucose and therefore catabolite regulation of antibiotic biosynthesis does not occur. To prevent such carbon catabolite regulation, gradual feeding of glucose during the total fermentation cycle yields high antibiotic levels.

The effect of nitrogen sources on the regulation of antibiotic biosynthesis has been reported. The addition of soybean meal to erythromycin fermentation medium markedly repressed antibiotic production without regard to the carbon source. The addition of glycine or NH_4Cl also caused repression (Smith et al., 1962). Usually the effect of the nitrogen source depends on its composition, when soybean meal was used as a nitrogen source rather than other nitrogen compounds, magnamycin production by *S. halstedii* increased remarkably (Abou-Zeid et al., 1977).

It is well known that the addition of metal ions is important for the production of antibiotics or secondary metabolites. The production of actinomycin, monensin, neomycin, candicidin, grisein, streptomycin, chloramphenicol, and mitomycin requires $1 \times 10^{-5}-100 \times 10^{-5}$ M manganese or $0.1 \times 10^{-5}-10 \times 10^{-5}$ M zinc (Weinberg, 1970). Addition of magnesium phosphate and related insoluble materials to the fermentation medium of *S. kitasatoensis* KA-429 (a mutant strain of the original leucomycin producer NRRL 2486), *S. ambofaciens* ATCC 23877 (a spiramycin producer) and *S. fradiae* KA427 (a tylosin producer) increased each antibiotic titer from 2 to 100 times over the titer of the control fermentation (Table 11). Since the leucomycin fermentation is suppressed by

Table 11 Stimulation of the Production of the Macrolide Antibiotics Leucomycin, Spiramycin, and Tylosin by Magnesium Phosphate and Related Materials

Antibiotic	Addition Compound	Amount (%)	Antibiotic produced (µg/ml) No addition	Addition
Leucomycin	$Mg_3(PO_4)_2 \cdot 8H_2O$ (=MgP)	1	270	900
	$NH_4MgPO_4 \cdot 6H_2O$	1	85	840
	$Mg_3(PO_4)_2 \cdot 8H_2O$	1	480	2600
	$Mg_3(PO_4)_2 \cdot 8H_2O$	1	<10	960
Spiramycin	$Mg_3(PO_4)_2 \cdot 8H_2O$	3	<10	150
	$Ca_3(PO_4)_2$	1	<10	165
	$NH_4MgPO_4 \cdot 6H_2O$	1	<10	270
	$Ca_3(PO_4)_2$	0.5	200	400
	$Na_3P_2W_{12}O_{43} \cdot 18H_2O$	1	200	340
Tylosin	$NH_4MgPO_4 \cdot 6H_2O$	1	86	340

Source: Ōmura et al. (1980).

a high concentration of ammonium ions in the medium, Ōmura et al. (1980) suggested that a limited supply of ammonium ions caused by magnesium phosphate leads to the high titers of leucomycin as well as those of spiramycin and tylosin. This finding indicates that a high antibiotic production level requires regulation of the concentration and the consumption rate of ammonium ions. Related phenomena were also observed in the maridomycin fermentation. A limited supply of ammonium ions has led to the economical production of many other antibiotics.

A precursor may be supplied to the medium to increase the antibiotic titer. In industrial macrolide antibiotic production, the procedure is rarely used because of the higher costs. The addition of n-propanol increased the titer of erythromycin fermentation (Bošnjak et al., 1978), the addition of isoleucine and methionine increased maridomycin III in maridomycin fermentation (Miyagawa et al., 1979a), and the addition of leucine increased leucomycin A₃ in leucomycin fermentation (Vezina et al., 1979).

The interrelation between antibiotic production and primary metabolism has been previously described (Miyagawa et al., 1979a): The regulation of intracellular adenosine triphosphate (ATP) levels is important for antibiotic production. The level of ATP of *Streptomyces griseus* (candicidin and polyene antibiotic producer) decreased rapidly before antibiotic biosynthesis began; it was concluded that a low level of intracellular ATP during idiophase was necessary to allow antibiotic synthesis. Addition of a high concentration of inorganic phosphate resulted in a rapid increase in the intracellular ATP concentration accompanied by inhibition of antibiotic production (Martin et al., 1978). The relation between the level of the intracellular adenylates (the concentration of ATP, adenosine diphosphate, and adenosine monophosphate) and the biosynthesis of tylosin by *S. fradiae* NRRL 2702 was investigated by Vu-Trong et al. (1980). The intracellular concentration of ATP and adenosine diphosphate reached maximal values during the period of rapid growth (2 days) and dropped rapidly before the onset of tylosin synthesis (Fig. 17). The total sum of adenyl-

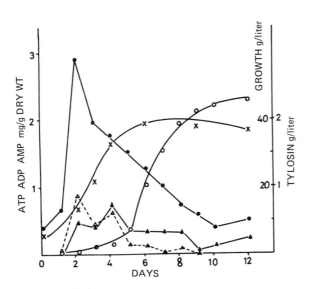

Figure 17 Intracellular concentration of adenylate during the batch fermentation of tylosin. (From Vu-Trong et al., 1980.)

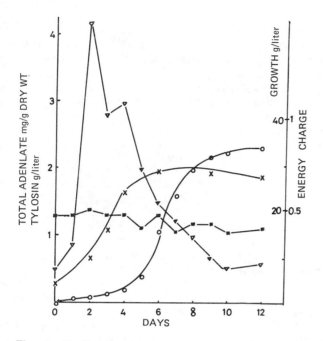

Figure 18 Energy charge and total intracellular adenylate during the fermentation of tylosin. (From Vu-Trong et al., 1980.)

ates decreased rapidly before antibiotic synthesis began, but energy charge was almost constant (0.44–0.5) throughout the fermentation (Fig. 18). The addition of glucose during the idiophase has no effect on the energy charge, but causes a sharp decrease in the level of total adenylate and suppression of tylosin biosynthesis. Propionyl-CoA carboxylase and methyl malonyl-CoA carboxyltransferase in the culture showed highest activities after 5–6 days of fermentation. The activity of the former was about two orders of magnitude higher. These data suggest that the ATP concentration, the level of the adenylate pool, or both might be the intracellular effectors controlling the onset of tylosin biosynthesis. Carboxyltransferase might be the major enzyme and it was found that the α-ketoglutarate pool and succinyl-CoA might be important in tylosin biosynthesis.

Physical factors such as pH, temperature, viscosity, and dissolved oxygen are also important in antibiotic biosynthesis. In the case of macrolide antibiotic production, the optimal initial pH of the fermentation medium is usually weakly acidic to neutral, while the optimum pH during the idiophase differs according to the properties of the antibiotic and its producer. For magnamycin fermentation by S. halstedii NRRL 2331, the optimum initial pH for production was studied in the pH range 3.0–9.05. The titers were 173.8, 302, and 199.5 µg/ml at pH 5.4, 6.2, and 6.64, respectively. The optimum pH for growth was 7.5–8.5 (Abou-Zeid et al., 1977). The optimum pH for maridomycin biosynthesis during the idiophase was 6.0 (Miyagawa et al., 1979a). The maintenance of an optimum pH for production during the idiophase is one of the crucial factors for increasing antibiotic titers. In addition, the optimum temperature for growth is usually different from that for

antibiotic production. In the maridomycin fermentation, the optimum temperature for growth was 28–35°C (Ono et al., 1973), while for maridomycin biosynthesis it was 25°C (Asai et al., 1978a). Fermentation carried out at 28°C during trophophase (0–24 hr) and at 25°C during idiophase (24–130 hr) yielded
a higher titer.

The concentration of dissolved oxygen in the fermentation broth is also
an important factor for obtaining high antibiotic titers. Actinomycetes grow
even at low concentrations of dissolved oxygen, but antibiotic biosynthesis
usually requires a high concentration. The critical oxygen levels for growth
of *S. hygroscopicus* B-5050 and for biosynthesis of maridomycin are 15 and
60 mmHg, respectively. In the erythromycin fermentation, the inoculum quality, which influenced final erythromycin yield, was clearly dependent on the
unit power input consumed during inoculum cultivation (Ettler and Greger,
1978). The relation between oxygen transfer and antibiotic biosynthesis has
also been investigated. An erythromycin fermentation was carried out in a
fermenter agitated at a power input of 1.85 kW per 1000 liters of medium and
aerated at 0.4 vol/vol per minute from 0 to 12 hr and at 0.9 vol/vol per minute
from 12 hr till harvest. Carbon dioxide was added to the incoming air at 0.1
vol/vol per minute or 11% of the inlet air. The supplement of carbon dioxide
did not affect pH and growth, but the erythromycin titer decreased to 40%
of the control. The latter effect might result from feed-back inhibition of
methylmalonic acid formation (Nasch, 1974).

Scale-up for industrial macrolide fermentation may be achieved by taking
into consideration the factors mentioned above. Mechanical differences between
a shaking culture and a tank culture cause physiological problems for growth
as well as for antibiotic biosynthesis. In the maridomycin fermentation, the
amount of inorganic phosphate required in a shaking culture was different
from that needed in a tank culture. It was assumed that this physiological
difference resulted from mechanical damage to the mycelium caused by the
impeller in the tank fermentation. Using 200-ml Erlenmeyer flasks containing
various types of beans (B flask), Asai et al. (1978b) were able to determine
the amount of inorganic phosphate required for tank fermentation. Young
(1979) presented an example of the scale-up/scale-down of an oleandomycin
fermentation from a 50-liter lab fermenter to a 200-m^3 plant fermenter. In
this example, the fermenter environments (glucose, phosphate, nitrogen,
and oxygen concentration/uptake rate curves) were similar. Intermediate
keto acid, carbon dioxide, cell mass (DNA), and the product concentration/
formation rate curve were also similar in both systems.

Recently, the development of computers for fermentation systems has
progressed. At present, at least pH, temperature, airflow rate, and O_2 and
CO_2 concentrations can be determined and controlled; carbon and nitrogen
substrate addition as well as dissolved oxygen can be monitored (Humphrey,
1977).

VII. PRODUCT: ASSAY AND ISOLATION

A. Assay

Usually the amount of an antibiotic is determined by microbioassay. Quantification of macrolide antibiotic is carried out by the diffusion assay method (cup
method, paper disk method, or agar well method) with *Bacillus subtilis* as
a test organism. In industrial production, a chemical assay method is used
for rapid and accurate determination. Tylosin, extracted with chloroform

from a culture broth, was determined spectrophotometrically at 283 nm (Caltrider and Hayes, 1969). The amount of erythromycin can be determined by colorimetry (Kakemi et al., 1956; Korchagin et al., 1961). *Streptomyces hygroscopicus* B5050 produces maridomycins together with leucomycin group and carbomycin group antibiotics as coproducts. Maridomycin has an epoxy group in its structure, and a colorimetric method using picric acid can be selectively applied to detect it. This method determined the amount of maridomycin in an ethyl acetate extract (Uchida et al., 1979).

A gas-liquid chromatographic method was established for the determination of each component of the macrolide antibiotics (Tsuji and Robertson, 1974). The method is inconvenient because it requires at least 24 hr of derivatization. High-performance liquid chromatography (HPLC) finds its uses in the pharmaceutical and fermentation industries. Because of its ability to combine fast chromatographic separation with selective determination of an antibiotic complex from a small sample, HPLC has become one of the most powerful tools for industrial analysis. HPLC of the following macrolide antibiotics has been reported: leucomucin (Ōmura et al., 1973), tylosin (Bhuwapathapum and Gray, 1977), spiramycin (Mourot et al., 1978), erythromycin (Tsuji and Goetz, 1978a–c; Tsuji, 1978; Sugden et al., 1978), and maridomycin (Kondo, 1979). A few examples of chromatograms are shown in Figure 19.

B. Isolation

Almost all macrolides are produced extracellularly in the fermentation broth. Prior to isolation of the antibiotic, the mycelium should be separated from the broth. The procedure of mycelium should be separated from the broth. The procedure of mycelium separation is performed by filter press, centrifuge, or drum filter with a filter aid. In acidic conditions mycelium is separated easily from the broth.

The basic macrolide antibiotics can then be extracted with a water-immiscible organic solvent such as methylisobutyl ketone or ethyl acetate and transferred to acidic water adjusted with HCl, H_3PO_4, acetic acid, or citric acid. For the separation a weak acidic ion exchange resin such as Amberlite IRC-50 is available. After such purification and concentration, the antibiotic is precipitated by adding diethyl ether or n-hexane.

It is further purified by chromatography on silica gel or an aluminum column. Suzuki et al. (1973) established an economical industrial procedure for purification of macrolide antibiotics composed of many analogous components. Resin of the nonionic type (Amberlie XAD-2, Rhom & Haas Co.; Diaion HP-20, Mitsubishi Chem. Ind. Ltd.) gives a high yield and excellent separation. The antibiotic absorbed on the resin is eluted by a mixture composed of an organic solvent and water pH 3–8. As for purification of maridomycin, the antibiotic is obtained from the culture broth as a mixture, dissolved in methanol, diluted with water, and absorbed on a column of Amberlite XAD-2. The antibiotics absorbed are then eluted with 1/50 M acetate buffer (pH 5.0) containing 7% butanol, and maridomycins VI, V, IV, III, II, and I were separated. The isolation and purification of leucomycin, spiramycin, cirramycin, and the narbomycin–pikromycin group were reported by Suzuki et al. (1973).

Generally, the fermentation and purification of industrial products are controlled by strict quality design, as antibiotics produced commercially are prepared in various formulations for injection or oral administration. In addition, certain macrolide antibiotics are transformed into derivatives that display better therapeutical effectiveness (acetylspiramycin, triacetyloleandomycin, propionyljosamycin, and propionylmaridomycin etc.).

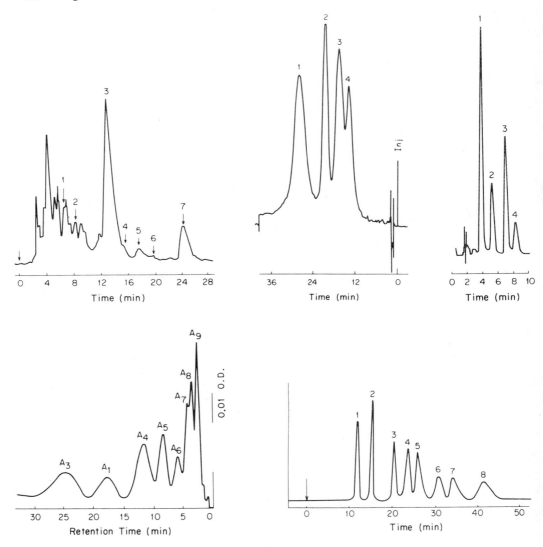

Figure 19 HPLC chromatograms of erythromycin, tylosin, and spiramycin. (From Tsuji and Goetz, 1978a, Vu-Trong and Gray, 1982, and Mourot et al., 1978.) HPLC of leucomycin and maridomycin. (From Ōmura et al., 1973, and Kondo, 1979.)

VIII. MODE OF ACTION

Almost all of the macrolide antibiotics show strong activity, especially against gram-positive bacteria and some gram-negative bacteria, but weak activity against gram-negative rods and acid-fast bacteria. They also exhibit strong activity against mycoplasmas; in this respect, 16-membered macrolides (tylosin, spiramycin, leucomycin, and maridomycin) and the 17-membered macrolides (lankacidin [T-2636 C], bundlin B [T-2636 A]) (Table 12) have stronger activity than 14-membered macrolides (erythromycin and oleandomycin).

Table 12 Antimycoplasma Activity of Macrolide Antibiotics

	Minimum inhibitory concentration (μg/ml)					
	M. hominis type 1-C	M. salivarium C Hup	M. pneumoniae CL, FH	M. gallisepticum KP-13	M. gallisepticum S-6	Ureaplasma
Amaromycin	>100	>100	0.78	25	100	—
Erythromycin A	100	100	0.19	100	1.56	0.0696
Lankamycin	>100	>100	>100	>100	>100	—
Megalomycin A	>100	100	0.39	>100	>100	—
Oleandomycin	100	100	0.19	100	100	0.137
Albocycline	100	>100	<0.19	>100	25	—
Chalcomycin	3.13	<0.19	<0.19	>100	>100	—
Spiramycin	50	1.56	<0.19	1.56	>100	—
Cirramycin A	<0.19	0.39	<0.19	<0.39	1.56	0.456
Tylosin	6.25	6.25	<0.19	<0.19	<0.19	—
Leucomycin A3	<0.19	<0.19	0.19	<0.19	<0.19	0.0812
Maridomycin	—	—	—	0.01	0.1	0.0512
Lankacidin (T-2636C)	—	—	—	0.01	0.01	0.01
Nystatin	>100	>100	25	>100	>100	0.0533
Amphotericin B	>100	>100	12.5	>100	>100	—

Source: Ōmura et al. (1972), Kuraeki et al. (1973), and Kishima and Hashimoto (1979).

Inhibition of protein synthesis in intact bacteria or in cell-free systems by macrolide antibiotics has been reported for erythromycin (Brock and Block, 1959; Taubman et al., 1963; Wolfe and Hahn, 1964), spiramycin (Ahmed, 1968; Vazquez, 1966), Chalcomycin, tylosin, niddamycin (Jordan, 1963; Mao and Wiegand, 1968), leucomycin (Tago and Nagano, 1970), kujimycin (Sawada et al., 1974), midecamycin (Nojiri et al., 1976), maridomycin, and lankacidin (T-2636 C) (Imada and Nozaki, personal communication; Tsuchiya et al., 1971). It is widely recognized that all these macrolides are inhibitors of protein synthesis and of amino acid incorporation directed by synthetic polynucleotides. T-2636 inhibited the growth of leukemia L-1210, lymphosarcoma 6C3HED/OG, and melanoma B-16 in mice (Ootsu and Matsumoto, 1973).

Macrolides of the carbomycin and spiramycin groups (Table 13) are moderately active as inhibitors of proline and lysine incorporation, directed, respectively, by poly C, poly A, and poly U, but erythromycin, methymycin, and the lankamycin group have little or not activity on poly U-dependent phenylalanine incorporation (Mao and Wiegand, 1968). Direct evidence for macrolide binding (1:1) on the 50-S subunit of the bacterial ribosome has been obtained with erythromycin (Taubman et al., 1966; Oleinick and Corcoran, 1969; Teroaka and Tanaka, 1971), spiramycin III (Vazquez, 1967), leucomycin (Tago and Nagano, 1970), midecamycin (Nojiri et al., 1976), and maridomycin (Imada and Nozaki, personal communication). Ribosomes from gram-positive bacteria bind erythromycin with a higher affinity than those from gram-negative bacteria. Ribosomes from erythromycin-resistant mutants or macrolide-resistant *Staphylococcus aureus* (macrolide resistance group A) have a decreased affinity for erythromycin or maridomycin (Oleinick and Corcoran, 1969; Imada and Nozaki, personal communication). About 20% of lankacidin (T-2636) combined the 50-S subunit of both bacterial ribosomes (Macrolide sensitive and resistant) and a difference in affinity was not observed (Imada and Nozaki, personal communication).

Macrolide antibiotics show weak toxicity to mammalian tissue and erythromycin and spiramycin do not bind to cytoplasmic ribosomes from rabbit reticulocyte and rat liver (Mao et al., 1970; Rodriguez-Lopez and Vazquez, 1968).

Antibiotics of the carbomycin and spiramycin groups block peptide bond formation, whereas macrolides of the erythromycin, methymycin, and lankamycin groups do not (Monro and Vazquez, 1967). The antibiotics of the carbomycin and spiramycin groups inhibit not only formation of the peptide bond, but polypeptide synthesis as well (Cundliffe, 1969). The target of the action of midecamycin seems to be the peptidyltransferase reaction and it strongly inhibits the synthesis of both polylysine and polyphenylalanine. Inhibition

Table 13 The Macrolide Groups of Antibiotics

Spiramycin	Spiramycin, angolamycin, relomycin, tylosin
Carbomycin	Carbomycins, leucomycins, niddamycins, maridomycin, midecamycin, josamycin
Erythromycin	Erythromycin, neospiramycin, oleandomycin, megalomycins
Methymycin	Methymycins, narbomycins, neomethycin, pikromycin
Lankamycin	Chalcomycin, neutramycin, lankamycin, kujimycin A

of the synthesis of the latter by erythromycin A is rather slight. The mechanism of action of midecamycin may resemble that of spiramycin (Ahmed, 1968; Nojiri et al., 1974). The mechanism of action of the neutral macrolides kujimycin and chalcomycin is similar to that of erythromycin (Sawada et al., 1974).

Otaka and Kaji (1975) reported that erythromycin A released peptidyl-tRNA in the in vitro polypeptide synthesis system. This result shows that this antibiotic inhibits translocation by preventing the proper location of oligopeptidyl tRNA at the donor site on ribosomes. In a recent study on the binding of erythromycin and 50-S ribosome from *Escherichia coli*, total reconstitution experiments with the 50-S subunit demonstrated an absolute requirement for L-15 and L-16 (a component of the 50-S ribosomal protein of *S. coli*, Kaltschmidt and Wittmann, 1970) with respect to both drug binding and peptidyl-transferase activity (Teraoka and Nierhans, 1978).

Three groups of macrolide resistance in *S. aureus* have been reported (Garrod, 1957; Mitsuhasi, 1967). That in group A is constitutive and is exhibited by all macrolide antibiotics. Macrolide resistance can be induced in both groups B and C, and erythromycin and oleandomycin show inducer activity. Ono et al. (1975) investigated the antibacterial and inducer activities of 32 erythromycin, oleandomycin, and other macrolide antibiotic derivatives. The macrolides were classified into five groups (A, B, C, D, and E) according to their inducer activity. Allen (1977) showed that modification of erythromycin at the C4" position of cladinose could destroy the inducer property, but did not affect the inhibitory properties of the antibiotic. Thus the inducer and inhibitor activities can be dissociated (Allen, 1977). The detailed elucidation of the mechanism of action of macrolide antibiotics will eventually result in the discovery of new derivatives of microbial products or chemical synthesis of macrolides having improved activity against gram-negative and acid-fast bacteria as well as against gram-positive organism and macrolide-resistant bacteria.

REFERENCES

Abou-Zeid, A., El-Diway, I., and Salem, M. H. (1977). The fermentative production of magnamycin by *Streptomyces halstedii*. *J. Appl. Chem. Biotechnol.* 27:318–325.

Achenbach, H., and Karl, W. (1975a). Untersuchungen an Stoffwechselprodukten von Mikroorganismen. VI. Zur Struktur des Antibiotikums Aldgamycin E. *Chem. Ber.* 108:759–771.

Achenbach, H., and Karl, W. (1975b). Untersuchungen an Stoffwechselprodukten von Mikroorganismen. VIII. Aldgemycin F, ein neues Antibiotikum aus *Streptomyces lavendulae*. *Chem. Ber.* 108:780–789.

Ahmed, A. (1968). Mechanism of inhibition of protein synthesis by spiramycin. *Biochim. Biophys. Acta* 166:205–212.

Aizawa, W., Chikaike, N., Kodama, A., Iwasaki, T., Suguro, S., Takenaka, A., and Hiruumi, T. (1975a). Preservation and improvement of leucomycin producing organism on the industrial production. I. Study on the preservation of the organism. Paper presented at the 198th Meeting of the Japan Antibiotics Research Association, Tokyo, May 16.

Aizawa, W., Chikaike, N., Kodama, A., Iwasaki, T., Suguro, S., Takenaka, A., and Hiruumi, T. (1975b). Preservation and improvement of leucomycin producing organism on the industrial production. II. Study on the improvement of the organism. Paper presented at the 198th Meeting of the Japan Antibiotics Research Association, Tokyo, May 16.

Albers-Schonberg, G., Burg, R. W., Miller, T. W., Ormond, R. E., and
Wallick, H. (1977). A new insecticide and the procedure for the production. Jpn. Kokai Tokkyo Koho 77 151,197, Appl. March 19, 1976.

Allen, N. E. (1977). Macrolide resistance in *Staphylococcus aureus* inducers
of macrolide resistance. *Antimicrob. Agents Chemother.* 11:669–674.

Aoki, A., Fukuda, R., Nakayama, T., Ishibashi, K., Takeuchi, C., and
Ishida, M. (1974). Insecticide and tickicide. Jpn. Kokai Tokkyo Koho
74 14, 624, Appl. June 8, 1972.

Arnox, B., Pascaerd, C., Raynord, L., and Lunel, J. (1980). 23627RP a new
macrolide antibiotic from *Streptomyces chryseus*. *J. Am. Chem. Soc.*
102:3605–3608.

Asai, T., Sawada, H., Yamaguchi, T., Higashide, E., Kono, T., and Uchida,
M. (1978a). Kinetics study on the production of maridomycin by *Streptomyces hygroscopicus*. *J. Ferment. Technol.* 56:369–373.

Asai, T., Sawada, H., Yamaguchi, T., Suzuki, M., Higashide, E., and Uchida, M. (1978b). The effect of mechanical damage on the production of
maridomycin by *Streptomyces hygroscopicus*. *J. Ferment. Technol.* 56:
374–379.

Bhuwapathapum, S., and Gray, P. (1977). High pressure liquid chromatography of the macrolide antibiotic tylosin. *J. Antibiot.* 30:673–674.

Bibb, J. M., Ward, J. M., and Hopwood, D. A. (1978). Transduction of plasmid DNA into *Streptomyces* at high frequency. *Nature* 274:398–400.

Bibikova, M. V., Ivanotskaya, L. P., and Tikhonova, A. S. (1980). Resistance of *Micromonospora* to definite antibiotics and their capacity for production of structurally analogous antibiotics. *Antibiotiki (Moscow)* 25:9–12.

Birch, J., Djerassi, C., Ducher, J. D., Mejer, J., Perlman, D., Pride, E.,
Richard, R. W., and Thompson, P. J. (1964). Studies in relation to bio-
synthesis. Part XXXV. Macrolide antibiotics. Part XII. Methymycin.
J. Chem. Soc. 1964:5274–5278.

Bošnjak, M., Tporovec, V., and Vrana, M. (1978). Growth kinetics of *Streptomyces erythreus* during erythromycin biosynthesis. *J. Appl. Chem.
Biotechnol.* 28:791–798.

Brock, T. D., and Block, M. L. (1959). Similarity in mode of action of chloramphenicol and erythromycin. *Biochim. Biophys. Acta* 33:274–279.

Brockmann, H., and Henkel, W. (1950). Pikromycin, ein neues Antibiotikum
aus Actinomyceten. *Naturwissenschaften* 37:138–139.

Brockmann, H., and Henkel, W. (1951). Pikromycin, ein bitter schmerckendes
Antibiotikum aus Actinomyceten. *Chem. Ber.* 84:284–288.

Bu'Lock, J. D., Hamilton, D., Hulme, M. A., Powell, A. J., Smalley, H. M.,
Shepherd, D., and Smith, G. N. (1975). Metabolic development and
secondary biosynthesis in *Penicillium urticae*. *Can. J. Microbiol.* 11:
765–778.

Burg, R. W., Miller, B. M., Baker, E. E., Birnbaum, J., Currier, S. A.,
Hartman, R., Yu-Lin Kong, Monaglan, R. L., Olson, G., Putter, I.,
Tunac, J. B., Wallnick, H., Stapley, E. O., Ōiwa, R., and Ōmura, S.
(1979). Avermectins, new family of potent anthelmintic agents: Producing organism and fermentation. *Antimicrob. Agents Chemother.* 15:361–
367.

Caltrider, P. G., and Hayes, H. B. (1969). Antibiotic production and thereof.
U.S. Patent No. 3,433,711.

Celmer, D. (1965a). Macrolide stereochemistry. I. The total absolute configuration of oleandomycin. *J. Am. Chem. Soc.* 87:1797–1799.

Celmer, D. (1965b). Macrolide stereochemistry. III. A configurational model
for macrolide antibiotics. *J. Am. Chem. Soc.* 87:1801–1821.

Celmer, D. (1971). Stereochemical problems in macrolide antibiotics. *Pure Appl. Chem. 28*:413–453.

Celmer, W. D., Else, H., and Murai, K. (1958). Olenadomycin derivatives. Preparation and characterization. *Antibiot. Annu. 1957/58*:476–483.

Corbaz, R., Ettlinger, L., Gäumann, E., Keller, W., Kladorfer, F., Kyburz, Z., Neipp, L., Prelog, V., Reusser, R., and Zähner, H. (1955). Stoffwechselprodukte von Actinomyceten. Narbomycin. *Helv. Chim. Acta 38*:935–942.

Corcoran, J. W. (1975). S-Adenosyl methionine: Erythromycin C O-methyl transferase. *Methods Enzymol. 43*:487–498.

Corcoran, J. W., and Majer, J. (1975). Erythromycin D, a key intermediate in the biogenesis of the erythromycins. *Fed. Proc. Fed. Am. Soc. Exp. Biol. 34*:589.

Cundliffe, E. (1969). Antibiotics and polyribosomes. II. Some effects lincomycin, spiramycin and streptogramin A in vitro. *Biochemistry 8*:2063–2068.

Delic, V., and Pigac, J. (1969). Detection and study of cosynthesis of tetracycline antibiotics by an agar method. *J. Gen. Microbiol. 55*:103–108.

Djerassi, C., and Halpern, D. (1957). The structure of the antibiotic neomethymycin. *J. Am. Chem. Soc. 79*:2022–2023.

Djerassi, C., and Zderic, J. A. (1956). The structure of antibiotic methymycin. *J. Am. Chem. Soc. 78*:5390–6395.

Djerassi, C., Halpern, O., Wilkinson, D. I., and Eisenbraun, E. J. (1958). Macrolide antibiotics VIII. The absolute configuration of certain centers in neomethymycin, erythromycin and related antibiotics. *Tetrahedron 4*:369–381.

Donin, N. N., Pagano, J., Dutcher, J. D., and McKee, C. M. (1953). Methymycin, a new crystalline antibiotic. (1953). *Antibiot. Annu. 1953/54*:179–185.

Dulaney, E. L. (1955). Induced mutation and strain selection in some industrially important microorganisms. *Ann. N.Y. Acad. Sci. 60*:155–163.

Els, H., Celmer, W. D., and Murai, K. (1958). Oleandomycin(PA 105) II. Chemical characterization(1). *J. Am. Chem. Soc. 80*:3777–3782.

Ettler, J. P., and Greger, V. (1978). Oxygen transfer rate in media used for erythromycin biosynthesis. *J. Ferment. Technol. 56*:144–151.

Forberg, W., Bradler, G., and Knöll, H. (1979). Turimycin-ein Antibiotikum der Makrolidgruppe. *Pharmazie 34*:338.

Forhardt, R. P., Piffillo, R. P., and Ehrlich, J. (1962). Chalcomycin and its fermentation production. U.S. Patent No. 3,065,137.

Fricke, H., Ihn, W., Heinrich, T., and Drewell, S. (1979). Chemische Charakterisierung des Turimycin-Komplexes. *Pharmazie 34*:339–340.

Friedman, S. M., Kaneda, T., and Corcoran, J. W. (1964). Antibiotic glycosides. V. A comparison of 2-methylmalonate and propionate as precursors of the C_{21}-branched chain lactone in erythromycin. *J. Biol. Chem. 239*:2386–2391.

Fugono, T., Higashide, E., Suzuki, T., Yamamoto, H., Harada, S., and Kishi, T. (1970). Interconversion of T-2636 antibiotics produced by *Streptomyces rochei* var. *volubilis. Experientia 26*:36.

Furumai, T., and Suzuki, M. (1975a). Studies on the biosynthesis of basic 16-membered macrolide antibiotics, platenomycins. I. Selection of cosynthesis by non-platenomycin-producing mutants. *J. Antibiot. 28*:770–774.

Furumai, T., and Suzuki, M. (1975b). Studies on the biosynthesis of basic 16-membered macrolide antibiotics, platenomycins. II. Production, isolation, and structures of 3-O-propionyl-5-O-mycarosylplatenolides I and II,

9-dehydro demycarosyl platenomycin and demycarosyl platenomycin. *J. Antibiot. 28*:775–782.

Furumai, T., and Suzuki, M. (1975c). Studies on the biosynthesis of basic 16-membered macrolide antibiotics, platenomycins. III. Production, isolation and structures of platenolides I and II. *J. Antibiot. 28*:783–788.

Furumai, T., Nagahama, N., and Okuda, T. (1967). Studies on a new antibiotic. Albocycline. II. Taxonomic studies on albocycline producing strains. *J. Antibiot. 20*:84–90.

Furumai, T., Seki, Y., Takeda, K., Kinumaki, A., and Suzuki, M. (1973). An approach to the biosynthesis of macrolide antibiotic platenomycin. *J. Antibiot. 26*:708–710.

Furumai, T., Shimizu, Y., Takeda, K., Matsuzawa, N., Tani, K., and Okuda, T. (1974). Studies on the macrolide antibiotic YL-704 complex. I. Taxonomy of the producing strain and production of the complex. *J. Antibiot. 27*:95–101.

Furumai, T., Maezawa, I., Matsuzawa, N., Yano, S., Yamaguchi, T., Takeda, T., and Okuda, T. (1977). Macrolide antibiotics M-4365 produced by Micromonospora. I. Taxonomy, production, isolation, characterization and properties. *J. Antibiot. 30*:443–449.

Ganguly, A. K., Lee, B. K., Branbilla, R., Condon, R., and Scare, O. (1976). Biosynthesis of rosamicin. *J. Antibiot. 29*:976–977.

Garrod, L. P. (1957). The erythromycin group of antibiotics. *Br. Med. J. 2*:57–63.

Gäumann, E., Hütter, R., Keller-Schierlein, W., Neipp, L., Prelog, V., and Zähner, H. (1960). Stoffwechselprodukte von Actinomyceten. 21. Lankamycin und Lankacidin. *Helv. Chim. Acta 43*:601–606.

Gersch, D. (1980). Metabolic regulation by cyclic AMP in macrolide antibiotic-producing strains of *Streptomyces hygroscopicus*. *Proc. Biochem. 1980*: 21–25.

Gersch, D., Backer, H., and Thrum, H. (1977). Biosynthetic studies on the macrolide antibiotic turimycin using ^{14}C-labeled precursors. *J. Antibiot. 30*:488–493.

Gersch, D., Skurk, A., and Römer, W. (1979). Phosphate inhibition of secondary metabolism in *Streptomyces hygroscopicus* and its reversal by cyclic AMP. *Arch. Microbiol. 121*:91–96.

Grisebach, H. (1978). Biosynthesis of macrolide antibiotics. In *Antibiotics and Other Secondary Metabolites. Biosynthesis and Production.* R. Hütter, T. Leisinger, J. Nuesch, and W. Wehrli (Eds.). Academic, London, pp. 113–139.

Hamill, R. L., and Stark, W. M. (1964). Macrocin, a new antibiotic, and lactenocin, an active degradation product. *J. Antibiot. 17*:133–139.

Hamill, R. L., Haney, M. E., Jr., Wiley, P. E., and Stamper, M. E. (1959). Tylosin, a new antibiotic, and its microbiologically active degradation product, desmycosin. Abstract of papers, 136th Meeting of American Chemical Society, 1959, p. 16c.

Hamill, R. L., Haney, M. E., Stamper, M., and Wiley, P. E. (1961). Tylosin, a new antibiotic. II. Isolation, properties, and preparation of desmycosin, microbiologically active degradation product. *Antibiot. Chemother. 11*:328–334.

Harada, S., Higashide, E., Fugono, T., and Kishi, T. (1969). Isolation and structure of T-2636 antibiotics. *Tetrahedron Lett. 27*:2239–2244.

Harris, D. R., McGeachin, S. G., and Mills, H. H. (1965). The structure and stereochemistry of erythromycin A. *Tetrahedron Lett. 11*:679–685.

Hata, T., and Sano, Y. (1955). Method for isolation of leucomycin, a new antibiotic. Jpn. Tokkyo Koho 55 2,299, Appl January 21, 1953.

Hata, T., Sano, Y., Ohki, N., Yokoyama, Y., Matsumae, A., and Ito, S. (1953). Leucomycin, a new antibiotic. J. Antibiot. Ser. A 6:87–89.

Hatano, K., Harada, S., and Kishi, T. (1975). Studies on lankacidin group antibiotics(T-2636). IX. Preparation of ^{14}C-labeled lankacin C-14-propionate. J. Antibiot. 28:15–20.

Hatano, K., Higashide, E., and Shibata, M. (1976). Studies on juvenimicin, a new antibiotic. I. Taxonomy, fermentation and antimicrobial properties. J. Antibiot. 29:1163–1170.

Hayakawa, T., Tanaka, T., Sakaguchi, K., Ōtake, N., and Yonehara, H. (1979). A linear plasmid-like DNA in Streptomyces sp. producing lankacidin group antibiotics. J. Gen. Appl. Microbiol. 25:255–260.

Higashide, E., Hasegawa, T., Shibata, M., Mizuno, K., Imanishi, M., and Miyake, A. (1965). Studies on Streptomycetes. Simultaneous production of oleandomycin and chromomycins by Streptomyces olivochromogenes No. 69895. J. Antibiot. 18:26–37.

Higashide, E., Hasegawa, T., Ono, H., Asai, M., Muroi, M., and Kishi, T. (1972). Antibiotics and the production procedure thereof. Jpn. Tokkyo Koho 72 7,531, Appl. August 13, 1969.

Higashide, E., Fugono, T., Hatano, T., and Shibata, M. (1971). Studies on T-2636 antibiotics. I. Taxonomy of Streptomyces rochei var. volubilis, and production of the antibiotics and an esterase. J. Antibiot. 24:1–24.

Higashide, E., Asai, M., Ootsu, K., Tanida, S., Hasegawa, T., Kishi, T., Sugino, Y., and Yoneda, M. (1977). Ansamitocin, a group of novel maytansinoid antibiotics with antitumor properties from Nocardia. Nature 270:721–722.

Hochstein, F. A., and Murai, K. (1954). Magnamycin B, a second antibiotic from Streptomyces halstedii. J. Am. Chem. Soc. 76:5080–5083.

Hopwood, D. A. (1978). Extrachromosomally determined antibiotic production. Annu. Rev. Microbiol. 32:373–392.

Hopwood, D. A., Wright, H. M., Bibb, M. J., and Cohen, S. N. (1977). Genetic recombination through protoplast fusion in Streptomyces. Nature 268:171–174.

Hori, T., Maezawa, T., Nagahama, N., and Suzuki, M. (1970). The biosynthesis of narbomycin and picromycin. Abst. Paper of 14th Symposium on the Chemistry of Natural Products, pp. 35–42.

Hori, T., Maezawa, I., Nagahama, N., and Suzuki, M. (1971). Isolation and structure of narbonolide, narbomycin aglycon, from Streptomyces venezuelae and its biological transformation into picromycin via narbomycin. Chem. Commun. 1971:304–305.

Huber, G., Wallhäusser, K. H., Fries, L., Steigler, A., and Weidenmüller, H. (1962). Niddamycin, a new macrolide antibiotic. Arneim. Forsch. 12:1191–1195.

Humphrey, A. E. (1977). The use of computers in fermentation systems. Proc. Biochem. 15:19–25.

Hütter, R., Keller-Schierlein, W., and Zähner, H. (1961). Zur Systematik der Actinomyceten. 6. Die Produzenten von Makrolid-Antibiotika. Arch. Mikrobiol. 39:158–194.

Iijima, T., and Sakane, T. (1973). A method for preservation of bacteria and bacteriophage by drying in vacuo. Cryobiology 10:379–385.

Imada, T., and Nozaki, U. The mode of action of lankacidin(T-2636). Personal communication.

Imada, A., Kitano, K., Kintaka, K., Muroi, M., and Asai, M. (1981). Sulfazecin and isosulfazecin, novel β-lactam antibiotics of bacterial origin. *Nature 289*:590–591.

Inouye, S. (1973). Recent advances in the chemistry of macrolides and related antibiotics. *Sci. Rep. Meiji Seika Kaisha 13*:100–142.

Inouye, S. (1974). Recent advances in the chemistry of macrolides and related antibiotics. II. *Sci. Rep. Meiji Seika Kaisha 14*:28–78.

Inouye, S. (1976a). Recent advances in the chemistry of macrolides and related antibiotics. III. *Sci. Rep. Meiji Seika Kaisha 15*:48–99.

Inouye, W. (1976b). Recent advances in the chemistry of macrolides and related antibiotics. IV. *Sci. Rep. Jeiji Seika Kaisha 16*:50–118.

Inouye, W., Tsuruoka, T., Shomura, T., Omoto, S., and Niida, T. (1971). Studies on antibiotic, SF-837, a new antibiotic. II. Chemical structure of antibiotic, SF-837. *J. Antibiot. 24*:460–475.

Jensen, A. L., Darken, M. A., Schultz, J. S., and Shay, A. J. (1964). Relomycin: Flasks and tank fermentation studies. *Antimicrob. Agents Chemother. 1963*:49–53.

Jordan, D. C. (1963). Effect of chalcomycin on protein synthesis by *Staphylococcus aureus. Can. J. Microbiol. 9*:129–132.

Kakemi, K., Uno, T., and Yamashina, H. (1956). Chemical method for determination of antibiotics. VI. Colorimetric determination of erythromycin. *J. Pharm. Soc. Jpn. 76*:1116–1117.

Kaltschmidt, E. and Wittmann, H. G. (1970). Ribosomal proteins, XII. Number of proteins in small and large ribosomal subunits of *Escherichia coli* as determined by two-dimensional gel electrophoresis. *Proc. Nat. Acad. Sci. U.S.A. 67*:1276–1282.

Kanda, N., and Abe, K. (1972). Procedure of the production of a new antibiotic, K-231-1. Jpn. Kokai Tokkyo Koho 72 25,391, Appl March 28, 1969.

Kaneda, T., Butte, J. C., Taubman, S. B., and Corcoran, J. W. (1962). Actinomycete antibiotics. III. The biogenesis of erythronolide. The C_{21}-branched chain lactone in erythromycin. *J. Biol. Chem. 237*:322–328.

Kawamoto, I., Okachi, R., Kato, H., Takahashi, I., Takasawa, S., and Nara, T. (1974). The antibiotic XK-41 complex. I. Production, isolation and characterization. *J. Antibiot. 27*:493–501.

Kenhne, M. E., and Benson, B. W. (1965). The structure of the spiramycin and magnamycin. *J. Am. Chem. Soc. 87*:4660–4662.

Kikuchi, M. (1980). Application of genetics for strain improvement in industrial microorganisms. *Biotechnol. Bioeng. 22*:195–208.

Kinumaki, A., and Suzuki, M. (1972). The structure of a new antibiotic, YC-17. *J. Chem. Soc. Chem. Commun. 1972*:744–745.

Kinumaki, A., Takamori, I., Sugawara, Y., Nagahama, N., Suzuki, M., Egawa, Y., Sakurazawa, M., and Okuda, T. (1974a). Isolation, and physicochemical properties of YL-704 components. *J. Antibiot. 27*:102–106.

Kinumaki, A., Takamori, I., Sugawara, Y., Seki, Y., Suzuki, M., and Okuda, T. (1974b). Studies on the macrolide antibiotic YL-704 complex IV. The structure of minor components. *J. Antibiot. 27*:117–122.

Kinumaki, A., Harada, K., Suzuki, T., Suzuki, M., and Okuda, T. (1977). Macrolide antibiotics M-4365 produced by *Micromonospora*. II. Chemical structure. *J. Antibiot. 30*:450–454.

Kishi, T., Harada, S., Yamana, Y., and Miyake, A. (1976). Studies on juvenimicin, a new antibiotic. II. Isolation, chemical characterization and structures. *J. Antibiot. 29*:1171–1181.

Kishima, H., and Hashimoto, K. (1979). *In vivo* insensitivities to antimicrobial drugs of ureaplasms isolated from the bovine respiratory tract, genital tract and eye. *Res. Vet. Sci.* 27:218–222.

Kitano, K., Kintaka, K., Suzuki, S., Katamoto, K., Nara, K., and Nakao, Y. (1976a). Screening of microorganisms capable of producing β-lactam antibiotics from n-paraffins. *J. Ferment. Technol.* 54:683–695.

Kitano, K., Kintaka, K., and Nakao, Y. (1976b). Some characteristics of a β-lactam antibiotic-hypersensitive mutant derived from a strain of *Pseudomonas aeruginosa*. *J. Ferment. Technol.* 54:696–704.

Kitao, C., Ikeda, H., Hamada, H., and Ōmura, S. (1979a). Bioconversion and biosynthesis of 16-membered macrolide antibiotics. XIII. Regulation of spiramycin I 3-hydroxyacylase formulation by glucose, butyrate and cerulenin. *J. Antibiot.* 32:593–599.

Kitao, C., Hamada, H., Ikeda, H., and Ōmura, S. (1979b). Bioconversion and biosynthesis of 16-membered macrolide antibiotics. XV. Final steps in the biosynthesis of leucomycins. *J. Antibiot.* 32:1055–1057.

Kondo, K. (1979). Analytical studies of maridomycin. I. High-performance liquid chromatography of maridomycins and some macrolide antibiotics. *J. Chromatogr.* 169:329–336.

Korchagin, V. B., Semenov, S. M., and Savushkima, L. M. (1961). Colorimetric method of determination of erythromycin. *Antibiotiki (Moscow)* 6:311–314.

Koshiyama, H., Okanishi, M., Ohmori, T., Miyaki, T., Tsukiura, H., Matsuzaki, M., and Kawaguchi, H. (1963). Cirramycin, a new antibiotic. *J. Antibiot.* 16:59–66.

Kunstmann, L. A., Mitscher, A., and Patterson, E. L. (1964). Aldgamycin E, a new neutral macrolide antibiotic. *Antimicrob. Agents Chemother.* 1964:87–90.

Kuraeki, S., Tamura, H., Ooe, I., Kishi, T., Tsuchiya, K., and Kanekiyo, T. (1973). Therapeutic agent for mycoplasma infections diseases. Jpn. Kokai Tokkyo Koho 73 15,604.

Kusaka, T., Yamamoto, H., and Suzuki, T. (1970). Studies on antibiotic B-58941. *Streptomyces fradiae* var. *acinicolor* B-58941 and its product belonging to macrolide antibiotic. *Annu. Rep. Takeda Res. Lab.* 29:239–245.

Lefemine, D. V., Barbartsch, F., Dann, M., Thomas, S. O., Kunstman, M. P., Mitschev, L. A., and Bohonos, N. (1963). Neutramycin, a new neutral macrolide antibiotic. *Antimicrob. Agents Chemother.* 1963:41–44.

LeMahieu, R. A., Ax, H. A., Blonunt, J. F., Carson, M., Despreuaux, C., Pruss, D. L., Scannell, J. P., Weiss, F., and Kierstread, R. W. (1976). A new semisynthetic macrolide antibiotic, 3-O-oleandrosyl-5-O-desosaminylerythronolide A oxime. *J. Antibiot.* 29:728–734.

Lindenbein, W. (1952). Über einige chemische interessente Actinomyceten-Stamm und ihre Klassifizierung. *Arch. Mikrobiol.* 17:361–383.

McGuire, J. M., Bunch, R. L., Anderson, R. C., Boaz, H. E., Flyan, E. H., Powell, H. M., and Smith, J. W. (1952). "Ilotycin" a new antibiotic. *Antibiot. Chemother.* 2:281–283.

McGuire, J. M., Boniece, W. S., Higgins, C. E., Hoehn, M. M., Stark, W. M., Westhead, J., and Wolf, R. N. (1961). Tylosin, a new antibiotic: I. Microbiological studies. *Antibiot. Chemother.* 11:320–327.

Maezawa, I., Hori, T., and Suzuki, M. (1974a). Accumulation of narbonolide caused by the addition of organic acids. *Agric. Biol. Chem.* 38:91–96.

Maezawa, I., Hori, T., and Suzuki, M. (1974b). Accumulation of narbonolide by the addition of sodium arsenite. *Agric. Biol. Chem.* 38:539–542.

Mallams, A. K., Jaret, R. S., and Reimann, H. (1969). The megalomicins. II. The structure of megalomicin A. *J. Am. Chem. Soc.* 91:7506–7508.

Majer, J., Martin, J. R., Egan, R. S., and Corcoran, J. W. (1977). Antibiotic glycosides. 8. Erythromycin D, a new macrolide antibiotic. *J. Am. Chem. Soc.* 99:1620–1622.

Mao, J. C. H., and Wiegand, R. G. (1968). Mode of action of macrolides. *Biochim, Biophys. Acta* 157:404–413.

Mao, J. C. H., Putterman, M., and Wiegand, R. G. (1970). Biochemical basis for the selective toxicity of erythromycin. *Biochem. Pharmacol.* 19:391–399.

Marquez, J., Murauski, A., and Wagman, G. H. (1969). Isolation, purification and preliminary characterization of megalomicin. *J. Antibiot.* 22:259–264.

Martin, J. F. (1977). Control of antibiotic synthesis by phosphate. *Adv. Biochem. Eng.* 6:105–127.

Martin, J. F., and Demain, A. L. (1980). Control of antibiotic biosynthesis. *Microbiol. Rev.* 44:230–251.

Martin, J. F., Liras, P., and Demain, A. L. (1978). ATP and adenylate energy charge during phosphate-mediated control of antibiotic synthesis. *Biochem. Biophys. Res. Commun.* 83:822–828.

Martin, J. R., and Goldstein, A. W. (1970). Final steps in erythromycin biosynthesis. *Prog. Antimicrob. Anticancer Chemother.* 2:1112–1116.

Martin, J. R., and Rosenbrook, W. (1968). Studies on the biosynthesis of the erythromycins. III. Isolation and structure of 5-deoxy-5-erythronolide B, a shunt metabolite of erythromycin biosynthesis. *Biochemistry* 7:1728–1733.

Martin, J. R., Egan, R. S., Goldstein, A. W., and Collum, D. (1975). Extention of the erythromycin biosynthetic pathway. Isolation and structure of erythromycin E. *Tetrahadron 31*:1985–1989.

Matern, H., Brillinger, G. U., and Pape, H. (1973). Stoffwechselprodukte von Mikroorganismen. 114. Mitteilung. Thymidindiphopho-D-Glucose-Oxidoreductase aus *Streptomyces rimosus*. *Arch. Mikrobiol.* 88:37–48.

Miller, T. W., Chaiet, L., Cole, D. J., Cole, L. J., Flor, J. E., Goegelman, R. T., Gullo, V. P., Joshua, H., Kempf, A., Krellwitz, W. R., Monaghan, R. L., Ormond, R. E., Wilson, K. E., Albers-Schonberg, G., and Putter, I. (1979). Avermectins, new family of potent agents: Isolation and chromatographic properties. *Antimicrob. Agents Chemother. 15*:368–371.

Mitscher, L. A., and Kunstmann, M. P. (1969). The structure of neutramycin. *Experientia 25*:12–13.

Mitsuhashi, S. (1967). Epideminological and genetical studies of drug resistance in *Staphylococcus aureua*. *Jpn. J. Microbiol. 11*:49–68.

Miyagawa, K., Suzuki, M., Higashide, E., and Uchida, M. (1979a). Effect of aspartic acid family amino acids on production of maridomycin. III. *Agric. Biol. Chem. 43*:1103–1109.

Miyagawa, K., Suzuki, M., and Uchida, M. (1979b). Predominant accumulation of maridomycin III by a valine resistant mutant of *Streptomyces hygroscopicus* No. B-5050-HA. *Agric. Biol. Chem. 43*:1111–1116.

Miyairi, N., Takashima, M., Shimozu, K., and Sakai, H. (1966). Studies on new antibiotics, cineromycins A and B. *J. Antibiot. Ser. A 19*:56–62.

Miyoshi, T., Iseki, M., Konomi, T., and Imanaka, H. (1980). Biosynthesis of bicyclomycin. I. Appearence of aerial mycelia negative strains(am⁻). *J. Antibiot. 33*:480–487.

Monro, R. E., and Vazquez, D. (1967). Ribosome-catalysed peptidyl transfer: Effects of some inhibitors of protein synthesis. *J. Mol. Biol. 28*:161–165.

Morin, R. B., Gorman, H., Hamill, R. L., and Demarco, P. V. (1970). The structure of tylosin. *Tetrahedron Lett. 54*:4737–4740.

Mourot, D., Delepire, B., Biosseau, J., and Gayot, G. (1978). Reverse phase high pressure liquid chromatography of spiramycin. *J. Chromatog. 161*: 386–388.

Muntwyler, R., and Keller-Schierlein, W. (1972). Stoffwechsel-produkte von Mikroorganismen. 103. Mitteilung. Zur Stereochemie des Lankamycins. *Helv. Chim, Acta 55*:460–467.

Muroi, M., Izawa, H., Asai, M., Kishi, T., and Mizuno, K. (1973). Maridomycin, a new macrolide antibiotic. II. Isolation and characterization. *J. Antibiot. 26*:199–205.

Muxfeldt, H., Srader, S., Hasen, P., and Brockmann, H. (1968). The structure of pikromycin. *J. Am. Chem. Soc. 90*:4748–4749.

Nagahama, N., Suzuki, M., Awataguchi, S., and Okuda, T. (1967). Studies on a new antibiotic, albocycline. I. Isolation, purification and properties. *J. Antibiot. 20*:261–266.

Nagahama, N., Takamori, I., and Suzuki, M. (1971). Studies on an antibiotic, albocycline. V. High pressure-high temperature hydrogenation products and structure proof of albocycline. *Chem. Pharm. Bull. 19*:660–666.

Nakazawa, K. (1955). Studies on *Streptomycetes*. On *Streptomyces albireticuli* nov. sp. *J. Agric. Chem. Soc. Jpn. 29*:647–649.

Namiki, S., Ōmura, S., Nakayoshi, H., and Sawada, J. (1969). Studies on antibiotic from *Streptomyces spinichromogenes* var. *kujimyceticus*. I. Taxonomic and fermentation studies with *Streptomyces spinichromogenes* var. *kujimyceticus*. *J. Antibiot. 22*:494–499.

Nara, T., Takazawa, S., Okachi, R., Kato, H., and Yamamoto, S. (1973). Antibiotic XK-41-B$_2$ and the production procedure. Jpn. Kokai Tokkyo Koho 73 91,281.

Nach, C. H., III. (1974). Effect of carbon dioxide on synthesis of erythromycin. *Antimicrob. Agents Chemother. 5*:544–545.

Nash, C. H., III, Coch, K. F., and Hoehn, M. M. (1980). Antibiotic A-6888C and A-6999X. Jpn. Kokai Tokkyo Koho 80 154,994.

Niida, T., Tsuruoka, T., Ezaki, N., Shomura, T., Akita, E., and Inouye, S. (1971). A new antibiotic, SF-837. *J. Antibiot. 24*:319–320.

Nishimura, N., Kumagai, K., Ishida, N., Saito, K., Kato, F., and Azumi, M. (1965). Isolation and characterization of shinocymycin A. *J. Antibiot. Ser. A 18*:251–258.

Nojiri, C., Goi, H., and Yamada, Y. (1976). Mode of action of a antibiotic SF-837. *Sci. Rep. Meiji Seika Kaisha 16*:40–49.

Okachi, R., Kawamoto, I., Kato, H., Yamamoto, S., Takazawa, S., and Nara, T. (1974). Isolation, purification and biological properties of antibiotic XK-41. *Abst. Annual Meeting of Japan Agricultural Chemical Society 1974*, pp. 429.

Okuda, T., and Awata, S. (1971). Procedure of YL-704, a new macrolide antibiotic. Jpn. Tokkyo Koho 71 28,836, Appl. May 19, 1969.

Oleinick, N. L. (1975). The erythromycins. In *Antibiotics*, Vol. 3. J. W. Corcoran and T. E. Huhn (Eds.). Springer-Verlag, Berlin, pp. 396–419.

Oleinick, N. L., and Corcoran, J. W. (1969). Two type of binding of erythromycin to ribosomes from antibiotic-sensitive and -resistant *Bacillus subtilis* 168. *J. Biol. Chem. 244*:727–735.

Okami, Y., Okazaki, T., Kitahara, T., and Umezawa, H. (1976). Studies on marine microorganisms. V. A new antibiotic, aplasmomycin, produced by a streptomycete isolated from shallow sea mud. *J. Antibiot.* 29:1019–1025.

Ōmura, S., and Kitao, C. (1979). Biosynthesis of macrolide antibiotics. *Hakko To Kogyo* 37:749–764.

Ōmura, S., and Nakagawa, A. (1975). Chemical and biological studies on 16-membered macrolide antibiotics. *J. Antibiot.* 28:401–433.

Ōmura, S., and Takeshima, H. (1975a). Biosynthesis of macrolide antibiotics. *Kagaku To Seibutsu* 15:309–315.

Ōmura, S., and Takeshima, H. (1975b). Biosynthesis of macrolide antibiotics. III. *Kagaku To Seibutsu* 15:381–453.

Ōmura, S., and Takeshima, H. (1975c). Biosynthesis of macrolide antibiotics. III. *Kagaku To Seibutsu* 15:447–452.

Ōmura, S., Hironaka, Y., and Hata, T. (1970). Chemistry of leucomycin. XI. Identification of leucomycin A_3 with josamycin. *J. Antibiot.* 23:511–513.

Ōmura, S., Hironaka, Y., Nakagawa, A., Umezawa, I., and Hata, T. (1972). Antimycoplasma activities of macrolide antibiotics. *J. Antibiot.* 25:105–108.

Ōmura, S., Suzuki, Y., Nakagawa, A., and Hata, T. (1973). Fast liquid chromatography of macrolide antibiotics. *J. Antibiot.* 26:794–796.

Ōmura, S., Nakagawa, A., Takeshima, H., Atsumi, K., and Miyazawa, J. (1975a). Biosynthetic studies using ^{13}C-enriched precursors on the 16-membered macrolide antibiotic laucomycin A_3. *J. Am. Chem. Soc.* 97:6600–6602.

Ōmura, S., Nakagawa, A., Takeshima, H., Miyazawa, J., and Kitao, C. (1975b). A ^{13}C nuclear magnetic resonance study of the biosynthesis of the 16-membered macrolide antibiotic tylosin. *Tetrahedron Lett.* 50:4503–4506.

Ōmura, S., Takeshima, H., Nakagawa, A., and Miyazawa, J. (1976). The biosynthesis of pikromycin using ^{13}C enriched precursors. *J. Antibiot.* 29:316–317.

Ōmura, S., Kitao, C., Miyazawa, J., Imai, H., and Takeshima, H. (1978). Bioconversion and biosynthesis of 16-membered macrolide antibiotic, tylosin, using enzyme. *J. Antibiot.* 31:254–256.

Ōmura, S., Kitao, C., Hamada, H., and Ikeda, H. (1979a). Bioconversion and biosynthesis of 16-membered macrolide antibiotics. X. Final steps in the biosynthesis of spiramycin, using enzyme inhibitor: Cerulenin. *Chem. Pharm. Bull.* 27:176–182.

Ōmura, S., Ikeda, H., and Kitao, C. (1979b). The detection of a plasmid in *Streptomyces ambofaciens* KA-1028 and its possible involvement in spiramycin production. *J. Antibiot.* 32:1058–1060.

Ōmura, S., Tanaka, Y., Tanaka, H., Takahashi, Y., and Imai, Y. (1980). Stimulation of the production of macrolide antibiotics by magnesium phosphate and related insoluble materials. *J. Antibiot.* 33:1568–1569.

Ōmura, S., Ikeda, H., and Tanaka, H. (1981). Extraction and characterization of plasmid from macrolide antibiotic-producing streptomycetes. *J. Antibiot.* 34:478–482.

Ono, H., Hasegawa, T., Higashide, E., and Shibata, M. (1973). Maridomycin, a new macrolide antibiotic. I. Taxonomy and fermentation. *J. Antibiot.* 26:191–198.

Ono, H., Harada, S., and Kishi, T. (1974). Maridomycin, a new macrolide antibiotic. VII. Incorporation of labeled precursors into maridomycin and preparation of ^{14}C-labeled 9-propionyl maridomycin. *J. Antibiot.* 27:442–448.

Ono, H., Inoue, M., Mao, J. C. H., and Mitsuhashi, S. (1975). Drug resistance in *Staphylococcus aureus*. Induction of macrolide resistance by erythromycin, oleandomycin and their derivatives. *Jpn. J. Microbiol.* 19:343–347.

Ootsu, K., and Matsumoto, T. (1973). Effects of lankacidin group (T-2636) antibiotics on the tumor growth and immune response against sheep erythrocytes in mice. *Gann* 64:481–492.

Osono, T., Oka, Y., Watanabe, S., Numazaki, Y., Moriyama, K., Ishida, H., Suzaki, K., Okami, Y., and Umezawa, H. (1967). A new antibiotic, josamycin. I. Isolation and physicochemical characteristics. *J. Antibiot. Ser. A* 20:174–180.

Otaka, T., and Kaji, A. (1975). Release of (oligo)peptidyl-tRNA from ribosomes by erythromycin A. *Proc. Nat. Acad. Sci. U.S.A.* 72:2649–2652.

Pape, H., and Brillinger, G. (1973). Stoffwechselprodukte von Mikroorganismen. 113. Mitt. Biosynthese von Thymidin-diphosphomycarose durch ein zellfreis System aus *Streptomyces rimosus*. *Arch. Mikrobiol.* 88:25–35.

Pettinga, C. W., Stark, W. M., and Van Abbeele, F. R. (1954). The isolation of a second crystalline antibiotic from *Streptomyces erythreus*. *J. Am. Chem. Soc.* 76:569–571.

Pinnert-Sindico, S. (1954). Une novelle espèce. De Streptomyces productrice d'antibiotiques: *Streptomyces ambofaciens* n. sp. caractères culturaux. *Ann. Inst. Pasteur* 87:702–707.

Pinnert-Sindico, S., Ninet, L., Preud'homme, J., and Coar, C. (1955). A new antibiotic—Spiramycin. *Antibiot. Annu.* 1954/55:724–727.

Puar, M. S., Brambilla, R., Bartner, P., Schumacher, D., and Jaret, R. S. (1979). Sch 23831, a new macrolide from *Micromonospora rosaria*. *Tetrahedron Lett.* 30:2767–2770.

Puar, M. S., Lee, B. K., Munayyer, H., Brambilla, R., and Waitz, J. A. (1981). The biosynthesis of AR-5(mycinamicins) antibiotics. *J. Antibiot.* 34:619–620.

Queener, S. W., Sebek, O. K., and Vezine, C. (1978). Mutants blocked in antibiotic synthesis. *Annu. Rev. Microbiol.* 32:593–636.

Raczynska-Bojanowska, K., Ruezaj, Z., Swanor-Korozynska, D., and Rafaski, A. (1973). Limiting reactions in activation of acyl units in biosynthesis of macrolide antibiotics. *Antimicrob. Agents Chemother.* 3:162–167.

Reiman, H., and Jaret, R. S. (1972). Structure of rosamicin, a new macrolide from *Micromonospora rosaria*. *J. Chem. Soc. Chem. Commun.* 1972:1270.

Rodoriquez-Lopez, M., and Vazquez, D. (1968). Comparative studies on cytoplasmic ribosomes from algae. *Life Sci. Part II*, 7:327–336.

Sakamoto, J. M. J., Kondo, S., Yumoto, H., and Arishima, M. (1962). Bundlins A and B, two antibiotics produced by *Streptomyces griseofuscus* nov. sp. *J. Antibiot. Ser. A* 15:98–102.

Satoi, S., Muto, N., Hayashi, N., Fujii, T., and Otani, M. (1980). Mycinamicins, a new macrolide antibiotics. I. Taxonomy, production, isolation, characterization and properties. *J. Antibiot.* 33:364–376.

Sawada, J., Namiki, S., Onodera, M., Ōmura, S., and Tanaka, I. (1974). Mode of action of kujimycins, neutral macrolide antibiotics. *J. Antibiot.* 27:639–641.

Seino, A. (1977). The preservation method for actinomycetes. In *The Preservation Method for Microorganism*. T. Nei (Ed.). Tokyo University Press, Tokyo, pp. 225–238.

Seno, E. T., Pieper, R. L., and Huber, F. M. (1977). Terminal stage in the biosynthesis of tylosin. *Antimicrob. Agents Chemother. 11*:455–461.

Sensi, P., and Thieman, J. E. (1965). Production of rifamycins. *Prog. Ind. Microbiol. 6*:21–60.

Shibata, M., and Uyeda, M. (1978). Microbial transformations of antibiotics. *Annu. Rep. Ferment. Proc. 2*:267–303.

Shimauchi, Y., Okamura, K., Koki, A., Hasegawa, M., Date, M., Fukagawa, Y., Kouno, K., Ishikura, T., and Lein, J. (1978a). Deltamycins, new macrolide antibiotics. I. Producing organism and fermentation. *J. Antibiot. 31*:261–269.

Shimauchi, Y., Kubo, K., Osumi, K., Okamura, K., Fukagawa, Y., Ishikura, T., and Lein, J. (1978b). Deltamycins, new macrolide antibiotics. II. Isolation and physicochemical properties. *J. Antibiot. 31*:270–275.

Shimauchi, Y., Kubo, K., Osumi, K., Okamura, K., Fukagawa, Y., and Ishikura, T. (1979). Deltamycin, new macrolide antibiotics. III. Chemical structures. *J. Antibiot. 32*:878–883.

Shimi, I. R., Shonkry, S., and Ali, F. T. (1979). Staphcoccomycin, a new basic macrolide antibiotic. *J. Antibiot. 32*:1248–1255.

Shomura, T., Yoshida, J., Amano, S., Kojima, M., Inouye, S., and Niida, T. (1979). Studies on *Actinomycetales* producing antibiotics only on agar culture. I. Screening, taxonomy and morphology–productivity relationship of *Streptomyces halstedii*, strain SF-1993. *J. Antibiot. 32*:427–435.

Smith, R. L., Bungay, H. R., and Pittenger, R. C. (1962). Growth–biosynthesis relationships in erythromycin fermentation. *Appl. Microbiol. 10*:293–296.

Sobin, B. A., English, A. R., and Celmer, W. D. (1955). PA 105, a new antibiotic. *Antibiot. Annu. 1955/56*:827–830.

Srinivasan, D., and Srinivasan, P. R. (1967). Studies on the biosynthesis of magnamycin. *Biochemistry 6*:3111–3118.

Stokes, J. L., and Woodward, C. R., Jr. (1942). The isolation from soil of spore-forming bacteria which produce bactericidal substances. *J. Bacteriol. 43*:253–263.

Stuttard, C. (1979). Transduction of auxotrophic markers in chloramphenicol-producing strain of Streptomyces. *J. Gen. Microbiol. 110*:479–482.

Sugawara, R., and Takeda, T. (1955). Studies on the improvement of leucomycin producing strain. *J. Antibiot. Ser. A 8*:139–144.

Sugden, K., Cox, G. B., and Loscombe, C. R. (1978). Chromatographic behaviour of basic amino compounds on silics and ODS-silica using aqueous mobile phases. *J. Chromatogr. 149*:377–390.

Suzuki, T. (1970). The structure of an antibiotic, B-58941. *Bull. Chem. Soc. Jpn. 43*:292.

Suzuki, T., and Aota, T. (1969). Studies on antibiotics B-62169. II. The chemical structures of macrolide antibiotics B-62169A and B produced by *Streptomyces fradiae* No. B-62169. *Annu. Rep. Takeda Res. Lab. 28*:61–68.

Suzuki, M., and Nakao, Y. (1979). Improvement of antibiotic producing organism. *Hakko To Kogyo 37*:712–722.

Suzuki, T., Asai, M., and Sugita, N. (1973). The procedure of the antibiotic production. Jpn. Tokkyo Koho 73 76,880.

Suzuki, M., Miyagawa, K., and Uchida, M. (1975). Procedure of the antibiotic production. Jpn. Kokai Tokkyo Koho 75 64,495.

Suzuki, M., Takamaki, T., Miyagawa, K., Ono, H., Higashide, E., and Uchida, M. (1979). Interconversion among 16-membered macrolide antibiotics belonging to leucomycin-maridomycin group. *Agric. Biol. Chem.* *43*:1331-1336.

Tago, K., and Nagano, M. (1970). Mechanism of inhibition of protein synthesis by leucomycin. In *Progress in Antimicrobial and Anticancer Chemotherapy*, Vol. 1. University of Tokyo Press, pp. 199-205.

Takeshima, H., Kitao, C., and Ōmura, S. (1977). Inhibition of the biosynthesis of leucomycin, a macrolide antibiotic, by cerulenin. *J. Biochem.* *81*:1127-1132.

Takiguchi, Yo., Mishima, H., Okuda, M., Terao, M., Aoki, A., and Fukuda, R. (1980). Milbemycins, a new family of macrolide antibiotics. Fermentation, isolation, and physiocochemical properties. *J. Antibiot.* *33*:1120-1127.

Tanner, F. W., Jr., English, A. R., Lee, T. M., and Routien, J. B. (1952). Some properties of magnamycin, a new antibiotic. *Antibiot. Chemother.* *2*:441-443.

Taubman, S. B., Young, F. E., and Corcoran, J. W. (1963). Antibiotic glycosides. IV. Studies on the mechanism of erythromycin resistance in *Bacillus subtilis*. *Proc. Nat. Acad. Sci. U.S.A.* *50*:955-962.

Taubman, S., Jones, N., Young, F., and Corcoran, J. (1966). Sensitivity and resistance to erythromycin in *Bacillus subtilis* 168: The ribosomal binding of erythromycin and chloramphenicol. *Biochim, Biophys. Acta* *123*:438-440.

Teroaka, H., and Nierhaus, K. H. (1978). Proteins from *Escherichia coli* ribosomes involved in the binding erythromycin. *J. Mol. Biol.* *126*:185-193.

Teraoka, H., and Tanaka, K. (1971). An alteration in ribosome function caused by equimolar binding of erythromycin. *Biochim. Biophys. Acta* *232*:509-513.

Thompson, R. M., and Strong, F. M. (1971). Identification of erythromycin A in cultures of *Streptomyces griseoplanus*. *Biochem. Biophys. Res. Commun.* *43*:313-316.

Torboshikina, L. I., and Dormidoshina, T. A. (1964). Mechanism of glucose dissimilation in the oleandomycin producing *Actinomyces antibioticus*. *Mikrobiologiya* *33*:325-331.

Torbochikina, L. I., Dormidoshina, T. A., and Saizeva, L. P. (1964). Carbohydrate metabolism in the oleandomycin producing *Actinomyces antibioticus*. *Mikrobiologiya* *33*:162-166.

Tsuji, K. (1978). Fluorometric determination of erythromycin and erythromycin ethylsuccinate in serum by a high-performance liquid chromatographic post-column, on-stream derivation and extraction method. *J. Chromatogr.* *158*:337-348.

Tsuji, K., and Goetz, J. F. (1978a). High-performance liquid chromatographic determination of erythromycin. *J. Chromatogr.* *147*:359-367.

Tsuji, K., and Goetz, J. F. (1978b). Elevated column temperature for the high-performance liquid chromatographic determination of erythromycin and erythromycin ethylsuccinate. *J. Chromatogr.* *157*:185-196.

Tsuji, K., and Goetz, J. F. (1978c). HPLC as a rapid means of monitoring erthromycin and tetracycline fermentation process. *J. Antibiot.* *31*:302-308.

Tsuji, K., and Robertson, J. H. (1974). Determination of erythromycin and its derivatives by gas–liquid chromatography. *Anal. Chem. 43*:818–821.

Tsuchiya, T., Yamazaki, T., Takeuchi, Y., and Oishi, T. (1971). Studies on T. 2636 antibiotics. IV. In vitro and in vivo antibacterial activity of T-2636 antibiotics. *J. Antibiot. 24*:29–41.

Tsukiura, H., Konishi, M., Saka, M., Naito, T., and Kawaguchi, H. (1969). Studies on cirramycin A₁. III. Structure of cirramycin A₁. *J. Antibiot. 22*:89–99.

Tsuruoka, T., Shomura, T., Ezaki, N., Akita, E., Inouye, S., Fukatsu, S., Amano, S., Watanabe, H., and Niida, T. (1971). New antibiotic and the procedure of the production. Jpn. Tokkyo Koho 71 28,834, Appl. February 6, 1969.

Tsuruoka, T., Shomura, T., Ezaki, N., Watanabe, H., Akita, E., Inouye, S., and Niida, T. (1971). Studies on antibiotic, SF-837, a new antibiotic. I. The producing organism, and isolation and characterization of the antibiotic. *J. Antibiot. 24*:452–459.

Uchida, M., Suzuki, M., Sugita, N., Kondo, K., Muroi, M., and Kishi, M. (1975a). The procedure of the antibiotic production. Jpn. Kokai Tokkyo Koho 75 19,989.

Uchida, M., Suzuki, M., Miyagawa, K., Sugita, N., Sawada, H., and Higashide, E. (1975b). Studies on maridomycin III fermentation by the mutant improved. *Abst. Paper of Ann. Meeting of Japan Agricultural Biological Chemical Society*, pp. 86.

Uchida, M., Suzuki, M., Takayama, T., and Sugita, N. (1979). Colorimetric determination of maridomycin. *Agric. Biol. Chem. 43*:847–852.

Umezawa, S., Machida, I., Shiozu, S., Yokota, K., Makino, S., Kawaguchi, T., and Honda, K. (1971). Espinomycin, a new antibiotic, and the production procedure. Jpn. Tokkyo Koho 71 30,796, Appl. August 9, 1969.

Uramoto, M., Ōtake, N., Ogawa, Y., Yonahara, H., Marumo, F., and Sato, Y. (1969). The structure of bundlin A(lankacidin) and bundlin B. *Tetrahadren Lett. 1969*:2249.

Uramoto, M., Ōtake, N., Cary, L., and Tanabe, M. (1978). Biosynthetic studies with carbon-13 lankacidin group of antibiotics. *J. Am. Chem. Soc. 100*:3616–3617.

Vazquez, D. (1966). Antibiotics affecting chloramphenicol uptake by bacteria. Their effect on amino acid incorporation in a cell-free system. *Biochim. Biophys. Acta 114*:289–295.

Vazquez, D. (1967). Binding to ribosomes and inhibitory effect on protein synthesis of the spiramycin antibiotics. *Life Sci. 6*:845–853.

Vazquez, D. (1975). The macrolide antibiotics. In *Antibiotics*, Vol. 3. J. W. Corcoran, and F. E. Hahn (Eds.). Springer-Verlag, Berlin, pp. 459–479.

VEB Jenapharm. (1970). Werkwiejze voor het winnen van een macrolide antibioticum. Netherlands Patent No. 70 03,002, Appl. March 5, 1969.

Vezina, C., Bolduc, C., Kudelski, A., and Audet, P. (1979). Biosynthesis of kitasamycin(leucomycin) by leucine analogue resistant mutants of *Streptomyces kitasatoensis*. *Antimicrob. Agents Chemother. 15*:738–746.

Vu-Trong, K., and Gray, P. P. (1982). Continuous culture studies on the regulation of tylosin biosynthesis. *Biotechnol. Bioeng. 24*:1093–1104.

Vu-Trong, K., Bhuwapathanapun, S., and Gray, P. P. (1980). Metabolic regulation in tylosin-producing *Streptomyces fradiae*: Regulatory role of adenylate nucleotide pool and enzyme involved in biosynthesis of tylonolide precursors. *Antimicrob. Agents Chemother. 17*:519–525.

Wagman, G. H., and Weinstein, M. J. (1980). Antibiotics from *Micromonospora*. *Annu. Rev. Microbiol.* *34*:537-557.

Wagman, G. H., Waitz, J. A., Marquez, J., Muracushi, A., Oden, E. M., Testa, R. T., and Weinstein, M. J. (1972). A new micromonospora produced macrolide antibiotic, rosamicin. *J. Antibiot.* *25*:641-646.

Walter, A. M., and Heilmeyer, L. (Begündet). (1975). Antibiotika mit begrenztem Wirkungsbereich Indikation vorwiegend bei Infectionen mit grampositiven Keiman. Die Makrolid-Antibiotika. In *Antibiotika-Fibel. Antibiotika und Chemotherapeutika Therapie mikrobieller Infektionen*. Georg Thieme Verlag, Stuttgart, pp. 473-497.

Weinberg, E. D. (1970). Biosynthesis of secondary metabolites: Role of trace metals. *Adv. Microb. Physiol.* *4*:1-44.

Weinstein, M. J., Wagman, G. H., Marquez, J. A., Testa, R. T., Oden, E., and Waitz, J. A. (1969). Megalomicin, a new macrolide antibiotic complex produced by *Micromonospora*. *J. Antibiot.* *22*:253-258.

Whaley, H. A., Patterson, E. L., Dornbush, A. C., Backus, E. J., and Bohonus, N. (1963). Isolation and characterization of relomycin, a new antibiotic. *Antimicrob. Agents Chemother.* *1963*:45-48.

Wiley, P. F., Gale, P., Pittinga, C. W., and Gezon, K. (1957). Erythromycin XII. The isolation, properties and partial structure of erythromycin C. *J. Am. Chem. Soc.* *79*:6074-6077.

Wolfe, A. D., and Hahn, F. E. (1964). Erythromycin: Mode of action. *Science* *143*:1445-1446.

Woo, P. W. K., Dion, H. W., and Bartz, Q. R. (1964). The structure of chalcomycin. *J. Am. Chem. Soc.* *86*:2726-2727.

Woodward, R. B. (1957). Struktur und Biogenese der Makrolide. Eine neue Klasse von Naturstoffen. *Angew. Chem.* *69*:50-58.

Yamamoto, H., Kusaka, T., Higashide, E., and Suzuki, T. (1969). Studies on antibiotic B-62169. I. *Streptomyces fradiae* B-62169 and its products, macrolide antibiotic B-62169A and B. *Ann. Rep. Takeda Res. Lab.* *28*:50-60.

Young, T. B., III. (1979). Fermentation scale-up: Industrial experience with a total environment approach. *Ann. N.Y. Acad. Sci.* *326*:165-180.

Antibiotics Used in Medical or
Agricultural Practice:
Antibacterial Antibiotics

16

THE POLYENES: PROPERTIES, BIOSYNTHESIS, AND FERMENTATION

JOSÉ A. GIL, PALOMA LIRAS, AND JUAN F. MARTIN *University of León, León, Spain*

I. INTRODUCTION

The first member of the group of polyene macrolide antifungal antibiotics was discovered by Hazen and Brown (1950), and it was tentatively named fungicidin (later renamed nystatin). It was effective in vitro against a large number of nonpathogenic and pathogenic fungi and had no effect against some of the common bacteria.

Since 1950 more than 90 different members of this group have been described, and more are being discovered each year, although many of the new isolates are no longer characterized or reported in the literature (Martin and McDaniel, 1977). Most of the polyenes isolated up until 1981 are produced by soil actinomycetes, mainly of the genus *Streptomyces*, although the production of polyenes by species of the genera *Streptoverticillum, Actinosporangium,* and *Chainia* has been reported.

The polyenes are usually formed as mixtures of polyenic and nonpolyenic products; this situation, common in secondary metabolism, has given rise to great difficulties in separating the different polyenic entities and in determining their precise molecular formulas. The identities of many of the polyene antibiotics remain questionable, and caution should be taken in considering all of them as different.

Although purification of polyene macrolide antibiotics is difficult, considerable progress has been achieved in their purification by high-pressure liquid chromatography and determination of the complex chemical structure of these compounds by sensitive analytical methods, such as proton magnetic resonance, x-ray, and mass spectrometry (Rinehart et al., 1974; Mechlinsky et al., 1970; Hansen and Thomsen, 1976).

II. CHEMICAL STRUCTURE

The polyene macrolides form a subdivision of the macrolide antibiotics containing hydroxylated macrocyclic lactone rings and usually one or more sugars. From the biosynthetic point of view, the macrolides are a homogeneous group, being synthesized from acetate, propionate, and other short-chain acids via the oligoketide pathway (Martín, 1975, 1976). The macrolide antibiotics are divided into two subgroups: (1) polyene macrolides (antifungal) and (2) nonpolyene macrolides (antibacterial) antibiotics. The polyene subgroup has a chromophore formed by a system of three to seven conjugated double bonds in the macrolactone ring. Polyene macrolides are amphipatic molecules containing both a rigid planar lipophilic portion and a flexible hydrophilic polyhydroxylated region. The chromophore accounts for some of the characteristic physical and chemical properties of the polyenes (strong light absorption, photolability, and poor solubility in water) and appears to be responsible for the differences in the modes of action of the polyene and nonpolyene macrolide subgroups.

The chromophore gives a typical multipeak ultraviolet visible light absorption spectrum. Polyene macrolides are subdivided into trienes, tetraenes, pentaenes, hexaenes, and heptaenes according to the number of conjugated double bonds in the chromophore. A list of the polyene macrolide antibiotics that are better characterized or more medically relevant is included in Table 1, together with the producer strains (Martín, 1977a). The polyene macrolides have lactone rings of 26–38 carbon atoms, which are much larger than those of the nonpolyene macrolides, except for the axenomycins, which have 34 atoms in the macrolide ring (Bianchi et al., 1974).

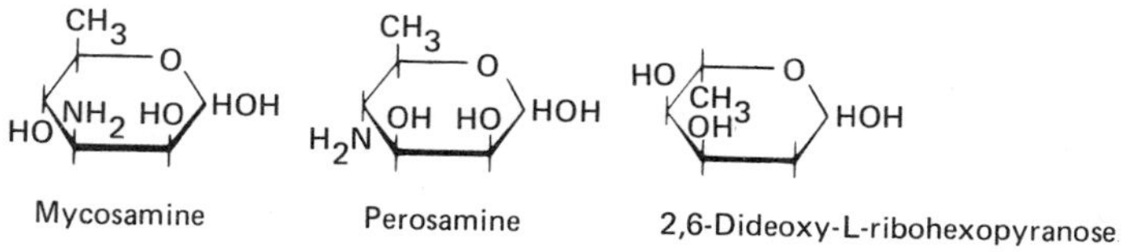

Figure 1 Chemical structures of the sugar moieties of polyene macrolide antibiotics. (From Martín and Gil, 1979.)

III. SUGAR AND AMINO SUGAR COMPONENTS

A large number of polyene macrolide antibiotics have an amino sugar (or neutral sugar) moiety that is linked to the macrolide ring by a glycoside bond. In all cases, the amino sugar is attached to a carbon adjacent to the chromophore. Two different amino sugars have been described as components of polyene macrolide antibiotics, namely, mycosamine (3-amino-3,6-dideoxy-D-mannose) and its isomer perosamine (4-amino-4,6-dideoxy-D-mannose) (Fig. 1).

Mycosamine is present in all the amino sugar-containing polyene macrolides, with the exception of the heptaene perimycin (fungimycin), which contains perosamine. Until 1979 it was believed that neutral sugar did not occur in polyene macrolides, but Zielinski et al. (1979) isolated the first neutral sugar from nystatin A_3, candidinin, and polifungin B, and it was identified as 2,6-dideoxy-L-ribohexopyranose. It is not clear, however, whether this neutral sugar is present in minor components of other polyene antibiotics.

IV. AROMATIC MOIETIES

Several polyenes belonging to the heptaene subgroup (ascosin, aureofungin, ayfactin, candicidin, heptamycin, levorin, trichomycin, DJ 400 B2, vacidin, etc.) have an aromatic p-aminoacetophenone moiety. Other heptaenes, including candimycin, DJ 400 B1, and perimycin, have an aromatic N-methyl-p-aminoacethophenone moiety (Fig. 2).

V. STRAIN IMPROVEMENT AND SCREENING PROGRAMS

Selection of high-production mutants after empirical mutagenesis treatment is one of the first steps prior to and during industrial production of an anti-

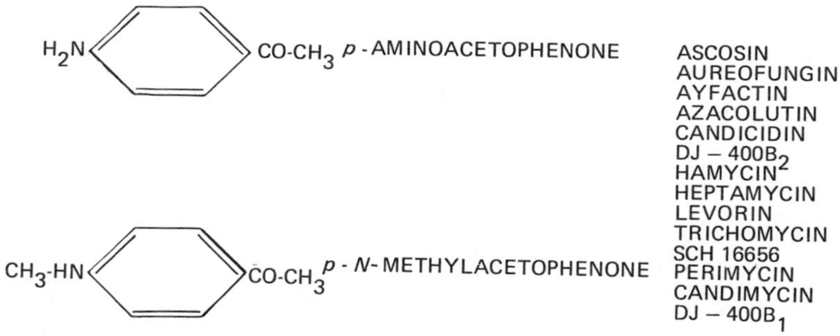

Figure 2 Aromatic moieties present in polyene macrolide antibiotics. (From Martín and McDaniel, 1977).

Table 1 Some of the More Representative Polyene Macrolide Antibiotics

Name (alternate name)	Producer strain	Chromophore subgroup	Aminosugar moiety	Aromatic moiety
Etruscomycin (lucensomycin)	*Streptomyces lucensis*	Tetraene	Mycosamine[a]	None
Pimaricin (tennecetin)	*S. natalensis* *S. chattanoogensis* *S. gilveosporus*	Tetraene	Mycosamine	None
Tetrin A,B	*Streptomyces* sp.	Tetraene	Mycosamine	None
Rimocidin	*S. rimosus*	Tetraene	Mycosamine	None
Nystatin (polifungin)	*S. noursei* *S. albulus*	Tetraene	Mycosamine	None
Chainin	*Chainia* sp.	Methylpentaene	None	None
Filipin (Durhamycin)	*S. filipensis*	Methylpentaene	None	None
Fungichromin (moldcidin B) (pentamycin) (lagosin)	*S. cinnamoneus* *S. roseoluteus* *S. cellulosae* *S. pentaticus*	Methylpentaene	None	None
Eurocidin A,B	*S. eurocidicus* *S. albireticuli*	Pentaene	Mycosamine	None
Flavofungin (mycoticin A)	*S. ruber* *S. flavofungini*	Carbonylpentaene	None	None
Mycoticin B	*S. flavofungini*	Carbonylpentaene	None	None

Candihexin A,B	*S. viridoflavus*	Hexaene	Mycosamine	None
Candihexin E,F	*S. viridoflavus*	Hexaene	None	None
Flavacid	*S. flavus*	Hexaene	None	None
Dermostatin	*S. viridogriseus*	Carbonylhexaene	None	None
Ascosin	*S. canescus*	Aromatic heptaene	Mycosamine	p-Aminoacetophenone
Ayfactin (aureofacin)	*S. aureofaciens*	Aromatic heptaene	Mycosamine	p-Aminoacetophenone
	S. viridofaciens	Aromatic heptaene	Mycosamine	p-Aminoacetophenone
Aureofungin	*S. cinnamoneus*	Aromatic heptaene	Mycosamine	p-Aminoacetophenone
Candicidin	*S. griseus*	Aromatic heptaene	Mycosamine	p-Aminoacetophenone
Hamycin	*S. primprina*	Aromatic heptaene	Mycosamine	p-Aminoacetophenone
Heptafungin A	*S. longisporolavendulae*	Aromatic heptaene	Mycosamine	p-Aminoacetophenone
Levorin	*S. levoris*	Aromatic heptaene	Mycosamine	p-Aminoacetophenone
Trichomycin	*S. hachijoensis*	Aromatic heptaene	Mycosamine	p-Aminoacetophenone
DJ-400 B2	*S. surinam*	Aromatic heptaene	Mycosamine	p-Aminoacetophenone
DJ-400 B1	*S. surinam*	Aromatic heptaene	Mycosamine	N-methyl-p-aminoacetophenone
Perimycin (Fungimycin)	*S. coelicolor*	Aromatic heptaene	Perosamine[b]	N-methyl-p-aminoacetophenone
Amphotericin B	*S. nodosus*	Nonaromatic heptaene	Mycosamine	None
Candidin	*S. viridoflavus*	Nonaromatic heptaene	Mycosamine	None
Mycoheptin	*Streptoverticillium mycoheptinicum*	Nonaromatic heptaene	Mycosamine	None
	S. netropsis			

[a] 3-Amino-3,6-dideoxy-D-mannopyranose.
[b] 4-Amino-4,6-dideoxy-D-mannopyranose.

Source: Reproduced, with permission, from the *Annual Review of Microbiology*, Volume 31. c 1979. Annual Reviews Inc.

biotic. Such work has been carried out with the producer strains of amphotericin B and nystatin (Thoma, 1971), levorin (Tereshin, 1976), and most probably with other polyene producers in several industries. Soil isolates of polyene producers showed a large population variability. High spontaneous variability in the shape of spores and sporophores and antibiotic titers was found in the amphotericin B-producing organism *Streptomyces nodosus*. Zhukova et al. (1970) proposed that the high variability is due to the heterogeneous nature of the strain, consisting probably of several genotypes. Such differences in genotype have been found among sporeforming and spore-minus variants of the macrolide producer *Streptomyces reticuli* (Schrempf, 1982). They are probably due to chromosomal rearrangement. Mutants with abundant sporulation and sporophores in the form of short tight spirals were highly active in amphotericin C production, whereas those with poor sporulation were low or nonproducers. This correlation is frequently observed in *Streptomyces* species which produce antibiotics (Kern et al., 1980). It possibly indicates that sporulation and antibiotic production are regulated by similar intracellular effectors (Sarkar and Paulus, 1972).

Sometimes a strain produces more than one antibiotic, and it is convenient to eliminate the production of one or several of them. The original soil isolate of *Streptomyces noursei* produced cycloheximide in addition to the polyene macrolide nystatin (Spizek et al., 1965a). Strains that do not produce any compound of the cycloheximide series synthesize about double the amount of nystatin (Dolezilova et al., 1965). Russian workers claimed that selection of high-production levorin variants may be carried out simply by screening for high-production clones plated from submerged cultures (Tereshin, 1976).

VI. BIOSYNTHESIS

The three different moieties of polyene macrolides (i.e., the macrolide ring, the aromatic moiety, and the amino sugar moiety, are synthesized by three different pathways (Martin, 1977a).

A. Biosynthesis of the Macrolide Aglycone Ring

The macrolide ring of polyene macrolide antibiotics is formed basically from acetate and propionate. Initial studies by Birch et al. (1964) showed that [1-, 2-, and 3-^{14}C]propionate and [1-^{14}C]acetate were incorporated into the nystatin aglycone (nystatinolide), whereas negative results were obtained with [2-^{14}C]mevalonic acid and [methyl-^{14}C]methionine, indicating the absence of terpenoid or C1 precursor units. Similar precursor incorporations were observed in the biosynthesis of the polyene macrolides lucensomycin (Manwaring et al., 1969), amphotericin B (Linke et al., 1974; Perlman and Semar, 1965), candicidin (Liu et al., 1972a; Martín and Liras, 1976), levorin (Belousova et al., 1971), and fungimycin (Liu et al., 1972b). The lack of incorporation of methyl groups was also described by Perlman and Semar (1965) in amphotericin B and by Liu et al. (1972a) in candicidin.

Table 2 shows the proposed precursor units of some polyene macrolide antibiotics. As observed, straight side chains of four to six carbon atoms exist in the polyenes filipin and chainin, which might represent direct introduction of six to eight preformed carbon fatty acids, although such fatty acids would be formed by previous condensation of acetate units (Martín, 1977a).

The biosynthesis of the macrolide rings of the polyene antibiotics occurs via the oligoketide pathway by repeated head-to-tail condensation of acetate

Table 2 Biosynthetic Units Required for Formation of the Macrolide Rings of Polyene Antibiotics

Polyene macrolide antibiotics	Units required	Reference
Nystatin (polifungin)	Three propionate and 16 acetate units	Birch et al. (1964)
Lucensomycin (etruscomycin)	Two propionate and 12 acetate units	Manwaring et al. (1969)
Amphotericin B	Three propionate and 16 acetate units	Linke et al. (1974)
Pimaricin (tennecetin)	One propionate and 12 acetate units	Martin (1977a)
Filipin	One eight-carbon unit, one propionate, and 12 acetate units	Martin (1977a)
Mycoticin A (flavofungin)	Three propionate and 14 acetate units	Martin (1977a)
Chainin	One six-carbon unit, one propionate, and 12 acetate units	Martin (1977a)

Source: With permission from J. F. Martin, "Polyene macrolide antibiotic," in *Economic Microbiology*, Vol. 3, A. H. Rose (Ed.). © 1979 Academic Press.

and propionate units. The term *polyketide* (polyacetate) *pathway* was introduced to indicate the formation of a β-polyketo chain by condensation of acetate units, which is involved in the biosynthesis of a large number of plant
and microbial metabolites (Light, 1970; Turner, 1971). However, the term
oligoketide rather than *polyketide* should be used, because of the limited number of units which are polymerized. Fatty acids are synthesized by a specific
pathway in which a single acetate (or propionate) starter unit and a small
number of malonate units are joined together. Oligoketides are formed by a
nonspecific pathway in which one starter unit (acetate, propionate, or p-
aminobenzoyl coenzyme A) and a highly variable number of malonate or methylmalonate units are similarly attached together. In fatty acid synthesis, the
β-keto group introduced in each condensing step is modified by reduction,
dehydration and hydrogenation prior to the next condensation step (Volpe
and Vagelos, 1973). In oligoketide synthesis, the β-keto group formed at
each chain-extending step may or may not be retained. Successive condensation steps result in the formation of the oligoketide chain (Martín, 1976)
(Fig. 3).

In fatty acid synthesis reduced NADPH is required as a cofactor in the
reduction and hydrogenation steps. The requirement for NADPH appears
to be lower in the biosynthesis of the polyenes, since some of them are simply
reduced to hydroxyl groups without undergoing subsequent dehydration and
hydrogenation (Martín, 1977a).

Since both fatty acids and macrolides are synthesized by the same cell,
an interesting question is whether the fatty acid synthetase and macrolide
synthetase are one single enzyme or two different entities. Studies on the
biosynthesis of 6-methylsalicylic acid have demonstrated that the oligoketide
synthetase involved is a multienzyme complex similar to but physically different from fatty acid synthetase. The synthetase has been separated by
gel filtration and sucrose density gradient centrifugation (Light and Hager,
1968). Preliminary studies in *Streptomyces erythreus*, which synthesizes
the nonpolyene macrolide erythromycin, indicates that this macrolide is also
synthesized by a multienzyme synthetase complex (Corcoran, 1974). Cerulenin,
an inhibitor of fatty acid synthesis, inhibits specifically the biosynthesis of
the polyene macrolide antibiotic candicidin by *Streptomyces griseus* (Martín
and McDaniel, 1975a). Since cerulenin is known to inhibit the condensation
of malonyl coenzyme A (CoA) subunits in the formation of fatty acids, it is
concluded that polyenes are synthesized via the oligoketide pathway by condensation steps similar to those occurring in fatty acid biosynthesis. Preliminary results in our laboratory on in vitro biosynthesis of candicidin indicate that both malonyl-CoA and methylmalonyl-CoA seem to be incorporated
into a candicidin-like molecule (Naharro and Martín, unpublished results).

B. Biosynthesis of the Aromatic Moiety

The p-aminoacetophenone and N-methyl-p-aminoacetophenone moieties of the
polyene (heptaenes) macrolides are synthesized from glucose via the aromatic
pathway. p-Aminobenzoic acid (PABA) has been identified as the immediate
precursor of the aromatic moiety of candicidin. The intact ring and the carboxyl group of PABA are incorporated into candicidin, as concluded from the
quantitative incorporation of [ring UL-^{14}C]PABA and [7-^{14}C]PABA (Liu et
al., 1972a; Martín and Liras, 1976). The branch-point compound of the aromatic pathway giving rise to PABA in bacteria is chorismic acid (Gil and
Martín, in press). Neither tryptophan, phenylalanine, tyrosin, nor anthra-

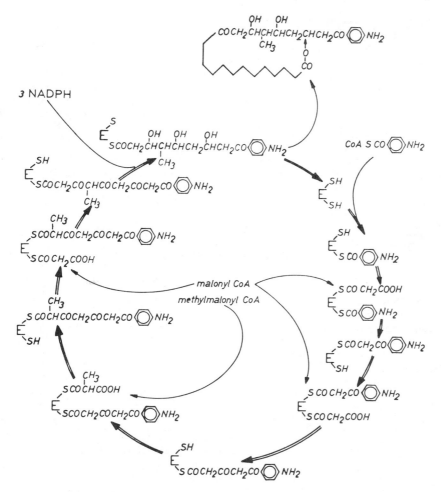

Figure 3 Proposed pathway for the formation of the oligoketide chain of aromatic polyene macrolide antibiotics. (Reproduced, with permission, from the *Annual Review of Microbiology*, Volume 31. © 1977a. Annual Reviews Inc.)

nilic acid are incorporated into candicidin. In *Escherichia coli*, *Enterobacter aerogenes*, *Neurospora crassa*, and *Bacillus subtilis*, chorismic acid is converted into PABA in the presence of L-glutamine as the amino donor (Gibson et al., 1964; Huang and Gibson, 1970; Altendorf et al., 1971; Kane and O'Brien, 1975).

The aromatic pathway is a complex, multibranched, diverging pathway leading to the biosynthesis of the aromatic amino acids and to other minor metabolic products such as PABA, 4-hydroxybenzoic acid, and 2,3-dehydroxybenzoic acid (Gibson and Pittard, 1968).

In strains producing aromatic polyene macrolides under conditions of active antibiotic synthesis, the PABA synthetase is very active (Gil and Martín, unpublished results), and this enzyme is subject to regulation by aromatic amino acids (Gil et al., 1980). It is very interesting to note that PABA synthethase has not been detected in several *Streptomyces* strains which do not produce aromatic polyene macrolides. In *Streptomyces* this enzyme seems specific for PABA biosynthesis, which is used for production of secondary metabolites.

There is evidence from biosynthetic studies of several polyketide-derived products with aromatic moieties which indicates that aromatic rings are used as primer units in polyketide formation. Light (1970) described several examples in which an aromatic cinnamoyl starter unit produced several compounds of the chalcone or stilbene series by addition of malonyl-CoA units. Another well-known example of an aromatic moiety acting as a starter in the synthesis of aromatic antibiotics is the C_7N unit involved in the biosynthesis of the macrolactam antibiotic rifamycin. Recent studies at the Ciba-Geigy Laboratories (Basel, Switzerland) have identified the 3-amino-5-hydroxybenzoic acid as a direct precursor of the seven-carbon amino acid starter unit for the biosynthesis of ansamycin (Ghisalba and Nüesch, 1981). By chemical and biochemical studies they demonstrate that this compound is derived from an intermediate of the shikimate pathway between 3-deoxy-D-arabinoheptulosonic acid-7-phosphate and shikimate (Ghisalba and Nüesch, 1978a,b).

The chemical structure of the aromatic heptaenes shows that the aromatic moiety is covalently linked to the carbon atom of the macrolide ring that supports the lactone bond.

This suggests that cyclization occurs when the polyketide chain formed by the polymerization of C2 and C3 units, using as starter the aromatic ring, has achieved the adecuate length. Supporting a starter function of the aromatic moiety in polyene macrolide biosynthesis is the fact that cerulenin rapidly inhibits incorporation of PABA into candicidin (Martin and McDaniel, 1975a). Rapid inhibition of PABA incorporation into candicidin would not be expected until the pool of macrolide ring is depleted if PABA is attached after the macrolide ring is formed.

B. Biosynthesis of the Macrolide Amino Sugars

No in vitro studies have been carried out specifically on the biosynthesis of the amino sugar moieties of polyene macrolide antibiotics; however, the similarity with the biosynthesis of amino sugars of nonpolyene macrolides and bacterial lipopolysaccharides provides some data on the probable biosynthetic pathway. It has been known that the amino sugars of macrolides are formed from hexoses without cleavage of the sugar chain (Corcoran and Chick, 1966). All biosynthetic transformations of these sugars occur while they are in the form of nucleoside diphosphate derivatives. In all cases, the corresponding 4-keto-6-deoxysugar nucleoside diphosphate has been described as an intermediate (Lüderitz et al., 1966; Pape and Brillinger, 1973; Ortman et al., 1974). The formation of 4-keto-6-deoxihexose nucleoside dephosphate from deoxythimidine diphosphate (dTDP) glucose is catalized by a NAD-dependent dTDP-glucose oxidoreductase (Melo et al., 1968; Wang and Gabriel, 1969; Glasser and Zarkowski, 1971). This enzyme has been purified from a tylosin-producing strain of *Streptomyces rimosus* and its activity increases during the stationary phase together with the tylosin-synthesizing system (Mattern et al., 1973).

The following step seems to be the conversion of TDP-4-keto-6-deoxy-D-glucose to TDP-4-amino-4,6-dideoxy-D-glucose. It is carried out by a pyridoxal phosphate-requiring transaminating enzyme that catalyzes the transfer of an amino group from L-glutamate (Matsuhashy and Strominger, 1966). The intermediate TDP-4-amino-4,6-dideoxy-D-glucose may be converted into TDP-perosamine by epimerization. A similar biosynthetic process seems probable

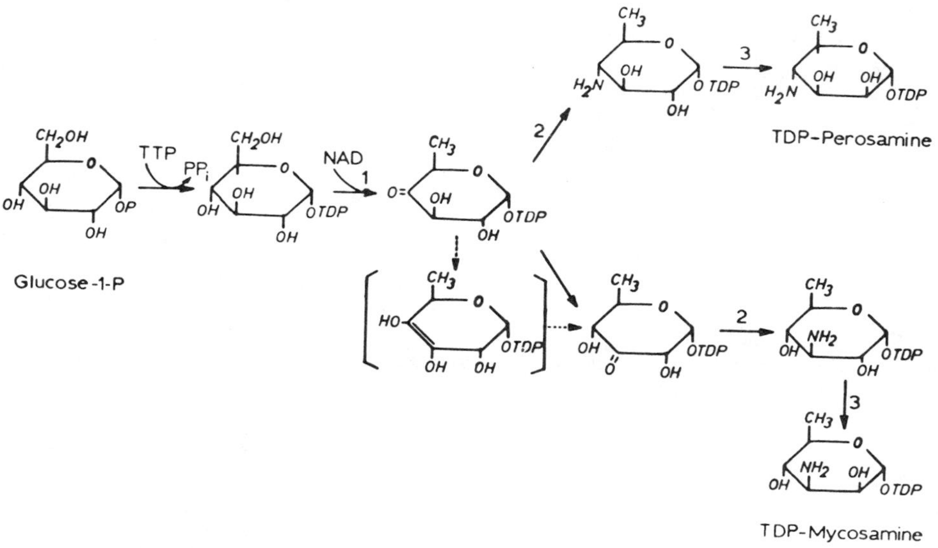

Figure 4 Proposed pathway of biosynthesis of the aminosugar moieties of polyene macrolide antibiotics. (Reproduced, with permission, from the *Annual Review of Microbiology*, Volume 31, 1977. Annual Reviews Inc.)

in the formation of the polyene amino sugar mycosamine, except that isomerization of the keto group is required. In the light of this evidence, Martin (1977a) proposed a pathway for the synthesis of the amino sugar moieties of polyene macrolides (Fig. 4).

The attachment of the amino sugar moiety to the macrolide ring seems to take place during the secretion of the polyene. In the biosynthesis of candihexin it was observed that the intracellular antibiotic consists exclusively of nonglycosylated inactive components, whereas the excreted antibiotic contains a mixture of glycosylated and nonglycosylated components (Martin and McDaniel, 1975b). This evidence is consistent with the theory that the macrolide ring acts as a lipid-soluble carrier of the amino sugar during the secretion process (Martin, 1977a). Sugar attachment during biosynthesis of the macrolide antibiotic tylosin is also a late step, since mutants blocked in either sugar biosynthesis or in glycosyltransferase reactions are able to form the corresponding aglycone tylonolide (Baltz and Seno, 1981).

VII. NUTRITIONAL AND ENVIRONMENTAL FACTORS AFFECTING POLYENE MACROLIDE FERMENTATION

Nutritional studies of polyene-producing *Streptomyces* have been carried out with batch cultures, usually in shake flasks, with occasional scale-up to pilot-plant or large-scale fermenters with those polyenes that are produced industrially. In most studies several nutritional factors or physical parameters are studied and the combinations giving the highest yields are selected for further uses (McDaniel et al., 1976).

Empirical medium development studies have not substantially contributed to an understanding of the regulation and control of the metabolism of the producing strain.

A. Carbohydrates as Carbon Sources

Most polyene antibiotic fermentations have been carried out using glucose as a carbon source. Concentrations of glucose as high as 7–9.5% have been routinely used (Martin and McDaniel, 1977). However, in pilot-plant fermentations of candidin it was observed that the high initial concentration of glucose retarded growth and resulted in abnormal fermentations patterns. Slow feeding of glucose, initially described in the penicillin fermentation (Soltero and Johnson, 1954) to bypass the negative effect of high glucose concentrations, resulted in increased synthesis of the polyenes candidin and candihexin (Martín and McDaniel, 1974). Intermittent addition of glucose in the amphotericin B fermentation also resulted in a slight increase in the titer of amphotericin B (Brewer and Frazier, 1962). The higher antibiotic synthesis under slow feeding of glucose suggests regulation of the polyene macrolide synthetases by the metabolites of rapid glucose utilization, which is bypassed when glucose is fed slowly.

Strong carbon catabolite regulation occurs in the biosynthesis of several antibiotics (penicillin, actinomycin, streptomycin, siomycin, bacitracin, etc.) (for a review see, Martin and Demain, 1980); however, there are significant differences between catabolite regulation of the biosynthesis of these antibiotics and the effect of glucose in polyene macrolide synthesis. For example, actinomycin formation occurred only after glucose was exhausted from the culture medium, whereas polyene macrolide synthesis takes places better at low concentrations of glucose (5–15 mg/ml), but it stops when glucose is depleted (Martín and McDaniel, 1977). The evidence existing on catabolite regulation of polyene macrolide biosynthesis by glucose is obscured by the fact that glucose is required for synthesis of the product.

Disaccharides, especially lactose, were poor substrates for polyene production. In the amphotericin fermentation, British gum (a dextrin) was used as a carbon source because it gives large amounts of amphotericin B relative to amphotericin A. The same effect is achieved by certain alcohols (see below).

B. Short-Chain Fatty Acids as Substrates for Polyene Biosynthesis

Acetate and propionate as their biologically active units, acetyl-CoA and propionyl-CoA, and their carboxylated derivatives, malonyl-CoA and methyl-malonyl-CoA, are the building blocks used in the head-to-tail condensation leading to the formation of the macrolide antibiotics. Acetate and propionate do not support antibiotic synthesis by themselves, but supplementation of a glucose basal medium with acetate, propionate, or malonate significantly stimulates candicidin synthesis in batch cultures (Martin and McDaniel, 1976). Addition of acetate also stimulates fungimycin biosynthesis (Mohan et al., 1963).

The inability of acetate, propionate, and malonate to serve as a better carbon source than glucose is explained by the fact that glucose does not provide acetate and propionate and it is involved in the biosynthesis of the aromatic moiety (p-aminoacetophenone) through the pentose phosphate cycle and provides NADPH for the reductive steps of the highly oxidized polyketide chain.

Citrate is a positive effector of the acetyl-CoA carboxylase (Volpe and Vagelos, 1973), the first enzyme involved in the branching of acetyl-CoA for polyketide biosynthesis. However, high levels of citrate were clearly inhibitory, apparently due to its metal-chelating properties (Martin and McDaniel, 1976).

Early studies (McCarthy et al., 1955, Brock, 1956) demonstrated that oils and fatty acids stimulated the production of the polyene macrolide antibiotics fungichromin and filipin but not amphotericin B (Brewer and Frazier, 1962) or candicidin (Ethiraj, 1969).

Although oils and fatty acids contain more energy per unit weight than glucose, the increased polyene macrolide production with oils cannot be attributed to the additional available energy, since the yields are far greater than would be expected from the energy supplied. The stimulation of polyene antibiotic biosynthesis by oils might be a simple precursor effect. Catabolism of fatty acids results in a increased pool of acetyl-CoA, which is subsequently used for polyene biosynthesis.

Martín and McDaniel, (1975a) found that exogenous fatty acids were unable to support candicidin biosynthesis when endogenous condensation of acetyl-CoA units was prevented by cerulenin. It is not clear at present why oils stimulate the biosynthesis of some polyene macrolides (fungichromin and filipin) and have an inhibitory effect on the biosynthesis of others.

C. Nitrogen Sources

Complex nitrogen sources are the choice for large-scale antibiotic production. A large variety of complex nitrogen sources, including plant (and some animal) proteins have been used. These include yeast extract, soy peptone, cottonseed meal, soybean meal, corn meal, casein, corn-steep liquor, and distiller's solubles. Soybean meal is a good nitrogen source for producing many polyene macrolides (Brewer and Frazier, 1962; Mohan et al., 1963; Martín and McDaniel, 1974), probably because of its balance of nutrients and slow hydrolysis, which create a physiological condition during trophophase that favors antibiotic production during idiophase (Nefelova and Pozmogova, 1960). Considerable differences in the ability of different amino acids to support candicidin biosynthesis were founded by Acker and Lechevalier (1954). L-Asparagine was the best nitrogen source. L-Histidine, glycine, L-glutamic acid, and L-aspartic acid were next in effectiveness as nitrogen sources. The D-isomers of the amino acids supported negligible growth and no candicidin production.

Its interesting to note that β-alanine supported a 10-fold higher yield than α-alanine and was the best amino acid source for the production of ayfactin. This result is interesting, as β-alanine is a component of the pantetheinyl moieties of coenzyme A and pantetheinyl−protein of fatty acid and macrolide synthetases (Abou-Zeid and Abou-el-Atta, 1971).

Inorganic salts are generally poor nitrogen sources for polyene production; however, ammonium salts were acceptable in the case of mycoheptin (Tereshin, 1976).

In the last few years a new regulatory mechanism, similar to carbon catabolite regulation, which controls utilization of nitrogen sources by the cell has been discovered. Ammonia represses the formation of the many enzymes involved in the utilization of other nitrogen sources (Aharanowitz, 1980). There have been specific studies on nitrogen catabolite regulation of polyene antibiotic synthesis, but there are several reports in the literature which suggest that antibiotic synthesis may be repressed by ammonia and other rapidly utilized nitrogen sources. Martín and McDaniel (1974) described that addition of soy bean meal extracts produced an immediate increase in respiration and decreased the rate of candidin and candihexin synthesis.

Aromatic polyenes macrolide antibiotics, at least candicidin, are subject to regulation by aromatic amino acids. The biosynthesis of candicidin was inhibited by L-tryptophan, L-phenylalanine, and, to a lesser degree, L-tyrosine. A mixture of the three aromatic amino acids inhibited candicidin biosynthesis to a greater extent than did each amino acid separately. The inhibitory effect of tryptophan was partially reverted by exogenous PABA, suggesting that this effect is exerted at the PABA synthetase level (Gil et al., 1980). In vitro experiments have clearly shown that PABA synthetase is the main target enzyme in the feed-back regulation by aromatic amino acids of the biosynthesis of candicidin (Gil et al., 1980). This is a fine example of a regulatory system in which the biosynthesis of a secondary metabolite (candicidin) is regulated by the mechanisms controlling the biosynthesis of primary metabolites, such as aromatic amino acids.

D. Phosphate and Adenylate Charge Regulation

The biosynthesis of many antibiotics is inhibited by phosphate. Antibiotics are synthesized only at concentrations of inorganic phosphate that are suboptimal for growth. The lowest concentration of inorganic phosphate inhibitory for antibiotic production ranges from 1 to 50 mM for different antibiotics. Phosphate in the range 0.3–500 mM permits excellent cell growth, whereas 10 mM phosphate often suppresses biosynthesis of antibiotics. The same inhibitory effect is produced by deoxyribonucleotides, because they are cleaved during uptake by *S. griseus* cells (Martín and Demain, 1977; Martín, 1977b).

The molecular mechanism of the phosphate effect is largely unknown. Several mechanisms have been proposed to explain the regulatory effect of phosphate (Martín, 1977b): Phosphate (1) favors primary over secondary metabolism, (2) shifts carbohydrate catabolic pathways, (3) limits the synthesis of inducers of antibiotic biosynthetic pathways, (4) inhibits the formation of antibiotic precursors, and (5) inhibits or represses the phosphatases necessary for antibiotic production. The different mechanisms may well be distinct aspects of a common underlying mechanism: the regulation of the intracellular ATP pool levels by exogenous extracellular phosphate. Intracellular ATP levels or the adenylate energy charge of the cell (Atkinson, 1969) might well be the intracellular effector carrying the nutritional message to the gene expression machinery. Cyclic adenosine 5'-monophosphate does not seem to be the intracellular effector mediating phosphate control of antibiotic biosynthesis (Vining et al., 1982). The phosphate effect seems to be exerted at two different expression levels: It (1) represses the formation of enzymes involved in antibiotic biosynthesis during the growth phase (Liras et al., 1977) and (2) inhibits the activity of the same enzymes once they have been formed (Martín et al., 1977). Recent results in our laboratory indicate that phosphate control of candicidin biosynthesis is exerted by repressing PABA synthetase and at the same time by indirectly inhibiting the activity of the DAHP synthetase. Phosphate stimulates the formation of the sugar phosphates, which inhibit the activity of DAHP synthetase (G. Naharro, unpublished results).

E. Effect of Metal Ions

Weinberg (1962, 1970) has reviewed the role of trace metals in secondary metabolism and has concluded that iron and zinc are "key metals" for actinomycetes and fungi. The concentrations of metal required for secondary metabolism are one or more log units higher than the minimum required for growth (about 10^{-7} M). Moreover, primary metabolism tolerates up to 10^{-3} M of each metal, about 2 log units higher than the level that is inhibitory for secondary metabolism.

Early studies indicated that iron and zinc were essential for growth or polyene antibiotic production (Acker and Lechevalier, 1954). Stimulation of the production of the tetraene antimycin by calcium and magnesium ions was reported by Schaffner et al. (1958). Brewer and Frazier (1962) described stimulatory effect of manganous, nickel, zinc, and cobalt salts on the production of amphotericin B in soybean meal media. Stimulatory effects of iron, zinc, and magnesium on candicidin biosynthesis were reported by Liu et al. (1975). Addition of zinc at any time during the fermentation resulted in increased candicidin synthesis, reduced dry weight, and an increased rate of glucose utilization.

A probable role of metal ions in the biosynthesis of polyene macrolides is either the activation of polyene synthetases or the stabilization of the intermediate substrates in polyketide biosynthesis, as suggested by Bu'lock (1967). β-Polyketide intermediates have a great tendency to form coordination complexes with metal ions. The stimulatory effect of magnesium ions on polyene macrolide biosynthesis might be related to the activating effect on enzymes involved in polyene biosynthesis. A magnesium requirement has been found in in vitro studies on acetyl-CoA carboxylase (Rasmussen and Klein, 1968). We have found recently that PABA synthetase activity is greatly stimulated by Mg^{2+}, but this enzyme is inhibited by Zn^{2+} (Gil and Martin, unpublished results).

Recently, it was reported (Tanaka et al., 1980) that microbial conversion of glycine to L-serine was stimulated by magnesium phosphate. The relevance of the stimulation to the nitrogen catabolite regulation was postulated, based upon the fact that the ammonium ion concentration in the culture broth was depressed in the presence of magnesium phosphate. Removal of nitrogen regulation might explain why the production of macrolide antibiotic was increased 2- to 100-fold in the presence of magnesium phosphate (Omura et al., 1980).

REFERENCES

Abou-Zeid, A. A., and Abou-el-Atta, A. Y. (1971). *Zentralbl. Bakteriol. Parasitenkd. Infectidnskr. Hyg. Abt.* 2:126,371.

Acker, R. F., and Lechevalier, H. (1954). *Appl. Microbiol.* 2:152–157.

Aharonowitz, Y. (1980). *Annu. Rev. Microbiol.* 34:209–234.

Altendorf, K. H., Gilch, B., and Lingens, F. (1971). *FEBS Lett.* 16:95–98.

Atkinson, D. E. (1969). *Annu. Rev. Microbiol.* 23:47–68.

Baltz, R. H., and Seno, E. T. (1981). *Antimicrob. Agents Chemother.* 20: 214–225.

Belousova, I. I., Lishnevskaya, E. B., and Elgart, R. E. (1971). *Antibiotiki* 16:684–687.

Bianchi, M., Cotta, E., Ferni, G., Grein, A., Inlita, P., Mazzoleni, R., and Spalla, C. (1974). *Arch. Microbiol.* 98:289–299.

Birch, A. J., Holzapfel, C. W., Rickards, R. W., Djerassi, C., Suzuki, M., Westley, J., Dutcher, J. D., and Thomas, R. (1964). *Tetrahedron Lett.* 1964:1485–1490.

Brewer, G. A., and Frazier, W. R. (1962). *Antimicrob. Agents Chemother.* 1961:212–217.

Brock, T. D. (1956). *Appl. Microbiol.* 4:131–133.

Bu'lock, J. D. (1967). *Essays in Biosynthesis and Microbial Development.* Wiley, New York.

Chatterjee, S., and Vining, L. C. (1982). *Can. J. Microbiol. 28*:311–317.

Corcoran, J. W. (1974). In *Genetics of Industrial Microorganisms*, Vol. 2. Z. Vanek, Z. Hostalek, and J. Cudlin (Eds.). Prague Academja, Prague, pp. 339–351.

Corcoran, J. W., and Chick, M. (1966). *Biosynth. Antibiot. 1*:159–201.

Dolezilova, L., Spizek, J., Vondracek, M., Paleckova, K. and Vanek, Z. (1965). *J. Gen. Microbiol. 39*:305–309.

Ethiraj, S. (1969). Ph.D. thesis, Rutgers University, New Brunswick, New Jersey.

Ghisalba, O., and Nüesch, J. (1978a). *J. Antibiot. 31*:202–214.

Ghisalba, O., and Nüesch, J. (1978b). *J. Antibiot. 31*:215–225.

Ghisalba, O., and Nüesch, J. (1981). *J. Antibiot. 34*:58–63.

Gibson, F., and Pittard, J. (1968). *Bacteriol. Rev. 32*:465–492.

Gibson, F., Gibson, M., and Cox, G. B. (1964). *Bioshim, Biophys. Acta. 82*:637–638.

Gil, J. A., Liras, P., Naharro, G., Villanueva, J. R., and Martin, J. F. (1980). *J. Gen. Microbiol. 118*:189–195.

Glasser, L., and Zarkowski, H. (1971). In *The Enzymes*. P. D. Boyer (Ed.). Academic, London, p. 465.

Hansen, S. H., and Thomsen, M. (1976). *J. Chromatogr. 123*:205–211.

Hazen, E. L., and Brown, R. (1950). *Science 112*:423.

Huang, M., and Gibson, F. (1970). *J. Bacteriol. 102*:797–773.

Kane, J. F., and O'Brien, H. D. (1975). *J. Bacteriol. 123*:1131–1138.

Kern, B. A., Hendlin, D., and Inamine, E. (1980). *Antimicrob. Agents Chemother. 17*:679–685.

Light, R. J. (1970). *J. Agric. Food. Chem. 18*:260–267.

Light, R. J., and Hager, L. P. (1968). *Arch. Biochem. Biophys. 125*:326–333.

Linke, H. A. B., Mechlinski, W., and Schaffner, C. P. (1974). *J. Antibiot. 27*:155–160.

Liras, P., Villanueva, J. R., and J. F. Martín. (1977). *J. Gen. Microbiol. 102*:269–277.

Liu, C. M., McDaniel, L. E., and Schaffner, C. P. (1972a). *J. Antibiot. 25*:116–121.

Liu, C. M., McDaniel, L. E., and Schaffner, C. P. (1972b). *J. Antibiot. 25*:187–188.

Liu, C. M., McDaniel, L. E., and Schaffner, C. P. (1975). *Antimicrob. Agents Chemother. 7*:196–202.

Lüderitz, O., Stanb, A. M., and Westphal, O. (1966). *Bacteriol. Rev. 30*:192–255.

McCarthy, F. J., Fisher, W. P., Charney, J., and Tytell, A. (1955). *Antibiot. Annu. 1954*:719–723.

McDaniel, L. E., Bailey, E. G., Ethirag, S., and Andrews, H. P. (1976). *Dev. Ind. Microbiol. 17*:91–98.

Manwaring, D. G., Rickards, R. W., Gaudiano, G., and Nicolella, V. (1969). *J. Antibiot. 22*:545–550.

Martin, J. F. (1976). *Dev. Ind. Microbiol. 17*:223–231.

Martin, J. F. (1977a). *Annu. Rev. Microbiol. 31*:13–38.

Martin, J. F. (1977b). In *Advances in Biochemical Engineering*, Vol. 6. Ed. T. K. Ghose, A. Fiechter, and N. Blakebrough (Eds.). Springer-Verlag, Berlin, pp. 105–127.

Martin, J. F., and Demain, A. L. (1977). *Can. J. Microbiol. 23*:1334–1339.

Martin, J. F., and Demain, A. L. (1980). *Microbiol. Rev. 44*:230–251.

Martin, J. F., and Liras, P. (1976). *J. Antibiot.* 29:1306–1309.

Martin, J. F., and McDaniel, L. E. (1974). *Dev. Ind. Microbiol.* 15:324–337.

Martín, J. F., and Mcdaniel, L. E. (1975a). *Biochim, Biophys. Acta.* 411: 186–194.

Martin, J. F., and McDaniel, L. E. (1975b). *Antimicrob. Agent. Chemother.* 8:200–208.

Martín, J. F., and McDaniel, L. E. (1976). *Eur. J. Appl. Microbiol.* 3:135–144.

Martín, J. F., and McDaniel, L. E. (1977). *Adv. Appl. Microbiol.* 21:2–52.

Martin, J. F., Liras, P., and Demain, A. L. (1977). *FEMS Microbiol. Lett.* 2:173–176.

Matsuhashy, M., and Strominger, J. L. (1966). *J. Biol. Chem.* 241:4738–4744.

Mattern, U., Brillinger, G. U., and Pape, H. (1973). *Arch. Mikrobiol.* 88: 37–48.

Mechlinsky, W., Schaffner, C. P., Ganis, P., and Avitabile, G. (1970). *Tetrahedron Lett.* 1970:3873–3876.

Melo, A., Elliot, W. H., and Glasser, L. (1968). *J. Biol. Chem.* 243:1467–1474.

Mohan, R. R., Pianotti, R. S., Martin, J. F., Ringel, S. M., Schwartz, B. S., Bailey, E. G., McDaniel, L. E., and Schaffner, C. P. (1963). *Antimicrob. Agents Chemother.* 1963:462.

Nefelova, M. W., and Pozmogova, I. N. (1960). *Mikrobiologiya* 29:856–861.

Omura, S., Tanaka, Y., Tanaka, H., Takahashi, Y., and Iwai, Y. (1980). *J. Antibiot.* 33:1568–1569.

Ortman, R., Mattern, U., Grisebach, H., Stadler, P., Sinnwell, V., and Paulsen, H. (1974). *Eur. J. Biochem.* 43:265–271.

Pape, H., and Brillinger, G. U. (1973). *Arch. Mikrobiol.* 88:254–264.

Perlman, D., and Semar, J. B. (1965). *Biotechnol. Bioeng.* 7:133–137.

Rasmussen, R. K., and Klein, H. P. (1968). *J. Bacteriol.* 95:727.

Rinehart, K. L., Cook, J. C., Maurer, K. H., and Rapp, U. (1974). *J. Antibiot.* 271–13.

Sarkar, N., and Paulus, H. (1972). *Nature (London) New Biol.* 239:228–230.

Schaffner, C. P., Steinman, I. D., Safferman, R. S., and Lechevalier, H. (1958). *Antibiot. Annu.* 1958:869.

Schrempf, H. (1982). *J. Chem. Technol. Biotechnol.* 32:292–295.

Soltero, F. V., and Johnson, M. J. (1954). *Appl. Microbiol.* 2:41–44.

Spizek, J., Malek, I., Dolezilova, L., Vondracek, M., and Vanek, Z. (1965a). *Folia Microbiol. (Prague)* 10:259–263.

Spizek, J., Malek, I., Suchy, J., Vondracek, M., and Vanek, Z. (1965b). *Folia Microbiol. (Prague)* 10:263–266.

Tanaka, Y., Araki, K., and Nakayamu, K. (1980). *J. Ferment. Technol.* 58: 189–196.

Tereshin, I. M. (1976). *Polyene Antibiotics. Present and Future.* University of Tokyo Press, Tokyo.

Thoma, R. W. (1971). *Folia Microbiol. (Prague)* 16:197.

Turner, W. B. (1971). *Fungal Metabolites.* London, Academic, p. 446.

Volpe, J. J., and Vagelos, P. R. (1973). *Annu. Rev. Biochem.* 42:21–60.

Wang, S. F., and Gabriel, O. (1969). *J. Biol. Chem.* 244:3430.

Weinberg, E. D. (1962). *Perspect. Biol. Med.* 5:432–445.

Weinberg, E. D. (1970). *Adva. Microb. Physiol.* 4:1–44.

Zhukova, R. A., Zhuravlera, M. P., and Morozov, V. M. (1970). *Antibiotiki* 15:758.

Zielinski, J., Jereczek, E., Sowinski, P., Falkowski, L., Rudowski, A., and Borowski, E. (1979). *J. Antibiot.* 32:565–568.

17

CYCLOHEXIMIDE: PROPERTIES, BIOSYNTHESIS, AND FERMENTATION

JOHN L. JOST* AND LEO A. KOMINEK *The Upjohn Company, Kalamazoo, Michigan*

GREGORY S. HYATT AND HENRY Y. WANG *University of Michigan, Ann Arbor, Michigan*

I. INTRODUCTION

Cycloheximide (also known as Actidione and naramycin A) was discovered by Whiffen et al. (1946) in a streptomycin-yielding culture of *Streptomyces griseus*. Cycloheximide is produced by several other strains of *Streptomyces* including *Streptomyces noursei, Streptomyces naraensis, Streptomyces pulveraeus, Streptomyces albus*, and *Streptomyces ornatus*. Cycloheximide was first isolated in crystalline form, from the culture filtrate of *S. griseus*, by Leach et al. (1947). The chemical structure was determined by Kornfeld et al. (1949) as 3-[2-(3-5-dimethyl-2-oxocyclohexyl)-2-hydroxyethyl]-glutarimide, which is illustrated in Figure 1.

Present affiliation: Genentech, Inc., South San Francisco, California.

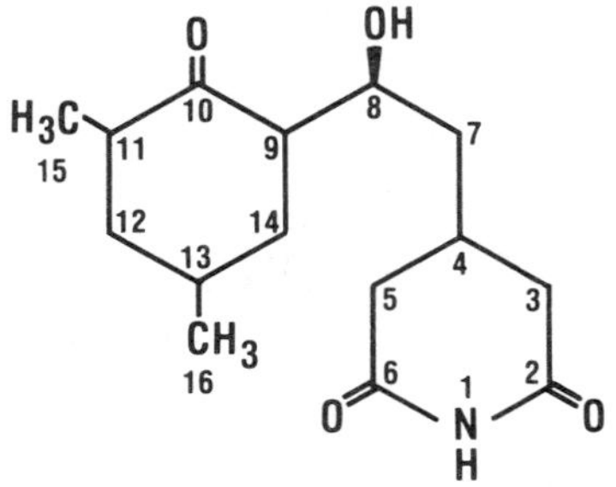

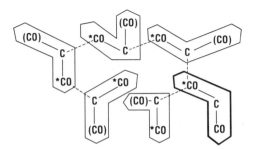

Figure 1 Structure of cycloheximide with asymmetric centers at positions 8, 9, 11, and 13.

Kerridge (1958) showed that cycloheximide was an effective protein synthesis inhibitor. This property has been exploited in the research laboratory, where cycloheximide has been used extensively for culture selection and isolation because of its relatively bacteriostatic characteristics. Cycloheximide is used as an agricultural fungicide and is an abscission agent.

Cycloheximide was shown to be synthesized from malonyl coenzyme A units. Control mechanisms exist for both production and degradation of cycloheximide. Other relevant publications on cycloheximide are listed in the References, but only the most pertinent articles were cited in the text. The other publications were included in the References for the sake of convenience and completeness.

The various methods of isolation and purification of cycloheximide produced by fermentation are discussed.

II. CYCLOHEXIMIDE PRODUCTION

A. Biosynthesis

A molecule of cycloheximide has 4 asymmetric centers, and therefore 16 stereo-isomeric forms are possible (Fig. 1). Three of these stereoisomers have been described and include cycloheximide, isocycloheximide, and naramycin B (Ume-

Figure 2 Condensation of malonate units into the cycloheximide skeleton. The carbons in parentheses are those lost by decarboxylation. The malonate unit in heavy type does not undergo decarboxylation during condensation. (From Vanek et al., 1967.)

zawa, 1964; Vondracek and Vanek, 1964). Radioactive labeling studies have led to a description of the mode of cycloheximide biosynthesis. The antibiotic cycloheximide is formed from the condensation of six malonate units, five of which undergo decarboxylation as shown in Figure 2. This was demonstrated in a number of publications using *Streptomyces noursei* as the producing organism (Vanek et al., 1967, 1969). Degradation studies using ^{14}C-labeled substrates ([Me-^{14}C]L-methionine, [1-^{14}C]acetic acid, [1,2-^{14}C]malonic acid, and [$^{14}CO_2$]) indicated that six acetic acid units were incorporated into cycloheximide after carboxylation to malonyl coenzyme A. This was followed by decarboxylation of five of the malonate units, with the initiating malonate unit remaining unchanged. Other degradation studies showed that two methyl groups of the dimethylcyclohexanone nucleus were derived from carbon dioxide via L-methionine.

B. Control Mechanisms

Studies with *S. griseus* by Kominek (1975a) showed that the synthetic rate of cycloheximide production was inversely proportional to the concentration of cycloheximide in the medium, suggesting that feed-back inhibition was operating (Fig. 3). The same effect was found using isocycloheximide or anhydrocycloheximide in place of cycloheximide. The production of cycloheximide in relation to glucose utilization during a standard fermentation and a fermentation which is glucose fed is illustrated in Figure 4. Figure 4 shows that glucose utilization starts about 1 day after inoculation and corresponds to the initiation of cycloheximide accumulation. When glucose is exhausted, cessation of antibiotic accumulation also occurs, followed by a rapid degradation of cycloheximide. When a glucose feed is used, degradation of cycloheximide is not apparent and titers are increased by approximately 20%. However, the cycloheximide titers still level off, even in the presence of excess glucose and an extended fermentation time.

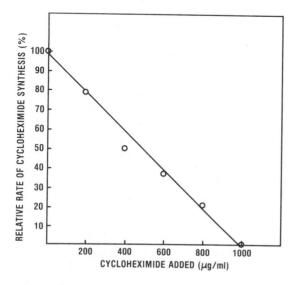

Figure 3 The rate of cycloheximide biosynthesis in the presence of added cycloheximide of increasing concentration.

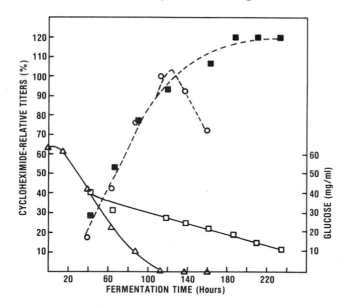

Figure 4 Cycloheximide production and glucose utilization during a standard and a glucose-fed fermentation. The fermentation was run in a tank containing 250 liters of medium at 25°C with an aeration and agitation of 250 standard liters/min and 300 rpm, respectively. Cycloheximide titer was determined by the colorimetric assay. The glucose feed was initiated at 42 hr at a rate of 0.24 g/hr per liter [standard fermentation: cycloheximide (○) and glucose (△); glucose-fed fermentation: cycloheximide (■) and glucose (□)].

Washed cells of *S. griseus* were used to determine the ability of cells of increasing age to synthesize or degrade cycloheximide. The rate of cycloheximide synthesis in the system utilizing washed cells was linear with respect to cell concentration (Fig. 5). Figure 6 shows a washed cell system in which cycloheximide biosynthesis was determined both by chemical assay (Takeshita et al., 1962) and by the incorporation of [14C]glucose. This system shows that the addition of cycloheximide results in a greatly reduced production of cycloheximide when determined by chemical assay, approximately 14% of the control. However, [14C]glucose incorporation is about 30% of the control, indicating that the synthetic process is functioning faster than indicated by chemical assay. This can be explained by the fact that feed-back inhibition reduces the effectiveness of the synthetic enzyme sufficiently so that the rate of cycloheximide biosynthesis approximates the rate of cycloheximide degradation.

Chloramphenicol was found to severely inhibit cycloheximide synthesis, and after an initial lag this inhibition was complete. Chloramphenicol did not affect the rate of glucose utilization, so it appears that this inhibition is specific for cycloheximide biosynthesis and unrelated to protein synthesis.

Glucose prevents cycloheximide degradation as shown in Figure 7. This is not due solely to synthesis in its presence, since the addition of chloramphenicol which prevents biosynthesis does not increase the rate of degradation. The addition of chloramphenicol in the absence of glucose has no effect of cycloheximide degradation. These results permit washed cells to be used

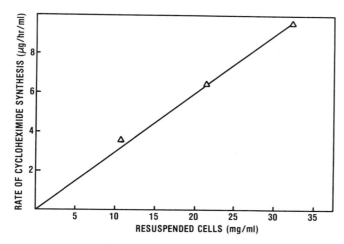

Figure 5 Cycloheximide biosynthesis in a washed cell system with respect to cell concentration. Resuspension medium consisted of glucose, 60 mg/ml; tris(hydroxymethyl)aminomethane buffer, 0.5 M (pH 7.0); and washed cells as indicated.

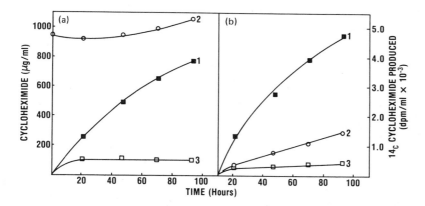

Figure 6 The effect of cycloheximide and chloramphenicol on cycloheximide biosynthesis in a washed cell system: (a) cycloheximide titer as determined by the colorimetric assay and (b) incorporation of [^{14}C]glucose (UL) into cycloheximide. The resuspension medium consisted of the following: (1) [^{14}C]glucose (UL), (2.2 × 10^5 dpm/ml), 60 mg/ml; tris(hydroxymethyl)amino-methane buffer, 0.05 M (pH 7.0); and washed cells, 21.5 mg/ml; (2) same as (1) plus cycloheximide (1 mg/ml); and (3) same as (1) plus chloramphenicol (0.1 mg/ml). Incorporation of [C^{14}]glucose into cycloheximide was determined by extracting a portion of the washed cell resuspension with an equal volume of chloroform and placing a suitable portion of the solvent in Diotol counting solution. The radioactivity was determined in a Beckman scintillation counter.

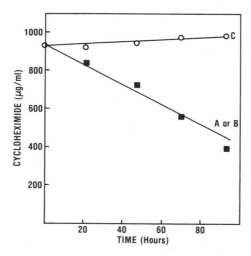

Figure 7 Degradation of cycloheximide in a washed cell system. Cycloheximide concentration was determined by the colorimetric assay. The resuspension medium consisted of the following: (a) cycloheximide, 1 mg/ml; tris(hydroxymethyl)aminomethane buffer, 0.05 M (pH 7.0); and washed cells, 21.5
mg/ml; (b) same as (a) plus chloramphenicol, (0.1 mg/ml); and (c) same as
(a) plus chloramphenicol (0.1 mg/ml) and glucose (60 mg/ml).

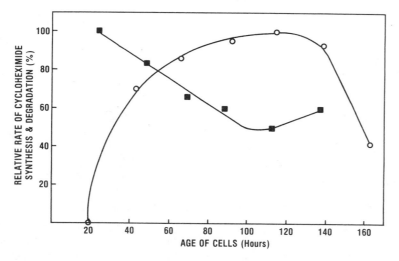

Figure 8 Rate of cycloheximide biosynthesis and degradation by cells of increasing age. To determine the rate of synthesis (○), 200 ml of whole beer
was harvested at the indicated time interval, washed, and resuspended in
100 ml of a medium containing glucose (60 mg/ml) and tris(hydroxymethyl)-
aminomethane buffer (0.05 M; pH 7.0). To determine the rate of degradation
(■), 200 ml of whole beer was harvested at the indicated time interval,
washed, and resuspended in 100 ml of a medium containing cycloheximide (1
mg/ml) and tris(hydroxymethyl aminomethane)buffer (0.05 M; pH 7.0). In
both cases cycloheximide was determined by chemical assay.

to determine the ability of cells of increasing age to synthesize or degrade cycloheximide. The rate of cycloheximide biosynthesis and degradation of cyclo- heximide by cells of increasing age is illustrated in Figure 8. The ability to synthesize cycloheximide appears to be inducible and reaches a maximum between 3 and 5 days of fermentation. After 5 days the ability to produce cycloheximide drops off significantly at a point that corresponds to glucose exhaustion in the fermentation medium from which these cells were harvested. The ability to degrade cycloheximide decreases from the beginning of the fer- mentation and reaches a minimal value between 4 and 5 days. Degradative ability increases from this time, which again corresponds to exhaustion of glucose from the fermentation medium.

The accumulation of cycloheximide in a fermentation medium by *S. griseus* is a dynamic process involving both synthesis and degradation of this anti- biotic. The biosynthetic system responsible for cycloheximide formation ap- pears to be inducible and retains a high level of activity throughout the fer- mentation, provided that glucose is present. The system responsible for de- gradation of cycloheximide appears to be repressible and subject to catabolite inhibition in the presence of glucose. Cycloheximide has been found to inter- fere with its own production, probably owing to feed-back inhibition. There- fore the accumulation of cycloheximide is a product of the balance between synthesis and degradation. When product inhibition becomes great enough to balance the rate of synthesis to that of degradation, accumulation of the antibiotic ceases.

C. Fermentation Process

Fermentations producing cycloheximide titers greater than 1000 μg/ml have been described for *S. griseus* by Churchill (1959) and for *S. noursei* by Abou-Zeid and El-Sherbini (1974).

Streptomyces griseus was maintained on agar slants which consist of the following: glucose, 10 g; Torula yeast, 10 g; distiller's solubles, 5 g; Kaysoy, 4 g; calcium carbonate, 1 g; agar (Difco), 15 g; tap water to 1000 ml. The seed medium consisted of glucose, 10 g; beef extract, 5 g; peptone, 5 g; sodium chloride, 5 g; and tap water to 1000 ml.

The fermentation medium was composed of glucose, 60 g; defatted soy- bean flour, 15 g; yeast, 2.5 g; ammonium sulfate, 5 g; calcium carbonate, 8 g; sodium chloride, 5 g; KH_2PO_4, 0.2 g; and tap water to 1000 ml.

The fermentation medium which supported greater than 1000 μg/ml of cycloheximide production by *S. noursei* contained the following ingredients: glucose, 10 g; peptone, 2 g; yeast extract (Difco), 1 g; corn-steep liquor, 5 g; KH_2PO_4, 5 g; sodium chloride, 5 g; $MgSO_4 \cdot 7H_2O$, 0.5 g; $MgCO_3$, 1 g; and 1000 ml of tap water.

In both cases all media and media components were sterilized by auto- claving at 121°C at 15 psi(gauge) for 25 min.

A typical fermentation using *S. griseus* as a producing organism con- sisted of 250 liters of medium aerated at a rate of 250 standard liters of air per minute and agitated at 300 rpm. The temperature was maintained at 25°C throughout the fermentation. Maximum titers were achieved between 5 and 7 days.

The medium described for *S. noursei* has been used only in shake flasks. Erlenmeyer flasks (500 ml) each containing 100 ml of fermentation medium were inoculated with 2% of the vegetative seed. The seed medium consisted of starch, 10 g; glucose, 10 g; peptone, 10 g; sodium chloride, 5 g; KH_2PO_4,

5 g; and 1000 ml of tap water. The fermentation flasks were inoculated with 2% of the seed volume and incubated at a temperature of 27°C on a rotary shaker. Titers of 1000 µg/ml were achieved after 5 days of fermentation.

III. RECOVERY AND PURIFICATION OF CYCLOHEXIMIDE

Cycloheximide was first found to have antifungal properties by Whiffen et al. at the Upjohn Co. in Kalamazoo, Michigan, early in 1946. Since that time productivity through biosynthesis and recovery methods has been greatly improved. The following is a review of published material on the recovery and purification methods used for the isolation of cycloheximide.

The earliest methods of cycloheximide recovery in crystalline form involved techniques and solvents still in use today. The basic procedure was one of filtration, adsorption by activated carbon, elution, liquid–liquid extraction, and crystallization. The detailed block diagram shown in Figure 9 illustrates the method used by Ford and Leach in 1947 and is a good example of the recovery method.

The discovery and preliminary research on cycloheximide brought about recognition of some physical properties very different from the valuable co-product of S. griseus, streptomycin. Cycloheximide is soluble in chloroform, ether, and other polar solvents, and streptomycin is not. In addition, cycloheximide is acid stable whereas streptomycin is unstable in acid. These properties have aided in the development of recovery and purification procedures for cycloheximide.

The recognized acid stability of cycloheximide resulted in sulfuric acid treatment of the beer as an initial isolation step (Whiffen et al., 1946). Filtration of fermentation beer is common to many recovery processes to remove mycelia and was also used in most cycloheximide recovery schemes. Inert diatomaceous earth filters served the purpose nicely and could be used before or after acidification of the beer.

Crude cycloheximide was first isolated from fermentation beer, which also contained streptomycin, by adsorption on activated carbon. Cycloheximide was recovered by eluting with 80% acetone, distilling the acetone, and extracting the remaining aqueous solution with chloroform. The chloroform extracts were either orange-brown or intense green in color, depending on the culture medium employed. Most of the coloring compounds could be eliminated by passage over a carbon column. The chloroform extracts were then vacuum distilled. An orange-brown sticky oil having a moldy odor usually resulted with a purity of 30–60% based on a pure standard (Leach et al., 1947). This material is further purified by several dissolution–cyrstallization steps using amyl acetate. The pure cycloheximide crystals have a melting point of 115–116.5°C.

A further development in the purification found in 1947 utilized a final extraction with warmed water after the crystallization from amyl acetate. The warm water aided in dissolving the crystals resulting in a slightly purer product.

Alternatively, purification of the product could be achieved using countercurrent distribution between benzene and water (Craig et al., 1945). The use of this process for purifying cycloheximide has not been reported for applications larger than bench scale.

Genetic manipulation and complex medium development resulted in a strain of S. griseus that could produce over 100 µg/ml of cycloheximide. In

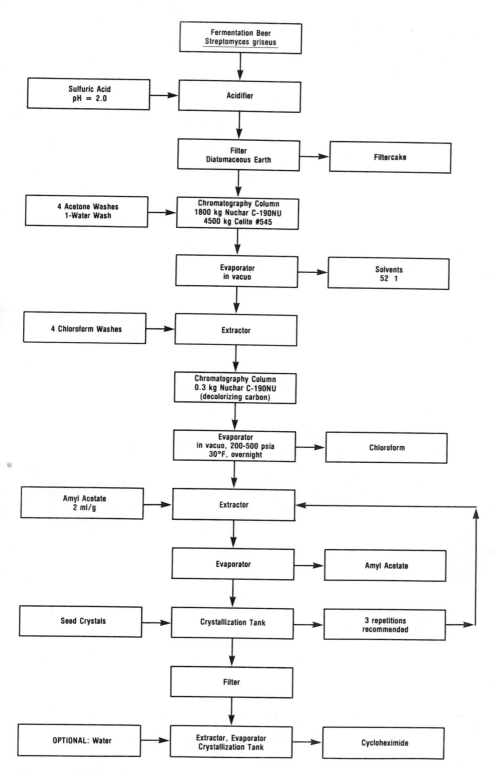

Figure 9 Block diagram I: the cycloheximide recovery method used by Ford and Leach in 1947.

conjunction with these developments, a simpler procedure was developed by Churchill in 1959 to isolate and purify the cycloheximide from the fermentation beer.

The beer is acidified to pH 3.5–5.5 with 60% sulfuric acid and heated to 60°C for 10 min, cooled to 30°C, and clarified in a filter press with distomaceous earth as the filter aid. The clarified beer is then extracted in a Podbielniak extractor with methylene chloride (0.2 liter methylene chloride/ liter of beer). The solution obtained is clarified further in a centrifugal separator and decolorized with activated carbon. It is then concentrated 20- to 40-fold and decolorized again. The methylene chloride extract is concentrated further until there is a solid content on 0.5 kg/liter and diluted with 3.0 liters of amyl acetate/kg solids. The remaining methylene chloride is then removed by vacuum distillation at 70°C or below. Cycloheximide is crystallized from the amyl acetate solution, washed with technical hexane, and recrystallized from amyl acetate. By following this procedure as illustrated in Figure 10, Churchill (1959) claimed that the following specifications of the pure product could be met: (1) a melting point of 105–120°C, (2) weight loss on drying not exceeding 0.3%, and (3) a potency of 800 μg/mg as tested with *Streptomyces pastorianus*.

Brown and Hazen (1955) published a procedure for the recovery of cycloheximide from *S. noursei* fermentation beer. Virtually the same procedure was used as that shown in Figure 9. The major difference was that the acidification step occurred after the cells were filtered off and was used to eliminate the coproduct fungicidin rather than streptomycin.

The discovery of a family of antitumor and antifungal agents called streptoviticins also produced by *S. griseus* aided further development of recovery methods for cycloheximide. A gradient partition chromatography procedure was employed for the separation. The five streptoviticins are similar in structure to cycloheximide and have differing polarities greater than that of cycloheximide.

The method developed by Eble and his associates at the Upjohn Co. in Kalamazoo, Michigan (Eble et al., 1960), is illustrated in Figure 11. The chromatography column used is one of 600 g of Celite 545, slurried in 5.0 liters of the upper phase of solvent system I (Table 1) and 240 ml of the lower phase of solvent system I. The agitated mixture is packed to 66 in. under 5 psi- (gauge) air pressure.

A 20-g sample of acidified and filtered fermentation beer is dissolved in 25 ml of solvent system I lower phase. To the solution is added 150 ml of upper-phase solvent system I and 75 g of Celite 545. The whole mixture is homogenized and added to the prepared column. Fresh upper phase of solvent system I is added to the column for elution. The column is developed at a rate of 5 ml/min in 20-ml fractions. Cycloheximide is collected in the first fractions totaling 6.51 and is 4.8% of the total starting material (0.952 g). The pooled fractions are concentrated to dryness, crystallized with amyl acetate, and recrystallized. This method provides for the subsequent recovery of five streptoviticins and isocycloheximide using two additional solvent systems of increasing polarity (Eble et al., 1960).

The isocycloheximide was found to elute with the cycloheximide and required further separation steps. The isolation of isocycloheximide from the cycloheximide is performed as follows: The cycloheximide product from above is partition chromatographed over Celite 545 using solvent system II (Table 1). The column is developed at a rate of 1.0 ml/min per 2 g of sample and 15 liters are collected. Isocycloheximide elutes in the 2.02–2.38 liter fraction.

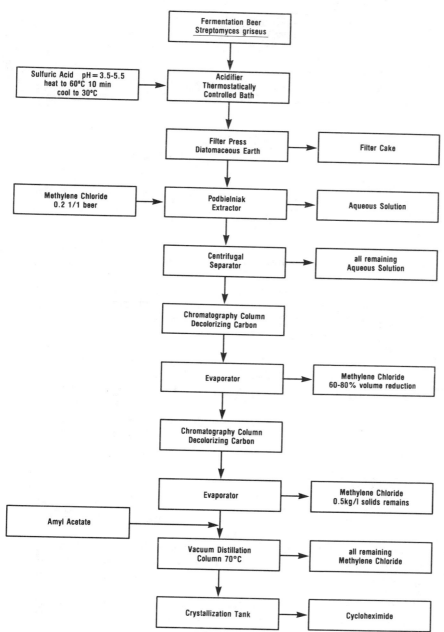

Figure 10 Block diagram II: the cycloheximide purification procedure described by Churchill (1959).

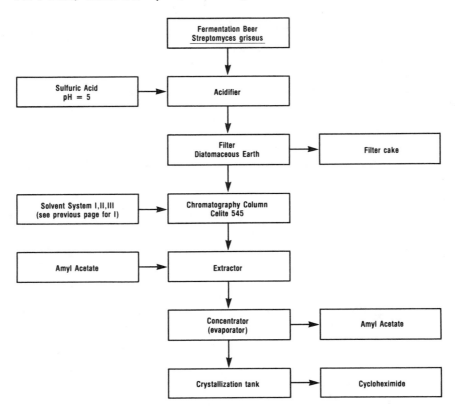

Figure 11 Block diagram III: the purification and isolation of cycloheximide by gradient partition chromatography described by Eble et al. (1960).

The remaining fractions contain cycloheximide, which is crystallized from amyl acetate (Eble et al., 1960).

Okuda et al. (1963) reported finding another cycloheximide-like compound, naramycin B, in 1963. Naramycin B (an isomer) appeared in the product of a procedure published by Eble et al. in 1960. Naramycin B was successfully separated from cycloheximide and isocycloheximide by employing finer diatomaceous earth (Celite analytical filter aid) instead of Celite 545 in the final separation step. The order of elution in increasing polarity was found to be isocycloheximide, naramycin B, and cycloheximide.

The method investigated most recently for the isolation of cycloheximide-like compounds involves ion exchange chromatography. The structure of

Table 1 Isolation Solvent Systems

System	Ethyl acetate	Cyclohexane	McIlvanine's pH 5 buffer
Solvent system I	3 parts	7 parts	1 part
Solvent system II	1 part	9 parts	1 part

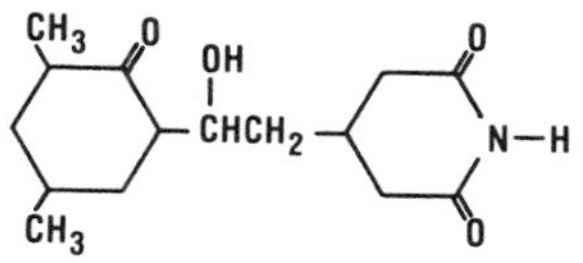

Figure 12 Cycloheximide with its hydantoin moiety which would allow the isolation of this antibiotic by ion exchange chromatography: (I) cycloheximide, (II) hydantoin, and (III) hydantoin ion.

cycloheximide [1] includes the chemical configuration of a hydantoin [2] (Blau and King, 1977), which can be isolated by ion exchange chromatography (Fig. 12). The separation of various hydantoins was investigated by Hirs et al. in 1955 and again by Stark and Smyth in 1963 for the determination of terminal amino acid groups. The ion exchange resins used in these cases were Dowex 50-X, Dowex 50-X2, and Dowex 1-X8 purchased from Technical Service and Development, the Dow Chemical Co., Midland, Michigan. The hydantoins were eluted using various concentrations of hydrochloric acid. The resins have a higher affinity for the smaller ion Cl^- versus the hydantoin ion [3]. Three characteristics determine the adsorption elution patterns of the ions: (1) differences in the extent of ionization of the individual compounds in their passage through the resin packed column, (2) differences in charge, and (3) adsorptive (Van der Waals) forces related to the hydrophobic portions of the resin structure. The publications cited do not deal directly with cycloheximide, but involve similar structures and the separation method deals directly with the identical functional group arrangement [3]. Therefore applications are likely to be suited to cycloheximide, its isomers, and related compounds.

A direct study of the use of nonionic resins for the isolation of cycloheximide has been performed by the joint efforts of Wang of the University of Michigan and Kominek and Jost of the Upjohn Co. in 1980. Several neutral resins were investigated: Rohm and Haas Amberlite XAD-2, XAD-4, XAD-7, XE-348, and Mitsubishi's BACM. The resins were tested in the fermentation broth which eliminates end-product inhibition and allows for a higher final antibiotic titer similar to the dialysis extraction fermentation developed by Kominek (1975b,c). The activities of the resins were found to be dependent on their surface area and chemical configuration (Fig. 13).

After fermentation, the cycloheximide is recovered from the resin using a butyl acetate extraction. Purification of the resulting cycloheximide could be performed using the method discussed earlier.

A novel isolation method has been reported by Kominek (1975b,c). This method consists of an on-line dialysis of the fermentation broth. The isolation scheme is unique in that the dialysate is further extracted with a solvent and the spent dialysate is then recycled. This approach minimizes the volume of dialysate used and makes the method practical.

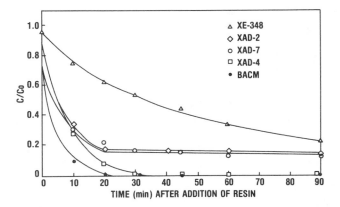

Figure 13 Cycloheximide uptake by various resins versus time: C is the concentration of cycloheximide (μg/ml) in solution; C_0 is the concentration of added cycloheximide (1000 μg/ml); resin was added at t_0.

The apparatus used for this on-line recovery process is illustrated in Figure 14. Dialysate was circulated through cellulose dialysis tubing inside a standard fermenter (Fig. 14a). The circulation rate Kominek (1975b) reported was about 0.8 volume dialysate per volume fermentation medium per hour. The dialysate was then sparged through a column (Fig. 14b) containing solvent to accomplish the extraction of the cycloheximide. The spent dialysate was aerated to remove the residual solvent and pumped back to the dialysis tubing. The solvent volume was kept constant by manual replacement once per day while the dialysate volume was automatically controlled by water addition.

The dialysis–extraction eliminated the effects of product inhibition and eliminated the need for a large dialysate reservoir. The reported yield doubled owing to the use of this method. An added advantage was that a separa-

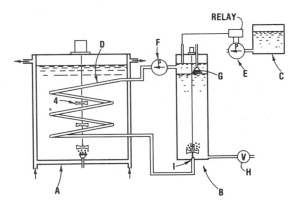

Figure 14 Equipment for continuous dialysis extraction: (a) fermenter, (b) extraction column, (c) makeup water reservoir, (d) dialysis tubing, (e) solenoid valve, (f) dialysate pump, (g) air sparger of spent dialysate, (h) solvent inlet valve, and (i) dialysate sparger.

tion-purification also occurred. It was reported that the solids in the solvent phase were 85% cycloheximide.

The advantages of this method on a production scale are numerous. Further purification steps would probably only include decolorizing, distillation, and then a crystallization. This procedure would then eliminate, at a minimum, the filtration step.

While the dialysis-extraction method has only been reported on a pilot scale, it showed promise of increasing the fermentation yield while simultaneously simplifying the isolation process; however, a production-scale dialysis system has not been reported in the literature.

The nature of fermentation processes is such that it would be impossible to predict the technological advances over the next few years. Advances are likely to occur in the productivity of the fermentation process as well as the recovery processes. Good product purity will probably result from improved isolation steps resulting in fewer, less costly steps in the process.

IV. APPLICATIONS FOR CYCLOHEXIMIDE

The major uses of cycloheximide include its application as a laboratory research tool and as an agricultural fungicide. It has been used as an abscission agent and a human topical fungicide. A huge volume of literature exists documenting the use of cycloheximide in the research laboratory. Studies of biosynthetic pathways have exploited its ability to inhibit protein synthesis. The antifungal properties and relatively bacteriostatic (Whiffen, 1948) characteristics of cycloheximide have led to its use in culture selection and isolation.

Cycloheximide is best known as a protein synthesis inhibitor. This was first proven by Kerridge in 1958. Later Siegel and Sisler (1963) showed that cycloheximide prevented the incorporation of amino acids into protein. An excellent review of the mode of action of cycloheximide has been presented by Sisler and Siegel (1967). While the majority of the literature on cycloheximide deals with its use as a laboratory research tool, this application will not be considered further here, since it is of little commercial importance.

A. Cycloheximide as a Fungicide

The uses of cycloheximide as an agricultural fungicide have been reported for treatment of a wide variety of plant diseases. One major area is the treatment of fungal diseases of turf grasses. Cycloheximide is effective against many phytopathogens of turf grasses, including *Sclerotinia homoeocarpa* (Massie et al., 1968), *Helminthosporium* and *Curvularia* (Meyer et al., 1971).

Several studies have shown that the use of cycloheximide over longer periods of time does not alter the ecosystems in the soil. Partially this is due to the fact that the antibiotic is not significantly active against bacteria (Whiffen, 1948). Smiley and Craven (1979) found that cycloheximide did not affect the numbers of bacteria, actinomycetes, or fungi in the soil and thatch, even after 3 years of repeated application. Meyer et al. (1971) also found that the fungal populations of the soil were not altered after 4 years of fungicide treatment. They attributed this partially to the chemical and microbial degradation of cycloheximide in the soil and thatch.

Miller (1968) has reported that cycloheximide is an effective agent for treatment of powdery mildew of roses. Cycloheximide eradicated even heavy infections of the mildew.

Cycloheximide effectively controlled the many fungal infestations of forests and nurseries. Burdekin and Phillips (1971) reported the control of needle blight (by *Didymascella thujina*) in cedar trees by cycloheximide. It was found that both the rate of application and the timing of the application were important variables. Three applications per year at 50 gal/acre of an 85 ppm solution resulted in about one-quarter the amount of infection as compared to a 200 gal/acre test applied only twice. Increasing the dosage rate by about a factor of 3 to 227 ppm lowered the infection rate by another factor of 2. Here the infection was measured by enumerating the number of fungal apothecia per plant. It was concluded that no chemicals tested other than cycloheximide showed promise of controlling the fungus *D. thujina*. The use of cycloheximide in this instance eliminated the need for the previous control schemes of nursery rotation and isolation.

In contrast to the above study, Klingstrom and Lundeberg (1978) reported that only a single application of cycloheximide at 5 ppm was adequate to control various fungal infestations in pine tree nurseries and forests. The effectiveness of cycloheximide depends also upon the fungal species or type. Harvey and Grasham (1979) reported that the antibiotic was only fungistatic when tested against the causative agent of white pine blister rust.

Cycloheximide has been cited for other applications as a fungicide. It has been successfully used as means of preventing decay of fruit in transit to the consumer (Sharma and Wahab, 1971). The fruit was dipped in a cycloheximide solution of 10 µg/ml or less. This treatment prevented the growth of the fungus *Pythium apharnidermatum* for up to 70 hr. Cycloheximide has been reported as an effective fungicide in controlling mal secco of lemon plants. Solel et al. (1972) used cycloheximide as a systemic fungicide. The antibiotic was applied to the soil and leaves of the lemon trees. While it prevented germination of *Deuterophoma tracheiphila* spores, little fungicide accumulated in the leaves of the lemon plants. Thus the treatment did not completely prevent leaf infection.

B. Cycloheximide as an Abscission Agent

Cycloheximide was also introduced as an abscission agent. The use of abscission agents can allow for faster handpicking of fruits or the use of mechanical harvesters (Whitney, 1972). The exact mode of action of cycloheximide in abscission is complicated. Addicott (1965) showed that abscission can be caused by the enzymatic destruction of materials in plant cell walls. That this process can be accelerated by ethylene was shown by Crocker et al. (1935). Rubinstein and Abeles (1965) demonstrated that many defoliants increased the ethylene concentration near the abscission zone. Abeles and Holm (1967) explain the role of ethylene as stimulating the production of enzymes involved in the separation of cells. They then could show that the protein synthesis inhibitor, cycloheximide, could prevent abscission by preventing the synthesis of these enzymes. Thus the effectiveness of cycloheximide as an abscission agent would be a balance between its role in stimulating ethylene production and inhibiting the synthesis of the enzymes which cause abscission. A critical factor in abscission would then be the relative diffusion rates of ethylene and cycloheximide to the abscission zone. Therefore the effectiveness of cycloheximide as an abscission agent would be expected to be dependent upon its concentration in the abscission layer. Cooper et al. (1968) demonstrated that application of cycloheximide to Valencia oranges stimulates ethylene production and fruit loosening. Concentrations higher than 32 µmol/liter did not increase ethylene production in proportion to the increased cyclohexi-mide levels.

When Fisher (1971) treated oranges with [^{14}C]cycloheximide, less than 2% of the radioactivity was found in the edible portion of the fruits; most of the radioactivity was found on the peel. From 80 to 83% of the radiocarbon could be removed by washing. Only a single degradation product of cyclo-heximide could be identified—anhydrocycloheximide. Since most of the radio-activity could be either washed off or found in the peel, it was concluded that the abscission property of cycloheximide is associated with the rind.

Davies et al. (1975) found that the important variables of an abscission agent are the concentration of the abscission agent and the general condition of the orange grove. Holm and Wilson (1976) reported reduced effectiveness of abscission agents during a certain period of orange growth (the regreening period). After this period, oranges again responded normally to the applica-tion of abscission agents.

REFERENCES

Abeles, F. B., and Holm, R. E. (1967). Abscission: Role of protein synthe-sis. *Ann. N.Y. Acad. Sci. 114*:367–373.

Abou-Zeid, A. A., and El-Sherbini, S. H. (1974). Fermentation production of cycloheximide by *Streptomyces griseus* and *Streptomyces noursei. J. Appl. Chem. Biotechnol. 24*:283–291.

Addicott, F. T. (1965). Physiology of abscission. In *Encyclopedia of Plant Physiol.* W. Ruhland (Ed.). Springer-Verlag, Heidelburg, pp. 1094–1126.

Blau, K., and King, C. (1977). In *Handbook of Derivatives for Chromatogra-phy.* K. Brau and G. King (Eds.). Heyden and Sons, London, p. 282.

Brown, R., and Hazen, E. (1955). Production of actidione by *Streptomyces noursei. Antibiot. Annu. 1955*:245.

Burdekin, D. A., and Phillips, D. H. (1971). Chemical control of *Didymas-cella thujina* on Western red cedar in forest nurseries. *Ann. Appl. Biol. 67*:131–136.

Churchill, B. W. (1959). Production of cycloheximide. U.S. Patent No. 2,885,326. U.S. Patent Office, Washington, D.C.

Cooper, W. C., Rasmussen, G. K., Rogers, B. J., Reece, P. C., and Henry, W. H. (1968). Control of abscission in agricultural crops and its phys-iological basis. *Plant Physiol. 43*:1560–1576.

Corrado, M. L., Weitzman, I., Stanek, A., Goetz, R., and Agyare, E. (1980). Subcultaneous infection with *Phialophora richardsiae* and its suscepti-bility to 5-flourocytosine, amphotericin B and miconazole. *Subouraudia 18*:97–104.

Craig, L. C., Golumdie, C., Mighton, H., and Titus, E. (1945). Identifica-tion of small amounts of organic compounds by distribution studies. *J. Biol. Chem. 161*:321.

Crocker, W., Hitchcock, A. E., and Zimmerman, P. W. (1935). Similarities in the effects of the ethylene and plant auxins. *Boyce Thompson Inst. Contrib. 7*:231–248.

Darbe, A. (1977). Cyclization. In *Handbook of Derivatives for Chromatogra-phy.* K. Brau and G. King (Eds.). Heyden and Sons, London, pp. 279–282.

Davies, F. S., Cooper, W. C., and Galena, F. E. (1975). A comparison of four abscission compounds for use on 'Hamlin', 'Pineapple' and 'Valencia' oranges. *Proc. Fla. State Hortic. Soc. 88*:107–113.

Dolezilova, L., Spizek, J., Vondracek, M., Paleckova, F., and Vanek, Z. (1955). Cycloheximide-producing and fungicidin-producing mutants of *Streptomyces noursei. J. Gen. Microbiol. 39*:305–309.

Eble, T., Bergy, R., Herr, J., and Fox, J. (1960). The separation and properties of the streptoviticins. In *Antimicrobial Agents and Chemotherapy*. American Society for Microbiology, pp. 479–483.

Fisher, J. F. (1971). Distribution of radiocarbon in Valencia oranges after treatment with ^{14}C-cycloheximide. *J. Agric. Food Chem. 19*:1162–1164.

Ford, J., and Leach, B. (1947). Actidione, an antibiotic from *Streptomyces griseus. J. Am. Chem. Soc. 69*:1223.

Hartmann, H. T., El-Hamaday, M., and Whisler, J. (1972). Abscission induction in the olive by cycloheximide. *J. Am. Soc. Hortic. Sci. 97*:781–785.

Harvey, A. E., and Grasham, J. L. (1979). The effects of selected systemic fungicides on the growth of *Cronartium ribicola in vitro. Plant Dis. Rep. 63*:354–358.

Hirs, C., Moore, S., and Stein, W. (1955). Peptides obtained by tryptic-hydrolysis of performic acid-oxidized ribonuclease. *J. Biol. Chem. 219*: 623.

Holm, R. E., and Wilson, W. C. (1976). Loss in the capacity of 'Valencia' oranges treated with abscission chemicals to produce ethylene and fruit loosening during the regreening period. *Proc. Fla. State Hortic. Soc. 89*:35–38.

Holm, R. E., and Wilson, W. C. (1977). Ethylene and fruit loosening from combinations of citrus abscission chemicals. *J. Am. Soc. Hortic. Sci. 102*:576–579.

Kerridge, D. (1958). The effect of actidione and other antifungal agents on nucleic acid and protein synthesis in *Saccharomyces carlsbergensis. J. Gen. Microbiol. 19*:497–506.

Khartyan, S., Puza, M., Spizek, J., Dolezilova, L., Vanek, Z., Vondracek, M., and Rickards, R. W. (1963). Biogenesis of cycloheximide and related compounds. *Chem. Ind. 25*:1038–1039.

Klingstrom, A., and Lundeburg, G. (1978). Control of *Lophodermium* and *Phacidium* needle cast and *Scleroderris* canker in *Pinus sylvestris. Environ. J. Forest Pathol. 8*:20–25.

Kominek, L. A. (1975a). Cycloheximide production by *Streptomyces griseus*: Control mechanisms of cycloheximide biosynthesis. *Antimicrob. Agents Chemother. 7*:856–860.

Kominek, L. A. (1975b). Cycloheximide production by *Streptomyces griseus*: Alleviation of end-product inhibition by dialysis–extraction fermentation. *Antimicrob. Agents Chemother. 7*:861–863.

Kominek, L. A. (1975c). Dialysis process and apparatus. U.S. Patent 3,915,802. U.S. Patent Office, Washington, D.C.

Kornfield, E. C., Jones, R. J., and Parke, T. V. (1949). The structure and chemistry of actidione, an antibiotic from *Streptomyces griseus. J. Am. Chem. Soc. 71*:150–159.

Leach, B. E. (1952). Extraction of cycloheximide. U.S. Patent 2,612,502. U.S. Patent Office, Washington, D.C.

Leach, B., Ford, J., and Whiffen, A. (1947). Actidione, an antibiotic from *Streptomyces griseus. J. Am. Chem. Soc. 69*:474.

Ledre, E., and Ledre, M. (1957). Amino acid ion exchange chromatography. In *Chromatography*. Elsevier, New York, pp. 292–303.

Lemin, A. J., and Ford, J. H. (1960). Isocycloheximide. *J. Org. Chem. 25*: 344–346.

Massie, L. B., Cole, H., and Duick, J. (1968). Pathogen variation in relation to disease severity and control of *Sclerotinia* dollar spot of turfgrass by fungicides. *Phytopathology 58*:1616–1619.

Meyer, W. A., Britton, M. P., Gray, L. E., and Sinclair, J. B. (1971). Fungicide effects on fungal ecology in creeping bentgrass turf. *Myco-pathol. Mycol. Appl. 43*:309–315.

Miller, N. H. (1968). Powdery mildew of roses. *Sunshine State Agric. Res. Rep. 13*:8–9.

Okuda, T., Suzuki, M., Furumai, M., and Takahashi, H. (1963). Studies on *Streptomyces* antibiotic, cycloheximide. *Chem. Pharm. Bull. 11*:730.

Rasmussen, G. K. (1976). Effect of abscission-inducing chemicals on mandarin oranges. *Proc. Fla. State Hortic. Soc. 89*:39–41.

Rubinstein, B., and Abeles, F. B. (1965). Relationship between ethylene evolution and leaf abscission. *Bot. Gaz. 126*:255–259.

Sharma, B. B., and Wahab, S. (1971). Efficacy of actidione and aureofungin in the control of post-harvest decay of some cucurbitaceous fruits due to *Pythium apharnidermatum*. *Hind. Antibiot. Bull. 13*:8–13.

Siegel, M. R., and Sisler, H. D. (1963). Inhibition of protein synthesis in vitro by cycloheximide. *Nature 200*:675–676.

Sisler, H. D., and Siegel, M. R. (1967). Cycloheximide and other glutarimide antibiotics. In *Antibiotics,* Vol. 1. D. Gottlieb and P. D. Shaw (Eds.). Springer-Verlag, Heidelberg, pp. 283–307.

Smiley, R. W., and Craven, M. M. (1979). Microflora of turfgrass treated with fungicides. *Soil Biol. Biochem. 11*:349–353.

Solel, Z., Pinkas, K., and Loebenstein, G. (1972). Evaluation of systemic fungicides and mineral oil adjuvants for the control of mal secco disease of lemon plants. *Phytopathology 62*:1007–1013.

Spizek, J., Malek, I., Dolezilova, L., Vondracek, M., and Vanek, Z. (1965). Metabolites of *Streptomyces noursei* IV. Formation of secondary metabolites by producing mutants. *Folia Microbiol. 10*:259–262.

Stark, G., and Smyth, D. (1963). The use of cyanide for the determination of NH_2-terminal residues in proteins. *J. Biol. Chem. 238*:214.

Takeshita, M., Takahashi, H., and Ikuda, T. (1962). Studies of *Streptomyces* antibiotic cycloheximide. XIII. New spectrophotometric determination of cycloheximide. *Chem. Pharm. Bull. 10*:304–308.

Umezewa, H. (1964). Glutarimide antibiotics. In *Recent Advances in Chemistry and Biochemistry of Antibiotics*. Microbial Chemistry Research Foundation, Tokyo, p. 40.

Vanek, Z., and Vondracek, M. (1965). Biogenesis of cycloheximide and of related compounds. In *Antimicrobial Agents and Chemotherapy*. G. L. Hobby (Ed.). American Society for Microbiology, pp. 982–991.

Vanek, Z., Puza, M., Cudlin, J., Dolezilova, L., and Vondracek, M. (1964). Metabolites of *Streptomyces noursei* III. Incorporation of the carbon-dioxide into cycloheximide. *Biochem. Biophys. Research Commun. 17*: 532–535.

Vanek, Z., Cudlin, J., and Vondracek, M. (1967). Cycloheximide and other glutarimide antibiotics. In *Antibiotics,* Vol. 2. D. Gottleib and P. D. Shaw (Eds.). Springer-Verlag, Heidelberg, p. 222.

Vanek, Z., Puza, M., Cudlin, J., Vondracek, M., and Rickards, R. W. (1969). Metabolites of *Streptomyces noursei* X. Biogenesis of cycloheximide. *Folia Microbiol. 14*:388–397.

Vondracek, M., and Vanek, Z. (1964). Metabolites of *Streptomyces noursei*. Some new metabolites and the structure of albonoursin. *Chem. Ind. 26*: 1686–1687.

Wang, H. Y., Kominek, L. A., and Jost, J. L. (1981). On-line extraction fermentation. In *Advances in Biotechnology*. M. Moo-Young (Ed.). Pergamon Press, Toronto, pp. 601–607.

Wardowski, W. F., and Wilson, W. C. (1972). Observations on early and mid-season orange abscission demonstrations using cycloheximide. *Proc. Fla. State Hortic. Soc. 84*:81–84.

Whiffen, A. J. (1948). The production, assay, and antibiotic activity of actidione, an antibiotic from *Streptomyces griseus*. *J. Bacteriol. 56*:283–291.

Whiffen, A. J., Bohonos, N., and Emerson, R. L. (1946). The production of an antifungal antibiotic by *Streptomyces griseus*. *J. Bacteriol. 52*:610–611.

Whiffen, A. J., Emerson, R. L., and Bohonos, N. (1951). Cycloheximide and process for its production. U.S. Patent 2,574,519. U.S. Patent Office, Washington, D.C.

Whitney, J. D. (1972). Citrus harvest results with the air shaker concept. *Proc. Fla. State Hortic. Soc., 85*:250–254.

18

GRISEOFULVIN: PROPERTIES, BIOSYNTHESIS, AND FERMENTATION

FLOYD M. HUBER AND ANTHONY J. TIETZ *Biochemical Development Division, Eli Lilly and Company, Indianapolis, Indiana*

I. INTRODUCTION

In 1939 Oxford et al. reported the isolation of griseofulvin from mycelia of *Penicillium griseofulvum* grown in defined medium. This substance was given the empirical formula of $C_{17}H_{17}O_6Cl$ and had a melting point of 218–219°C. Griseofulvin was also characterized as being a colorless, crystalline, neutral compound giving no color with $FeCl_3$ and containing no free hydroxyl or carboxyl groups.

In 1946 Brian et al. described a substance that would curl or gnarl the mycelia of the fungus *Botrytis allii*. The substance, which was named the "curling factor," was produced by *Penicillium janczewskii* Zal. Griseofulvin and the "curling factor" were later shown to be both chemically and biologically the same substance (Grove and McGowan, 1947; Brian et al, 1949). Grove et al. determined the structure of griseofulvin [1] in 1952. Both dechlorogriseofulvin and bromogriseofulvin have also been isolated from fungal cultures (Macmillan, 1951, 1954).

Perlman (1978) surveyed the antibiotic industry and found that griseofulvin was either manufactured or sold by several companies. Thus this substance has achieved great economical significance.

II. FERMENTATION SYSTEMS

Fermentation conditions for griseofulvin production have appeared infrequently in the scientific literature. Since this substance is of economic importance, many developments relating to this antibiotic can be found in the patent literature.

Table 1 lists numerous organisms that have been shown to produce griseofulvin. It is important to note that this is an antifungal antibiotic produced by fungi. The fermentation media and culture conditions for most of the producing organisms appear to be quite similar.

A. Preservation and Maintenance of Cultures

Bayan et al. (1962) described procedures for sporulation of *P. griseofulvum*, *Penicillium nigricans*, and *Penicillium patulum* on a medium of cracked corn (750 g/liter) and distilled water. Permanent stocks were prepared by drying the spores grown on corn and storing the spores in sand. A medium suitable for maintaining these fungi on agar plates was also described (Table 2). The Czapek-Dox agar medium has also been successfully employed by Oxford et al. (1939) and in a Glaxo patent (U.S. Patent No. 2,843,527).

B. Inoculum Development

Both spore suspensions and vegetative cells have been reported as suitable inocula for laboratory-scale fermentations. For example, the Glaxo group

Table 1 Griseofulvin-Producing Organisms

Organism	Reference
Penicillium griseofulvum	Oxford et al. (1939)
Penicillium janczewskii	Brian et al. (1946)
Penicillium nigricans	Brian et al. (1955)
Penicillium urticae	Brian et al. (1955)
Penicillium raistrickii	Brian et al. (1955)
Penicillium albidum	U.S. Patent No. 2,843,527
Penicillium raciborskii	U.S. Patent No. 2,843,527
Penicillium melinii	U.S. Patent No. 2,843,527
Penicillium patulum	U.S. Patent No. 2.843,527
Aspergillus versicolor	Kingston et al. (1976)
Carpenteles brefeldianum	Dean et al. (1957)
Khauskia oryzae	German Patent No. 1,813,572
Nigrospora musae	U.S. Patent No. 3,616,247
Nigrospora oryzae	Japanese Patent No. 7,213,717
Nigrospora saccharii	Japanese Patent No. 1,813,572
Nigrospora splaerica	U.S. Patent No. 3,616,247

(U.S. Patent No. 2,843,527) described the sporulation medium (Table 3) which provided inoculum suitable for direct inoculation of shaken-flask cultures. Bayan et al. (1962) employed a germination medium, described in Table 4, to provide inoculum for their shake-flask medium studies. Larger-scale development and production fermentations have typically utilized a seed stage. An example of this approach is listed in Table 5.

C. Fermentation Media and Conditions

Oxford et al. (1939) described a chemically defined medium for production of griseofulvin in unagitated liquid culture (Table 6). The incubation period for this technique lasted 65–85 days, and a final mass of approximately 11 g of dry mycelium per liter was present when the culture flasks were harvested. This medium contains the same ingredients as standard Czapek-Dox

Table 2 Solid Medium for Maintenance of Griseofulvin-Producing Cultures

Ingredient	Amount (g/liter)
Difco Peptone	5.00
Glycerol	7.50
Molasses	7.50
NaCl	40.00
$MgSO_4 \cdot 7H_2O$	0.05
K_2HPO_4	0.06
Agar	25.00

Source: Bayan et al. (1962).

Table 3 Medium for the Preparation of
Spore Inoculum (U.S. Patent No.
2,843,527)[a]

Ingredient	Amount (g/liter)
Whey powder lactose	30.0
Whey powder nitrogen	50.5
KH_2PO_4	4.0
KCl	0.5
Corn-steep liquor solids	3.8

[a]A total of 600 ml of medium per 2-liter
conical flask, 25°C, 7 days.

Table 4 Germination Medium for Preparation of Vegetative Inoculum[a]

Ingredient	Amount (g/liter)
Protopeptone #159	20.0
Malted cereal extract	10.0
Glucose	40.0
Soluble starch	20.0
$NaNO_3$	3.0
KH_2PO_4	1.0
$MgSO_4 \cdot 7H_2O$	0.5
$FeSO_4 \cdot 7H_2O$	0.02

[a]A total of 50 ml of medium per 250-ml
Erlenmeyer flask, 25°C, 48 hr, 280 rpm
on a shaker with a two-in. throw.
Source:Bayan et al. (1962).

Table 5 Vegetative Seed Stage Medium
(U.S. Patent No. 3,038,839)[a]

Ingredient	Amount (g/liter)
Corn-steep liquor—nitrogen	3.0
Brown sugar	20.0
Chalk	10.0
Maize oil	10.0
Hodag MF	0.3

[a]One liter of inoculum ($3 \times 10^7 - 5 \times 10^7$
spores/ml) per 100 gal, 25°C, 40 hr or
until mycelial volume equals 25%.

Table 6 Defined Fermentation Medium[a]

Ingredient	Amount (g/liter)
Glucose	80.0
NaNO$_3$	2.5
KH$_2$PO$_4$	1.0
KCl	0.5
MgSO$_4$·7H$_2$O	0.5
FeSO$_4$·7H$_2$O	0.02

[a]A total of 350 ml of medium per 1-liter conical flask, 30°C for 65–85 days or until residual glucose equals 6–8 g/liter.
Source: Oxford et al. (1939).

broth, but at increased concentration of the components. Brian et al. (1946) reported a modified fermentation medium related to Czapek-Dox, but with additional trace elements (Table 7). Bayan et al. (1962) employed the medium of Brian et al. in shake-flask cultures and reported a yield of 200 μg/ml with *P. griseofulvum*. This medium was also suitable for production of griseofulvin by *P. nigricans* and *P. patulum*, although the yields with the latter two organisms were somewhat lower than that of *P. griseofulvum*.

Typically, high-yield industrial fermentations in submerged cultures have employed complex media with corn-steep liquor as the nitrogen source. These processes were highly reminiscent of the penicillin cultural techniques.

Table 7 Defined Fermentation Medium for Surface and Submerged Culture[a]

Ingredient	Amount (g/liter)
Glucose	75.0
KNO$_3$	2.3
KH$_2$PO$_4$	1.0
MgSO$_4$·7H$_2$O	0.5
FeSO$_4$·7H$_2$O	0.001
CuSO$_4$·5H$_2$O	0.00015
ZnSO$_4$·7H$_2$O	0.001
MnSO$_4$·4H$_2$O	0.0001
KMoO$_4$	0.0001

[a]Other conditions were the following: surface culture, 25°C, 24 days; shake flasks, 50 ml of medium per 250-ml flask, 5 ml of inoculum, 25°C, 5–7 days; all media adjusted to pH 4.4 with HCl prior to autoclaving.
Source: Bayan et al. (1962) and Brian et al. (1946).

Table 8 450-Liter Batch Fermentation
Medium (U.S. Patent No. 2,843,527)[a]

Ingredient	Amount (g/liter)
Corn-steep nitrogen	2.0
Lactose	70.0
Limestone	8.0
K_2HPO_4	4.0
KCl	1.0

[a]Other conditions were 10% inoculum,
25°C, 350 rpm, 10 ft^3 of air per minute
0–8 hr, then 20 c.f.m.

Yields exceeding 1500 µg/ml on a 450-liter scale with the medium described
in Table 8 and a mutant strain of *P. patulum* have been reported (U.S. Patent
No. 2,843,527).

A significant improvement in the fermentation technology of griseofulvin
production was realized with the introduction of the fed-batch process described in a Glaxo patent (U.S. Patent No. 3,069,328). The initial medium
(Table 9) was deliberately formulated to be deficient in carbohydrate. When
the pH of the culture rose after inoculation, repeated additions of glucose
were used to maintain the pH at or near neutrality. A yield of 6 g of griseofulvin per liter was obtained after 220 hr of cultivation. Utilization of the
fed-batch technique permitted an increase in the medium's nutrient concentration. When the nitrogen level (corn-steep liquor) was increased to 4–5 g
per liter, antibiotic yields improved to 11 g per liter after 260 hr (U.S. Patent
No. 3,038,839). Partial replacement of corn-steep liquor nitrogen by ammonium sulfate was shown to be advantageous. This modification gave a yield
of 14 g per liter in a 1000-gal process described in another Glaxo patent (U.S.

Table 9 Fed-Batch Medium 400-Gal
Scale (U.S. Patent No. 3,069,328)[a]

Ingredient	Amount (g/liter)
Corn-steep nitrogen	1.75
Limestone	4.00
KH_2PO_4	4.00
KCl	1.50
Antifoam	0.55

[a]Other conditions were 10% inoculum
from vegetative seed stage, 25°C, 25
c.f.m. 0–5 hr, 50 ft^3/min 5–10 hr, 80
c.f.m. after 10 hr, glucose solution
(50%) fed 0.5–5.0 liters/hr as required
to maintain pH 6.8–7.2.

Table 10 Fed-Batch Medium 100-Gal
Scale (U.S. Patent No. 3,069,329)[a]

Ingredient	Amount (g/liter)
Corn-steep nitrogen	3.50
$(NH_4)_2SO_4$	0.50
KH_2PO_4	4.00
KCl	1.00
$CaCO_3$	4.00
H_2SO_4	0.125
Mobilpar S	0.275
White mineral oil	0.275

[a]Other conditions were 800-gal preinoculation volume, 10% inoculum, 25°C, 40
c.f.m. 0–5 hr, 80 ft^3/min 5–10 hr, 125
c.f.m. after 10 hr, 220 rpm, glucose
(50%) fed 0.5–6 liter/hr as required
to maintain pH 6.8–7.2.

Patent No. 3,069,329). The medium and fermentation conditions for this process are described in Table 10.

In still another Glaxo patent (U.S. Patent No. 3,095,360) it was claimed
that the corn-steep requirement could be partially replaced by choline. In
shaken flasks using the medium described in Table 8, the corn-steep solids
could be reduced to 1 g (as nitrogen) per liter if 1 g/liter choline was added,
and the resulting griseofulvin yields were superior to that produced in the
conventional medium. Other "methyl donor" compounds were also claimed to
be of similar benefit.

The use of hydrocarbons as carbon and energy sources for griseofulvin
fermentation has been reported (U.S. Patent No. 3,607,656). The fermentation, as reported in the patent assigned to the British Petroleum Company,
was conducted in two stages. If hydrocarbons were used in the first stage,
other microorganisms, particularly the yeasts *Candida tropicalis* and *Candida
lipolytica*, were employed to initiate catabolism of the alkanes. Alternately,
the first stage employed carbohydrate as the carbon source for either *P.
griseofulvum*, *P. patulum*, or *P. nigricans*. The hydrocarbons were used
exclusively as the carbon source for the second stage in which the bulk of
the antibiotic synthesis occurred. Daily additions of hydrocarbons were made
to the initial second-stage medium described in Table 11; resulting in a yield
of 2.55 g of griseofulvin per liter after 4 days.

An unusual reactor design has been disclosed in a patent assigned to
Biochemie GmbH (British Patent No. 1,159,695) and it relates to the production of griseofulvin in high yield by surface culture. Cultivation occurred
in stacked fermentation dishes contained within a closed vessel. Humidified
air was supplied to the vessel so that the airflow was directed away from the
surface mycelial mats in the dishes. After culturing *Penicillium urticae* for
21–24 days, a yield of 7–10 g/liter was observed. The medium used in this
process is described in Table 12. Although the maximum yields and productivity appear to be lower than for submerged fermentations, the energy requirements are greatly reduced.

Table 11 Hydrocarbon Medium for Griseofulvin Production (U.S. Patent No. 3,607,656)

Ingredient	Amount (g/liter)
Gas oil	200.00
$NaNO_3$	2.00
KH_2PO_4	4.00
KCl	1.00
$MgSO_4$	0.500
$FeSO_4$	0.005
$CuSO_4$	0.0075
$ZnSO_4$	0.005
$MnSO_4$	0.005
$KMoO_4$	0.005

D. Fermentation Control Strategy

Optimization of fermentation yield appears to be significantly related to control of carbon limitation. Hence the use of lactose in batch fermentation presumably is a mechanism to provide a slow release of the hexose moieties through the enzymatic hydrolysis of this disaccharide. Such a reaction would control the rate of catabolism of the carbohydrate. The replacement of lactose by a glucose feed, as described in the various patents, probably provided a more direct control over the carbon consumption rate. Similar fermentation strategies have been widely employed for industrial penicillin fermentations. A Soviet patent assigned to Leningrad Antibiotics (Soviet Patent No. 637,647) has specifically claimed a method of improving griseofulvin fermentation yields by controlling the growth rate with a glucose feed. The Soviet scientists reported a 60% improvement in antibiotic synthetic rate when the growth rate was maintained between 0.004 and 0.007 hr^{-1}, as compared with a control fermentation's growth rate of 0.0017–0.0025 hr^{-1}.

A survey of the current literature has not revealed any published reports of on-line computer control being applied to the griseofulvin fermentation process. However, since many aspects of the griseofulvin control strategy seem to parallel those for penicillin fermentation, interested investigators would

Table 12 Surface Culture Production Medium (British Patent No. 1,159,695)

Ingredient	Amount (g/liter)
Sucrose	210.0
Corn-steep nitrogen	2.5
KCl	1.5
Na_2SO_4	2.0
KH_2PO_4	8.0

be advised to explore the process control approaches published for penicillin. Presumably many of these strategies could be adapted to the griseofulvin fermentation.

E. Fermentation of Griseofulvin Analogs and Bioconversions of Griseofulvin-Like Materials

Dechlorogriseofulvin is produced by *P. griseofulvum* and *P. nigricans* when cultured in Czapek-Dox medium and under the same conditions used to produce griseofulvin (MacMillan, 1951). The bromo analog has been reported to be readily produced by simply substituting potassium bromide for potassium chloride in the Czapek-Dox medium (MacMillan, 1954). Deuterated griseofulvin has been synthesized by *P. janczewskii* when the water in the culture medium was replaced by deuterium oxide (Nona et al., 1968).

The conversion of dehydrogriseofulvin to (+)-5'-hydroxygriseofulvin and of dehydro-1-thiogriseofulvin to (+)-1-thiogriseofulvin and (+)-5'-hydroxy-1-thiogriseofulvin has been demonstrated by researchers at American Cyanamide (U.S. Patents Nos. 3,532,714, 3,557,151, 3,616,237, and 3,616,238). These conversions were carried out aerobically by *Streptomyces cinereocrocatus* NRRL 3443. The substrates were added to the cultures after a period of growth had been observed.

III. STRAIN DEVELOPMENT

Brian et al. (1955) surveyed several fungal cultures of the *Penicillium raistrickii* series for their ability to produce griseofulvin. The authors observed great variations in yield between different strains on various media. When their morphological characters were compared, the strains with conidia produced the greatest amount of griseofulvin. The cultures that produced sclerotia only synthesized negligible quantities of the antibiotic. Those forms which were intermediate in morphological character were also intermediate producers of griseofulvin. The best yield reported by these authors was 102 µg/ml of fermenter broth. In British Patent No. 1,132,217, a process was disclosed that resulted in titers of 11 mg/ml. This yield was obtained by mutating spores of *P. urticae* with uv light and ethyl methane sulfonate and then culturing the mutants in a corn-steep liquor medium. Mutants of *P. nigricans* have been derived by treatment with nitrosomethyl urea, ultraviolet irradiation and sulfur isotopes (Soviet Patents Nos. 412-787 and 458-577). The use of the first two mutagens resulted in a culture producing 3.6 mg griseofulvin per milliliter. The sulfur isotope procedure produced a culture that was reported to yield 3.5 mg griseofulvin per milliliter, or 60–70% more than its parent.

IV. PURIFICATION

A review of purification schemes for griseofulvin should be prefaced by a recognition of the physical properties of the compound. According to *The Merck Index* (Windholz, 1976), griseofulvin is practically insoluble in water or nonpolar solvents such as petroleum ether, but somewhat soluble in the polar solvents ethanol, chloroform, acetone, and ethyl acetate.

In the original griseofulvin publication, Oxford et al. (1939) isolated the antibiotic from the mycelial cake. After drying, the mycelium was extracted with light petroleum and then extracted with ether. The evaporated residue from the ether extract was dissolved into hot benzene and, after cooling, another microbial product containing nitrogen crystallized. This compound was probably mycelianamide. Bayan et al. (1962) showed this compound to be produced by this organism in a similar medium. Griseofulvin was then isolated by evaporation of the benzene and recrystallization from ethanol. An estimation of the yield is not possible, since no standard assay existed for determination of the antibiotic concentration in the surface culture.

In the first Glaxo patent (British Patent No. 784,618), a purification scheme was disclosed for larger-scale (300-400 liters) extractions. The wet mycelium was collected from a rotary string discharge filter and extracted three times with butyl acetate. The combined ester extracts were clarified by centrifugations, concentrated under reduced pressure at 50°C, and then cooled to give a crude material. Further purification was accomplished by washing with chloroform and recrystallization from aqueous acetone, and the final product was washed with butyl acetate. Pure griseofulvin was obtained with a 58% yield from the fermentation medium. No attempt was made to extract the antibiotic not associated with the mycelium, although the authors indicated that approximately 10% of the total griseofulvin remained in the medium.

In Soviet Patent No. 135,187 the mycelium was extracted three times by homogenization in the presence of methylene chloride. The solvent extracts were decanted, concentrated two-thirds under reduced pressure below 50°C, and then centrifuged to separate mycelial residues. The extract was decolorized with iron-free charcoal, filtered, and evaporated to 1/15 of the original volume. Cooling to 8°C produced crystals which after a final acetone wash resulted in 88% pure griseofulvin, with a 95% yield from the mycelial content. Two subsequent recrystallizations from acetone improved the purity to 99% but reduced the overall yield to about 72%.

Further improvements were reported in a patent assigned to Leningrad Antibiotics (French Patent No. 1,462,217) in which the above scheme was modified to eliminate acetone recrystallizations. In this case, the methylene chloride extracts were concentrated to 1/10 of the original volume and cooled to 4°C to remove resinous impurities. After decolorization with charcoal, crystallization was accomplished by evaporation at 50°C to 1/15 the original volume and subsequent cooling to 0°C. After washing with cold acetone, an overall yield of 95% with 95% pure griseofulvin was claimed. In addition to the improved yield, this patent also claimed the process to be simpler and safer than other schemes using solvents more toxic and more combustible than methylene chloride.

V. BIOSYNTHESIS

As suggested in earlier reviews on the biosynthesis of griseofulvin (Grove, 1967; Rhodes, 1963), a complication arises in that several different producing organisms have been employed in biosynthetic studies. Consequently the producing organism as well as the biochemistry involved will be described.

The carbon skeleton of the griseofulvin rings has been shown by Birch et al. (1958, 1962) to arise from seven acetate units. Radioactive [1-14C]-acetate was incorporated into griseofulvin by *P. griseofulvum,* as shown in Structure [2]. An identical pattern of acetate incorporation was reported by Tanube and Detre (1966). In this case the incorporation of 2-13C by *P. urticae* was shown by nuclear magnetic resonance (NMR) spectroscopy to occur at alternate carbon sites to those reported by Birch et al. (1958). More recently Simpson and Holker (1977) have similarly shown alternating site incorporation of [1-^{13}C] or [2-^{13}C]acetate into griseofulvin by *P. patulum*. In still other studies, Sato et al. (1975, 1976a) have used [2-^{2}H] and [2-^{3}H]acetate incorporation by *P. urticae* and shown deuterium and tritium label incorporation consistent with the carbon labeling studies.

The isolation of three metabolites, griseophenones A, B, and C [3],

GRISEOPHENONE A $R_1 = CH_3$ $R_2 = Cl$
GRISEOPHENONE B $R_1 = H$ $R_2 = Cl$
GRISEOPHENONE C $R_1 = H$ $R_2 = H$

from a *P. patulum* culture was reported by Rhodes et al. (1961) and a biosynthetic scheme of griseophenone C → B → A was proposed. This progression was confirmed by preparation of the ^{36}Cl-labeled griseophenones A and B and ^{14}C-labeled griseophenone C by Rhodes et al. (1963) and incubating the labeled compounds with the same *P. patulum* culture. However, label from C and B only could be incorporated into griseofulvin, suggesting that only C and B are precursors of griseofulvin.

Birch et al. (1958) proposed that the ring structure originated from the folding of a single heptaketide formed by head-to-tail linkage of the seven acetate units. Other suggested mechanisms have included condensation of two smaller polyketide chains (Vanek and Sousek, 1962; Dean et al., 1975) and ring closure followed by ring fission (Whalley, 1961). Sato et al. (1976b) have confirmed the simple heptaketide cyclization route in *P. urticae* by employing [1,2-^{13}C]acetate-labeled griseofulvin. NMR analysis in pentadeuteriopyridine of ^{13}C-^{13}C coupling indicated two different results for positioning of the acetate units in ring A. Sato et al. (1976b) reasoned that chlorination of griseophenone C could occur at either of two positions in ring A (5 and 7 are equivalent), but that ring B closure would only occur if the chlorinated carbon was adjacent to C7a, thus resulting in the two arrangements of acetate units in ring A [4]. Neither of the other two proposed mechanisms of Vanek and Sousek (1962) or Whalley (1961) are consistent with the ^{13}C-^{13}C coupling patterns arising from 1,2-^{13}C-labeled griseofulvin.

Additional evidence for the polyketide precursor was provided by Simpson and Holker (1977). Incorporation of [1-^{13}C], [2-^{13}C], and [1,2-^{13}C]-acetate into griseofulvin was achieved using *P. patulum*. In this work, the

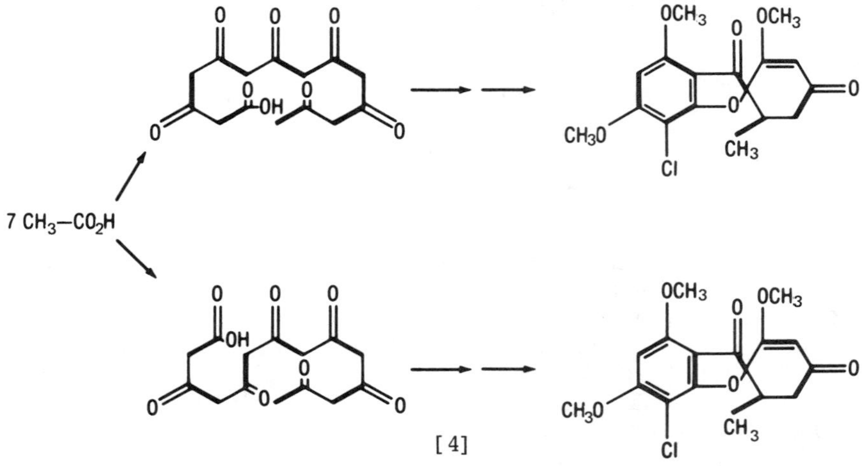

[4]

incorporation patterns were consistent with those reported by Sato et al. (1976b). Simpson and Holker also reported that the 6' and 6'-methyl carbons were relatively more enriched with label than the other 12 carbons of the gri-san ring. This was attributed to a starter acetyl coenzyme A (CoA) unit, which was then sequentially attached to the other six acetate units derived from malonyl CoA units. These authors reasoned that the first acetate unit bore greater enrichment, since it may be directly incorporated. The next six acetate units must first have been converted to malonyl-CoA and thus were diluted by endogenous malonyl-CoA.

The o-methyl groups were shown to be derived from typical C1 pools by Hockenhull and Faulds (1955), who demonstrated that [^{14}C]methyl choline effectively labeled the methoxy groups of griseofulvin in a culture of *P. patulum*. It has also been shown by Sato et al. (1976b), Tanube and Derre (1966), and Simpson and Holker (1977) that [2-^{13}C]acetate, but not [1-^{13}C]acetate, is incorporated into the methoxy groups. Simpson and Holder (1977) proposed that the acetate is metabolized via the Krebs cycle to malate and subsequently via pyruvate and serine into the C1 pool.

The sequence of methylations has been partially elucidated by Harris et al. (1976). First, a proposed benzophenone intermediate (II) [5] was chem-

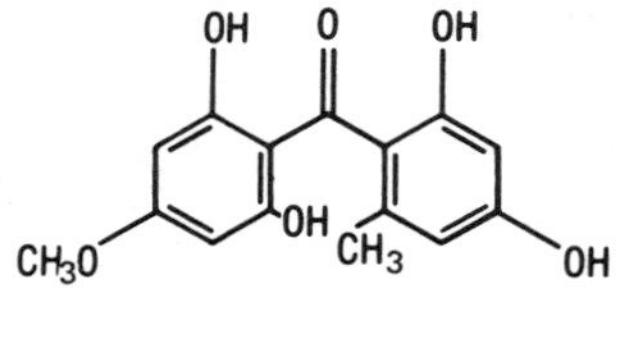

[5] 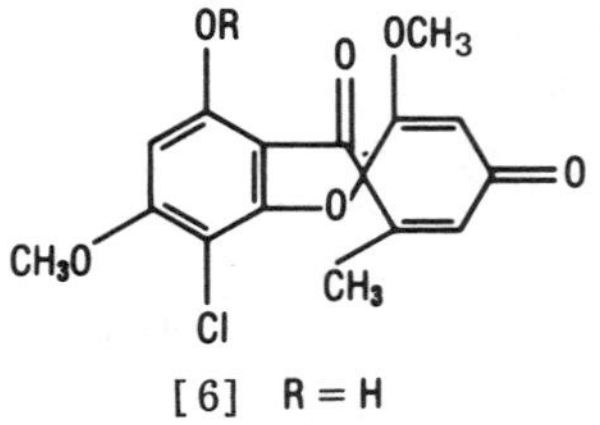   [6] R = H

[7] R = CH$_3$

ically synthesized with tritium label in the 6-methoxy group. In a culture of *P. griseofulvum*, label from II was incorporated into griseofulvin, with all radioactivity maintained in the methoxy group; thus [5] was proposed as a precursor of griseophenone C. In the same publication, Harris and co-workers also provided evidence that the third and final methylation occurred after closure of ring B. 4-Demethyldehydrogriseofulvin (III) [6] was synthesized chemically from [1-^{14}C]acetate-labeled griseofulvin. Incubation of the grisan

III with *P. griseofulvum* resulted in the reincorporation of the ^{14}C label into griseofulvin. The authors concluded that [6] was derived by ring closure of griseophenone B, which preceded methylation of [6], affording dehydro-griseofulvin [7]. The methylation and reduction of [6] to griseofulvin has also been reported by Sato et al. (1978). In this study, [6] deuterated at the 5'-position was prepared and the label was incorporated into griseo-fulvin by a culture of *P. urticae*. NMR studies showed that the 5'-deuterium label resided exclusively in the α-position, indicating the stereospecificity of the final hydrogenation.

The putative terminal step, reduction of IV to griseofulvin, has been demonstrated by Birch (1962). ^{14}C-labeled IV was incorporated into griseo-fulvin by *P. griseofulvum*. Sato et al. (1977) have also reported the reduc-tion of IV to griseofulvin by *S. cinereocrocatus* with the same stereospecificity for 5' addition of hydrogen as in the fungus *P. griseofulvum*.

In summary, there now exists considerable evidence that the griseofulvin molecule arises from the cyclization of a heptaketide derived from acetate and malonate. No conclusive evidence exists describing the order of closure of rings A and C and the methylation of the 6-hydroxyl group. The most recent evidence suggests that the terminal biosynthetic pathway involves methylation of the 2'-hydroxyl group giving griseophenone C, which is then chlorinated to griseophenone B. Closure of ring B occurs next, resulting in the tricyclic grisan [6], which is then methylated to dehydrogriseofulvin [7] and finally hydrogenated to griseofulvin.

It should be noted that not every step described here has been observed with all griseofulvin-producing organisms. Indeed, most biosynthetic studies have been limited to either *P. griseofulvum*, *P. patulum*, or *P. urticae*. How-ever, considerable species overlap does exist in these pathways, and there have been no obvious differences between producers. Thus all reports to date point to a common biosynthetic pathway for griseofulvin.

VI. DETERMINATION OF GRISEOFULVIN

A. Biological Determination

A precise description of a biological assay for the determination of griseofulvin has been reported by Nona et al. (1968). In this procedure the test organ-ism, *Microsporum gypseum*, is cultured on a medium containing dextrose (40 g), peptone (10 g), agar (1.5 g), chloramphenicol (0.05 g), and distilled water to 1 liter. After sufficient growth had occurred at 25°C, a spore sus-pension was prepared by dislodging the surface growth in distilled water. To prepare test plates, a base layer of the above medium was added to the plate and then overlaid with the same medium containing cycloheximide and spore suspension. Solutions to be assayed were dispensed into stainless steel cylinders placed directly on the surface of the medium. The exact concentra-tion of the antibiotic is determined by relating the zone of inhibition to a stan-dard curve prepared in a like manner.

B. Spectrophotometric Determination

A method for spectrophotometrically determining griseofulvin in fermentation broths has been reported by Ashton and Brown (1956). These authors ex-tracted whole fermentation broth with butyl acetate. The extract was then filtered and diluted further with the same solvent and the ultraviolet absorption determined at 288, 290, 292, 294, 298, and 300 µm. The resulting data were mathematically manipulated to exclude any other compounds absorbing in that area and the concentration of the antibiotic calculated.

VII. MECHANISM OF ACTION

A review covering most aspects concerned with the mechanism of action of griseofulvin was published in 1974 by Huber. Since that time only a minimal number of studies have been reported as to how this antibiotic controls fungal growth. Of interest is the study of Gull and Trinci (1973) using the fungus *Basidiobolus ranarum* as the test system. These investigators reported that griseofulvin produced multinucleation. In such multinucleated cells the nuclei were observed to be always in close proximity. It was suggested that griseofulvin may act by inhibiting the postulated sliding of microtubules.

Some studies that have been published describe basic differences in the mode of action of griseofulvin and other mitotic inhibitors (colchicine, etc.). Malawista (1971) tested several mitotic agents in a system that employed the rapid reversible darkening of frog skin under the influence of melanocyte-stimulating hormone. Unlike vinblastine, vincristine, colcemid, and colchicine, preincubation of frog skin with griseofulvin followed by washing had no subsequent effects on darkening or lightening. In using spindle birefringance as a reflection of the concentration of spindle microtubules, Malawista et al. (1976) reported that this index of activity decreased linearly as a function of time at high concentrations of griseofulvin. The biological test system for their experiments were oocytes of the starfish *Pisaster ochraceous*. Grisham et al. (1973) examined the effect of various chemical agents on the polymerization reaction needed to form microtubules in brain extracts. Although colchicine and other substances prevented microtubule formation, griseofulvin did not. Microscopic examination of griseofulvin-inhibited HeLa cells showed them to have normal microtubules. Steroidogenesis in rate adrenal cells has been linked to microtubule formation by Crivello and Jefcoate (1978), because the antibiotic inhibited microtubule formation. A minireview of the action of these drugs on microtubules has been published by Wilson (1975).

REFERENCES

Ashton, G. C., and Brown, A. D. (1956). Determination of griseofulvin in fermentation samples. *Analyst 81*:220.

Bayan, A. P., Unger, U. F., and Brown, W. E. (1962). Factors affecting the biosynthesis of griseofulvin. *Antimicrob. Agents Chemother. 1962*: 669.

Bekker, Z. E., Gorlenko, M. V., Lisina, E. S., Rodionova, E. G., and Voronina, E. V. (1963). Antiphytopathogenic properties of some antibiotics of fungal origin. *Chem. Abstr. 58*:14465.

Birch, A. J. (1962). Some pathways in biosynthesis. *Proc. Chem. Soc. 1962*:3.

Birch, A. J., Massey-Westrop, R. A., Richards, R. N., and Smith, H. (1958). Studies in relation to biosynthesis. Part XIII, Griseofulvin. *J. Chem. Soc. 1958*:360.

Brian, P. W., Curtis, P. J., and Hemming, H. G. (1946). A substance causing abnormal development of fungal hyphae produced by *Penicillium janczewskii* Zal. *Trans. Br. Mycol. Soc. 29*:173.

Brian, P. W., Curtis, P. J., and Hemming, H. G. (1949). A substance causing abnormal development of fungal hyphae produced by *Penicillium janczewskii* Zal. III. Identity of curling factor with griseofulvin. *Trans. Br. Mycol. Soc. 32*:20.

Brian, P. W., Curtis, P. J., and Hemming, H. G. (1955). Production of griseofulvin by *Penicillium raistrickii*. *Trans. Br. Mycol. Soc. 38*:305.

Crivello, J. F., and Jefcoate, C. R. (1978). Mechanism of corticotropin action in rat adrenal cells. I. The effects of inhibitors of protein synthesis and microfilament formation on corticosterone synthesis. *Biochim. Biophys. Acta. 542*:315.

Dean, F. M., Eade, R. A., Moubasher, R., and Robertson, A. (1957). Fulvic acid: Its structure and relationship to citromycetin and fusarubin. *Nature 179*:366.

Grisham, L. M., Wilson, L., and Bensch, K. G. (1973). Antimitotic action of griseofulvin does not involve disruption of microtubules. *Nature 244*: 294.

Grove, J. F. (1967). Griseofulvin. In *Antibiotic*, Vol. s. D. Gottlieb and P. D. Shaw (Eds.). Springer-Verlag, New York, p. 123.

Grove, J. F., and McGowan, J. C. (1947). Identity of griseofulvin and "curling factor." *Nature 160*:574.

Grove, J. F., MacMillan, J., Mulholland, T. P. C., and Rogers, M. A. T. (1952). Griseofulvin. Part IV. Structure. *J. Chem. Soc. 1952*:3977.

Gull, K., and Trinci, A. P. J. (1973). Griseofulvin inhibits fungal mitosis. *Nature 244*:292.

Harris, C., Roberson, J. S., and Harris, T. M. (1976). Biosynthesis of griseofulvin. *J. Am. Chem. Soc. 98*:5380.

Hockenhull, D. J. D., and Faulds, W. F. (1955). Origin of the methoxy groups of griseofulvin. *Chem. Ind. 1955*:1390.

Huber, F. M. (1974). Griseofulvin. In *Antibiotics*, Vol. 3. J. W. Corcoran and F. E. Hahn (Eds.). Springer-Verlag, Berlin, p. 606.

Kingston, D. G. I., Chen, P. N., and Percellotte, J. R. (1976). Metabolites of *aspergillus versicolor*: 6,8-Di-o-methylnidurufin, griseofulvin, dechlorgriseofulvin and 3,8-dihydroxy-6-methoxy-1-methylxanthrone. *Phytochemistry 15*:1037.

MacMillan, J. (1951). Dechlorogriseofulvin, a metabolic product of *Penicillium griseofulvum* Dierckx and *Penicillium janczewskii* Zal. *Chem. Ind. (London) 1951*:179.

MacMillan, J. (1954). Griseofulvin. Part 9. Isolation of the bromoanalogue from *Penicillium griseofulvum* and *Penicillium nigricans*. *J. Chem. Soc. 1954*:2585.

Malawista, S. E. (1971). The melanocyte model. *J. Cell. Biol. 49*:848–855.

Malawista, S. E., Sato, H., and Creasey, W. A. (1976). Of the mitotic spindle in oocytes exposed to griseofulvin and vinblastine. *Exp. Cell Res. 99*: 193.

Nona, D. A., Blake, M. I., Crespi, H. L., and Katy, J. J. (1968). Effect of deuterium oxide on the culturing of *Penicillium janczewskii*. III. Antifungal activity of fully deuterated griseofulvin. *J. Pharm. Sci. 57*: 1993.

Oxford, A. E., Raistrick, H., and Simonart, P. (1939). Studies on the biochemistry of microorganisms 60. Griseofulvin, $C_{17}H_{17}O_6Cl$, a metabolic product of *Penicillium griseofulvum* Dierckx. *Biochem. J. 33*:240.

Perlman, D. (1978). Fermentation products. 1977. In *The Fermentor*. Newsletter of the Microbial Chemistry section of the American Chemical Society.

Rhodes, A. (1963). Griseofulvin. Production and biosynthesis. *Prog. Ind. Microbiol.* 4:167.

Rhodes, A., Boothroyd, B., McGonagle, M. P., and Somerfield, G. A. (1961). Biosynthesis of griseofulvin: The methylated benzophenone intermediates. *Biochem. J.* 81:28.

Rhodes, A., Somerfield, G. A., and McGonagle, M. P. (1963). Biosynthesis of griseofulvin: Observations on the incorporation of C^{14}-griseophenone C and Cl^{36}-griseophenone B and A. *Biochem. J.* 88:349.

Sato, Y., Machida, T., and Oda, T. (1975). Incorporation of [2-$^3H_3$] and [2-$^2H_3$]-acetate by *Penicillium urticae* into griseofulvum and determination of the stereochemistry of isotopes at C-5'. *Tetrahedron Lett.* 51:4571.

Sato, Y., Oda, T., and Saito, H. (1976a). A novel biosynthetic study of griseofulvin by ^2H nuclear magnetic resonance: Determination of deuterium incorporation from [2-$^2H_3$] acetate by *Penicillium urticae*. *Tetrahedron Lett.* 31:2695.

Sato, Y., Oda, T., and Uvano, S. (1976b). Griseofulvin biosynthesis: New evidence of two acetate-dispositions in the ring A from ^{13}C nuclear magnetic resonance studies. *Tetrahedron Lett.* 44:3971.

Sato, Y., Oda, T., and Saito, H. (1977). Microbial transformation of dehydrogriseofulvin and griseofulvin: ^2H NMR and mass spectrometric studies of stereochemical courses of microbial hydrogenation and hydroxylation. *J. Chem. Soc. Chem. Commun.* 1977:415.

Sato, Y., Oda, T., and Saito, H. (1978). H^2 nuclear magnetic resonance studies on biosynthesis. Stereochemistry of the 5'-hydrogen atoms of griseofulvin derived from griseophenone B and 4-demethyldehydroxy griseofulvin. *J. Chem. Soc. Chem. Commun.* 1978:135.

Simpson, T. J., and Holker, J. S. E. (1977). C^{13}-NMR studies on griseofulvin biosynthesis and acetate metabolism in *Penicillium patulum*. *Phytochemistry* 16:229.

Tanube, M., and Detre, G. (1966). The use of C^{13}-labelled acetate in biosynthetic studies. *J. Am. Chem. Soc.* 88:4515.

Vanek, Z., and Sousek, M. (1962). Factors determining the biosynthesis of griseofulvin and similar substances. *Folia Microbiol.* 7:262.

Whalley, W. B. (1961). Some structural and biogenic relationships. In *Plant Phenolics*. W. D. Oleis (Ed.). Pergamon, New York, p. 20.

Wilson, L. (1975). Action of drugs on microtubules. *Life Sci.* 17:303.

Windholz, M. (Ed.) (1976). *The Merck Index*. Merck, Rayway, New Jersey.

Antibiotics Used in Medical Practice:
Antitumor and Antiviral Antibiotics

19

DAUNORUBICIN AND ADRIAMYCIN: PROPERTIES, BIOSYNTHESIS, AND FERMENTATION

RICHARD J. WHITE* AND RONALD M. STROSHANE† *FCRC Fermentation Program, Frederick Cancer Research Center, Frederick, Maryland*

Present affiliation: Medical Research Division, Lederle Laboratories, Pearl River, New York
†*Present affiliation:* Sterling-Winthrop Research Institute, Rensselaer, New York

I. INTRODUCTION

Daunorubicin and adriamycin are closely related anthracycline antibiotics with potent anticancer activity. As a group, the anthracyclines are characterized by the possession of an anthraquinone chromophore that, in most cases, is substituted with one or more sugars. Daunorubicin and adriamycin have the amino sugar daunosamine as a substituent (Fig. 1). The first anthracycline to be described was β-rhodomycin I (Brockmann and Bauer, 1950) and this was followed by cinerubins A and B (Ettlinger et al., 1959). The high toxicity of these early anthracyclines precluded any clinical application of their antibacterial, antifungal, or anticancer activity. The first clinically effective anthracycline to be isolated was daunorubicin. It was discovered independently in 1963 by groups working at Farmitalia in Italy (Grein et al., 1963) and at Rhône-Poulenc in France (Dubost et al., 1963). The Italian group isolated daunorubicin (originally named daunomycin) from fermentations of new species of streptomyces which they designated *Streptomyces peucetius*. In France the antibiotic was produced by an isolate of the known species *Streptomyces coeruleorubidus* and was called rubidomycin. Daunorubicin is a contraction of the original Italian and French names daunomycin and rubidomycin. Although highly toxic, daunorubicin showed interesting activity against several experimental cancers in animals, and has progressed to clinical trials as an anticancer antibiotic. Adriamycin (also referred to as doxorubicin was discovered several years later as a result of a programmed attempt to find an improved daunorubicin. Structurally it is closely related to daunorubicin and is produced by a mutant strain of *S. peucetius* referred to as variant *caesius* (Arcamone et al., 1969). Adriamycin was reported to have an improved therapeutic index and to be active against a wider spectrum of

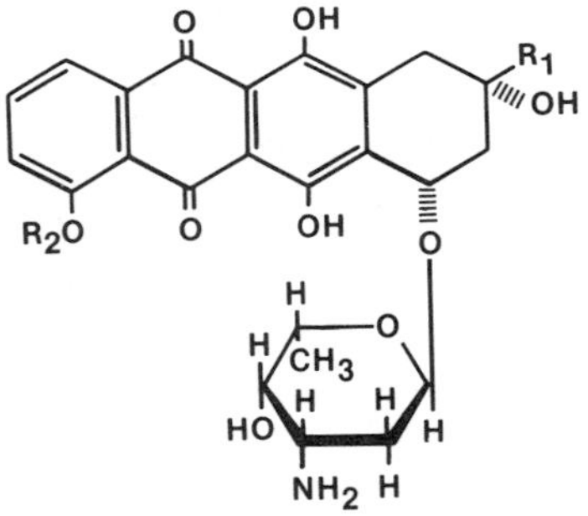

	R₁	R₂
DAUNORUBICIN	CH_3CO	CH_3
ADRIAMYCIN	CH_2OHCO	CH_3
CARMINOMYCIN	CH_3CO	H

Figure 1 Structures of daunorubicin, adriamycin, and carminomycin.

tumors than daunorubicin (Bonadonna et al., 1970). Adriamycin rapidly replaced use of the older drug (daunorubicin) in most clinical trials and is now probably the most important anticancer antibiotic in current usage. Figure 1 also gives the structure of carminomycin, a third anthracycline antibiotic closely related to daunorubicin and adriamycin. It was discovered by Gause et al. (1974) in Russia and is produced by *Actinomadura carminata*.

II. DISCOVERY

It is not clear exactly what screening method was used to discover daunorubicin, but both the groups involved stressed its remarkable cytotoxicity; at concentrations ranging from 0.01 to 0.1 μg/ml it completely inhibits the mitotic activity of normal and neoplastic cells in vitro (Di Marco et al., 1964). Adriamycin, on the other hand, was discovered during the course of screening mutants of the daunorubicin producer *S. peucetius* in a deliberate attempt to find new structurally related metabolites with improved biological properties (Arcamone et al., 1969). It is interesting to note that daunorubicin frequently occurs in fermentations in the form of a somewhat higher glycoside that liberates daunorubicin on simple acid hydrolysis (McGuire et al., 1980b). Thus it is not clear whether it was daunorubicin itself or a higher glycoside that was detected initially, as they both have similar biological activities. A number of different isolates have now been categorized as producing daunorubicin and related anthracyclines; these are listed in Table 1.

Table 1 Organisms Identified as Producing Daunorubicin and Related Anthracyclines

Anthracycline produced	Species	Origin of Culture
Daunorubicin	*Streptomyces bifurcus* 23219	Rhône-Poulenc
	S. coeruleorubidus 8899 and 31723	Rhône-Poulenc
	S. coeruleorubidus ME 130-A	Sanraku-Ocean[a]
	S. coeruleorubidus	Gause
	S. griseoruber	Takeda
	S. griseus 32041	Rhône-Poulenc
	S. griseus IMET SA, 3933, 5570, 10086, 10431	Strauss Jena
	S. peucetius	Farmitalia
	S. viridochromogenes	Pfizer
Adriamycin	*S. peucetius* var. *caesius*	Farmitalia
Carminomycin	*Actinomadura carminata*	Gause

[a]J. Lunel, personal communication.
Source: Adapted from Lunel et al. (1977).

III. STRAIN IMPROVEMENT

Very little information has been published on the approaches used and the results obtained for improving productivity of the daunorubicin and adriamycin producer organisms. Fermentation titers quoted in the patent literature are in the range of 60—70 µg/ml for daunorubicin (Di Marco et al., 1977) and 5—15 µg/ml for adriamycin (Arcamone et al., 1968). For propriety reasons current titers achieved in production-scale fermentations are not released by Rhône-Poulenc and Farmitalia, but it seems reasonable to estimate that results of at least 500—1000 µg/ml have now been achieved. The present authors will report in some detail their own experiences with the daunorubicin fermentation at the Frederick Cancer Research Center in Maryland acquired over a 3-year period while producing material for clinical trials at the National Cancer Institute (NCI). The producer organism used at Frederick is an as yet unspeciated streptomycete. Initial attempts at classification have shown that it is not *S. peucetius* or *S. coeruleorubidus* but have yet to establish it as a new species. This organism was originally obtained from another NCI fermentation contractor (Kern et al., 1977). Initially shake-flask titers of around 50 µg/ml were achieved. Based on knowledge of the biosynthetic origin of the aglycone moiety of daunorubicin (see Sec. IV) a rational attempt was made to devise a selective screen for improved mutants of the producer organism. Daunorubicinone (the aglycone of daunorubicin) has a polyketide-derived carbon skeleton resulting from the condensation of one propionate and nine acetate units. The formation of linear polyketide chains is thought to proceed in a manner analogous to that of fatty acid biosynthesis. Once formed, the polyketide chain is modified and cyclized to yield the anthraquinone chromophore. Cerulenin (Fig. 2) is an antibiotic produced by *Cephalosporium caerulens* (Omura, 1976) that specifically inhibits fatty acid and polyketide biosynthesis. In the case of fatty acids it has been shown that the target for this inhibition is the enzyme β-ketoacylacyl carrier protein synthetase with which it forms an irreversible complex (Omura, 1976). The biosynthesis of polyketides has been reported to be 5 to 10 times more sensitive to inhibition by cerulenin than is growth of the producer organism (Omura and Takeshimi, 1974).

On this basis we have used cerulenin to select for mutants altered in the production of daunorubicin. Normally the daunorubicin producer forms red-pigmented colonies on agar due to the accumulation of anthracyclines. Conditions were first established which resulted in cerulenin-suppressed pigment formation by colonies on minimal agar medium. When mutagenized (ultraviolet irradiated) spores of the daunorubicin producer were plated on agar containing ∿50 µM cerulenin, the majority of the colonies formed were not pigmented (McGuire et al., 1980a). Rare pigmented colonies were evident and readily isolated for testing in shake-flask fermentations. In Figure 3 the spread of titers produced by different mutants isolated using the cerulenin screen is shown. The mutants obtained could be divided into three approximately equal groups: those producing the same titer as the parent strain,

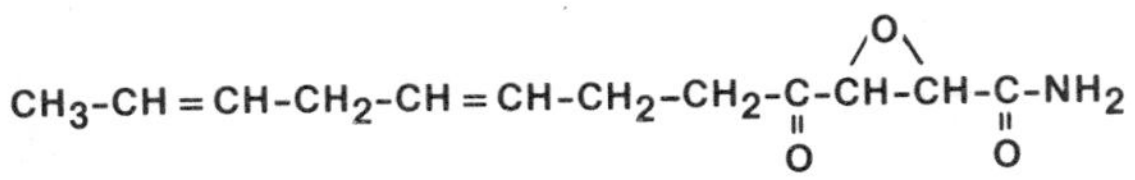

$$CH_3-CH=CH-CH_2-CH=CH-CH_2-CH_2-\overset{O}{\underset{\parallel}{C}}-CH-CH-\overset{O}{\underset{\parallel}{C}}-NH_2$$

Figure 2 Structure of the antibiotic cerulenin.

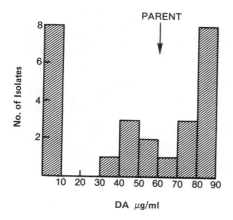

Figure 3 Daunorubicin (DA) titers achieved in shake-flask fermentations by mutants isolated in the cerulenin screen. The parent was strain V8. (From McGuire et al., 1980a.)

those producing substantially more, and those producing substantially less. The very low producers (less than 10 μg/ml) were found to accumulate high concentrations of the anthracycline aglycone, ε-rhodomycinone (see Fig. 6), which was responsible for the red pigmentation of their colonies on agar. The mutant (C5) with the highest titer in the overproducing group was substituted for the parent strain (V8) in production-scale fermentations. Although the initial titers observed in shake flasks were about twice those obtained with C5, production scale (10,000-liter vessels) gave three- to fivefold higher titers (Hamilton et al., 1981). The reason for the increased yields of antibiotic pro-

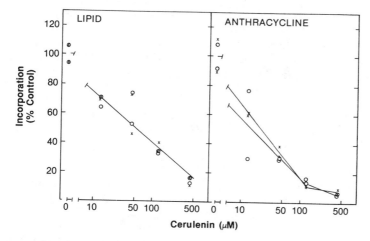

Figure 4 Effects of cerulenin on incorporation of [14C]acetate into lipid and anthracycline by parent strain V8 (×) and cerulenin mutant strain C5 (○). Incorporation is expressed relative to that observed in control flasks untreated with cerulenin. Each symbol represents incorporation by a separate shake-flask culture. (From McGuire et al., 1980a.)

duced by C5 has not been completely elucidated. In an experiment in which
the effect of different concentrations of cerulenin on the incorporation of labeled acetate into anthracyclines and lipid was compared, no significant differences were observed between the parent strain and mutant C5 (Fig. 4).
It appears that anthracycline synthesis is more sensitive to cerulenin inhibition than lipid synthesis. It may be that inhibition of anthracycline synthesis by cerulenin, while permitting the growth of colonies, allowed the facile
identification of all overproducers (colored colonies) irrespective of the mutation responsible, that is, it need not have involved a mutation in the β-ketoacylacyl carrier protein synthetase.

IV. BIOSYNTHESIS

A. Origin of the Carbon Skeleton

As a result of studies using ^{13}C-enriched acetate (Paulick et al., 1976) and
propionate (Shaw et al., 1979) it is known that the anthracyclinone moiety
of daunorubicin is derived from a linear condensation of a single propionate
starter unit and nine sequential acetate units, with subsequent loss of the
terminal carboxyl (see Fig. 5). Although the data are not as complete for
adriamycin, one can be fairly confident that an analogous biosynthetic scheme
is involved. Evidence that the methoxyl group of daunorubicin and adriamycin
(see Fig. 1) is derived from methionine comes from experiments with methylation inhibitors (Blumauerova et al., 1979) and blocked mutants (Casinelli et
al., 1978). The amino sugar moiety (daunosamine) of daunorubicin is apparently derived directly from glucose (Pavanosenkova and Karpov, 1976).

B. Biosynthetic Interrelationships

Anthracycline antibiotics are generally produced as complex mixtures of closely
related metabolites that may differ with respect to their chromophores or carbohydrate moieties. Daunorubicin is no exception: *S. peucetius* and *S. coeruleorubidus* produce a total of at least 20 daunorubicin-related compounds
(Blumauerova et al., 1977). The biosynthetic interrelationships between these
cometabolites have been investigated by the use of blocked mutants and biotransformation (Blumauerova et al., 1979a; McGuire et al., 1981; Yoshimoto
et al., 1980b). There is still some dispute as to the roles of ε-rhodomycinone
and daunorubicinone in the biosynthetic pathway leading to daunorubicin.
The conflicting results that have been reported may be due to the use of different producer organisms and mutants, and this may reflect that there is
not a unique sequence of biosynthetic steps leading to daunorubicin. ε-Rhodomycinone is frequently a major and sometimes the principal anthracycline present in daunorubicin fermentations (McGuire et al., 1980b). Thus, understanding its relationships to the desired end product, daunorubicin, could
have important implications with respect to strategies adopted for increasing
yields. Two groups maintain that ε-rhodomycinone can be biotransformed
to daunorubicin by a variety of blocked mutants and the parent producer
strain itself (McGuire et al., 1980b; Yoshimoto et al., 1980a). A third group
was unable to demonstrate the transformation of ε-rhodomycinone to daunorubicin (Blumauerova et al., 1978). A second point of dispute is whether daunorubicinone itself is an intermediate in the biosynthesis. Umezawa and co-
workers were unable to demonstrate conversion of this aglycone to daunorubicin (Yoshimoto et al., 1980a) and suggested that glycoslation occurs at an

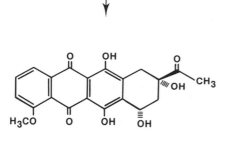

● ─── Acetate Unit   ●──/ Propionate Unit

9 Acetates + 1 Propionate

COOH

OH

Hypothetical Intermediate

CH₃

H₃CO

DAUNORUBICINONE

Figure 5 Biosynthetic origin of the carbon skeleton of daunorubicinone. The single arrows correspond to an undetermined number of enzymatic steps.

earlier stage of the biosynthesis. In Figure 6 a proposed biosynthetic inter-relationship for this group of compounds is shown. Baumycins are shown as the higher glycosides of daunorubicin (Takahashi et al., 1977).

V. FERMENTATION AND RECOVERY

The fermentation processes for production of daunorubicin and adriamycin differ little from those of other secondary metabolites from streptomyces. Basically, the cultures are revived from frozen stocks, progress through various inoculum buildup stages, and are used to inoculate production fermenters. The fermenters are operated under aerobic conditions for up to 10 days.

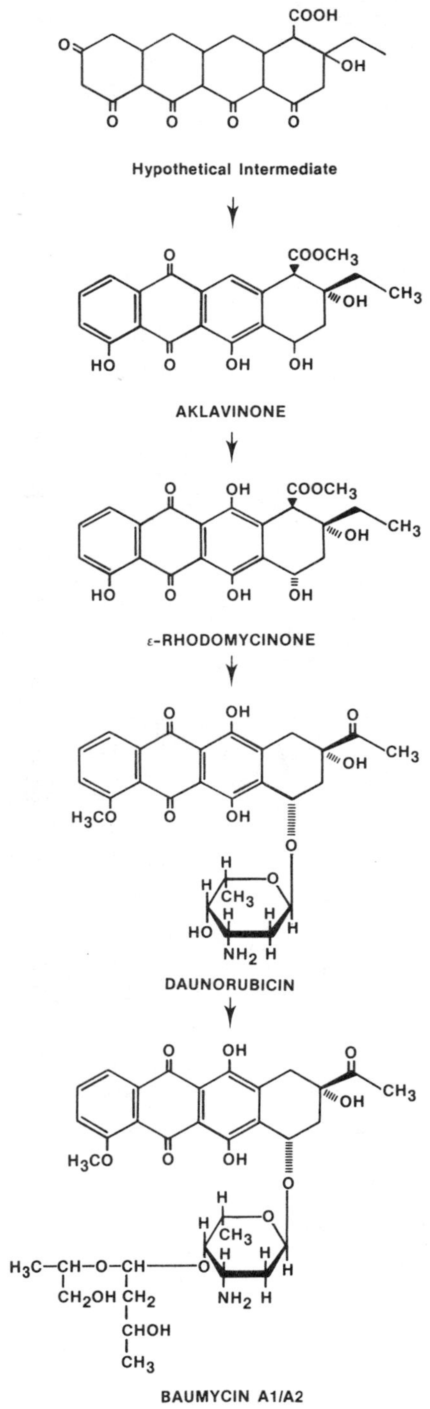

Figure 6 Proposed biosynthetic interrelationship among daunorubicin-related anthracyclines. The single arrows between compounds represent an undetermined number of enzymatic steps. This scheme does not necessarily represent a unique pathway of daunorubicin biosynthesis but is consistent with biotransformation data.

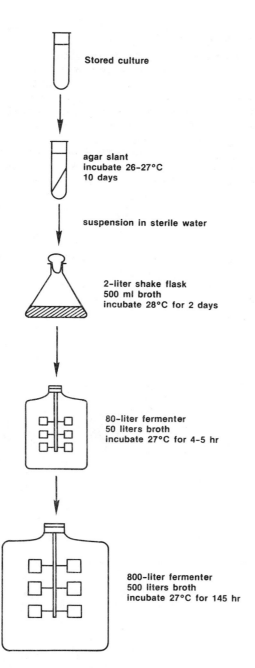

Stored culture

agar slant
incubate 26-27°C
10 days

suspension in sterile water

2-liter shake flask
500 ml broth
incubate 28°C for 2 days

80-liter fermenter
50 liters broth
incubate 27°C for 4-5 hr

800-liter fermenter
500 liters broth
incubate 27°C for 145 hr

Figure 7   Process flow sheet for production of daunorubicin (Rhône-Poulenc).

A. Inoculum Preparation and Production of Daunorubicin

Two methods of inoculum preparation have been reported. In one (DiMarco et al., 1977) the frozen stock culture is used to inoculate agar slants (Fig. 7). After 10 days of growth, a mycelial suspension in sterile water from slants serves as inoculum for shake-flask cultures. A 170-liter fermenter is batched with 100 liters of seed medium (Pinnert et al., 1976) (Table 2) and inoculated with 200 ml (0.17%, vol/vol) of culture medium from the shake flasks. The seed culture is grown for 27 hr at 26–27°C with agitation and aeration (0.69 vol/vol per minute). The daunorubicin production is carried out in an 800-liter fermenter containing 500 liters of sterilized production medium (Table 2). This fermenter is inoculated with 50 liters (10% vol/vol) of the 27-hr seed culture, and the fermentation proceeds at 28°C with agitation and aeration (0.5 vol/vol per minute) for 67 hr.

Table 2 Media Used in Rhône-Poulenc Fermentation of Daunorubicin

	Amount (%)
Medium for slant culture	
Saccharose	2
Dry yeast	0.1
Potassium hydrogen phosphate	0.2
Sodium nitrate	0.2
Magnesium sulfate	0.2
Agar	2
Tap water	q.s.
Medium for shake flask	
Peptone	0.6
Dry yeast	0.3
Calcium nitrate, hydrate	0.05
Tap water	q.s.
pH after sterilization: 7.2	
Vegetative medium	
Corn steep	2.4[a]
Sucrose	3.6
Calcium carbonate	0.9
Ammonium sulfate	0.24
Water	q.s.
pH after sterilization: 7.2	
Production medium	
Soy flour	4
Distiller's solubles	0.5
Starch	4
Soy oil	0.5
Sodium chloride	1
Water	q.s.
pH is adjusted to 7.2 with sodium hydroxide solution.	

[a]As batched; volume increases by 20% on sterilization.
Source: Adapted from Pinnert et al. (1976).

A second method (McGuire et al., 1979) involves direct inoculation of the shake-flask medium with frozen culture stock (Fig. 8). The shake flask is incubated at 28°C for 2.5 days and is used to inoculate 75 liters (1.3%, vol/vol) of seed medium (Table 3) contained in a 100-liter fermenter. Incubation in the 100-liter fermenter continues at 30°C, 0.5 vol/vol per minute aeration, 125 rpm (750 ft/min impeller tip speed) for 24 hr. The seed from the 100-

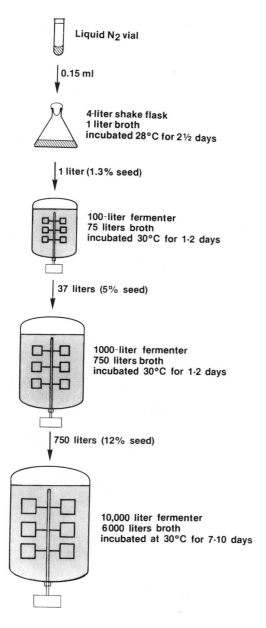

Figure 8 Process flow sheet for production of daunorubicin (Frederick Cancer Research Center). (From Hamilton et al., 1981.)

Table 3 Media Used in the Fermentation of Daunorubicin at the Frederick
Cancer Research Center

	Concentration (wt/vol) (%)
Seed medium	
Corn starch	1.5
Cottonseed flour	0.5
Defatted soy flour	0.5
Autolyzed nutritional yeast	0.1
NaCl	0.25
CaCO$_3$	0.5
Prochem #51, antifoam	0.8[a]
Tap water	q.s.
Production medium	
Glucose	5.0
Herring meal	1.2
Baker's nutrisoy	1.25
Autolyzed nutritional yeast	0.75
NaCl	0.33
CaCO$_3$	1.0
Prochem #51, antifoam	1.0[b]
Tap water	q.s.

[a]Initial concentration; no antifoam addition during seed fermentation is necessary.
[b]Initial concentration; antifoam addition during production fermentation is
necessary to control foam.
Source: McGuire et al. (1979).

liter fermenter serves as inoculum (5%) for a 1000-liter fermenter, which is
operated under exactly the same conditions as the 100-liter fermenter for 1–2
days. At this point, 750 liters (12%) of the seed culture is used to inoculate
6000 liters of production medium contained in a 10,000-liter fermenter. The
operational parameters of this fermenter are exactly the same as for the 100-
and 1000-liter seed fermentations. The production fermentation continues
for 7–10 days (Fig. 9).

Substitution of maltose or malt extract for glucose monohydrate in the
production medium gave up to twofold increases in titer in shake-flask fermentations (Hamilton et al., 1981). These results have yet to be confirmed in
production fermenters.

B. Isolation and Purification of Daunorubicin

Daunorubicin and many of its cometabolites are cardiotoxic (Von Hoff et al.,
1977; Lenaz and Page, 1976) or mutagenic (Benedict et al., 1977; Marquart
et al., 1976). As such, special handling and precautions are required when
isolating and purifying daunorubicin from fermentation broth. In addition,
fermentation broth contains very little daunorubicin per se, since the antibiotic is present in the mycelium and broth primarily as higher glycosides,

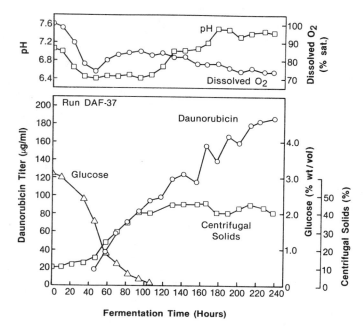

Figure 9 Time course of a typical daunorubicin production fermentation run in the 10,000-liter fermenter. (From Hamilton et al., 1981.)

such as the baumycins (Pandey et al., 1979). The recovery process for daunorubicin then requires a mild hydrolysis step to convert higher glycosides to the desired product. Excess hydrolysis must be avoided lest the aglycone be generated.

Two approaches have been taken to solve these problems. At Rhône-Poulenc, (Pinnert et al., 1976), the fermentation broths are immediately treated with excess oxalic acid and heated at 50°C for 60-90 min (Fig. 10). This process both lyses the cells and hydrolyzes the higher glycosides to daunorubicin. This makes the process somewhat safer, since daunorubicin is less active than baumycins (Komiyama et al., 1977). The hydrolyzed broth is then clarified by filtration and the cake is washed with water. Large-scale liquid-liquid extractions, which unless rigorously contained tend to expose operators to undesirable quantities of toxic solvents and solutes, are obviated by passing the filtrate and washings (adjusted to pH 4.5) through ion exchange columns of IRC-50 (acid cycle). After loading, the columns are washed exhaustively with water, 50% aqueous methanol, and methanol-water (9:1). These washings are discarded and the column is eluted with a methanol solution containing 1% sodium chloride and 10% water. The broth is concentrated fivefold in vacuo and then adjusted to pH 8.5 and extracted with a half volume of chloroform. The chloroform extract is concentrated 100-fold and 10 volumes of hexane are added to precipitate the crude daunorubicin. The crude product is acidified with aqueous hydrochloric acid and crystallized by acetone addition. Recrystallization from acetone, from methanol-chloroform, and from n-butanol gives 88% pure daunorubicin hydrochloride.

In the process (Fig. 11) used at the Frederick Cancer Research Center (Pandey et al., 1979) acidification of the whole broth is also the first process

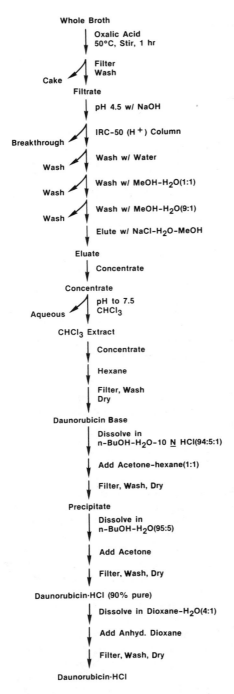

Whole Broth

Oxalic Acid
50°C, Stir, 1 hr

Filter
Wash

Cake

Filtrate

pH 4.5 w/ NaOH

IRC-50 (H⁺) Column

Breakthrough

Wash w/ Water

Wash

Wash w/ MeOH-H₂O(1:1)

Wash

Wash w/ MeOH-H₂O(9:1)

Wash

Elute w/ NaCl-H₂O-MeOH

Eluate

Concentrate

Concentrate

pH to 7.5
CHCl₃

Aqueous

CHCl₃ Extract

Concentrate

Hexane

Filter, Wash
Dry

Daunorubicin Base

Dissolve in
n-BuOH-H₂O-10 N HCl(94:5:1)

Add Acetone-hexane(1:1)

Filter, Wash, Dry

Precipitate

Dissolve in
n-BuOH-H₂O(95:5)

Add Acetone

Filter, Wash, Dry

Daunorubicin·HCl (90% pure)

Dissolve in Dioxane-H₂O(4:1)

Add Anhyd. Dioxane

Filter, Wash, Dry

Daunorubicin·HCl

Figure 10 Rhône-Poulenc process for the purification of daunorubicin.

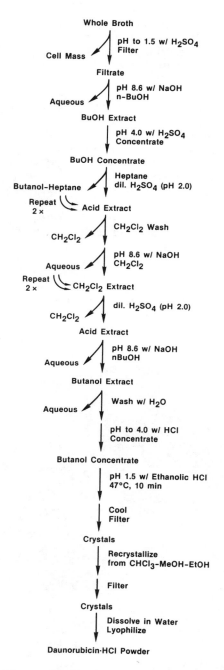

Figure 11 Frederick Cancer Research Center process for the purification of daunorubicin.

step. Here, however, sulfuric acid is used and the pH is taken to 1.5. The broth is not heated at this time, so this step only serves to solubilize the derivable daunorubicin and lyse the cells. After clarifying the broth by filtration, the filtrate and washings are pooled and extracted with n-butanol at pH 8.6. The extracts are adjusted to pH 4.5 to stabilize the daunorubicin and concentrated 20-fold. Heptane is added and the broth is extracted with dilute sulfuric acid solution (pH 2.0). The aqueous extracts are pooled, washed with methylene dichloride, and then adjusted to pH 8.6 and extracted with methylene dichloride. The methylene dichloride extracts are extracted with dilute aqueous acid and these extracts are adjusted to pH 8.6 and extracted with n-butanol, which, after washing with neutral water, completes the countercurrent extraction process. These countercurrent extraction steps can be safely performed when the operators are suitably protected with supplied-air hoods or adequate respirators, nitrile gloves, plasticized coveralls, and rubber boots which protect the workers from exposure to antibiotic and solvent (Figure 12) and the operations are done in ventilated tanks equipped with fixed manifold piping (Figure 13). The remaining process steps involve hydrolysis of higher gycosides to daunorubicin using ethanolic HCl at 45°C for 10 min, followed by chilling and filtration of the crude crystals. Recrystallization from a mixture of chloroform−methanol−ethanol gives a product containing some trapped solvent. This is removed by dissolving the product

Figure 12 Recovery operator equipped with supplied air hood, nitrile gloves, disposable plasticized coveralls, and steel-toed rubber boots.

Figure 13 Fixed piping system used to contain recovery process streams.

in water and lyophilizing the solution to give daunorubicin hydrochloride of approximately 95% purity (with 4–6% water as the major "impurity").

Waste treatment (Hamilton et al., 1981) involves incineration of the filter cake, although normal disposal is also practiced following chemical decontamination by adjustment of a cake slurry to pH 12 and heating at 70°C for 4 hr or at 100°C for 1 hr. Cakes so treated show no mutagenicity by the Ames test, while prior to treatment they were highly active. Contaminated solvent streams are disposed of by concentrating the waste and transferring the solvent to drums for burial at approved disposal sites. Disposal of contaminated solvents by incineration is a more effective and less costly alternative.

C. Inoculum Preparation and Production of Adriamycin

Streptomyces peucetius var. *caesius*, the producing organism for adriamycin, was obtained through chemical mutagenic treatment of *S. peucetius* (Arcamone

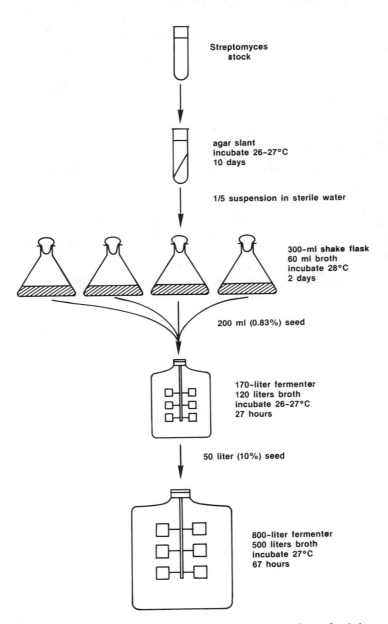

Figure 14 Process flow sheet for the production of adriamycin.

et al., 1968, 1969), the daunorubicin-producing culture. The steps for fermentation of this organism (Fig. 14) involve growth on agar slants at 28°C for 10 days, followed by removal of the deep red vegetative mycelium, which is homogenized in a tissue grinder, suspended in sterile distilled water, and serves as inoculum for 500 ml of growth medium in a 2-liter baffled round-bottom flask. After 48 hr of rotary shaking at 28°C (120 rpm, 70-mm stroke) the medium is transferred to an 80-liter stainless steel fermenter containing 50 liters of medium. This culture then serves as inoculum for 500 liters of production medium in an 800-liter fermenter. Fermentation continues for 145 hr.

D. Isolation and Purification of Adriamycin

Although adiramycin is found primarily in the mycelium, it can be recovered from both the filter cakes and filtrate following filtration of the whole broth (Fig. 15). The cake is suspended in a 4:1 mixture of acetone and 0.1 N aqueous sulfuric acid. The solids are removed by filtration and the filtrate neutralized and concentrated in vacuo. Lipophilic materials are removed by a

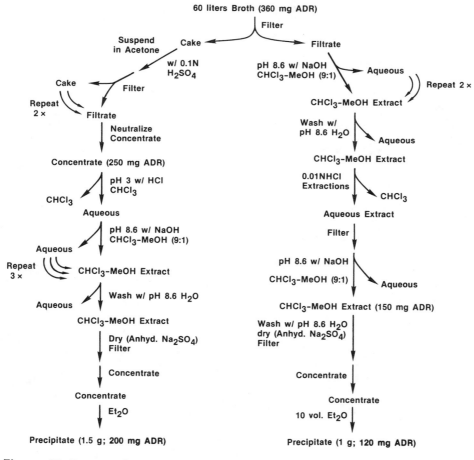

Figure 15 Process for the recovery of crude adriamycin.

chloroform wash at pH 3 and the aqueous phase is made alkaline and then extracted with a mixture of chloroform and methanol. This extract is dried over anhydrous sodium sulfate, filtered, and then concentrated. Addition of ether gives a crude precipitate containing 13–15% adriamycin. The broth filtrate is analogously treated to give a precipitate containing 12–15% adriamycin. The crude material is purified (Fig. 16) by partition chromatography on cellulose powder at pH 5.4. Elution with n-butanol (previously saturated with pH 5.4 phosphate buffer) or a propanol–ethyl acetate–water mixture (7:1:2) gives fractions rich in adriamycin. These are pooled, concentrated, and extracted into chloroform. After aqueous washing and drying over anhydrous sodium sulfate, the chloroform extract is concentrated, treated with methanolic hydrogen chloride, and chilled. The resulting adiramycin hydrochloride crystals are recovered and recrystallized from ethanol.

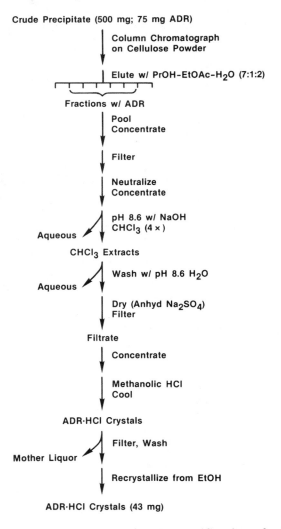

Figure 16 Process for the purification of adriamycin.

VI. ASSAY

Numerous assays have been developed for analysis of daunorubicin and its metabolites in complex biological systems. These assays include fluorescence monitoring (Formelli et al., 1979; Schwartz, 1973; Bachur et al., 1976), radio-immunoassay (Bachur et al., 1977; Van Vunakis et al., 1974), detection of biological activity (Wright and Matsen, 1980), paper chromatography with quantitative ultraviolet detection (Belloc et al., 1968), induction of prophage (Anderson et al., 1980), thin-layer chromatography analysis (Issaq et al., 1979), light scattering techniques (Woolley et al., 1978), and high-pressure liquid chromatography (HPLC) quantitation (Hulhoven and Desager, 1976; Baurain et al., 1978, 1979; Strauss et al., 1980; Quattrone and Ranney, 1980; Paul et al., 1980). In most cases, the assay involves extraction of the material from a biological fluid and subsequent quantitation by analytical methods.

The problems associated with analysis of daunorubicin in fermentation broths are equally complex, but are complicated by factors different from those of metabolism. First, the amount of daunorubicin per se in the fermentation broth is only a fraction of the total derivable daunorubicin present after mild hydrolysis. Fermentation broths contain most of the derivable daunorubicin as higher glycosides attached to the 4'-hydroxyl of the daunosamine moiety. The hydrolytic conversion to daunorubicin is further complicated, because overhydrolysis cleaves the anthracycline–aminoglycoside linkage, giving rise to daunorubicinone. The second complication involves the distribution of the derivable daunorubicin within the fermentation broth. Typically 20-30% of the total can be found in the supernatant after centrifugation (or the filtrate after filtration); the remainder is closely associated with the mycelial pellet or filter cake. Whether this material is intracellular or merely bound to the cell membrane has not been established. This distribution makes representative sampling of fermentation broths vital to the accuracy of the analysis: Samples containing more mycelia than is truly representative of the whole will give artificially high titer values. Great pains must be taken to assure that samples submitted for assay truly reflect the composition of the whole fermenter. These include adequate agitation before sampling, acquisition of a large sample and subsequent subsampling, and use of large-bore pipets for sample removal. The final assay complication involves the presence of structurally related fermentation cometabolites. Compounds such as daunomycinone, ε-rhodomycinone, baumycins, and 7-deoxydaunorubicinol aglycone, which can be present in fermentation broth either naturally or through under- or overhydrolysis, must be distinguishable from daunorubicin in the assay procedure. Simply reading the ultraviolet spectrum of a clarified broth or single emission wavelength scanning by fluorescence will give erroneous results, necessitating multiple countercurrent extractions or chromatography. Simple bioassays must also involve prior purification, since cometabolites are also bioactive.

An HPLC assay which addresses all of these problems has been reported (Stroshane et al., 1979). The first two complications, conversion of higher glycosides to daunorubicin and release of mycelial components to the clarified broth, are overcome by mildly heating the whole broth in the presence of oxalic acid. Oxalic acid is preferred to mineral acids because it can be added as a solid, thus avoiding volume corrections, and it is self-buffering at pH 1.3, obviating tedious titrations and preventing aglycone production. Reverse-phase HPLC analysis on an analytical octadecasilyl column allows direct injection of the centrifuged and filtered posthydrolysis sample, eliminating the need for extraction into organic solvent and allowing for short run times,

DAUNORUBICIN

COLUMN: μBONDAPAK C$_{18}$
3.9 mm x 30 cm

SOLVENT: Methanol-water-PIC B-7
(65:35:1.8 vol/vol/vol

FLOW: 2.0 ml/min

DETECTION: UV, 254 nm, 0.05 AUFS

Figure 17 High-performance liquid chromatography conditions for analysis of daunorubicin. Sample pretreatment involves adding 300 mg of oxalic acid to 10 ml of whole broth in a 15-ml centrifuge tube, incubation at 50°C for 45 min, centrifugation, and decantation and filtration of the supernatant.

since the polar contaminants of the fermentation broth are eluted in the void volume (Fig. 17). The chromatographic power of these high-performance columns and subsequent ultraviolet detection give resolution of the cometabolites and direct quantitation of the daunorubicin. With an automatic sample injector and computer interfacing more than 100 samples can be assayed daily on a single instrument by one technician.

VII. MECHANISM OF ACTION

Daunorubicin and adriamycin have both been demonstrated to interact with double-stranded helical DNA. Binding of these antibiotics to DNA results in inhibition of DNA and RNA synthesis and in the usual features associated with intercalation, for example, the complex formed is metachromatic and sensitive to increases in ionic strength, the thermal denaturation temperature is raised, and viscosity is enhanced (Di Marco et al., 1975). In addition, single-strand scission of DNA has also been reported in vitro in the presence of reducing agents (Lown et al., 1977) and in vivo (Schwarz, 1975).

VIII. BIOLOGICAL ACTIVITY

In addition to their previously mentioned cytotoxicity, daunorubicin and adriamycin are moderately active against gram-positive and acid-fast bacteria, with little or no effect on the gram negative ones. As a consequence of their interaction with and damage to DNA, daunorubicin and adriamycin cause induction of prophage in lysogenic bacteria (Anderson et al., 1980) and are mutagenic in the Ames test (Benedict et al., 1977). The antitumor activity of these anthracyclines will be discussed in a separate section.

IX. APPLICATIONS AND ECONOMY

Daunorubicin and adriamycin have made a major impact on cancer chemotherapy. They have been used alone and in combination with other agents in the treatment of a variety of cancers, but their most successful application has been in the treatment of acute leukemia (Davis and Davis, 1979). Adriamycin is probably the most "broad-spectrum" single agent in cancer therapy and is used in the treatment of a number of solid tumors, for example, breast and lung cancers. The spectrum of activity of daunorubicin is believed to be narrow and confined mainly to acute leukemia (Von Hoff et al., 1978); however, investigations in solid tumors have been few and difficult to evaluate (Von Hoff et al., 1976). In common with all other chemotherapeutic agents used to treat cancer, the major limitation to the use of daunorubicin and adriamycin is their toxicity. In addition to the more usual side effects associated with cancer chemotherapy, such as bone marrow suppression, these anthracyclines both demonstrate major cumulative cardiotoxicity. This poses limitations on the total dose that can be administered. Apparently, there is no guarantee that limiting total dosage will avoid cardiotoxicity altogether, as there is a continuum of risk with no absolutely safe level (Von Hoff et al., 1977; Lenaz and Page, 1976). The prevention of cumulative cardiotoxicity remains the key goal in anthracycline chemotherapy and major semisynthetic analog programs are underway in an attempt to find a less toxic derivative.

The wholesale price of daunorubicin in the United States is $920,000/kg, while that for adriamycin is $1,370,000/kg. Although total synthesis of both antibiotics has been reported (Arcamone, 1978), none of the methods described thus far are competitive on a cost basis with the fermentation process. The semisynthesis of adriamycin from fermentation-derived daunorubicin would be commercially interesting if attainable adriamycin fermentation titers were significantly lower than those obtained for daunorubicin. The total free world market for adriamycin has been estimated to be $109,000,000 for 1980 and is much larger than that for daunorubicin.

REFERENCES

Anderson, W. A., Moreau, P. L., Devoret, R., and Maral, R. (1980). Induction of prophage-λ by daunorubicin and derivatives correlation with antineoplastic activity. *Mutat. Res.* 77197–208.

Arcamone, F. (1978). Daunomycin and related antibiotics. In *Topics in Antibiotic Chemistry*, Vol. 2. P. G. Sammes (Ed.). Wiley, New York, pp. 102–239.

Arcamone, F., Cassinelli, G., Di Marco, A., and Gaetani, M. (1968). Adriamycin, a new antibiotic substance. South African Patent No. 6802378.

Arcamone, F., Cassinelli, G., Fantini, G., Grein, A., Orezzi, P., Pol, C., and Spalla, C. (1969). Adriamycin, 14-hydroxydaunomycin, a new antitumor antibiotic from *S. peucetius* var. caesius. *Biotechnol. Bioeng.* 11: 1101–1110.

Bachur, N. R., Riggs, C. E., Green, M. R., Langone, J. J., Van Vunakis, H., and Levine, L. (1977). Plasma adriamycin and daunorubicin levels by fluorescence and radioimmunoassay. *Clin. Pharmacol. Ther.* 21:70–77.

Baurain, R., Zenebregh, A., and Trouet, A. (1978). Cellular uptake and metabolism of daunorubicin as determined by high-performance liquid chromatography: Application to L1210 cells. *J. Chromatogr.* 157:331–336.

Baurain, R., Deprez-De Campeneere, D., and Trouet, A. (1979). Rapid determination of doxorubicin and its fluorescent metabolites by high pressure liquid chromatography. *Anal. Biochem.* 94:112–116.

Belloc, A., Charpentie, Y., Lunel, J., and Preud'homme, J. (1968). Process for the production of rubiodomycin. South African Patent No. 6801487.

Benedict, W. F., Baker, M. S., Haroun, L., Choi, E., and Ames, B. N. (1977). Mutagenicity of cancer chemotherapeutic agents in the salmonella/ microsome test. Cancer Res. 37:2209–2213.

Blumauerova, M., Mateju, J., Stajner, K., and Vanek, Z. (1977). Studies on the production of daunomycinone-derived glycosides and related metabolites in Streptomyces coeruleorubidus and Streptomyces peucetius. Folia Microbiol. 22:275–285.

Blumauerova, M., Kralovcova, E., Mateju, J., Hostalek, Z., and Vanek, Z. (1978). Genetic approaches to the biosynthesis of anthracyclines. In Genetics of Industrial Microorganisms. O. K. Sebek and A. I. Laskin (Eds.). American Society for Microbiology, Washington, D.C., pp. 90–96.

Blumauerova, M., Kralovkova, E., Mateju, J., Jizba, J., and Vanek, Z. (1979a). Biotransformation of anthracyclinones in Streptomyces coeruleorubidus and Streptomyces galilaeus. Folia Microbiol. 24:117–127.

Blumauerova, M., Jizba, J., Stajner, K., and Vanek, Z. (1979b). Effect of DL-ethionine on the biosynthesis of anthracyclines in Streptomyces coeruleorubidus. Biotechnol. Lett. 1:471–476.

Bondonna, G., Monfardini, S., De Lena, M., Fossati-Bellani, F., and Beretta, G. (1970). Phase I and preliminary Phase II evaluation of adriamycin (NSC 123127). Cancer Res. 30:2572–2582.

Brockmann, H., and Bauer, K. (1950). Rhodomycin, ein rotes Antibioticum aus Actinomyceten. Naturwissenschaften 37:492–493.

Casinelli, G., Grein, A., Masi, P., Suarato, A., Bernardi, L., Arcamone, F., Di Marco, A., Casazza, A. M., Pratesi, G., and Sorano, C. (1978). Preparation and biological evaluation of 4-O-demethyl daunorubicin (carminomycin 1) and its 13-dihydro derivative. J. Antibiot. 31:178–184.

Davis, H. L., and Davis, T. E. (1979). Daunorubicin and adriamycin in cancer treatment: An analysis of their roles and limitations. Cancer Treat. Rep. 63:809–815.

Di Marco, A., Gaetani, M., Orezzi, P., Scarpinato, B. M., Silvestrini, R., Soldati, M., Dasdia, T., and Valentini, L. (1964). Daunomycin, new antibiotic of the rhodomycin group. Nature 201:706–707.

Di Marco, A., Arcamone, F., and Zunino, F. (1975). Daunomycin (daunorubicin) and adriamycin and structural analogs: Biological activity and mechanism of action. In Antibiotics, Vol. 3. J. W. Corcoran and F. E. Hahn (Eds.). Springer-Verlag, New York, pp. 101–128.

Di Marco, A., Canevazzi, G., Grein, A., Orezzi, P., and Gaetani, M. (1977). Process for preparation of antibiotics F.1. 1762 derivatives. U.S. Patent No. 4012284.

Dubost, M., Ganter, P., Maral, R., Ninet, L., Pinnert, S., Preud'homme, J., and Werner, G. H. (1963). Un nouvel antibiotique à propriétés cytostatiques: La rubidomycin. C. R. Acad. Sci. 257:1813–1815.

Ettlinger, L., Gaumann, E., Hutter, R., Keller-Schierlein, W., Kradolfer, F., Neipp, L., Prelog, V., Reusser, P., and Zahner, H. (1959). Metabolic products from actinomycetes. XVI. Cereulins. Chem. Ber. 92:1867–1879.

Formelli, F., Casazza, A. M., Di Marco, A., Mariani, A., and Pollini, C. (1979). Fluorescence assay of tissue distribution of 4-demethoxydaunorubicin and 4-demethoxydoxorubicin in mice bearing solid tumors. Cancer Chemother. Pharmacol. 3:261–269.

Gausse, G. F., Brazhnikova, M. G., and Shorin, V. A. (1974). A new anti-tumor antibiotic, carminomycin. *Cancer Chemother. Rep.* 58:255–256.

Grein, A., Spalla, C., Di Marco, A., and Canevazzi, G. (1963). Descrizione e classificazione di un attinomicete (*Streptomyces peucetius* sp. nova) producttore di una sostanza ad attivita antitumorale: La daunomicina. *G. Microbiol.* 11:109–118.

Hamilton, B. K., White, R. J., McGuire, J., Montgomery, P., Stroshane, R., Kalita, C., and Pandey, R. (1981). Improvement of the daunorubicin fermentation realized at the 10,000 L scale. In *Advances in Biotechnology,* Vol. 1, M. Moo-Young (Ed.). Pergamon Press, New York, pp. 63–68.

Hulhoven, R., and Desager, J. P. (1976). Quantitative determination of low levels of daunomycin and daunomycinol in plasma by high-performance liquid chromatography. *J. Chromatogr.* 125:369–374.

Issaq, H. J., Risser, N. H., and Aszalos, A. A. (1979). Thin-layer chromatographic separation and quantitation of the anti-tumor agent daunorubicin in fermentation media. *J. Liquid Chromatogr.* 2:533–538.

Kern, D. L., Bunge, R. H., French, J. C., and Dion, H. W. (1977). The identification of ε-rhodomycinone and 7-deoxydaunorubicin aglycone in daunorubicin beers. *J. Antibiot.* 30:432–434.

Komiyama, T., Matsuzawa, Y., Oki, T., and Inui, T. (1977). Baumycins, new antitumor antibiotics related to daunorubicin. *J. Antibiot.* 30:619–621.

Lenaz, L., and Page, J. A. (1976). Cardiotoxicity of adriamycin and related anthracyclines. *Cancer Treat. Rev.* 3:111–120.

Lown, J. W., Sim, S. K., Majumdar, K. C., and Chang, R. Y. (1977). Strand scission of DNA by bound adriamycin and daunorubicin in the presence of reducing agents. *Biochem. Biophys. Res. Commun.* 76:705–710.

Lunel, J., Florent, J., Mancy, D., and Renaut, J. (1977). Production of anthracycline antibiotics by fermentation. Abstr. 174th National Meeting Am. Chem. Soc., MICR 041.

McGuire, J. C., Hamilton, B. K., and White, R. J. (1979). Approaches to development of the daunorubicin fermentation. *Proc. Biochem* 14:2–5.

McGuire, J. C., Glotfelty, G., and White, R. J. (1980a). Use of cerulenin in strain improvement of the daunorubicin fermentation. *FEMS Microbiol. Lett.* 9:141–143.

McGuire, J., Thomas, M. C., Pandey, R. C., Toussaint, M., and White, R. J. (1981b). Biosynthesis of daunorubicin glycosides: Analysis with blocked mutants. In *Advances in Biotechnology,* Vol. 3, M. Moo-Young (Ed.). Pergamon Press, New York, pp. 117–122.

McGuire, J. C., Thomas, M. C., Stroshane, R. M., Hamilton, B. K., and White, R. J. (1980b). Biosynthesis of daunorubicin glycosides: Role of ε-rhodomycinone. *Antimicrob. Agents Chemother.* 18:454–464.

Marquart, H., Philips, F. S., and Sternberg, S. S. (1976). Tumorigenicity in vivo and induction of malignant transformation and mutagenesis in cell cultures by adriamycin and daunomycin. *Cancer Res.* 36:2065–2069.

Omura, S. (1976). The antibiotic cerulenin, a novel tool for biochemistry as an inhibitor of fatty acid biosynthesis. *Bacteriol. Rev.* 40:681–697.

Omura, S., and Takeshima, H. (1974). Inhibition of the biosynthesis of leucomycin, a macrolide antibiotic, by cerulenin. *J. Biochem.* 75:193–195.

Pandey, R. C., Kalita, C. C., White, R. J., and Toussaint, M. W. (1979). Process development in the purification of daunorubicin from fermentation broths. *Proc. Biochem.* 14:6–13.

Paul, C., Baurain, R., Gahrton, G., and Peterson, C. (1980). Determination of daunorubicin and its main metabolites in plasma, urine and leukemic cells in patients with acute myeloblastic leukaemia. *Cancer Lett.* 9:263–269.

Paulick, R. C., Casey, M. L., and Whitlock, H. W. (1976). Carbon-13 nuclear magnetic resonance study of the biosynthesis of daunomycin from $^{13}CH_3^{13}CO_2Na$. *J. Am. Chem. Soc. 98*:3370—3371.

Pavanosenkova, V. I., and Karpov, V. L. (1976). Studies on biosynthesis of carbohydrate fragment of rubomycin. *Antibiotiki 21*:299—301.

Pinnert, S., Ninet, L., and Preud'Homme, J. (1976). Antibiotics and their preparation. U.S. Patent No. 3997662.

Quattrone, A. J., and Ranney, D. F. (1980). Simplified toxicologic monitoring of adriamycin, its major metabolites and nogalamycin by reverse-phase high pressure liquid chromatography, Part 1: Analytical techniques for isolated human plasma. *J. Anal. Toxicol. 4*:12—15.

Schwartz, H. S. (1973). A fluorometric assay for daunomycin and adriamycin in animal tissues. *Biochem. Med. 7*:396—404.

Schwartz, H. S. (1975). DNA breaks in P388 tumor cells in mice after treatment with daunorubicin and adriamycin. *Res. Commun. Chem. Pathol. Pharmacol. 10*:51—64.

Shaw, G. J., Milne, G. W. A., and Minghetti, A. (1979). Propionate precursors in the biosynthesis of daunomycin and adriamycin: A ^{13}C nuclear magnetic resonance study. *Phytochemistry 18*:178—179.

Strauss, J. F., Kitchens, R. L., Patrizi, V. W. and Frenkel, E. P. (1980). Extraction and quantification of daunomycin and doxorubicin in tissues. *J. Chromatogr. Biomed. Appl. 221*:139—144.

Stroshane, R. M., Guenther, E. C., Piontek, J. L., and Aszalos, A. A. (1979). A rapid high performance liquid chromatographic assay for total derivable daunorubicin in fermentation broth. Abstr., American Chemical Society Annual Meeting, Washington, D.C., Anal. 044.

Takahashi, Y., Naganawa, H., Takeuchi, T., Umezawa, H., Komiyama, T., Oki, T., and Inui, T. (1977). The structure of baumycins A1, A2, B1, B2, C1 and C2. *J. Antibiot. 30*:622—624.

Van Vunakis, H., Langone, J. J., Riceberg, L. J., and Levine, L. (1974). Radioimmunoassays for adriamycin and daunomycin. *Cancer Res. 34*: 2546—2552.

Von Hoff, D. D., Rozencweig, M., Slavik, M., and Muggia, F. M. (1976). Activity of daunomycin in solid tumors. *J. Am. Med. Assoc. 236*:1693.

Von Hoff, D. D., Rozencweig, M., Layard, M., and Slavik, M. (1977). Daunomycin induced cardiotoxicity in children and adults: A review of 110 cases. *Am. Med. J. 62*:200—208.

Von Hoff, D. D., Rozensweig, M., and Slavik, M. (1978). Daunomycin: An anthracycline antibiotic effective in acute leukemia. In *Advances in Pharmacology and Chemotherapy*, Vol. 15. S. Gavattini, A. Goldin, F. Hawking, and I. J. Koplin (Eds.). Academic, New York, pp. 2—50.

Woolley, C., Mellett, L. B., and Wyatt, P. J. (1978). Laser differential light scattering bioassays for selected antitumor agents. *Res. Commun. Chem. Pathol. Pharmacol. 21*:531—537.

Wright, D. N., and Matsen, J. M. (1980). Bioassay of antibiotics in body fluids from patients receiving cancer chemotherapeutic agents. *Antimicrob. Agents Chemother. 17*:417—422.

Yoshimoto, A., Oki, T., Takeuchi, T., and Umezawa, H. (1980a). Microbial conversion of anthracyclinones to daunomycin by blocked mutants of *Streptomyces coeruleorubidus*. *J. Antibiot. 33*:1158—1166.

Yoshimoto, A., Oki, T., and Umezawa, H. (1980b). Biosynthesis of daunomycinone from aklavinone and ε-rhodomycinone. *J. Antibiot. 33*:1199—1201.

20

THE BLEOMYCINS: PROPERTIES, BIOSYNTHESIS, AND FERMENTATION

TOMOHISA TAKITA *Institute of Microbial Chemistry*, Tokyo, Japan

I. INTRODUCTION

In 1956, Umezawa and his associates discovered phleomycin during the course of a study of water-soluble basic antibiotics (Maeda et al., 1956). This antibiotic was found to inhibit Ehrlich carcinoma with a high therapeutic index, but it showed strong renal toxicity toward dogs precluding its clinical study.

Bleomycin (BLM) was discovered by Umezawa and his associates after an intensive search for phleomycin-like antibiotics (Umezawa et al., 1966). The BLM did not cause renal damage in dogs, and the clinical study started in 1965. BLM is currently used in the treatment of squamous cell carcinomas, Hodgkin's lymphomas, and testis tumors.

BLM is produced by *Streptomyces verticillus* ATCC 15003. The antibiotic is a mixture of congeners and is obtained as the blue-colored copper complex

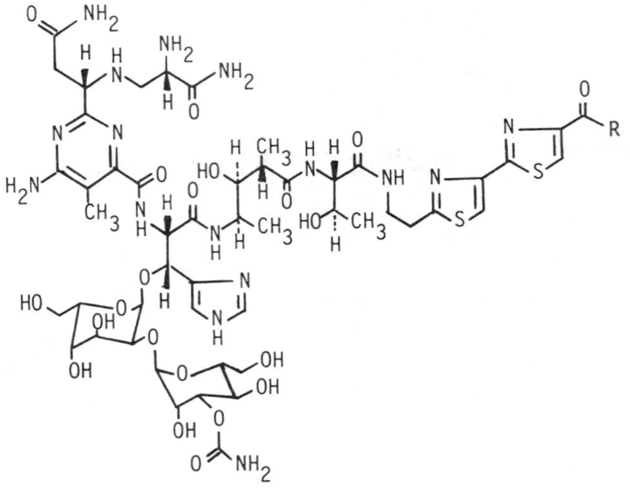

R:

A1 $-NH-(CH_2)_3SOCH_3$

Demethyl-A2 $-NH-(CH_2)_3-S-CH_3$

A2 $-NH-(CH_2)_3-\overset{+}{\underset{\underset{CH_3}{|}}{S}}-CH_3X^-$

A2'-a $-NH-(CH_2)_4-NH_2$

A2'-b $-NH-(CH_2)_3-NH_2$

A2'-c $-NH-(CH_2)_2\text{-imidazole}$

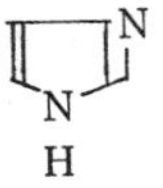

A5 $-NH-(CH_2)_3-NH-(CH_2)_4-NH_2$

A6 $-NH-(CH_2)_3-NH-(CH_2)_4-NH-(CH_2)_3-NH_3$

B2 $-NH-(CH_2)_4-NH-\underset{\underset{NH}{\|}}{C}-NH_2$

B4 $-NH-(CH_2)_4-NH-\underset{\underset{NH}{\|}}{C}-NH-(CH_2)_4-NH-\underset{\underset{NH}{\|}}{C}-NH_2$

Figure 1   Structure of bleomycins

The structure was determined as shown in Figure 1 (Takita et al., 1978a). The congeners are different from each other in the carboxy terminal amine moiety.

BLM consists of a linear hexapeptide and an O-carbamoyl disaccharide. The peptide part is composed from pyrimidoblamic acid (PBA) [1], erythro-β-hydroxy-L-histidine (OH-His) [2], (2S,3S,4R)-4-amino-3-hydroxy-2-methyl-pentanoic acid (AHM) [3], L-threonine (Thr), 2'-(2-aminoethyl)-2,4'-bithia-zole-4-carboxylic acid (ABC) [4], and the terminal amine specific to each bleo-mycin.

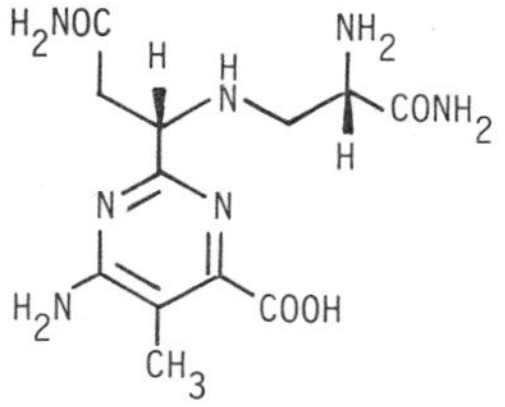

(PBA)[1]

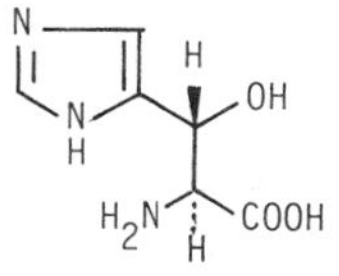

(OH-His)[2]

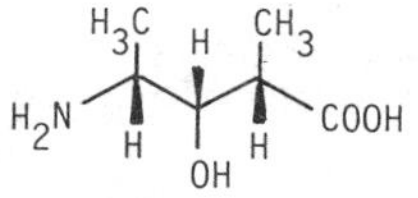

(AHM)[3]

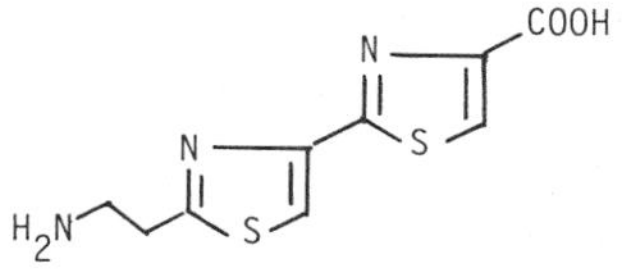

(ABC)[4]

A mixture containing metal-free BLM A2 (55–70%) and B2 (25–32%, as the major components has been used clinically. Recently peplomycin, a new biosynthetic BLM (see Sec. II.A), has been used clinically in Japan as a second-generation bleomycin.

II. BIOSYNTHESIS

The BLM-producing strain (*Streptomyces verticillus* ATCC15003) was isolated from a soil sample collected at a coal mine in Fukuoka Prefecture, Japan. The productivity of BLM has been improved by selection of the highly productive strain, which was obtained by conventional physical and chemical methods of mutation. A CM Sephadex C-25 chromatographic separation of BLM mixture produced by an improved strain under an ordinary culture is shown in Figure 2a.

A. Incorporation of the Terminal Amine

The incorporation of a specific amine into the carboxy terminal amine moiety of BLM was first proved by a culture experiment, in which radioactive 3-amino-propyl-dimethylsulfonium chloride ([14C]methyl) was added to the culture medium and the labeled BLM A2 was produced (Fig. 2b) (Fujii et al., 1974). By this experiment, it became further evident that the addition of the amine of A2 suppressed the production of the other natural BLMs. This precursor effect was also found in the terminal amines of the other natural BLMs, except for those of BLMs A1, demethyl-A2, and A6. The results are summarized in Table 1. The terminal amines of BLMs A1 and demethyl-A2 were not incorporated. These BLMs are derived from BLM A2 by spontaneous demethylation

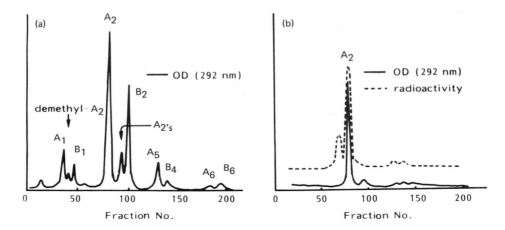

Figure 2 Analysis of bleomycin mixture produced by fermentations by CM Sephadex C-25 chromatography: (a) produced without addition of amine and (b) produced by addition of 3-aminopropyldimethylsulfonium chloride ($^{14}CH_3$).

and oxidation. Addition of A6 amine (spermine) caused the selective production of BLM A5, whose terminal amine is spermidine.

The above results suggested that amino alkyl derivatives having at least one more additional basic function such as amine, guanidine, imidazole, sulfonium, and so on, could be incorporated into the terminal amine moiety of BLM. In fact, the unnatural amines shown in Table 2 were well incorporated into the terminal amine moiety of BLM, and new biosynthetic BLMs have been prepared. Among them, peplomycin, whose terminal amine is N-[(S)-1-phenylethyl]-1,3-diaminopropane, was selected as a second-generation BLM and recently has been used clinically in Japan.

Table 1 Bleomycins Produced by Addition of Terminal Amines of Natural Bleomycins

Terminal amine added	Produced BLMs (%)	Control (%)[a]
A1	Almost same as control	9.2
Demethyl-A2	Almost same as control	2.4
A2	A2 82	54.5
A2'-a	A2'-a 25	
A2'-b	A2'-b 85	3.6
A2'-c	A2'-c 93	
B2	B2 45, new BLM 34[b]	26.7
B4	B4 38, new BLM 40[b]	2.4
A5	A5 ∿ 100	1.3
A6	A5 ∿ 100, A6 trace	Trace

[a]Composition of BLMs produced without addition of amine.
[b]Deaminated products of the terminal guanido group of BLMs B2 and B4, respectively.

Table 2 Unnatural Amines Well Incorporated into the Terminal Amine Moiety of Bleomycin

$H_2N-CH_2-CH_2-NH_2$	$H_2N-(CH_2)_3-N\langle\square$
$H_2N-CH_2-\underset{\underset{CH_3}{\mid}}{CH}-NH_2$	$H_2N-(CH_2)_3-N\langle\bigcirc$
$H_2N-(CH_2)_3-NH-CH_3$	$H_2N-(CH_2)_3-N\langle\bigcirc O$
$H_2N-(CH_2)_3N(CH_3)_2$	$H_2N-(CH_2)_2-N\langle\bigcirc NH$
$H_2N-(CH_2)_3-\overset{+}{N}(CH_3)_3X^-$	$H_2N-(CH_2)_3-NH-CH_2-\langle\bigcirc\rangle$
$H_2N-(CH_2)_3-NH-(CH_2)_3-N(CH_3)_2$	$*H_2N-(CH_2)_3-NH-\underset{\underset{CH_3}{\mid}}{CH}-\langle\bigcirc\rangle$
$H_2N-(CH_2)_3-\underset{\underset{CH_3}{\mid}}{N}-(CH_2)_3-NH_2$	
$H_2N-(CH_2)_3-NH-\underset{\underset{CH_3}{\mid}}{CH}-(CH_2)_2-NH_2$	$H_2N-CH_2-\langle\bigcirc\rangle-CH_2-NH_2$
$H_2N-(CH_2)_3-NH-(CH_2)_3-OH$	$H_2N-(CH_2)_3-NH-\langle\bigcirc\rangle$
$H_2N-(CH_2)_3-NH-(CH_2)_3-OCH_3$	

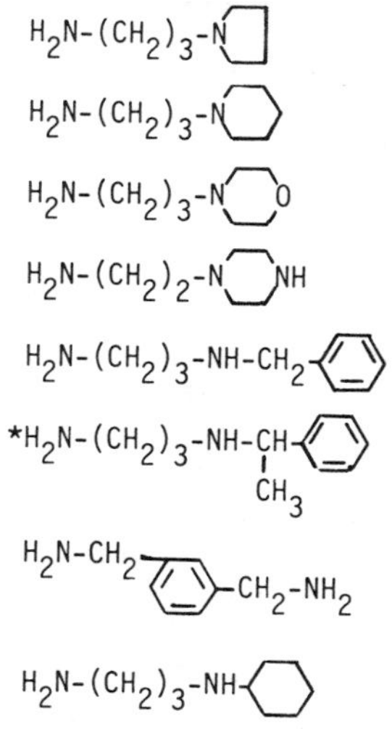

[a]Terminal amine of peplomycin.

B. Biosynthesis of the Peptide Part of Bleomycin

BLM contains three novel amino acids, PBA [1], AHM [3], and ABC [4], which have unusual skeletons. The biosyntheses of these amino acids have been studied by incorporation of ^{14}C-labeled compounds. The results are shown in Table 3. The PBA moiety of BLM is decomposed into a pyrimidine-containing amino acid [5] and β-aminoalanine [6] by competitive β-elimination under

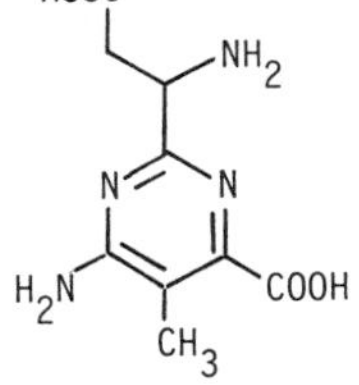

[5]

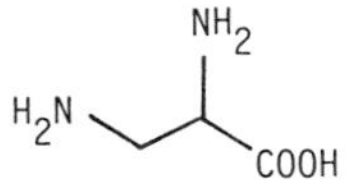

[6]

total acid hydrolysis condition. Thus it was found that serine was incorporated into [6] and asparagine (but not aspartic acid) and methionine−CH_3 were incorporated into [5]. Therefore PBA appears to be biosynthesized from 2 mol of asparagine and 1 mol each of serine and methionine−CH_3 (Fig. 3). AHM was found to be biosynthesized from alanine, acetic acid, and methionine−CH_3,

Table 3 Incorporation of ^{14}C-Labeled Compounds into Novel Amino Acid Constituents of Bleomycin

Novel amino acid	Significantly incorporated labeled compound
PBA	L-Asn (U)[a], L-Met (CH$_3$), L-Ser (U)
AHM	L-Ala (U), L-Met (CH$_3$), acetic acid (2)
ABC	β-Ala (1), DL-Cys (3)

[a]Position of labeling; U, uniformly labeled.

and ABC was synthesized from 1 mol of β-alanine and 2 mol of cysteine (Fig. 3).

The incorporation of methionine−CH$_3$ into the methyl on the pyrimidine of PBA and the α-methyl of AHM was confirmed further by the incorporation study using [^{13}C]L-methionine−methyl analyzed by ^{13}C-nuclear magnetic resonance spectrometry. The incorporation of alanine into the C3−C5 moiety of AHM was also confirmed by the incorporation of [^{13}C]DL-alanine-3 (Nakatani et al., 1980). The ^{3}H-labeled ABC was not incorporated into BLM. This suggested that ABC is formed after incorporation of β-alanine and cysteines into the peptide chain.

The nine biosynthetic intermediates (P3 ∿ deglyco-BLM; see Fig. 4) have been isolated from the fermentation broth of BLM (Fujii, 1979). The isolation of a series of the intermediary peptides indicates that (1) the peptide chain is elongated from the N-terminal amino acid by stepwise incorporation of common amino acids and acetic acid, (2) the novel amino acids are formed by modification of these acids after incorporation into the peptide chain, and (3) the formation of AHM during the peptide chain elongation suggests that the peptide part of BLM is biosynthesized by transpeptidation and transthiolation like gramicidin S, more similar to fatty acid synthesis rather than the ordinary peptide synthesis (Lipmann, 1971).

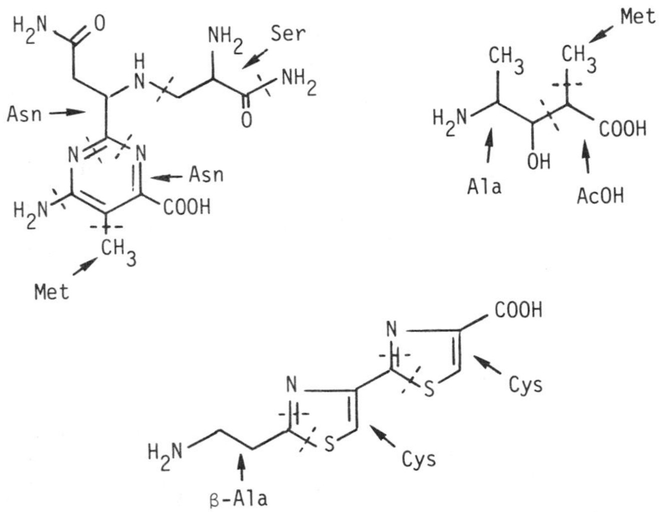

Figure 3 Biosyntheses of PBA [1], AHM [3], and ABC [4].

P-3   DM-PBA—His   (DM: demethyl)

P-3A   DM-PBA—His—Ala
                          *
P-4   DM-PBA—His—AHM

P-5   DM-PBA—His—AHM—Thr
                          *
P-5m   PBA—His—AHM—Thr

P-5mB   PBA—His—AHM—Thr—β-Ala
                              *
P-6m   PBA—His—AHM—Thr—ABC
                              *
P-6mo   PBA—OH-His—AHM—Thr—ABC

Deglyco-BLM PBA—OH-His—AHM—Thr—ABC—Terminal amine

Figure 4 Biosynthetic intermediates of bleomycin. The tilted arrow indicates elongation of peptide chain. Modification occurs at the step marked by an asterisk.

III. FERMENTATION

A highly productive strain of BLM is inoculated on steamed barley grain and kept at 27°C for about 2 weeks. The well-sporulated grain is dried at room temperature under reduced pressure and stored at 0°C until use.

The sporulated grain is well suspended in sterilized water and inoculated to the seed culture fermenter (6 m³) containing the following medium: 1% glucose, 1% starch, 0.75% peptone (Kyokuto Seiyaku Co., Tokyo), 0.75% meat extract (Mikuni Kagaku Co., Tokyo), and 0.3% NaCl at pH 7.2 before sterilization. After culturing for 30–36 hr at 27–28°C under agitation and aeration, the cultured broth is transferred to the production fermenter (125 m³) containing the following medium: 6.4% millet jelly (Asadaame Shokuhin Co., Kanagawa-ken, Japan), 0.5% glucose, 3.5% soybean meal, 0.75% corn-steep liquor (Ajinomoto Co., Tokyo), 0.3% NaCl, 0.1% K_2HPO_4, 0.05% $ZnSO_4 \cdot 7H_2O$, 0.01% $CuSO_4 \cdot 5H_2O$, and 0.2% $NaNO_3$ at pH 6.5 before sterilization. It is cultured at 27–28°C under agitation and aeration keeping the pH at 7.4 or below by addition of sulfuric acid. The culture is stopped when the production of reaches a plateau (7–8 days).

For the production of peplomycin, the precursor, N-[(S)-1-phenylethyl]-1,3-diaminopropane dihydrochloride, is added to the production medium at the initial stage of the fermentation.

IV. ISOLATION AND PURIFICATION

The culture broth is filtered with a filter aid after adjusting pH to 6.0 with sulfuric acid. The filtrate is passed through a column of Amberlite IRC-50 (carboxylic acid ion exchange resin) to adsorb the BLM. The elution is made with 0.3 M HCl. The BLM fraction is neutralized with dilute sodium hydroxide and passed through a column of active carbon to adsorb the BLM. The elution is made with acetone-0.2% M HCl (1:1). The BLM fraction is neutralized with Dowex 44 (a weakly basic ion exchange resin) and concentrated under reduced pressure. The crude powder thus obtained is further purified by alumina chromatography using 80% aqueous methanol as the developing solvent followed by Sephadex G-25 chromatography. Purified BLM mixture thus obtained is separated into each component by CM Sephadex C-25 column chromatography developed with 0.1–0.5 M NH_4Cl.

The copper of the BLM-Cu(II) complex can be removed as follows. The BLM-Cu(II) complex is adsorbed on a column of Amberlite XAD-2. The column is washed with 4% EDTA neutral solution followed with 5% sodium sulfate solution. Elution with 50% methanol gives the colorless metal-free BLM sulfate.

V. ASSAY METHOD AND MODE OF ACTION

The potency of BLM is determined by a cylinder-plate method using *Mycobacterium smegmatis* ATCC 607 as the test organism. The potency of BLM is referred to that of metal-free BLM A2 free base; 30 μg (potency)/ml and 15 μg (potency)/ml solutions of the BLM working standard are used as the reference. The BLM working standard is obtained from National Institute of Health of Japan, Tokyo.

The primary target of BLM appears to be DNA. BLM is a bifunctional compound. The bithiazole chromophore and the terminal amine are involved in the binding to DNA, and the metal complex formation part is involved in the reaction with DNA (Fig. 5) (Takita et al., 1978b). The DNA degradation by BLM is mediated by an active oxygen species formed from the BLM-Fe(II)-O_2 complex (Sausville et al., 1976). The active oxygen species was studied further, and the redox system between BLM-Fe(II) and BLM-Fe(III) involved in DNA cleavage was proposed (Fig. 6) (Kuramochi et al., 1981).

VI. APPLICATIONS AND ECONOMY

BLM is now used in the treatment of squamous cell carcinoma, Hodgkin's lymphoma, and testis tumor, alone or in combination with other chemotherapeutic agents or therapies. The annual sale was about 1,500,000 vials in terms of 15 mg (potency) in all the world in 1980 (Europe, 35%; Asia, 32%; North America, 21%; other 11%). BLM is used as an injection (90%), an ointment (8%), or in oil (2%).

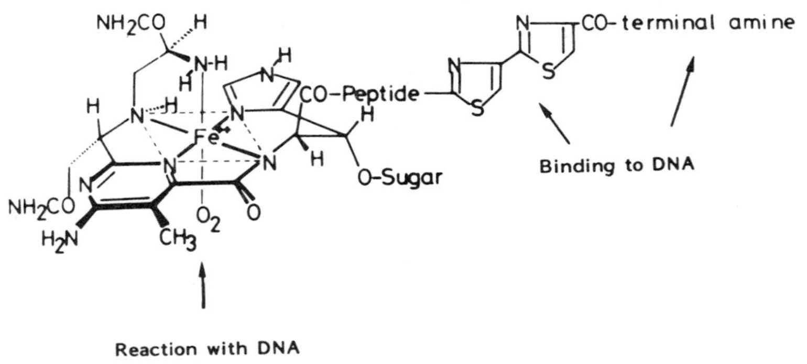

Figure 5 Interaction of bleomycin with DNA.

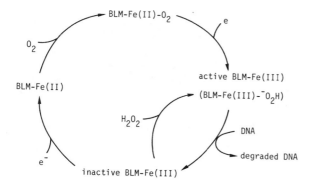

Figure 6 Redox and O_2 activation process of the bleomycin–iron complex. react with DNA.

REFERENCES

Fujii, A. (1979). Biogenetic aspects of bleomycin–phleomycin group antibiotics. In *Bleomycin: Chemical, Biochemical and Biological Aspects.* S. Hecht (Ed.). Springer-Verlag, New York, pp. 75–91.

Fujii, A., Takita, T., Shimada, N., and Umezawa, H. (1974). Biosyntheses of new bleomycins. *J. Antibiot. 27*:73–77.

Kuramochi, H., Takahashi, K., Takita, T., and Umezawa, H. (1981). An active intermediate formed in the reaction of bleomycin–Fe(II) complex with oxygen. *J. Antibiot. 34*:576–582.

Lipmann, F. (1971). Attempts to map a process evolution of peptide biosynthesis. Synthesis of peptide antibiotics from thiol-linked amino acids paralles fatty acid synthesis. *Science 173*:875–884.

Maeda, K., Kosaka, H., Yagishita, K., and Umezawa, H. (1956). A new antibiotic, phleomycin. *J. Antibiot. A9*:82–85.

Nakatani, T., Fujii, A., Naganawa, H., Takita, T., and Umezawa, H. (1980). Biosynthetic study using ^{13}C-enriched precursors. *J. Antibiot. 33*: 717–721.

Sausville, E. A., Peisach, J., and Horwitz, S. B. (1976). A role of ferrous ion and oxygen in the degradation of DNA by bleomycin. *Biochem. Biophys. Res. Commun. 73*:814–822.

Takita, T., Muraoka, Y., Nakatani, T., Fujii, A., Umezawa, Y., Naganawa, H., and Umezawa, H. (1978a). Revised structures of bleomycin and phleomycin. *J. Antibiot. 31*:801–804.

Takita, T., Muraoka, Y., Nakatani, T., Fujii, A., Iitaka, Y., and Umezawa, H. (1978b). Metal-complex of bleomycin and its implication for the mechanism of bleomycin action. *J. Antibiot. 31*:1073–1077.

Umezawa, H., Maeda, K., Takeuchi, T., and Okami, Y. (1966). New antibiotics, bleomycin A and B. *J. Antibiot. A19*:200–209.

Antibiotics Used in Agricultural Practice

21

NISIN: PROPERTIES, BIOSYNTHESIS, AND FERMENTATION

KHALIL RAYMAN AND A. HURST *Microbiology Research Division, Health and Welfare Canada, Ottawa, Ontario, Canada*

I. INTRODUCTION

Nisin, known for over four decades, is in a unique position in that it is the only antibiotic-like substance presently used as a preservative in foods intended for human consumption. The polypeptide is produced by certain strains of *Streptococcus lactis* belonging to the serological group N. This classification and the observation that the metabolite is an inhibitory substance were used by Mattick and Hirsch (1947) to coin the name nisin.

In previous publications, nisin has been referred to as an antibiotic. This term is questionable for the following reasons: As a therapeutic agent in human and veterinary medicine, nisin is ineffective (Gowans et al., 1952); it is not used as an additive to animal feeds, nor is it used for growth promotion; nisinlike substances are commonly produced by lactic streptococci (Hurst, 1967), for example, an inhibitor named diplococcin is produced by *Streptococcus cremoris*, which also belongs to the serological group N (Oxford, 1944), and streptococcin A-FF22 is produced by *Streptococcus pyogenes*, the most pathogenic member of the genus (Tagg and Wannamaker, 1978). These inhibitors appear to be bacteriocins and they have been recently reviewed along with other bacteriocins produced by gram-positive bacteria (Tagg et al., 1976); finally, nisin-producing streptococci have been often found in milk and have been reported in farmhouse cheese (Chevalier et al., 1957). It is therefore likely that nisin has been unwittingly consumed for a long period of time without apparent ill effects.

In view of the foregoing, it would appear that nisin is more closely related to a bacteriocin than an antibiotic and should be placed with the other biological inhibitors produced by the lactic acid bacteria. The variety of inhibitors produced by this group of organisms makes them excellent antagonists to pathogenic and spoilage bacteria in foods (Hurst and Collins-Thompson, 1979).

A number of reviews have been published dealing with various aspects of nisin (Berridge, 1953; Hawley, 1957, 1962; Marth, 1966, Jarvis and Morisetti, 1969; Hurst, 1978, 1981; Lipinska, 1977).

In this article, we shall attempt a comprehensive coverage of all aspects of nisin, with an emphasis on its biosynthesis and production. However, the large number of publications on this subject precludes complete coverage of all the literature; omissions or partial treatment of references should therefore not be construed as a slight to the authors.

II. HISTORY AND DISCOVERY

Nisin has a long history punctuated by periods during which research on the biological inhibitor was at a standstill. The first published reports on an inhibitor produced by lactic streptococci was by Rogers in 1928 and in the same year by Rogers and Whittier. They reported the production of a substance diffusible through a collodion membrane which inhibited growth of other streptococci and prevented acid production by *Lactobacillus bulgaricus*. Five years later, Whitehead and collaborators in New Zealand (Whitehead, 1933; Whitehead and Riddete, 1938), while investigating slow acid production in cheese making, isolated from the milk a lactic streptococcus which produced an inhibitory substance. Not much importance was attached to this observation, because phage was soon discovered to be the major problem in starter culture failure. However, Whitehead (1933) isolated the inhibitor and showed that it was a polypeptide. It was not until 1943 that interest in the inhibitory streptococci was rekindled when Meanwell ascribed slowness in cheese manufacture to the presence of these organisms. In 1944, Mattick and Hirsch showed that the substance produced by the lactic streptococci isolated by Meanwell (1943) was inhibitory to a number of gram-positive organisms. The biological inhibitor was later named nisin (Mattick and Hirsch, 1947). It is estimated that about one-third of all strains of S. lactis isolated from milk produce nisin (Hoyle and Nichols, 1948).

The generally accepted belief that the usefulness of antibiotics or antibiotic-like substances lies in their medicinal application dictated the area in which research on nisin was initially concentrated. Mattick and Hirsch (1946, 1947) and Hirsch and Mattick (1949) investigated the potential of nisin for medical use and obtained encouraging in vitro results on the susceptibility of a number of important pathogenic organisms to nisin; however, because of its low solubility at physiological pH (Taylor et al., 1949; Gowans et al., 1952) and its destruction by certain enzymes of the digestive system (Jarvis and Mahoney, 1969), the potential for nisin as a therapeutic agent lost ground. Furthermore, antibiotics of fungal and actinomycete origin were available and of greater therapeutic value than nisin (Gowans et al., 1952).

The lack of medicinal application (Gowans et al., 1952), low toxicity (Frazer et al., 1962), absence of growth stimulation (Barber et al., 1952), and its inhibitory action on clostridia were admirable qualities for nisin to be used as a food preservative. This was first explored by Hirsch et al. (1951), who used a nisin-producing starter culture to prevent "blowing" in Swiss-type cheese caused by growth of *Clostridium butyricum* and *Clostridium tyrobutyricum*. McClintock et al. (1952) later extended this application to processed cheese with even greater success, and in the Netherlands, Kooy and Pette (1952) also reported inhibition of butyric acid fermentation in cheese by using a nisin-producing starter culture of S. lactis. It is ironic that a substance originally conceived as a problem in cheese manufacture should be first used as a preservative in none other than cheese. Interest in nisin as a food preservative gained momentum, and in 1953 the first commercial preparation of nisin was marketed.

III. UNIT OF NISIN ACTIVITY

An International Reference Preparation of Nisin was established in compliance with a request from the World Health Organization Expert Committee on Biological Standardization. This preparation is maintained at the Central Veterinary Laboratory, Weybridge, Surrey, England (Tramer and Fowler, 1964). The International Reference Preparation is identical to the commercial preparation which is marketed by Aplin and Barrett under the trade name Nisaplin. One gram of Nisaplin is standardized to contain 10^6 international units (IU) and 1 g of pure nisin is equivalent to 40×10^6 IU. One IU is equivalent to one Reading unit, which was the former designation of nisin activity; the Reading unit was chosen because much of the original research on nisin was done at the University of Reading in England.

IV. SELECTION AND STRAIN IMPROVEMENT

Work done on selection of strains for improved nisin production is minimal, probably because of the extreme susceptibility of nisin-producing strains to phage attack. Unfortunately, most of the phage-resistant strains of S. lactis are poor nisin producers. Mattick and Hirsch (1947) observed a considerable variation in nisin production among different strains. In broth culture, they noted that nisin production was stable for about 2 years, after which further subculturing resulted in a decline in production. A similar observation was made by Cheeseman and Berridge (1957), who selected from a 10-year-old freeze-dried preparation, a culture which produced nisin but which after a number of subcultures lost this ability and showed a slight sensitivity to the inhibitor. From this culture, they recovered a nisin producer by using a 10% inoculum in broth containing 500 IU of added nisin per milliliter and, after overnight incubation, plating the culture on agar containing 2000 IU nisin per milliliter. A colony selected from the plate yielded 700 IU/ml nisin, whereas, the parent strain produced only 250 IU/ml. Hirsch (1951b) described a similar phenomenon, but his isolate was selected from a nisin-free medium and produced a much lower level of nisin. He also noted that maximal nisin production was attained after several serial transfers of the culture, but the number of transfers necessary varied from experiment to experiment.

Csiszar and Pulay (1956) isolated high nisin-producing mutants by ultraviolet irradiation of cultures with doses that reduced the viable population to less than 1%. Three of their isolates produced 10–15 times the amount of nisin produced by the parent strain, and this property remained stable for at least 1 year. Similar work was done by Kalra et al. (1973). Lipinska (1977) summarized the work of others in which γ- and x-rays were used. She reported that stable mutants have been selected which are also capable of producing 10 times as much nisin as the parent strain.

V. GENETIC CONTROL OF NISIN SYNTHESIS

The possibility that nisin synthesis might be plasmid controlled was first suggested by Kozak et al. in 1973. Using proflavine and ethidium bromide, agents known to "cure" cells of plasmids, they noted a significant increase in the frequency of conversion of nisin-producing strains (nis+) to nisin-negative strains (nis-) when compared to the frequency of spontaneous conversion in untreated cultures. This was very pronounced in one strain where spontaneous conversion was 0.82%, but after treatment with ethidium bromide and proflavine, the frequency of conversion increased to 9 and 77%, respectively. In these experiments, the nis- strains could not have been preferentially selected, because the susceptibility of the nis+ and nis- strains to the curing agents was essentially the same. Furthermore, treatment of the nis- isolates with nitrosoguanidine did not reverse the effect of the curing agents, indicating that the ability to produce nisin was permanently lost by the "cured" strains and that nisin synthesis was most likely plasmid controlled (Kozak et al., 1974). In further experiments, Fuchs et al. (1975) were unable to prove unequivocally that nisin production was plasmid regulated. Among eight nis+ strains, two were found by CsCl gradient not to contain covalently closed circular (ccc) DNA, which is characteristic of plasmid DNA; also, of two nis- isolates, only one had lost its ccc DNA. Plausible explanations were offered by the authors for these apparent discrepancies, but it is evident that more work is necessary to resolve this uncertainty using more up-to-date methods, such as agarose electrophoresis.

VI. BIOSYNTHESIS

Nisin has a molecular weight of 7000 (Cheeseman and Berridge, 1959) and occurs naturally as a dimer (Jarvis et al., 1968); the molecular weight of each

subunit is approximately 3500 (Gross and Morell, 1967). The polypeptide
contains two unusual sulfur-containing amino acids, lanthionine and β-methyllanthionine (Newton et al., 1953), as well as two unsaturated amino acids,
dehydroalanine and dehydrobutyrine (β-methyldehydroalanine) (Gross and
Morell, 1968). A diagrammatic representation of the structure of nisin is shown
in Figure 1 (Gross, 1977). The unusual and unsaturated amino acids are not
unique to nisin, for they are also found in subtilin (Gross et al., 1969; Gross
and Kiltz, 1973).

The biosynthesis of nisin has yet to be fully elucidated. Using radioactive tracers to follow the biochemical activity of the cells, and a bioassay
to follow nisin production, Hurst (1966) showed that synthesis of nisin in
a chemically defined medium was unaffected by penicillin and mitomycin, antibiotics to which the organism was sensitive. Actinomycin D had an immediate
inhibitory effect on RNA synthesis, but inhibition of nisin synthesis was delayed for 60 min. The insensitivity to mitomycin and the delayed effect of
actinomycin D suggested that nisin synthesis was independent of newly formed
DNA, but could be dependent on messenger RNA, which appeared to have
an unusually long half-life.

Antibiotics which inhibit protein synthesis such as chloramphenicol, puro-
mycin, and tetracycline also inhibited nisin synthesis. This inhibition was
immediate and was independent of the time of addition of the antibiotics in
the growth cycle of the organism. An example of this is shown in Figure 2,
where addition of chloramphenicol before initiation of nisin synthesis prevented
production of the polypeptide, whereas, addition of chloramphenicol after
the commencement of nisin synthesis inhibited further production of nisin
by the organism. Nisin synthesis was also observed to be more sensitive to
chloramphenicol than was protein synthesis (Fig. 3). These results were
unexpected, because synthesis of peptides containing unusual amino acids
was generally thought to be insensitive to inhibitors of protein synthesis (Bo-
dansky and Perlman, 1964). The observations of Hurst (1966) were confirmed by
Ingram (1969, 1970). Using the chemically defined medium of Hurst (1966), Ingram followed de novo synthesis of nisin by the incorporation of radiolabeled amino

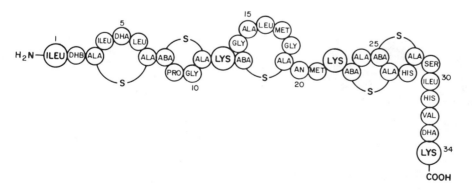

Figure 1 Structure of nisin: ABA, aminobutyric acid; DHA, dehydroalanine;
DHB, dehydrobutyrine (β-methyldehydroalanine); ALA-S-ALA, lanthionine;
ABA-S-ALA, β-methyllanthionine. (After Gross, 1977.)

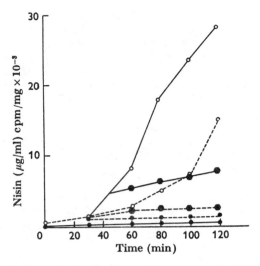

Figure 2 The effect of chloramphenicol (20 µg/ml) on nisin synthesis and incorporation of [U-^{14}C]L-glutamic acid by *Streptococcus lactis*: radioactivity (—), nisin (---); control (○), chloramplenicol added at time zero (●), at 50 min (⬡). (After Hurst, 1966.)

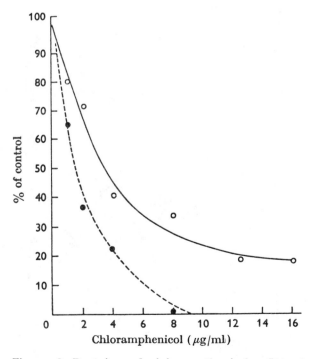

Figure 3 Protein and nisin synthesis by *Streptococcus lactis* in the presence of various concentrations of chloramphenicol as a percentage of the control without antibiotic: protein (○—), nisin (●---). (After Hurst, 1966.)

acids into material which behaved as nisin after purification. He observed that chloramphenicol or tetracycline did indeed inhibit incorporation of ^{35}S-labeled cysteine into lanthionine and β-methyllanthionine, the two thioether amino acids found in nisin. In addition, this inhibition was greater than the inhibition of incorporation of cysteine into protein, which confirmed the observations of Hurst (1966). In the absence of chloramphenicol, Ingram followed the incorporation of labeled amino acids and noted that of several tested, only serine, cysteine, and threonine were incorporated into lanthionine and β-methyllanthionine, and the molar ratios incorporated were close to the theoretically anticipated values. This led Ingram to conclude that serine, cysteine, and threonine were precursors of the thioether amino acids and a pathway for the synthesis of lanthionine and β-methyllanthionine was proposed. Ingram postulated that serine, cysteine, and threonine were first incorporated into the peptide, then serine and threonine were dehydrated to the dehydro-amino acids, which in turn condensed with cysteine to form lanthionine and β-methyllanthionine. The incorporation of label derived from threonine and cysteine into nisin, which contains neither of these amino acids, supported the biosynthetic pathway proposed by Ingram (1970).

Because an invariant genetic code would not permit synthesis of poly-peptides containing unusual amino acids (Bodanszky and Perlman, 1964), nisin production should not have been inhibited by antibiotics which affect protein synthesis at the ribosomal level. To explain this anomaly, Ingram (1969, 1970) suggested that the polypeptide chain containing cysteine, serine, and threonine was first synthesized via the ribosomal mechanism, and these amino acids were subsequently modified nonribosomally to lanthionine and β-methyllanthionine, ultimately resulting in the production of nisin. This suggestion was confirmed and extended by the work of Hurst and Paterson (1971). Before this study, Hurst (1967) had isolated a strain of *S. lactis* that did not produce nisin but which did produce a low molecular weight protein which, by Sephadex chromatography and polyacrylamide gel electrophoresis, resembled nisin. Preparations of this noninhibitory protein when incubated with a cell extract from a nisin-producing strain of *S. lactis* generated an inhibitory substance that behaved like nisin. The enzyme responsible for the conversion of the low molecular weight protein (pronisin) to the inhibitory nisinlike substance was made only briefly during the exponential growth phase of the organism. It appeared just before the start of detectable nisin synthesis and attained maximum specific activity in about 2.5 hr, after which it declined rapidly and was undetectable 3 hr after its first appearance. The enzyme was heat labile and its activity decreased after 30 min of incubation at 30°C (Hurst and Paterson, 1971).

The enzyme was readily obtained by blending cells of nisin-producing *S. lactis* with microbeads for 45 sec, a treatment that had no effect on either the phase-contrast appearance or the gram reaction of the cells. This indicated that the location of the enzyme and most probably the site of conversion of pronisin to nisin were at the cell surface. The observations of White and Hurst (1968) were consistent with this postulate. They showed by chemical and physical fractionation of the producer organism that nisin was associated mainly with the cell envelope. These observations and those of Ingram (1969, 1970) suggested a two-stage process for nisin biosynthesis: Firstly, the nisin precursor or pronisin was synthesized via the conventional protein-synthesizing mechanism involving transcription and translation, and was sensitive to protein inhibitors, and secondly, the pronisin was transported to the cell surface, where it was converted enzymatically to the active biological inhibitor.

Precursor proteins are not uncommon; for example, insulin, which is similar to nisin in molecular weight, is now known to be synthesized via not only a proinsulin but a preproinsulin (Steiner, 1979).

VII. FERMENTATION

A. General

The extreme susceptibility of nisin-producing strains of S. lactis to bacterio-phage attack is one of the major problems encountered in producing nisin on an industrial scale. Another common problem is infection of the culture by "wild" yeasts, which could have disastrous effects at any stage of manufac-ture. Additionally, Hawley (1955) pointed out that if the nisin preparation is intended for use against clostridia, as in most cases, care has to be taken to avoid contamination of the fermentation menstruum by these organisms, because nisin-resistant strains could develop under these conditions and be carried over to the product. This is a very real danger if milk is a component of the medium. For these reasons, special efforts have to be made to destroy clostridia present in the milk and to ensure freedom from bacteriophage con-tamination. Therefore sterile media have to be used and the entire manufac-turing and drying process has to be done in a sterile plant.

B. Inoculum

Information on the preparation of inocula for large-scale production of nisin is limited. Cheeseman and Berridge (1957) started their inoculum from a freeze-dried culture which was grown in 10 ml of broth at 30°C for 24 hr and then transferred to 100 ml of broth and grown for an additional 24 hr. At this point, the culture was examined for purity and it was then transferred to 100 ml of broth containing 500−1000 IU/ml of added nisin. Incubation was continued for 24 hr at 30°C, after which the culture was used to inoculate 7 × 100 ml of broth. This was grown for 8 hr and the entire volume was added to 7 liters of broth, which was again incubated for 8 hr. After purity of the culture was established by microscopic examination, the entire 7 liters were used to inoculate a 150-liter batch fermentation broth.

   Bailey and Hurst (1971) started their inoculum from slopes 3 days before the fermentation. The cultures were grown in a glucose-rich medium contain-ing meat extract, yeast extract, and mineral salts adjusted to pH 6.8. By daily subculturing to larger volumes, they obtained several 1-liter cultures which were used to inoculate each 20-liter fermentation medium.

C. Medium

Hirsch (1951b) noted that in an unbuffered medium, a nisin-producing strain of S. lactis yielded about 80 IU/ml and the pH of the medium fell from its origi-nal value of 7.0 to 4.6 in 24 hr. The lower pH, brought about by an accumu-lation of lactic acid, arrested growth of the organism and the nisin synthe-sized was distributed between the medium and the cells. By periodically ad-justing the pH back to neutrality, twice as much nisin was obtained from a correspondingly higher cell mass, and the nisin was almost entirely associated with the cells. However, Hirsch noted that glucose, which was at a concen-tration of 1%, became limiting after 12 hr of incubation. By increasing the glucose concentration to 2.5% and periodically adjusting the pH, the nisin yield was increased and this could be boosted to 620 IU/ml by adding liver extract to the medium. Calcium pantothenate was found to be more effective than liver extract or a mixture of B vitamins, but it was necessary to have an excess of pantothenate (0.6 µg/ml), because microbiological assays showed that in the glucose-rich medium, pantothenate was rapidly utilized and became the limiting factor. Hirsch further improved the nisin yield to 1700 IU/ml by replacing the periodic pH adjustment with a three-component buffer system comprising acetate, citrate, and phosphate.

A simple medium for nisin production was devised by Egorov et al. (1971). They formulated a medium consisting of optimum concentrations of corn extract, casein hydrolysate, glucose, KH_2PO_4, and $MgSO_4$ which yielded about 2200 IU/ml. In a later report, Egorov et al. (1976) claimed that addition of thymine, adenine, and hypoxanthine to the growth medium stimulated nisin and biomass production. However, the nisin yield was lower than that reported earlier, but a medium of totally different composition was used. Neither Hirsch (1951b) nor Kozak and Dobrzanski (1977) observed any stimulatory effects by incorporating purine or pyrimidine bases into their respective growth media. Removal of $MgSO_4$ did not significantly alter the nisin yield, but Kozlova et al. (1979) noted that lowering the concentration of KH_2PO_4 while maintaining the pH near neutrality with NaOH caused a 35–40% reduction in nisin yield. These authors suggested that KH_2PO_4 was essential not only as a buffering agent, but also as a source of K^+, which was required for growth. Baranova et al. (1980) subsequently replaced the more expensive casein hydrolysate with a protein- and vitamin-rich nitrogenous base obtained from enzymatically hydrolyzed microbial biomass (mainly yeast). The yield of nisin was slightly reduced, but the lower cost of the enzymic hydrolysate more than compensated for this.

Kozak and Dobrzanski (1977) described a synthetic medium containing a minimum of organic components necessary for growth and nisin production. It consisted of nine amino acids, four vitamins from the B group, glucose, and mineral salts. In addition to the nine amino acids, six of which were present in the nisin molecule, Kozak and Dobrzanski noted that addition of serine, proline, cysteine, and cystine either singly or in combination increased the yield of nisin but had no effect on either the growth rate or the final biomass yield. Biotin stimulated growth and marginally increased the yield of nisin. Unfortunately, their yield could not be compared with others because they expressed their results as nisin per colony-forming unit or nisin per unit optical density.

In the patent held by Aplin and Barrett (Hawley and Hall, 1957), sterilized whole, skimmed, or reconstituted milk was used for the commercial preparation of nisin, but in a later patent (Hall, 1963), a modified milk medium was used. This medium was prepared by treating milk with papain, pepsin, or trypsin to hydrolyze part of the milk protein and then heating to destroy the enzyme. Precipitated protein was then removed and the liquid protein was sterilized at 115°C and used to cultivate the organism. For optimum activity of papain, pepsin, and trypsin, the milk was adjusted to pH 6.5, 1.5, and 7.8, respectively.

D. Culture Conditions

The growth environment has a marked effect on the yield of nisin and its antibacterial spectrum. Hirsch (1951a) reported that a single strain of *S. lactis* could produce different types of nisin, depending on the method and conditions of growth. Berridge et al. (1952), and Willimowska-Pelc et al. (1976) also observed that a single preparation of nisin consisted of a number of close-ly related polypeptides, the ratios of which could be altered by a variety of conditions. For instance, Berridge (1949) noted that nisin produced by S. lactis grown at 30°C was inhibitory to S. cremoris and Streptococcus agalac-tiae, whereas, when grown at 20°C, the nisin produced was active against S. cremoris but not S. agalactiae.

As pointed out by Hirsch (1951b), highest yields of nisin were obtained by continuously adjusting the pH close to neutrality. He noted that nisin production was directly related to cell mass and that aeration was antagonistic to growth and nisin production. Hawley (1955) also observed that optimal yields could be realized if the environment was adjusted to maintain the or-ganism in single or diplococcal forms during active growth; however, he cau-tioned that nisin produced under conditions favoring maximum yield must not be assumed to have the same antibacterial properties as nisin produced sub-maximally at, say, a different pH.

Typical fermentation conditions would be incubation at 28–30°C for 16–18 hr, continuous or periodic adjustments of the pH in the range 6.0–6.8, and no agitation except during pH adjustments or in an atmosphere devoid of oxy-gen.

E. Product Recovery and Purification

Mattick and Hirsch (1947) extracted nisin from the fermentation liquor by creating a gel-like emulsion with chloroform. The culture was acidified to pH 2.0 to release the cell-associated nisin, and then the cell-free supernatant adjusted to pH 4.5 was vigorously stirred after adding 5% chloroform and 0.1% secondary octyl alcohol. The gel-like layer was removed and cold absolute ethanol was added to it to solubilize the chloroform and precipitate the nisin. The precipitated material was collected by centrifugation, dissolved in 0.05 N HCl at 50°C, and residual chloroform removed by blowing hot air over the liquid while stirring. A series of pH adjustments and ethanol extraction fol-lowed and the resulting material was freeze dried. This material had an ac-tivity of approximately 2.0×10^6 IU/g and it could be further purified by repeated solubilizing in 0.01 N HCl and salting out with NaCl. Berridge (1947) crystallized the preparation of Mattick and Hirsch (1947) by dissolving it in absolute ethanol and slowly cooling the solution. The crystals were in the form of long filaments.

Hirsch (1951b) also tried recovering nisin by adsorbing it to columns of synthetic resins and eluting with acidified 80% ethanol. Although most of the nisin was recovered, the purity of the eluate was disappointing. The activity of the dried material was only 1.0×10^5 IU/g, which was lower than that obtained by the chloroform gel procedure (Mattick and Hirsch, 1947). The effect of frothing was then investigated. Nitrogen containing 5% CO_2 was bubbled into the culture fluid and the froth formed was allowed to creep up to 1-in.-diameter glass tube set above the medium at a 30° angle. The froth collected in this manner contained 90% of the nisin and was contained in 1/10 the original volume of the growth medium.

Cheeseman and Berridge (1957) used a combination of n-propanol and NaCl to recover nisin from the fermentation liquor. The propanol was first dissolved in the acidified broth, and then NaCl was added, which decreased the solubility of both the propanol and nisin. The propanolic layer containing most of the nisin separated from the broth and was siphoned off. The pro-cedure was repeated and the combined propanolic layers contained 85–100%

of the nisin. By dissolving the propanolic layer in 0.1 N HCl and salting out with an excess of NaCl, two layers of relatively small volumes formed; the upper layer rich in propanol contained one-half of the nisin and a layer of grey precipitated material sandwiched between the upper layer and the aqueous NaCl layer contained the other half. This interfacial precipitate was fractionated with butanol and finally precipitated with acetone. The dried powder had an activity of approximately 3.0×10^7 IU/g. The procedure for the propanol-rich layer was more complicated, involving partitioning in a butanol-acetic acid system and a number of acetone precipitation steps. The activity of this preparation was 2.8×10^7 IU/g.

In the patent specifications submitted by Aplin and Barrett (Hawley and Hall, 1957), nisin was recovered by the frothing technique (Hirsch, 1951b). Concentration was achieved by physical and chemical separation techniques and the concentrate spray dried and standardized by addition of NaCl.

F. Assay

As early as 1934, a methylene blue reduction test was used by Cox for the detection of "inhibitory streptococci" in milk. In the test, organisms present in the milk reduced the methylene blue to a colorless form, but in the presence of nisin-producing organisms, the time to reduction was significantly increased. Chevalier et al. (1957) also described a test for detecting inhibitory streptococci in milk. They took advantage of the fact that these organisms were capable of growth at relatively low temperatures. A rich agar medium was inoculated simultaneously with the milk under test and with a nisin-susceptible thermophilic L. lactis. The plates were incubated for 1 day at 20°C, which allowed the nisin producers to develop, incubation was then shifted to 30°C, and as the susceptible strain grew it was inhibited in the vicinity of nisin-producing organisms, which were identified by clear zones surrounding the colonies.

Mattick and Hirsch (1947) developed an extinction dilution method to quantitate nisin. To a series of tubes inoculated with a hemolytic strain of S. agalactiae, dilutions of the material under test were added, and the tubes were incubated at 37°C for 16–20 hr. The tubes were then examined for growth, and the titer was calculated as the reciprocal of the dilution added to the last tube in which no growth occurred. An extension of this method was described by Hirsch (1950) in which S. cremoris IP5 was used as the test organism and reduction of methylene blue as the growth indicator. In the same report, Hirsch also described a more complex lag-phase method using S. agalactiae as the test organism. In the latter assay, the length of the lag phase was linearly related to nisin concentration in the range 5–10 IU/ml.

The turbidimetric method of Berridge and Barrett (1952) was modified by Hurst (1966). The test organism was S. cremoris, which was inoculated into a series of tubes containing various dilutions of nisin. The tubes were incubated for 2.5–3 hr, and then growth was stopped by adding a disinfectant. The OD_{600} was measured and plotted against the log_{10} of the nisin concentration to establish a standard durve. The sensitivity of the test was 0.04–0.4 IU/ml, but only clear solutions could be assayed.

The conventional horizontal agar diffusion technique was only of limited usefulness, because of the difficulty with which nisin diffused through agar. To overcome this problem, Mocquot and Lefebvre (1956) incorporated Tween 80 into the agar medium and refrigerated the seeded plates overnight to allow nisin to diffuse through the medium. The plates were then incubated and the zones of inhibition of the test organism were measured. Tramer and Fowler

(1964) used 1% Tween 20 in the assay medium, which increased the rate of
diffusion of nisin, thereby obviating the need for a prediffusion step. The
method which used *Micrococcus flavus* as the test organism was capable of
detecting between 0.5 and 10 IU of nisin per milliliter. Cloudy samples presented no problem, making the assay especially useful for quantitative determination of nisin in foods. Suitable controls were built into the method to
compensate for interfering substances likely to be present in food extracts.
Tramer and Fowler (1964) also described a simple semiquantitative reverse-
phase procedure that utilized heat-shocked spores of *Bacillus stearothermophilus* and which could be used for routine screening of samples.

Stumbo et al. (1964) developed a sensitive micromethod for assaying residual nisin in canned food products. The test is based on the nisin sensitivity
of heat-damaged spores of *B. stearothermophilus*. The authors claimed that
as little as 0.3 IU of nisin per gram of food could be determined with excellent
precision.

VIII. MODE OF ACTION

With the exception of three strains of *Neisseria* (Mattick and Hirsch, 1947),
gram-negative bacteria are not inhibited by nisin; nor are yeasts and molds;
however, many of the gram-positive bacteria are inhibited. Among these,
the streptococci are especially sensitive and a species of the genus, *S. cremoris* strain IP5, is used in the biological assay of nanogram quantities of
nisin (Hurst, 1966). Mattick and Hirsch (1947) also reported nisin sensitivity
of six strains of *Mycobacterium tuberculosis*, but these organisms were about
100 times more resistant than the streptococci. Nisin has no effect on viruses
(Carlson and Bauer, 1957).

The importance of nisin as a food preservative stemmed from the observation that bacterial spores are generally sensitive. The degree of sensitivity
appears to be related to the manner in which the spore coat ruptures. Organisms such as *Bacillus subtilis* which produce small spores and depend on
mechanical pressure to rupture their coats (M spores) prior to outgrowth have
been found to be more sensitive to nisin than the large spore producers such
as *Bacillus cereus* which rely on a lytic enzyme to open their spore coats (L
spores). These phenomena could be explained by the work of Gould and Hurst
(1962), who reported that L spores contained an extractable nisin inactivator
which was either absent or produced in very low concentrations in M spores.
These observations were confirmed by Jarvis (1967), who demonstrated that
the inactivator was an enzyme which was active against both nisin and subtilin.
Campbell and O'Brien (1955), Boone (1966), and Hurst (1972) have reviewed
the effect of nisin on sporeformers.

It was first believed that nisin decreased the heat resistance of bacterial
spores (Lewis et al., 1954; O'Brien et al., 1956), but it became evident from
later studies that the apparent decrease in heat resistance was due to inhibition of spore outgrowth by nisin carried over from the heating menstruum
to the recovery medium (Campbell and Sniff, 1959; Denny et al., 1961).
Campbell and O'Brien (1955) showed that as little as 1.2 IU of nisin in the
recovery medium was sufficient to inhibit outgrowth of severely heated spores
of clostridia and bacilli. It would appear that nisin is sporostatic and not
sporocidal, because the diminished counts obtained by heating spores in the
presence of nisin could be restored to control levels when trypsin was incorporated into the plating medium (Campbell and Sniff, 1959). Thorpe (1960)
obtained similar results with *B. stearothermophilus* and concluded that nisin

was carried over by adsorption to the spore coat. This was consistent with the observation of Hawley (1962), who reported that nisin could not be washed off spores of *Clostridium botulinum*. On the other hand, Tramer (1964) found no evidence for adsorption to spores, because the inhibitory effect of carried over nisin could be diluted out. It is not clear why trypsin, which does not inactivate nisin (Jarvis and Mahoney, 1969), should neutralize the effect of the inhibitor. It may be speculated that nisin does attach to the spore coat through ester or amide bonds which are susceptible to trypsin hydrolysis.

The data presented by Hawley (1962) strongly suggested that spores of *C. botulinum* types A and B were injured by heat. He noted that without heating, nisin-resistant spores germinated and grew profusely in the presence of high concentrations of nisin; however, if the spores were heated, they became nisin sensitive. Direct evidence for injury of spores by heat was provided by Heinemann et al. (1965), who showed that the sensitivity to nisin of four different types of spores increased as the severity of heating increased. Rayman et al. (1981) also noted that the sensitivity to nisin of spores of *Clostridium sporogenes* PA3679 was greater in pork slurries heated to 70°C than in unheated slurries. Conversely, Tramer (1964) observed no difference between the nisin sensitivity of spores of *B. stearothermophilus* heated in the presence of nisin and those to which an equal concentration of nisin was added after heating.

Tramer (1964) reported that germination of spores of *B. stearothermophilus* was inhibited by nisin. However, the concensus appears to be that nisin does not affect germination, but affects the postgermination stages of spore development. Ramsier (1960) reported that spores of *C. butyricum* were unaffected by nisin until after germination, when further development was inhibited. Hitchins et al. (1963) observed phase darkening and loss of heat resistance of spores of *B. cereus* and *B. subtilis* in the presence of nisin, indicating that germination occurred, but they noted that outgrowth was inhibited. Gould (1964) similarly observed that germination of 56 strains of *Bacillus* spp. was unaffected by nisin but that postgermination swelling and subsequent stages of spore development were inhibited.

The mechanism by which nisin exerts its effect is largely unexplained. Apart from the report of Tramer (1964), the available evidence indicates that a prerequisite for nisin activity is adsorption to the organism (Hirsch, 1954; Ramsier, 1960; Thorpe, 1960). Working with *S. agalactiae*, Hirsch (1954) noted that the organism was killed in 10–15 min by 150 IU nisin per milliliter, but if charcoal were added to the recovery medium, it took 20–30 min to obtain the same result. Hirsch suggested that a nisin–bacterium complex was formed which could be disrupted by the charcoal. Ramsier (1960) also reported strong adsorption of nisin to vegetative cells of *C. butyricum*. The nisin-treated cells leaked ultraviolet absorbing material and subsequently lysed. This observation and the finding that anionic soaps neutralized the effects of nisin led Ramsier to conclude that nisin behaved like a cationic surface-active detergent. Lysis caused by nisin was also observed by Hurst and Kruse (1972) when it was added to the producer organism during its nisin-sensitive phase of growth.

Using an in vitro system, Linnett and Strominger (1973) estimated that 40 μg/ml nisin caused a 50% inhibition of peptidoglycan synthesis catalyzed by particulate enzyme proteins derived from either *B. stearothermophilus* or *Escherichia coli*. Reisinger et al. (1980) reported similar results and showed that inhibition resulted from the formation of a complex between nisin and the lipid intermediate of the murein biosynthetic pathway. The concentration of nisin necessary to bring about this inhibition (1600 IU/ml) was approximately 1000-fold higher than the minimum inhibitory concentration for *B. stearothermophilus*, making it unlikely that inhibition of peptidoglycan synthesis was the primary site of nisin action.

IX. TOXICITY

The toxicity of nisin has been extensively investigated by Frazer et al. (1962) in Britain. The tests included both acute toxicity and chronic effects of long-term feeding of nisin to animals, at levels 1000 times greater than would be expected to be consumed. The results indicated that nisin was safe for use in foods at the levels that would be necessary to have a preservative effect. It was also pointed out that some nisin may be naturally present in milk and cheese as a result of chance contamination by nisin-producing streptococci, and has therefore been consumed over long periods of time without apparent ill effects. While this does not rule out the possibility that nisin might have some adverse effects, it does indicate, at least, that nisin has low toxicity, if any at all. Caserio et al. (1979a) reported investigative work done by Japanese researchers in which kittens and rats were fed nisin and the LD_{50} was found to be similar to that of common salt, that is, about 7 g/kg body weight. Shtenberg and Ignat'ev (1970) also reported nontoxicity of nisin, and Lipinska (1977) quoted a number of other studies done in the Soviet Union, all of which confirm this phenomenon. Additionally, nisin has no effect on gram-negative organisms which constitute the major intestinal microflora; its ingestion will therefore not be expected to alter the intestinal ecological balance.

Despite the general consensus that nisin is nontoxic, its use in foods is permitted in only 27 countries. A list of these countries was published by Hurst (1981). At the present time, the United States and Canada do not permit the use of nisin in foods. An application to the U.S. Food and Drug Administration for use of nisin in certain dairy products is presently being considered.

X. APPLICATION

A. Cheese

Long before nisin was first permitted as a food additive in Great Britain (Anonymous, 1959), Hirsch et al. (1951) suggested its use in low-salt cheese to combat gas production by clostridia. They used a nisin-producing starter culture to prevent "blowing" in Swiss-type cheese caused by lactate-utilizing clostridia. In 1952, McClintock et al. added milk soured by nisin-producing streptococci to cheese melt, and although clostridial growth was inhibited, the method of application was not industrially feasible. Hawley (1955) recommended instead that nisin be added as a mixture with skim milk powder, which to the present time is one of the most effective ways in which nisin is utilized. Kooy and Pette (1952), Kooy (1952), Lipinska (1956), and Lipinska et al. (1962) have successfully used nisin to manufacture a variety of cheeses which had been plagued with clostridial problems. However, Winkler and Fröhlich (1958), Galesloot and Pette (1957), and Pulay (1956) claimed that in addition to suppressing growth of clostridia, nisin also inhibited the growth of desirable microflora, resulting in abnormal ripening and poor eye formation in the cheeses.

To overcome the problem of inhibition of desirable organisms in cheese, research was directed toward development of nisin-resistant starter cultures. Attempts to "train" *Streptococcus thermophilus*, various lactobacilli, and propionic acid bacteria to grow in the presence of nisin met with only limited success. Nisin resistance was either not permanent or the cultures became phage sensitive. These investigations were reviewed by Lipinska (1977), who also reported some successful experiments in Poland and the Soviet Union, where a combination of nisin-producing and nisin-resistant starter cultures were used to control butyric acid fermentation in Edam and Gouda cheeses which were either naturally or artificially infected with clostridial spores. In the Western world, nitrate is used to control butyric acid fermentation in cheese. The nitrate is most likely reduced by bacterial action to nitrite, which is the inhibitor, but it can react with naturally occurring amines to form N-nitrosamines (Mirvish, 1977). Because many of these N-nitroso compounds are carcinogenic, the use of high concentrations of nitrate and nitrite in foods is being discouraged. The work reported by Lipinska (1977) offers an interesting alternative to nitrate in cheese manufacture.

There are no recommended nisin levels in milk to be used for cheese manufacture. Lipinska (1977) reported that good-quality cheeses made from nisin-producing and nisin-resistant starter cultures contained between 100 and 400 IU/g. Nisin is seldom added to the milk before fermentation, because most of it would be lost in the whey.

B. Canned Food

In commercially sterilized foods, spores of extremely heat-resistant bacteria such as *B. stearothermophilus* and *Clostridium thermosaccharolyticum* may be present and under suitable conditions could grow out and cause flat-sour spoilage. Because nisin is very stable in an acid environment, Gillespy (1957) used it to control spoilage of canned beans in tomato sauce which were inoculated with spores of *C. thermosaccharolyticum*. At a concentration of 200 IU/g, no spoilage occurred.

In the canning of low-acid products (pH > 4.6), a basic concept is to heat the product to a sufficiently high temperature to eliminate any risk from botulinum poisoning. This is termed the "minimum botulinum cook," which is a heat treatment equivalent to at least F = 2.5, F being the number of minutes that the center of the can is held at 121°C (250°F), assuming instantaneous heating and cooling, or an equivalent amount of heat. Because the heat resistance of spoilage organisms is higher than that of *C. botulinum*, heating in excess of F = 2.5 is practiced with low-acid products, and nisin was used with the "minimum botulinum cook" with the expectation that the lowered thermal processing will not compromise safety or increase the potential for spoilage. At the same time, this process should result in improving product quality and economy. Several successful applications of nisin in these products have been reported from East European countries such as the Soviet Union, Poland, and Hungary. Examples of canned foods in which nisin was successfully used include carrot purée, mushrooms (Vas, 1964; Heinemann et al., 1965), peas (Kiss et al., 1968), soups (Bardsley, 1962), and a number of others which are summarized by Lipinska (1977). In the report of Kiss et al. (1968), a pilot-plant experiment is described in which addition of 200 IU nisin per milliliter to 1 kg of peas in jars allowed a 60% reduction in the heat process, thereby resulting in improved appearance and taste.

In chicken chow mein and cream-style corn artificially contaminated with spores of *C. sporogenes* PA3679 at levels 10^3–10^4 times greater than would occur naturally, Heinemann et al. (1965) noted only partial control of spoilage. Gibbs and Hurst (1964) were also unable to prevent spoilage in heavily contaminated soups, even at a nisin concentration of 500 IU/g. These observations indicate that nisin cannot be used to mask poor-quality raw materials or poor manufacturing practices (Jarvis and Morisetti, 1969).

C. Semipreserved Meats

There are conflicting reports on the usefulness of nisin in semipreserved mildly heat-processed meats. The American and British studies done during 1955–1965 have been reviewed by Hawley (1957, 1962), Marth (1966), Boone (1966), and Jarvis and Morisetti (1969). The East European studies have been reviewed by Lipinska (1977).

It was first thought that nisin might be used as a preservative in canned ham, tongue, meat, and in fish-based sandwich spreads; however, its use in semipreserved meats was not approved in Great Britain because of insufficient evidence for its effectiveness (Anonymous, 1959). Other countries similarly approved the use of nisin in dairy products and certain canned products, but not for meat preservation.

Because of the risk of carcinogenic nitrosamines being produced from nitrite used to preserve meats, especially those that are cooked at high temperatures, such as bacon and some types of ham, researchers have recently been directing their attention toward finding a suitable alternative to nitrite for meat preservation. This has sparked renewed interest in nisin. Caserio et al. (1979b) reported that 75 or 150 ppm nitrite did not completely suppress growth of *Clostridium perfringens* in frankfurters. These were inoculated with 70–200 spores/g and heated for 1 hr at 72°C. However, 75 ppm of nitrite with 200 ppm (8000 IU/g) nisin was effective in preventing outgrowth of the spores. This nisin concentration was much higher than that used for vegetable and dairy products, but, as pointed out by Hurst (1981), previous failures of nisin as a preservative in cured cooked meats could have been attributed to the low concentrations used. Rayman et al. (1981) observed that 75 pp, (3000 IU/g) nisin was superior to 150 ppm nitrite in preventing outgrowth of spores of *C. sporogenes* PA3679 in meat slurries which had been heated to 70°C. Nisin–nitrite combinations were even more effective. Scott and Taylor (1981) noted that of type A, B, or E spores of *C. botulinum*, type A was most resistant to nisin, requiring 1000–2000 IU/ml to cause 50% inhibition of outgrowth on laboratory medium; type E spores were the least resistant, requiring 50–100 IU/ml; while type B spores were intermediate, requiring 500–1000 IU/ml for a 50% inhibition of outgrowth. They also noted that 5000 and 2000 IU/ml were insufficient to prevent spore outgrowth of type A and B spores, respectively, in medium containing cooked meat, and they suggested that the larger amounts of nisin necessary to prevent outgrowth in the meat medium was due to binding of nisin to the meat particles. We have also observed that at the normal pH of most meats (5.8–6.4), up to 12,000 IU of nisin per gram alone or in combination with 60 ppm nitrite did not prevent outgrowth of a mixture of type A or B spores in pork slurries which were heated to 70°C and then held at 25°C. Higher concentrations of nisin were not soluble (Rayman, Malik, and Hurst, unpublished results).

D. Chocolate Milk

Chocolate milk is a low-acid food. It contains cocoa powder which may be contaminated with large numbers of bacterial spores, so that processing at relatively high temperatures is necessary to prevent spoilage. The high-heat treatment required to obtain complete sterility could result in an over-cooked flavor and may be accompanied by stability problems (Fowler and Mc-Cann, 1971). Heinemann et al. (1965) showed that the processing temperature could be significantly reduced if nisin were added as a preservative. They monitored cans of chocolate milk stored at room temperature for 6 months and noted no spoilage in cans that contained as little as 80 IU/ml nisin and which received a heat treatment equivalent to F = 3. Without nisin, it was estimated that a heat treatment of F = 11 would be necessary to obtain the same shelf-life. The reduced heating resulted in a significant saving (center of the can held at 250° for 3 min as opposed to 11 min) and the product had a better flavor, with improved nutritional quality.

Fowler and McCann (1971) suggested that nisin could also be used in preparing milk powder to be shipped to those countries which use it to produce recombined and reconstituted milk. They suggested that a lower than normal heat process could then be used to produce secondary milk products of improved nutritional quality without compromising keeping quality.

REFERENCES

Anonymous. (1959). *Food Standards Committee Report on Preservatives in Food.* Her Majesty's Stationery Office, London, p. 49.

Bailey, F. J., and Hurst, A. (1971). Preparation of a highly active form of nisin from *Streptococcus lactis. Can. J. Microbiol.* 17:61–67.

Baranova, I. P., Yegorov, N. S., Golovkina, G. P., and Grigoryan, A. N. (1980). Use of enzymic hydrolysates of a microbial biomass in media for cultivating *Streptococcus lactis*, the producer of nisin. *Antibiotiki* 25: 735–738.

Barber, R. S., Braude, R., and Hirsch, A. (1952). Growth of pigs given skim milk soured with nisin-producing streptococci. *Nature* 169:200–201.

Bardsley, A. (1962). Antibiotics in food canning. *Food Technol. Aust.* 14: 532–537, 606–611.

Berridge, N. J. (1947). Further purification of nisin. *Lancet* 2:13.

Berridge, N. J. (1949). Preparation of the antibiotic nisin. *Biochem. J. 100:* 486–493.

Berridge, N. J. (1953). The antibiotic nisin and its use in the making and processing of cheese. *Chem. Ind. 1953:*1158–1161.

Berridge, N. J., and Barrett, J. (1952). A rapid method for the turbidimetric assay of antibiotics. *J. Gen. Microbiol.* 6:14–20.

Berridge, N. J., Newton, G. G. F., and Abraham, E. P. (1952). Purification and nature of the antibiotic nisin. *Biochem. J.* 52:529–535.

Bodanszky, M., and Perlman, D. (1964). Are peptide antibiotics small proteins? *Nature* 204:840–844.

Boone, P. (1966). Mode of action and applications of nisin. *Food Manuf. 41:* 49–51.

Campbell, L. L., and O'Brien, R. T. (1955). Antibiotics in food preservation. *Food Technol.* 9:461–465.

Campbell, L. L., and Sniff, E. E. (1959). Effect of subtilin and nisin on the spores of *Bacillus coagulans. J. Bacteriol.* 77:766–770.

Carlson, S., and Bauer, H. H. (1957). A study of problems associated with resistance to nisin, an antibiotic produced by *Streptococcus lactis. Arch. Hyg. Bakteriol.* 141:445–449.

Caserio, G., Ciampella, A., Gennari, M., and Barluzzi, A. M. (1979a). Research on the use of nisin in cooked pork products. *Ind. Aliment. 18:* 157–161.

Caserio, G., Stecchini, M., Pastore, M., and Gennari, M. (1979b). The individual and combined effects of nisin and nitrite on the spore germination of *Clostridium perfringens* in meat mixtures subjected to fermentation. *Ind. Aliment. 18:*894–898.

Cheeseman, G. C., and Berridge, N. J. (1957). An improved method of preparing nisin. *Biochem. J. 65:*603–608.

Cheeseman, G. C., and Berridge, N. J. (1959). Observations on the molecular weight and chemical composition of nisin A. *Biochem. J. 71:*185–194.

Chevalier, R., Fournaud, J., Lefebvre, E., and Mocquot, G. (1957). Mise en évidence des stréptocoques lactiques inhibiteurs et stimulants dans le lait et les fromages. *Ann. Technol. Agric. 2:*117–137.

Cox, G. A. (1934). A simple method for the detecting of "non-acid" milk. *N. Z. J. Agric. 49:*231–234.

Csiszar, J., and Pulay, G. (1956). Studies on *Streptococcus lactis* which produce antibiotics effective against clostridia. I. The activity spectrum of antibiotic-producing strains, and how to increase their antibiotic production. *Proc. 14th. Int. Dairy Congr. 2:*423–430.

Denny, C. B., Sharpe, L. E., and Bohrer, C. W. (1961). Effects of tylosin and nisin on canned-food-spoilage bacteria. *Appl. Microbiol. 9:*108–110.

Egorov, N. S., Baranova, I. P., and Kozlova, Y. I. (1971). Optimization of nutrient medium composition for the production of the antibiotic nisin by *Streptococcus lactis*. *Mikrobiologiya 40:*993–998.

Egorov, N. S., Baranova, I. P., and Kozlova, Y. I. (1976). Influence of purine and pyrimidine bases on the growth of *Streptococcus lactis* and the biosynthesis of nisin. *Mikrobiologiya 45:*100–103.

Fowler, G. G., and McCann, B. (1971). The growing use of nisin in the dairy industry. *Aust. J. Dairy Technol. 26:*44–46.

Frazer, A. C., Sharratt, M., and Hickmann, J. R. (1962). The biological effects of food additives. I. Nisin. *J. Sci. Food Agric. 13:*32–42.

Fuchs, P. G., Zajdel, J., and Dobrzanski, W. T. (1975). Possible plasmid nature of the determinant for production of the antibiotic nisin in some strains of *Streptococcus lactis*. *J. Gen. Microbiol. 88:*189–192.

Galesloot, T. E., and Pette, J. W. (1957). The formation of normal eyes in Edam cheese made by nisin-producing starters. *Neth. Milk Dairy J. 11:* 144–151.

Gibbs, B. M., and Hurst, A. (1964). Limitations of nisin as a preservative in non-dairy foods. In *Microbial Inhibitors in Foods*. N. Molin (Ed.). Almqvist and Wiksell, Stockholm, pp. 151–165.

Gillespy, T. G. (1957). Nisin trials. Fruit and vegetable canning and quick freezing research association. Leaflet No. 3. Chipping Campden, United Kingdom.

Gould G. W. (1964). Effect of food preservatives on the growth of bacterial spores. In *Microbial Inhibitors in Foods*. N. Molin (Ed.). Almqvist and Wiksell, Stockholm, pp. 17–24.

Gould, G. W., and Hurst, A. (1962). Inhibition of *Bacillus* spore development by nisin and subtilin. Proc. 8th. Int. Conf. Microbiol. Abstra. A2–11.

Gowans, J. L., Smith, N., and Florey, H. W. (1952). Some properties of nisin. *Br. J. Pharmacol. 7:*438–449.

Gross, E. (1977). α,β-Unsaturated and related amino acids in peptides and proteins. In *Protein Cross-Linking*, Vol. B. M. Friedman (Ed.). Plenum, New York, pp. 131–153.

Gross, E., and Kiltz, H. H. (1973). The number and nature of α,β-unsaturated amino acids in subtilin. *Biochem. Biophys. Res. Commun. 50*:559–565.

Gross, E., and Morell, J. L. (1967). The presence of dehydroalanine in the antibiotic nisin and its relationship to activity. *J. Am. Chem. Soc. 89*: 2791–2793.

Gross, E., and Morell, J. L. (1968). The number and nature of α,β-unsaturated amino acids in nisin. *FEBS Lett. 2*:61–64.

Gross, E., Morell, J. L., and Craig, L. C. (1969). Dehydroalanyllysine: Identical COOH-terminal structures in the peptide antibiotics nisin and subtilin. *Proc. Nat. Acad. Sci. U.S.A. 62*:952–956.

Hall, R. H. (1963). The production of nisin. Patent specification 916,351.

Hawley, H. B. (1955). The development and use of nisin. *J. Appl. Bacteriol. 18*:388–395.

Hawley, H. B. (1957). Nisin in food technology. *Food Manuf. 32*:370–376, 430–434.

Hawley, H. B. (1962). The uses of antibiotics in canning. In *Antibiotics in Agriculture*. M. Woodbine (Ed.). Butterworth, London, pp. 272–288.

Hawley, H. B., and Hall, R. H. (1957). The production of nisin. Patent specification 844,782.

Heinemann, B., Voris, L., and Stumbo, C. R. (1965). Use of nisin in processing food products. *Food Technol. 19*:592–596.

Hirsch, A. (1950). The assay of the antibiotic nisin. *J. Gen. Microbiol. 4*: 70–83.

Hirsch, A. (1951a). Various antibiotics from one strain of *Streptococcus lactis*. *Nature 167*:1031–1032.

Hirsch, A. (1951b). Growth and nisin production of a strain of *Streptococcus lactis*. *J. Gen. Microbiol. 5*:208–221.

Hirsch, A. (1954). Some polypeptide antibiotics. *J. Appl. Bacteriol. 17*: 108–115.

Hirsch, A., and Mattick, A. T. R. (1949). Some recent applications of nisin. *Lancet 2*:190–197.

Hirsch, A., Grinsted, E., Chapman, H. R., and Mattick, A. T. R. (1951). A note on the inhibition of an anaerobic sporeformer in Swiss-type cheese by a nisin-producing *Streptococcus*. *J. Dairy Res. 18*:205–206.

Hitchins, A. D., Gould, G. W., and Hurst, A. (1963). The swelling of bacterial spores during germination and outgrowth. *J. Gen. Microbiol. 30*: 445–453.

Hoyle, M., and Nichols, A. A. (1948). Inhibitory strains of lactic streptococci and their significance in the selection of cultures for starter. *J. Dairy Res. 15*:398–408.

Hurst, A. (1966). Biosynthesis of the antibiotic nisin by whole *Streptococcus lactis* organisms. *J. Gen. Microbiol. 44*:209–220.

Hurst, A. (1967). Function of nisin and nisin-like basic proteins in the growth cycle of *Streptococcus lactis*. *Nature 214*:1232–1234.

Hurst, A. (1972). Interactions of food-starter cultures and food-borne pathogens: The antagonism between *Streptococcus lactis* and sporeforming microbes. *J. Milk Food Technol. 35*:418–423.

Hurst, A. (1978). Nisin: Its preservative effect and function in the growth cycle of the producer organism. In *Streptococci*. F. A. Skinner and L. B. Quesnel (Eds.). Academic, London, pp. 297–314.

Hurst, A. (1981). Nisin. *Adv. Appl. Microbiol. 27*:85–123.

Hurst, A., and Collins-Thompson, D. (1979). Food as a bacterial habitat. *Adv. Microb. Ecol. 3*:79–133.

Hurst, A., and Kruse, H. (1972). Effect of secondary metabolites on the organisms producing them: Effect of nisin on *Streptococcus lactis* and enterotoxin B on *Staphylococcus aureus*. *Antimicrob. Agents Chemother.* 1:277–279.

Hurst, A., and Paterson, G. M. (1971). Observations on the conversion of an inactive precursor protein to the antibiotic nisin. *Can. J. Microbiol.* 17:1379–1384.

Ingram, L. (1969). Synthesis of the antibiotic nisin: Formation of lanthionine and β-methyl-lanthionine. *Biochim, Biophys. Acta 184*:216–219.

Ingram, L. (1970). A ribosomal mechanism for synthesis of peptides related to nisin. *Biochim. Biophys. Acta 224*:263–265.

Jarvis, B. (1967). Resistance to nisin and production of nisin-inactivating enzymes by several species of *Bacillus*. *J. Gen. Microbiol.* 47:33–48.

Jarvis, B., and Mahoney, R. R. (1969). Inactivation of nisin by alpha-chymotrypsin. *J. Dairy Sci.* 52:1448–1450.

Jarvis, B., and Morisetti, M. D. (1969). The use of antibiotics in food preservation. *Int. Biodeterior. Bull.* 5:39–61.

Jarvis, B., Jeffcoat, J., and Cheeseman, G. C. (1968). Molecular weight distribution of nisin. *Biochim. Biophys. Acta. 168*:153–155.

Kalra, M. S., Kuila, R. K., and Ranganathan, B. (1973). Activation of nisin production by UV-irradiation in a nisin-producing strain of *Streptococcus lactis*. *Experimentia 29*:624–625.

Kiss, I., Kiss, K. N., Farkas, J., Fabri, I., and Vas, K. (1968). Further data on the application of nisin in pea preservation. *Elelmiszertudomany* 2:51–57.

Kooy, J. S. (1952). Strains of *Lactobacillus plantarum* which inhibit the activity of the antibiotics produced by *Streptococcus lactis*. *Neth. Milk Dairy J. 6*:323–330.

Kooy, J. S., and Pette, J. W. (1952). The inhibition of butyric acid fermentation in cheese by using antibiotic-producing streptococci as a starter. *Neth. Milk Dairy J. 6*:317–322.

Kozak, W., and Dobrzanski, W. T. (1977). Growth requirements and the effect of organic components of the synthetic medium on the biosynthesis of the antibiotic nisin in *Streptococcus lactis* strain. *Acta Microbiol. Pol. 26*:361–368.

Kozak, W., Rajchert-Trzpil, M., and Dobrzanski, W. T. (1973). Preliminary observations on the influence of proflavin, ethidium bromide and elevated temperature on the production of the antibiotic nisin by *Streptococcus lactis* strains. *Bull. Acad. Pol. Sci. 21*:811–817.

Kozak, W., Rajchert-Trzpil, M., and Dobrzanski, W. T. (1974). The effect of proflavin, ethidium bromide and an elevated temperature on the appearance of nisin-negative clones in nisin-producing strains of *Streptococcus lactis*. *J. Gen. Microbiol. 83*:295–302.

Kozlova, Y. I., Golikova, T. I., Baranova, I. P., and Egorov, N. S. (1979). Investigation on the influence of KH_2PO_4 on the growth of *Streptococcus lactis* and nisin synthesis at constant pH values of the medium. *Mikrobiologiya 48*:443–446.

Lewis, J. C., Michener, H. D., Stumbo, C. R., and Titus, D. R. (1954). Antibiotics in food processing: Additives accelerating the death of spores by moist heat. *J. Agric. Food Chem.* 2:298–302.

Linnett, P. E., and Strominger, J. L. (1973). Additional antibiotic inhibitors of peptidoglycan synthesis. *Antimicrob. Agents Chemother.* 4:231–236.

Lipinska, E. (1956). Microbiological methods of control of butyric acid fermentation in cheese. *Acta Microbiol. Pol.* 5:271–275.

Lipinska, E. (1977). Nisin and its applications. In *Antibiotics and Antibiosis in Agriculture.* M. Woodbine (Ed.). Butterworth, London, pp. 103–130.

Lipinska, E., Strzalkowska, M., Goettlich, W., and Soltys, W. (1962). Application of nisin resistant starters to the production of Edam cheese. *Proc. 16th Int. Dairy Congr.* 2:849–860.

McClintock, M., Serres, L., Marzolf, J. J., Hirsch, A., and Mocquot, G. (1952). Action inhibitrice des streptocoques producteurs de nisine sur le développement des sporulés anaérobies dans le fromage de Gruyère fondu. *J. Dairy Res.* 19:187–193.

Marth, E. H. (1966). Antibiotics in foods—Naturally occurring, developed and added. *Residue Rev.* 12:65–161.

Mattick, A. T. R., and Hirsch, A. (1944). A powerful inhibitory substance produced by group N streptococci. *Nature 154*:551.

Mattick, A. T. R., and Hirsch, A. (1946). Sour milk and the tubercle bacillus. *Lancet 1*:417–418.

Mattick, A. T. R., and Hirsch, A. (1947). Further observations on an inhibitory substance (nisin) from lactic streptococci. *Lancet 2*:5–12.

Meanwell, L. J. (1943). The influence of raw milk quality on "slowness" in cheesemaking. *Proc. Soc. Agric. Bacteriol.* 19:21.

Mirvish, S. S. (1977). N-Nitroso compounds: Their chemical and in vivo formation and possible importance as environmental carcinogens. *J. Toxicol. Environ. Health* 2:1267–1277.

Mocquot, G., and Lefebvre, E. (1956). A simple procedure to detect nisin in cheese. *J. Appl. Bacteriol.* 19:322–323.

Newton, G. G. F., Abraham, E. P., and Berridge, N. J. (1953). Sulphur-containing amino-acids of nisin. *Nature 171*:606.

O'Brien, R. T., Titus, D. S., Devlin, K. A., Stumbo, C. R., and Lewis, J. C. (1956). Antibiotics in food preservation. II. The influence of subtilin and nisin on the thermal resistance of food spoilage organisms. *Food Technol.* 10:352–355.

Oxford, A. E. (1944). Diplococcin, an anti-bacterial protein elaborated by certain streptococci. *Biochem. J.* 38:178–182.

Pulay, G. (1956). Research on *Streptococcus lactis* strains producing antibiotics against clostridia. II. Unfavourable effects of nisin-type antibiotics on the quality of cheese. *Proc. 14th Int. Dairy Congr.* 2:432–433.

Ramsier, H. R. (1960). The action of nisin on *Clostridium botulinum. Arch. Mikrobiol.* 37:57–94.

Rayman, M. K., Aris, B., and Hurst, A. (1981). Nisin: A possible alternative or adjunct to nitrite in the preservation of meats. *Appl. Environ. Microbiol.* 41:375–380.

Reisinger, P., Seidel, H., Tschesche, H., and Hammes, W. P. (1980). The effect of nisin on murein synthesis. *Arch. Microbiol.* 127:187–193.

Rogers, L. A. (1928). The inhibitory effect of *Streptococcus lactis* on *Lactobacillus bulgaricus. J. Bacteriol.* 16:321–325.

Rogers, L. A., and Whittier, E. O. (1928). Limiting factors in lactic fermentation. *J. Bacteriol.* 16:211–214.

Scott, V. N., and Taylor, S. L. (1981). Effect of nisin on the outgrowth of *Clostridium botulinum* spores. *J. Food Sci.* 46:117–120.

Shtenberg, A. I., and Ignat'ev, A. D. (1970). Toxicological evaluation of some combinations of food preservatives. *Food Cosmet. Toxicol.* 8:369–380.

Steiner, D. F. (1979). Processing of protein precursors. *Nature 279*:674–675.

Stumbo, C. R., Voris, L., Skaggs, B. G., and Heinemann, B. (1964). A procedure for assaying residual nisin in food products. *J. Food Sci.* 29:859–861.

Tagg, J. R., and Wannamaker, L. W. (1978). Streptococcin A-FF22: Nisin-like antibiotic substance produced by group A streptococcus. *Antimicrob. Agents Chemother. 14*:36–39.

Tagg, J. R., Dajani, A. S., and Wannamaker, L. W. (1976). Bacteriocins of gram positive bacteria. *Bacteriol. Rev. 40*:722–756.

Taylor, J. I., Hirsch, A., and Mattick, A. T. R. (1949). The treatment of bovine streptococcal and staphylococcal mastitis with nisin. *Vet. Rec. 61*:197–198.

Thorpe, R. H. (1960). The action of nisin on spoilage bacteria. I. The effect of nisin on the heat resistance of *Bacillus stearothermophilus*. *J. Appl. Bacteriol. 23*:136–143.

Tramer, J. (1964). The inhibitory action of nisin on *Bacillus stearothermophilus*. In *Microbial Inhibitors in Foods*. N. Molin (Ed.). Almqvist and Wiksell, Stockholm, pp. 25–33.

Tramer, J., and Fowler, G. G. (1964). Estimation of nisin in foods. *J. Sci. Food Agric. 15*:522–528.

Vas, K. (1964). Nisin in the food industry. *Dtsch. Lebensm. Rundsch. 60*: 63–67.

White, R. J., and Hurst, A. (1968). The location of nisin in the producer organism, *Streptococcus lactis*. *J. Gen. Microbiol. 53*:171–179.

Whitehead, H. R. (1933). A substance inhibiting bacterial growth produced by certain strains of lactic streptococci. *Biochem. J. 27*:1793–1800.

Whitehead, H. R., and Riddet, W. (1938). Slow development of acidity in cheese manufacture. *N. Z. J. Agric. 46*:225–229.

Willimowska-Pelc, A., Olichwier, Z., Malicka-Blaszkiewicz, M., and Majbaum-Katzenellenbogen, W. (1976). The use of gel-filtration for the isolation of pure nisin from commercial products. *Acta Microbiol. Pol. 25*:71–77.

Winkler, S., and Frölich, M. (1958). Investigations of emmental cheese with nisin. *Milchwiss. Ber. 8*:87–96.

22

ANTIMYCIN A: PROPERTIES, BIOSYNTHESIS, AND FERMENTATION

CLAUDE VÉZINA* AND S. N. SEHGAL *Ayerst Research Laboratories, Princeton, New Jersey*

Present affiliation: Institut Arnand-Frappier, Ville de Laval, Quebec, Canada

I. INTRODUCTION

Antimycin A is an antifungal and piscicidal antibiotic. In 1948, Leben and Keitt isolated an unidentified streptomycete from soil and reported its high antifungal activity against plant pathogens. The active principle was isolated in crystalline form 1 year later by Dunshee et al. (1949) and named antimycin A. The minimum inhibitory concentration (MIC) against saprophytic and pathogenic fungi ranged from 0.02 to 1 µg/ml (Koaze et al., 1956; Lockwood et al., 1954; Nakayama et al., 1956; Sakagami et al., 1956; Watanabe et al., 1957). The MIC against *Venturia inaequalis*, the apple scab organism, and *Ceratocystis ulmi*, the causative agent of the Dutch elm disease, is about 0.125 µg/ml. Field trials (P. H. Derse, personal communication) showed that infected elms could be protected by antimycin A. The antibiotic is active against several yeasts and filamentous fungi, including the dermatophytes, pathogenic to man and animals. H. A. Baker and A. Sidorowicz (personal communication) conducted experiments with guinea pigs experimentally infected with *Trichophyton mentagrophytes* and found that antimycin A, administered topically, was extremely toxic to the animals. This unacceptable toxicity precluded its use as an antifungal antibiotic for use in man and animals. Antimycin A has very little or no activity against bacteria. Early studies were reviewed by Strong (1956).

Soon after its isolation, antimycin A was found by Ahmad et al. (1950) to inhibit succinoxidase at very low concentration (2×10^{-8} M) and it became an important biochemical tool which contributed significantly to the understanding of respiration (Rieske and Zaugg, 1962). This mode of action instigated Derse and Strong (1963) to evaluate the toxicity of antimycin A to fish: they found that the antibiotic was more toxic to fish (by immersion) by several orders of magnitude (LC_{50}, 0.7 ppb) than to mammals (intraperitoneal LD_{50}, 1–2 mg/kg; oral LD_{50}, 2–55 mg/kg) and birds (oral LD_{50}, 2–200 mg/kg).

The piscicidal properties of antimycin A were studied extensively at the Fish Control Laboratories in La Crosse, Wisconsin, where the antibiotic was developed into an exceptional tool in fish management. At the same time, high-production strains and improved fermentation conditions were selected and suitable formulations prepared at Ayerst Research Laboratories. Antimycin A was reviewed by Lennon and Vézina (1967, 1971).

II. CHEMISTRY AND ASSAY METHODS

A. Chemistry

Dunshee et al. (1949) reported the isolation of antimycin A as an apparently homogeneous crystalline antibiotic. Subsequent studies of Lockwood et al. (1954) recognized antimycin A to be a mixture of four cocrystallizable components which were designated as A_1–A_4 in order of their increasing polarities. The two major components of the mixture, antimycins A_1 and A_3, were isolated in pure form (Harada et al., 1958). One year later Liu and Strong (1959) further obtained pure fraction A_2. The structures of antimycins A_1 and A_3, the two major components of the complex, were determined by the degradation studies of Strong, van Tamelen, and their collaborators (van Tamelen et al., 1961; Dickie et al., 1963). The basic structure of antimycin A consists of a 3-formamidosalicyclic acid residue linked through an amide group to a dilactone ring bearing an acyl side chain and an alkyl side chain (Fig. 1). Under alkaline degradation, the molecule is broken into blastmycic acid and C_{14} or

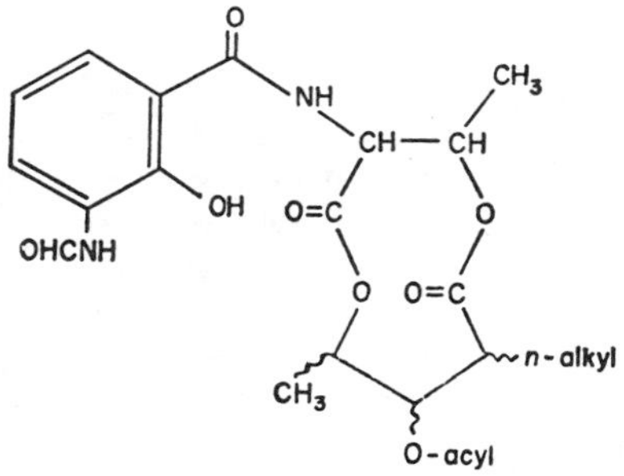

Component		n-alkyl	acyl
A_0			
	(a)	n-hexyl	hexanoyl
	(b)	n-butyl	heptanoyl
	(c)	octyl	butyryl
	(d)	heptyl	isovaleryl
A_1		n-hexyl	isovaleryl
A_2		n-hexyl	butyryl
A_3		n-butyl	isovaleryl
A_4		n-butyl	butyryl
A_5		ethyl	isovaleryl
A_6		ethyl	butyryl

Figure 1 Structures of antimycin A_0–A_6, according to Schilling et al. (1970).

C_{16} antimycin lactone. Blastmycic acid is further deformylated to antimycic acid. Kleupfel et al. (1970) used a countercurrent distribution technique to obtain gram quantities of the four major components, A_1, A_2, A_3, and A_4, of the antimycin complex. In addition, three other components, designated A_0, A_5, and A_6 in order of their increasing polarities and comprising less than 1% of the complex, were also detected. All seven components, A_0–A_6, were subjected to pyrolysis, and pyrolysates separated and collected by gas-liquid chromatography (Schilling et al., 1970). The pyrolytic fractions were submitted to mass spectrometry; on the basis of their mass number, infrared, ultraviolet, and nuclear magnetic resonance spectra, and microchemical analysis, previously published structures of components A_1 and A_3 were confirmed; the structures of components A_2 and A_4 were elucidated and the structures of A_0, A_5, and A_6, which were not available in pure form, were proposed (Fig. 1).

The analysis of antimycin A produced by various strains of *Streptomyces* sp. revealed that mutations which lead to increased production were not necessarily, but sometimes, accompanied by a change in the centesimal composition of the complex; however, the major component was always A_1 (Kluepfel et al., 1970). An antimycin complex consisting mainly of antimycin A_3 (blastmycin) was isolated from *Streptomyces blastmyceticus* (Watanabe et al., 1957).

Tappel (1960) reported that the major components of antimycin A complex have same activity as electron transport inhibitors. This was confirmed by Kluepfel et al. (1970), who reported that the four major components, A_1–A_4, exhibited identical fungicidal activity against *Saccharomyces cerevisiae* Y-30 and same teleocidal activity against goldfish. Differences in antifungal activity between antimycin A_1 and A_3 were reported by early workers (Liu and Strong, 1959; Harada et al., 1959), the A_3 component apparently being several times more active than the A_1 component. Kluepfel et al. (1970) explained this discrepancy in terms of different diffusion rates of components in the agar gel used in the cylinder plate method.

Kinoshita and Umezawa (1970) reported a total synthesis of the antimycin homolog, dehexyldeisovaleryloxy antimycin A_1, in which alkyl and acyl groups attached to the dilactone ring were replaced with hydrogens. They also made an antimycin analog (Kinoshita et al., 1971) with a 15-member dilactone ring in two stereoisomeric forms. Rieska (1980) in a recent review has compared the biological activities of antimycin homologs and analogs. The synthetic homologs have the same activity as antimycin A. The 15S epimer of 15-member ring analogs of antimycin A have the same activity as antimycin A, whereas the 15R epimer is devoid of activity.

B. Assay Methods

The antimycin A complex is a colorless crystalline lipophilic antibiotic. It is freely soluble in acetone, chloroform and other chlorinated solvents, and ethyl acetate; moderately soluble in benzene, methanol, and ethanol; sparingly soluble in petroleum ether and hexane; and practically insoluble in water. In alcohol solutions it has a typical ultraviolet spectrum with maxima at 320 and 225 nm.

Lockwood et al. (1954) described the microbiological assay of antimycin A complex using *Saccharomyces cerevisiae* Y-30 by the standard cylinder plate method. The range of sensitivity of this method is between 1 and 10 µg/ml. Murphy and Derse (Wisconsin Alumni Research Foundation, personal communication) modified the yeast inoculum to increase the sensitivity 100-fold and to permit determination of antimycin A in extremely dilute solutions. Schneider et al. (1952) employed turbidimetric and manometric assay procedures using *Saccharomyces cerevisiae* as test organism. The sensitivity of these methods is between 1 and 6 µg/ml. Because of the nature of the antimycin A complex, where even slight changes in the centesimal composition would alter the mean diffusion coefficient in agar gel, the microbiological assays of antimycin A are not satisfactory for fermentation broths.

An alcoholic solution of antimycin A is fluorescent with an excitation wavelength of 350 nm and an uncorrected emission wavelength of 420 nm. Sehgal et al. (1965) reported that maximal fluorescence is obtained when the alcoholic solution is at pH 8.0. They utilized this property to develop a spectrophotofluorometric assay of antimycin A. Sehgal and Vézina (1967) automated this fluorometric assay using standard Technicon Auto Analyzer components. They were able to assay fermentation broths at a rate of 30 samples per hour and the system could be adjusted to assay broth samples in the range of 50–1000 to 2000–8000 µg/ml of antimycin A. Finally, a specific and extremely sensitive assay (2–10 ppb) was developed by Walker et al. (1964) using goldfish as the test organism and the teleocidal property of antimycin A as indicator.

III. BIOSYNTHESIS

A. Growth Experiments

Kannan et al. (1968) used a complete medium and devised a synthetic medium containing glycine, L-tryptophan, and DL-alanine for *Streptomyces antibioticus* NRRL 2838, an actinomycin and antimycin A producer. They found that tryptophan added to the synthetic medium was well assimilated but not required for growth; no absolute relationship existed between antimycin A production and growth. Řeháček et al. (1968a) isolated from the same organism a new microbial metabolite, N-formylaminosalicyclic acid, which, they believed, is a precursor of the antibiotic; this metabolite was isolated and further characterized by Řeháček and Švarc (1968). Ramankutty et al. (1969) observed that 5 g of potassium phosphate per liter of medium led to optimal yields when added at 18 hr of incubation; the role of phosphate was found to lie in the conversion of poly-β-hydroxybutyrate into a readily assimilable source of carbon. Neft and Farley (1972a), using the chemically defined medium of Ramankutty et al. (1969), in which ferrous sulfate was replaced by zinc sulfate, manganese chloride, and cobalt chloride, reported that iron was highly inhibitory to *Streptomyces* sp. AY B-265 (strain M-106 in Fig. 5), and could not be added to the medium for maximal yields; tryptophan enhanced antimycin A production. A. Kudelski (Ayerst Research Laboratories, personal communication), using complete medium 4A (Table 1) for *Streptomyces* sp. M-406 (Fig. 5), found that among a large number of substances added separately to medium 4A only DL-isoleucine at 2 mg/ml yielded a 10% increase in antimycin A (2360 μg/ml for medium 4A); shikimic acid at the same concentration somewhat inhibited the production (1520 μg/ml).

Vézina et al. (1976) devised a chemically defined medium (medium 8 in Table 1) for strains M-506 and M-706 of *Streptomyces* sp. (Fig. 5). None of the ingredients listed could be deleted without reducing yields of antimycin A drastically. A. Kudelski (personal communication) tested a large number of substances, separately added to medium 8 or added in combination, with the following results: medium 8 (strain M-706), 1000 μg antimycin A/ml; medium 8 + methyl myristate (2.5 g/liter), 1800 μg/ml; medium 8 + shikimate (2.5 g/liter), 1600 μg/ml; medium 8 + DL-threonine (2.5 g/liter), 900 μg/ml; medium 8 + methyl myristate + shikimate, 2850 μg/ml; medium 8 + methyl myristate + DL-threonine, 1750 μg/ml; medium 8 + shikimate + DL-threonine, 1850 μg/ml; medium 8 + methyl myristate + shikimate + DL-threonine, 2650 μg/ml. Thus the addition of myristate and shikimate to medium 8 tripled the yields of antimycin A; these had no effect when added to the production medium described in the Secs. IV.B and C on fermentation. The time course of antimycin A production by *Streptomyces* sp. M-706 in medium 8 enriched with methyl myristate and shikimate is shown in Figure 2.

B. Studies with ^{14}C-Labeled Precursors

Incorporation studies with ^{14}C]formic, [1-14]C]acetic, and [2-14]C]pyruvic acids by Birch et al. (1962) led to the following conclusions (Fig. 3): Acetic acid added to a complete medium inoculated with *S. kitazawaensis* was incorporated into moieties A, B, C, and D, whereas pyruvic acid was incorporated into moieties B and C only. Therefore moiety A would originate through the acetate–malonate pathway, and moiety B from pyruvate directly. Moiety C (the acyl fragment) would be from the degradation of valine, leucine, and isoleucine; this is confirmed by the yield increase observed by A. Kudelski (personal communication) when isoleucine was added to a complex medium. Moiety D (threonine) originates from the aspartate pathway. Fragment E would be from shikimate, and moiety F from the C1 unit transfer mechanism; however, ^{14}C]formic acid was not incorporated. Řeháček et al. (1968a) found high incorporation of [2-14]C]tryptophan and [5-14]C]tryptophan into formamidosalicylic acid and fragment E of antimycin A (Fig. 3) by *S. antibioticus* NRRL 2838, thus confirming the hypothesis of Birch et al. (1962) that this fragment is formed from shikimic acid. Neft and Farley (1972b), using *Streptomyces* sp. AY B-265 (strain M-106 in Fig. 5), found that tryptophan was incorporated to a larger extent than phenylalanine and shikimate into moiety E, but it did not enhance yields of antimycin A; they also showed that C2 of the tryptophan ring was incorporated into the formamidocarbonyl (moiety F) of antimycin A.

S. N. Sehgal (unpublished data) inoculated synthetic medium 8 (Table 1) enriched with methyl myristate with *Streptomyces* sp. M-506 (Fig. 5). After 3 days of incubation [U-L-14]C]threonine and [U-DL-14]C]shikimic acid were added to separate flasks and incubation was continued for 3 days. Antimycin A was extracted and purified as usual, and then degraded according to the methods of Tener et al. (1953a,b). Incorporation of threonine and shikimic acid was 2.5 and 2% respectively; 90% of the radioactivity of added labeled threonine was recovered in the threonine fraction, and 89% of the

Table 1 Fermentation Media and Conditions

Ingredients (g/liter) and conditions	Medium and condition combination number								
	1	2	3	4	4A	5	6	7	8a
Soy flour (44% protein)b	40	40	40						
Special X soy flourc				20	40				
Nutrisoy 220 soy flourc									
Cerelosed						60	60	60	
Lactose	20	20	20	20	20	20	20	20	
Glucose							5	5	
Sodium citrate									30
Soybean oil									8.1
Sperm oil							15		
Fresh yeast								20	
(NH4)2SO4						5		5	
CaCO3	1.5	1.5	1.5	6	3	3	6	6	8.1
K2HPO4	1.5	1.5	1.5	1.5	1.5	1.5	3	3	1.0
									0.5

Trace elements solutions[e] (ml)				0.1				
$MgSO_4 \cdot 7H_2O$								0.25
$ZnSO_4 \cdot 7H_2O$								0.05
$MnSO_4 \cdot 7H_2O$								0.0025
$CuSO_4 \cdot 7H_2O$								0.003
Tap water to (liter)	1	1	1	1				
H_2O to (liter)					1	1	1	1
pH (after sterilization)	6.5	7.0	7.0	7.0	7.5	7.4	7.4	7.0
Temperature (°C)	28	25	25	25	25	25	25	25
Volume (ml/500-ml flask)	100	50	50	50	50	50	50	50

[a] Synthetic medium.

[b] Staley Manufacturing Co., Decatur, Illinois.

[c] Archer Daniels Midland Co., Minneapolis, Minnesota, Special X (50% protein, 6–8% fat), Nutrisoy 220 (43% protein, 22–23% fat).

[d] Corn Products Corporation, New York. Cerelose is a pharmaceutical grade of glucose.

[e] Trace elements solutions contained (mg/100 ml), respectively, $Na_2B_4O_7 \cdot 10H_2O$, 88 (10 mg B); $CuSO_4 \cdot 5H_2O$, 393 (100 mg Cu); $MnCl_2 \cdot 4H_2O$, 72 (20 mg Mn); $(NH_4)_6Mo_7O_{24} \cdot 4H_2O$, 37 (20 mg Mo); $ZnSO_4 \cdot 7H_2O$, 8807 (2000 mg Zn); and $FeNH_4(SO_4)_2 \cdot 12H_2O$, 1725 (200 mg Fe). Medium 4A contained 0.1 ml of each solution per liter.

Source: From Vézina et al. (1976).

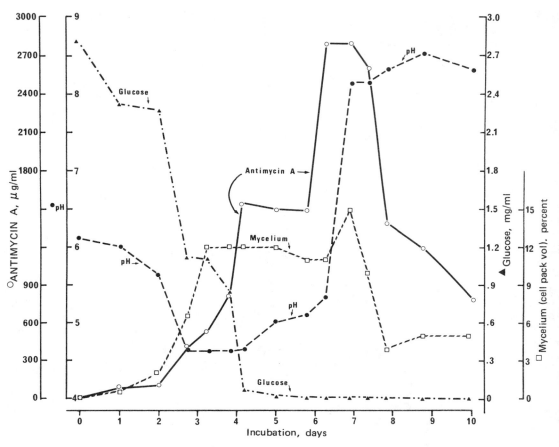

Figure 2 Time course of antimycin A fermentation with *Streptomyces* sp. M-706 in medium 8 (Table 1) enriched with 0.25% shikimic acid.

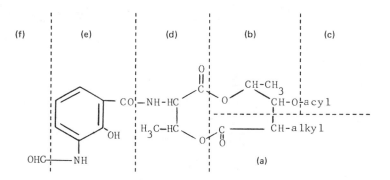

Figure 3 Biosynthesis of antimycin A. (From Birch et al., 1962.)

labeled shikimic acid in the 3-aminosalicylic acid moiety. These results support the hypothesis that threonine is incorporated as such in the dilactone, while shikimic acid is the precursor of the salicylic acid moiety.

To our knowledge no attempt has been made to isolate the enzymes of the antimycin A biosynthetic pathway.

C. Genetic Determinants of Antimycin A Biosynthesis

El-Kersh and Vézina (1981) have tried to establish the chromosomal or plasmidic nature of the genes involved in the biosynthesis of antimycin A and melanin. Their results are summarized in Figure 4. In prototroph M-506, the genes coding for antimycin A (*ant*) and melanin (*mel*) formation were completely resistant to all plasmid "curing" treatments tested: Thus they would be localized in the chromosome. In all the auxotrophs descended from strain M-506 (three are shown in Fig. 4: B-1069 *gua*, B-1138 *his*, and B-1086 *arg trp*), *ant* and *mel* were lost at high frequency, either spontaneously ($\sim$7%) or upon "curing" treatments (25—84%): In these auxotrophs *ant* and *mel* would exist autonomously as extrachromosomal elements. When the auxotrophs were reverted to prototrophy (*gua* in B-1069 probably resulted from a deletion, since no revertants could be isolated), *ant* and *mel* tended to be reintegrated in the chromosome; they could not be eliminated by "curing" treatments in re-

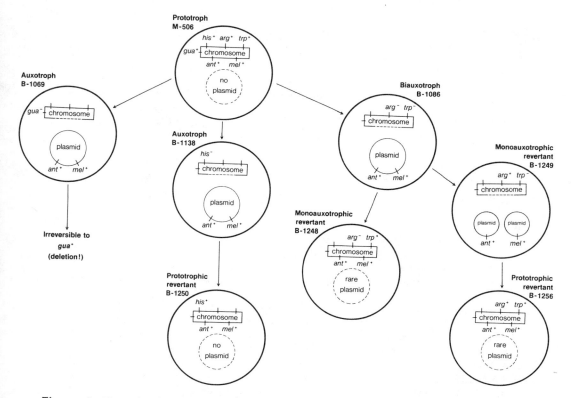

Figure 4 Hypothetical localization of antimycin A (*ant*) and melanin (*mel*) determinants in a family of auxotrophs and revertants descended from *Streptomyces* sp. M-506. (From El-Kersh and Vézina, 1981.)

vertants B-1250 and B-1256 (Fig. 4). In the monoauxotrophic revertants isolated from the biauxotroph B-1086, *ant* and *mel* either remained on plasmids (B-1249 *arg*+ *trp*-) or were relocalized in the chromosome (B-1248 *arg*- *trp*+). This episomal movement in and out of the chromosome could occur by translocation mechanisms. Since they are generally "cured" together, *ant* and *mel* would code for some common functions probably involving the bioconversion of tryptophan into the salicylic moiety of antimycin A and of tyrosine into melanin. Plasmid DNA (6×10^6–7×10^6 daltons) could be isolated from all Ant+ auxotrophs, never from the Ant+ prototroph or from Ant- strains; however, its biological activity could not be demonstrated.

IV. FERMENTATION, EXTRACTION, RECOVERY, AND FORMULATION

A. Strain Improvement

From 16 different antimycin A-producing streptomycetes, Vézina et al. (1976) selected *Streptomyces* sp. M-106 for the development of an improved antimycin A-producing strain. In the strain improvement program emphasis was placed on increasing the power of the selective system. First, they used a rapid and convenient automated fluorometric method of assay of antimycin A (Sehgal and Vézina, 1967). Second, at each round of mutagenesis, several improved strains were compared not only for productivity, but also for early dense conidiation, absence of pigment, and growth rate; only one strain which represented a distinct advantage over the parent was chosen for further selection. Third, since the fermentation environment plays an important role in the expression genetic determinants, fermentation media and conditions were periodically revised to best suit the organism selected for further improvement. The various media used are given in Table 1. The improved antimycin A production in shake flasks was evaluated in aerated–agitated fermenters. Throughout the antimycin improvement program, a rigid maintenance, propagation, and preservation scheme was followed. This program resulted in major yield improvement from 0.62 to 9.75 g/liter, an overall 125-fold increase. The program is summarized in Figure 5.

B. Fermentation in Shake Flasks

Vézina et al. (1976) have reported optimization of medium and fermentation conditions in shake flasks. The critical factors for the medium were absence of phosphate and organic salts, proper choice of the organic nitrogen source and the ratio of organic to inorganic nitrogen, addition of calcium carbonate, and the presence of a rapidly utilizable carbon source, glucose. Addition of a slowly utilizable carbon source, lactose, and a noncarbohydrate carbon source, soybean oil, were highly beneficial. The time course of a typical flask fermentation is illustrated in Figure 6. It shows three phases of fermentation. During the first phase, from inoculation to the third day, the organism grows to almost its maximal value, glucose and ammonia nitrogen are almost completely utilized, and pH drops to 4.3, antimycin A is produced at a rapid rate between the second and the third day, which corresponds to the late growth period of the organism. In the second phase, from the third to the seventh day, the mycelium reaches its maximal mass, lactose utilization begins at the depletion of glucose, ammonia nitrogen remains low, and pH does not change appreciably. During this phase there is a slow and steady increase in the titer of antimycin A. During the last phase, from the seventh day to the end

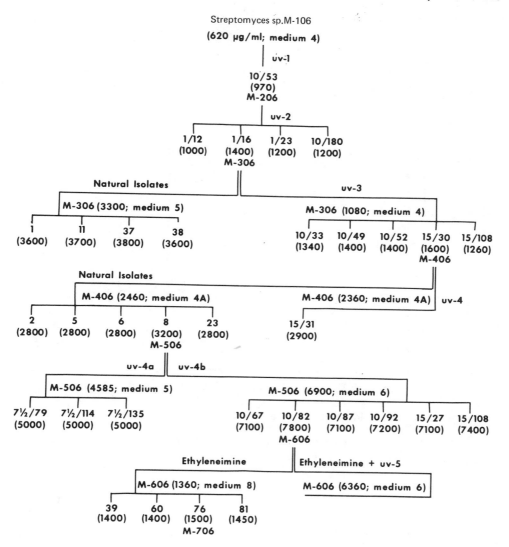

Figure 5 Family of antimycin A-producing strains of *Streptomyces* sp. M-106. Yields (µg/ml) are given in parentheses; refer to Table 1 for medium composition. (From Vézina et al., 1976.)

of fermentation, ammonia nitrogen is produced by the organism to reach, on the eleventh day, a value higher than that at the beginning of fermentation. This corresponds to an autolytic phase accompanied by the rise in the pH of the medium. The rate of antimycin A production remains constant and maximum titers are reached on the ninth day. The titers of antimycin A drop after the ninth day as the pH of the medium rises above 8, a value at which antimycin A is rapidly destroyed.

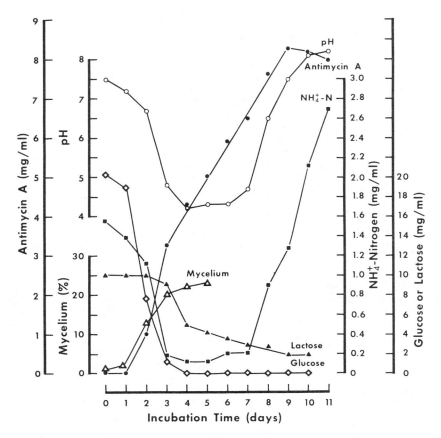

Figure 6 Time course of antimycin A fermentation by *Streptomyces* sp. M-506 in medium 6 (Table 1). (From Vézina et al., 1976.)

C. Fermentation in Aerated—Agitated Fermenters

Our laboratory has reported the optimization of antimycin A fermentation on a pilot-plant scale in aerated—agitated vessels (Sehgal et al., 1976; Sehgal, 1977). The fermenters employed were 250-liter New Brunswick fermenters, equipped with automatic antifoam addition systems and pH recorders/controllers. The fermenters were charged with 160 liters of production medium consisting of (g/liter tap water): Nutrisoy 220 soy flour (Archer Daniels Midland Co., Minneapolis, Minnesota), 60; Cerelose, 20; $(NH_4)_2SO_4$, 6; $CaCO_3$ (U.S.P.), 3; lard oil, 2 ml; pH 7.1–7.2. The role of lard oil was to prevent excessive foaming during sterilization, which was done at 121°C for 45 min under an agitation of 150 rpm. The lard oil was rapidly utilized after inoculation.

The fermenters were inoculated with 2% of 18-hr-old second-stage inoculum. The fermentation was done at 25°C, 250 rpm agitation, and 0.5 vol/vol per minute aeration. During fermentation, foaming was controlled by on-demand addition of antifoam agent DF-143PX (Mazer); only 0.3–0.4% was needed during the course of fermentation. A typical time course of fermentation in enriched production medium is shown in Figure 7. The available carbohy-

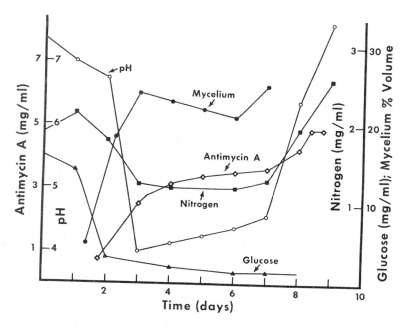

Figure 7 Time course of antimycin A fermentation with *Streptomyces* sp. M-506 in 250-liter fermenters. The production medium was enriched with 0.5% lactose and 1.5% soybean oil. (From Sehgal et al., 1976.)

drate was nearly exhausted in the first 48 hr, resulting in a sharp drop in pH (pH 4 at 72 hr). Antimycin A was produced rapidly. From the third to the sixth day, pH, mycelial mass, nonprotein nitrogen, and glucose remained constant, and antimycin A was produced at a very low rate. After 6 days, the pH started to rise sharply and increase in cell volume was obtained. During that period, another cycle of rapid antimycin A production was observed. On the ninth day, the pH reached 8.0 and the optimum titer of 4.5 g/liter was achieved. Alternately, feeding of glucose to production medium starting at 30 hr of fermentation kept the pH of broth at 4 for up to 12 days, at which time glucose addition was stopped and the pH rose slowly, reaching 7.6 on the fifteenth day. Antimycin A titers of 5.9 g/liter were obtained. Thus the effects of lactose and oil in enriched production medium could be reproduced by the continuous addition of glucose, but at the cost of an increase of 6−7 days in fermentation time. From these experiments in fermenters and shake flasks (Vézina et al., 1976), it was concluded that a satisfactory fermentation was always associated with a drop in pH, but that at a pH value below 5 the organism neither grew nor produced the antibiotic. Therefore it was deemed desirable to control the pH of the fermentation broth for optimal growth of the organism under continuous addition of a suitable carbon source. Of the various pH values tested, 6.0 was determined to give the best titers of the antibiotic.

The time course of antimycin A fermentation, with pH control set at 6.0 by on-demand addition of 10 N NH$_4$OH and a glucose feed rate of 0.7% vol/vol per day is shown in Figure 8. Optimum antimycin A titers of 6.8 g/liter were obtained in 7 days.

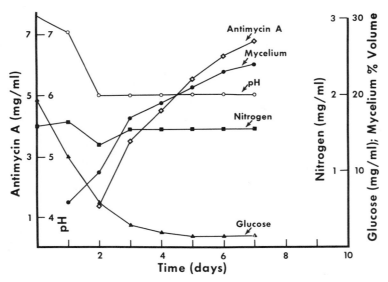

Figure 8 Time course of antimycin A fermentation with *Streptomyces* sp. M-506 in 250-liter fermenters. Glucose was added continuously to the production medium at 0.7%/day; pH was controlled at 6 by automatic addition of 10 N NH$_4$OH. (From Sehgal et al., 1976.)

In the process of medium improvement, Sehgal et al. (1976) substituted defatted Special X soy flour (50% protein, 6–8% fat) for Nutrisoy 220 soy flour (43% protein, 22–23% fat), and observed that the maximum titers of antimycin A dropped from 6.8 to only 4.5 g/liter. This indicated the desirability of using soybean oil as an energy source in the medium. The optimized conditions of antimycin A fermentation were as follows.

Fermenters of 250 liters were charged with 160 liters of production medium and inoculated with 2% of the second-stage inoculum; fermentation was continued at 25°C, 250 rpm agitation, and 0.5 vol/vol per minute aeration. At 30 hr after inoculation, by which time the initial drop of pH had taken place, sterile soybean oil was added continuously at the rate of 1.25% per day. The pH was maintained at 6.0 by intermittent (on-demand) addition of either ammonia gas of 10 N ammonium hydroxide. The addition of oil was continued until the pH started to rise above 6.0. The fermentation was terminated at that point (usually 6–7 days). The time course of a typical fermentation is shown in Figure 9.

Antimycin A accumulated almost linearly from the second day to reach a maximum of 9 g/liter in 6 days. The control of continuous addition of oil during antimycin A fermentation is very critical. The excess of oil would smother fermentation and lack of it would starve it. A close examination of the surface of the centrifuged broth used for cell volume determination showed if an excess of oil was being added. On the other hand, if demand for ammonium hydroxide solution ceased or was reduced significantly, the addition of oil was considered lower than optimal. The pH of fermentation broth was observed closely before pH control was started. A slight and temporary rise in pH after an initial drop to 6.0 (probably due to a lag in the utilization of oil) is a characteristic of good and normal antimycin A fermentation.

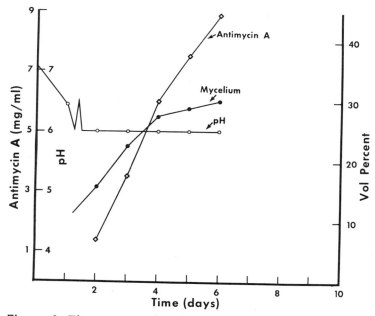

Figure 9 Time course of antimycin A fermentation with *Streptomyces* sp. M-506 in 250-liter fermenters. Soybean oil was added continuously to the production medium at 1.25%/day; pH was controlled at 6 by automatic addition of 10 N NH$_4$OH. (From Sehgal et al., 1976.)

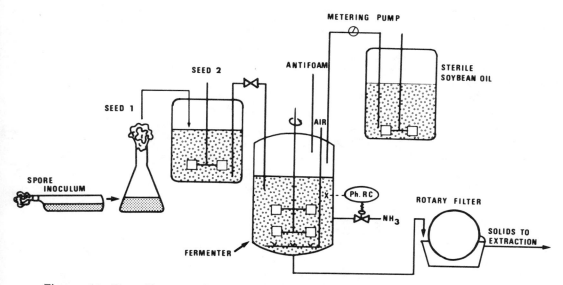

Figure 10 Flow diagram of antimycin A fermentation.

The pilot-plant fermentation was scaled up to production stage in 9000-gal fermenters at a charge of 6000 gallons. The agitation was maintained at 84 rpm and aeration at 400 ft^3/min. A flow diagram of antimycin A production is given in Figure 10. Antimycin A titers of 10 g/liter were obtained on the production scale.

D. Extraction from Fermentation Broth

Antimycin A is a lipophilic antibiotic which is produced intracellularly. Only about 2–5% is excreted into the medium during fermentation. Therefore it is economical to separate the fermentation solids from the liquid medium at the end of fermentation and extract the antibiotic from the solids. A flow diagram of antimycin A extraction and recovery is given in Figure 11.

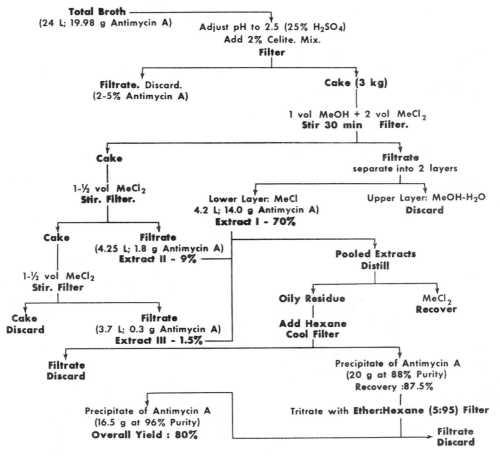

Figure 11 Flow diagram of antimycin A extraction from microbial broth. (From Sehgal et al., 1976.)

E. Formulation of Antimycin A

To exploit fully the advantages of antimycin A as a piscicide, considerable efforts were made to develop field formulations that would meet target needs and yet preserve the desirable features of the antibiotic (Lennon and Vézina, 1973). Fintrol-5 is a dry, granular formulation that penetrates the surface film of water readily, does not adhere to vegetation or debris, and releases the toxicant within 5 ft from the surface of water. It contains 1% antimycin A, 24% polyethylene glycol 6000 (Carbowax), and 75% sand by weight. The formulation is prepared as follows: Antimycin A is dissolved in acetone and the solution mixed with polyethylene glycol; the mixture is sprayed on sand (50–70 mesh) by spray column application technique. Fintrol-5 is easily and uniformly spread over the surface of water by means of hand-powered or motor-powered, grass-seed spreaders from shore, boats, or aircraft (Lennon et al., 1967). Fintrol-15 is similar to Fintrol-5, except that it contains 5% antimycin A and is designed to release the piscicide within the first 15 ft below the surface. Other granular, coated-sand formulations were prepared by Ayerst Laboratories and reviewed by Vézina (1967).

Fintrol-Concentrate contains 10% antimycin A in acetone and is used to inject antimycin A into streams and shallow lakes and nonaccessible bodies of water. Fintrol-BAR, containing 35 g of antimycin and 215 g of Pluronic F68, can be anchored below the surface of a stream and releases its antimycin A over a period of 7 hr at 10°C and 5 hr at 15°C. Once the concentration of antimycin A to be used has been determined by preliminary bioassays on site against target fish (Berger et al., 1969), a fraction of a bar or one or more bars are placed in the stream to release the desired concentration of the piscicide in parts per billion for a desired amount of time.

V. APPLICATIONS AND MODE OF ACTION

A. Applications

The main applications of antimycin A today relate to its use as an aid in modern fishery management. Prévost (1960) stated that the reclamation of waters by poisoning fish had become the best management tool available. Fish killing is beneficial for the control of a pest situation and the management and production of food or game fishes. In their early laboratory experiments at the Fish Control Laboratories in La Crosse, Wisconsin, Walker et al. (1964) and Berger et al. (1969) recognized several desirable properties of antimycin A: specificity to fish, toxicity to fish in extremely small concentrations (parts per billion or per trillion), irreversible toxicity, nonrepellency to fish, and rapid degradation in water (degradation products are inactive on fish and other forms of life); they also found that antimycin A is effective against target fish in warm or cold, soft or hard, acid or alkaline, and fresh or marine waters. The rate of degradation increases with increasing hardness, temperature, and pH. Degradation can be accelerated by detoxication with potassium permanganate. Field trials by Gilderhus et al. (1969) confirmed laboratory results and further showed that antimycin A is relatively harmless to nontarget aquatic life. With the formulations available, antimycin A is a precision tool, provided that on-site bioassays against target species in target water are conducted to determine the optimum concentration to be used. When properly treated with the piscicide, most waters can be restocked with fish 2–14 days after treatment. The application of antimycin A as a general and selective piscicide was reviewed by Lennon and Vézina (1973). Of the fish toxicants, antimycin A comes closest by far in meeting the needs of fishery managers and the requirements of regulatory agencies.

Treatment cost for antimycin A compares favorably with that for other piscicides. The rapid degradation of the antibiotic combined with the possibility of early restocking with fish represents a real saving in reclamation of waters. The excipients present in the formulation are inexpensive; the main cost of the constituted formulation is related to the fermentation and purification of the antibiotic which is produced at the yield of 10 g/liter. The cost of antimycin A can be reduced if a crude extract is used instead of the pure material (>92% purity); however, a crude extract could contain repellent substances. Therefore high purity of the active ingredient is desirable to preserve nonrepellency of antimycin A formulations.

B. Mode of Action

In reviewing the toxicity of antimycin A to various life forms, Rieske (1980) concluded that all the toxic effects of the antibiotic can be directly or indirectly traced to an inhibition of energy-yielding electron transport through a single, highly specific binding to a site associated with the cytochromes $b-c_1$ segment known as complex III.

Most bacteria, including the antimycin A-producing streptomycetes (Řeháček et al., 1968b), are insensitive to antimycin A, even though they possess a full complement of cytochromes a, b, and c. They may have no antimycin A-sensitive site or alternate electron transfer pathways. Rieske noted (1980) that S. antibioticus also lacks the BAL-sensitive Slater factor (BAL = British antilewisite = 2,3-dimercaptopropanol), coenzyme Q, and vitamin K and, interestingly, is sensitive to 2-N-heptyl-4-hydroxyguinoline-N-oxide (HQNO), a compound that mimics the effect of antimycin A on the respiratory chain of higher organisms. The fish, as a gilled animal, is more sensitive to antimycin A than any other life form perhaps because it is separated from its aquatic environment by a membrane only one cell layer thick; therefore all the cells would be exposed at the same time to antimycin A dissolved in the water.

VI. CONCLUSION

Antimycin A is produced in high yields by fermentation and recovered as a crystalline complex. High yields could be attained by mutagenizing the producing organism, randomly selecting high-production strains, and optimizing the production medium and cultural conditions. Continuous addition of soybean oil as a slowly utilizable carbon source and continuous adjustment of pH at 6 are necessary for full expression of productivity in the antimycin A producers and for reducing the fermentation time to 6 days. The quality of the antibiotic produced (>92% purity) is suitable for preparing formulations that make antimycin A a precision tool in fish management. Purity can be increased to 99% (with some loss) for use of antimycin A in the study of the mechanism of respiratory electron transport.

REFERENCES

Ahmad, K., Schneider, H. C., and Strong, F. M. (1950). Studies on the biological action of antimycin A. *Arch. Biochem. 28*:281–294.

Berger, B. L., Lennon, R. E., and Hogan, J. W. (1969). Investigations in fish control. Laboratory studies on antimycin A as a fish toxicant. *U.S. Bur. Sport Fish. Wildl. Resour. Publ. 26*:1–21.

Birch, A. J., Cameron, D. W., Harada, Y., and Rickards, R. W. (1962). Studies in relation to biosynthesis. Part XXV. A preliminary study of the antimycin A complex. *J. Chem. Soc. London* p. 303–305.

Derse, P. H., and Strong, F. M. (1963). Toxicity of antimycin A to fish. *Nature 200*:600–601.

Dickie, J. P., Loomans, M. E., Farley, T. M., and Strong, F. M. (1963). The chemistry of antimycin A. XI. N-substituted 3-formamidosalicylic amides. *J. Med. Chem. 6*:424–427.

Dunshee, B. R., Leben, C., Keitt, G. W., and Strong, F. M. (1949). The isolation and properties of antimycin A. *J. Am. Chem. Soc. 71*:2436–2437.

El-Kersh, T., and Vézina, C. (1981). Localization of antimycin A and melanin determinants in auxotrophs of *Streptomyces* sp. M-506 and in their prototrophic revertants. In *Advances in Biotechnology—Recent International Developments*, Vol. 3. C. Vézina and K. Singh (Eds.). pp. 37–42. Pergamon Press, New York, N.Y.

Gilderhus, P. A., Berger, B. L., and Lennon, R. E. (1969). Investigations in fish control. Field trials of antimycin A as a fish toxicant. *U.S. Bur. Sport Fish. Wildl. Resour. Publ. 27*:1–21.

Harada, Y., Uzu, K., and Asai, M. (1958). Separation of antimycin A. *J. Antibiot. Ser. A. 11*:32–36.

Harada, Y., Kumabe, K., Kagawa, T., and Sato, T. (1959). Antifungal activity of antipiricullin for several growing stages of *Piricularia oryzae*. *Nippon Shokubutsu Byori Gakkaiho 24*:247.

Kannan, L. V., Kozová, J., and Řeháček, Z. (1968). Biogenesis of peptide antibiotics. I. Dynamics of antimycin A biosynthesis by *Streptomyces antibioticus* NRRL 2838. *Folia Microbiol. Prague 13*:1–6.

Kinoshita, M., and Umezawa, S. (1970). The total synthesis of dehexyldeisovaleryloxy antimycin A. *Bull. Chem. Soc. Jpn. 43*:897–901.

Kinoshita, M., Ishii, K., and Umezawa, S. (1971). Synthesis of (3S,4R,15S)-4,15-dimethyl-1,5-dioxa-3-(3'-formamido salicylamido)-cyclopentadecane-2,6-dione and its (15R) epimer, new antimycin analogs. *Bull. Chem. Soc. Jpn. 44*:3395–3399.

Kluepfel, D., Sehgal, S. N., and Vézina, C. (1970). Antimycin A components. I. Isolation and biological activity. *J. Antibiot. Jpn. Ser. A 23*:75–80.

Koaze, Y., Sakai, H., Yonehara, H., Asakawa, M., and Misato, T. (1956). Studies on the activities of antibiotics against plant pathogenic microorganisms. *J. Antibiot. Jpn. Ser. A 9*:89–96.

Leben, C., and Keitt, G. W. (1948). An antibiotic substance active against certain phytopathogens. *Phytopathology 38*:899–906.

Lennon, R. E., and Vézina, C. (1973). Antimycin A, a piscicidal antibiotic. *Adv. Appl. Microbiol. 16*:55–96.

Lennon, R. E., Berger, B. L., and Gilderhus, P. A. (1967). A powered spreader for antimycin A. *Progr. Fish Cult. 29*:110–113.

Liu, W. C., and Strong, F. M. (1959). The chemistry of antimycin A. VI. Separation and properties of antimycin A subcomponents. *J. Am. Chem. Soc. 81*:4387–4390.

Lockwood, J. L., Leben, C., and Keitt, G. W. (1954). Production and properties of antimycin A from a new *Streptomyces* isolate. *Phytopathology* 14:438–446.

Nakayama, K., Okamoto, F., and Harada, Y. (1956). Antimycin A: Isolation from *Streptomyces kitazawaensis* and its activity against rice plant blast fungi. *J. Antibiot. Jpn. Ser. A* 9:63–66.

Neft, N., and Farley, T. M. (1972a). Conditions influencing antimycin production by a *Streptomyces* species grown in chemically defined medium. *Antimicrob. Agents Chemother.* 1:274–276.

Neft, N., and Farley, T. M. (1972b). Studies on the biosynthesis of antimycin A. I. Incorporation of ^{14}C-labeled metabolites into the 3-formamidosalicyl moiety. *J. Antibiot.* 25:298–303.

Prévost, G. (1960). Use of fish toxicants in the Province of Quebec. *Can. Fish Cult.* 28:13–36.

Ramankutty, M., Kannan, L. V., and Řeháček, Z. (1969). Effect of phosphate on biosynthesis of antimycin A and production and utilization of poly-β-hydroxybutyrate by *Streptomyces antibioticus*. *Indian J. Biochem.* 6: 230–231.

Řeháček, Z., and Švarc, S. (1968). 3-(N-Formylamino)-salicylic acid, a new metabolite of *Streptomyces antibioticus*. *Chem. Ind. London* p. 1523.

Řeháček, Z., Kannan, L. V., Ramankutty, M., and Puza, M. (1968a). Relation of tryptophan metabolism to antimycin A biosynthesis in *Streptomyces antibioticus*. *Hind. Antibiot. Bull.* 10:280–286.

Řeháček, Z., Ramankutty, M., and Kozová. (1968b). Respiratory chain of antimycin A-producing *Streptomyces antibioticus*. *Appl. Microbiol.* 16: 29–32.

Rieske, J. S. (1967). Antimycin A. In *Antibiotics*, Vol. 1. D. Gottlieb and P. D. Shaw (Eds.). Springer-Verlag, Berlin, pp. 542–584.

Rieske, J. S. (1976). Composition, structure, and function of complex III of the respiratory chain. *Biochim. Biophys. Acta* 456:195–247.

Rieske, J. S. (1980). Inhibitors of respiration at energy-coupling site 2 of the respiratory chain. *Pharmacol. Ther.* 11:415–450.

Rieske, J. S., and Zaugg, W. S. (1962). The inhibition by antimycin A of the cleavage of one of the complexes of the respiration chain. *Biochem. Biophys. Res. Commun.* 8:421–426.

Sakagami, Y., Takeuchi, S., Sakai, H., and Takashima, M. (1956). Antifungal substances no. 720A and no. 720B and the probable identity of no. 720A with antimycin A, virosin, and antipiricullin. *J. Antibiot. Jpn. Ser. A* 9:1–5.

Schilling, G., Berti, D., and Kluepfel, D. (1970). Antimycin A components. II. Identification and analysis of antimycin A fractions by pyrolysis-gas liquid chromatography. *J. Antibiot. Jpn. Ser. A* 23:81–90.

Schneider, H. G., Tener, G. H., and Strong, F. M. (1952). Separation and determination of antimycin. *Arch. Biochem.* 12:191–195.

Sehgal, S. N. (1977). Antimycin A fermentation. In *Symposium on "Following On-Going Fermentations."* Annual Meeting of the American Society for Microbiology, New Orleans, May 9, 1977. American Society for Microbiology, Washington, D.C.

Sehgal, S. N., and Vézina, C. (1967). Automated fluorometric assay of antimycin A. *Anal. Biochem.* 21:266–272.

Sehgal, S. N., Singh, K., and Vézina, C. (1965). Spectrophotofluorometric assay of antimycin A. *Anal. Biochem.* 12:191–195.

Sehgal, S. N., Saucier, R., and Vézina, C. (1976). Antimycin A fermentation. II. Fermentation in aerated-agitated fermentors. *J. Antibiot.* 29:265–274.

Strong, F. M. (1956). Isolation, structure and properties of antimycin A. In *Topics in Microbial Chemistry.* Wiley, New York, pp. 1–43.

Tamelen, E. E. van, Dickie, J. P., Loomans, M. E., Dewey, R. S., and Strong, F. M. (1961). The chemistry of antimycin A. X. Structure of the antimycins. *J. Am. Chem. Soc.* 83:1639–1646.

Tappel, A. L. (1960). Inhibition of electron transport by antimycin A, alkyl hydroxy naphthoquinones and metal co-ordination compounds. *Biochem. Pharmacol.* 3:389–396.

Tener, G. M., Bumpus, F. M., Dunshee, B. R., and Strong, F. M. (1953a). The chemistry of antimycin A. II. Degradation studies. *J. Am. Chem. Soc.* 75:1100–1104.

Tener, G. M., Tamelen, E. E. van, and Strong, F. M. (1953b). The chemistry of antimycin A. III. The structure of antimycic acid. *J. Am. Chem. Soc.* 75:3623–3625.

Vézina, C. (1967). Antimycin A, a teleocidal antibiotic. *Antimicrob. Agents Chemother.* 1966:757–766.

Vézina, C. (1971). Antibiotics for non-human uses. *Pure Appl. Chem.* 28:681–698.

Vézina, C., Bolduc, C., Kudelski, A., and Sehgal, S. N. (1976). Antimycin A fermentation. 1. Production and selection of strains. *J. Antibiot.* 29:248–264.

Walker, C. R., Lennon, R. E., and Berger, B. L. (1964). Investigations in fish control. 2. Preliminary observations on the toxicity of antimycin A to fish and other aquatic animals. *U.S. Bur. Sport Fish. Wildl. Resour. Publ.* 186:1–18.

Watanabe, K., Tanaka, T., Fukuhara, K., Miyairi, N., Yonehara, H., and Umezawa, H. (1957). Blastmycin, a new antibiotic from *Streptomyces* sp. *J. Antibiot. Jpn. Ser. A* 10:39–45.

23

BLASTICIDIN S: PROPERTIES, BIOSYNTHESIS, AND FERMENTATION

HIROSHI YONEHARA *Kaken Pharmaceutical Co., Tokyo, Japan*

I. INTRODUCTION

In 1952, a large-scale screening program was started for new antibiotics especially effective against *Mycobacterium tuberculosum.* More than 6000 species of actinomycetes were isolated from soil samples collected from all over Japan and their antibiotic activities were tested against several types of microorganisms, including *Bacillus subtilis, Escherichia coli, Mycobacterium* sp. 607, *Aspergillus oryzae,* and *Saccharomyces cerevisiae* (Table 1).

In the autumn of that year Dr. Fukunaga and his collaborators from the National Institute of Agricultural Sciences asked us to make a joint effort to search for a new antibiotic effective against *Piricularia oryzae,* which is the

Table 1 Effective Strains Tested at the First Screening in 1952

	Number of strains tested	Effective against[a]				
		Myco.	G+.	G−.	F.	Y.
Actinomycetes	6705	782 (11.7%)	1709 (25.5%)	284 (3.5%)	561 (8.4%)	837 (12.5%)
Fungi	1472	102 (6.9%)	369 (25.3%)	65 (4.4%)	187 (12.7%)	135 (9.2%)
Bacteria	820	37 (4.5%)	214 (26.1%)	82 (10.0%)	15 (1.8%)	11 (1.3%)

[a]Myco., *Mycobacterium* sp. 607 and *Mycobacterium phlei*; G+., *Bacillus subtilis* PCI 219; G−., *Escherichia coli*; F., *Aspergillus oryzae* 0-8-1; Y., *Saccharomyces cerevisiae*.

most destructive pathogen of rice plants in Japan. We sent them the 561 streptomyces strains, which displayed antifungal activity against *A. oryzae* in our screening tests.

Using these strains, they tested the antifungal activities in vitro and in vivo against *P. oryzae*. In the case of in vivo tests, the culture broths were directly sprayed on young rice plants, inoculated with the pathogen, and grown in a greenhouse. About 3 weeks after the spray treatment, the infected spots were counted and the degree of infection was observed.

At that time in Japan the phenyl acetate mercuric compounds had been widely used to prevent rice blast and the Japanese government worried much about their influence on public health and promoted the search for nonmercuric pesticides effective against rice blast and other plant pathogens.

Fukunaga's group selected about 10 effective streptomyces strains after repeated greenhouse tests against *P. oryzae*.

We then tried to isolate the antifungal principles from cultures of these streptomyces. First, we extracted an antifungal substance from the cultured mycelium of a streptomyces, designated as *Streptomyces griseochromogenes*, and named it blasticidin A (Misato et al., 1959, 1961a,b).

Blasticidin A showed antifungal activity against *P. oryzae* at a concentration of 0.1 µg/ml in vitro, but about 100−200 ppm were needed to show the protective activities obtained with phenyl mercuric compounds for rice blast in greenhouses. Blasticidin A proved to be an unstable compound under ultraviolet rays and sunlight. When the culture broth was checked for its effectiveness against rice blast, a second antibacterial factor effective against *B. subtilis* was detected in the culture broth of *S. griseochromogenes* (Takeuchi et al., 1958).

Consequently, we isolated the new antibiotic—blasticidin S—in crystalline form from the culture of *S. griseochromogenes* by ion exchange method and recrystallization from water. The antibiotic showed antibacterial and antifungal activity, and the minimum inhibitory concentration against *P. oryzae* was 1−5 µg/ml (Table 2).

Table 2 Therapeutic Effect of Blasticidin S Against Rice Blast Disease[a]

| Fungicides | Concentration (μg/ml) | Average number of blast spots per pot | | Chlorosis |
		Sprayed 4 days before inoculation (preventive effect)	Sprayed 1 day after inoculation (therapeutic effect)	
Blasticidin S	2	96	19.3	−
	5	64	3.7	−
	10	100	1.5	−
	20	76	0.2	±
	50	42	0	+
	100	11	0	‡
	200	10	0	+++
Phenyl mercuric	10	48	10.0	−
Acetate (per mg)	20	28	7.8	−
Control		86	76.3	−

[a]Each pot contained 10 plants having 4 leaves.

II. PHYSICAL AND CHEMICAL PROPERTIES OF BLASTICIDIN S

Blasticidin S appears as white needlelike crystals, melting at 235–236°C under decomposition. The stability of blasticidin S in aqueous solutions of various pH values was examined at 100°C. It was most stable at pH 5.0–7.0 and less stable at pH 4.0 than at pH 2.0 and 8.0–9.0.

The free base of blasticidin S is soluble in water and acetic acid, but insoluble in methanol, ethanol, acetone, benzene, ether, butyl acetate, chloroform, carbon tetrachloride, methyl ethyl ketone, cyclohexane, xylene, tetrahydrofuran, methyl isobutyl ketone, pyridine, and dioxane. It is negative in ferric chloride, Fehling, Tollens, sodium nitroprusside, triphenyltetrazonium chloride, bromonitroso, maltol, Millon, Ehrlich, Sakaguchi, Molisch, Biuret, xanthoprotein, Grafs aldehyde, ammoniacal silver nitrate, and ninhydrin tests.

The decomposition point of blasticidin S hydrochloride is 224–225°C. Optical rotation of the free base is $[\alpha]_D^{11°} = +108.4$ (C = 1% in water). Elementary analysis of blasticidin S gave a molecular formula of $C_{17}H_{26}O_5N_8$.

In the ultraviolet absorption spectrum, the absorption maximum in $E_{1\,cm}^{1\%}$ =349 at 275 μm in 10 N hydrochloric acid and $E_{1\,cm}^{1\%}$ = 266 at 266–270 μm in 10 N sodium hydroxide.

Typical infrared absorption bands are 3318, 3130, 2860, 1675, 1614, 1600, 1557, 1492, 1422, 1400, 1353, 1300, 1282, 1235, 1200, 1111, 1067, 1042, 1002, 943, 860, 822, 775, and 714 cm^{-1}.

III. CHEMICAL STRUCTURE OF BLASTICIDIN S

The chemical structure of blasticidin S is represented in Figure 1 (see also Structure [1]). Blasticidin S ($C_{17}H_{26}O_5N_8$) [(c, 1.0 in H_2O), $\lambda_{max}^{0.1\,N\,HCl}$, 274 μm (ε 13,400); $\lambda_{max}^{0.1\,N\,NaOH}$, 266 μm) (ε 8850); pK$_a$, 2.4, 4.6, 8.0, and above 12.5] yields corresponding monohydrochloride, [$C_{17}H_{26}O_5N_8 \cdot HCl$; m.p., 229°C (dec.)], dihydrochloride [$C_{17}H_{26}O_5N_8 \cdot 2HCl$; m.p., 195°C (dec.)], and monomethyl ester trihydrochloride [$C_{17}H_{25}O_4N_8$)OCH_3)$\cdot 3HCl$; m.p., 206–208.5°C (dec.)] derivatives. Therefore it was characterized as a monoacidic tribasic compound. Functional group analysis demonstrated the presence of one N—CH$_3$ group, two ethylenic groups, and two NH$_2$ groups (Van Slyke determination) (Otake et al., 1966; Yonehara and Otake, 1966).

Under mild alkaline hydrolysis, liberation of 1 mol of ammonia took place to yield a crystalline compound designated cytomycin [2] [$C_{17}H_{23}O_5N_7$; m.p., 237–239°C (dec.); $\lambda_{max}^{0.1\,N\,HCl}$, 274 μm (ε 12,000); $\lambda_{max}^{0.1\,N\,NaOH}$, 266 μm (ε 6930); pK$_a$, 2.4, 4.6, and above 12.5]. Further alkaline hydrolysis resulted in cleavage of cytomycin into cytosinine [3], ($C_{10}H_{12}O_4N_4$), whose structure was elucidated by Otake et al. (1966), and into blastidone [4] [$C_7H_{13}O_2N_3$; m.p., 209–210°C (dec.)]. The structure of blastidone was pre-

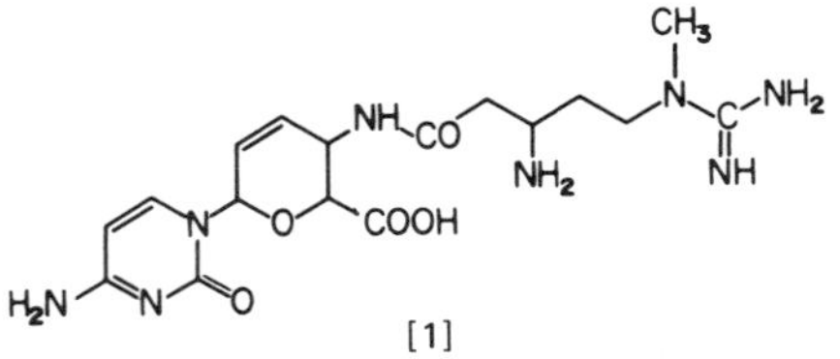

Figure 1 Formula of blasticidin S.

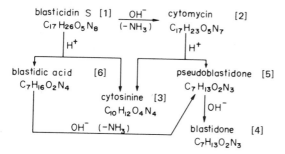

Figure 2 Blasticidin S derivatives.

viously elucidated as 4-ureido-N-methyl-2-piperidone (Fig. 2). The interpretation of the origin of the piperidone ring (whether it is natively contained in blasticidin S or arises during degradation) became a key problem.

On acid hydrolysis, cytomycin gave cytosinine and pseudoblastidone [5] [$C_7H_{13}O_2N_3$; m.p., 282–283°C (dec.); $\lambda_{max}^{H_2O}$, only end absorption; IR (Nujol, 1720 (shoulder), 1640, and 1610 cm^{-1}; NMR (in D_2O) ppm; 1.8 (2H m), 2.50 (2H d, J = 7.5 cps), 3.05 (3H s, N—CH$_3$), 3.40 (2H t), and 3.65 (1H m)], which was characterized as a monoacidic monobasic compound with pK_a values 3.75 and 12.5.

Blasticidin S could be cleaved with acid under selected conditions to yield cytosinine and a new amino acid designated blastidic acid [6] as dihydrochloride [$C_{17}H_{16}O_2N_4$·2HCl; m.p., 192–192.5°C (dec.); $[\alpha]_D^{15}$, +25.0°C (c, 1.0 in H_2O); pK_a, 3.2, 8.0, and above 12.5; containing one amino group (Van Slyke determination)]. Blastidic acid showed nuclear magnetic resonance (NMR) peaks (in D_2O + DCl) at ppm: 1.8 (2H m), 2.50 (2H d, J = 7.5 cps), 2.95 (3H s, N—CH$_3$), 3.15 (2H t), and 3.50 (1H m). Spin decoupling data established a relationship between 10 protons and provided proof for the partial structure [7] in blastidic acid (Fig. 3).

Blastidic acid was transformed almost quantitatively to pseudoblastidone with liberation of 1 mol of ammonia by the action of dilute alkali , similar to blasticidin S yielding cytomycin.

Potassium permanganate oxidation of blastidic acid yielded N-methylguanidine, characterized as picrate (m.p., 200–201°C), which was identified unequivocally by comparison of infrared (IR) spectra and mixed melting point with the authentic sample.

These evidences allowed the assignment of structure [6] for blastidic acid, and the transformation of structure [6] to [5] could be reasonably explained by Figure 4.

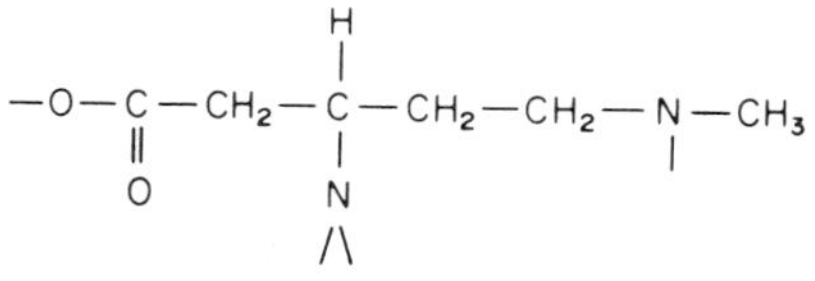

$$-O-\underset{\underset{O}{\|}}{C}-CH_2-\underset{\underset{\underset{/\backslash}{N}}{|}}{\overset{\overset{H}{|}}{C}}-CH_2-CH_2-\underset{\underset{|}{}}{N}-CH_3$$

[7]

Figure 3 Partial structure of blastidic acid.

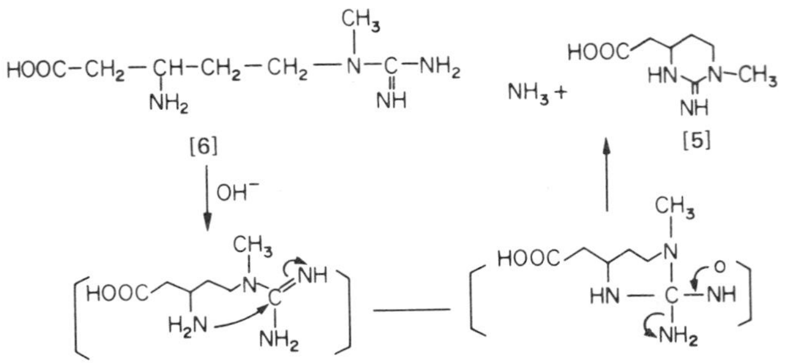

Figure 4 Transformation of blastidic acid to pseudoblastidone.

Indeed, pseudoblastidone gave N-methylguanidine characterized as pi-crate by the oxidation with potassium permanganate. This result, in addition to NMR data, supported the assignment of Structure [5] for pseudoblastidone.

An approach to the complete elucidation of the structure of blasticidin S was provided by spectral evidence, namely, spin decoupling experiments on blasticidin S showed the intact presence of the partial structure [7] in the antibiotic.

If one takes account of the information obtained so far, Structures [1], [8], and [9] were possible, but data on dissociation constants excluded Structure [9] (Fig. 5).

The following experiments gave unequivocal support to Structure [1]. Blasticidin S or cytomycin was hydrogenated with PtO_2 in acetic acid to yield the product designated SC_{13} compound [10] [$C_{13}H_{25}O_4N_5$; m.p., 208–209°C (dec.); pK_a, 3.2, 8.0, and above 12.5] or a 13-carbon compound [11] ($C_{13}H_{22}O_4N_4$; m.p., above 320°C; pK_a, 3.2 and above 12.5), respectively. Upon acid hydrolysis of [10], 3-aminotetrahydropyran-2-carboxylic acid [12] and blastidic acid were obtained; similarly, [11] yielded [12] and pseudo-blastidone (Fig. 6).

Under selective conditions Structure [11] was partially hydrolyzed by alkali into a compound designated pHC_{13} [13] [$C_{13}H_{24}O_5N_4$; m.p., 177–177.5°C; pK_a, 3.0 and 10.8; IR (Nujol), 1595, 1650, and 2760 cm^{-1}; NMR (D_2O) ppm: 2.0–1.4 (6H m), 2.29 (2H d, J =6.5 cps), 2.54 (3H s, N—CH$_3$), 2.90 (2H t), 3.49 (1H d, J = 8.5 cps), and 3.7 (3H m)]. Structure [13] gave corresponding crystalline ethyl ester hydrochloride, [$C_{13}H_{23}O_3N_4(OC_2H_5)$ ·HCl; m.p., 222–223°C (dec.); IR (Nujol), 1740, 1650, and 1250 cm^{-1}] and

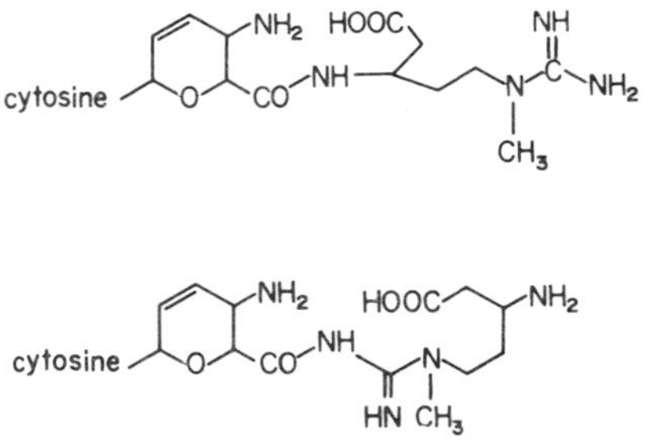

Figure 5 Hypothetical structures proposed for Blasticidin S.

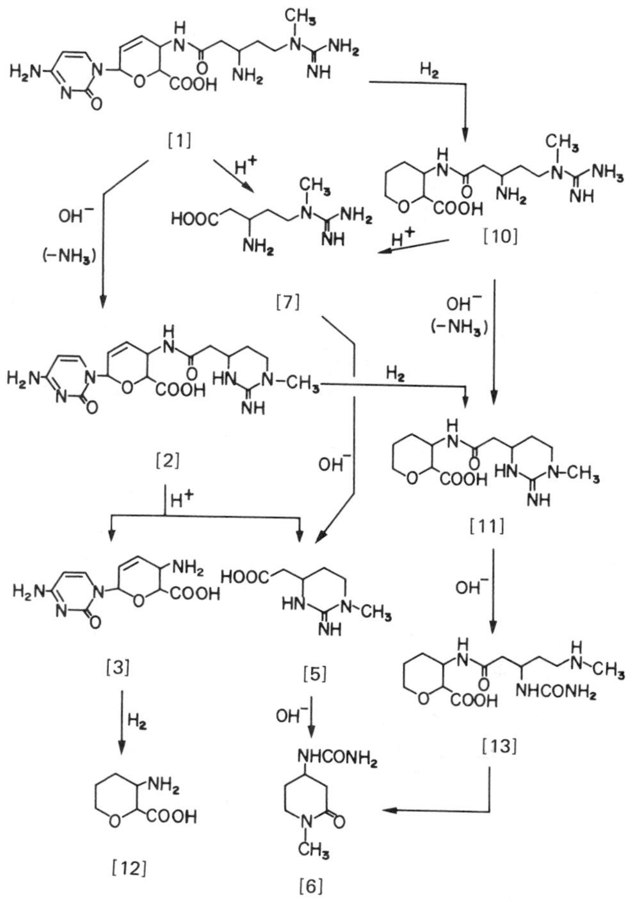

Figure 6 Chemical derivatives of blasticidin S.

also yielded a monodinitrobenzene derivative, $C_{13}H_{21}O_4N_4(C_6H_3O_4N_2)$ (m.p., 188.5–190°C), which on drastic acid hydrolysis yielded N-methylaniline and [12] as characteristic products. These data demonstrated the presence of one free N-methylamino group and one free carboxyl group in [13]. Two amino groups in Structure [13] could be detected by Van Slyke determination; nevertheless, they did not show basicity and therefore were believed to be ureido functions. Indeed, PHC_{13} exhibited the characteristic color reaction of diphenylcarbohydrazide for urea.

These results permitted the assignment of Structure [13] for PHC_{13} compound. Clearly, the piperidone ring was formed by recyclization, since by either acid or alkaline hydrolysis, PHC_{13} gave rise to blastidone and 3-amino-tetrahydropyran-2-carboxylic acid. Simultaneous formation of terminal N-methyl and ureido groups could be explained by the rupture of the guanidino ring [11] → [13].

On the basis of these accumulated evidences, the structure [11] could be assigned unequivocally to C_{13} compound, [10] to SC_{13} compound, [2] to cytomycin, and [1] to blasticidin S, respectively.

Structural elucidation by x-ray analysis using blasticidin S monohydro-bromide was investigated in parallel with the chemical work, and reached complete agreement with Structure [1].

IV. BIOGENESIS OF BLASTICIDIN S

Blasticidin S is an antibiotic produced by *S. griseochromogenes* and its structure has been established as shown in Figure 1. It consists of a pyrimidine nucleoside, a β-amino acid named cytosinine, and blastidic acid.

Hitherto several reports have been published on the biogenesis of purine nucleoside antibiotics such as cordycepin and angustimycin C (Psicofuranine), but very few reports are concerned with pyrimidine nucleoside antibiotics. On the other hand, the biogenesis of β-alanine and α-methyl-β-alanine are known, while these β-amino acids should be regarded as ω-amino acids rather than as β-amino acids from the point of view of their formation mechanisms.

The metabolic pathways of the unique nucleoside and anomalous amino acid will be discussed here in view of the biogenesis of blasticidin S (Seto et al., 1966, 1968).

For incorporation tests of ^{14}C-labeled compounds (except [^{14}C]glucose) into blasticidin S, a synthetic medium consisting of sucrose, glucose, and mineral salts was used. When [^{14}C]glucose was used, the medium consisted of 5% sucrose, 2% soybean meal, and 0.5% sodium chloride. Each ^{14}C-labeled compound was added to the 48-hr-old culture broths of *S. griseochromogenes*. After 72 hr of cultivation, radioactive blasticidin S was isolated by cation exchange procedures.

The incorporation ratio of ^{14}C-labeled compounds into the antibiotic are presented in Table 3. According to these results, it became evident that [U-^{14}C]D-glucose, [1-^{14}C]D-glucose, [6-^{14}C]D-glucose, [2-^{14}C]cytosine, [U-^{14}C]cytidine, [methyl-^{14}C]L-methionine, [guanidino-^{14}C]L-arginine, and [U-^{14}C]L-arginine were incorporated into blasticidin S in higher yield as compared to other compounds.

The isolated blasticidin S was degraded by the scheme presented in Figure 7 and radioactivity of the degradation products was measured by liquid scintillation.

The results presented in Table 4 indicated the following facts: (1) Cytosine was incorporated intact into the cytosine nucleus of blasticidin S. (2) The sugar part of cytosinine was derived from D-glucose. Levulinic acid,

Table 3   Incorporation Ratio of ^{14}C-Labeled Compounds into Blasticidin S

	Percentage		Percentage
[U-^{14}C]D-Glucose	3.7	[U-^{14}C]L-Aspartic acid	0.5
[1-^{14}C]D-Glucose	4.0	[1-^{14}C]β-Alanine	0.6
[6-^{14}C]D-Glucose	4.9	[U-^{14}C]Acetic acid	0.5
[2-^{14}C]Cytosine	95.1	[U-^{14}C]Glycine	1.1
[U-^{14}C]Cytidine	15.3	[U-^{14}C]L-Alanine	0.5
[Methyl-^{14}C]L-methionine	38.3		
[Guanidino-^{14}C]L-arginine	51.2		
[U-^{14}C]L-Arginine	30.3		

aUnlabeled cytosine was added simultaneously.

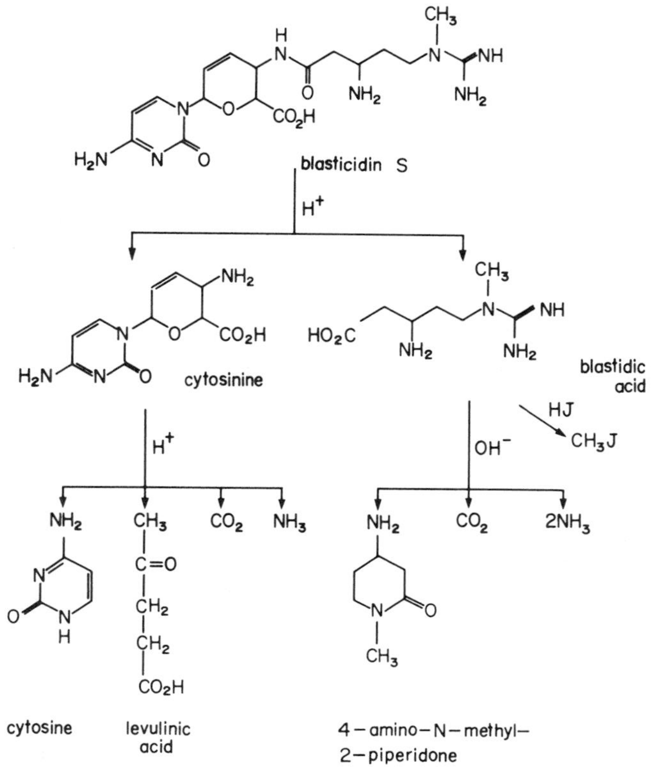

Figure 7 Degradation scheme for blasticidin S.

ammonia, and carbon dioxide were formed from the sugar part of cytosinine, as shown in Figure 8, and the formation mechanism was supported by experiments using ^{14}C-labeled glucose. (3) When [U-^{14}C]cytidine was added, almost all of the radioactivity occurred in the cytosine nucleus and was almost negligible in the sugar portion. Therefore only the cytosine nucleus is incorporated into blasticidin S after cleavage of the C—N linkage of cytidine. (4) The N-methyl group of blastidic acid was derived from methionine in a manner similar to the formation of creatine from guanidino acetic acid. (5) The whole molecule of L-arginine, except the α-amino group, was incorporated into the skeleton of blastidic acid. The mechanism of β-amino acid formation from α-amino acids remains to be disclosed.

From the results mentioned above, the biogenesis of blasticidin S can be depicted as follows (Fig. 9) (Yonehara et al., 1963; Endo et al., 1969).

V. PRODUCTION OF BLASTICIDIN S

The production of blasticidin S was studied in shake flask cultures at 27°C. The basal medium contained 1% soybean meal, 0.4% potassium chloride, 1.4% calcium carbonate, and 0.03% potassium phosphate (dibasic). Among carbon sources added at 1.0%, molasses was found to be superior to sucrose, lactose, starch, dextrin, and dextrose for production (Yonehara et al., 1963).

Table 4 Percentage of Radioactivity in the Degraded Products of Blasticidin S

Compound	Precursor							
	[2-14C]-Cytosine	[U-14C]-Cytidine	[U-14C]-D-Glucose	[1-14C]-D-Glucose	[6-14C]-D-Glucose	[Methyl-14C]-L-methionine	[U-14C]-L-arginine	[Guanidino-14C]-L-arginine
Blasticidin S	100	100	100	100	100	100	100	100
Blastidic acid	0	0.5	16.9	14.8	26.6	98.3	98.0	97.9
CO_2	—	—	—	—	—	—	*24.5	99.5
4-Amino-N-methyl-2-piperidone	—	—	—	—	—	—	—	—
CH_3I	—	—	—	—	—	101.6	*69.2	0
Cytosinine	97.1	98.2	83.6	89.2	74.6	0.1	0	0
Cytosine	96.1	97.9	22.8	21.6	23.8	0.1	1.5	2.7
Levulinic	0	0.2	52.0	62.9	0	—	1.1	2.6
CO_2	0	0.3	11.3	6.7	45.7	—	—	—

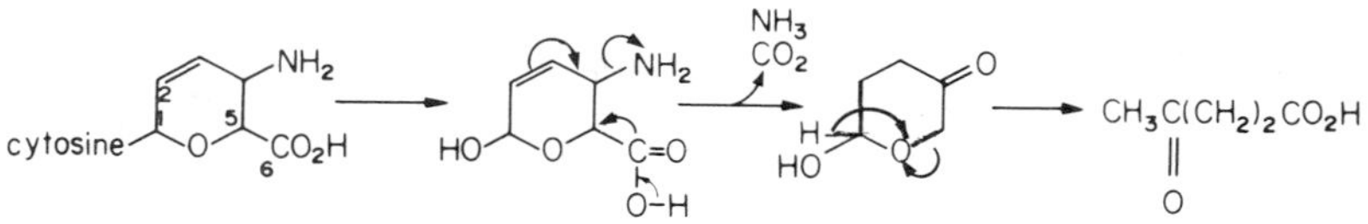

Figure 8 Formation of levulinic acid.

In the case of tank cultures, the 48-hr shake cultured broth of 1000 ml was inoculated into 200 liters of the medium and incubated under aeration and agitation at 27°C for 72–120 hr. The cultured broth showed an antibiotic activity of about 150 μg/ml (Endo et al., 1964).

VI. RECOVERY AND PURIFICATION

After filtering the culture broth, the clear supernatant (150 liter) was adjusted to pH 4.0 with hydrochloric acid, 0.5% active carbon was added, and the mixture was stirred for several minutes. The antibiotic was adsorbed completely onto the carbon. After washing the carbon with water, the antibiotic was repeatedly eluted with 50% aqueous acetone. The combined eluates were concentrated to 1/10 their volume in vacuo at 40°C. The solution was passed through a column of IR 4B, an anion exchange resin of the hydroxy type. The antibiotic was then adsorbed onto a column of IRC 50 cation exchange resin of the hydrogen type. After washing with water and 50% aqueous acetone, the antibiotic was eluted with acid aqueous acetone (acetone 1:1N hydrochloric acid 1). Five volumes of acetone were added to the active eluate and the antibiotic was precipitated.

The crude powder of the antibiotic hydrochloride was dissolved in 90% aqueous methanol and adsorbed on alumina column (neutral alumina charged with absolute acetone). The elution was performed by increasing the water content of aqueous acetone. The chromatogram was developed by 90–70% aqueous acetone and the antibiotic appeared in the eluate with 50% aqueous acetone. To the active eluate, absolute acetone was added and the antibiotic precipitated in crystalline form.

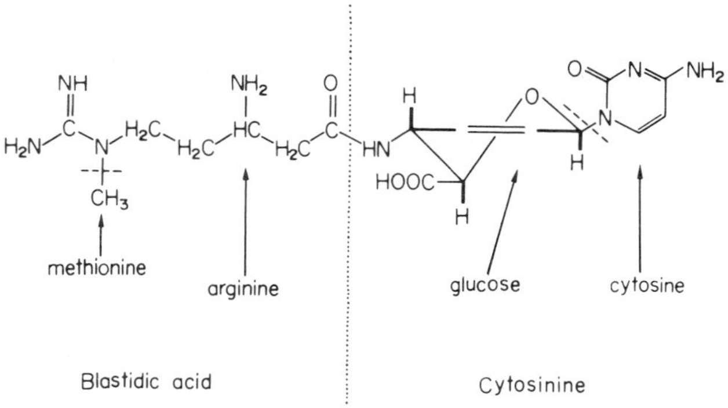

Figure 9 Biogenesis of blasticidin S.

Upon recrystallization from aqueous acetone, 8.5 g of white needlelike crystals of blasticidin S hydrochloride were obtained. The yield was 37.7%. The crystalline hydrochloride was dissolved in water, passed through the column of the free type of IR4B, and concentrated in vacuo. Recrystallization of the residue from the water yielded the needle crystal of free base antibiotic blasticidin S.

VII. ASSAY, BIOLOGICAL ACTIVITIES, AND TOXICITY OF BLASTICIDIN S

A cylinder agar plate method was applied using *B. subtilis* PCI 219 as a test organism. The assay medium contained 0.5% peptone and 2.0% agar and the pH was adjusted to 7.0. A linear relation between the logarithm of concentrations of the antibiotic and the diameters of inhibition zones was obtained in a range of 50–1000 µg/ml.

Biological activities of blasticidin S were examined by the agar streak dilution method and the broth dilution method. The results are shown Tables 5 and 6. It is highly effective against *Pseudomonus* species and *P. oryzae*.

Blasticidin S is toxic to mice; the LD_{50} by intravenous injection is 2.82 mg/kg.

Table 5 Antimicrobial Spectrum of Blasticidin S

Test organisms	Growth inhibition (µg/ml)
Micrococcus pyogenes var. *aureus* 209 P	50
Bacillus subtilis	50
Bacillus lactis	>100
Bacillus agri	100
Sarcina lutea	50
Escherichia coli	50
Pseudomonas fluorescens	5
Mycobacterium tuberculosis ATCC 607	50
Mycobacterium phlei	50
Mycobacterium tuberculosis $H_{37}RA$	10
Penicillium notatum	>100
Penicillium chrysogenum Q_{176}	>100
Aspergillus oryzae	>100
Trichophyton purpureum	>100
Candida albicans	>100
Pseudosaccharomyces santacruzensis	>100
Saccharomyces cerevisiae	>100
Torula albida	>100
Torula utilis	>100

Table 6 Antiphytopathogenic Spectrum of Blasticidin S

Test organisms	Growth inhibition (μg/ml)
Alternaria kikuchiana	50
Cladosporium fulvum	100
Cladosporium sphaerosporum	>100
Corticium centrifugum	>100
Fusarium lini	100
Fusarium roseum	>100
Fusarium oxysporium	>100
Glomerella cingulata	>100
Gibberella fujikuroi	>100
Gibberella saubinetii	100
Gloeosporium kaki	50
Gloeosporium lacticolor	50
Helminthosporium sesanum	>100
Mactosporium bataticola	>100
Ophiobolus miyabeanus	>100
Piricularia oryzae	5-10
Sclerotinia mali	10
Sclerotinia arachidis	>100
Sclerotium rolfsii	>100
Bacterium citri	5
Bacterium aroideae	5
Erwinia aroidae	50
Pseudomonas tabaci	5
Pseudomonas solanacearum	50
Xanthomonas citri	50

REFERENCES

Endo, T., Ōtake, N., Takeuchi, S., and Yonehara, H. (1964). *J. Antibiot.* *17A*:172–173.

Misato, T., Ishii, I., Asakawa, M., Okimoto, Y., and Fukunaga, K. (1959). *Ann. Phytopathol. Soc. Jpn.* *24*:302–306.

Misato, T., Ishii, I., Asakawa, M., Okimoto, Y., and Fukinaga, K. (1961a). *Ann. Phytopath. Soc. Jpn.* *26*:19–24.

Misato, T., Ishii, I., Okimoto, Y., Asakawa, M., and Fukunaga, K. (1961b). *Ann. Phytopathol. Soc. Jpn.* *26*:25–30.

Ōtake, N., Takeuchi, S., Endo, T., and Yonehara, H. (1966). *Agric. Biol. Chem.* *30*:132–141.

Seto, H., Yamaguchi, I., Ōtake, N., and Yonehara, H. (1966). *Tetrahedron Lett.* 3793–3799.

Seto, H., Yamaguchi, I., Ōtake, N., and Yonehara, H. (1968). *Agric. Biol. Chem.* *32*:1292–1298.

Takeuchi, S., Hirayama, K., Ueda, K., Sakai, H., and Yonehara, H. (1958). *J. Antibiot.* *11A*:1–5.

Yonehara, H., and Ōtake, N. (1966). *Tetrahedron Lett.* p. 3785–3791.

Yonehara, H., Takeuchi, S., Ōtake, N., Endo, T., Sakagami, Y., and Sumiki, Y. (1963). *J. Antibiot.* *16A*:195–202.

24

THE BACITRACINS: PROPERTIES, BIOSYNTHESIS, AND FERMENTATION

ØYSTEIN FRØYSHOV *A/S Apothekernes Laboratorium for Specialpraeparater, Oslo, Norway*

I. INTRODUCTION

Bacitracin consists of a group of closely related peptides which are produced by certain strains of *Bacillus licheniformis* and *Bacillus subtilis*. The original producing strain was first isolated in 1943 by Johnson and co-workers from the wound of a 7-year-old girl, Margaret Tracy—thus the name bacitracin (Johnson et al., 1945; Hickey, 1964).

Bacitracin is most active against gram-positive bacteria in a concentration of 50–500 ppm. It seems to affect protein synthesis, cell wall synthesis, and membrane functions. Bacitracin is almost exclusively used as a topical antibiotic in the treatment of infections (Brunner, 1965). Bacitracin is often supplemented with antibiotics such as polymyxin B and neomycin to make a broad-spectrum combination preparation.

Bacitracin is an important feed supplement for a number of animal species. It improves weight gain and feed efficiency when added in a concentration range of 5–100 ppm (Council Directive European Economic Communities (EEC), 1970).

Improvement of the producer strains and the fermentation media as well as the purification procedure has been an important goal for the bacitracin maufacturers.

In this review, a summary is given of the research leading to the elucidation of the structure of bacitracin, its chemical properties, and biosynthetic pathway. Furthermore, its production, commercial use, and mechanism of action are discussed.

II. CHEMISTRY

A. Isolation

The bacitracins are readily isolated from the fermentation broth at pH 7.0 by 1-butanol extraction and are subsequently transferred to an aqueous phase at pH 3.0 (Johnson et al., 1945). A number of different techniques have been used in order to separate and identify the individual bacitracins. The techniques include the following:

    Countercurrent distribution (Craig et al., 1952; Newton and Abraham, 1953a)
    Zone electrophoresis (Porath, 1954)
    Carboxylmethyl cellulose chromatography (Konigsberg and Craig, 1959)
    Disk electrophoresis (Dubost and Pascal, 1970; Coombe, 1972; Fink, 1978)
    Isoelectric focusing in gel (Frøyshov, 1978a)
    Reversed-phase chromatography (Tsuji et al., 1974; Tsuji and Robertson, 1975; Frøyshov and Mathiesen, 1983)

By countercurrent distribution, the bacitracins were separated into at least 10 different components. By high-performance liquid chromatography (reversed phase), Tsuji et al. (1974) reported further separation into 22 components. However, some of these components are probably degradation products of the microbiologically active bacitracins.

B. Structure

The structure of bacitracin A, the most potent and main bacitracin component, was initially studied during the 1950s by Craig and co-workers at The Rockefeller Institute for Medical Research, New York, and by Newton and co-workers at Oxford University.

Newton and Abraham (1953a) reported that bacitracin A is a peptide containing the amino acids isoleucine, cysteine, leucine, glutamic acid, lysine, ornithine, phenylalanine, histidine, and aspartic acid. The ionizable functions were shown to be one free amino group, the δ-amino group of ornithine, two free carboxyl groups, and the imidazole group of histidine. It was suggested that bacitracin A has a cyclic structure in which one carboxyl group is condensed with the ε-amino group of lysine. By partial hydrolysis, Lockhart and Abraham (1954) were able to isolate the following amino acid sequence:

$$-S$$
$$|$$
$$Ile-Cys-Leu-Glu$$

Continued structural studies by Hausmann et al. (1955) indicated that bacitracin A contained three isoleucine residues and the following structure was suggested:

$$
\begin{array}{c}
Ile \to Cys \to Leu \to Glu \to Ile \to Lys \to Orn \to Ile \\
\uparrow \qquad\qquad\qquad \downarrow \\
Asp \leftarrow His \leftarrow Phe \\
Asp
\end{array}
$$

Here the arrows denote the peptide bonds and the amide bond between the ε-amino group of lysine and the free carboxyl group of aspartic acid. Later it was shown by Ressler and Kashelikar (1966) and by Galardy et al. (1971) that the internal peptide ring was a heptapeptide ring, where the free α-carboxyl group of the C-terminal asparagine residue was amide bound to the ε-amino group of lysine as indicated below:

$$
\begin{array}{c}
Ile \to Cys \to Leu \to Glu \to Ile \to Lys \to Orn \to Ile \\
\qquad\qquad\qquad\qquad \downarrow \\
Asn \leftarrow Asp \leftarrow His \leftarrow Phe
\end{array}
$$

The presence of D-amino acids in bacitracin was initially reported by Sharp et al. (1949). It was concluded from experiments with D-amino acid oxidase that some of the amino acids in crude bacitracin had the D configuration. Later it was shown (Lockhart and Abraham, 1954; Ressler and Kashelikar, 1966; Galardy et al., 1971) that glutamic acid, ornithine, phenylanine, and asparagine occur as D-amino acids as shown:

$$
\begin{array}{c}
L-Ile-L-Cys \to L-Leu \to D-Glu \to L-Ile \to L-Lys \to D-Orn \to L-Ile \\
\qquad\qquad\qquad\qquad\qquad \downarrow \\
L-Asn \leftarrow D-Asp \leftarrow L-His \leftarrow D-Phe
\end{array}
$$

Experiments performed by Newton and Abraham (1953b) suggested that bacitracin contained a 2-thiazoline ring:

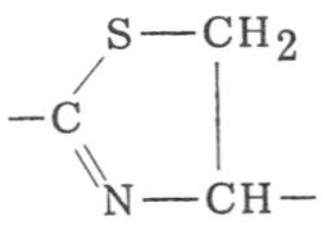

$$\overset{S—CH_2}{\underset{N—CH—}{-C}}$$

Later it was shown that the thiazoline ring was a condensation product of the N-terminal isoleucine residue and the cysteine residue (Lockhart et al., 1955; Weisiger et al., 1955). Based on the above findings, a primary structure of bacitracin A was proposed, as represented in Figure 1. In Figure 2 further details of the structure is shown. The empirical formula of bacitracin A is $C_{63}H_{103}O_{16}N_{17}S$. The molecular weight of bacitracin A is 1423.

In order to verify the primary structure of bacitracin A, Munekata et al. (1973) tried to synthesize the molecule chemically. Owing to extremely low yields of intermediate products and the instability of the thiazoline ring, total synthesis of bacitracin A was not achieved. Later the group succeeded in synthesizing bacitracin F, which contains the thiazole ring (Hirotsu et al., 1979).

Galardy et al. (1971) suggested on the basis of thin-film dialysis studies that bacitracin A has a compact structure with the peptide tail folded over the ring. By electron microscopy studies Ottensmeyer et al. (1978) were able to observe the configurations of two circles of slightly different sizes. The structure had a size of about 15 × 20 Å. In accordance with the studies of Hausmann et al. (1955), which indicated the existence of an interaction between the N-terminal isoleucine residue and phenylalanine, a space-filling model of bacitracin was built. In this model the five N-terminal amino acids loop back to phenylalanine on the seven-membered cyclopeptide portion. This forms a slightly larger circle of nine amino acids.

The nature of the binding between isoleucine and phenylalanine is not known. Carbon-13 nuclear magnetic resonance (NMR) studies (Lyerla and Freedman, 1972) show that the isoleucine resonances in the peptide ring and side chain deviated somewhat from their expected values. This might reflect interactions between the peptide ring and the side chain which restrain the isoleucine methyl groups in specific orientations.

In bacitracin B, one of the L-isoleucine residues is substituted with L-valine (Craig and Konigsberg, 1957). The substitution of one of the amino acids in the bacitracin molecule reflects the substrate unspecificity of the multienzyme complex in *B. licheniformis* which carries out the formation of bacitracin (Frøyshov and Laland, 1974; Frøyshov, 1978b). There may be

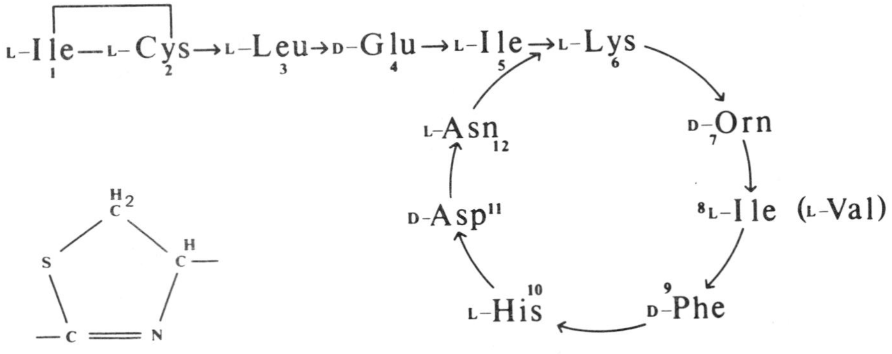

Figure 1 Amino acid sequence of bacitracin A.

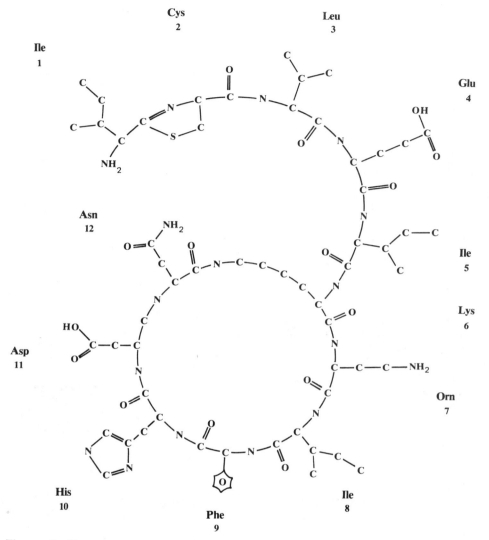

Figure 2 Structure of bacitracin A. For convenience, most of the H atoms in the formula are omitted.

other structural differences between bacitracins A and B, but this has not been further investigated. In general, bacitracins A and B represent about 96% of the total microbiological activity in commercial products.

It has been observed that bacitracin A is transformed to bacitracin F in neutral or slightly alkaline solutions (Craig and Konigsberg, 1957). The ketothiazole in bacitracin F is shown in Figure 3. In commercial products, bacitracin F (oxidized bacitracins A and B) usually makes up less than 2% by weight of the total bacitracin present. A more rapid oxidation of bacitracin is observed in the presence of Cu^{2+} ions (Newton and Abraham, 1953b).

By carboxymethyl cellulose chromatography, Konigsberg and Craig (1959) separated and identified a minor component from bacitracin A. The component

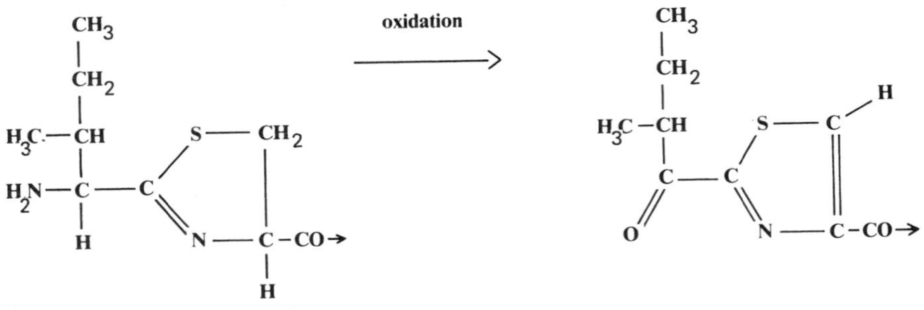

Figure 3 Formation of bacitracin F.

which accumulated under acetic acidic conditions was identified as D-alloiso-leucylbacitracin A, where the N-terminal amino acid L-isoleucine was epimer-ized to D-alloisoleucine. The epimerization product, which possessed a re-duced antimicrobial activity, was named "low-potency bacitracin A." In addi-tion, it was found that bacitracin B was also composed of the same type of optical isomers (Craig et al., 1960).

Other transformation products of the bacitracins are the microbiologically inactive deamidobacitracins. Under alkaline conditions (NaOH) the amide group of L-asparagine is lost (Newton and Abraham, 1953a; Ressler and Kashelikar, 1966).

In commercial lots of bacitracin the presence of the minor components bacitracins C, D, E, and G has been reported (Newton and Abraham, 1953b). However, further investigations on their structure have not been performed.

C. Properties

Purified bacitracin is a white, amorphous, hygroscopic powder that is readily soluble in polar solvents. The antimicrobial potency of commercial bacitracin (free base) is on the order of 70–74 IU/mg (Lightbown et al., 1964).

The maximum stability of bacitracin in solutions is in the pH range 5–6, with decreased stability outside this range. The isoelectric point for bacitra-cin A is 7.1 and that for bacitracin F is 6.0 (Frøyshov, 1978a). The corre-sponding values for bacitracins B and F are similar.

The thiazoline ring in the microbiologically active peptides exhibits an absorption maximum at 253 nm (Newton and Abraham, 1953b; Craig et al., 1969), while the thiazole ring of bacitracin F gives a maximum at 288 nm. Fur-thermore, the thiazoline and the thiazole rings provide two differential pulse polarograms (Jacobsen et al., 1977). Bacitracin shows remarkable resistance against the action of proteolytic enzymes like trypsin, pepsin, ficin, and ficus protease (Hickey, 1964). However, some minor effect of papain has been ob-served. Bacitracin is readily hydrolyzed by acid (Craig et al., 1952). Under mild alkaline conditions the thiazoline ring of bacitracin is oxidized to a keto-thiazole, and bacitracin F is formed (Craig and Konigsberg, 1957). In 0.1 N NaOH the amide group of the C-terminal amino acid asparagine is readily lost (Ressler and Kashelikar, 1966).

The solubility of bacitracin, zinc bacitracin, and oxidized bacitracin (bacitracin F) are shown in Table 1.

Table 1 Solubility of Bacitracin

Solvent	Solubility (mg/ml)		
	Bacitracin	Zinc bacitracin	Bacitracin F
Petrolether (b.p., 100−120)	—	—	0.002
Petrolether (b.p., 40−60)	—	—	0.007
Carbotetrachloride	0.19	0.009	0.37
Toluene	—	—	—
Diethylether	—	0.002	—
Chloroform	0.006	0.07	0.24
Pyridine	17.0	16.1	11.4
1-Butanol	4.2	0.68	>15
Acetone	0.08	0.06	0.09
Methanol	>40	17.3	>15
Acetonitrile	0.005	0.005	0.07
Water	>40	10.1	11.2

The solubility of bacitracin (pharmaceutical grade), zinc bacitracin (pharmaceutical grade), and oxidized bacitracin were tested at 25°C. At our laboratory, the detection limit was 0.002 mg/ml. The microbiological activities, when tested with *M. luteus* against the Second International Standard, were 67.0, 67.3, and 1.8 IU/mg, respectively.

Under proper conditions, bacitracin may be precipitated by the agents listed in Table 2 (Hickey, 1964).

The thiazoline ring of bacitracin is reduced to a thiazolidine ring with diborane (Atassi and Rosenthal, 1969) or sodium borohydride (Shipchandler, 1976).

Table 2 Some Precipitating Agents for Bacitracin Resulting in Salts or Other Derivatives

Ammonium molybdate
Ammonium rhodanilate
Azobenzene p-sulfonic acid
Benzoic acid
Divalent metal ions
Furoic acid
Methylene disalicyclic acid
Molybdic acid
Phosphomolybdic acid
Phosphotungstic acid
Picric acid
Picrolonic acid
Salicylic acid

D. Metal Binding Properties

Bacitracin precipitates from aqueous solutions by the addition of divalent metal ions. Under proper conditions bacitracin is also precipitated by molybdate, methylene disalicylate, and other agents (Baron, 1956; Gollaher and Honohan, 1956; Chornock, 1957; Zinn and Chornock, 1958; Zorn, 1959, 1961, 1962; Ziffer and Cairney, 1962; Lewis et al., 1965; Hodge and Riddick, 1975). Divalent metal ions are essential for antimicrobial activity of bacitracin. The most active ions were found to be Cd^{2+}, Mn^{2+}, and Zn^{2+} (Weinberg, 1958, 1960; Adler and Snoke, 1962). The interaction between bacitracin and divalent metal ions has been demonstrated by a number of techniques, including potentiometric titration, optical absorption, optical rotatory dispersion, and NMR studies. By titration studies Garbutt et al. (1961) found the following apparent order of metal binding:

$$Cu^{2+} > Ni^{2+} > Co^{2+} = Zn^{2+} > Mn^{2+}$$

Bacitracin has been found to be most stable as its zinc salt. The formation of the zinc bacitracin complex was studied by Craig et al. (1969) and Cornell and Guiney (1970). Their NMR studies indicate that the thiazoline ring and the nitrogen in the 3-position in the imidazole ring of the histidine residue provide two coordination sites for the metal ion. Other coordination sites may also be involved (Konigsberg and Craig, 1959; Scogin et al., 1980; Mosberg et al., 1980). The association constant of zinc bacitracin was estimated to be 2.5 × 10^3 at pH 6.34. Commercial pharmaceutical zinc bacitracin has a zinc content of 4–6%. The binding properties of Ni^{2+}, Cu^{2+}, and Mn^{2+} to bacitracin are similar to that of Zn^{2+} (Weinberg, 1964; Washylishen and Graham, 1975; Scogin et al., 1980; Mosberg et al., 1980).

III. BIOCHEMISTRY

A. Formation in Whole Cells

The production of bacitracin is frequently reported to be initiated in the late growth phase. During the last years it has been apparent that bacitracin can normally be produced during active growth (Haavik, 1974a; Frøyshov, 1977). The growth/production relationship may be pH dependent (Haavik, 1974a,b). Similar observations have been made for the peptide antibiotics polymyxin E, penicillin, tyrothricin, and gramicidin S (Haavik and Frøyshov, 1975; Vandamme, 1981; Vandamme et al., 1981).

In Figure 4, bacitracin production in relation to growth is shown for a defined medium. The pH development in the culture and the bacitracin synthetase activity (see Sec. V.B) have also been illustrated.

B. Mechanism of Biosynthesis

Most of the studies on the biosynthesis of bacitracin have been performed with strain ATCC 10716 of B. licheniformis. An industrial high-production strain and nonproducers of bacitracin have been used in some experiments (Haavik and Frøyshov, 1975; Haavik, 1979a,b).

A cell-free system for de novo bacitracin formation has been reported by Ishihara et al. (1968), Pfaender et al. (1973), and Frøyshov and Laland

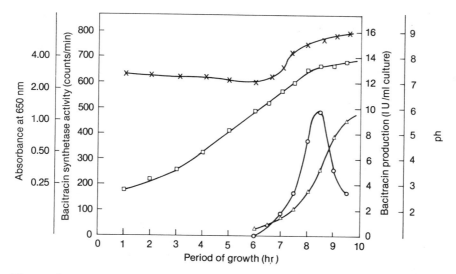

Figure 4 The production of bacitracin and bacitracin synthetase activity in relation to growth and pH. *Bacillus licheniformis* ATCC 10716 was grown in a glutamate–citrate–starch medium. The following parameters were measured: Cell growth (□), as the absorbance at 650 nm; pH (×); bacitracin (△) and bacitracin synthetase activity (○). (From Frøyshov and Laland, 1974.)

(1974). Both the L isomers of the constituent amino acids and the D-amino acids, D-glutamic acid, D-phenylalanine, and D-asparatic acid, which occur in bacitracin, support its synthesis in the presence of adenosine triphosphate (ATP) and a divalent cation like Mn^{2+}, Mg^{2+}, Co^{2+}, or Fe^{2+} (Frøyshov, 1978b; Frøyshov et al., 1980).

Bacitracin is synthesized by an enzyme complex, bacitracin synthetase. The L isomers of the constituent amino acids are first activated by the enzyme complex as aminoacyl adenylates as measured by the [^{32}P]PP_i–ATP exchange reaction. The D-amino acids, glutamic acid, phenylalanine, and ornithine, which occur in the peptide, also promote the exchange reaction. After activation, the amino acids are transferred and covalently bound as thioesters to separate thiol sites on the synthetase (Ishihara and Shimura, 1974; Frøyshov and Laland, 1974). The most likely activation scheme for one single amino acid in the biosynthesis of bacitracin is shown in Figure 5.

It was of importance to demonstrate that the thioester bound amino acids really are intermediates in the biosynthesis of bacitracin. This was done by isolation of the ^{14}C-labeled aminoacyl–enzyme complex and reincubation with

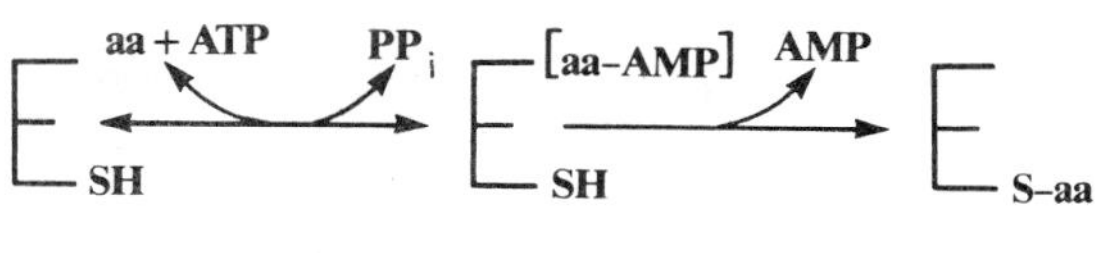

$$\underset{\text{SH}}{\text{E}} \xrightarrow[\text{PP}_i]{\text{aa + ATP}} \underset{\text{SH}}{\text{E}} [\text{aa–AMP}] \xrightarrow{\text{AMP}} \underset{\text{S–aa}}{\text{E}}$$

Figure 5 Schematic representation of the activation of a single amino acid (aa) in the biosynthesis of bacitracin. Brackets indicates that the aa–AMP complex is noncovalently bound to the synthetase.

unlabeled substrate amino acids. A significant portion of the thio ester bound amino acids was incorporated into bacitracin (Frøyshov and Laland, 1974).

During bacitracin formation, intermediate peptides are thio ester linked to the synthetase. The labeled peptides were liberated from bacitracin synthetase by performic acid oxidation and identified by thin-layer chromatography and radioautography. The peptides were

Ile—Cys

Ile—Cys—Leu

Ile—Cys—Leu—Glu

Ile—Cys—Leu—Glu—Ile

Ile—Cys—Leu—Glu—Ile—Lys

Ile—Cys—Leu—Glu—Ile—Lys—Orn

Ile—Cys—Leu—Glu—Ile—Lys—Orn—Ile

Ile—Cys—Leu—Glu—Ile—Lys—Orn—Ile—Phe

Ile—Cys—Leu—Glu—Ile—Lys—Orn—Ile—Phe—His

Ile—Cys—Leu—Glu—Ile—Lys—Orn—Ile—Phe—His—Asp

The radioactivity of the undecapeptide spot was low and any dodecapeptide spot containing all the amino acids of bacitracin could not be detected. This might be due to the reaction kinetics of the different peptidization steps. By reincubation experiments it was shown that significant amounts of the thio ester bound peptides were transferred to bacitracin (Frøyshov, 1975).

Bacitracin synthetase is an enzyme complex consisting of three multifunctional enzymes (Frøyshov, 1974). The three enzymes A, B, and C occur simultaneously in the cytoplasm during the growth phase. Later bacitracin synthetase associates with the membranes or cell walls. This association does not seem to have any negative effect on bacitracin formation and may represent a natural immobilization of an enzyme complex protecting it against increasing proteolytic activities (Frøyshov, 1977). A schematic representation of bacitracin synthetase is given in Figure 6.

Enzymes A, B, and C activate amino acids 1–5, 6–7, and 8–12, respectively, of bacitracin. It is believed that the enzymes A, B, and C contain one thiol site for each amino acid, as shown in Figure 6 (Frøyshov, 1974; Ishihara et al., 1975).

Bacitracin synthetase is less specific than the ribosomal system. In addition to the amino acids normally activated by enzyme A, amino acid analogs and nonconstituent amino acids of bacitracin like D-alloisoleucine, S-methylcysteine, D-leucine, L-methionine, and L-serine are activated. However, it has not been shown that these amino acids are incorporated into bacitracin. Both enzymes A and C activate L-valine, one of the amino acids of bacitracin B. The experiments suggest that valine is activated mostly at the leucine site, and to a smaller extent at one or both of the isoleucine sites on enzyme A. On enzyme C, which does not activate leucine, valine seems to be activated at the isoleucine site (Frøyshov, 1978b). This provides a possibility for directed control of the bacitracin composition in minimal media, for instance, by adding larger amounts of a single amino acid to the media, as described for the peptide antibiotic tyrocidine (Mach and Tatum, 1964).

The enzymes of bacitracin synthetase all contain the cofactor phosphopantothenic acid. It is probably present as phosphopantetheine, which may

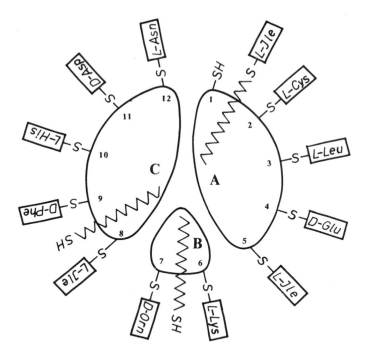

Figure 6 Schematic representation of bacitracin synthetase. A, B, and C represent the three complementary enzymes of bacitracin synthetase. The activation sites for the constituent amino acids (numbered 1–12 from the N-terminal end) in bacitracin are indicated. The zigzag line illustrated covalently linked phosphopantetheine. (© O. Frøyshov, Ph.D. thesis.)

function as a carrier of intermediates in the biosynthesis of bacitracin (Frøyshov and Laland, 1974; Ishihara et al., 1975; Roland et al., 1977).

The molecular weights of enzymes A, B, and C have been estimated to be 140,000, 210,000, and 380,000, respectively, by sodium dodecylsulfate (SDS) polyacrylamide gel electrophoresis (Ogawa et al., 1981), and 335,000, 200,000, and 325,000 by gel filtration (Frøyshov and Mathiesen, 1982).

It has been observed that trypsin treatment of bacitracin synthetase may reduce the bacitracin formation while the amino acid activation reactions still occur (Frøyshov and Mathiesen, 1979). It was suggested that the multifunctional enzymes A, B, and C are built up of smaller subunits or domains. This has been demonstrated by limited trypsin digestion of the individual enzymes of bacitracin synthetase, followed by gel filtration separation of the cleavage products. Smaller domains with molecular weights of 70,000–75,000 were isolated (Frøyshov and Mathiesen, 1979, 1982).

Other types of domains which are not detectable in this test system may also be present, for instance, smaller proteins like the pantetheine carrier protein of tyrocidine synthetase (Lee and Lipmann, 1975) which was split off by a factor in the growth medium.

In Figure 7 the functioning of bacitracin synthetase is suggested. After activation as aminoacyl adenylates, the constituent amino acids are bound as thioesters to specific thiol groups on the surface of the synthetase (Fig. 7a). Chain growth is initiated on enzyme A. L-isoleucine is transferred to

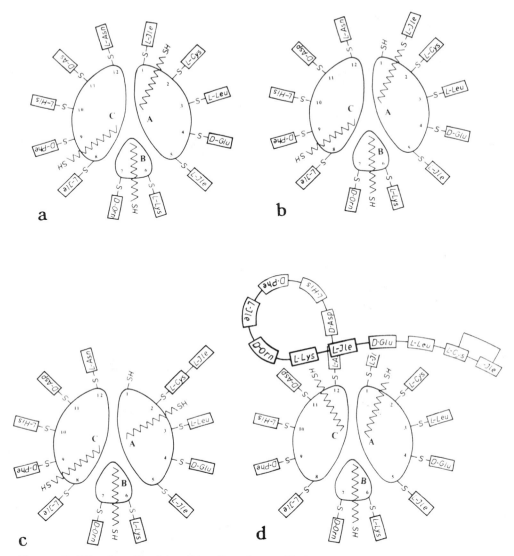

Figure 7 The functioning of bacitracin synthetase.

the phosphopantetheine arm which is covalently linked to this enzyme component (Fig. 7b). By movement of the arm, L-isoleucine is brought sufficiently
close to thioester bound cysteine for reaction to occur. The resulting dipeptide
Ile—Cys Fig. 7c) is transferred to the same phosphopantetheine arm. By further
movement of the arm, the dipeptide is brought close to the thioester bound
leucine. The formed tripeptide then joins onto glutamic acid and the tetrapep-
tide joins onto isoleucine in a similar way. The resulting pentapeptide is
transferred to enzyme B, where leucine and ornithine are added. The resulting heptapeptide is transferred to enzyme C, where the linear dodecapeptide L-Ile—L-Cys—L-Leu—D-Glu—L-Ile—L-Lys—D-Orn—L-Ile—D-Phe—L-His—D-
Asp—L-Asn is formed. The termination of bacitracin biosynthesis probably

takes place in the following way. After formation of the dodecapeptide, the C-terminal amino acid is thio ester linked to enzyme C. An attack by the ε-amino group of lysine on this bond results in the liberation of the molecule from the enzyme with the simultaneous formation of the amide bond in the hepta-peptide ring. In Figure 7d the peptide ring of bacitracin is about to form. The thiazoline ring, which is illustrated in Figure 7 is probably formed at the dipeptide stage (Ishihara and Shimura, 1979).

This mechanism of formation is similar to that described for the other *Bacillus* peptides gramicidin S, tyrocidine, linear gramicidin, polymyxin, and edeine, and it has been characterized as the thio template mechanism of forma-tion by Laland and Zimmer (1973). The about 53 catalytical functions of baci-tracin synthetase hitherto described can by summarized in the following way:

1. Intradomain reactions
    Amino acid activation reactions   12 reactions
    Amino acid thiolization reactions   12 reactions
    Racemization reactions   4 reactions
    Formation of the thiazoline ring   1 reaction

2. Interdomain reactions
    Chain initiation   1 reaction
    Elongation reactions   11 reactions
    Trans-thiolization reactions   9 reactions
    Termination reaction (probably)   1 reaction

3. Interenzymatic reactions
    Peptidyl transfer from A to B   1 reaction
    Peptidyl transfer from B to C   1 reaction

C. Regulation of Biosynthesis

At present, much interest is focused on possible control mechanisms for anti-biotic formation. Synthesis of several antibiotics may be controlled by cata-bolite regulation and phosphate regulation (Demain, 1972; Martin, 1977). How-ever, in the case of bacitracin production by *B. licheniformis* ATCC 10716, both the glucose effect (catabolite regulation) and the phosphate effect may be due to a more general response of the cells to environmental factors (Haa-vik, 1974a,b,c). Several antibiotics exert feed-back effects on their own synthesis (Demain, 1976) and trace metals also seem to have some regulatory functions (Weinberg, 1970). The effect of bacitracin and divalent metal ions on bacitracin formation by the bacitracin synthetase have therefore been in-vestigated.

Addition of bacitracin to bacitracin synthetase inhibited the enzyme ac-tivity in the presence of several different metal ions and it was suggested that product—metal ion complexes exert feed-back inhibition on the enzyme activity (Frøyshov et al., 1980).

The amino acid pool in cells of *B. licheniformis* seems to be of critical importance to the amount of bacitracin produced. (Haavik and Vessia, 1978; Haavik, 1979a,b, 1981). Further studies indicate that the substrate amino acid L-leucine plays an inducer role in the regulation of bacitracin synthetase (Haa-vik and Frøyshov, 1982). Also other amino acids may play a certain inducer role (Haavik, 1979a, 1981).

IV. PRODUCTION

A. Strains

The original bacitracin-producing organism was called *Bacillus subtilis* strain Tracy I. Later it was classified as a variant of *B. licheniformis* (Johnson et al., 1945; Meleney and Johnson, 1949) and named *B. licheniformis* ATCC 10716 (Snoke, 1960). Ayfivin, which is produced by *B. licheniformis* strain A-5 (Arriagada, et al., 1949) was later found to be identical to bacitracin (Newton and Abraham, 1950). Other bacitracin-producing strains are *B. licheniformis* ATCC 9945, 11945, 11946, and 14580 and *B. subtilis* 14593.

Strain improvement and media development are key factors in antibiotic production. Traditional strain improvement involves mutation and selection of mutants. To induce mutants, ultraviolet irradiation, alkylating agents, and the use of acridines have frequently been used. Interesting mutants may be those which are:

> High producers of a certain metabolite
> Nonproducers of a certain metabolite
> Nonproducers of bacitracin
> Auxotrophs
> Resistant to bacteriophages
> Displaying better growth characteristics than the parent strain

For each interesting mutant isolated, a parallel study of the fermentation conditions (medium composition, pH, aeration, agitation, incubation temperature) is undertaken in order to optimize the conditions for growth and bacitracin production.

Many results from the intensive studies on mutants and growth conditions of the bacitracin-producing bacilli which have been carried out in industrial research laboratories are considered proprietary knowledge and are not being published.

Recently recombinant DNA techniques have been adapted for strain improvement. For the antibiotic-producing bacilli these new "genetic engineering" techniques look very promising. In several species of the genus Bacillus, genetic transformation has been demonstrated (Erickson and Young, 1978). Since this mechanism of genetic exchange involved nondiscriminating uptake of DNA as regards the source of DNA, foreign DNA can enter such cells. Thus the antibiotic-producing bacilli should be suitable organisms for "genetic engineering" purposes.

B. Media

For bacitracin production different types of media have been used. Among those are synthetic media containing amino acids as C and N sources (Anker et al., 1948; Haavik, 1981), rich media containing beef extracts or tryptone broth (Anker et al., 1948), and high-producing media with soybean meal and starch or dextrines/sugars as nutrients (Darker, 1951; Keko et al., 1953; Cohen, 1957; Freaney and Allen, 1958; Zorn, 1961; Zorn et al., 1961; Smekal et al., 1979).

Reported levels of bacitracin production are 408 IU/ml (Tyc, et al., 1976), and 570–590 IU/ml (Istvan et al., 1978). As defined by Johnson et al. (1945), 74 IU (international units) corresponds to 1 mg of the Second International Standard of bacitracin (Lightbown et al., 1964).

C. Fermentation

The bacitracin-producing strains of B. *licheniformis* are kept on agar slants or in the spore form in dry sterile sand. A suitable amount of the culture is used to inoculate shake flasks of tryptone or peptone broth. An intermediate fermenter is usually used as a seed tank for the main fermenter. The fermenters are equipped with an agitator and a jacket for steam or water. The medium is sterilized either by batch or continuous sterilization. The fermentation process is usually monitored by continuous pH and O_2 measurements.

The fermentation of bacitracin is carried out at 30–37°C, for 24–30 hr (Cohen, 1957; Freaney and Allen, 1958; Zorn et al., 1961).

Recently cells of B. *licheniformis*, producing bacitracin, have been immobilized in polyacrylamide gel (Morikawa et al., 1979; Stepanov and Rudenskaya, 1978). This provides a new technique for the production of pure bacitracin and bacitracin analogs.

D. Recovery

Bacitracin is processed/purified in several ways, according to its application. Different principles for the isolation of bacitracin are shown in Figure 8.

1. *Feed-Grade Bacitracin*

For the preparation of feed-grade products, direct precipitation of bacitracin in the fermentation broth by salts of divalent cations is common. Zinc bacitracin (Chornock, 1957; Zinn and Chornock, 1958), nickel bacitracin (Zorn, 1959), manganese bacitracin (Zorn, 1961; Zorn et al., 1961, and cobalt bacitracin (Zorn, 1962) are examples.

Bacitracin can also be precipitated after fermentation as salts of methylene disalicylic acid (Baron, 1956), lignin sulfonic acid (Ziffer and Cairney, 1962), hydroxymethane sulfonic acid (Lewis et al., 1965), and alkylbenzene sulfonic acid (Hodge and Riddick, 1975).

Furthermore, bacitracin can be adsorbed/precipitated on solid carriers like bentonite (Charney, 1952; Wehrmeister, 1956) and celite (Monroe and Ward, 1967).

Before final drying, precipitated bacitracin is concentrated by filtration, centrifugation (Chornock, 1957), or evaporation (Baron, 1956). The product

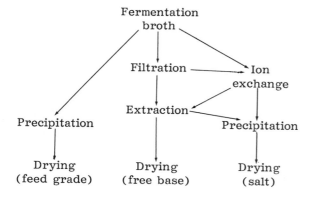

Figure 8 Principles of bacitracin preparation.

is dried either by drum drying (Chornock, 1957; Zorn et al., 1961) or spray drying (Baron, 1956; Øystese, 1978, 1979).

At present, feed-grade bacitracin is mostly used in the form of zinc bacitracin (ZB) and bacitracin-methylene disalicylate (BMD).

A typical analysis of a final fermentation product is shown here:

Crude protein	40%
Ether extract	5%
Fiber	5%
Ash	23%
Moisture	4%

The product may be standardized with calcium carbonate, soybean meal, or starch to give, for instance, 100 or 150 g of bacitracin activity per kilogram of product (100 g bacitracin activity corresponds to 4,200,000 international units). The product is mixed into feeds in the concentrations 5–100 ppm as shown in Table 3. See definition under V.B.

2. Pharmaceutical-Grade Bacitracin

Different processes are used for the purification of bacitracin:

Solvent extraction of bacitracin with butanol from clarified fermentation broth (Senkus and Markunas, 1952; Chornock, 1957; Miescher, 1974; Kindraka and Gallaher, 1978)

Removal of bacitracin from the fermentate on ion exchanger, followed

Table 3 Proposed Concentrations in Feedstuff[a]

Animal species	Content in complete feedstuff (ppm)		Remarks and maximum age
	Minimum	Maximum	
Turkeys	5	50	4 weeks
	5	20	5–26 weeks
Other poultry (excluding ducks, geese, laying hens, pigeons)	5	50	4 weeks
	5	20	5–16 weeks
Calves, lambs, kids	5	50	16 weeks
	5	20	17–26 weeks
	5	80	Milk feeds
Piglets	5	50	4 months
	5	80	Milk feeds
Pigs	5	20	4–6 months
Animals bred for fur	5	20	
Laying hens	15	100	

[a]EEC Council Directive 70/5241EEC with later amending directives lists the above doses for zinc bacitracin in feeds. The bacitracin content is given in parts per million. 100 ppm corresponds to 4200 IU/kg feed.

by solvent extraction (Shortridge, 1957; Chaiet and Cochrane, 1959; Hoff, 1972)

Final precipitation of bacitracin with divalent metal ions

Bacitracin (free base) and zinc bacitracin, pharmaceutical grade, are purified dried products containing about 70 and 65 IU/mg, respectively.

V. ASSAY

A. Microbiological Assays

The microbiological assay of bacitracin is usually carried out by a plate agar diffusion technique (Craig, 1967; Grynne and Hoff, 1968; Grynne, 1971; Council Directive EEC, 1978; A. L. Zinc Bacitracin assay methods, 1978). *Micrococcus luteus* is used as a test organism. A turbidimetric assay has also been described (Darker et al., 1948; Ragheb et al., 1976).

The microbiological assay of bacitracin may be disturbed by the presence of substances which influence the biological activities of zinc bacitracin. This is especially a problem in the feed industry. There are several possible ways in which the assay of bacitracin may be influenced:

The samples may contain other substances with microbiological activity.
Zinc bacitracin may bind to substances in the feed, rendering it more difficult to extract.
The feed may contain substances which react chemically with zinc bacitracin during the extraction and dilution steps.
The feed may contain substances which influence the diffusion of zinc bacitracin in the agar plates.
The feed may contain substances which influence the test organism used in the zinc bacitracin assay.
The handling of the feed during processing may influence the distribution of zinc bacitracin particles in the finished feed.
Incorrect sampling and assay procedure should also be considered when an assay result does not turn out according to the claimed values.

The A. L. Zinc Bacitracin Assay Methods (1978) contains a survey of principles and techniques which can be used in handling samples containing interfering substances.

In order to clarify the definition of the microbiological activity of bacitracin, a short summary of the different bacitracin standards is made. One unit of bacitracin was originally defined as the amount of bacitracin which, when diluted 1:1042 in a series of twofold dilutions in 2 ml of beef infusion broth, completely inhibits the growth of a stock strain (Chanin) of a group A hemolytic streptococcus under closely defined conditions (Anker, et al., 1945). This material was defined to contain 38.5 units/mg and was officially designated as the Food and Drug Administration Bacitracin Working Standard in 1949.

In 1950 The FDA Master Standard with a potency of 42 units/mg was established. The definition of 42 units/mg is still used by the feed industry to express the "content" of bacitracin in animal feeds on a weight basis (Code of Federal Regulations, 1970). In 1952 the World Health Organization (WHO) First International Standard with 55 units/mg was established (Humprey et al., 1953). It was gradually used up and a new standard containing zinc bacitracin, which is more stable than bacitracin, was prepared. The WHO Second International Standard contains 74 units/mg (Lightbown et al., 1964).

B. Alternative Methods

The determination of bacitracin is usually carried out by microbiological assays. However, alternative methods are often needed in cases where the microbiological assay results are uncertain, the purpose of the investigations is to look for biological transformation or degradation products, or a rapid answer is wanted (process control).

Paper and thin-layer chromatography has been used to separate bacitracin from other antibiotics. Coloring reagents like ninhydrin, ponceau C, and chlortoluidin, ultraviolet detection and overlaying techniques with microorganisms have been used for the identification. The techniques have been described by several authors (Aszalos and Frost, 1975; Betina, 1975; Aszalos and Issaq, 1980).

By high-voltage electrophoresis on agar, similar separations can be carried out (Lightbown and Rossi, 1965; Grynne, 1973; Smither and Vaughan, 1978; Smither et al., 1980).

Polyacrylamide gel electrophoresis (PAGE) may be used for the separation of bacitracin from other antibiotics in pharmaceuticals (Coombe, 1972). The technique has been further developed for the determination of zinc bacitracin in feeds (Fink, 1978; Kaemmerer and Fink, 1980a,b).

Isoelectric focusing in gel has been used for the separation of microbiologically active and inactive bacitracins (Frøyshov, 1978a).

Pulse polarographic techniques are currently used for rapid quantitative determination of bacitracin in the fermentation broth and during the purification process (Jacobsen et al., 1977). In Figure 9, the polarogram of a bacitracin fermentate is shown.

High-performance liquid chromatography (HPLC) is a very important chemical technique for bacitracin analyses owing to its speed, high resolution and high detection sensitivity, good prospects for quantitative analysis, possibility for detection of degradation and transformation products, and good prospects for instrumentation and automation.

In 1974 Tsuji et al. and in 1975 Tsjui and Robertson determined bacitracin by using a reversed-phase column and a programmed gradient elution.

At our laboratories an isocratic system was developed for bacitracin de-

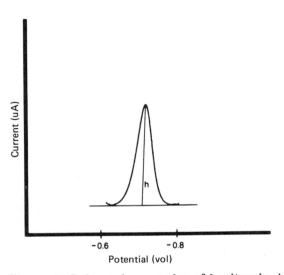

Figure 9 Pulse polarography of bacitracin A + F.

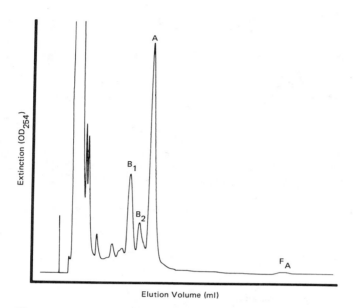

Figure 10 HPLC of purified bacitracin. The bacitracins are detected by a spectro photometer at 254 nm. Peak A represents bacitracin A, peaks B1 and B2 represent bacitracins where one of the isoleucine residues is subsituted with valine, and peak F3 represents bacitracin F (oxidized bacitracin A).

termination in the fermentation broth and during the purification of bacitracin (Frøyshov and Mathiesen, 1983). In Figure 10, an HPLC chromatogram of bacitracin fermentate is shown.

VI. MODES OF ACTION

A. Function in the Producer Organism

Peptide antibiotics in the producer organisms have been ascribed a wide variety of roles, ranging from evolutionary relics and waste products to elevated control in metabolism (Weinberg, 1970).

In the case of bacitracin it has been shown that the antibiotic is normally produced during active growth and may also have a function during growth (Haavik 1974a; Haavik and Frøyshov, 1975). The normal function of bacitracin during growth of B. licheniformis may be to promote the uptake of essential divalent metal ions from soils poor in that element. Bacitracin probably acts as a "helper molecule" which modifies the cell membrane in some way so that it takes up the metals more readily.

Haavik (1976) found that relatively high (10–30 IU/ml) concentrations of bacitracin were needed to inhibit growth of the producer organism, but in mineral salt solutions of sublethal levels much lower antibiotic concentrations (5 IU/ml) inhibited growth. Manganese, cobalt, nickel, copper, and zinc all had their toxic effects magnified by bacitracin. This effect was antagonized by magnesium. It was suggested that bacitracin promotes an increase of the uptake of several divalent cations from anionic "waiting sites" on the cell walls. Bacitracin, which also attaches to the membrane, was suggested to mediate cation transfer to the cell membrane and further into the cell by an "inte-trated cation exchange system."

The role of bacitracin as a trigger for sporulation was ruled out by the isolation of a zero producer of bacitracin which sporulates normally (Haavik and Thomassen, 1973). The absence of bacitracin production was due to a defect in bacitracin synthetase (Haavik and Frøyshov, 1975; Frøyshov, 1975).

B. Antimicrobial Activity

Bacitracin is mainly active against gram-positive bacteria, including the genera *Micrococcus*, *Staphylococcus*, *Sarcina*, *Diplococcus*, *Streptococcus*, *Coryne-bacterium*, *Clostridium*, *Actinomyces*, and *Treponema*. The gram-negative bacteria are generally considered resistant to zinc bacitracin, but some species in the genera *Vibrio*, *Neisseria*, *Hemophilus*, and *Bordetella* have been reported to be sensitive (Johnson et al., 1945; Hickey, 1964; Brunner, 1965; Weinberg, 1967).

Bacitracin affects both membrane functions, cell wall synthesis, and protein synthesis in the cells.

1. Membrane Permeability

Bacitracin seems to affect the membrane permeability of gram-positive bacteria. Protoplasts of *B. licheniformis* and *Micrococcus lysodeikticus* are rapidly lysed in the presence of bacitracin and zinc or cadmium ions at concentrations which also inhibit the growth of whole cells (Snoke and Cornell, 1965). The growth of protoplasts does not require peptidoglycan biosynthesis and, as expected, bacitracin, in contrast to penicillin, affects protoplast growth (Shockman and Lampen, 1962). By electron microscopic studies, morphological changes of plasma membranes are observed after bacitracin treatment. This may be due to the surface-active properties of bacitracin. However, the mechanism of bacitracin action on membranes is not yet clear.

2. Cell Wall Synthesis

Bacitracin inhibits cell wall synthesis. Concentrations of bacitracin which have no effect on protein synthesis cause the accumulation of peptidoglycan precursors (Park, 1958). Siewert and Strominger (1967) demonstrated that bacitracin inhibits the dephosphorylation of the pyrophosphate derivative of the 55-carbon isoprenol lipid, which is an intermediate in the peptidoglycan synthesis. In the presence of bacitracin and magnesium, isoprenyl pyrophos-phate accumulated at higher levels.

The precise way in which bacitracin binds to its target is not known. It is most likely to be via a divalent metal ion bridge in addition to secondary hydrophobic and van der Waals interactions (Storm, 1974). There is a rea-sonable correlation between antibiotic activity and binding constants for the different bacitracins (Storm and Toscano, 1979). Maximum binding required a divalent cation and the greatest affinity for the pyrophosphate function was seen in the presence of Zn^{2+}, Cd^{2+}, or Mg^{2+} in decreasing order. Simi-larly, it is known that a divalent cation is required for antibiotic activity, with Zn^{2+} and Cd^{2+} being most effective (Weinberg, 1958).

Bacitracin also interacts with the biosynthesis of squalene and other sterols which proceed through the formation of C-farnesyl pyrophosphate and the biosynthesis of rat liver sterols, which proceeds through the forma-tion of isopentenyl or farnesyl pyrophosphate (Stone and Strominger, 1972).

C. Function as a Feed Additive

The mechanism of growth promotion and feed efficiency by bacitracin for domestic animals is largely unknown. It is generally accepted that the addition of low levels of antibiotics to feeds can improve animal growth and production; however, no general conclusion on the mechanism of the improved animal performance can be drawn. Bacitracin may slightly affect the gut microflora in a positive way, (Barnes et al., 1978; Smith and Tucker, 1980; Fuller, 1981; Henderickx et al., 1981; Walton, 1981).

It has also been shown that zinc bacitracin causes lesions in the cell wall of intestinal *Escherichia coli* (Walton and Bird, 1975; Walton, 1977). The authors postulated an increase in sensitivity of these damaged cells to the action of the immunological and cellular body defense mechanisms.

The thinner intestinal wall in germ-free animals may account for the increased weight gain and feed utilization. It has been observed that the thickening of the intestinal wall observed in conventionally reared animals can be prevented by the addition of zinc bacitracin to the feed (King, 1976). It was concluded that this may be one of the ways in which bacitracin improves growth.

Addition of bacitracin to the feed may affect the activity and synthesis of certain liver enzymes (Rybinska, 1977) and increase the level of proteases and amylases in the digestive tract of laying hens (Makarova, 1976).

D. Other Interactions

Bacitracin interacts with several enzyme systems. It affects the biosynthesis of β-galactosidases (Gale and Folkes, 1955) and the α-hemolysin in bacteria (Hinton and Orr, 1960). Bacitracin inhibits the biodegradation of β-endorphin by exopeptidases (Patthy et al., 1977).

Furthermore, bacitracin inhibits the enzymatic inactivation of glucagon (Desbuquois and Cuatrecasa, 1972), the degradation of the thyrotropin and luteinizing hormone releasing factor, (McKelvey et al., 1976), respiration of rat liver mitochondria (Nohl and Hegner, 1977), carboxylic proteinases (Stephanov et al., 1978), and the high-affinity binding of α_2-macroglobulin to plasma membranes (Dickson et al., 1981).

The presence of bacitracin resulted in high extracellular protein production in *Bacillus brevis* strains (Udaka, 1979).

VII. APPLICATION

A. Feed Additive

Zinc bacitracin and bacitracin methyldisalicylate (feed grade) are widely used for growth promotion. The use of up to 100 ppm bacitracin in the feed (which corresponds to 4200 IU/kg feed) improves weight gain and feed efficiency for growing animals like pigs, calves, rabbits, turkeys, and broilers, and improves the egg production of layers. Zinc bacitracin is also one of the most tested feed antibiotics. There seems not to be any problems with residues, bacterial resistance, or effects on the environment when bacitracin is used at nutritional levels up to 100 ppm (Rosen, 1980; Rosen et al., 1976, 1978).

The proposed concentrations in feeding stuffs in the EEC Council Directives are shown in Table 3.

B. Other Applications

In veterinary medicine, bacitracin may be used for treatment of enteritis. In human medicine, bacitracin is almost exclusively used as a topical antibiotic in the treatment of infections. Bacitracin is often supplemented by other anti-biotics such as polymyxin B or neomycin to make broad-spectrum combination preparations (Brunner, 1965; Otten et al., 1975). Furthermore, bacitracin has been used for the grouping of β-meolytic *Streptococci* (Arvilommi, 1976) and for protease purification by using immobilized bacitracin (Stepanou and Rudenskaya, 1978).

VIII. REVIEWS

During the last few years reviews have been written on the chemistry (Wein-berg, 1967), biochemistry (Frøyshov et al., 1978; Laland et al., 1978; Kura-hashi, 1981), manufacture (Hickey, 1964), assay (Brewer, 1980), function (Weinberg, 1960; Storm and Toscano, 1979; Katz and Demain, 1977), and ap-plication (Hickey, 1964; Brunner, 1965) of bacitracin.

ACKNOWLEDGMENTS

The author wishes to thank T. Høyland, Director of the Department of Re-search and Development, B Øystese, and Dr. S. Pedersen, for discussing the manuscript.

REFERENCES

Adler, R. H., and Snoke, J. E. (1962). Requirement of divalent metal ions for bacitracin activity. *J. Bacteriol. 83*:1315–1317.

A. L. Zinc Bacitracin Assay Methods (1978). A/S Spothekernes Laboratorium, Oslo.

Anker, H. S., Johnson, B. A., Goldberg, J., and Meleney, F. L. (1948). Bacitracin: Methods of production, concentration, and partial purifica-tion, with a summary of the chemical properties of crude bacitracin. *J. Bacteriol. 55*:249–255.

Arriagada, A., Savage, M. C., Abraham, E. P., Heatley, N. G., and Sharp, A. E. (1949). Ayfivin: An antibiotic from *B. licheniformis;* production in potato–dextrose medium. *Br. J. Exp. Pathol. 30*:425–427.

Arvilommi, H. (1976). Grouping of beta-haemolytic streptococci by using co-agglutination, precipitation or bacitracin sensitivity. *Acta Pathol. Mi-crobiol. Scand. Sect. B 84*:79–84.

Aszalos, A., and Frost, D. (1975). Thin layer chromatography of antibiotics. *Methods Enzymol. 18*:172–213.

Aszalos, A., and Issaq, H. J. (1980). Thin layer chromatographic systems for the classification of antibiotics. *J. Liquid Chromatogr. 3*:867–883.

Atassi, M. Z., and Rosenthal, A. F. (1969). Specific reduction of caroxyl groups in peptides and protein by diborane. *Biochem. J. 111*:593–601.

Barnes, E. M., Mead, G. C., Impev, C. S., and Adams, B. W. (1978). The effect of dietary bacitracin on the incidence of *Streptococcus faecalis* sub-species *liquefaciens* and related streptococci in the intestines of young chicks. *Br. Poult. Sci. 19*:713–723.

Baron, A. L. (1956). Bacitracin compound and recovery of bacitracin. U.S. Patent No. 2,774,712.

Betina, V. (1975). Paper chromatography of antibiotics. *Methods Enzymol.* *18*:100–172.

Brewer, G. A. (1980). Bacitracin. *Anal. Profiles Drug Subst.* *9*:1–69.

Brunner, R. (1965). Polypeptide. In *Die Antibiotica.* R. Brunner and G. Machek (Eds.). Verlag Carl, Nürnberg, pp. 167–214, 702–707.

Chaiet, L., and Cochrane, T. J. (1959). Recovery and concentration of bacitracin. U.S. Patent No. 2,915,432.

Charney, J. (1952). Purification of bacitracin by adsorption. U.S. Patent No. 2,582,921.

Chornock, F. W. (1957). Zinc bacitracin feed supplement. U.S. Patent No. 2,809,892.

Code of Federal Regulations (1970). Zinc bacitracin, tests and methods of assay. *Code Fed. Reg.* *21*:6, 168.

Cohen, I. R. (1957). Process for producing bacitracin. U.S. Patent No. 2,789,941.

Coombe, R. G. (1972). Pharmaceutical analysis II. Separation and determination of bacitracin, polymyxin and neomycin by acrylamide disc gel electrophoresis. *Aust. J. Pharm. Sci. NS1*:6–8.

Cornell, N. W., and Guiney, D. G. (1970). Binding sites for zinc (II) in bacitracin. *Biochem. Biophys. Res. Commun.* *40*:530–536.

Council Directive 70/524/EEC (1970). 14.12.70.

Council Directive 78/633/EEC (1978). Determination of zinc bacitracin by diffusion in an agar medium. 29.7.78. *Off. J.E.E.C. L 206*:44–55.

Craig, G. H. (1967). Assay for bacitracin in feed. U.S. Patent No. 3,306,827.

Craig, L. C., and Konigsberg, W. (1957). Further studies with the bacitracin polypeptides. *J. Org. Chem.* *22*:1345–1353.

Craig, L. C., Weisiger, J. R., Hausmann, W., and Harfenist, E. J. (1952). The separation and characterization of bacitracin popypeptides. *J. Biol. Chem.* *199*:259–266.

Craig, L. C., King, T. P., and Konigsberg, W. (1960). Homogeneity studies with insulin and related substances. *Ann. N.Y. Acad. Sci.* *88*:571–585.

Craig, L. C., Phillips, W., and Burachik, M. (1969). Bacitracin A. Isolation by counter double-current distribution and characterization. *Biochemistry 8*:2348–2356.

Darker, G. D. (1951). Production of bacitracin. U.S. Patent No. 2,567,698.

Darker, G. D., Brown, H. B., Free, A. H., Biro, B., and Goorley, J. T. (1948). The assay of bacitracin. *J. Am. Pharm. Assoc.* *37*:156–160.

Demain, A. L. (1972). Cellular and environmental factors affecting the synthesis and excretion of metamolites. *J. Appl. Chem. Biotechnol.* *22*:345–362.

Demain, A. L. (1976). Genetic regulation of fermentation organisms. In *Stadler Genetic Symposium*, Vol. 8. University of Missouri, Columbia, Missouri, pp. 41–55.

Desbuquois, B., and Cuatrecasa, P. (1972). Independence of glucagon receptors and glucagon inactivation in liver cell membranes. *Nature New Biol.* *237*:202–204.

Dickson, R. B., Willingham, M. C., Gallo, M., and Pastan, I. (1981). Inhibition by bacitracin of high affinity binding of $125I$-α_2-M to plasma membranes. *FEBS Lett.* *126*:265–268.

Dubost, M., and Pascal, C. (1970). Determination of antibiotics in food. Electrophoretic method. *Ann. Falsif. Expert. Chim.* *63*:189–202.

Erickson, R. J., and Young, F. E. (1978). Mechanisms of interspecific genetic exchange in the bacilli and their potential application in the industrial laboratory. *Dev. Ind. Microbiol.* 19:245–257.

Fink, J. (1978). Analytische Trennung von Zincbacitracin durch Polyacrylamid-Gelelectrophorese. *Draftfutter* 61:126–128.

Freaney, T. C., and Allen, L. P. (1958). Production of bacitracin. U.S. Patent No. 2,828,246.

Frøyshov, Ø. (1974). Bacitracin biosynthesis by three complementary fractions from *Bacillus licheniformis*. *FEBS Lett.* 44:75–78.

Frøyshov, Ø. (1975). Enzyme-bound intermediates in the biosynthesis of bacitracin. *Eur. J. Biochem.* 59:201–206.

Frøyshov, Ø. (1977). The production of bacitracin synthetase by *Bacillus licheniformis* ATCC 10716. *FEBS Lett.* 81:315–318.

Frøyshov, Ø. (1978a). Separation of bacitracin A and bacitracin F by isoelectric focusing in gel. *Anal. Chim, Acta* 98:137–139.

Frøyshov, Ø. (1978b). Studies on the substrate specificity of bacitracin synthetase. Twelfth FEBS Meeting, Dresden. *Abstr. Commun.* 2922.

Frøyshov, Ø., and Laland, S. G. (1974). On the biosynthesis of bacitracin by a soluble enzyme complex from *Bacillus licheniformis*. *Eur. J. Biochem.* 46:235–242.

Frøyshov, Ø., and Mathiesen, A. (1979). Tryptic cleavage of enzyme A in bacitracin synthetase. *FEBS Lett.* 106:275–278.

Frøyshov, Ø., and Mathiesen, A. (1982). Bacitracin synthetase, tryptic cleavage of enzymes B and C. In *Enzymatic Biosynthesis of Peptides*. H. Kleinkauf and H. von Döhren (Eds.). Walter De Gruyter, Berlin, pp. 307–314.

Frøyshov, Ø., and Mathiesen, A. (1983). High performance liquid chromatography of bacitracin in fermentation broth. To be published.

Frøyshov, Ø., Zimmer, T. L., and Laland, S. G. (1978). Biosynthesis of microbial peptides by the thiotemplate mechanism. In *International Review of Biochemistry, Amino Acid and Protein Biosynthesis II*, Vol. 18. H. R. V. Arnstein (Ed.). University Park Press, Baltimore, Maryland, pp. 49–77.

Frøyshov, Ø., Mathiesen, A., and Haavik, H. I. (1980). Regulation of bacitracin synthetase by divalent metal ions in *Bacillus licheniformis*. *J. Gen. Microbiol.* 117:163–167.

Fuller, R. (1981). Development and dynamics of the aerobic gut flora in gnotobiotic and conventional animals. *Fortschr. Veterinaermed.* 33:7–15.

Galardy, R. E., Printz, M. P., and Craig, L. C. (1971). Tritium–hydrogen exchange of bacitracin A. Evidence for an intramolecular hydrogen bond. *Biochemistry* 10:2429–2436.

Gale, E. F., and Folkes, J. P. (1955). The assimilation of amino acids by bacteria. *Biochem. J.* 59:675–684.

Garbutt, J. T., Morehouse, A. L., and Hanson, A. M. (1961). Metal binding properties of bacitracin. *J. Agric. Food Chem.* 9:285–288.

Gollaher, M. G., and Honohan, E. J. (1956). Bacitracin recovery. U.S. Patent No. 2,763,590.

Grynne, B. (1971). An improved method for the determination of bacitracin in animal feeds. *Analyst* 96:338–342.

Grynne, B. (1973). Identification of small amounts of antibiotics by electrophoresis and bioautography. *Acta Pathol. Microbiol. Scand. Sec. B 81*: 583–588.

Grynne, B., and Hoff, E. (1968). Verfaren zur quantitativen Bestimmung geringer Mengen von Bacitracin. German Patent No. 1,817,652.

Haavik, H. I. (1974a). Studies on the formation of bacitracin by *Bacillus licheniformis*: Effect of glucose. *J. Gen. Microbiol. 81*:383–390.

Haavik, H. I. (1974b). Studies on the formation of bacitracin by *Bacillus licheniformis*: Effect of inorganic phosphate. *J. Gen. Microbiol. 84*: 226–230.

Haavik, H. I. (1974c). Studies on the formation of bacitracin by *Bacillus licheniformis*: Role of catabolite repression and organic acids. *J. Gen. Microbiol. 84*:321–326.

Haavik, H. I. (1976). Possible functions of peptide antibiotics during growth of producer organisms: Bacitracin and metal (II) ion transport. *Acta Pathol. Microbiol. Scand. Sect. B 84*:117–124.

Haavik, H. I. (1979a). Amino acid control mechanism for bacitracin formation by *Bacillus licheniformis*. *Folia Microbiol. 24*:234–239.

Haavik, H. I. (1979b). On the physiological meaning of secondary metabolisms. *Folia Microbiol. 24*:365–367.

Haavik, H. I. (1981). Effects of amino acids upon bacitracin production by *Bacillus licheniformis*. *FEMS Microbiol. Lett. 10*:111–114.

Haavik, H. I., and Frøyshov, Ø. (1975). Function of peptide antibiotics in producer organisms. *Nature 254*:79–82.

Haavik, H. I., and Frøyshov, Ø. (1982). On the role of L-leucine in the control of bacitracin formation by *Bacillus licheniformis*. In *Enzymatic Biosynthesis of Peptides*. H. Kleinkauf and H. van Döhren (Eds.). Walter De Gruyter, Berlin, pp. 155–159.

Haavik, H. I., and Thomassen, S. (1973). A bacitracin-negative mutant of *Bacillus licheniformis* which is able to sporulate. *J. Gen. Microbiol. 76*: 451–454.

Haavik, H. I., and Vessia, B. (1978). Bacitracin production by the high-yielding mutant *Bacillus licheniformis* strain AL: Stimulatory effect of L-leucine. *Acta Pathol. Microbiol. Scand. Sect. B 86*:67–70.

Hausmann, W., Weisiger, J. R., and Craig, L. C. (1955). Bacitracin A. Further studies on the composition. *J. Am. Chem. Soc. 77*:721–722.

Henderickx, H. K., Vervaeke, I. J., Decuypere, J. A., and Dierick, N. A. (1981). Effect of growth promoting agents on the intestinal gut flora. *Fortschr. Veterniaermed. 33*:56–63.

Hickey, R. J. (1964). Bacitracin, its manufacture and uses. *Prog. Ind. Microbiol. 5*:93–150.

Hinton, N., and Orr, J. H. (1960). The effect of antibiotics on the toxin production of *Staphylococcus aureus*. *Antibiot. Chemother. 10*:758–765.

Hirotsu, Y., Nishiuchi, P., and Shiba, T. (1979). Total synthesis of bacitracin F. In *Peptide Chemistry 1978*. N. Izumiya (Ed.). Protein Research Foundation, Osaka, Japan, pp. 171–176.

Hodge, E. B., and Riddick, J. A. (1975). Recovery of bacitracin as the calcium or magnesium complex of an alkylbenzenesulfonic acid. U.S. Patent No. 3,891,615.

Hoff, E. (1972). Sorption of antibiotic from unfiltered liquids. U.S. Patent No. 3,660,279.

Humprey, J. H., Lightbown, J. W., Mussett, M. V., and Perry, W. L. M. (1953). The international Standard for bacitracin. *Bull. WHO 9*:861–869.

Ishihara, H., and Shimura, K. (1974). Biosynthesis of bacitracin III. Partial purification of a bacitracin-synthesizing enzyme system from *Bacillus licheniformis*. *Biochim. Biophy. Acta 338*:588–600.

Ishihara, H., and Shimura, K. (1979). Thiazoline ring formation in bacitracin biosynthesis. *FEBS Lett. 99*:109–112.

Ishihara, H., Sasaki, T., and Shimura, K. (1968). Biosynthesis of bacitracin II. Incorporation of ^{14}C-labelled amino acids into bacitracin by a cell-free preparation from *Bacillus licheniformis*. *Biochim. Biophys. Acta 166*: 496–504.

Ishihara, H., Endo, Y., Abe, S., and Shimura, K. (1975). The presence of 4'-phosphopantetheine in the bacitracin synthetase. *FEBS Lett. 50*:43–46.

Istvan, S., Jarai, M., Piukovich, S., Horvath, I., and Inczeffy, I. (1978). Zinc-bacitracin. *Hung. Teljes 15*:483.

Jacobsen, E., Pederstad, J. H., and Øystese, B. (1977). Determination of bacitracin by differential pulse polarography. *Anal. Chim. Acta 91*:121–128.

Johnson, B. A., Anker, S. H., and Meleney, F. L. (1945). Bacitracin: A new antibiotic produced by a member of the *B. subtilis* group. *Science 102*: 376–377.

Kaemmerer, K., and Fink, J. (1980a). Thermische Einflüsse auf den Futter-zusatzstoff Zincbacitracin. *Dtsch. Tieraertzl. Wochenschr. 86*:405–409.

Kaemmerer, K., and Fink, J. (1980b). Prüfung der Gewebespiegel nach Zink-bacitracingabe an Ratten mittels der Polyacrylgelelektrophorese. *Berl. Muench. Tieraerztl. Wochenschr. 93*:193–195.

Katz, E., and Demain, A. L. (1977). The peptide antibiotics of *Bacillus*: Chemistry, Biogenesis and possible functions. *Bacteriol. Rev. 41*:449–474.

Keko, W. L., Bennett, R. E., and Arzberger, F. C. (1953). Production of bacitracin. U.S. Patent No. 2,627,494.

Kindraka, J. A., and Gallagher, J. B. (1978). Bacitracin recovery process. U.S. Patent No. 4,101,539.

King, J. O. L. (1976). The feeding of zinc bacitracin to growing rabbits. *Vet. Rec. 99*:507–508.

Konigsberg, W., and Craig, L. C. (1959). Cellulose ion exchange and rotatory dispersion studies with the bacitracin polypeptides. *J. Am. Chem. Soc. 81*:3452–3458.

Kurahashi, K. (1981). Biosynthesis of peptide antibiotics. In *Antibiotics IV. Biosynthesis*. J. W. Corcoran (Ed.). Springer-Verlag, Heidelberg, pp. 325–352.

Laland, S. G., and Zimmer, T. L. (1973). The protein thiotemplate mechanism of synthesis for the peptide antibiotics produced by *Bacillus brevis*. *Assays Biochem. 9*:31–57.

Laland, S. G., Zimmer, T. L., and Frøyshov, Ø. (1978). Biosynthesis of bio-active peptides produced by microorganisms. In *Bioactive Peptides Produced by Microorganisms*. H. Umezawa, T. Takita, and T. Shiba (Eds.). Kodansha, Tokyo, pp. 7–34.

Lee, S. G., and Lipmann, F. (1975). Tyrocidine synthetase system. *Methods Enzymol. 43*:585–602.

Lewis, A. D., Niager, F. C., and Pattison, I. (1965). Bacitracin derivatives. U.S. Patent No. 3,205,137.

Lightbown, J. W., and Rossi, P. (1965). The identification and assay of mixtures of antibiotics by electrophoresis in agar gels. *Analyst 90*:89–98.

Lightbown, J. W., Kogut, M., and Uemura, K. (1964). The second international standard for bacitracin. *Bull. WHO 31*:101–109.

Lockhart, I. M., and Abraham, E. P. (1954). The amino acid sequence in bacitracin A. *Biochem. J. 58*:633–647.

Lockhart, I. M., Abraham, E. P., and Newton, G. G. F. (1955). The N-terminal and sulphur-containing residues of bacitracin A. *Biochem. J. 61*: 534–544.

Lyerla, J. R., and Freedman, M. H. (1972). Spectral assignment and conformational analysis of cyclic peptides by carbon-13 nuclear magnetic resonance. *J. Biol. Chem. 247*:8183–8192.

Mach, B., and Tatum, E. L. (1964). Environmental control of amino acid substitutions in the biosynthesis of the antibiotic polypeptide tyrocidine. *Proc. Nat. Acad. Sci. U.S.A. 52*:876–884.

McKelvy, J. F., LeBlanc, P., Laudes, C., Perrie, S., Grimm-Jorgensen, Y., and Kordon, C. (1976). The use of bacitracin as an inhibitor of the degradation of thyrotropin releasing factor and luteinizing hormone releasing factor. *Biochem. Biophys. Res. Commun. 73*:507–515.

Makarova, V. I. (1976). Effect of zinc bacitracin on the hydrolysis of glycosides and the activity of digestive enzymes in laying hens. *Tr. Vses. Nauchno Issled. Tekhnol. Inst. Ptitsevod. 42*:99–103.

Martin, J. F. (1977). Control of antibiotic synthesis by phosphate. *Adv. Biochem. Eng. 6*:105–127.

Meleney, F. L., and Johnson, B. A. (1949). Bacitracin. *Am. J. Med. 7*:794–806.

Miescher, G. M. (1974). Recovery of bacitracin. U.S. Patent No. 3,795,663.

Monroe, C. H., and Ward, G. E. (1967). Precipitation of bacitracin upon an inert insoluble inorganic support for use in animal feeds. U.S. Patent No. 3,345,178.

Morikawa, Y., Ochiai, K., Karube, I., and Suzuki, S. (1979). Bacitracin production by whole cells immobilized in polyacrylamide gel. *Antimicrob. Agents Chemother. 15*:126–130.

Mosberg, H. J., Scogin, D. A., Storm, D. R., and Gennis, R. B. (1980). Proton nuclear magnetic resonance studies on bacitracin A and its interaction with zinc ions. *Biochemistry 19*:3353–3357.

Munekata, E., Shiba, T., and Kaneko, T. (1973). Synthetic studies of bacitracin. IX. Synthesis of peptide fragments for bacitracin of cycloheptapeptide formula. *Bull. Chem. Soc. Jpn. 46*:3835–3839.

Newton, G. G. F., and Abraham, E. P. (1950). Ayfivin and bacitracin: Resolution of crude products into similar series of peptides. *Biochem. J. 47*: 257–267.

Newton, G. G. F., and Abraham, E. P. (1953a). Some properties of the bacitracin polypeptides. *Biochem. J. 53*:597–604.

Newton, G. G. F., and Abraham, E. P. (1953b). Observations on the nature of bacitracin A. *Biochem. J. 53*:604–613.

Nohl, H., and Hegner, D. (1977). The effects of some nutritive antibiotics on the respiration of rat liver mitochondria. *Biochem. Pharmacol. 26*:433–437.

Ogawa, I., Ishihara, H., and Shimura, K. (1981). Component I protein of bacitracin synthetase: A multifunctional protein. *FEBS Lett. 124*:197–201.

Otten, H., Plempel, M., and Siegenthaler, W. (1975). In *Antibiotika Fibel*, 4. Ed. Thieme Verlag, Stuttgart, pp. 542–545.

Ottensmeyer, F. P., Bazett-Jones, D. P., Hewitt, J., and Price, G. B. (1978). Structure analysis of small proteins by electron microscopy: valinomycin, bacitracin and low molecular weight cell growth stimulators. *Ultramicroscopy 3*:303–313.

Øystese, B. (1978). Zinc bacitracin composition for use as a feed supplement and method for making the same. U.S. Patent No. 4,906,246.

Øystese, B. (1979). Zinc bacitracin composition for use as a feed supplement and method for making the same. U.S. Patent No. 4,164,572.

Park, J. T. (1958). Inhibition of cell wall synthesis in *Staphylococcus aureus* by chemicals which cause accumulation of wall precursors. *Biochem. J.* 70:2P.

Patthy, A., Gráf, L., Kenessey, A., Székely, J. I., and Bajusz, S. (1977). Effect of bacitracin on the biodegradation of β-endorphin. *Biochem. Biophys. Res. Commun.* 79:254–259.

Pfaender, P., Specht, D., Heinrich, G., Schwarz, E., Kuhnle, E., and Simlot, M. M. (1973). Enzymes of *Bacillus licheniformis* in the biosynthesis of bacitracin A. *FEBS Lett.* 32:100–104.

Porath, J. (1954). Purification of bacitracin polypeptides by charcoal chromatography and zone electrophoresis. *Acta Chem. Scand.* 8:1813–1826.

Ragheb, H. S., Black, L., and Graham, S. (1976). Turbidimetric and diffusion assay of bacitracin in feeds. *J. Assoc. Off. Anal. Chem.* 59:526–535.

Ressler, C., and Kashelikar, V. V. (1966). Identification of asparaginyl and glutaminyl residues in endo position in peptides by dehydration–reduction. *J. Am. Chem. Soc.* 88:2025–2035.

Roland, I., Frøyshov, Ø., and Laland, S. G. (1977). A rapid method for the preparation of the three enzymes of bacitracin synthetase essentially free from other proteins. *FEBS Lett.* 84:22–24.

Rosen, G. D. (1980). Multi-factorial models for antibacterials in broiler nutrition. Sixth European Poultry Conference, Hamburg. *Conf. Proc. III*:302–309.

Rosen, G. D., Roberts, P., and Widdowson, V. M. (1976). An algebraic model for bacitracin in laying hen nutrition. Fifth European Poultry Conference, Malta. *Conf. Proc. I*:201–212.

Rosen, G. D., Roberts, P., and Widdowson, V. M. (1978). Algebraic models for zinc bacitracin in pig nutrition. Third World Congress on Animal Feeding, Madrid. *Congr. Proc. VIII*:120.

Rybinska, K. (1977). Effect of zinc-bacitracin on the activity of certain enzymes in rats. Part I. Determination of aspartate and alanine aminotransferases in the liver and kidney of test animals. *Rocz. Panstw. Zakl. Hig.* 28:133–140.

Scogin, D. A., Mosberg, H. I., Storm, D. R., and Gennis, R. B. (1980). Binding of nickel and zinc ions to bacitracin A. *Biochemistry* 19:3348–3352.

Senkus, M., and Markunas, P. C. (1952). Recovery of bacitracin. U.S. Patent No. 2,609,324.

Sharp, V. E., Arrigada, A., Newton, G. G. F., and Abraham, E. P. (1949). Ayfivin: Extraction, purification, and chemical properties. *Br. J. Exp. Pathol.* 30:444–457.

Shipchandler, M. T. (1976). Reduced bacitracin. U.S. Patent No. 3,966,699.

Shockman, G. D., and Lampen, J. O. (1962). Inhibition by antibiotics of the growth of bacterial and yeast protoplasts. *J. Bacteriol.* 84:508–512.

Shortridge, R. W. (1957). Recovery and purification of bacitracin. U.S. Patent No. 2,776,240.

Siewert, G., and Strominger, J. L. (1967). Bacitracin: An inhibitor of the dephosphorylation of lipid pyrophosphate, an intermediate in biosynthesis of the peptidoglycan of bacterial cell walls. *Proc. Nat. Acad. Sci. USA* 57:767–773.

Smekal, F., Paleckova, F., and Vinter, V. (1979). Fermentation production of bacitracin. Czechoslovak Patent No. 175,992.

Smith, H. W., and Tucker, J. F. (1980). Further observations on the effect of feeding diets containing avoparcin, bacitracin, and sodium arsenilate on the colonization of the alimentary trace of poultry by *Salmonella* organisms. *J. Hyg. 84*:137-150.

Smither, R., and Vaughan, D. R. (1978). An improved electrophoretic method for identifying antibiotics with special reference to animal tissues and animal feeding stuffs. *J. Appl. Bacteriol. 44*:421-429.

Smither, R., Lott, A. F., Dalziel, R. W., and Ostler, D. C. (1980). Antibiotic residues in meat in the United Kingdom; and assessment of specific tests to detect and identify antibiotic residues. *J. Hyg. 85*:359-369.

Snoke, J. E. (1960). Formation of bacitracin by washed cell suspensions of *Bacillus licheniformis*. *J. Bacteriol. 80*:552-557.

Snoke, J. E., and Cornell, N. (1965). Protoplast lysis and inhibition of growth of *Bacillus licheniformis* by bacitracin. *J. Bacteriol. 89*:415-420.

Stepanov, V. M., and Rudenskaya, G. N. (1978). Purification of proteolytic enzymes. U.S. Patent No. 4,100,028.

Stepanov, V. M., Gonchar, M. V., and Rudenskaya, G. N. (1978). Bacitracin and gramicidin S as inhibitors of carboxylic proteinases. *Khim. Prir. Soedin. 3*:385-389.

Stone, K. J., and Strominger, J. L. (1972). Preparation of C_{55}-isoprenyl pyrophosphate. *Methods Enzymol. 27*:306.

Storm, D. R. (1974). Mechanism of bacitracin action: A specific lipid-peptide interaction. *N.Y. Acad. Sci. Ann. 233*:387-398.

Storm, D. R., and Toscano, W. A., Jr. (1979). *Bacitracin in Antibiotics*, Vol. 1. F. E. Hahn (Ed.). Springer-Verlag, Heidelberg, pp. 1-17.

Tsuji, K., and Robertson, J. H. (1975). Improved high-performance liquid chromatographic method for polypeptide antibiotics and its application to study the effects of treatments to reduce microbial levels in bacitracin powder. *J. Chromatogr. 112*:663-672.

Tsuji, K., Robertson, J. H., and Bach. J. A. (1974). Quantitative high-pressure liquid chromatographic analysis of bacitracin, a polypeptide antibiotic. *J. Chromatogr. 99*:597-608.

Tyc, M., Kadzikiewicz, T., Karabin, L., Wurzue, J., Taflinska, K., Michalak, D., and Komorowska, C. (1976). Bacitracin. Polish Patent No. 79,336.

Udaka, S. (1979). Extracellular production of proteins by bacteria. *Dev. Food. Sci. 2*:285-289.

Vandamme, E. J. (1981). Properties, biogenesis, and fermentation of the cyclic decapeptide antibiotic, gramicidin S. *Top. Enr. Ferment. Biotechnol. 5*:185-261.

Vandamme, E. J., Leyman, D., Visscher, P. de, Buyser, D. de, and Steenkiste, G. van. (1981). Effect of aeration and pH on gramicidin S production by *Bacillus brevis*. *J. Chem. Technol. Biotechnol. 31*:247-257.

Walton, J. R. (1977). Antibiotics and antibiosis in agriculture. A mechanism of growth promotion: Non-lethal feed antibiotic induced, cell wall lesions in enteric bacteria. *Easter Sch. Agric. Sci. 25*:259-264.

Walton, J. R. (1981). Modes of action of growth promoting agents. *Fortschr. Veterinaermed. 33*:77-82.

Walton, J. R., and Bird, R. G. (1975). A possible mechanism to explain the growth promotion effect of feed antibiotics in farm animals: Zinc bacitracin induced cell wall damage in *Escherichia coli* in vitro. *Zentralbl. Bet. Med. B. 22*:318-325.

Washylishen, R. E., and Graham, M. R. (1975). A nuclear magnetic resonance study of the metal binding sites in bacitracin. *Can. J. Biochem.* 53: 1250–1254.

Wehrmeister, H. L. (1956). Bacitracin feed supplement from crude fermentation liquors. U.S. Patent No. 2,739,063.

Weinberg, E. D. (1958). Enhancement of bacitracin by the metalic ions of group II B. *Antibiot. Annu. 1958–1959*:924–929.

Weinberg, E. D. (1960). The relation of metal-binding to antimicrobial action. In *Metal Binding in Medicine.* M. Seven (Ed.). Lippincott, Philadelphia, pp. 329–334.

Weinberg, E. D. (1964). Manganese requirement for sporulation and other secondary biosynthetic processes of Bacillus. *Appl. Microbiol. 12*:436–441.

Weinberg, E. D. (1967). Bacitracin. In *Antibiotics,* Vol. 1. D. Gottlieb and P. D. Shaw (Eds.). Springer-Verlag, Heidelberg, pp. 90–101.

Weinberg, E. D. (1970). Biosynthesis of secondary metabolites; role of trace metals. *Adv. Microbiol. Physiol. 4*:1–44.

Weisiger, J. R., Hausman, W., and Craig, L. C. (1955). Bacitracin A. The nature of the linkages surrounding the sulfur. *J. Am. Chem. Soc.* 77: 3123–3127.

Ziffer, J., and Cairney, T. J. (1962). Liquor–bacitracin complex as growth stimulant and bacitracin purific. U.S. Patent No. 3,035,919.

Zinn, E., and Chornock, F. W. (1958). Production of bacitracin. U.S. Patent No. 2,834,711.

Zorn, R. A. (1959). Bacitracin product and processes utilizing nickel compounds. U.S. Patent No. 2,903,357.

Zorn, R. A. (1961). Bacitracin product and processes utilizing manganese compounds. U.S. Patent No. 2,985,533.

Zorn, R. A. (1962). Bacitracin product and processes utilizing cobalt compounds. U.S. Patent No. 3,021,217.

Zorn, R. A., Malzahn, R. C., and Hanson, A. M. (1961). Bacitracin product and processes. U.S. Patent No. 2,985,534.

25

VIRGINIAMYCIN: PROPERTIES, BIOSYNTHESIS, AND FERMENTATION

ANDRÉ M. BIOT *Smith Kline—RIT, Rixensart, Belgium*

I. INTRODUCTION

The compound antibiotic virginiamycin, produced by *Streptomyces virginiae*, is representative of the homogeneous and original group of antibiotics which contain synergistic components. The properties and uses of these protein synthesis inhibitors have been recently reviewed (Cocito, 1979).

Virginiamycin, a natural mixture of two cyclic lactone peptolides, called factors M and S (Fig. 1) (Crooy and de Neys, 1972), is soluble in organic solvents (methanol, ethanol, chloroform, dimethyl formamide), poorly soluble in water (55 mg in 100 ml) (de Somer and van Dijck, 1955), and practically

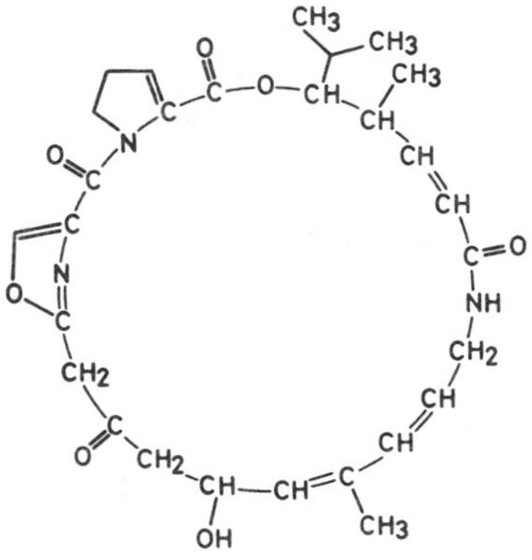

Figure 1 Chemical structure of factors M and S, components of virginiamycin.

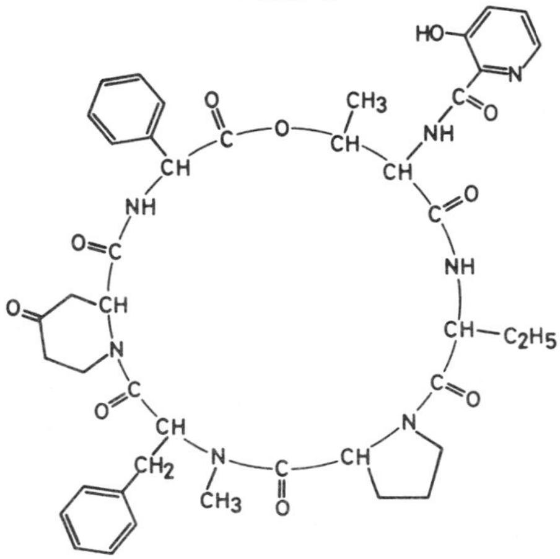

insoluble in hexane and petroleum ether. Factor M has a molecular weight of 525 ($C_{28}H_{35}N_3O_7$) and factor S a molecular weight of 823 ($C_{43}H_{49}N_7O_{10}$).

The antibiotic activity of virginiamycin is measured by plate diffusion assay using *Corynebacterium xerosis* NCTC 9755 (Boon and Dewart, 1974) or *Micrococcus luteus* ATCC 9341 (European Communities, 1972) as the susceptible microorganism. A respirometric automated method, using *Micrococcus pyogenes var. aureus*, ATCC 6538P, has also been described (Dewart et al., 1965). Although the bactericidal action of virginiamycin is selective against gram-positive microorganisms (Table 1) (van Dijck, 1969), it has been shown that the ribosomes of gram-negative organisms are also sensitive in cell-free systems (Cocito, 1979). The indifference of the gram-negative cells is explained by the impermeability of their envelope for the antibiotic. Since virginiamycin is not active against the gram-negative Enterobacteriaceae, for example, *Escherichia coli* or *Salmonella*, it plays no part in the gram-negative organism R-factor transfer problem.

The minimum inhibitory concentration values of the separate factors are higher than those of virginiamycin (van Dijck, 1969) (Table 1), not only

Table 1 Minimum Inhibitory Concentrations of Virginiamycin and Single Components

Microorganism	Virginiamycin (μg/ml)	Factor M (μg/ml)	Factor S (μg/ml)
Bacillus subtilis			
NCTC 8226	0.04	0.5	0.4
Brucella abortus	75	75	>100
Candida albicans	>100	>100	>100
Corynebacterium xerosis			
NCTC 9755	0.03	0.2	2
Escherichia coli 637	>100	>100	>100
Mycobacterium tuberculosis			
H37Rv	1	20	>100
Neisseria catarrhalis			
NCTC 3622	0.3	0.7	15
Proteus mirabilis	>100	>100	>100
Pseudomonas fluorescens	15	30	>100
Sarcina lutea	0.03	0.05	0.5
Shigella flexneri	>100	>100	>100
Staphylococcus aureus			
6538 P	0.2	0.6	1
Streptococcus faecalis			
NCTC 779	15	75	>100
Streptococcus pneumoniae	0.07	1	20
Streptococcus pyogenes			
ATCC 8668	0.07	0.3	10
Yersinia pestis	3	30	50

Source: van Dijck (1969).

against highly susceptible strains, but also against moderately susceptible organisms (Biot, 1979). Factor S per se has little or no activity, representing, for example, for Staphylococcus aureus only about 3.5% of the activity of the same amount of factor M. However, it potentiates the activity of factor M so that the resulting activities of mixtures of both factors may be represented by a bell-shaped synergy curve (Fig. 2). A solely additive effect would have been represented by a straight line joining the activities of the single components. A maximum synergistic action is observed when 25–40% factor S is present in virginiamycin, increasing the activity of factor M by about 3.8 times. Knowing precisely the concentrations of factors M and S in a given product, one can calculate, using the synergy curve for S. aureus, the corresponding microbiological activity. For a maximum antibacterial effect only a minimum amount of product, in weight, is needed. If 1.4 mg of factor M produces 1000 units of activity, it can be calculated that 31.2 mg of factor S or only 0.37 mg of a mixture containing 70% factor M and 30% factor S will produce the same activity. This is obviously an advantage when safety of use is concerned. The unique feature of synergistic interaction is most important not only in explaining the absence of significant resistance induction after many years of intensive use as an aid to livestock production (Devriese, 1980; Schäfer et al., 1980), but also in understanding that the production

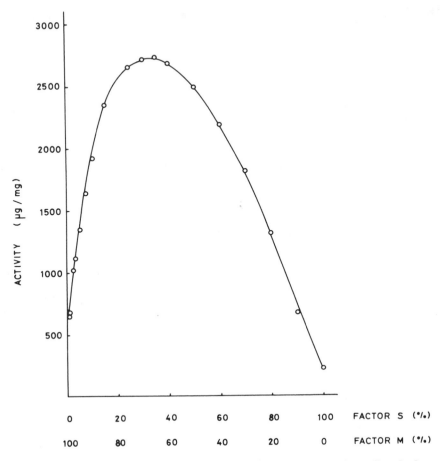

Figure 2 Microbiological activity of virginiamycin against *Staphylococcus aureus* as a function of M and S factor ratios.

and particularly the fermentation of this type of antibiotic presents specific characteristics. The virginiamycin-producing streptomycete elaborates the two components together in the same fermentation medium at levels corresponding to the maximum values of synergistic activity. Both factors are extracted from the fermentation broth using the same solvent system, purified, and stabilized by incorporation into a patented formulation, marketed as Stafac 500 (or Eskalin 500 in the United Kingdom), which is used as a very effective performance promoter for animals, increasing daily live weight gain and improving feed conversion efficiency.

Virginiamycin corresponds to the profile of a modern performance promoter: activity confined to the gram-positive microorganisms; stability at acidic pH; low solubility and low absorption through the intestinal wall; and presents as well high safety standards related to synergistic activity, that is, lowered probability of resistance induction, and absence of toxicity or residue problems. Although many studies of the properties and uses of virginiamycin have been published, little information has been made available on the technical aspects of the production of this industrially important antibiotic. It is the purpose of this chapter to convey information on more than 20 years of collaborative, mostly unpublished work performed in optimizing the production process of virginiamycin.

II. HISTORICAL ASPECTS

The producing microorganism, *S. virginiae*, was isolated in 1954 from a sample of Belgian soil taken in Kessel-Lo, near Leuven, and has been given the ATCC number 13161. The antibiotic produced by this microorganism was first described as antibiotic number 899 (de Somer and van Dijck, 1955) before being known as virginiamycin. Its antibacterial action, principally against staphylococci resistant to penicillin, was considered interesting enough to justify further development work together with more basic research on its composition and properties. The first industrial production, at low yields, started in 1958 and the pure product, under the trade name Staphylomycin, at once raised interest in its external therapeutic use against staphylococci. Ten years of use in human therapy were able to demonstrate the efficacy and safety of Staphylomycin.

During the same period intensive development work increased the productivity of the strain and optimized the composition of the culture medium, the fermentation conditions, and the purification process. However, it was soon realized that virginiamycin was best suited for use as a performance promoter in animal husbandry rather than as a human therapeutic agent and therefore the medical applications were discontinued.

The first stabilized formulation, Stafac 500, which does not include the dried mycelial cake and which has an activity corresponding to 50% virginiamycin, came out in Belgium in 1966. Since then its use has been accepted by the official regulatory bodies both in the United States and in Europe.

III. THE ORGANISM

The original strain number 899, sensitive to its own antibiotics and to phages, was subjected in a sequence of selection operations which led, via several steps representative of important changes in the controlled biosynthesis of virginiamycin, to the isolation of different successive strains with new characteristics (Table 2). The overall result was an increase of more than 1000 times the production of the original strain. The major improvements obtained during the development stages were mainly due to the following approaches:

Blind selection on the basis of yield, morphology, and stability of the strains with mutagenic treatments using ultraviolet rays or nitrogen mustard

Resistance to virginiamycin by successive transfers on increasing concentrations of the antibiotics either in liquid or on solid media and also by continuous culture

Elimination of two phages and a colicin-like substance (Roelants and Naudts, 1964)

Lowering of the inhibition of virginiamycin production by factor S.

Global optimization of strain, culture medium composition, and fermentation conditions

Selection of mutants phenotypically and genetically stable on unfavorable seed medium.

Table 2 Representative Selected Strains of *Streptomyces virginiae*

Number	Year of isolation	Characteristics
899	1954	Original strain, ATCC 13161
PDT 30	1958	Moderately resistant to its own secondary metabolites
PDT 1830	1961	Pigment producer; phage sensitive
R 81	1962	Pigmentless; phage insensitive
R 341	1964	Lowered production inhibition by factor S
R 1081	1965	Optimal strain versus optimal culture medium
SV 32	1973	Highly resistant to its secondary metabolites; stable on unfavorable seed medium
SV 422	1974	Revertant from tryptophane auxotroph
SV 3582	1976	Phenylalanine analog resistant
SV 6282	1979	Methionine analog resistant

Resistance to antimetabolites of assumed precursors
Revertants from auxotrophs and DNA transformation

These techniques were found, among others (Elander et al., 1977), to be most suitable for obtaining improved mutants. More than 90,000 strains were tested for their production in Erlenmeyer flasks on rotary shaking tables and the best ones were progressively transferred to the industrial process. A description of the characteristics of the most typical selected strains will provide a better understanding of some regulatory mechanisms and strain behavior during virginiamycin production.

The original strain number 899 produced very low amounts of virginiamycin during fermentation on a liquid corn steep–glucose medium. The organism started its antibiotic production during the growth phase and was inevitably exposed to protein synthesis inhibition by its own secondary metabolites, resulting in a rapid stop to growth and production. Mutagen treatment, combined with an increase of resistance to 250 µg/ml virginiamycin, optimization of the production medium composition, and choice of a modified and better adapted inoculation procedure yielded a new strain with increased productivity, PDT 30, the first one used for industrial production.

The level of resistance was further increased to 1000 µg/ml virginiamycin. A better producing strain, PDT 1830, was then isolated but was found to form a red-brown pigment during fermentation, which was possibly related to a modification process of virginiamycin. In addition, in some well-defined conditions, growth stopped after 30 hr of fermentation in deep cultures and the mycelium underwent lysis with a consequently low production level. The lysing factor, passing through a Millipore 0.45-µm filter, thermostable (80°C during 12 min) and unstable at pH values of $\leqslant 2.0$ or $\geqslant 8.0$, was identified with phage, ϕS_1 (Roelants et al., 1968). Phage ϕS_1 was very specific for *S. virginiae*, since it was not active against other tested *Streptomycetes* (*antibioticus, aureofaciens, erythreus, fradiae, graminofaciens, griseolus, griseus, lavendulae, madurae, olivaceus, pelletieri, rimosus*) and *Nocardiae* (*lutea, minima, parafinae, polychromogenes, rubra,* and *uniformis*). The addition of phages at different times to deep cultures affected differently the production of virginiamycin: Whereas the addition of 10^7 phages/ml before 12 hr of fermentation reduced drastically the production, no influence was observed when the addition was made after 12 hr of fermentation. After successive transfers on solid media in the presence of increasing concentrations of ϕS_1, strains were obtained which were considered phage resistant when no lysis of the cells occurred at phage concentrations of 10^8/ml.

However, these strains underwent lysis again after 30 hr growth in the same defined conditions as previously established for the sensitive strains. It was therefore assumed that the resistance was not complete (pseudolysogeny) or that a second phage was present in the culture medium. The presence of a second phage, ϕS_2, was indeed revealed. It was either a mutant of phage ϕS_1 or a second phage whose activity was covered by the first one. Stable strains of S. virginiae, resistant to ϕS_2, were easily obtained after isolation of colonies growing inside the lytic zones produced on inoculated solid agar media. Nevertheless, the mycelium was still affected, in deep cultures, by a production inhibitor which was not virginiamycin. After investigation of its properties, the inhibitor could be identified with a bacteriocin-like substance (Roelants and Naudts, 1964): It was not dialyzable, was precipitated with ammonium sulfate, and was inactivated by proteolytic enzymes. Only active against Streptomycetes, particularly Streptomyces aureofaciens, and specifically adsorbed on sensitive viable cells, its synthesis could be increased by ultraviolet irradiation. Resistance to the inhibitor, obtained after growth of S. virginiae in the presence of high concentrations of the isolated product, was obtained in several hundreds of strains, only two of which were found to be nonproducers of the inhibitor. Concomitant increase of resistance to virginiamycin yielded finally a stable strain, R 81, free of pigment production, devoid of any impairing ability of virginiamycin, insensitive both to phage infection and to the action of the bacteriocin-like substance, but with an unimproved production level. The strain remained sensitive to production inhibition by virginiamycin. Reduction or neutralization of this inhibition was the next target in strain improvement. It was known from previous experiments that small amounts of virginiamycin, added to the production medium just after inoculation, reduced drastically the biosynthesis of virginiamycin without affecting growth. A similar action was observed in the presence of factor S alone, but not in the presence of factor M (Table 3). The inhibiting action was released if the addition took place after 20 hr of fermentation.

Two approaches were designed and used to obtain strains whose production was not influenced by the early addition of factor S to the culture medium: selection of strains resistant to the influence of factor S in deep cultures and selection on agar plates supplemented with factor S and covered, after growth of the separated colonies of S. virginiae, with a S. aureus agar culture, made resistant to factor S but remaining susceptible to factor M and the synergizing action of factor S. One mutagen-treated strain out of 8000, R 341, produced 60% more virginiamycin than the mother strain R81. It was found to be much less sensitive to the inhibiting action of factor S (Table 4). Interestingly, the inhibition of virginiamycin production of strain R 81 could also be neutralized upon addition to the starting culture of 10^{-4} M nickel sulfate (Table 5). Further study of this phenomenon demonstrated that factor S, unlike factor M, was reversibly complexed with nickel ions to produce a compound having characteristic ultraviolet absorption properties and a strongly modified infrared spectrum. Nickel affected the 3-hydroxypicolinic acid moiety of factor S, responsible for the synergistic activity, as was confirmed after preparation

Table 3 Influence of Virginiamycin or Virginiamycin Components, Added to the Culture Medium After Inoculation, on the Production of Virginiamycin (in Percent of the Control) by Strain R 81

Addition (μg/ml)	Virginiamycin	Factor M	Factor S
0	100	100	100
10	15	96	80
20	10	96	56
40	10	100	32
80	15	94	<10

Source: P. Roelants (unpublished results).

Table 4 Influence of the Addition of Factor S to the Culture Medium After Inoculation on the Virginiamycin Production (in Percent) by Strain R 341 and Its Mother Strain R 81

Addition (μg/ml factor S)	Strain R 81	Strain R 341
0	100	160
10	38	150
25	15	115
50	<10	108
100	<10	58

Source: P. Roelants (unpublished results).

Table 5 Influence of the Presence of 10^{-4} M Nickel Sulfate and Factor S, Added in Starting Cultures, on the Production of Virginiamycin (in Percent) by Strain R 81

Addition (μg/ml)	Factor S	Factor S + NiSO$_4$
0	100	100
10	76	106
25	32	96
50	<10	100

Source: P. Roelants (unpublished results).

of a derivative of factor S where the hydroxyl function of the acid was blocked (Rondelet, unpublished results, 1965). Traces of nickel introduced into a culture of *S. aureus* also strongly inhibited the action of virginiamycin on that organism.

It was finally also possible to reduce the influence of the early addition of factor S on production by changing the composition of the culture medium, namely, by addition of other carbon sources.

A combination of mutagenic treatments and better-defined environmental conditions and media compositions led to the improved strain R 1081.

A normal population of strain R 1081, grown on agar medium, always contained about 2% white variants. The number of these variants increased to about 20% after treatment with acriflavine or ethidium bromide, without affecting the overall production in deep cultures. Similar white mutants had already been described for other antibiotic-producing streptomycetes.

On the other hand, numerous variants or mutants were obtained by simple serial transfer in liquid seed media containing low concentrations of proteins and carbohydrates. The variations affected morphological aspect, pigmentation, aerial mycelium formation, sporulation, and production capacity. The selection properties of these poor seed media were used to isolate strain SV 32, which was at the same time rendered highly resistant to its own antibiotics. In an attempt to modify or suppress additional regulatory systems of the virginiamycin biosynthesis, based on some known metabolic pathways, the strain SV 32 was selected for auxotrophy. Auxotrophs were obtained on minimal supplemented medium after treatment of the strain with N-methyl-N'-nitro-N-nitrosoguanidine (100–500 µg), heat (1–45 min at 100–130°C), or near-ultraviolet rays (90 min at 365 nm) in combination with 100 µg/ml of 8-methoxypsoralen.

Stable revertants were selected from tryptophane auxotrophs. The most representative of these revertants was strain SV 422, stable during storage and after serial transfers on liquid media. High recombinant levels were obtained on potato–dextrose–agar medium after crossing of the auxotrophs: trp × ura and trp × leu (Table 6).

Finally, after it was clearly established that *S. virginiae* was able to take up 0.5–2% of homologous exogenous donor DNA during the early stationary growth phase (Roelants et al., 1976), a property confirmed by transfection experiments (Konvalinkova et al., 1977), other sophisticated selection methods were employed, including transformation, analog resistance, and protoplast fusion. The best strains isolated were a phenylalanine analog-resistant strain, SV 3582, and a methionine analog-resistant strain, SV 6282.

The carrying out of the strain selection program for the improved production of virginiamycin used classic microbiological procedures either dictated by the actual behavior of the selected strains or suggested by newly available techniques.

The more than 1000-fold increase of production which resulted from applying these procedures was accompanied, for every new mutant, by a noteworthy constancy of the ratio of the two virginiamycin components produced. The ratio corresponded always, within the analytical limits, to the maximum of possible synergy. Regulatory mechanisms, whereby the cells defend themselves from becoming overproducers of virginiamycin, included the inhibition of the production by factor S. Although little is known about the mechanisms protecting the mutant strains of *S. virginiae* from the action of factor S, there is some indication of a modification and desensitizing of the target site for the antibiotic, the 50-S ribosomal subunit (Vining, 1979).

Table 6 Recombination of *Streptomyces virginiae* Auxotrophs on Potato–Dextrose–Agar Medium

Auxotrophs	Minimal medium (number of colonies)	Complete medium (number of colonies)	Frequency of prototrophs	Ratio of recombinants to revertants
gua	7.0×10^4	6.0×10^{10}	10^{-6}	—
trp 1	0	1.4×10^{10}	$>10^{-10}$	—
ura	6.0×10^2	5.0×10^{11}	10^{-9}	—
trp 2	6.0×10^5	7.5×10^{10}	8.0×10^{-5}	—
leu	0	8.5×10^{10}	$>10^{-10}$	—
gua + ura	3.7×10^5	1.1×10^{11}	3.0×10^{-6}	1
gua + trp 2	1.4×10^5	3.7×10^{10}	3.8×10^{-6}	1
trp 1 + leu	6.5×10^3	5.0×10^{11}	1.3×10^{-8}	10^2
trp 1 + ura	5.0×10^6	5.0×10^{11}	10^{-5}	10^4

Source: P. Roelants (unpublished results).

The antibiotic is probably prevented from binding to the ribosome by adenine dimethylation, as found in a streptogramin type B (similar to factor S) producer, *Streptomyces diastaticus* NRRL 2560 (Fujisawa and Weisblum, 1981).

IV. BIOSYNTHESIS

The almost complete inhibition of virginiamycin production, brought about by a small amount of factor S added initially to the culture, without any effect on the growth of *S. virginiae*, suggested the instability of the virginiamycin-synthesizing system (Yanagimoto and Terui, 1971a). Results of additional experiments indicated that a substance effective in inducing virginiamycin production was formed just prior to the initiation of production and that factor S inhibited not only its formation, but also its inducing activity (Yanagimoto and Terui, 1971b). The inducing material (IM), found in the culture filtrate, was given the molecular formula of $C_{12}H_{22}O_4$ and contained a γ- or δ-lactone group, three hydroxy groups, and two methyl groups. IM was judged to be a carboxylic acid and was effective at concentrations under 0.02 μg/ml. Other substances, including γ-undecalactone, γ-nonalactone, and γ-octalactone, were also found to have some ability to induce virginiamycin production (Yanagimoto et al., 1979). IM was not consumed during the fermentation, implying that it might not be one of the precursors of virginiamycin.

While IM seems to activate the transcription of an operon having a production function, without influencing the cellular functions which are essential to the organism, factor S exerts a repressing activity, even when exogenous IM is supplied, and prevents the "operator" from initiating synthesis.

Not only an understanding of the regulatory mechanisms controlling antibiotic production but also knowledge of the biosynthetic pathways will provide basic elements for determining the optimum physiological and growth conditions and genetic modifications leading to increased antibiotic production.

Preliminary results establishing the overall biosynthetic origin of the major portion of factor M were obtained after isolation of labeled factor M from cultures of strain PDT 30, inoculated on a nutrient medium containing potential [14]C- or ^{13}C-labeled precursors. D-Glucose as well as sodium acetate, methionine, serine, proline, valine, and glycine were incorporated into the factor M molecule. These findings enabled a tentative pathway for the biosynthesis of factor M to be made (Dicuollo, 1979; Kingston and Kolpak, 1980).

Work is in progress on more detailed pathways by which the dehydroproline double bond and the unusual oxazole ring are formed and on the biosynthesis of factor S (Kingston and Kolpak, 1980). Factor S contains a sequence of amino acids, including threonine, aminobutyric acid, proline, phenylalanine, and phenylglycine, which were used as markers for the isolation of auxotrophs. Supplementation with certain amino acids enhanced antibiotic production, whereas inorganic nitrogen sources such as nitrates, nitrite, and several ammonium salts resulted in a much lower virginiamycin production. Some kind of ammonium repression of production occurs when ammonium is supplied in excess.

V. FERMENTATION

Both components of virginiamycin are produced by *S. virginiae* at a constant ratio in the same fermentation medium. The operations are conducted accord-

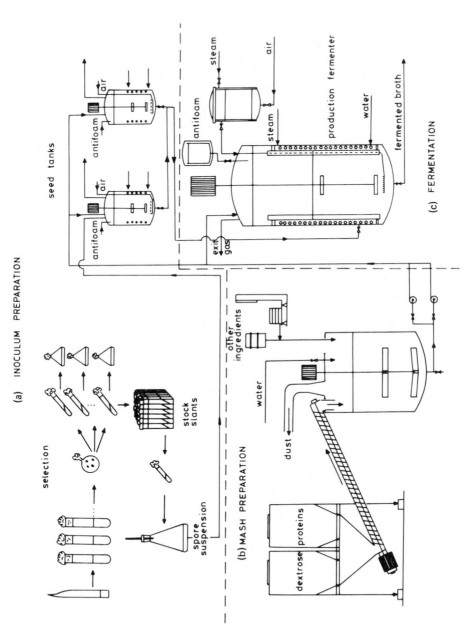

Figure 3 Production of virginiamycin. The operations include (a) inoculum preparation, (b) mash preparation, and (c) fermentation.

ing to the generally described procedures for the production of antibiotics by aerobic fermentation. However, the media compositions and fermentation conditions were designed to meet the optimal parameter conditions for the production of both compounds at a maximal rate.

A. Inoculum Preparation

Only well-grown and checked production strains are used for the operations. Each new stock of strains is prepared from a natural selection of existing strains and after checking of sterility and productivity (Fig. 3a). Spores collected from a previously selected slant or from a lyophilized tube are diluted in sterile physiological saline and plated on Kolle bottles to obtain separated colonies. A number of colonies, having a regular shape and color, are picked up and inoculated on slants to obtain spores. The medium for sporulation contains (amounts expressed as g/liter) soluble starch, 10; ammonium sulfate, 2; K_2HPO_4, 1; $MgSO_4$, 1; NaCl, 1; $CaCO_3$, 3; and agar, 20. Moisture and pH are known to affect sporulation on slants. After incubation a spore suspension is made from each slant, inoculated into Erlenmeyer flasks containing a seed medium, and incubated for 2 days on a rotary shaking table at 25°C. The seed medium is then transferred into a production medium, incubated for 3 days at 22°C, and analyzed for virginiamycin content. From the best-yielding colony a stock of sporulated slants is prepared and stored in the refrigerator until use.

Vegetative growth of S. virginiae occurs as a thin moist pellicle upon which, after 3–4 days, a light purple and dry aerial mycelium is superimposed. The aerial hyphae, approximately 1 µm in diameter, are long, many branched, and always lightly curved. Sporulation occurs by fragmentation of the hyphae and elaboration of spores from the fragments. The spores, when released, are hydrophobic and adhere to dry glass because of a large amount of lipids in their walls. They are cylindrical with curved ends. When inoculated in submerged shaken cultures, the spores swell and give rise to germ tubes after a lag period of about 20 hr. Short hyphae are formed which finally elaborate the mycelial network.

The preparation of the industrial inoculum is an essential step in obtaining good yields in the production fermenter, where growth must proceed very rapidly after inoculation.

A two-step inoculation procedure is carried out. Firstly, a spore suspension from agar slants is inoculated, using a special inoculation device, into a liquid vegetative medium designed for sporulation. The medium is well aerated for 48 hr. After exhaustion of the glucose and formation of threads of sporogenous mycelium, polymorphous segmentation spores (arthrospores) are separated from the network. In addition, a strictly vegetative reproduction system is observed with the appearance of mycelial threads divided into fragments of 8–10 µm in length. At the same time the sporulating culture releases a typical earthy-smelling odor and a black pigment is secreted into the medium, the intensity of which is greatest with maximum aeration. Conditions allowing a high proportion of the cells to sporulate seem to inhibit the production of virginiamycin.

The sporulated first vegetative medium is further inoculated into a second vegetative medium containing (amounts expressed as g/liter) corn-steep solids, 20; peanut oil cake, 15, yeast hydrolysate, 5; soy meal, 10; glucose, 40; and calcium carbonate, 5.

A normal growth is needed after 48 hr incubation for inoculating the production tank. A poor inoculum gives bad results and cultures which are too old also have a low quality. A poor inoculum is never compensated by an increased volume of inoculation.

The fermentations of the seed media are carried out with well-defined parameter values: 0.7–1.0 vol air/vol medium per minute, 25°C, and an agitation speed of 150–300 rpm. Foaming is controlled by automatic addition of anti-foam, and samples are taken daily to monitor sterility, pH, growth, and carbohydrate consumption (cryoscopic determination).

B. Fermenter Operations

The raw materials needed for the growth of the organism and for the biosynthesis of virginiamycin in the production tanks include proteins, dextrose, calcium carbonate, and antifoaming agents. A typical production medium consists of (amounts expressed as g/liter) corn-steep solids, 20; peanut oil cake, 10; yeast hydrolysate, 5; glucose, 5; glycerol, 25; linseed oil, 10; and calcium carbonate, 5. The correct amounts are weighed, mixed with water, and pumped into the empty fermenter (Fig. 3b). Sterilization is conducted for 30 min at 110°C and the sterile mash is cooled down to 23°C. After inoculation with the growth of the seed tank, sterile air is admitted to the fermenter and maintained between 0.4 and 0.6 vol/vol medium per minute, with a head pressure on the tank of 0.1 bar. Agitator speed is adapted as a function of the dissolved oxygen concentration, which is kept between 15–20% oxygen saturation. During the course of the fermentation, which lasts about 72 hr, temperature is maintained at 23°C and pH at between 6.8 and 7.0. Foaming is controlled, and regularly taken samples provide information about sterility, pH, growth, and carbohydrate consumption.

When oxygen is no longer consumed by the strain, the fermented broth is acidified at pH 6.5, analyzed for its virginiamycin content, and sent to the extraction area.

Virginiamycin activity is measured by microbiological diffusion tests on agar plates inoculated with a suitable susceptible microorganism, such as *M. luteus*.

Factor M may be determined by a colorimetric method using Ehrlich's reagent, and factor S with a spectrofluorimetric method.

Growth during fermentation may be estimated by a rapid density measurement carried out in graduated centrifuge tubes (PMV or packed mycelial vol-

Table 7 Influence of the Impeller Type, Impeller Size, and Baffle Size on the production of Virginiamycin

Impeller type	Impeller size[a]	Baffle size[b]	Speed range (rpm)	Maximum yield (%)
Curved blade	0.40	0.16	150–300	100
Flat blade	0.55	0.16	100–200	108.4
Flat blade	0.55	0.32	100–200	101.7
Flat blade	0.79	0.16	75–150	99.6

[a]Ratio of the diameter of the impeller to that of the fermenter.
[b]Ratio of the width of the baffle to the diameter of the fermenter.

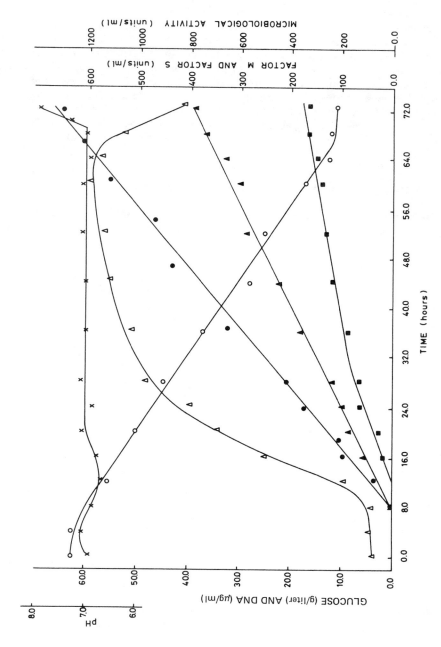

Figure 4 Raw data of virginiamycin fermentation: DNA (△), pH (×), glucose concentration (○), factor M (▲), factor S (■), and microbiological activity (●). Factors M and S and the microbiological activity are expressed in arbitrary units per milliliter.

ume), or more accurately measured by DNA determinations. Dry weight measurements are not reliable.

The fermenters (Fig. 3c) used for the production of virginiamycin are very similar in shape and relative dimensions to the 100-m^3 fermenter described by Aiba et al. (1973) for antibiotic production. The height-to-diameter ratio reaches 2.4, the maximum agitation speed 120 rpm and the airflow rate ranges between 0.3 and 0.6 vol air/vol medium per minute. The power uptake has a maximum value of 3.1 hp/m^3 or 2.3 kW/m^3 and is sufficient to achieve the optimum productivity of the strain. Power uptake and impeller design were optimized on pilot and industrial fermenters. A summary of results obtained with different impeller types, impeller sizes, and baffle sizes (Table 7) illustrates the superiority of flat-blade impellers and normal-sized baffles over curved-blade turbines and/or oversized baffles or impellers.

Data analysis of virginiamycin fermentations enabled the fermentation to be divided into four phases (Fig. 4):

A lag phase, lasting about 6–10 hr, with a slight increase of pH and formation of very few mycelial threads. Production of virginiamycin is not detectable.

A logarithmic growth phase between about 10 and 20 hr, with a decrease of pH to a value of between 6.8 and 7.0. Rapid mycelial growth is associated with a linear dextrose consumption and a high linear rate

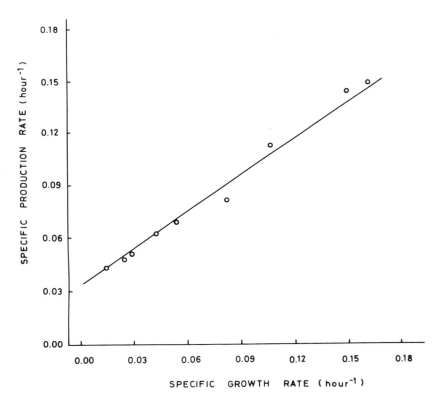

Figure 5 Relationship between the specific growth rate and the specific production rate of virginiamycin.

of virginiamycin production. Large crystals, identified as factor M, are visible in the fermentation broth under the microscope. The crystallization of factor M excreted in the medium depends on its concentration and on temperature. Simple dilution with hot water removed the crystals. At 35°C the broth no longer contained visible crystals.

A production phase, between 20 and 65 hr, follows the logarithmic growth phase without modification of cell volume or pH. Dextrose consumption still proceeds linearly, as well as virginiamycin production. However, the rate of production is not as high as the maximum obtained in the previous period.

The postproduction phase is sharply defined and starts between 65 and 75 hr of fermentation. pH increases abruptly when glucose consumption is stopped. Cell volume and DNA content decrease and virginiamycin is no longer produced.

Analysis of the data collected during fermentation gives a better insight into the function of the important variables.

The linear relationship existing between the specific growth rate and the specific production rate (Fig. 5) suggests that the most rapid fermentations will have the highest yields and that the production of virginiamycin is growth associated. This was confirmed when the final virginiamycin content also

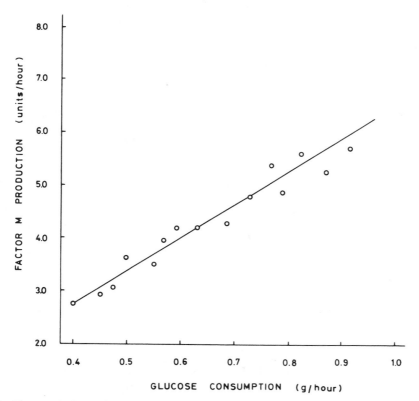

Figure 6 Relationship between factor M production and glucose consumption. Factor M is expressed in arbitrary units.

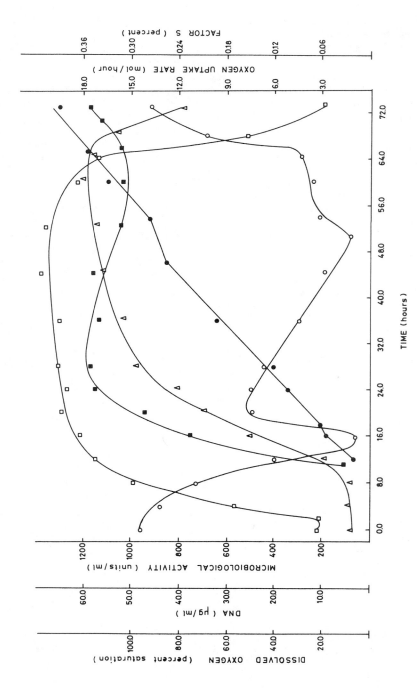

Figure 7 Evolution of parameter values during fermentation of virginiamycin: DNA (△), dissolved oxygen (○), microbiological activity expressed in arbitrary units per milliliter (●), factor S (■), and oxygen up-take rate (□).

showed a linear relationship with the virginiamycin production rate. The ratio of factors M and S at the end of the fermentation is constant thanks to the media composition, the fermentation conditions, and the strain. The percentage of factor S produced ranges between 30 and 35%, which corresponds to the maximum value of the synergy. In cases where the biosynthesis of one of the two factors is increased by using modified conditions, the production of the second factor is lowered in the same proportions, resulting in an important drop of the overall microbiological activity.

The production of factor M is directly dependent on glycolysis, as observed when glucose consumption is plotted against factor M production (Fig. 6). Factor S, whose production rate is highest during the logarithmic growth phase (Fig. 7), derives only indirectly from glycolysis via the biosynthetic pathways of the amino acids. Production of factor S is therefore dependent on amino acid biosynthesis and in particular their regulation systems.

Growth-associated production presupposes that the dissolved oxygen levels during fermentation are maintained at a sufficient level so as not to harm the growth rate of the microorganism. An insufficient dissolved oxygen concentration results in a lowered microbiological activity production rate, followed immediately by a recovery of production increase at the same rate as before when the dissolved oxygen concentration is again increased (Fig. 7).

The end of the fermentation is sharply defined and corresponds to a stop of dextrose consumption and virginiamycin production. Different parameters of the fermentation react within minutes: increase of pH, dissolved oxygen and oxygen measured in the exit gas, decrease of the oxygen uptake rate, carbon dioxide measured in the exit gas, cell volume, and DNA content.

The production of virginiamycin was found to be related to the total number of cells and the specific growth rate. Therefore not only cell mass concentration but also the microbiological activity may be estimated from the oxygen uptake rate data, provided that a model adequately describes the oxygen utilization in terms of cell maintenance, cell growth, and product formation. The model proposed by Luedeking and Piret (1959) establishing the relationship between the cell growth rate and the product formation rate has been useful in finding a close agreement between the predicted values and the measured results (Armiger, unpublished data, 1978).

VI. EXTRACTION

Unlike the other antibiotics used as performance promoters in animals, virginiamycin is not marketed as a dried mycelial cake but as a formulation containing the purified product extracted from the fermentation broth, thus avoiding the presence of inactive and unknown impurities in the preparation. The fermented broth, already at pH 6.5, is pumped into holding tanks and further acidified (Fig. 8).

Virginiamycin is extracted from the acidified broth by mixing with methyl isobutyl ketone (MIBK). The organic phase or extract is collected and the aqueous phase is submitted to a second and third extraction. Thereafter the spent broth is submitted to waste treatment. The mixed extract fractions are concentrated under vacuum until a light paste is obtained. The recovered MIBK is stored for later extractions.

Addition of hexane to the concentrated MIBK precipitates virginiamycin, which is further separated from the mother liquor by centrifugation under CO_2 pressure to avoid fire hazard.

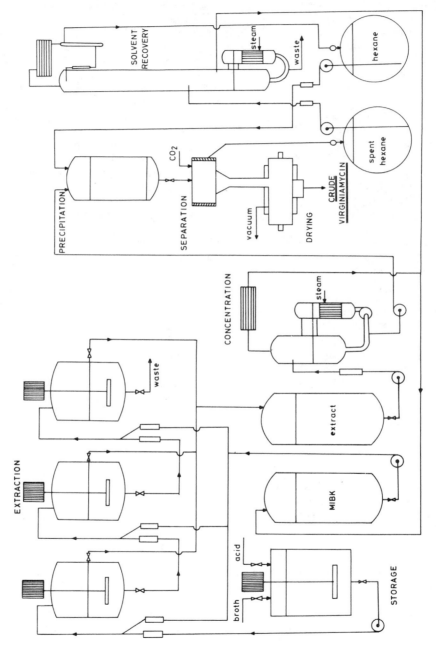

Figure 8 Production of virginiamycin: extraction operations of the fermented broth.

Finally, crude virginiamycin is collected after drying the centrifugate under vacuum at 40°C in a rotary dryer. The spent precipitation liquor, containing hexane and MIBK, is distilled for recovery and reuse of the solvents, while the bottom of the distillation column is discharged for waste treatment.

The waste treatment consists in concentrating the nonutilized fractions of the extraction process by reverse osmosis, using tubular cellulose acetate membranes. The permeate produced by reverse osmosis satisfies the requirements for discharging waste into the municipal main sewers. Crude virginiamycin, as obtained at the end of the extraction process is a very stable powder. No reduction of activity has been observed after more than 5 years storage at ambient temperature, protected from moisture. It is used as a raw material for the manufacture of the commercially patented formulation Stafac 500.

VII. STAFAC MANUFACTURE

The final destination of virginiamycin is incorporation at low levels, ranging between 0.0005 and 0.008%, in animal feeds, via intermediate premixes. Premixes may contain high levels of minerals and vitamins, and feeds are complex mixtures with variable granulation.

Experiments have demonstrated that in unfavorable environmental conditions some aggressive components of the feeds or premixes may be responsible for physicochemical alteration of crude virginiamycin, factor M being more sensitive than factor S. To avoid these drawbacks, the crude virginiamycin is protected by incorporation into granules containing in addition a binding agent and an inert filling agent.

The granulometry of the end product is further adapted to meet several other requirements: first, analytical requirements which are a function of the quality of distribution of the antibiotic preparation in normal feeds. Confidence limits of microbiological analyses made on virginiamycin in feeds are generally accepted by regulatory bodies as the "permitted analytical variations" and range between ±40% for 50-g sample sizes and ±50% for 20-g sample sizes for feeds supplemented with 10 ppm virginiamycin.

These values are based on results of collaborative studies, the most recent one, performed in 1980, involving 12 laboratories in five European countries and implying 201 determinations on the same pig feed at 10 ppm virginiamycin (Table 8).

Second, the granulometry has to meet physiological requirements whereby the standard deviation between daily food uptakes by newborn chicks, eating

Table 8 Relationship Between the Relative Standard Deviation (σ) or Confidence Limits (2σ) of Analytical Results and Sample Size: Collaborative Study on a Pig Feed Supplemented with 10 ppm Virginiamycin as Stafac 500

	Sample size		
	10g	20g	50g
σ	34.2	25.5	20.1
2σ	68.4	51.0	40.2

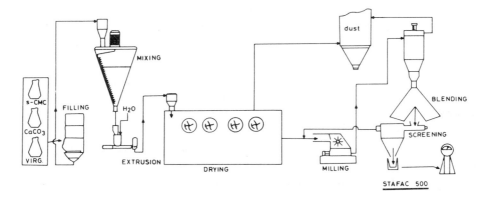

Figure 9 Manufacture operations of Stafac 500.

10 g of feed per day, is, by convention, below ±50%. These requirements
are met by the granulation process applied for the manufacture of Stafac 500.

Crude virginiamycin is blended with sodium carboxymethylcellulose and
milled chalk (Fig. 9). After addition of a measured amount of deionized water,
a paste is obtained which is extruded. The wet strands produced by the ex-
truder are dried in an air dryer at 60°C until a residual humidity of less than
7% has been obtained. The dry product is then milled, homogenized in a
blender, and sieved to retain particles larger than 1 mm which are again milled.
The final powder is transferred to polyethylene bags contained in cardboard
drums. Samples are taken and after analytical approval the bags are soldered
and the drums are sealed and labeled before shipment.

VIII. MODE OF ACTION

The efficacy of virginiamycin on growth and feed conversion has been demonstrated in many animal species: pigs, poultry, calves, rabbits, trout, and
so on. While the beneficial effects of virginiamycin have been recognized,
the mechanism of activity has remained, until recently, obscure. Numerous
theories have been advanced having as a common factor some modification of
the microbial action of the intestinal flora.

Recent research has elucidated the quantitative mode of action of virginiamycin, used as a performance promoter, from the purely nutritional point
of view and not associated with subclinical disease (Vervaeke et al., 1979).
Following the development of static, kinetic, and continuous fermentation techniques using young pigs on an artificial milk powder diet as the standard experimental unit, the quantitative effect of virginiamycin was found to be threefold and related to bacterial flora, biochemical metabolites and physiological
performance of the intestinal tract. Under experimental conditions and as
a result of the use of virginiamycin, a positive nutritional effect of 10% increased growth rate and 7% improved feed conversion were recorded and attributed to the following measured effects:

Changes in the composition and topographical distribution of the intestinal flora by a direct action to give a new equilibrium of the gram-positive flora (Decuypere et al., 1973).

Decrease in the microbial production of lactic acid, volatile fatty acids, ammonia, and amines reducing the toxic effects of these breakdown compounds (Henderickx and Decuypere, 1973).

Saving and optimal availability for the animal of glucose and amino acids—principally, lysine, serine, proline, alanine, glycine, and histidine (Vervaeke et al., 1979; Dierick et al., 1980), leading to a reduced waste of energy.

Reduction in the rate of passage through the intestine accompanied by improved absorption of protein, fat, and carbohydrates. The slower gut motility, probably due to the reduced production of lactic acid, means more time available for absorption (Fausch, 1981).

Direct, mostly beneficial, influence on the permeability of the intestinal mucosa (Dierick et al., 1980).

The experiments substantiated the regulatory effect of virginiamycin as a growth promoter. Virginiamycin did not destroy the intestinal flora, but changed the ratio between the different bacterial species and generated a new equilibrium.

In these new conditions bacterial metabolism is reduced so that fewer carbohydrates and amino acids are wasted (Henderickx et al., 1981). Similar studies in poultry appear to confirm that the mode of action of virginiamycin is identical in that species.

Gaining nutrients and energy has led to the useful "cost-saving concept" whereby virginiamycin allows the successful use of low-protein diets (−10%) with similar outcomes as control groups fed diets rich in protein without virginiamycin (Lesecq and Fremont, 1981).

IX. CONCLUSION

Several aspects of the properties and the production of virginiamycin have been described in the foregoing.

Virginiamycin belongs to the distinct and restricted group of antibiotics containing synergistic components of dissimilar chemical structure. Active against gram-positive microorganisms and not absorbed from the intestine, it presents outstanding advantages for use as a performance promoter in animals, thanks to its safety and efficacy. Additionally, it is used as a tool for scientific research, for example, in the elucidation of protein synthesis and inhibition mechanisms (Chinali et al., 1981) or transport of cations through cell membranes (Oberbäumer, 1979). It is known that the transportation of potassium ions through the cell membranes is greatly facilitated by factor S. The presence of potassium ions in the cells is important for the fixation of factor M on the ribosomes.

Both components of virginiamycin are produced by *S. virginiae* in the same fermentation broth, extracted together with the same solvent system and precipitated before being incorporated into a granulated formulation Stafac 500 (Massart and Naudts, 1981). The low production of the early fermentations has been outstandingly increased after modification first of the strain, by using blind mutation, directed selection, or genetic techniques, and secondly, of the growth conditions by improving the seed stage, the culture media, and the fermentation operations. The production ratio of factor M to factor S has been remarkably stable throughout all the development stages. More insight in the biosynthetic pathways and regulatory mechanisms is needed and is under actual investigation.

It is clear that the continued and even growing interest of many researchers and users in the very special properties and applications of virginiamycin will provide a better understanding of biologically important phenomena in the field, among others, of animal nutrition and of protein synthesis inhibition.

ACKNOWLEDGMENTS

The author is deeply indebted to F. Naudts and B. Boon for their guidance and stimulating interest, and to Y. Massart and G. Tasset for criticism and appraisal on reading the manuscript. Development of the strain at the crucial stages was performed by P. Roelants while the whole production team also contributed to improving the process. Part of the research work was supported by the Institut pour l'Encouragement de la Recherche Scientifique dans l'Industrie et l'Agriculture.

REFERENCES

Aiba, S., Humphrey, A. E., and Millis, N. S. (1973) *Biochemical Engineering*, 2nd ed. Academic, New York

Biot, A. (1979). The influence of virginiamycin on the bacterial cell. In *Performance in Animal Production*. Minerva Medica, Milano, pp. 295–310.

Boon, B., and Dewart, R. (1974). Methods for identification and assay of virginiamycin in animal feeds. *Analyst 99*:19–25.

Chinali, G., Moureau, P., and Cocito, C. G. (1981). The mechanism of action of virginiamycin M on the binding of aminoacyl-tRNA to ribosomes directed by elongation factor Tu. *Eur. J. Biochem. 118*:577–583.

Cocito, C. (1979). Antibiotics of the virginiamycin family, inhibitors which contain synergistic components. *Microbiol. Rev. 43*:2, 145–198.

Crooy, P., and Neys, R. de. (1972). Virginiamycin: Nomenclature. *J. Antibiot. 25*:6, 371.

Decuypere, J., Heyde, H. van der, and Henderickx, H. (1973). Study of the gastrointestinal microflora of suckling and early weaned piglets using different feeding systems and feed additives. In *Germfree Research. Biological Effect of Gnotobiotic Environments*. J. B. Heneghan (Ed.). Academic, New York, pp. 369–377.

Devriese, L. A. (1980). Sensitivity of staphylococci from farm animals to antibacterial agents used for growth promotion and therapy. A ten year study. *Ann. Rech. Vet. 11*:4, 399–408.

Dewart, R., Naudts, F., Lhoest, W. (1965). Automation in the microbiological assays of antibiotics. *Ann. N.Y. Acad. Sci. 130*:686–696.

Dicuollo, C. J. (1979). Establishment of human safety for virginiamycin when employed as performance promoter and virginiamycin's impact on the environment. In *Performance in Animal Production*. Minerva Medica, Milano, pp. 321–338.

Dierick, N. A., Decuypere, J. A., Vervaeke, I. J., and Henderickx, H. K. (1980). Resorption of amino acids from an isolated loop of the pig's small intestine in vivo: Influence of a nutritional dose of virginiamycin. In *Proceedings of the IPVS Congress*. Copenhagen, p. 302.

Dijck, P. van. (1969). Further bacteriological evaluation of virginiamycin. *Chemotherapy 14*:322–332.

Elander, R. P., Chang, L. T., and Vaughan, R. W. (1977). Genetics of industrial microorganisms. *Ann. Rep. Ferment. Proc. 1*:1–40.

European Communities (1972). Animal feeding stuffs: Determination of virginiamycin. *Directive 72/199/EEC.* Her Majesty's Stationery Office, London, pp. 40–43.

Fausch, H. D. (1981). The effect of virginiamycin on rate of passage of ingesta in growing, finishing pigs. In *Proceedings of the Smith Kline AHP Symposium.* Kansas City, Missouri, pp. 55–61.

Fujisawa, Y., and Weisblum, B. (1981). A family of r-determinants in *Streptomyces* spp. that specifies inducible resistance to macrolide, lincosamide and streptogramin type B antibiotics. *J. Bacteriol. 146*:2, 621–631.

Henderickx, H., and Decuypere, J. (1973). Influence of nutritional levels of spiramycin and virginiamycin on the bacterial metabolites in the gastrointestinal tract and urine of artificially reared early weaned piglets. In *Germfree Research. Biological Effect of Gnotobiotic Environments.* J. B. Heneghan (Ed.). Academic, New York, pp. 361–368.

Henderickx, H. K., Vervaeke, I. J., Decuypere, J. A., and Dierick, N. A. (1981). Mode of action of growth promotion drugs. In *Proceedings of the Smith Kline AHP Symposium.* Kansas City, Missouri, pp. 1–9.

Kingston, D. G. I., and Kolpak, M. X. (1980). Biosynthesis of antibiotics of the virginiamycin family. 1. Biosynthesis of virginiamycin M_1: Determination of the labeling pattern by the use of stable isotope techniques. *J. Am. Chem. Soc. 102*:5964–5966.

Konvalinkova, V., Roelants, P., and Mergeay, M. (1977). Transfection in *Streptomyces virginiae. Biochem. Soc. Trans. 5*:941–943.

Lesecq, P., and Fremont, Y. (1981). Effects of virginiamycin on the performance of chickens and pigs with diets at different protein levels. In *Proceedings of the Conference on Feed Additives.* Budapest, pp. 49–56.

Luedeking, R., and Piret, E. L. (1959). A kinetic study of the lactic acid fermentation. Batch process at controlled pH. *J. Biochem. Microbiol. Technol. Eng. 1*:393.

Massart, Y., and Naudts, F. (1981). Production process of virginiamycin and Stafac. *Proc. 2nd World Congr. Chem. Eng. 1*:281.

Oberbäumer, I. (1979). Streptogramine als modellsysteme für den Kationentransport durch Membranen. Ph.D. thesis, Georg-August University, Göttingen.

Roelants, P., Boon, B., and Lhoest, W. (1968). Evaluation of a commercial air filter for removal of viruses from the air. *Appl. Microbiol. 16*:10, 1465–1467.

Roelants, P., and Naudts, F. (1964). Properties of a bacteriocin-like substance produced by *Streptomyces virginiae. Antonie van Leeuwenhoek J. Microbiol. Serol. 30*:1, 45–53.

Roelants, P., Konvalinkova, V., Mergeay, M., Lurquin, P., and Ledoux, L. (1976). DNA uptake by *Streptomyces* species. *Biochim. Biophys. Acta 442*:117–122.

Schäfer, V., Knothe, H., and Lenz, W. (1980). Zur gegenwärtigen Resistenzsituation von *Staphylococcus aureus*—Stämmen von Broilern, Betriebsangehörigen und städtischer Bevölkerung. In *Sixth European Poultry Conference.* WPSA Congress, Hamburg, pp. 289–296.

Somer, P. de, and Dijck, P. van (1955). A preliminary report on antibiotic number 899, a streptogramin-like substance. *Antibiot. Chemother. 5*: 11, 632–639.

Vervaeke, J., Decuypere, J. A., Dierick, N. A., and Henderickx, H. K. (1979). Quantitative in vitro evaluation of the energy metabolism influenced by virginiamycin and spiramycin used as growth promoters in pig nutrition. *J. Anim. Sci. 49*:3, 846–856.

Vining, L. C. (1979). Antibiotic tolerance in producer organisms. *Adv. Appl. Microbiol. 25*:147–168.

Yanagimoto, M., and Terui, G. (1971a). Physiological studies on Staphylomycin production. I. Product inhibition. *Hakko Kogaku Zasshi 49*:7, 604–610.

Yanagimoto, M., and Terui, G. (1971b). Physiological studies on Staphylomycin production. II. Formation of a substance effective in inducing Staphylomycin production. *Hakko Kogaku Zasshi 49*:7, 611–618.

Yanagimoto, M., Yamada, Y., and Terui, G. (1979). Physiological study on the production of Staphylomycin. III. Extraction and purification of inducing material produced in Staphylomycin fermentation. *Hakko Kogaku Zasshi 57*:1, 6–14.

26

SALINOMYCIN: PROPERTIES, BIOSYNTHESIS, AND FERMENTATION

NOBORU ŌTAKE *Institute of Applied Microbiology, University of Tokyo, Tokyo, Japan*

I. INTRODUCTION

During the past decade, considerable effort has been directed to the study of the polyether antibiotics, a class of naturally occurring inophores produced predominantly by *Streptomyces* species. These antibiotics are known for their unique structure and for the fact that they possess a wide range of biological activities. Developmental research in recent years has shown their usefulness as cocciodiostats for chickens and as agents for improving the fattening of ruminants.

Coccidiosis, one of the most serious infections of poultry, is an infectious disease spread by the parasite *Eimeria protozoa*, belonging to the sporozoa. Poultry raising has attained during the past 20 years increasingly larger scales, and this has been accompanied by a rapid increase in the occurrence of poultry diseases. Among them, coccidiosis has been especially damaging. As a result, interest has been expressed in the development of coccidiostats, and many chemotherapeutic agents have been discovered. Nevertheless, this disease still continues to break out ceaselessly even today, and with the appearance of drug-resistant strains, there has been a new emergence of this old disease. Salinomycin displays anticoccidial activity. Its structure is represented in Figure 1.

Salinomycin was discovered by Dr. Miyazaki and associates of Kaken Chemicals; the structure and biochemistry were elucidated in collaboration with the author's group at the Institute of Applied Microbiology (Miyazaki, 1974).

II. DISCOVERY AND STRAIN CHARACTERIZATION

The salinomycin-producing organism is a *Streptomyces* species designated as strain 80614, isolated from a soil sample collected at Fuji city, Shizuoka Prefecture, Japan. According to taxonomic studies described below, this organism was identified as a strain of *Streptomyces albus*.

A. Morphological Characteristics

The morphology of the culture was microscopically observed on inorganic salts–starch agar, and oatmeal agar at 27°C for 10–14 days. The aerial mycelium of the strain 80614 branches and produces spirals of two or three volutions in sporophores. Formation of true whorls was not observed on the electron microscope; the spores were ellipsoidal and cylindrical 0.5–1.0 and 1.0–1.5 μm. The surface of the spores was smooth.

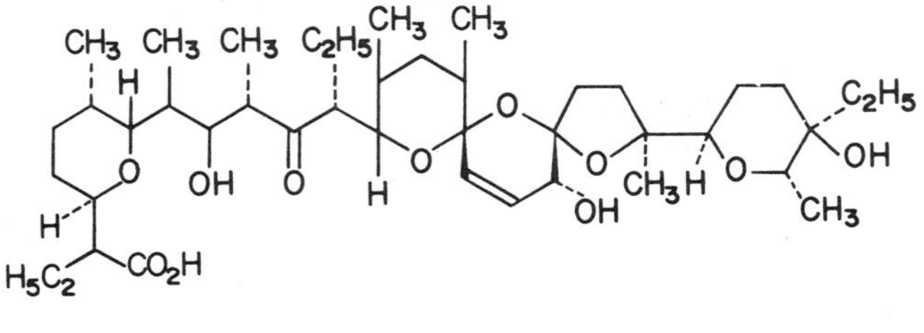

Salinomycin

Figure 1 Structure of salinomycin.

Table 1 Cultural Characteristics of Strain 80614

Medium	Growth	Aerial mycelium	Soluble pigment
Czapek's agar	Good, raised, white to pale yellow	Poor, white	None
Starch–inorganic salts agar	Poor or moderate, white to tan	Moderate, powdery, white to whitish gray	None
Glucose asparagine agar	Poor, thin, white to tan	Moderate to poor, velvety, white to whitish gray	None
Glycerol asparagine agar	Poor to moderate, tan	Poor to moderate, white	None
Calcium malate agar	Good, raised, white to pale tan	None or poor, white	None
Tyrosine agar	Poor, thin, white to pale brown	None	None or faint brown
Nutrient agar	Poor, thin, golden yellow	None or poor, white	None
Yeast malt extract agar	Good, yellowish brown	Good, white to yellowish white	Pale brown
Oatmeal agar	Poor, colorless	Poor to moderate, white to whitish gray	None or pale brown
Glucose nutrient agar	Poor, thin, yellowish white	None or poor, white	None
Glucose peptone agar	Poor, thin, yellowish white	None or poor, white	None or pale brown
Glycerol Czapek's agar	Good, raised, white to pale tan	Poor or none, white	None or pale brown
Potato plug	Poor, thin, brown	None or scanty, white to whitish gray	None
Cellulose	Scant, thin, colorless	None	None
Litmus milk	Ring growth in medium, colorless	None	Faint red
Egg	Poor, thin, yellowish white to white	None	None
Loeffler's serum medium (27°C, 10 days)	Good, raised, brownish yellow	None	None

Table 2 Physiological Properties of
Streptomyces—Strain 80614

Temperature for growth	21 ∿ 37°C
pH range for growth	5.5 ∿ 8.5
Tyrosine reaction	Negative
Melanoid pigment	Negative
Reduction of nitrate	Doubtful
Liquefaction of gelatin	Positive
Coagulation of milk	Negative
Peptonization of milk	Negative
Hydrolysis of starch	Positive
Cellulose decomposition	Negative
Product	Salinomycin

B. Cultural and Physiological Characteristics

The cultural characteristics and physiological properties of strain 80614 are
shown in Tables 1–3. All media used in this study were prepared according
to the recommendations of the International Streptomyces Project (ISP) (Shirling and Gottlieb, 1966).

Cultural and physiological studies were carried out at 27°C and the results were observed after 21 days. Utilization of carbon sources was investigated by the method of Pridham and Gottlieb (1948).

C. Comparison of Strain 80614 with Related *Streptomycetes*

As a result of these observations, the following characteristics were noticed
as distinctive features of strain 80614.

1. Spores are born in short spirals and the spore surface is smooth.
2. The color of aerial mycelia is white to whitish gray. The color of
   vegetative growth is noncharacteristic, colorless to pale yellow.
   No soluble pigment is produced on synthetic agar.
3. Melanoid pigment is not formed.

Table 3 Carbon Utilization Pattern of Strain 80614

Carbon source	Response[a]	Carbon source	Response[a]
Arabinose	±	Sucrose	±
Glucose	++	Trehalose	+
Galactose	++	Xylose	++
Lactose	+	Salicin	±
Levulose	++	Inulin	++
Mannose	±	Adonitol	−
Maltose	±	Dulcitol	−
Melezitose	−	i-Inositol	−
Melibiose	++	Mannitol	++
Raffinose	−	Sorbitol	−
Rhamnose	−	Cellobiose	++

[a]++, strongly positive utilization; +, positive utilization; ±, doubtful utilization; −, negative utilization.

Strain 80614 was checked for identity with known species described in *Bergey's Manual* (Breed et al., 1974). As a result of the above research, strain 80614 was found to be closely related to *S. albus* (Rossi-Doria) Waksman and Henrici (1943). Therefore the salinomycin-producing strain 80614 was identified as a strain of *S. albus* (Rossi-Doria) Waksman and Henrici.

III. BIOSYNTHESIS OF SALINOMYCIN

Thanks to the development of ^{13}C nuclear magnetic resonance (NMR), ^{13}C compounds have recently become quite popular, and it has now become possible to elucidate, within a relatively short period of time, the pattern of biosynthesis, even of compounds which are so complex that they cannot even by studied by the ^{14}C-labeling method. Polyether antibiotics in particular are very complex compounds (Fig. 1). Detailed studies have been made of the biosynthesis of such compounds as salinomycin (Seto et al., 1977), lysocellin (Ōtake et al., 1978), monensin (Day et al., 1973), and lasalocid A (Westley et al., 1972).

The ^{13}C-NMR spectrum of salinomycin has been investigated thoroughly and the complete assignment of each carbon atom has been accomplished as shown in Figure 2 (Seto et al., 1977).

On the basis of the chemical structure of salinomycin, biosynthesis of a polyketide chain is anticipated. For this reason, experimental proof was obtained by making use of ^{13}C-NMR. About 500 µg/ml of salinomycin is produced in *S. albus* shake cultures after 90 hr at 30°C in a medium containing glucose (5%), soybean four (1%), brewer's yeast (0.2%), table salt (0.2%), potassium chloride (0.2%), and calcium carbonate (0.02%). During the fermentation process under these conditions, [1-^{13}C] or [1,2-^{13}C]acetic acid, propionic acid, or butyric acid were added as precursors. Salinomycin enriched with each of these labels was refined and isolated. It became clear by studies of the ^{13}C-NMR spectra of these compounds and analysis of their $^{13}C-^{13}C$ coup-

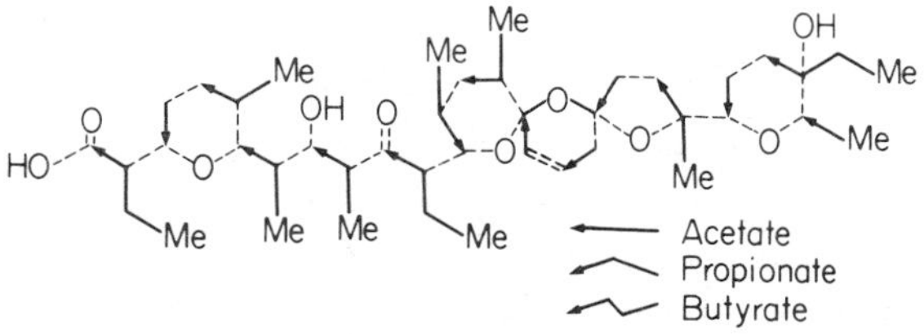

Acetate
Propionate
Butyrate

Assignment of ^{13}C-nmr(ppm,CDCl$_3$)

Carbon No.		Carbon No.		Carbon No.		Carbon No.	
1	177.2	11	214.5	21	106.4	32	6.3
2	48.9	12	56.5	22	36.2	33	25.8
3	74.9	13	71.7	23	30.2	34	17.9
4	20.1	14	36.5	24	88.5	35	15.6
5	26.4	15	38.6	25	73.7	36	22.7
6	28.0	16	40.7	26	21.9	37	11.9
7	75.2	17	99.2	27	29.3	38	12.8
8	32.6	18	121.6	28	70.9	39	7.0
9	68.7	19	132.4	29	77.2	40	11.2
10	49.2	20	67.2	30	14.5	41	16.6
				31	30.6	42	13.2

Figure 2 Biosynthesis of salinomycin.

ling that the carbon skeleton of salinomycin is synthesized according to the label patterns shown in Figure 2.

It was proven, based on these experimental results, that salinomycin is biosynthesized through a polyketide chain with condensation of 6 mol of acetic acid, 6 mol of propionic acid, and 3 mol of butyric acid. It is interesting to note that salinomycin and the other polyether antibiotics have similar pathways of biosynthesis; they share in common the fact that two- and four-carbon fatty acids are incorporated unchanged. Furthermore, productivity of the antibiotic can be increased by regulating the fatty acid metabolic systems of the antibiotic-producing organisms and by introducing vegetable or animal oils. In the salinomycin fermentation, productivity was largely improved based on these biochemical findings.

IV. DEVELOPMENT OF THE TECHNOLOGY FOR SALINOMYCIN PRODUCTION BY FERMENTATION

Since the usefulness of salinomycin against coccidiosis of poultry as well as a fattening agent for ruminants was recognized, research has been directed toward methods for manufacturing this antibiotic industrially with good efficiency.

Based on findings during studies on the biosynthesis of salinomycin, improvements were made by manipulating the production medium, chiefly by using oil as the carbon source. The oil had poor assimilation properties in the initial production medium; however, the utilization rate of the oil increased greatly as a result of the addition of the ammonium salts of various inorganic or organic acids, and it became possible to use vegetable or animal oils as the main carbon sources. As a result, the productivity of this antibiotic improved considerably.

In an attempt to increase the productivity of salinomycin and to shorten the fermentation cycle, studies were made concerning the optimum culturing conditions. The effect of the inoculum size, the aeration level, stirring conditions and culture temperature were also studied in view of improving the salinomycin yield.

As a result of these improvements, it became easy to determine significant differences between several producer strains. Attempts were also made to improve the strains starting from single spores, or by applying various techniques of mutation and selection. As a result, SLS-K, a strain with an excellent oil—salinomycin conversion rate, was obtained. With this improved strain SLS-K, the conversion rate from soybean oil into salinomycin rose 1.5-fold as compared with the previous strains, and as a result the production titers also increased considerably.

As a result of culture and strain manipulation, the productivity of salinomycin has been improved phenomenally in comparison with the early product yield derived from the original strain.

The production of salinomycin is carried out by tank fermentation. *Streptomyces albus* was grown in a 500-ml shake flask containing 100 ml of the following culture medium; 4% glucose, 1% soybean meal, 1% brewer's yeast, and 0.2% $CaCO_3$. The culture was kept for 48 hr on a rotary shaker at 28°C and the culture (1 liter) was seeded into a 200-liter tank fermenter containing 100 liters of the following medium: 2% glucose, 1% starch, 2.5% soybean meal, 0.4% beer yeast, 0.1% meat extract, and 0.2% NaCl. The pH was adjusted to 8.5 with 1 N NaOH before sterilization. Incubation was carried out at 28°C for 90 hr under stardard conditions of aeration and agitation.

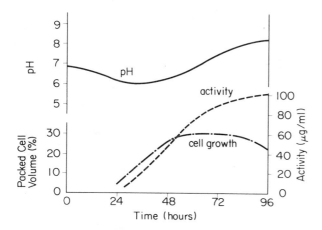

Figure 3 Time course of salinomycin production.

The formation of salinomycin was followed by testing the antibiotic activity at various times during the fermentation cycle. Each sample was mixed with an equal volume of methanol and the mixture was shaken for 2 hr and filtered. The filrate was assayed by paper disk method using *Bacillus subtilis* as a test organism.

A typical time course of fermentation is shown in Figure 3. The active substance was isolated from the mycelium as well as from the filtered broth. The fermentation broth was adjusted to pH 9.0 with 5 N NaOH and was filtered after the addition of a filter aid. The separated mycelium was extracted with 80% aqueous acetone. The extract was concentrated in vacuo and the aqueous residue was adjusted to pH 9.0 with 3 N NaOH and extracted with butyl ace-- tate. The filtered broth was extracted with butyl acetate. The solvent extracts were combined and concentrated to an oily residue under reduced pressure.

The antibiotic was further purified by column chromatography on alumina and silica gel. The residue was applied on top of an alumina column packed with an ethyl acetate–hexane mixture (3:1), and the column was washed with the same solvent system and then developed with an ethyl acetate–methanol mixture (3:1). The latter eluate, which contained the antibiotic, was concentrated in vacuo. The residue was dissolved in a small amount of chloroform and applied to a column of silica gel, which was developed with chloroform containing 4% methanol. The fractions containing the antibiotic were combined and concentrated to dryness. The dry residue was dissolved in a small amount of an acetone–water mixture and chilled in a refrigerator until crystallization was complete. Crude crystals of salinomycin were filtered, washed with water, and recrystallized from an acetone–water mixture and dried in vacuo. By the above-described isolation procedure, salinomycin was obtained in the form of colorless prisms of the sodium salt.

The free acid of salinomycin could be obtained as a white amorphous powder by adjusting the fermentation broth to pH 3.0 with 3 N HCl and following the previously described isolation procedure, or by shaking the ethyl acetate solution of salinomycin sodium salt with 0.1 N HCl. Both salinomycin and its

Table 4 Properties of Salinomycin and Sodium Salt

Property	Acid		Na salt	
Melting point (°C)	112.5 ∿ 113.5		140 ∿ 142	
$[\alpha]_{25}^{D}$(c = 1, ethanol)	−63°		−37°	
Ultraviolet absorption	285 nm (ε 108)		285 nm (ε 108)	
Infrared absorption	5.85		5.85	
(carbonyl)			6.40	
Molecular weight	750 (mass)		772 (mass)	
Molecular formula	$C_{42}H_{70}O_{11}$		$C_{42}H_{69}O_{11}Na$	
Elementary analysis	Calculated	Found	Calculated	Found
C	67.17	67.18	65.26	65.01
H	9.39	9.55	9.00	8.95
O	23.43	23.13	22.76	22.39
Na			2.97	2.80
pK_a'	6.40			

sodium salt are readily converted into each other and extractable from water into organic solvents.

Salinomycin free acid is a white amorphous powder. It is weakly acidic with pK_a' value of 6.4, and melts at 112.5–113.5°C. Salinomycin is levorotatory, having an optical rotation of −63° in ethanol solution. It is soluble in lower alcohols, acetone, ethyl acetate, benzene, chloroform, carbon tetrachloride, ether, petroleum ether, and hexane, and insoluble in water. It is stable as the sodium salt, but gradually loses antimicrobial activity under acidic conditions. Some physicochemical properties of salinomycin free acid and sodium salt are summarized in Table 4.

The molecular weight of salinomycin was determined from elementary analysis and its mass spectrum. The mass spectrum of salinomycin showed the molecular ion peak at m/e 750 and a dehydration peak at m/e 732, which is in good agreement with the molecular formula $C_{42}H_{70}O_{11}$; the mass spectrum of its sodium salt exhibited the molecular ion peak at m/e 772 corresponding to the molecular formula $C_{42}H_{69}O_{11}Na$.

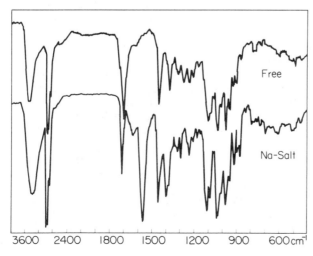

Figure 4 Infrared spectra of salinomycin.

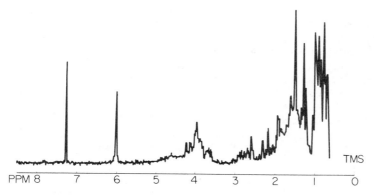

Figure 5   NMR spectrum of salinomycin.

Esterification of salinomycin with an excess of etherial diazomethane gave a crystalline monomethyl ester (m.p., 99–101°C; C, 67.35%, H, 9.28%; O, 22.85%; calculated for $C_{43}H_{72}O_{11}$: C, 67.54%, H, 9.42%; O, 23.03%). Its mass spectrum showed a distinct molecular ion peak at m/e 764.

Treatment of salinomycin with acetic anhydride–pyridine gave a crystalline monoacetyl derivative (m.p., 148–150°C; C, 66.45%; H, 9.14%; O, 24.45%; calculated for $C_{44}H_{72}O_{12}$: C, 66.66%; H, 9.09%; O, 24.24%). Its mass spectrum gave 792 as the molecular weight.

The molecular weight and formula of salinomycin methyl ester and acetyl derivative, as determined by elementary analysis and mass spectrometry, are consistent with the formula of salinomycin.

The ultraviolet absorption spectrum of salinomycin showed maxima of low intensity at 285 nm (ε 108), corresponding to a carbonyl group, in methanol solution, and at 285 nm (ε 218) in alkaline methanol solution, respectively. As shown in Figure 4, the infrared absorption spectra of salinomycin and its sodium salt showed bands at 3300, 3500, 1710, and 1560 cm^{-1} (in sodium salt), indicating the presence of hydroxyl, ketone, and carboxyl groups.

The NMR spectrum of salinomycin taken in deuterochloroform at 100 MHz is illustrated in Figure 5. The spectrum suggested the presence of many C-methyl groups at δ 0.7 and 1.5, and two unsplit vinyl protons at δ 6.0. In contrast to other polyether antibiotics, the absence of a methoxyl signal was characteristic of salinomycin.

V.   STRUCTURE AND REACTIONS OF SALINOMYCIN

A.   Elucidation of Salinomycin Structure by X-ray Analysis (Kinashi et al., 1973, 1975).

As with other polyether antibiotics, the structure of salinomycin was also elucidated by x-ray analysis, owing to extensive decomposition into unidentifiable products when chemical procedures are used.

Unlike other polyether antibiotics, metal salts of salinomycin could not form crystals suitable for x-ray analysis. Therefore the p-iodophenacyl ester of salinomycin was subjected to x-ray analysis in order to establish the entire molecular structure and the absolute configuration of the antibiotic. The p-iodophenacyl ester of salinomycin was prepared by the reaction of salinomycin with p-iodo-α-diazoacetophenone in dioxane in the presence of cupric chloride as a catalyst, and purified by thin-layer chromatography on solica gel using a mixture of chloroform and methanol (40:1). The crystals were grown by allowing the ethanol solution to evaporate in the dark. The crystals are color-less prisms elongated along the c axis. The unit cell dimensions and space group were determined from Weissenberg photographs. The former were re-fined on a diffractometer; crystal data are summarized below:

Salinomycin p-iodophenacyl ester, $C_{50}H_{75}O_{12}I$, molecular weight, 995.1, m.p. 191.5–192.5°C. Orthorhombic, a = 20.981(2), b = 22.761(2), c = 10.493(1) Å, U = 5010.7 Å^3, Dm = 1.32 g/cm^3, by flotation in an aqueous solu-tion of sodium potassium tartrate, Z = 4, D_x = 1.32 g/cm^3, F(000) = 2096, linear absorption coefficient for Mo K$_\alpha$ (λ = 0.7107 Å), μ = 7.27 cm^{-1}, space group, $P2_12_12_1$, from absent reflexions.

A crystal approximately 0.35 × 0.35 × 0.40 mm^3 was used to collect data on the Rigaku automated four-circle diffractometer. Integrated intensities were measured for 20 < 50° by the θ-2 scan technique with Mo K radiation at a scan speed of 2θ 4°/min. Backgrounds were counted at the beginning and the end of each scan for 10 sec. In this way 4936 reflexions were record-ed, of which 3288 gave intensities greater than three times their standard deviations. Three reflexions were measured as references every 50 reflexions: The net count of these reflexions reduced by about 4% over the period of data collection. All the data were corrected for this effect. They were also cor-rected for Lorentz and polarization factors, but not for absorption.

The structure was solved by the heavy-atom method. The position of the iodine atom was easily deduced from a three-dimensional Patterson map. A set of structure factors was calculated for 1708 reflexions up to 2θ =35°, and the electron density map was computed with the phases based on the iodine atom alone. The initial R value was 0.43 and the positions of 16 lighter atoms were located from this map.

Subsequent calculation of structure factors and electron density maps yielded 43 atomic positions and the least-squares plane through the other four atoms of the ring. The other three six-membered rings A, D, and F adopt chair conformations, while ring C takes the half-chair conformation owing to the double bond in the ring. For ring A, C(35) and C(61) are equatorial, whereas C(59) and O(60) are axial. For ring C, C(26), C(37), and O(57) are pseudoequatorial, while O(24) and O(34) are pseudoaxial. For ring D all the substituents are equatorial, except for C(33), which is axial. For ring E, C(18) is equatorial and C(11) and C(48) are axial.

A projection of the crystal structure viewed along the a axis is presented in Figure 6.

The chemical structure and the absolute configuration of salinomycin are therefore as shown in Figure 1. It is now well established that salinomycin is a new member of the polyether antibiotics containing four six-membered rings (A, C, D, and E) and one five-membered ring B. Rings B, C, and D form a unique tricyclic spiroketal ring system. Salinomycin is also the first example of a polyether antibiotic containing an unsaturated C ring in the mole-cule.

B. Chemical Transformation of Salinomycin (Ōtake et al., 1976)

The elucidation of the structure-activity relationship of salinomycin is of great interest.

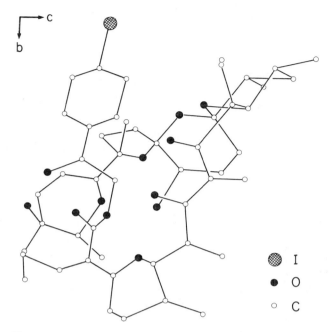

Figure 6 X-ray structure of salinomycin p-iodophenacyl ester (viewed along the a axis).

In order to clarify the contribution of the active sites to the biological activity of the molecule, chemical modification of the salinomycin molecule was undertaken. This modification was mainly focused on the oxygen functions such as hydroxyl, carboxylic acid, and ketone groups which might involve in ligand formation with cations.

Physiochemical properties of the compounds thus far prepared are listed in Table 5.

C. Mass Spectrometry of Salinomycin and Its Derivatives (Kinashi et al., 1975)

It is of particular interest that the structure of the molecule contains five ether ring systems of six members (rings A, B, C, and E), one of five members (ring D), and allylic alcohol and β-hydroxyketone groups, and that the rings B, C, and D form a unique tricyclic spiroketal system. It should be emphasized that this tricyclic spiroketal ring system was found here for the first time among natural products.

As part of our studies on ionophore antibiotics, we were interested in the mass spectrometry of this antibiotic and its derivatives in connection with the structural elucidation of minor components and transformation products. Therefore the fragment ions derived from the methyl ester, sodium salt, and free acid of salinomycin have been extensively investigated (Figs. 7, 8, and 9).

1. The Methyl Ester of Salinomycin

The mass spectrum of the methyl ester is discussed first, since it yielded more well-defined crystals with more sufficient volatility than the free acid and the sodium salt.

Table 5 Properties of Salinomycin and Its Derivatives

Compounds	R_1	R_2	R_3	Melting point (°C)	$[\alpha]_D^{25}$ (c=1.0, MeOH)	Formula[a]	Mass (M+, m/c)	NMR ($\delta_{TMS}^{CDCl_3}$)[b]
I	CO_2H	=O	OH	112~113	−63°	$C_{42}H_{70}O_{11}$	750	5.98 (2H, s)
II	CO_2CH_3	=O	OH	99~101	−50°	$C_{43}H_{72}O_{11}$	764	6.20 (1H, dd), 6.02 (1H, dd)
III	$CO_2CH_2COC_6H_4Br$	=O	OH	208~210	−51°	$C_{50}H_{75}O_{12}Br$	948	8.18 (2H, d), 7.62 (2H, d)
IV	CO_2CH_3	=O	$OCOCH_3$	192~193	−61°	$C_{45}H_{74}O_{12}$	806	6.16 (1H, d), 6.00 (2H,s), 5.43 (1H, d)
V	CO_2H	=O	OCHO	123~124	−70°	$C_{43}H_{70}O_{12}$	778	6.23 (1H, dd), 5.82 (1H, dd), 5.27 (1H, m)
VI	CO_2H	=O	$OCOCH_3$	148~150	−73°	$C_{44}H_{72}O_{12}$	792	8.10 (1H, s), 6.20 (1H, dd), 5.81 (1H, brd)
VII	CO_2H	=O	$OCOCH_2CH_3$	184~186	−85°	$C_{45}H_{74}O_{12}$	806	6.17 (1H, dd), 5.80 (1H, dd), 5.30 (1H,m)
VIII	CO_2H	=O	$OCO(CH_2)_2CH_3$	115~117	−45°	$C_{46}H_{76}O_{12}$	820	6.16 (1H, dd), 5.76 (1H, brd), 5.30 (1H, m)
IX	CO_2H	=O	$OCO(CH_2)_3CH_3$	120~122	−33°	$C_{47}H_{78}O_{12}$	834	6.23 (1H, dd), 5.86 (1H, dd), 5.39 (1H, m)
X	CO_2H	=O	OH	111~114	−23°	$C_{42}H_{72}O_{11}$	752	6.23 (1H, dd), 5.85 (1H, dd), 5.39 (1H, m)
XI	CO_2H	=O	=O	104~107	+10°	$C_{42}H_{68}O_{11}$	748	no vinylic protons
XIIa	CO_2H	H, OH	OH	203~204	− 5°	$C_{42}H_{72}O_{11}$	752	7.15 (1H, d), 6.23 (1H, d)
XIIb	CO_2H	H, OH	OH	240~241	−20°	$C_{42}H_{72}O_{11}$	752	6.13 (1H, dd), 5.94 (1H, brd)
XIIIa	CH_2OH	H, OH	OH	192~193	−18°	$C_{42}H_{74}O_{10}$	738	6.18 (1H, brd), 5.99 (1H, brd); 6.15 (1H, brd), 5.95 (1H, brd)
XIIIb	CH_2OH	H, OH	OH	241~243	−10°	$C_{42}H_{74}O_{10}$	738	6.07 (1H, brd), 5.91 (1H, brd)

[a] The results of elementary analysis of these compounds were consistent with the corresponding molecular formula within 0.05% errors.

[b] The abbreviation: s, d, dd, brd and m mean singlet, doublet, double-doublet, broad doublet and multiplet, respectively.

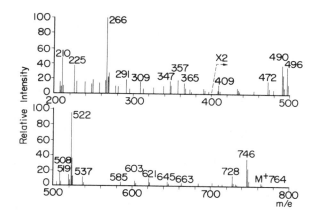

Figure 7 Mass spectrum of the methyl ester of salinomycin.

The mass spectrum of the methyl ester is shown in Figure 7; the molecular ion is observed at m/e 764 as a distinct peak accompanied by three successive dehydration peaks at m/e 764 ($C_{43}H_{70}O_{10}$), m/e 728 ($C_{43}H_{68}O_9$), and m/e 710 (low intensity). The base peak in the spectrum of the methyl ester occurs at m/e 266 ($C_{16}H_{26}O_3$).

Features of fragmentation observed in the spectrum of the methyl ester can be classified into three types, according to the origin of the ions; those formed by cleavage at α-positions of the cyclic ether rings or the nine-carbon hydroxyl group (α cleavage), those formed from the β-hydroxyketone group by McLafferty rearrangement (Ōtake et al., 1978) (M cleavage), and those formed by the cleavage of the ring C (C cleavage).

The ions so far observed in the spectrum of the methyl ester can be classified into the following groups:

α-Cleavage. The cleavage at the α-position of the ether rings yielding cyclic oxonium ions is the most important fragmentation exhibited by the α-alkylated tetrahydrofuran derivatives. Indeed, this type of cleavage was reported to play a significant role in the mass spectrometry of the other polyether antibiotics.

The ions with low or moderate intensity which appear at m/e 663, m/e 621, ($C_{35}H_{57}O_9$), m/e 537($C_{30}H_{49}O_8$), m/e 409($C_{23}H_{37}O_6$), and m/e 199 ($C_{11}H_{19}O_3$) are assigned to those derived by the α-cleavage as depicted in Scheme 1.

C Cleavage. It is noteworthy that ions which originate from the C cleavage are accompanied in most cases by rather intense dehydration peaks. Given the uniqueness of the tricyclic spiroketal ring system, the fragmentation pattern of this function under electron impact is of particular interest.

Cleavage of the parent ion at the ring C as illustrated in Scheme 1b yields a prominent ion at m/e 522 ($C_{30}H_{22}O_4$ along with a neutral lactone, with $C_{13}H_{22}O_4$ as empirical formula.

Relevant to the formation of the m/e 522 ion, the cleavage of ring C may occur with great ease, involving the allylic alcohol group by a concerted reaction; the resulting ion may eventually be formulated as a conjugated oxonium ion. Support for this explanation is provided by characterization of the carbon-13 lactone isolated from the alkaline hydrolysis product of the free acid in a reasonable yield. The formation mechanism seemed to be similar, an attack of

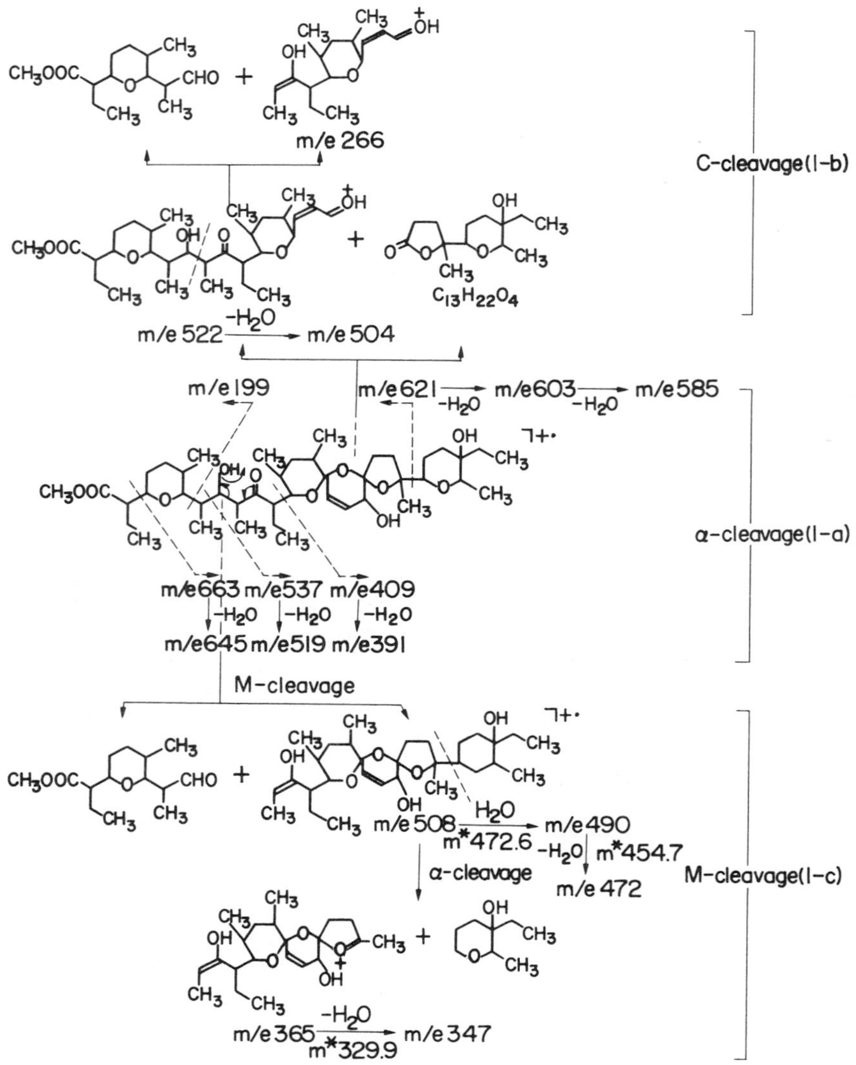

Scheme 1 Fragmentation pattern of the methyl ester of salinomycin.

OH⁻ on the allylic alcohol group resulting in cleavage of ring C to yield the 13-carbon lactone by a concerted reaction.

Further fragmentation of the m/e 522 ion by M cleavage gives rise to a base peak of m/e 266 ($C_{16}H_{26}O_3$).

It should be emphasized that C cleavage appears to be a characteristic feature in the mass spectrometry of salinomycin and its derivatives and that the ions derived from this cleavage can be used as diagnostic peaks for identification of salinomycin homologs.

M Cleavage. The ion at m/e 508 ($C_{29}H_{48}O_7$) which appears as a peak of moderate intensity is due to a cleavage of the methyl ester by the McLafferty rearrangement. It is observed that this cleavage is feasible at the β-hydroxyketone portion and the resulted fragments may be formulated as the charged enol and the neutral aldehyde.

Subsequent losses of two molecules of H_2O from the ion at m/e 508 yield peaks of appreciable intensity at m/e 490 ($C_{29}H_{46}O_6$) and m/e 472 ($C_{29}H_{44}O_5$). Metastable peaks corresponding to these dehydrations are present at m/e 472.6 and m/e 454.7, respectively.

An alternative pathway of fragmentation is observed in which the loss of the ring E portion by α cleavage from the m/e 508 ion gives rise to a peak at m/e 365 ($C_{21}H_{33}O_5$), and further loss of H_2O produces an ion at m/e 347 ($C_{21}H_{31}O_4$).

The ions derived from M cleavage are summarized in Scheme 1c.

2. The Sodium Salt of Salinomycin

As with other polyether antibiotics, the sodium salt has a characteristic low solubility in water, but is soluble in most organic solvents. This behavior can be explained in terms of its cyclic conformation, namely, the hydrophobic functions being located at the exterior surface and the metal ion in the center of a cavity of the molecule.

The mass spectrum of the sodium salt is shown in Figure 8. The molecular ion appears as a distinct peak at m/e 772, the (M-1) ion being of an appreciable intensity; in contrast to the methyl esters, however, the sodium salt shows no dehydration peaks derived from the molecular ion.

A prominent ion at m/e 728 ($C_{41}H_{69}O_9Na$) which is due to the loss of CO_2 indicates that decarboxylation of the sodium salt readily occurs.

The ions which are derived from α cleavage appear at m/e 685 ($C_{38}H_{62}O_9$ Na), m/e 629, m/e 559 ($C_{30}H_{48}O_8Na$), and m/e 431 ($C_{23}H_{36}O_6Na$), respectively (Scheme 2).

Inspection of these ions and their empirical formulas indicates that the sodium salt yields ions analogous to those derived from α cleavage of the free acid (see below) and the methyl ester, respectively; they differ, however, in that Na is substituted for H.

For example, the prominent peak at m/e 559, which is attributable to cleavage of the C_8–C_9 bond, includes the four-ring portion (rings B, C, D,

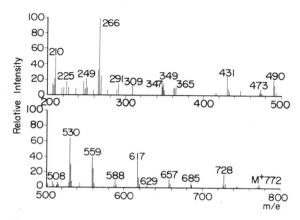

Figure 8 Mass spectrum of the sodium salt of salinomycin.

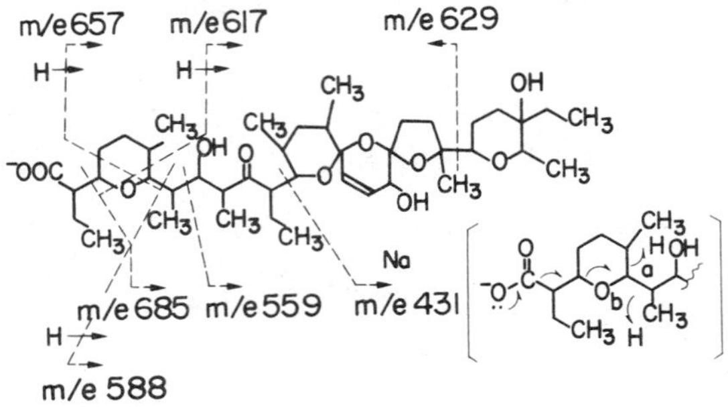

Scheme 2 Fragmentation pattern of the sodium salt of salinomycin.

and E), but not the carboxyl group or ring A. An ion of the same origin is observed at m/e 537 in the spectra of both the free acid and the methyl ester; it differs, however, by 22 mass units from the sodium salt, and a similar relationship is observed with the ions of m/e 685 and m/e 431, respectively.

Clearly, in spite of a loss of the carboxylic group moiety in the molecule, these fragment ions still contain sodium. This implies, significantly, that the sodium ion is not localized in the carboxylate group, but is complexed firmly with the residual oxygen functions of the molecules.

It is of particular interest that these results obtained by mass spectrometry provide direct evidence of the complexation nature of salinomycin with the metal ion.

C cleavage of the sodium salt yields a prominent peak at m/e 530 ($C_{29}H_{47}O_7Na$), whose composition is identical with that derived from M cleavage of the β-hydroxyketone group. In addition, these cleavages (C and M cleavages) afford two ions at m/e 508 with the same composition, $C_{29}H_{48}O_7$. This suggests that the foregoing cleavages induce the breakdown of the sodium—salinomycin complex expelling the metal ion from the molecule, and that the subsequent losses of two molecules of H_2O afford peaks of appreciable intensity at m/e 490 ($C_{29}H_{46}O_6$) and m/e 472 ($C_{29}H_{44}O_5$), respectively.

In addition, three characteristic ions of significant intensity which appear at m/e 657, m/e 617 ($C_{33}H_{54}O_9Na$), and m/e 588 ($C_{32}H_{53}O_8Na$), respectively, are observed exclusively in the spectrum of the sodium salt.

The origin of the ion at m/e 617 may be rationalized by a concerted mechanism with decarboxylation illustrated in Scheme 2; cleavage by route (a) may give rise to the ion at m/e 617, that by route (b) giving rise to the ion at m/e 588.

On the other hand, the peak at m/e 657, although its formation mechanism is uncertain, may be explained by a cleavage at ring A, illustrated by a dotted line in Scheme 2.

3. The Free Acid of Salinomycin

The mass spectrum of the free acid is shown in Figure 9. It is, however, accompanied by fragment ions derived from a trace of contaminating material.

A molecular ion is observed at m/e 750 accompanied by two dehydration peaks of considerable intensity at m/e 732 ($C_{42}H_{68}O_{10}$) and m/e 714, respectively.

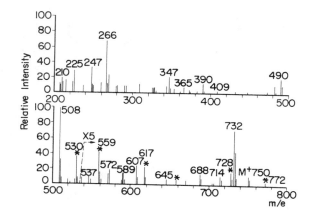

Figure 9 Mass spectrum of the free acid of salinomycin.

The ion at m/e 706 is attributed to the loss of CO_2 by decarboxylation, but appears as a weaker peak than that observed in the sodium salt. A dehydration peak is observed at m/e 688.

It should be mentioned here that the free acid gives rise to an ion of m/e 508 with the composition $C_{29}H_{48}O_7$ in both C and M cleavage; therefore it appears as an intensive base peak. Two dehydration peaks are observed at m/e 490 ($C_{29}H_{46}O_6$) and m/e 472 ($C_{29}H_{44}O_5$).

The ion at m/e 508 is further cleaved into a prominent peak at m/e 266 by M cleavage, as in the fragmentation pattern observed in the methyl ester.

By analogy with the cases of the sodium salt and the methyl ester, the ions due to α cleavage of the free acid are observed as peaks of moderate intensity at m/e 645 (663-18), m/e 607, m/e 537, and m/e 409, respectively, and these are further accompanied by respective dehydration peaks.

VI. BIOLOGICAL ACTIVITIES OF SALINOMYCIN

A. Antimicrobial Activity of Salinomycin

At the outset of the discovery of salinomycin, it was found to be highly effective against staphylococci with resistance to several antibiotics; it showed a limited activity against a wide range of mycobacteria and was also weakly active against some fungi.

The antimicrobial activity of salinomycin was determined by the agar dilution method (Miyazaki et al., 1974). and the results are presented in Table 6.

B. Coccidiostatic Effects of Salinomycin

The symptoms of coccidiosis in chickens differ, depending upon the types of *Eimeria protozoa* and the location of the parasites. The disease is extremely infectious and spreads within a short time in large-scale chicken breeding plants.

In 1973, the effect of salinomycin on coccidium protozoa was studied in in vitro cell systems, and it was confirmed to have an extremely superior coccidiostatic effect (Itagaki et al., 1974). In vivo tests followed, and it was

Table 6 Antimicrobial Activity of Salinomycin Free Acid

Test organisms	Minimum inhibitory concentration (μg/ml)
Bacillus subtilis PCI 219	3.12
Bacillus cereus IFO 3466	0.78
Bacillus circulans IFO 3329	3.12
Bacillus megaterium IFO 3003	>100
Staphylococcus aureus FDA 209P	1.56
Staphylococcus aureus[a]	3.12
Staphylococcus epidermidis IFO 3762	3.12
Sarcina lutea NIHJ	3.12
Micrococcus flavis IFO 3242	3.12
Micrococcus luteus IFO 2763	1.56
Mycobacterium smegmatis ATCC 607	25.0
Mycobacterium phlei IPCR	12.5
Mycobacterium avium IFO 3153	12.5
Escherichia coli NIHJ, P-17	>100
Klebsiella pneumoniae PCI 602	>100
Proteus vulgaris OX-19	>100
Proteus morganii CCM 680	>100
Xanthomonas oryzae	50.0
Xanthomonas citri NIAS	>100
Pseudomonas aeruginosa IFO 3445	>100
Pseudomonas aureofaciens IFO 3756	>100
Candida albicans YU 1200	>100
Piricularia oryzae	25.0
Alternaria kikuchiana NIAS, A-14	50.0
Alternaria tenuis IFO 4026	>100
Ophiobolus miyabeanus	>100
Diaphorthe citri	>100
Pellicularia filamentosa NIAS, C-37	>100
Penicillium chrysogenum	>100
Aspergillus niger IFO 6341	>100
Aspergillus fumigatus IMA	>100

[a]A resistant strain isolated from a patient (resistant to streptomycin, erythromycin, chloramphenicol, and tetracycline).

soon ascertained to have similar superior effects in practice (Danforth et al., 1977a,b). As a result of large-scale outdoor tests including battery tests and natural infection, salinomycin displayed pronounced effects against *Eimeria tenella*, one of the most tenacious pathogens, as well as against *Eimeria necatrix*, *Eimeria acervulina*, *Eimeria maxima*, and *Eimeria hagani* strains, including their resistant variants. Improvements in the symptoms, prevention of death, and an increase in body weight were noted.

The rates of penetration of coccidium sporozoites into cultured cells in the presence of salinomycin and various other coccidiostats were investigated. Salinomycin displayed a pronounced effect in preventing their penetration and results were superior to those of any other drug. The sporozoites which succeeded in penetrating into the cells displayed various morphological abnor-

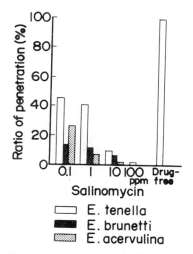

Figure 10 Coccidiostatic activity of salinomycin.

malities, such as swelling, transparency, and marginal irregularity. Even after the drugs were removed and new cells were inoculated, the sporozoites were still unable to penetrate into the cells.

Figure 10 shows the results of an investigation on the penetration of sporozoites into cells treated with salinomycin at various concentrations. At concentrations of 100 ppm, both drugs suppressed remarkably the penetration of *Eimeria tenella*, *E. brunetti*, and *E. acervulina* into the cells. In addition, they displayed a penetration-inhibiting effect even at concentrations of 1 and 0.1 ppm. Thus salinomycin has an excellent coccidiostatic effect at relatively low concentrations. It is effective against several types of protozia, including strains with drug resistance. As for concentrations to be used in practical applications, excellent preventive effects will be obtained when salinomycin is added to the feed in concentrations of 50 ppm.

Table 7 Feeding Effect of Salinomycin on Oxen[a]

Salinomycin (ppm)	0	20	30
Number of oxen	5	5	5
Average increase in body weights (kg/head)	262.0(100)	275.9(105.3)	283.8(108.3)
Increase of body weight per day (kg/head)	1.092	1.150	1.183
Total of consumed fodder (kg/head)	2.598(100)	2.119(81.6)	2.110(80.8)
Savings of feeding effect (%)		28.7	33.7

[a]Five heads of oxen (approximately 8 months old) were tested during 240 days.

C. Fattening Effects of Salinomycin on Ruminants

It is of great interest that salinomycin displayed also a unique biological activity apart from its antimicrobial or antiprotozoal action, for example, improvement of the fattening of ruminants.

For example, the effect on fattening of oxen was very remarkable, as shown in Table 7; a supplementation of 20–30 ppm salinomycin into combined feed showed a marked increase (5–10%) in body weight within 240 days. This means that a savings of 28.7–33.7% in feedings was obtained without any side effects during the period of supplementation. Today considerable amounts of salinomycin are used in practice all over the world. Similar effects have also been observed in hog raising.

VII. MECHANISM OF ACTION OF SALINOMYCIN

Following its steric structure, salinomycin has a cyclic conformation by fastening of the molecule through a hydrogen bond between the carboxylic and hydroxyl groups at both ends. It captures alkali metal ions in a cavity at the center of the molecule and forms an ion complex. This capturing action is different, depending on the type of ion. Its mechanism of action has been studied in detail by Mitani et al. (1975). It became clear that salinomycin acts as an ionophore across biomembranes and enhances the permeability of essential cations through the lipid barriers of the membranes; by disturbing the ion distribution inside and outside the membrane, it has an influence on the function of ion-dependent membranes. In the study of the selective binding affinity of this antibiotic with alkali metal ions, the following tests were used: Measurements of the binding values with respect to various radioactive metal

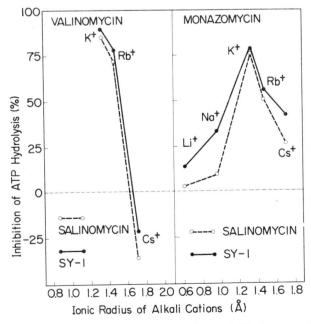

Figure 11 Effect of salinomycin and 20-deoxysalinomycin (SY-1) on the transport of various metal cations into mitochondria.

Table 8 Selective Bonding Properties of the Polyether Antibiotics with Ions

Antibiotic	Ion selectivity	
Salinomycin	$K^+ > Rb^+ > Na^+ > Cs^+ > Li^+$	
Monensin	$Na^+ \gg K^+ > Rb^+ > Li^+ > Cs^+$	
Nigericin	$K^+ > Rb^+ > Na^+ > Cs^+ > Li^+$	
Lasalocid A	$Cs^+ > Nb^+, K^+ > Na^+ > Li^+$	$Ba^{2+} > Ca^{2+} > Mg^{2+}$
Lysocellin	$Rb^+ > K^+ > Na^+ > NH_4^+$	$Ba^{2+} > Mg^{2+} > Ca^{2+}$
A-23817	$Li^+ > Na^+ > K^+$	$Ca^{2+} > Mg^{2+} > Ba^{2+}$
Antibiotic 6016	$NH_4^+, K^+ > Rb^+, Na^+ > Li^+, Ca^+$	$Mg^{2+} > Ca^{2+} > Ba^{2+}$

ions, the ion transport capacity in water–carbon tetrachloride–water model transport systems, and the functioning (respiration, ATPase activity, etc.), as well as morphological changes (swelling, contraction) of liver mitochondria in rats.

The ionophore antibiotics valinomycin (Pressman et al., 1967) and monazomycin (Nakayama et al., 1981) are known to be uncouplers that have the property of selectively transporting ions into mitochondria and which cause loss of coupling between electron transport and oxidative phosphorylation. On the other hand, several polyether antibiotics, including salinomycin, function to prevent ion transport into mitochondria or to transport ions accumulated inside mitochondria selectively out of the mitochondria.

Figure 11 demonstrates the transport of various alkali metal ions into mitochondria, using valinomycin and monazomycin in the presence of ATP, with salinomycin applied during this reaction (Mitani et al., 1978). In this manner, the selective ion binding properties of salinomycin and its transport capacities could be demonstrated (Mitani et al., 1975, 1976; Ōtake et al., 1976), using morphological changes of the mitochondria as indicators. The results indicate that salinomycin transports ions out of the mitochondria in the order of intensity K^+, Rb^+, and Na^+, but that it has a weak transport capacity for Cs^+ and Li^+, as shown in Table 8.

REFERENCES

Breed, R. S., Murray, E. G. D., and Smith, N. R. (1974). *Bergey's Manual of Determinative Bacteriology*, 8th ed. Williams and Wilkins, Baltimore.

Danforth, H. D., Ruff, M. D., Reid, W. M., and Miller, R. L. (1977a). Anticoccidial activity of salinomycin in battery raised broiler chickens. *Poult. Sci.* 56:926–932.

Danforth, H. D., Ruff, M. D., Reid, W. M., and Miller, R. L. (1977b). Anticoccidial activity of salinomycin in floor-pen experiments with broilers. *Poult. Sci.* 56:933–1938.

Day, L. E., Chamberlin, J. W., Gordee, E. Z., Chen, S., Gorman, M., Hamill, R. L., Ness, T., Weeks, R. E., and Stroshane, R. (1973). Biosynthesis of monensin. *Antimicrob. Agents Chemother.* 4:410–414.

Itagaki, K., Tsubokura, M., and Otsuki, K. (1974). Studies on methods for evaluation of anticoccidial drugs in vitro. *Jpn. J. Vet. Sci.* 36:195–202.

Kinashi, H., Ōtake, N., Yonehara, H., Sato, S., and Saito, Y. (1973). The structure of salinomycin, a new member of polyether antibiotics. *Tetrahedron Lett.* 49:4955–4958.

Kinashi, H., Ōtake, N., Yonehara, H., Sato, S., and Saito, Y. (1975). Studies on the ionophorous antibiotics. I. The crystal and molecular structure of salinomycin p-iodophenacyl ester. *Acta Crystallogr.* B31:2411–2415.

Mitani, M., Yamanishi, T., and Miyazaki, Y. (1975). Salinomycin: A new monovalent cation ionophore. *Biochem. Biophys. Res. Commun.* 66:1231–1236.

Mitani, M., Yamanishi, T., Miyazaki, Y., and Ōtake, N. (1976). Salinomycin effects on mitochondrial ion translocation and respiration. *Antimicrob. Agents Chemother.* 9:655–660.

Miyazaki, Y., Shibuya, M., Sugawara, H., Kawaguchi, O., Hirose, C., Nagatsu, J., and Esumi, S. (1974). Salinomycin, a new polyether antibiotic. *J. Antibiot.* 27:814–821.

Nakayama, H., Furihata, K., Seto, H., and Ōtake, N. (1981). The structure of monazomycin, a new ionophorous antibiotic. *Tetrahedron Lett.* 22:5217–5220.

Ōtake, N., Miyazaki, Y., Kinashi, H., Mitani, M., and Yamanishi, T. (1976). Chemical modification and structure–activity correlation of salinomycin. *Agric. Biol. Chem.* 40:1633–1640.

Ōtake, N., Seto, H., and Koenuma, M. (1978). The assignment of the ^{13}C-NMR spectrum of lysocellin and its biosynthesis. *Agric. Biol. Chem.* 42:1879–1886.

Pressman, B. C., Harris, E. J., Jagger, W. S., and Johnson, J. H. (1967). Antibiotic-mediated transport of alkali metal ions across lipid barriers. *Proc. Nat. Acad. Sci. U.S.A.* 58:1949–1956.

Pridham, T. G., and Gottlieb, D. (1948). The utilization of carbon compounds by some actinomycetales as an aid for species determination. *J. Bacteriol.* 56:107–114.

Seto, H., Miyazaki, Y., Fujita, K., and Ōtake, N. (1977). Studies on the ionophorous antibiotics X. The assignment of ^{13}C-NMR of salinomycin. *Tetrahedron Lett.* 28:2417–2420.

Shirling, E. B., and Gottlieb, D. (1966). Method for characterization of *Streptomyces* species. *Int. J. Syst. Bacteriol.* 16:313–340.

Westley, J. W., Preuss, D. L., and Pitcher, R. G. (1972). Incorporation of [I-^{13}C]-Butyrate into antibiotic X-537A: [13]C nuclear magnetic resonance study. *J. Chem. Soc. Chem. Commun.*, 161–162.

27

TYLOSIN: PROPERTIES, BIOSYNTHESIS, AND FERMENTATION

PETER P. GRAY AND S. BHUWAPATHANAPUN* *School of Biotechnology, University of New South Wales, Sydney, Australia*

Present affiliation: Kasetsart University, Bangkok, Thailand

I. INTRODUCTION

Tylosin is a macrolide antibiotic first isolated in the laboratories of Eli Lilly and Company, Indianapolis, Indiana. It became a commercial product in 1962 (Stark, 1977) and is marketed under the trade name Tylan, which is a tylosin mixture containing about 92% tylosin or tylosin A and small amounts of tylosin-like components (Debackere and Baeten, 1971).

Tylosin is produced commercially by *Streptomyces fradiae* (McGuire et al., 1961). It had been reported that strains of *Streptomyces rimosus* (Pape and Brillinger, 1973) and *Streptomyces hygroscopius* (Jensen et al., 1964) can also produce tylosin.

II. HISTORY AND DISCOVERY

The production of tylosin was first reported by McGuire et al. in 1961. Two tylosin-producing strains were isolated from soil samples obtained from Nong-kai, in the northeastern part of Thailand. The organisms have been permanently deposited with the culture collection of the Northern Regional Research Laboratories, Peoria, Illinois, and assigned the culture numbers NRRL 2702 and 2703 (Hamill et al., 1965). The organisms were classified as new strains of the species *S. fradiae* (McGuire et al., 1961). The organisms were similar to the Waksman strain of *S. fradiae* ATCC 10745 in the following characteristics: spore chain morphology, spore color, carbon utilization on 15 carbohydrates, positive nitrate reduction, absence of soluble pigment, and negative test for sulfide production. The tylosin-producing strains differed from the Waksman *S. fradiae* in that they utilized sucrose, produced tylosin rather than neomycin, and failed to grow on potato plugs (McGuire et al., 1961).

III. BIOSYNTHESIS MECHANISM

A. Structure of Tylosin

Tylosin is a macrolide antibiotic with a 16-membered lactone ring and three sugars, mycarose, mycaminose, and mycinose (Morin and Gorman, 1964). The structure of tylosin is shown in Figure 1. Mycinose is projected from C23 of the lactone ring, while mycarose is attached to the basic sugar mycaminose, the resulting disaccharide being attached to the oxygen at the C5 position of the lactone ring (Morin et al., 1970; Omura and Nakagawa, 1975). Tylosin

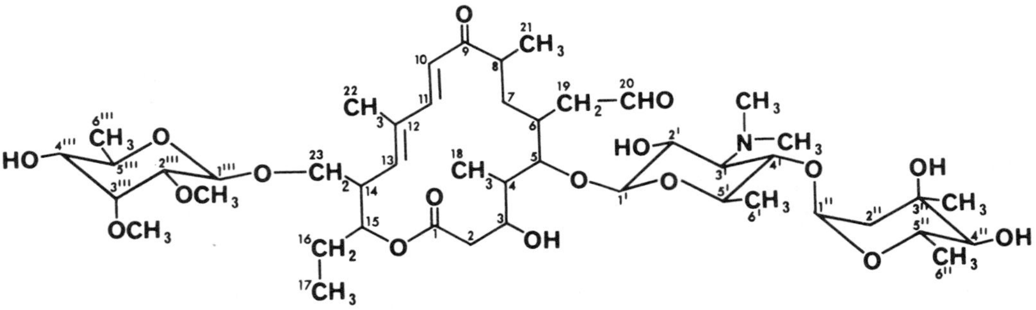

Figure 1 Structure of tylosin.

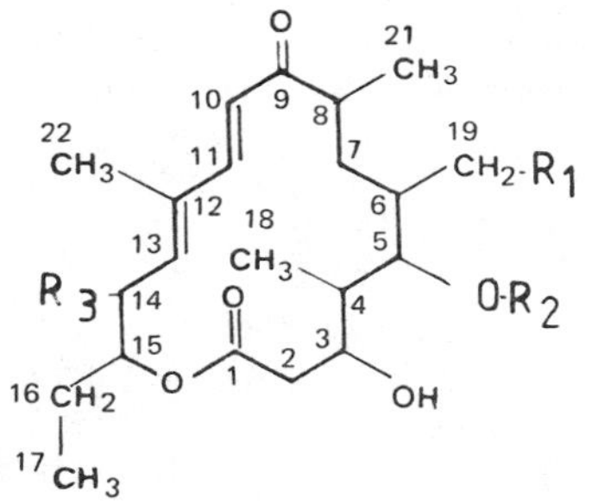

		R_1	R_2	R_3
TYLACTONE	[1]	CH_3	H	CH_3
O-MYCAMINOSYL TYLACTONE	[2]	CH_3	MYCAMINOSE	CH_3
O-MYCAMINOSYL TYLONOLIDE	[3]	CHO	MYCAMINOSE	CH_2OH
DEMYCINOSYL	[4]	CHO	MYCAMINOSYL	CH_2OH

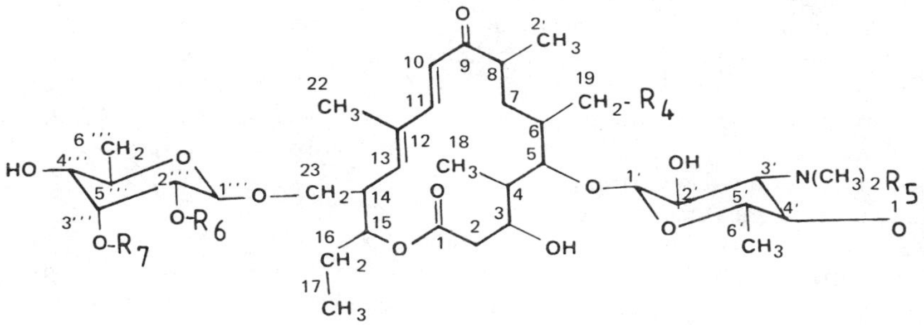

		R_4	R_5	R_6	R_7
DEMETHYL LACTENOCIN	[5]	CHO	H	H	H
LACTENOCIN	[6]	CHO	H	CH_3	H
DEMETHYL MACROCIN	[7]	CHO	MYCAROSE	H	H
MACROCIN	[8]	CHO	MYCAROSE	CH_3	H
DESMYCOSIN	[9]	CHO	H	CH_3	CH_3
TYLOSIN	[10]	CHO	MYCAROSE	CH_3	CH_3
RELOMYCIN	[11]	CH_2OH	MYCAROSE	CH_3	CH_3

Figure 2 Structures of tylactone and related compounds (top) and of tylosin and related compounds (bottom).

has been classified in the tylosin chalcomycin group (Omura and Nakagawa, 1975), other members of the group including cirramycin A, rosamicin, angolamycin, and B-58941.

B. Biosynthetic Pathway

The formation of the tylosin lactone ring (tylactone) is the initial step in the biosynthesis of tylosin followed by the sequential addition of the sugars.

It has been proposed (Masamune et al., 1977; Martin, 1977) that the formation of macrolide lactone rings is similar to the synthesis of saturated long-chain fatty acids, since they are formed from common intermediates like acetate, propionate, malonate, 2-methylmalonate, and butyrate or 2-ethyl-malonate. Corcoran (1974) and Rossi and Corcoran (1973) studied the biosynthesis of the lactone ring of the 14-membered macrolide erythromycin and found that it was synthesized by a multienzyme synthase complex. Based on the evidence for the synthesis of the erythromycin lactone ring, Masamune et al. (1977) proposed that tylactone and the aglycone of other macrolides are formed by "general fatty acid synthases."

Studies on the incorporation of ^{13}C precursors into tylosin indicated that the carbon skeleton of tylactone [1] (Fig. 2) is derived from five propionates, two acetates, and one butyrate (Omura et al., 1975, 1977). Two enzymes responsible for the production of propionyl coenzyme A (CoA) have been assayed in tylosin-producing S. fradiae. It was found that the activity of methyl malonyl-CoA carboxyltransferase (E.C. 2.1.3.1) was two orders of magnitude higher than that of propionyl-CoA carboxylase (E.C.6.4.1.3.) (Vu-trong et al., 1980), suggesting that the main route of propionyl-CoA synthesis is via the carboxyltransferase reaction.

Of the three sugars occurring in tylosin, the formation of mycarose is best documented. In studies on tylosin-producing S. rimosus it was shown (Pape and Brillinger, 1973) that TDP-mycarose is synthesized from TDP-D glucose and s-adenosyl-1-methionine. The proposed mechanism is shown in Figure 3. The reaction required NADPH and has TDP-4-keto-6-deoxy-D-glucose as an intermediate together with a second methylated TDP-sugar, the structure of which is unknown. Formation of TDP-4-keto-6-deoxy-D-glucose is catalyzed by the enzyme TDP-D-glucose oxidoreductase. The methyl

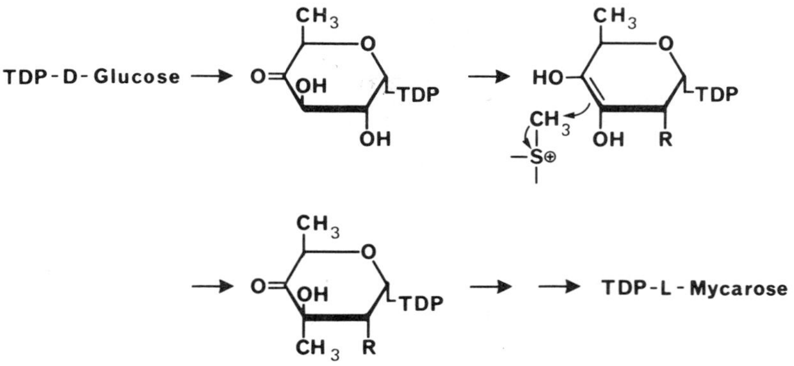

Figure 3 Proposed mechanism of mycarose synthesis. (From Pape and Brillinger, 1973.)

group at C3 of mycarose is transferred from methionine (Pape et al., 1969). The other two sugars, mycaminose and mycinose, are also derived from glucose and the methyl groups transferred from methionine (Achenbach and Grisebach, 1964; Brisebach, 1978). The synthesis of mycaminose is suggested to be similar to that of mycarose and a pathway has been proposed (Martin, 1979).

A pathway for the terminal stages of tylosin biosynthesis has been proposed by Seno et al. (1977). Subsequent work has further elucidated the pathway and the results of these studies are shown in Figure 4. In Figure 3 the structures of the relevant intermediates are listed. Following the formation of the tylactone, the first step is the addition of mycaminose to C5, fol-

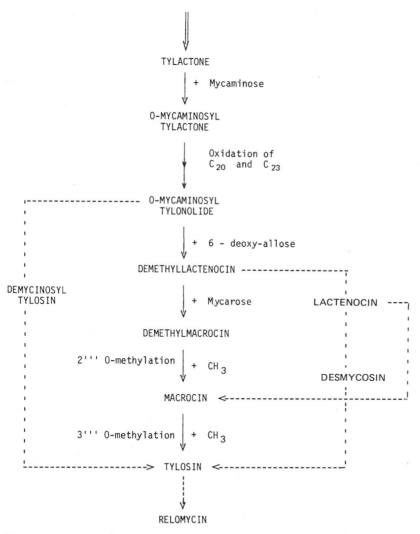

Figure 4 Proposed pathway for the terminal stages of tylosin biosynthesis (——→, major route for biosynthesis; ---→, possible bioconversions). (Based on Seno et al., 1977; Seno and Baltz, 1978; Baltz and Seno, 1980; Baltz et al., 1980.)

lowed by oxidation reactions of C20 and C23. The 6-deoxy-D-allose is then added to produce demethyllactenocin [5] which is then glycosylated with the amino sugar mycarose to form demethylmacrocin [7]. Demethylmacrocin is methylated to macrocin [8], which is subsequently methylated to tylosin [10]. Two different O-methyltransferases have been shown to be involved in the last two steps (Baltz et al., 1980). Demethylmacrocin [7] is methylated by the enzyme 2^{111}-O-methyltransferase to produce macrocin [8] and is not a substrate for the 3^{111}-O-methyltransferase, which converts macrocin [8] to tylosin.

Other conversions have been described as not being on the main pathway for tylosin biosynthesis and these are shown as secondary pathways in Figure 4. The work of Baltz et al. (1980) showed that mycarose could be glycosidated to O-mycaminosyltylanolide before addition of 6-deoxy-D-allose to produce demycinosyltylosin [4], but this compound is poorly converted to tylosin. Lactenocin [6] is produced by the addition of $2\text{-}O^{111}$-methyl to demethyllactenocin. It has been suggested (Seno et al., 1977) that lactenocin can then be either glycosidated with mycarosyl group at the C4 hydroxyl of mycaminose to produce macrocin, or methylated at the C3 hydroxyl of mycarose to produce desmycosin [9].

Hydrogenation of tylosin at C23 produces relomycin (Jensen et al., 1964), a compound which has been reported to accumulate during the tylosin fermentation together with tylosin and macrocin (Seno et al., 1977).

A recent paper by Baltz et al. (1982) describes the preferred biosynthetic pathway as: tylactone → O-mycaminosyl tylactone → 23-deoxy-20-O-mycaminosyl tylonolide → O-mycaminosyl tylonolide → demethyllactenosin → demethylmacrocin → macrocin → tylosin.

IV. FERMENTATION PROCESS

A. Inoculum Preparation

Tylosin-producing cultures of *S. fradiae* are usually stored as lyophilized pellets. The lyophilized pellets are used to plant first-generation agar slants. Both McGuire et al. (1961) and Stark et al. (1961) described the use of second-generation agar slants to inoculate vegetative medium. Several different formulations for agar slant media have been described (McGuire et al., 1961; Stark et al., 1861; Seno et al., 1977). The slant medium described by Seno et al. (1977) had the following composition (g/liter): glucose, 10.0; phytone, 10.0; agar, 25.0; biotin, 0.001; and sodium thiosulfate, 1.0. This medium was used for slants incubated at 28°C for 10 days and then stored at 4°C until used. Stark et al. (1961) reported that cultures held under refrigeration for extended periods showed decreases in antibiotic yields, and recommended that cultures refrigerated for less than 6 weeks be utilized. Spore suspensions obtained from the agar slants are used to inoculate liquid vegetative medium. The vegetative medium is formulated to provide consistent amounts of mycelial growth, a terminal pH near neutrality, and good yields of the antibiotic in the resulting fermentation (Stark et al., 1961). Several different vegetative media have been described by several workers (Caltrider and Hayes, 1969; Seno et al., 1977; Stark et al., 1961; McGuire et al., 1961) using a basic medium containing (g/liter) glucose, 15.0; corn-steep liquor, 10.0; yeast extract in the range 5.0–6.25; and calcium carbonate in the range 3.0–3.8. Aerobic growth of the vegetative medium was carried out for 48 hr and the resulting suspension of vegetative mycelia used to inoculate the production medium.

B. Medium Preparation and Substrates

There have been relatively few descriptions of media suitable for the production of tylosin. The influence on tylosin biosynthesis of media composition has been shown for complex media (Caltrider and Hayes, 1969; Hamill et al., 1965). It was shown that a complex medium containing (g/liter) fish meal (17.5), beet molasses (20.0), crude soybean oil (30.0), calcium carbonate (2.0), diammonium phosphate (0.4), and sodium chloride (1.0) could support higher yields of tylosin than a medium containing yeast extract (20.0), corn-steep liquor (5.0), beet molasses (20.0), and crude soybean oil (30.0), or one containing (g/liter); soybean meal (15.0), casein (1.0), crude glucose syrup (20.0 ml/l), calcium carbonate (2.5), and sodium nitrate (3.0), (4360, 3395, and 250 μg/ml of tylosin at 144 hr, respectively, for the three media). However, there have been no publications describing detailed media optimization for complex media. From the limited information published, it would seem that the following ingredients need to be present in a complex medium: a source of early assimilable carbohydrate, an insoluble protein source, a source of mineral salts, and a lipid source to supply energy and precursors during antibiotic synthesis. A detailed study has been published by Stark et al. (1961) on the biosynthesis of tylosin in synthetic media. These workers studied the effect of carbon source, amino acids, methylated fatty acids, and inorganic components on growth and tylosin biosynthesis.

The optimum medium developed by the workers contained (g/liter) NaCl, 2.0; $MgSO_4$, 5.0, $CoCl_2 \cdot 6H_2O$, 0.001; ferric ammonium citrate, 1.0; $MnCl_2 \cdot 4H_2O$, 1.0; $ZnSO_4 \cdot 7H_2O$, 0.01; $CaCO_3$, 3.0; glycine, 7.0; l-alanine, 2.0; l-valine, 1.0; betaine, 5.0; glucose, 35.0; methyloleate, 25.0; and K_2HPO_4, 2.3.

The substitution of calcium chloride for the calcium carbonate in the above medium and sodium glutamate instead of glycine, valine, and alanine resulted in a medium with all soluble ingredients (apart from the methyloleate) and a simplified amino acid composition (Gray and Bhuwapathanapun, 1980).

C. Fermentation Conditions

Information on the relationship between operating parameters and tylosin production is limited. Batch pilot-scale fermentations have been described (Hamill et al., 1965). Most studies (Hamill et al., 1965; Stark et al., 1961; Caltrider and Hayes, 1969; Seno et al., 1977) appear to use operating temperatures of 28 or 30°C and a pH near neutrality. The fermentation is highly aerobic and tylosin production will not occur during periods of oxygen limitation. A maximum specific oxygen uptake rate (qO_2) of 1.5 mmol O_2/g dry weight per hour has been reported during active growth falling to one-third of this value during the period of antibiotic synthesis for a batch fermentation with a classic trophophase–idiophase kinetic pattern (Gray and Bhuwapathanapun, 1980).

Batch and chemostat studies have shown that increased specific uptake rates of glucose and phosphate have a depressing effect on tylosin biosynthesis (Madry et al., 1979; Sprinkmeyer and Pape, 1978; Gray and Bhuwapathanapun, 1980; Vu-trong et al., 1980, 1981). It has also been shown in batch culture that increased uptake rates of glucose and phosphate depressed the levels of the enzymes methylmalonyl-CoA carboxyltransferase and propionyl-CoA carboxylase, enzymes involved in the formation of propionyl-CoA, a precursor for the synthesis of the lactone ring (Vu-trong et al., 1980, 1981). The specific activity of these two enzymes was shown to be closely correlated with the tylosin for a range of different metabolic states. Also in chemostat cultures it has been shown that as the dilution rate is increased and the value of $q_{tylosin}$ decreases, there is a corresponding drop in the specific activities of methylmalonyl-CoA carbosyltransferase and propionyl-CoA carboxylase (Vu-trong et al., 1981). Similarly, the level of macrocin 3'-O-methyltransferase (the enzyme catalyzing the conversion of macrocin to tylosin) has also been shown to be correlated with the value of $q_{tylosin}$ (Hughes, 1979).

These results show that under a range of environmental conditions enzymes involved in the provision of one of the main precursors for lactone ring synthesis, that is, at the beginning of the tylosin pathway, and the enzyme involved in the last step in the pathway appear to be under tight regulation, their specific activities coupled with the specific rate of tylosin production. The case regarding enzymes responsible for the three sugars in the tylosin molecule is not so clear-cut. Matern et al. (1973) showed that the activity of the enzyme dTDP-glucose 4,6-dehydratase increased during the stationary phase of *S. rimosus* together with tylosin production and the mycarose-synthesizing system. Sprinkmeyer and Pape (1978) showed that the enzyme system catalyzing the synthesis of TDP-mycarose is not affected in glucose-inhibited cultures where tylosin synthesis had been depressed. In chemostat cultures on a defined medium, it was found that increasing the specific uptake rate of sodium glutamate increased the $q_{tylosin}$ (Gray and Bhuwapathanapun, 1980); that is, there was no evidence of nitrogen catabolite repression (Drew and Demain, 1977) in tylosin biosynthesis. Stark et al. (1961) showed that tylosin synthesis in batch culture on defined medium followed a classic trophophase-idiophase kinetic pattern. In a medium containing increased levels of sodium glutamate and decreased levels of glucose, it was possible to obtain high values of $q_{tylosin}$ while the mycelia were still actively growing (Gray and Bhuwapathanapun, 1980), showing that tylosin production is not necessarily "non-growth-associated." It is possible then to obtain tylosin synthesis in a growth situation in both batch and continuous cultures if catabolite repression effects are taken into consideration. Fed-batch fermentations have been reported using this medium (Bhuwapathanapun and Gray, 1981).

The relationship between glucose uptake and the metabolism of the lipid source has been investigated. Madry et al. (1979) showed that increased levels of glucose in batch culture inhibited the oxidation of methyl oleate. Sprinkmeyer and Pape (1978) showed that oleyl-CoA inhibits the enzyme citrate synthase, and that long-chain fatty acids, as well as supplying subunits for tylosin synthesis, also led to increased levels of acetyl CoA by reducing its final oxidation through inhibition of citrate synthase. Chemostat studies have also shown that the specific uptake rates of methyl oleate were decreased when the glucose uptake rate exceeded 50 mg glucose/g dry weight per hour (Gray and Bhuwapathanapun, 1980). The possible role of intracellular adenylates and glucose-6-phosphate in mediating glucose, phosphate, and lipid effects on tylosin biosynthesis has been discussed (Madry et al., 1979; Vu-trong et al., 1980, 1981).

There have been few publications describing the influence of environmental and metabolic factors on the accumulation of tylosin-like components in the fermentation. Seno et al. (1977) reported that macrocin and relomycin also accumulated in the tylosin fermentation and they plotted a time course for the relative amounts of each antibiotic throughout the fermentation. Gray and Bhuwapathanapun (1980) observed that in batch culture tylosin, relomycin, and macrocin accumulated, but in chemostat cultures macrocin only accumulated at growth rates less than 0.01 hr^{-1}, only tylosin and relomycin being observed at growth rates greater than 0.01 hr^{-1}. Jensen et al. (1964) observed that tylosin accumulated initially in fermentations of *S. hygroscopicus*, being subsequently converted to relomycin.

There have been no publications describing equipment or operating conditions currently used for production-scale tylosin fermentations.

D. Tylosin Assay

Both chemical and bioassays have been reported for the assay of tylosin. McGuire et al. (1961) described a turbidometric assay procedure using *Staphylococcus aureus* (ATCC 9144) as the test organism; a paper disk agar diffusion plate assay was also described with *Sarcina lutea* (ATCC 9341) as the test organism.

A chemical assay method has been described by Caltrider and Hayes (1969) where the fermentation broth was adjusted to pH 5.0 and then extracted with chloroform. The concentration of tylosin-like components was then determined by measuring the absorbance reading at 283 nm. Both chemical and bioassays suffer from the problem that the presence of tylosin-like components will also contribute to the assay result. A high-performance liquid chromatographic method which will allow the quantitation of tylosin and related compounds has been described (Kennedy, 1978).

V. PRODUCT RECOVERY AND MODIFICATION

Tylosin can be removed from filtered or centrifuged broth by employing absorption or extraction techniques. Extractants which can be used include water-immiscible polar organic solvents such as ethyl acetate and amyl acetate, chlorinated hydrocarbons such as chloroform, water-immiscible alcohols, and ketones and ethers (Hamill et al., 1965). For recovery of tylosin by absorption, a range of absorbents and ion exchange resins can be used and the tylosin then eluted with an organic solvent. The organic solvent extract can then be either evaporated to dryness to provide crude tylosin or concentrated in vacuo and a precipitant added. The formation of various salts of tylosin has also been described (Hamill et al., 1965). Vasileva-Lukanova et al. (1980a) described the recovery of tylosin from a filtered fermentation liquor of *S. fradiae* by extraction into either chloroform or dichloromethane at pH 9.5 and 2–5°C with a 95% efficiency. Tylosin in the chloroform extract was concentrated by evaporation and then precipitated. The same workers also compared extraction into butylacetate, chloroform, and dichloromethane and found that increasing the temperature from 5 to 20–25°C increased the amount of tylosin recovered. Tylosin could be reextracted back into water from butyl acetate at a temperature of 5–15°C and at pH 4.0 (Vasileva-Lukanova et al., 1980b). The conversion of tylosin into desmycosin by dilute acid hydrolysis has been reported (Hamill et al., 1965), as has the preparation of macrocin from macrocin–tylosin mixtures and the production of lactenocin by dilute acid hydrolysis of macrocin (Hamill and Stark, 1967).

The production of novel compounds by the biochemical acylation of hydroxyl groups of tylosin has been reported (Okamoto et al., 1978) using species of *Streptomyces*. The selective cleavage of the mycarose sugar from tylosin has also been reported (Nagel and Vincent, 1979).

VI. MODE OF ACTION AND APPLICATIONS

Tylosin acts on microorganisms by inhibiting protein synthesis. The concentration of tylosin needed for inhibition of growth was found to be comparable to that needed to inhibit protein synthesis in cell-free extracts (Mao and Weigard, 1968).

Tylosin has been found to bind to the ribosome and inhibit the binding of the aminoacyl end of aminoacyl-tRNA (Pestka, 1970) and inhibit the formation of the mRNA—aminoacyl-tRNA—ribosome complex (Suzuki et al., 1970; Kageyama et al., 1971). The site of the binding of tylosin is on the 50-S subunit of the ribosome (Pestka, 1971).

Graham and Weisblum (1979) have shown that tylosin-producing S. fradiae contain N^6-methyl adenine in their 23-S ribosomal nucleic acid. Specific methylation of 23-S RNA mediates coresistance to the macrolides, lincosamide, and streptogramin antibiotics in clinical isolates of S. aureus. The presence of methylated RNA in S. fradiae could explain the mechanism by which the organism is resistant to its own antibiotic.

Macrocin, relomycin, and desmycosin are the structurally related compounds found accumulating during tylosin production by S. fradiae (Seno et al., 1977). Among these compounds, desmycosin has the greatest ability to bind to ribosomes (Corcoran et al., 1977), whereas little difference was observed in the binding properties of macrocin, relomycin, and tylosin. Although desmycosin has the highest binding ability, it is least active in inhibiting microbial growth (Corcoran et al., 1977). The differences in the ability to bind to ribosomes and inhibition of microbial growth of these structurally related compounds results from differences in their functional groups (Corcoran et al., 1977).

Studies of McGuire et al. (1961), Hamill and Stark (1964), Whaley et al., (1967), and Corcoran et al. (1977) demonstrate that tylosin and the other factors differ in their ability to inhibit microbial growth. Tylosin is most biologically active and macrocin has an almost comparable activity. The least active factor was found to be desmycosin, while relomycin has the intermediate activity.

The in vitro antimicrobial spectra of tylosin, macrocin, relomycin, and lactenocin are summarized in Table 1.

Tylosin is relatively nontoxic in vivo and is used exclusively for animal nutrition and veterinary medicine. As shown in Table 1, tylosin inhibits gram-positive bacteria and mycoplasma. It is used in the control of respiratory disease in poultry, improves weight gains in poultry and swine, and controls a number of other diseases in cattle, swine, poultry, dogs, and cats (Stark, 1977).

Many papers have been published describing the use of tylosin in animals. These will not be reviewed; instead, a few references describing the use of tylosin in a range of applications will be cited.

Tylosin has been found to reduce the incidence of liver abscesses, increase average daily weight gain, and improve feed conversion efficiency of feedlot cattle (Brown et al., 1973, 1975). The use of tylosin together with monensin has also been investigated in a high-energy diet for finishing steers (Pendulum et al., 1978; Heinemann et al., 1978). The use of tylosin to improve feed efficiency in broiler chicks has been described (De Schrijver, 1973), as has its use to improve feed efficiency in weaned pigs (Wahlstrom and Libal, 1975). Tylosin has also been used to treat mycoplasma in poultry (Wise and Fuller, 1975) and mycoplasma associated with bovine mastitis (Socci et al.,

Table 1 In Vitro Antimicrobial Spectrum of Tylosin, Macrocin, Relomycin, and Lactenocin[a]

Test organisms	Tylosin[b]	Macrocin[c]	Relomycin[d]	Lactenocin[c]
Bacteria				
Staphylococcus aureus (209P)	0.39	0.78	25	6.25
Staphylococcus albus	0.78	1.56	—	6.25
Bacillus subtilis	0.39	0.78	1.5	6.25
Sarcina lutea	0.2	0.1	—	1.56
Mycobacterium tuberculosis	—	0.4	—	50
Mycobacterium avium	3.13	0.4	—	6.25
Klebsiella pneumonia	50	50	—	100
Shigella paradysenteriae	100	25	—	100
Brucella brouchiseptica	100	100	—	100
Vibrio metschnikovi	—	50	—	100
Bacterial plant pathogens				
Corynebacterium michiganese	—	0.2	—	1.56
Erwinia amylovora	—	100	—	—
Fungi				
Trichophyton rubrum	>100	—	—	100
Trichophyton interidigitale	>100	—	—	100
Avium PPLO				
PPLO strain 699	—	0.25	—	—
PPLO strain 295	<0.09 (bd)[e]	4.0	—	—
PPLO strain 1991	—	8.0	—	—
PPLO strain 2453	<0.09 (bd)	8.0	—	—
PPLO strain 299	<0.09 (bd)	—	—	—
PPLO strain 455	<0.09 (bd)	—	—	—

[a]All the tests used the agar dilution technique unless otherwise specified.
[b]McGuire et al. (1961)
[c]Hamill and Stark (1964).
[d]Whaley et al. (1967).
[e]bd, broth dilution technique.

1970). Suzuki et al. (1970) reported the use of tylosin as a food preservative, but it would appear that this use is not widespread.

REFERENCES

Achenbach, H., and Grisebach, H. (1964). Zur biogenese der macrolide XII—Mitt Weitzf untersuchungen zer biogenese des magnamycin. *Z. Naturforsch. Teil B. 193*:561–568.

Baltz, R. H., and Seno, E. T. (1980). Biosynthesis of the macrolide antibiotic tylosin: cofermentation, bioconversion and in vitro enzymatic conversion studies with blocked mutants. In Abstracts of VIth International Fermentation Symposium, London, Ontario: F-4.2.2 (L).

Baltz, R. H., Seno, E. T., Stonesifer, P., Matusushima, P., and Wild, G. M. (1980). Genetics and biochemistry of tylosin production by *Streptomyces fradiae*. In *Microbiology—1980*. D. Schlessinger (Ed.). American Society for Microbiology, Washington, D.C., pp. 371–375.

Baltz, R. H., Seno, E. T., Stonesifer, P., Matsushima, P., and Wild, G. M. (1982). Genetics and biochemistry of tylosin production. *Trends in Antibiotic Research*, pp. 65–72.

Bhuwapathanapun, S., and Gray, P. P. (1981). Production of the macrolide antibiotic tylosin in feed-batch culture. *J. Ferment. Technol. 95(5)*:419–421.

Brown, H., Elliston, N. G., McAskill, J. W., Muenster, O. A., and Tonkinson, L. V. (1973). Tylosin phosphate (TP) and tylosin urea adduct (TUA) for the prevention of liver abscesses, improved weight gains and feed efficiency in feedlot cattle. *J. Anim. Sci. 37*:1085–1091.

Brown, H., Bing, R. F., Grueter, H. P., McCaskill, J. W., Cooley, C. O., and Rathmacher, R. P. (1975). Tylosin and chlortetracycline for the prevention of liver abscesses, improved weight gains and feed efficiency in feedlot cattle. *J. Anim. Sci. 40*:207–213.

Caltrider, P. G., and Hayes, H. B. (1969). Process for tylosin production. U.S. Patent No. 3,433,711.

Corcoran, J. W. (1974). Lipid and macrolide lactone biosynthesis in *Streptomyces erythreus*. *Dev. Ind. Microbiol. 15*:93–100.

Corcoran, J. W., Huber, M. L. B., and Huber, F. M. (1977). Relationship of ribosomal binding and antibacterial properties of tylosin-type antibiotics. *J. Antibiot. 30*:1012–1014.

Debackere, M., and Baeten, K. (1971). A thin-layer chromatographic method for the detection of tylosin in biological materials and feeds. *J. Chromatogr. 61*:112–132.

De Schrijver, R. (1973). Tylosin as a feed additive in a broiler chick's ration. *Meded. Fac. Landbouwwet. Rijksuniv. Gent 38*:396–401.

Drew, S. W., and Demain, A. L. (1977). Effects of primary metabolites on Secondary metabolism. *Annu. Rev. Microbiol. 31*:343–356.

Graham, M. Y., and Weisblum, B. (1979). 23 S Ribosomal ribonucleic acid of macrolide producing streptomycetes contains methylated adenine. *J. Bacteriol. 137*:1464–1467.

Gray, P. P., and Bhuwapathanapun, S. (1980). Production of the macrolide antibiotic tylosin in batch and chemostat cultures. *Biotechnol. Bioeng. 22*:1785–1804.

Grisebach, H. (1978). Biosynthesis of macrolide antibiotics. In *Antibiotics and Other Secondary Metabolites: Biosynthesis and Production*. R. Hütter, T. Leisinger, J. Nüesch, and W. Wehrl (Eds.). Academic, London, pp. 113–127.

Hamill, R. L., and Stark, W. M. (1964). Macrocin, a new antibiotic and lactenocin, an active degradation product. *J. Antibiot.* 17:133–139.

Hamill, R. L., and Stark, W. M. (1967). Antibiotics macrocin and lacterocin. U.S. Patent No. 3,326,759.

Hamill, R. L., Haney, M. E., McGuire, J. M., and Stamper, M. C. (1965). Antibiotics tylosin and desmycosin and derivatives thereof. U.S. Patent No. 3,178,341.

Heinemann, W. W., Hanks, E. B., and Young, D. C. (1978). Monensin and tylosin in a high energy diet for finishing steers. *J. Anim. Sci.* 47: 34–40.

Hughes, D. (1979). Honours thesis, "Studies on macrocin 3'-O-methyltransferase in cell-free extracts of *S. fradiae*." University of New South Wales, Sydney, New South Wales.

Jensen, A. L., Darken, M. A., Schultz, J. S., and Shay, A. J. (1964). Relomycin: Flask and tank fermentation studies. In *Antimicrobial Agents and Chemotherapy.* J. C. Sylvestor (Ed.). American Society for Microbiology, Washington, D.C., pp. 49–53.

Kageyama, B., Okazaki, M., and Shibasaki, I. (1971). Mode of action of tylosin (II). *J. Ferment. Technol.* 49:747–758.

Kennedy, J. H. (1978). High performance liquid chromatographic analysis of fermentation broths: Cephalosporin C and tylosin. *J. Chromatogr. Sci.* 16:492–495.

McGuire, J. M., Boniece, W. S., Higgens, C. E., Hoehn, M. M., Stark, W. M., Westhead, J., and Wolfe, R. N. (1961). Tylosin, a new antibiotic: I. Microbiological studies. *Antibiot. Chemother.* 2:320–327.

Madry, N., Sprinkmeyer, R., and Pape, H. (1979). Regulation of tylosin synthesis in *Streptomyces*: Effects of glucose analogs and inorganic phosphate. *Eur. J. Appl. Microbiol. Biotechnol.* 7:365–370.

Mao, J. C. H., and Weigard, R. G. (1968). Mode of action of macrolides. *Biochim. Biophys. Acta* 157:404–413.

Martin, J. F. (1977). Biosynthesis of polyene macrolide antibiotics. *Annu. Rev. Microbiol.* 31:13–38.

Martin, J. F. (1979). Nonpolyene macrolide antibiotics. In *Economic Microbiology,* Vol. 3. A. H. Rose (Ed.). Academic, New York, pp. 239–289.

Masamune, S., Bates, G. S., and Corcoran, J. W. (1977). Macrolides—Recent progress in chemistry and biochemistry. *Angew, Chem. Int. Ed. Engl.* 16:585–607.

Matern, U., Brillinger, G. U., and Pape, H. (1973). Metabolic products of microorganisms. Thymidine diphospho-D-Glucose oxidoreductase *Streptomyces rimosus. Arch. Mikrobiol.* 88:37–48.

Morin, R. B., and Gorman, M. (1964). The partial structure of tylosin, a macrolide antibiotic. *Tetrahedron Lett.* 34:2339–2345.

Morin, R. B., Gorman, M., Hamill, R. L., and Demarco, P. V. (1970). The structure of tylosin. *Tetrahedron Lett.* 54:4737–4740.

Nagel, A. A., and Vincent, L. (1979). Selective cleavage of the mycinose sugar from the macrolide antibiotic tylosin: An unique glycosidic scission. *J. Org. Chem.* 44:2050–2052.

Okamoto, R., Fukumoto, T., Takamatsu, A., and Takeuchi, T. (1978). Tylosin derivatives and their manufacturing process. U.S. Patent No. 4,092,473.

Omura, S., and Nakagawak, A. (1975). Chemical and biological studies on 16-membered macrolide antibiotics. *J. Antibiot.* 28:401–433.

Omura, S., Nakagawa, A., Takeshima, H., Miyazawa, J., and Kitao, C. (1975). A carbon-13 nuclear magnetic resonance study of the bio-synthesis of the 16-membered macrolide antibiotic tylosin. *Tetrahedron Lett.* *50*:4503–4506.

Omura, S., Takeshima, H., Nakagawa, A., Miyazawa, J., Piriou, F., and Lukacs, G. (1977). Studies on the biosynthesis of 16-membered macrolide antibiotics using carbon-13 nuclear magnetic resonance spectroscopy. *Biochemistry 16*:2860–2866.

Pape, H., and Brillinger, G. U. (1973). Metabolic products of microorganisms. Biosynthesis of thymidline diphospho mycarose in a cell-free system from *Streptomyces fradiae*. *Arch. Microbiol. 88*:25–35.

Pape, H., Schmid, R., and Grisebach, H. (1969). Ubertragung der intaken methylgruppe des methionins bei der biosynthese der l-mycarose. *Eur. J. Biochem. 11*:479–483.

Pendulum, L. C., Boling, J. A., and Bradley, N. W. (1978). Levels of monensin with and without tylosin for growing finishing steers. *J. Anim. Sci.* *47*:1–5.

Pestka, S. (1970). Studies on the formation of transfer ribonucleic acid–ribosome complexes. IX. *Arch. Biochem. Biophys. 136*:89–96.

Pestka, S. (1971). Inhibitors of ribosome functions. *Annu. Rev. Microbiol.* *25*:487–562.

Rossi, A., and Corcoran, J. W. (1973). Identification of a multienzyme complex synthesising fatty acids in the actinomycete *Streptomyces erythreus*. *Biochem. Biophys. Res. Commun. 50*:597–602.

Seno, E. T., and Baltz, R. H. (1978). Properties of mutants of *Streptomyces fradiae* blocked tylosin biosynthesis. In *Abstracts of the Third International Symposium on the Genetics of Industrial Microorganisms, Madison, Wisconsin.* American Society for Microbiology, Washington, D.C.

Seno, E. T., Pieper, R. L., and Huber, F. M. (1977). Terminal stages in the biosynthesis of tylosin. *Antimicrob. Agents Chemother. 11*:455–461.

Socci, A., Bertoldini, G., Marco, G., and Codazza, D. (1970). Antibiotics against *Mycoplasma* associated with bovine mastitis. *Arch. Vet. Ital. 21(4)*:235–243.

Sprinkmeyer, R., and Pape, H. (1978). Effects of glucose and fatty acids on the formation of the macrolide antibiotic tylosin by *Streptomyces*. In *Genetics of the Actinomycetales*. E. Freerksen, I. Tarnok, and J. H. Thumim (Eds.). Gustav Fischer Verlag, Berlin, pp. 51–58.

Stark, W. M. (1977). Lilly contribution to fermentation products. *Dev. Ind. Microbiol. 18*:1–8.

Stark, W. M., Daily, W. A., and McGuire, J. M. (1961). A fermentation study of the biosynthesis of tylosin in synthetic media. *Sci. Rep. 1st. Super. Sanita. 1*:340–354.

Suzuki, M., Okazaki, M., and Shibasaki, I. (1970). Mode of action of tylosin (I). *J. Ferment. Technol. 48*:525–532.

Vasileva-Lukanova, B., Atanasova, T., and Khlebarova, E. (1980a). Studies on the isolation of tylosin from the culture liquid of *Actinomyces fradiae*. I. Method for tylosin production. *Khim. Ind. (Sofia) 5*:200–202.

Vasileva-Lukanova, B., Atanasova, T., Gancheva, V., and Khlebarova, E. (1980b). Studies on the isolation of tylosin from the culture fluid of *Streptomyces fradiae*. II. Extraction of tylosin from a native solution. *Khim. Ind. (Sofia) 7*:300–302.

Vu-trong, K., Bhuwapathanapun, S., and Gray, P. P. (1980). Metabolic regulation in tylosin producing *Streptomyces fradiae*: Regulatory role of adenylate nucleotide pool and enzymes involved in biosynthesis of tylonolide precursors. *Antimicrob. Agents Chemother.* *17*:519–525.

Vu-trong, K., Bhuwapathanapun, S., and Gray, P. P. (1981). Metabolic regulation in tylosin producing *Streptomyces fradiae*: Control by phosphate of tylosin biosynthesis. *Antimicrob. Agents Chemother.* *19*:209–212.

Wahlstrom, R. C., and Libal, G. W. (1975). Effects of dietary antimicrobials during early growth and on subsequent swine performance. *J. Anim. Sci.* *40*:655–659.

Whaley, H. A., Patterson, E. L., Shay, A. J., and Tresner, H. D. (1967). Antibiotic AM-684 and methods of production. U.S. Patent No. 3,344,024.

Wise, D. R., and Fuller, M. K. (1975). Efficacy trials in turkey poults with tylosin tartrate against *Mycoplasma gallisepticum* and *Mycoplasma meleagridis*. *Res. Vet. Sci.* *19*:338–339.

Novel Trends in Microbial Antibiotic Production

28

ANTIBIOTIC PRODUCTION WITH IMMOBILIZED LIVING CELLS

ISAO KARUBE AND SHUICHI SUZUKI *Research Laboratory of Resources Utilization, Tokyo Institute of Technology, Yokohama, Japan*

ERICK J. VANDAMME *Laboratory of General and Industrial Microbiology, University of Ghent, Ghent, Belgium*

I. INTRODUCTION

Of the about 150 commercially available antibiotics, none are made by chemical synthesis (except for the relatively simple chloramfenicol and cellocidin molecules), despite the fact that chemical routes are known for many important antibiotics. The economics of the multistep chemical synthesis of such complex biological compounds are just too unfavorable. In this regard, the fact that the enormous development of new semisynthetic penicillins did not start from the moment the chemical synthesis route to the penicillin nucleus 6-aminopenicillanic acid (6-APA) was discovered, but only later, when it was found that under certain fermentation conditions the fungus *Penicillium chrysogenum* would excrete 6-APA which could be converted into new penicillins, is really illustrative (Vandamme, 1980; Savidge, 1982). As chemical large-scale synthesis of antibiotics is practically impossible, antibiotic synthesis by conventional fermentation also has its problems! Apart from the use of sophisticated production facilities, the impressive increase in antibiotic productivity in industrial fermentations is mainly a result of forcing the microorganism to overproduce the useful metabolite by mutation or mutasynthesis, by directly influencing cell metabolism and cell environment, for example, via nutritional control, precursor addition, or more recently by genetic engineering techniques. Most antibiotics are produced as secondary metabolites, that is, their synthesis is delayed until the growth of the cells declines or stops. Consequently, such classic fermentation processes always have a "nonproductive" phase. Once all the necessary enzymes are formed in the cell, theoretically a "linear" antibiotic production phase could be maintained over a long period if it were not for the fact that the involved enzymes are generally rapidly inactivated when growth stops. Conversion of sugar substrates into antibiotic is rather inefficient, owing to utilization for growth, maintenance, and the many side reactions occurring in intact growing cells. Furthermore, strain degeneration, that is, the selection of poorly producing strains which grow faster, causes a major problem in fermentation industry.

As a solution to such disadvantages, attempts have been made to replace antibiotic fermentation by (1) acellular processes, that is, total in vitro enzymatic synthesis (Demain et al., 1976; Demain and Wang, 1976) or (2) immobilized cell technology. Enzyme complexes in vivo localized on membranes or in cell organelles and vesicles have been found to be able to catalyze partial or even total antibiotic synthesis. In this respect, Kurzatkowski et al. (1982) immobilized vesicles (40–200 nm size), isolated from *Penicillium chrysogenum* PQ-96 protoplasts in alginate gels; and demonstrated penicillin G total synthesis.

In a total enzymatic synthesis, it is the ultimate aim to use isolated stabilized immobilized enzymes, which in sequential reactions perform the total synthesis of an antibiotic upon addition of its precursors, adenosine triphosphate (ATP) as an energy source, and cofactors. This concept can be seen as an extension of the already well-known and industrially applied simple enzymatic bioconversions of antibiotics (Vandamme, 1980; Sebek, 1980). This development—the acellular or cell-free total enzymatic synthesis of antibiotics— has yet to be exploited for the benefit of man. In vitro total enzymatic synthesis has been studied particularly at the Massachusetts Institute of Technology in the case of the oligopeptide antibiotics (gramicidin S and bacitracin) (Demain and Wang, 1976; Vandamme, 1981, 1983).

Despite the first successes already achieved with this method on a limited scale, there remain a number of major problems yet to be solved: fermentation of synthetic enzymes (ligases) on a large scale; the optimization and scale-up of cell disruption techniques to liberate active enzymes from the cells; the stabilization of these enzymes in vivo and in vitro (use of particulate enzyme fractions instead of soluble enzymes); the prevention of the in vitro inactivation of soluble, particle-, or membrane-bound enzymes; the large-scale immobilization of the enzymes; and continuous ATP regeneration. Only through a multidisciplinary approach involving the techniques of bioengineering, microbiology, and biochemistry will the efficient enzymatic total synthesis of antibiotics eventually become a reality. Biological effort must be applied to ensure the optimum genetic and cultural environment for maximal production of all the enzymes involved in the biosynthetic process. Chemical input would involve the isolation and kinetic and regulatory characterization of the relevant enzymes. Furthermore, modeling and computer studies should be used to optimize the process according to the kinetics of the system, so that rates much higher than those occurring in the living cell can be achieved. Engineering knowledge on the scale-up of the entire process is also essential to the success of such projects.

To circumvent the difficulties encountered at this moment with such complicated total enzymatic synthesis processes, it was thought that the use of whole immobilized living cells could lead to much greater success. Indeed, during the last decade, increased interest has been expressed in the use of whole microorganisms immobilized in solid supports (Abbott, 1975; Chibata and Tosa, 1977; Kierstan and Bucke, 1977; Mosbach and Larson, 1970; Shimizu et al., 1975; Vandamme, 1976, 1981; Vieth et al., 1973). In this respect, immobilized microorganisms—particularly their intracellular enzymes—have been used for the production of bioactive materials. Generally, utilization of immobilized whole living cells for total synthesis of fermentation or organic products is scarcely documented: Microbial electrodes using immobilized whole cells have been developed for the determination of biological oxygen demand (BOD) (Karube et al., 1977; Hikuma et al., 1980); laboratory-scale synthesis of α-amylase (Karube et al., 1978), glutamic acid (Slowinski and Charm, 1973), isoleucine, citric acid, hydrogen (Karube et al., 1976), ethanol, coenzyme A, ammonia, vitamin B_1 (Matsunaga et al., 1978), and a few antibiotics has recently been reported. Again the vast potential of whole immobilized cells in the field of antibiotic production is only just now being fully recognized. This concept would be particularly valuable for those antibiotics which are— or can be—excreted into the culture medium. The production of penicillin G, ampicillin, bacitracin, cephamycin (Freeman and Aharonowitz, 1980), and candicidin (Venkatasubramanian and Vieth, 1979) has been reported with immobilized living cells as catalysts. Compared to conventional fermentation, immobilized living cell systems can offer several important advantages, such as the possibility of continuous operation (at high dilution rates with no washout danger) or the plug-flow mode of action, "column fermentation," reduction of nonproductive growth phases, faster reaction rates at increased cell density, higher yields, easier rheological control, and control of cell reproduction (Vandamme, 1981). Advantages of using whole microorganisms compared to purified enzymes include the ability to catalyze a series of linked reactions, some of which may require cofactors (Chibata and Tosa, 1981; Klibanov, 1982; Fukui and Tahaka, 1982; Bucke, 1983).

II. PRODUCTION OF PENICILLIN G BY IMMOBILIZED *PENICILLIUM* MYCELIUM

Penicillin G is an important antibiotic and is also used as a raw material for 6-APA-production (Savidge, 1982). As a logical first choice, *P. chrysogenum* mycelium has been immobilized in polyacrylamide gel (PAA), collagen membranes, and Ca-alginate beads; the production of penicillin G from glucose with these catalysts is discussed here (Morikawa et al., 1979a).

A. Culturing of the Strain

The strain used in our studies was *P. chrysogenum* ATCC 12690. Fermentation was performed under aerobic conditions (reciprocal-type shaker: 120 strokes/min) at 25°C in the medium as described by Damin (1956). A total of 1 ml of 32% phenylacetate was added to the medium 24 hr after inoculation. Mycelium was harvested 70 hr after inoculation at its maximum rate of penicillin production (12 units/ml per hour). The penicillin concentration in the culture broth was between 200 and 300 units/ml.

B. Immobilization Procedure

The mycelium was collected by filtration and washed thoroughly with 0.05 M phosphate buffer (pH = 7.0). One gram of mycelium was suspended in 5 ml of 0.1 M phosphate buffer and 2 ml of 25% acrylamide monomer solution [85% acrylamide and 15% N,N'-methylene bisacrylamide (BIS)] were added to the suspension; the final volume was adjusted to 9.4 ml with water. Polymerization was initiated under nitrogen gas with 0.1 ml of 5% N,N,N',N'-tetramethylethylenediamine (TEMED) and 0.5 ml of 5% ammonium persulfate (APS), and allowed to proceed for 20 min at 5°C. The immobilized mycelium was then cut into small blocks (8–27 mm^3) and washed thoroughly with 0.05 M phosphate buffer (pH = 7.0). Collagen immobilization resulted in low activity, whereas Ca-alginate entrapment yielded the highest activity but did yield a gel too fragile to be used repeatedly under reactor conditions.

C. Penicillin Production by Immobilized Mycelium

Production of penicillin by the immobilized mycelium was performed as follows: The immobilized mycelium (containing 1 g of wet mycelium) was added to 25 ml of a medium containing 1% glucose, 0.5% $(NH_4)_2SO_4$, 0.01% phenylacetate, and 0.05 M phosphate buffer (pH = 7.0) in a 500-ml Sakaguchi flask. The mixture was incubated on a reciprocal shaker (120 strokes/min) for 5 hr at 25°C. Penicillin produced by the immobilized mycelium was assayed by the paper disk method using *Staphylococcus aureus* ATCC 6538P as the assay microorganism. Penicillin G (potassium salt, 1559 units/mg) was used as a standard.

The mycelium, immobilized in polyacrylamide gel, indeed produced penicillin G. However, the rate of penicillin production of immobilized mycelium was lower than that of washed mycelium. With the 5% PAA gel-entrapped mycelium, penicillin was produced (0.77 units/ml per hour) within 5 hr, but amounted to only 17% of that produced by washed mycelium; however, the half-life of immobilized mycelium activity was much longer. The reagents, such as acrylamide and TEMED, severely and adversely affected the penicillin-producing activity by the washed mycelium. These results suggested that the multienzyme system required for penicillin synthesis was inactivated by the reagents individually. In this respect, the amounts of acrylamide and TEMED were decreased to as low a concentration as possible: 5% of acrylamide and 0.05% TEMED were found to be ideal for polymerization of the gel. The total concentration and relative ratios of acrylamide and BIS determine both the pore size of the interstitial space within which whole cells are entrapped, and the physical properties of the complex.

The effect of monomer BIS content on penicillin production by the immobilized mycelium was also examined. The activity increased with increasing BIS content, probably owing to the increased pore size of the interstitial spaces. Since the gel became too fragile when the BIS content was higher than 20%, a BIS content of 15% was selected for further work; furthermore, it provided excellent mechanical rigidity.

D. Effect of Substrates on Penicillin Production

It is known that penicillin G is derived from three amino acids (L-α-aminoadipic acid, L-cysteine, and L-valine) and the side-chain precursor phenylacetate. Furthermore, penicillin acyltransferase, the enzyme presumably involved in the final step of penicillin G synthesis, catalyzes penicillin G synthesis from 6-aminopenicillanic acid (6-APA) and phenylacetate. In this respect, the effect of these substrates on penicillin production by the immobilized mycelium was expressed relative to that obtained with a glucose-$(NH_4)_2SO_4$ medium. The level of penicillin production in a medium containing the three amino acids was higher than that in a 6-APA medium. This might be caused by diffusional limitation of 6-APA through the cell membrane. The immobilized mycelium required oxygen for penicillin production.

E. Repeated Production of Penicillin by Immobilized Mycelium

Repeated penicillin production by immobilized mycelium was tested in a batch system. Experiments were performed as follows: About 1 g of washed mycelium or 10 g of immobilized mycelium (containing 1 g of wet mycelium) were incubated in a medium based on glucose-$(NH_4)_2SO_4$ for 5 hr at 25°C. After use, the immobilized mycelium was washed thoroughly with 0.05 M phosphate buffer and stored at 5°C. This experiment was repeated each day. Figure 1 shows the penicillin levels produced by the mycelium preparations. The level of penicillin production of the washed mycelium rapidly decreased with repeated use. On the other hand, the penicillin production level of the immobilized mycelium increased initially (up to 200% after the third run) and then gradually decreased. This initial activation of the immobilized mycelium activity might have been caused by growth of the mycelium within the polyacrylamide gel or by a change in the permeability of the mycelium cell wall. At least part of the immobilized mycelium remains viable during the entrapment and reaction procedure. However, the half-life of the immobilized mycelium

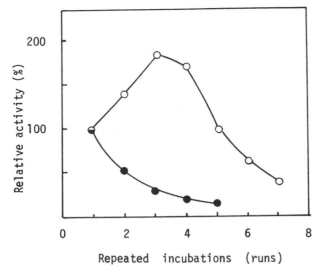

Figure 1 Penicillin production by the two forms of mycelium. Reactions were carried out under identical conditions as described in the text: immobilized mycelium (○) and washed mycelium (●).

activity was 6 days, compared to 1 day for the washed mycelium. The activity of the immobilized microorganisms decreased also during storage at low temperature.

The respiratory activity of the immobilized mycelium was examined with an oxygen electrode. The rate of oxygen uptake by the immobilized mycelium was about 30% that of the washed mycelium. Penicillin production by the immobilized mycelium was carried out in air as well as under a nitrogen gas atmosphere. Compared to the aerated mixture, only a small amount of penicillin was produced by the immobilized mycelium incubated under anaerobic conditions, indicating that oxygen is required for the immobilized system to synthesize penicillin.

Immobilized mycelium maintained respiratory activity. Furthermore, viable mycelium could be recovered from ground, immobilized mycelium in germ-free conditions. These results suggest that part of the mycelium is alive after being entrapped in the polyacrylamide gel and that this living mycelium is responsible for the production of penicillin from glucose and $(NH_4)_2SO_4$.

III. PRODUCTION OF CEPHAMYCIN BY IMMOBILIZED WHOLE CELLS

Morikawa et al. (1979a) obtained evidence that the PAA gel polymerization reagents individually inactivate the multienzyme systems in the mycelium. Such negative effects have also been encountered by Freeman and Aharonowitz (1980) when entrapping viable *Streptomyces* cells for cephalosporin (cephamycin) production. Indeed, the formation of PAA gels in the presence of viable cells results in a considerable loss of biological activity, mainly due to toxicity of the monomers and heat evolution during polymerization. Nevertheless, cross-linked PAA is one of the more common matrices used for entrapment of whole microbial cells. In order to retain the advantages of PAA and to avoid cell damage during preparative steps, Freeman and Aharonowitz (1980) proposed the use of preformed linear, water-soluble PAA chains (molecular weight $15 \times 10^4 - 17 \times 10^4$) partially substituted with acylhydrazide groups. This prepolymerized material was cross-linked in the presence of viable cells by stoichiometric amounts of bi- or multifunctional aldehydes or ketones (glyoxal, glutaraldehyde, polyvinylalcohol). Use of cross-linking agents of various chain lengths (glyoxal, glutardialdehyde, and periodate-oxidized polyvinylalcohol) allowed to control gel compactness. This cross-linking reaction, carried out in cold, neutral, physiological condition resulted in gel-entrapped cells with mechanical properties similar to those of conventional PAA gels. *Streptomyces clavuligerus* entrapped by this prepolymerization PAA gel procedure produced cephamycin C continuously for 96 hr with yields comparable to those of resting cells. Only 15% of the activity of washed *Penicillium* mycelium was obtained with the conventional PAA entrapment procedure (Morikawa et al., 1979a). The concept of using preformed linear, water-soluble PAA chains now deserves full attention and might considerably improve the performance of entrapped living cell systems.

IV. PRODUCTION OF BACITRACIN BY IMMOBILIZED WHOLE CELLS

The oligopeptide antibiotic bacitracin is synthesized according to the multienzyme thiotemplate mechanism present in *Bacillus* species (Katz and Demain, 1977; Frøyshov, 1982; Vandamme, 1981). Cell-free synthesis of bacitracin has also been studied in some detail (Katz and Demain, 1977; Frøyshov, 1982) and the antibiotic is excreted into the fermentation medium. This knowledge has contributed to the selection of this process as a model for synthesis of a secondary metabolite and of an economically important antibiotic by immobilized living bacterial cells.

A bacitracin-producing *Bacillus* species was immobilized within a polyacrylamide gel, and the production of bacitracin was tested out from a simple nutrient under varying cultural conditions. Furthermore, the growth of the bacteria in the gel was examined by electron microscopy.

A. Culture and Immobilization of Microorganisms

Bacillus sp. KY4515 was used for bacitracin production. The microorganism was cultured in a starch bouillon (SB) medium for 13 hr at 30°C. Cells were harvested from a batch fermentation when the rate of bacitracin production was maximal (40 units/ml per hour). Immobilization of the microorganism in polyacrylamide gel was performed by the method described in Section II. After immobilization in 5% PAA gel, the catalyst was cut into small blocks ($8-27$ mm^3).

B. Bacitracin Production by Immobilized Whole Cells

In a typical procedure, bacitracin production by immobilized whole cells was performed in a batch system as follows: 15 g of immobilized whole cells (contained 0.75 g of cells, wet weight) were added to 25 ml of fermentation medium (starch bouillon) in a 500-ml Sakaguchi flask; the mixture was incubated in a reciprocal shaker (150 strokes/min) for 4 hr at 30°C. For repeated batch production of bacitracin, the immobilized whole cells were washed thoroughly with saline after each incubation, resuspended in fresh medium, and incubated again for 4 hr. Bacitracin produced by the immobilized whole cells was assayed by the paper disk method, using *Micrococcus luteus* ATCC 10240 as the assay microorganism. Bacitracin (75 units/mg) was used as a standard. Samples for electron microscopy examination were fixed with 2.5% (wt/vol) glutaraldehyde in phosphate buffer (pH = 7.0). The wet gels were treated in a series of water–acetone, acetone, and isoamylacetate solutions. Thin sections were prepared, coated with gold, and examined in a JEOL (Japan Electron Optical Laboratory) scanning electron microscope (100CASID4D) operated at 40 KV.

Bacitracin was produced by washed cell preparations and the rate of production was usually 13–18 units/ml per hour. However, bacitracin productivity decreased with successive utilization of the washed cells. Harvesting washed cells from the reaction medium was also laborious. With immobilized whole cells, the level of bacitracin production was highest when the intact cells were immobilized in a gel prepared with 5% total acrylamide (95% acrylamide monomer and 5% BIS). On the other hand, no effect of BIS content was observed on the production level of bacitracin. However, highest productivity of bacitracin by immobilized whole cells was only 20–25% that of an equivalent amount of washed cells. These results suggest that the lower rate of bacitracin production is mainly caused by inactivation of intracellular enzymes by the polymerization reagents (especially acrylamide and APS, as shown in Table 1) and may be partly due to hindered diffusion of the substrates and/or product through the gel.

Table 1 Bacitracin Production by Whole Cells Entrapped in Various Acrylamide Gels

Acrylamide[a]	BIS[b]	APS	TEMED	% Bacitracin produced[c]
4	5	0.2	0.025	24
6	5	0.2	0.025	15
8	5	0.2	0.025	8
10	5	0.2	0.025	5
5	5	0.2	0.025	17
5	10	0.2	0.025	16
5	15	0.2	0.025	17
5	20	0.2	0.025	18
5	5	0.1	0.05	24
5	5	0.1	0.10	20
5	5	0.2	0.025	17
5	5	0.2	0.05	13
5	5	0.4	0.025	8
5	5	0.4	0.05	6

[a]Values indicate, in percent, the total weight of acrylamide per total volume used in the immobilization procedure.
[b]Values indicate, in percent, the BIS weight per total weight of the acrylamide monomer.
[c]Starch bouillon medium was used. Values indicate the production rate of immobilized whole cells, expressed as a percentage of the production rate of an equivalent amount of washed cells incubated under similar conditions.

C. Effect of Nutrients on Bacitracin Production

The effect of air (oxygen) on bacitracin production by immobilized whole cells was examined. Incubation of immobilized whole cells was carried out under normal aeration conditions and under nitrogen atmosphere. Bacitracin produced under anaerobic incubation conditions was only 30% that under aerobic conditions. Since air is required for an effective production of bacitracin by immobilized whole cells, it was concluded that whole cells entrapped within the gel display respiratory activity.

Preliminary experiments indicated that the level of bacitracin production by immobilized whole cells was markedly reduced upon repeated use when a starch bouillion fermentation medium was used as a substrate. Bacitracin productivity in a medium containing meat extract or peptone was higher than that in a medium based on carbohydrates. Furthermore, the productivity of immobilized whole cells was retained upon repeated use. This tendency was also observed in the case of bacitracin production with washed cells. Leakage of bacteria from the gel is not desirable for bacitracin production; external growth of leaked bacteria was observed, especially when the starch bouillion fermentation medium was employed for bacitracin production. Diluted peptone and meat extract were then selected as nutrients for the immobilized bacteria, as growth of leaded and contaminating bacteria was not observed under these conditions. Subsequently, 0.5% meat extract medium and especially 1% peptone medium (containing 0.25% saline and 1×10^{-5} M $MnSO_4$) were selected for repeated batch production of bacitracin. Repeated batch production of bacitracin was tested with immobilized whole cells as well as with washed cells (Fig. 2). Bacitracin produced by washed cells decreased gradually increased with successive utilizations, reaching a steady state at 30–90% of the activity of freshly washed cells. This increase in activity seems to be caused by active growth of the cells in and on the gel, as could be observed from electron microscopy examination. The bacitracin level produced in saline appeared to result from synthesis from whole cell nutrient pools.

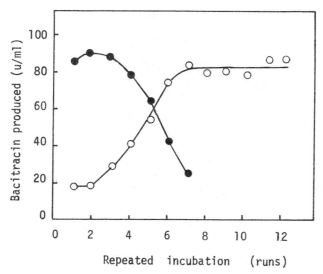

Figure 2 Repeated batch production of bacitracin; 1% peptone was used as a reaction medium: washed cells (●) and immobilized cells (○).

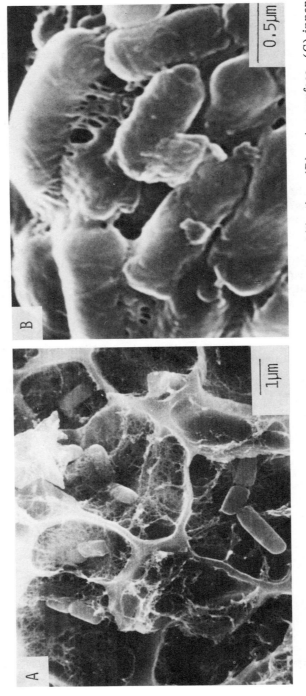

Figure 3 Electron micrographs of the gel: (A) gel immediately after immobilization, (B) outer surface, (C) inner layer near surface, (D) and center of the gel after 56 hr of incubation.

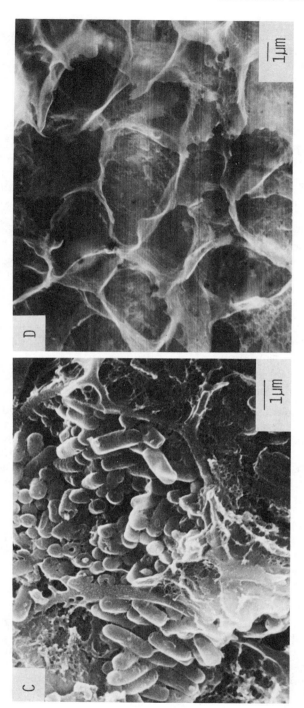

Figure 3 (continued)

The rate of bacitracin production increased with increasing amounts of cells in the gel, but upon repeated utilization an equal level of antibiotic was obtained. Bacitracin production from its constituent amino acids and ATP did not occur with these immobilized cell preparations as catalysts, probably owing to transport problems. The apparent half-life of this bacitracin synthesis catalyst was estimated to be at least 1 week.

D. Microscopic Observation of Immobilized Cells

Electron microscopy observation of cells entrapped in polyacrylamide gel revealed active growth to occur. Figure 3 shows the electron micrographs of whole cells immediately after immobilization and also after 56 hr of use for bacitracin production. Apparent growth of cells (especially near the surface layer of the gel) was observed after successive utilization, whereas the amount of cells evidently had decreased in the center of the gel matrix. The diffusion of oxygen through the gel matrix may be a limiting factor for bacitracin production. The active growth of the bacteria during incubation provides an obvious explanation for the increase in the level of bacitracin production with successive utilizations. A steady state of bacitracin production could mean that the interstitial space of the gel is completely filled with active bacteria, or that it is controlled by the diffusion rate of substrates and/or product through the gel. However, the detailed mechanism of a steady-state production of bacitracin is far from being fully understood. The overall rate of bacitracin production by immobilized whole cells was slightly higher than that of conventional fermentation.

E. Continuous Production of Bacitracin

Continuous production of bacitracin has recently been achieved in an "immobilized whole cell fermenter." A schematic diagram of the apparatus for continuous bacitracin production is illustrated in Figure 4. The water-jacketed reactor consists of an air sparger, inlet and outlet tubes, and an air exhaust tube. The flow rate of sterilized medium was controlled at 40 ml/hr. The medium level in the reactor was controlled by the level of an overflow tube. This outlet tube was covered with a nylon screen (100 mesh), so that the immobilized whole cells particles could not leak from the reactor.

Immobilized whole cells (30 g wet weight) washed thoroughly with sterilized saline were added to the sterilized reactor aseptically. After addition of the immobilized whole cells, the reactor volume was increased to 120 ml with fresh medium, which was then continuously fed to the reactor. Peptone (0.5%, pH = 6.5) containing 0.25% NaCl, 1×10^{-5} M $MnSO_4$, and 0.0025% Adekanol LG 126 (from Asahi Denka Co. Ltd., Japan, an antifoam agent) was used as a medium. The airflow rate was adjusted to 3.5 liters/min. The temperature of the reactor was maintained at 30°C. Maximum bacitracin levels were obtained after 1 day of operation, after which a gradual decrease in productivity was observed with increasing operation time. To screen for the most suitable conditions for the continuous production of bacitracin, changing the environmental cultural conditions (such as aeration rates, pH, components of the medium, and temperature) did not improve the productivity period of the immobilized whole cells. Bacterial growth in the effluent invariably began to increase when bacitracin productivity decreased. Furthermore, the richer the nutrient medium employed, the faster the rate of productivity loss. These results suggested that the decrease in bacitracin productivity was mainly caused by

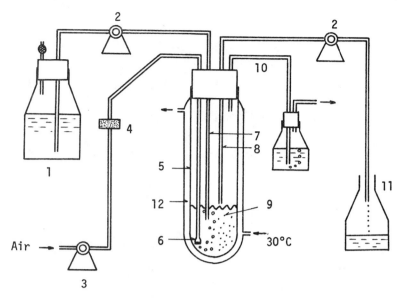

Figure 4 Schematic diagram of the continuous immobilized cell fermenter system: (1) substrate reservoir, (2) peristaltic pump, (3) air pump, (4) air filter, (5) reactor, (6) air sparger, (7) inlet tube, (8) outlet tube, (9) immobilized whole cells, (10) air exhaust tube, (11) product reservoir, (12) water jacket.

cell growth at the surface of the gel blocks. Such cell growth may prevent diffusion of oxygen and nutrients into the gel matrix, such that the turnover of cells within the gel became negligible. Indeed, it was obvious from electron microscopical observations that the surface of the gel blocks was completely covered with colonies after 50 hr of incubation. A washing treatment of the gel blocks was tried out to eliminate external growth: Sterilized saline was continuously fed to the reactor at a flow rate of 250 ml/hr during 2 hr once a day. After this interval washing, the reactor operated under ordinary conditions with improved production. Figure 5 shows the improved continuous production of bacitracin by immobilized whole cells. Under those operational conditions, high bacitracin productivity was obtained for at least 8 days and growth in the effluent was kept at low level. The half-life of the immobilized whole cells in this system was about 10 days.

F. Comparison of Bacitracin Productivity with That of Conventional Continuous Fermentation

Preliminary experiments showed that bacitracin productivity was proportional to the reciprocal of the flow rate within the range 30–100 ml/hr. In a conventional continuous fermentation bacitracin productivity rapidly decreased at high flow rates, because bacitracin fermentation is mainly non-growth-associated. A comparison of bacitracin productivity in the novel system described here with that in a conventional continuous fermentation was made. The results are given in Figure 6. In the immobilized cell fermenter system, the bacitracin concentration gradually decreased with increasing dilution rate,

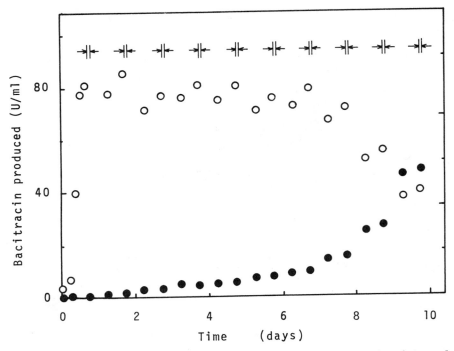

Figure 5 Improved continuous production of bacitracin. Time intervals between two arrows indicate the period where gels were washed with saline: bacitracin produced (○) and growth in the effluent (OD 660) (●).

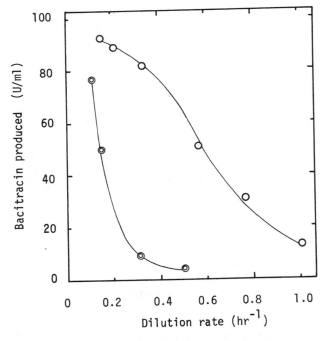

Figure 6 Effect of the dilution rate on bacitracin productivity. Continuous production of bacitracin was carried out under standard conditions. In continuous operation, sampling was performed after three residence times: immobilized whole cells (○) and conventional continuous fermentation (◎).

while in conventional continuous fermentation the bacitracin concentration decreased very rapidly with increasing dilution rate. For example, at a dilution rate of 0.33, the bacitracin concentration in the new system was 82 units/ml, whereas it was only 9 units/ml in the conventional continuous fermentation.

Continuous production of bacitracin by the immobilized whole cell fermenter seems to have several advantages over a conventional continuous fermentation process:

1. Production of bacitracin can be performed from a very simple nutrient such as peptone.
2. Purification of antibiotic from the broth is very simple because the broth does not contain the whole cells.
3. Bacitracin productivity in the immobilized whole cell reactor is higher than that in a conventional continuous fermentation at high dilution rates.

V. OTHER ANTIBIOTICS

Another antibiotic so far reported to be produced according to the immobilized cell principle is the polyene macrolide candicidin, produced by *Streptomyces griseus*. Immobilization of the streptomycete in collagen led to an antibiotic production capacity from glucose of 14% that obtained in the conventional batch fermentation process (Venkatasubramanian and Vieth, 1979). Nisin, an economical important polypeptide antibiotic (Hurst, Chap. 21, pp. 607–628), has been produced by *Streptococcus lactis* cells, immobilized in a polyacrylamide gel (Egorov et al., 1978). Also tylosin, colistin, and nikkomycin have been produced with immobilized cells and many other antibiotic compounds are bound to follow (Veelken and Pape, 1982; Vandamme, 1983).

VI. IMPROVED SYNTHESIS OF AMPICILLIN BY IMMOBILIZED WHOLE CELLS

As described above, several antibiotics can be produced de novo, from simple nutrients, by immobilized microorganisms (Morikawa et al., 1979a,b, 1980a). The authors have demonstrated that the cells remain alive during and after immobilization and were indeed growing. Immobilized cells are thus useful for de novo synthesis of secondary metabolites, but also for simple or multi-step enzymatic modifications of useful compounds. Indeed, increasing the cell number, resulting from cell growth within the gel matrix, may also result in global increase of the total enzymatic activity of the gel. Such immobilized growing whole cells might prove to be very effective for single-step enzyme modifications, such as penicillin acylation (Savidge, 1982; Vandamme, 1980, 1981).

Penicillin acylase (penicillin amidohydrolase E.C.3.5.1.11.) catalyzes the hydrolysis of penicillin into 6-aminopenicillanic acid (6-APA), but also catalyzes the reverse reaction, that is, the synthesis of penicillin from a side chain and 6-APA (Vandamme, 1980, 1981). Recently, there have been several studies on the continuous production of 6-APA using a column packed with immobilized penicillin acylase, but there are relatively few reports on penicillin acylation and/or deacylation by immobilized whole cells (Abbott, 1976; Vandamme, 1980, 1981; Savidge, 1982).

While cells of *Kluyvera citrophila*, a producer of penicillin acylase, were immobilized in polyacrylamide gel, and enhancement of penicillin acylase activity within the immobilized whole cells was attempted by continuous cultivation of the immobilized living cells in an aerated fermenter.

A. Immobilization Procedure of *Kluyvera citrophila* Cells and Enhancement of Penicillin Acylase Activity

Living cells of *K. citrophila* ATCC 14237 were immobilized in polyacrylamide gel (5%). The penicillin acylase activities of washed cells and immobilized cells were first compared. Washed cells harvested after 15 hr of growth produced essentially penicilloic acid and penicic acid when incubated with penicillin G. On the other hand, little penicilloic acid was produced by cells grown during 48 hr (Takasawa et al., 1972). Such cells, collected after 48 hr of cultivation, were immobilized in polyacrylamide gel. Penicillin G was rapidly hydrolyzed into 6-APA. It was then tried to synthesize ampicillin from 6-APA and D-phenylglycinemethylester or phenylacetate with immobilized cells.

The effect of the cell content of the matrix on ampicillin production was examined with washed cells and with immobilized cells; it was concluded that the immobilized cells retained approximately 60–70% of the ampicillin-producing activity of washed cells.

Using the aerated fermenter described above (Morikawa et al., 1980a), continuous cultivation of the immobilized cells was tested out to enhance penicillin acylase activity. A medium containing 1% peptone, 0.5% meat extract, and 0.25% NaCl was used for continuous cultivation (Morikawa et al., 1980b). The amount of 6-APA produced by immobilized cells cultivated continuously for 24 hr was clearly higher than that of the original cells. However, both penicilloic acid and penicic acid were also formed, as detected by thin-layer chromatography, probably owing to β-lactamase activity. Treatment of the cultivated immobilized cells with alkali could remove β-lactamase activity, as little penicilloic acid was produced. The immobilized cells thus treated produced 6-APA in higher yields than noncultivated ones. Alkali-treated, cultivated immobilized living cells produced twice the amount of ampicillin compared with noncultivated immobilized cells. These results indicate that the cultivation of the immobilized cells clearly increased their penicillin acylase activity.

B. Ampicillin Synthesis by an Immobilized Cell Column

Ampicillin production by immobilized cells was also investigated using a column system. The effect of the flow rate was examined on ampicillin production by cultivated and noncultivated cells. These experiments were carried out in peptone medium (1%) containing 0.25% NaCl and 0.05% phenylacetate or D-phenylglycinemethylester. The amount of ampicillin formed in continuous column operation by the cultivated immobilized cells was again higher than that produced by noncultivated cells. The maximal conversion ratio was 41% (based on 6-APA), and this ratio decreased when the flow rate became lower than 0.2. Low yields might be attributed to product or substrate inhibition, the reversibility of the reaction, and hydrolysis of the substrate (Vandamme, 1980, 1981; Savidge, 1982).

The operational stability of the immobilized, cultivated alkali-treated whole cells was investigated by continuously passing the substrate solution through the column at a flow rate of 0.5 for 9 days. As shown in Figure 7, the ampicillin concentration decreased slowly. The half-life of the catalyst was about 14 days. The observed enhancement of enzyme activity in immobilized cells by a cultivation stage as described here might provide for an effective technique for the continuous production of bioactive substances and should be tested out further with respect to other important bioconversion reactions.

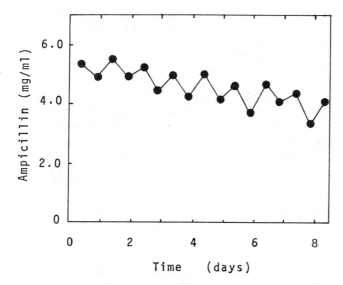

Figure 7 Operational stability of cultivated immobilized cells. The flow rate was 0.5 hr^{-1}.

VII. CONCLUSIONS

Except for ampicillin synthesis, these are all typical examples of synthesis by fixed viable cell systems of secondary metabolites, normally non-growth-associated complex fermentation products. Now, already, it seems that the culture age, physiological state, and viability of the cells in the immobilized reactor are of primordial importance for effecting such complex multienzyme reactions. It might indeed seem necessary to control cell growth at a fixed level within the immobilized matrix to obtain maximal productivity. In this perspective, new bioreactor configurations, such as the immobilized whole cell fermenter described by Morikawa et al. (1980a) for bacitracin production, should be designed combining fermenter and immobilized cell reactor characteristics, facilitating synthesis of both growth-associated as well as non-growth-associated products. Also, different cell species of types (spores) can be coimmobilized in such a fermenter and offer exciting perspectives for "artificial pathway" or multistep organic reactions. The same is true for systems where isolated enzymes, coenzymes, or organelles are combined with whole cells and coimmobilized (Hough and Lyons, 1972; Takasaki, 1979; Kierstan and Bucke, 1977; Lowe, 1981; Vandamme, 1976, 1981). Combination of such biocatalytic potential indicates that enzyme and whole cell immobilization can operate together successfully rather than compete with each other.

In contrast with immobilized monoenzyme systems, optimal catalysis by multistep enzyme complexes and by whole stabilized living microbial cells is quite complex and many basic aspects are yet to be well understood. Indeed, several typical microbiological, biochemical, physiological, and technical problems inherent to this immobilized cell development need to be examined further in order to arrive at optimal performance of immobilized cell reactors: for example, physiological state, viability, lysis or growth phenomena, quantification of growth and biomass, cell metabolism, and maintenance of energy at low or zero growth rates, cofactor utilization and regeneration, microbial contamination problems, and prevention of unwanted side reactions. Other important process design parameters such as the loading factor, stability of the cell, catalyst packing density, oxygen transfer, mass transport and diffusion efficiencies, and residence time distribution, which determine overall cell reactor productivity, need further study (Venkatasubramanian and Vieth, 1979; Vandamme, 1981; Pirt, 1983).

A comparison of immobilized cell systems reveals that very few immobilization methods were used relative to the large number that have been employed with cell-free enzymes. Especially absent are methods for the attachment of cells to insoluble supports by covalent bounds. Cells covalently immobilized should be more stable to dissociation than ionically bound cells, and less restricted by substrate and product diffusion than cells entrapped in a polymer, the usual technique applied so far. However, it is clear by now that immobilized cells can be considered as one of the tools of stereospecific and complex organic synthesis and bioproduction in the nearby future (Vandamme, 1983).

REFERENCES

Abbott, B. J. (1976). Preparation of pharmaceutical compounds by immobilized enzymes and cells. *Adv. Appl. Microbiol.* 20:203–257.

Bucke, C. (1983). Immobilized cells. *Phil. Trans. R. Soc. Lond.* B 300:369–389.

Chibata, I., and Tosa, T. (1977). Transformation of organic compounds by immobilized microbial cells. *Adv. Appl. Microbiol.* 22:1–27.

Chibata, I., and Tosa, T. (1981). Use of immobilized cells. *Ann. Rev. Biophys. Bioeng.* 10:197–216.

Demain, A. L. (1956). Inhibition of penicillin formation by amino acid analogues. *Arch. Biochem. Biophys.* 64:74–79.

Demain, A. L., and Wang, D. I. C. (1976). Enzymatic synthesis of gramicidin S. In *Second International Symposium on the Genetics of Industrial Microorganisms.* K. D. McDonald (Ed.). Academic, New York, pp. 115–128.

Egorov, N. S., Baranova, I. P., and Kozlova, Y. U. I. (1978). Nisin production by immobilized cells of the lactic acid bacterium *Streptococcus lactis. Antibiotiki (Moscow) USSR,* 23 (10):872–874.

Freeman, A., and Aharonowitz, Y. (1980). β-Lactam antibiotic biosynthesis and 17-ketosteroid reduction by bacteria immobilized in cross-linked prepolymerized linear polyacrylamide. Abstract VIth Internat. Ferment. Symposium, London, Ontario, Canada, p. 1211.

Frøyshov, O. (1984). The bacitracins: properties, biosynthesis, and fermentation. In *Biotechnology of Industrial Antibiotics.* E. J. Vandamme (Ed.) Marcel Dekker, New York, pp. 665–694

Fukui, S., and Tanaka, A. (1982). Immobilized microbial cells. *Ann. Rev. Microbial.* 36:145–172.

Hikuma, M., Suzuki, H., Yasuda, T., Karube, I., and Suzuki, S. (1980). A rapid electrochemical method for assimilation test of microorganisms. *Eur. J. Appl. Microbiol. Biotechnol.* 9:305–316.

Hough, J. S., and Lyons, T. P. (1972). Coupling of enzymes onto microorganisms. *Nature (London)* 235:389.

Karube, I., Matsunaga, T., Tsuru, S., and Suzuki, S. (1976). Continuous hydrogen production by immobilized whole cells of *Clostridium butyricum. Biochim. Biophys. Acta* 444:338–343.

Karube, I., Matsunaga, T., Mitsuda, S., and Suzuki, S. (1977). Microbial electrode BOD sensors. *Biotechnol. Bioeng.* *19*:1535–1547.

Katz, E., and Demain, A. L. (1977). The peptide antibiotics of *Bacillus*: Chemistry, biogenesis, and possible functions. *Bacteriol. Rev.* *41*:449–474.

Kierstan, M., and Bucke, C. (1977). The immobilization of microbial cells, subcellular organelles, and enzymes in calcium alginate gels. *Biotechnol. Bioeng.* *19*:387–397.

Klibanov, A. M. (1983). Immobilized enzymes and cells as bacterial catalysts. *Science 219*:722–727.

Kokubu, T., Karube, I., and Suzuki, S. (1978). α-Amylase production by immobilized whole cells of *Bacillus subtilis*. *Eur. J. Appl. Microbiol. Biotechnol.* *5*:233–240.

Kurzatowski, W., Kurylowics, W., and Paskiewics, A. (1982). Penicillin G production by immobilized fungal vesicles. *Eur. J. Appl. Microbiol. Biotechnol.* *5*:211–213.

Lowe, C. R. (1981). Immobilized coenzymes. *Top. Enzyme Ferment. Biotechnol.* *5*:13–146.

Matsunaga, T., Karube, I., and Suzuki, S. (1978). Electrochemical microbioassay of vitamin B_1. *Anal. Chim. Acta 98*:25–30.

Morikawa, Y., Karube, I., and Suzuki, S. (1979a). Penicillin G production by immobilized whole cells of *Penicillium crysogenum*. *Biotechnol. Bioeng.* *21*:261–270.

Morikawa, Y., Karube, I., and Suzuki, S. (1979b). Bacitracin production by whole cells immobilized in polyacrylamide gel. *Antimicrob. Agents Chemother.* *15*:125–130.

Morikawa, Y., Karube, I., and Suzuki, S. (1980a). Continuous production of bacitracin by immobilized whole cells of *Bacillus* sp. *Biotechnol. Bioeng.* *22*:1015–1023.

Morikawa, Y., Karube, I., and Suzuki, S. (1980b). Enhancement of penicillin acylase activity by cultivating immobilized *Kluyvera citrophilla*. *Eur. J. Appl. Microbiol. Biotechnol.* *10*:23–30.

Mosbach, K., and Larson, P. (1970). Preparation and application of polymer entrapped enzymes and microorganisms in microbial transformation process with special reference to steroid 11-β hydrosylation and Δ' dehydrogenation. *Biotechnol. Bioeng.* *12*:19–27.

Pirt, S. J. (1983). The role of microbial physiology in biotechnology. *J. Chem. Tech. Biotechnol.* *33B*:137–138.

Rayman, K., and Hurst, A. (1984). Nisin: properties, biosynthesis, and fermentation. In *Biotechnology of Industrial Antibiotics*. E. J. Vandamme (Ed.). Marcel Dekker, New York, pp. 607–628.

Savidge, T. (1984). Enzymatic conversions used in the production of penicillins and cephalosporins. In *Biotechnology of Industrial Antibiotics*. E. J. Vandamme (Ed.). Marcel Dekker, New York, pp. 171–224.

Sebek, O. K. (1980). Microbial transformation of antibiotics. In *Economic Microbiology*, Vol. 5. A. M. Rorel (Ed.). Academic, New York, pp. 576–612.

Shimizu, S., Morioka, H., Tani, Y., and Ogata, K. (1975). Synthesis of coenzyme A by immobilized microbial cells. *J. Ferment. Technol.* *53*:77–83.

Slowinski, W., and Charm, S. E. (1973). Glutamic acid production with gel-entrapped *Corynebacterium glutamicum*. *Biotechnol. Bioeng.* *15*:973–979.

Takasaki, Y. (1979). Binding of enzymes to microbial cells. *Agric. Biol. Chem.* *38*:1061-1082.

Takasawa, S., Okachi, R., Kawamoto, I., Yamamoto, M., and Nara, T. (1972). Some problems involved ampicillin formation by *Kluyvera's* penicillin acylase. *Agric. Biol. Chem. 36*:1701-1706.

Vandamme, E. J. (1976). Immobilized microbial cells as catalysts. *Chem. Ind. (London) 24*:1070-1072.

Vandamme, E. J. (1980). Penicillin acylases and β-lactamases. In *Economic Microbiology*, Vol. 5. A. H. Rose (Ed.). Academic, New York, pp. 467-522.

Vandamme, E. J. (1981). Use of microbial enzyme- and cell preparations to synthesize oligopeptide antibiotics. *J. Chem. Technol. Biotechnol. 31*: 637-659.

Vandamme, E. J. (1983). Peptide antibiotic production through immobilized biocatalyst technology. *Enz. Microbial Technol. 5*:403-416.

Veelken, M., and Pape, H. (1982). Production of tylosin and nikkomycin by immobilized *Streptomyces* cells. *Eur. J. Appl. Microbial. Biotechnol. 15*:206-210.

Venkatasubramanian, K., and Vieth, W. R. (1979). Immobilized microbial cells. *Proc. Ind. Microbiol. 16*:61-86.

Vieth, W. R., Wang, S., and Saini, R. (1973). Immobilization of whole cells in a membraneous form. *Biotechnol. Bioeng. 15*:565-569.

29

THE IMPACT OF THE NEW GENETICS ON ANTIBIOTIC PRODUCTION

ROLAND KURTH* AND ARNOLD L. DEMAIN   *Fermentation Microbiology Laboratory, Massachusetts Institute of Technology, Cambridge, Massachusetts*

There are no applied sciences. . . . There are only applications of science, and this is a very different matter . . . the study of the applications of science is easy to anyone who is master of the theory of it.

Louis Pasteur

I. INTRODUCTION

The general objective of genetic manipulation of industrially important antibiotic-producing microorganisms is to obtain a high degree of production of a desired product, thus keeping the costs for fermentation and isolation low. Furthermore, it is often desirable to change the proportion of natural antibiotics produced or to force the production of modified or new antibiotics in a given strain. Mutation and recombination are the tools to reach these goals.

*Present affiliation: Biotechnologie, BASF, Ludwigshafen, Federal Republic of Germany.

II. MUTATION

Throughout the antibiotic era, mutation has served admirably in insuring the viability of the antibiotic industry. Not only has mutation of fermentation organisms been extremely valuable in strain improvement, it has also been useful in eliminating undesirable antibiotics and in elucidating the biosynthetic pathways of natural products. An extended use of mutation is the practice of applying mutational biosynthesis to produce new antibiotic derivatives. In this technique antibiotic-negative mutants ("idiotrophs") are fed precursor analogs in order to produce antibiotic molecules never before formed in nature. Furthermore, even without feeding, certain blocked mutants have been obtained which produce shunt products with antibacterial activity (Kirst et al., 1983).

In the classic strain-development programs based on mutation, the following methods have been successfully applied:

Random screening for high-producing microorganisms
Screening for morphologically changed phenotypes
Selection of analog-resistant mutants that overproduce rate-limiting
    biosynthetic intermediates
Selection of auxotrophic and antibiotic-negative mutants followed by
    prototrophic reversion (or more properly, by "genetic suppression")

III. RECOMBINATION BY PROTOPLAST FUSION

Genetic recombination between parents of different sexes did not work well in industry during the 1950s and 1960s due to the extremely low frequency of genetic recombination in industrial microorganisms. For example, the frequency in streptomycetes is usually 10^{-6} or even less. Recently, however, there has been heightened interest in the application of genetic recombination to the production of important microbial products such as antibiotics.

Although knowledge of the genetics of antibiotic-producing organisms is still not extensive, new and widely applicable techniques for intraspecific recombination are becoming available. Complications arise in a variety of cases due to the lack of extensive information about synthetic pathways and regulation in the antibiotic-producing strain.

The knowledge of the genetics of *Escherichia coli, Bacillus subtilis*, and *Saccharomyces cerevisiae* and the techniques used with these organisms have greatly influenced the work on genetics of antibiotic-producing organisms. In addition, due to recent advances, well-developed experimental genetic systems are being developed for the antibiotic-producing actinomycetes (Hopwood, 1983).

The very low frequencies of intraspecific recombination among the actinomycetes required the introduction of genetic markers in order to detect recombinants. The introduction of such selectable markers was time consuming and resulted in a variety of negative pleiotropic effects on antibiotic production.

With the discovery of polyethyleneglycol-mediated protoplast fusion, a very powerful tool was introduced for genetic recombination in antibiotic producers. Protoplast fusion allows the genetic exchange between two or more whole chromosomes giving recombination frequencies as high as 10^{-1}. Protoplast fusion was first used with animal cells and plant cells and later with fungi and unicellular bacteria. Finally the work of Okanishi et al. (1974) on protoplast formation, fusion, and regeneration accelerated the use of this technique for gene manipulation in *Streptomyces*. Although one cannot obtain viable and stable recombinants by fusing completely unrelated species, successful interspecific protoplast fusion and recombination has been accomplished, for example, between *Penicillium chrysogenum* and *Penicillium cyaneofulvum*, *Aspergillus nidulans* and *Aspergillus rugolosus*, various streptomycetes, and even between *Candida* and *Endomycopsis*. This broadening of the recombination spectrum can probably be increased even further by the finding that ultraviolet irradiation of *Streptomyces* protoplasts before fusion can increase recombination frequencies after fusion by tenfold (Hopwood and Wright, 1981). Many excellent reviews have been written about the methodology of protoplast fusion (Sakaguchi et al., 1980; Ferenczy, 1981; Hopwood, 1981a; Baltz, 1980; Baltz and Matsushima, 1981; Wesseling, 1982).

Protoplast fusion offers the following advantages: (1) due to the very high recombination frequency, the introduction of auxotrophic markers is not necessary for the random screening of high-producing strains; (2) more than two different parents can be fused in a single recombination experiment; (3) large pieces of DNA have the possibility of being genetically exchanged; (4) the method is easily applicable to newly isolated industrially relevant strains and little time is needed for the optimization of the protoplasting, fusion and regeneration procedures.

Recombination is especially useful when combined with conventional mutation programs for two reasons. The first deals with the usual production of "sickly" organisms as a result of accumulated genetic damage over a series of mutated generations. By making a cross between a high-producing sick strain and a low-producing vigorously growing strain, it is possible to make a healthy high producer. For example, a cross via protoplast fusion has been carried out with strains of *Cephalosporium acremonium* from a commerical strain improvement program (Hamlyn and Ball, 1979). A low-titer, rapidly growing, spore-forming strain, which requires methionine to optimally produce cephalosporin C, was crossed with a high-titer, slow-growing, asporogeneous strain which could use the less expensive inorganic sulfate. The progeny included a recombinant which grew rapidly, sporulated, produced cephalosporin C from sulfate, and made 40% more antibiotic than the high-titer parent. The second application is the recombination of improved producers from a single mutagenesis treatment. In the conventional strain-improvement program, after considerable testing of survivors of a mutational step, the best producing mutant is retained for further mutagenesis and the rest of the improved producers are discarded. By recombination, one is able to combine the yield-increase mutations and obtain a superior producer before carrying out further mutagenesis. This has recently been demonstrated with *Nocardia lactamdurans*; two improved cephamycin-C producing strains from an industrial strain development program were fused and among the recombinants were two cultures which produced 10–15% more antibiotic than the best parent (Wesseling and Lago, 1981). Improvement in carbapenem antibiotic production by strains of *Streptomyces griseus* subsp. *cryophilus* has also been reported (Kitano et al., 1982).

Intergeneric recombination between *S. griseus* and *N. lactamdurans*, both producers of cephamycins, was successful and a recombinant was isolated which produced 7–11% more cephamycin C than the prototrophic *N. lactamdurans* grandparent (Wesseling and Lago, 1981).

A rather surprising development was the observation by Ikeda et al. (1983) that by mere protoplasting and regenerating a single macrolide-producing streptomycete, high-producing strains were obtained.

Another useful application of genetic recombination is the discovery of new antibiotics by fusing two producers of different, or even the same, antibiotics. Hopwood (1981b,c) has discussed two principal mechanisms: (1) The transfer of structural genes coding for the production of an antibiotic from one antibiotic-producing strain to another could lead to the production of a "hybrid" antibiotic. (2) The recombination of regulatory genes could positively influence the expression of unexpressed ("silent") genes.

Intraspecific genetic recombination in *N. mediterranei* has led to the discovery of two novel rifamycins (Traxler et al., 1982). Interspecific recombination has been achieved among *Streptomyces* species giving stable recombinants (Lomovskaya et al., 1977), some of which produce antibiotics not produced by either parent (Mazieres et al., 1981; Fleck, 1979). Although these examples were accomplished by conventional recombination, there is one recent example of the use of protoplast fusion. Idiotrophic strains of *S. griseus* SS-1198 (producing streptomycin) and *S. tenjimariensis* SS-939 (producing istamycin) were fused. Although most of the descendants showing antibiotic activity lost this property during subculture, an antibiotic produced by a stable strain was characterized and distinguished from the parental antibiotics (Yamashita et al., 1982).

Performance of unidirectional transfer of genetic information is possible by the liposome-mediated transformation of streptomycetes, in which isolated DNA of one strain is encapsulated in liposomes and fused with the protoplasts of another strain via polyethyleneglycol treatment (Makins and Holt, 1981). Another method is that of Hopwood and Wright (1981) in which ultraviolet-irradiated protoplasts can be used to eliminate a parental strain devoid of conventional counterselective markers. Other means include heat inactivation of one parental strain (Fodor et al., 1978) or chemical killing as described by Wright (1978). These methods are known as "dead donor" techniques.

IV. RECOMBINANT DNA

The low degree of success thus far reported for interspecific protoplast fusion between species of *Streptomyces* may be due to lack of DNA homology, different restriction systems, or cellular incompatibility. To overcome these problems, the development of in vitro recombinant DNA technology may be necessary.

Plasmid DNA has been detected in a variety of antibiotic-producing species and its exact function in connection to antibiotic production is under investigation. It has been shown that the genes of methylenomycin A biosynthesis are carried on plasmid SCP1 (Hopwood, 1979). In other cases, plasmid DNA may act more to regulate the expression of chromosomal structural genes of antibiotic biosynthesis. In *Streptomyces*, a number of plasmids have been found (Hopwood, 1981d), e.g., SCP1 and SCP2 in *S. coelicolor*, SRP1 in *S. rimosus*, and SLP1 in *S. lividans*. The first three are six plasmids and constitute an essential part of the sexual recombination process. Efforts are underway to try to exploit these discoveries, for instance, studies on plasmid gene amplification in streptomycetes aimed at increasing the dosage of genes coding for antibiotic production. Of course, bacterial viruses can also be used for gene transfer and gene amplification. Success with plasmids or phage could markedly reduce the cost of antibiotics as well as that of development of new antibiotics. According to Hopwood (1983), conjugal plasmids are probably involved in genetic recombination in *Streptomyces*. The best evidence for the role of plasmids has come from *S. lividans* 66. Plasmid pIJ101 was found to be self-transmissible by conjugation, to elicit "lethal zygosis" and to promote chromosomal recombination at high frequency (up to 1%) in both *S. lividans* 66 and *S. coelicolor* A3(2) (Kieser et al., 1982). It is a multicopy broad host range *Streptomyces* plasmid; 13 out of 18 strains, from diverse species, were successfully transformed. Moreover, several derivatives of pIJ101 suitable as DNA-cloning vectors were constructed. These were mostly designed to be nonconjugative and to carry pairs of resistance genes for selection. A bifunctional shuttle vector for *Escherichia coli* and *Streptomyces* has been constructed and a *Streptomyces* viomycin resistance gene on this shuttle vector was expressed in both hosts. That *S. lividans* is a useful host for foreign DNA was also shown by cloning resistance determinants from other antibiotic-producing streptomycetes (*S. azureus, S. erythreus, S. vinaceus,* and *S. fradiae*) into this organism (Thompson et al., 1982). Another example of the successful development of a cloning vehicle is the cloning of the neomycin resistance gene together with the entire *E. coli* pBR322 plasmid into *Streptomyces* plasmid pIJ103. The resulting *E. coli-Streptomyces* bifunctional vector (pIJ123) transformed both *E. coli* and *Streptomyces* (Richardson et al., 1982).

ϕC31 is an actinophage with a wide host range which has been used as a cloning vector (Lomovskaya et al., 1980; Chater, 1980). In addition to a broad host range, another advantage of using a phage-derived cloning vector is the low copy number. Often, high-copy number vectors (e.g., pIJ101) lead to physiological problems in the host due to unregulated gene expression (Chater, 1983). Excellent reviews about gene cloning in *Streptomyces* are those of Chater et al. (1982) and Okanishi et al. (1983).

At the moment, there is rapid progress in the development of recombinant DNA technology using antibiotic-producing *Streptomyces*. The methods have been established for cloning of genes using vectors with wide host ranges. Since in most antibiotic-producing organisms, knowledge about rate-limiting steps in the biosynthesis of antibiotics is only poorly known, direct approaches to gene amplification of certain steps to overcome this are not yet possible. In many of the strains examined thus far, at least partial clustering of antibiotic genes has been observed (Sermonti and Lanfaloni, 1982), [e.g., the biosynthetic structural genes of actinorhodin (Rudd and Hopwood, 1979) and the red prodiginine antibiotic in *S. coelicolor* (Rudd and Hopwood, 1980), of oxytetracycline in *S. rimosus* (Pigac and Alacevic, 1979; Rhodes et al., 1982) and of rifamycin in *N. mediterranei* (Schupp and Nüesch, 1979)]. Such clustering offers hope of incorporating chromosomal biosynthetic operons from actinomycetes into cloning vectors and transferring them to unicellular bacteria or to other actinomycetes.

Industrial fermentation problems might be solved by random cloning both for strain improvement and the development of new products. Self-shotgun cloning could result in clones which possess amplified genes for rate-limiting steps and therefore produce more antibiotic. Furthermore, interspecific shotgun cloning might lead to introduction of genes coding for the modification of a host antibiotic.

Since some enzymes of secondary metabolism are not highly specific, they often accept modified precursors. For example, it was recently shown that erythronolide B, an intermediate of erythromycin biosynthesis in *S. erythreus*, could be converted by an idiotroph of *S. antibioticus* (oleandomycin strain), to a hybrid antibiotic with a structure intermediate between erythromycin and oleandomycin (Spagnoli et al., 1983). By using interspecific "shotgun cloning" with a vector capable of being expressed and replicated in *S. antibioticus*, it should be possible to clone the genes for this biotransformation from a gene library of *S. erythreus* DNA. Use of an idiotrophic *S. antibioticus* strain would allow rapid screening for "hybrid" antibiotics.

Okamoto et al. (1979) described the biotransformation of tylosin by *S. thermotolerans* to 4"-acyltylosin derivatives which possess increased activity against macrolide-resistant Gram-positive bacteria. Again interspecific shotgun experiments could yield a tylosin-producing strain capable of carrying out the 4"-acylation.

V. THE FUTURE

The development of genetic methods such as protoplast fusion, plasmid-promoted conjugation, liposome-mediated transformation, and gene cloning bode well for the future of industrial genetics in antibiotic research and development (Hopwood and Chater, 1980). In addition to their use in producing new antibiotics, they might be useful in the transfer of antibiotic-producing operons from slow-growing streptomycetes to rapidly growing eubacteria (such as *E. coli* or *Bacillus subtilis*) in order to achieve rapid growth and more reproducible antibiotic production. Other advantages could be more rapid nutrient uptake due to a greater surface/volume ratio, better oxygen transfer, since filamentous organisms produce viscous non-Newtonian broths, better mixing and thus more reliable control of pO_2, pCO_2, and pH, and a better organism for mutagenesis. Another possibility is the transfer of such operons from one streptomycete to another in the hope that the structural genes might be better able to express themselves in another species. For example, a newly discovered aminoglycoside may be produced at very low levels, (e.g., 10 µg/ml), and a traditional strain-improvement program might take years to raise the titer to an economically feasible one. Transfer of a resistance gene(s) for a high kanamycin producer which already possesses resistance mechanisms to aminoglycoside antibiotics might yield a major increase in antibiotic titer.

Intraspecific protoplast fusion presents no problems today. On the other hand, with interspecific protoplast fusion, it has generally been observed that the hybrid strains are usually unstable. We probably should enlist the aid of taxonomists in order to make interspecific protoplast fusion more predictable. Also more knowledge is needed about the organization of the *Streptomyces* genome. The observation of repetitive DNA sequences in *Streptomyces* needs further exploration (Kirby et al., 1982; Robinson et al., 1981) and may contribute to the understanding of genetic instability in *Streptomyces* (Schrempf, 1983). The further development and optimization of vectors with a wide host range, low copy number, and sufficient cloning sites will be of great importance. We are confident that the new genetic methods will provide excellent tools for the improvement of existing antibiotic fermentations and will help to construct useful drugs.

ACKNOWLEDGMENT

The antibiotic research of A. L. D. is currently supported by the National Science Foundation and the National Institutes of Health. Funding for R. K. is provided by the Deutsche Forschungsgemeinschaft.

REFERENCES

Baltz, R. H. (1980). Genetic recombination by protoplast fusion in *Streptomyces*. *Devel. Ind. Microbiol.* *21*:43–54.

Baltz, R. H., and Matsushima, P. (1981). Protoplast fusion in *Streptomyces*: conditions for efficient genetic recombination and cell regeneration. *J. Gen. Microbiol.* *127*:137–146.

Chater, K. F. (1980). Actinophage DNA. *Devel. Ind. Microbiol.* *21*:65–74.

Chater, K. F. (1983). The deployment of *Streptomyces* vectors. In: *Proceedings of the Fourth International Symposium on Genetics of Industrial Microorganisms*. Y. Ikeda and T. Beppu (Eds.). Kodansha, Tokyo, pp. 71–75.

Chater, K. F., Hopwood, D. A., Kieser, T., and Thompson, D. J. (1982). Gene cloning in *Streptomyces*. *Curr. Top. Microbiol. Immunol.* *97*:69–95.

Ferenczy, L. (1981). Microbial protoplast fusion. *Symp. Soc. Gen. Microbiol.* *31*:1–34.

Fleck, W. F. (1979). Genetic approaches to new streptomycete products. In *Genetics of Industrial Microorganisms*. O. K. Sebek and A. I. Laskin (Eds.). American Society for Microbiology, Washington, D.C., pp. 117–122.

Fodor, K., Demiri, E., and Alfoldi, L. (1978). PEG-induced fusion of heat-irradiated and living protoplasts of *Bacillus megaterium*. *J. Bacteriol.* *135*:68–70.

Hamlyn, P. F., and Ball, C. (1979). Recombination studies with *Cephalosporium acremonium*. In *Genetics of Industrial Microorganisms*. O. K. Sebek and A. I. Laskin (Eds.). American Society for Microbiology, Washington, D.C., pp. 185–191.

Hopwood, D. A. (1979). Genetics of antibiotic production by actinomycetes. *J. Nat. Prod.* *42*:596–602.

Hopwood, D. A. (1981a). Genetic studies with bacterial protoplasts. *Ann. Rev. Microbiol.* *35*:237–272.

Hopwood, D. A. (1981b). Future possibilities for the discovery of new antibiotics by genetic engineering. In β-*Lactam Antibiotics: Mode of Action, New Developments and Future Prospects*. M. R. J. Salton and G. D. Shockman (Eds.). Academic Press, New York, pp. 585–598.

Hopwood, D. A. (1981c). Possible application of genetic recombination in the discovery of new antibiotics in actinomycetes. In *The Future of Antibiotherapy and Antibiotic Research*. L. Ninet, P. E. Bost, D. H. Bouanchaud, and J. Florent (Eds.). Academic Press, London, pp. 407–416.

Hopwood, D. A. (1981d). Genetic studies of antibiotics and other secondary metabolites. *Symp. Soc. Gen. Microbiol.* *31*:187–218.

Hopwood, D. A. (1983). Genetic manipulation in *Streptomyces*. In *Proceedings of the Fourth International Symposium on Genetics of Industrial Microorganisms*. Y. Ikeda and T. Beppu (Eds.). Kodansha, Tokyo, pp. 3–8.

Hopwood, D. A., and Chater, K. F. (1980). Fresh approaches to antibiotic production. *Phil. Trans. R. Soc. Lond. B 290*:313–328.

Hopwood, D. A., and Wright, H. M. (1981). Protoplast fusion in *Streptomyces*: Fusions involving ultraviolet-irradiated protoplasts. *J. Gen. Microbiol.* *126*:21–27.

Ikeda, H., Inoue, G., and Ōmura, S. (1983). Improvement of macrolide antibiotic-producing Streptomycete strains by the regeneration of protoplasts. *J. Antibiot.* *36*:283–288.

Kieser, T., Hopwood, D. A., Wright, H. M., and Thompson, A. J. (1982). pIJ101, a multi-copy broad host-range *Streptomyces* plasmid: Functional analysis and development of DNA cloning vectors. *Mol. Gen. Genet. 185*: 223–238.

Kirby, R., Gertsch, K., and Usdin, K. (1982). Repetitive DNA sequences in *Streptomyces*. *Abstracts of the Fourth International Symposium on the Genetics of Industrial Microorganisms* (Kyoto) p. 90.

Kirst, H. A., Wild, G. M., Baltz, R. H., Seno, E. T., Hamill, R. L., Paschal, J. W., and Dorman, D. E. (1983). Elucidation of structure of novel macrolide antibiotics produced by mutant strains of *Streptomyces fradiae*. *J. Antibiot. 36*:376–382.

Kitano, K., Nozaki, Y., and Imada, A. (1982). Strain improvement in carbapenem antibiotic production by *Streptomyces griseus* subsp. *cryophilus*. *Abstracts of the Fourth International Symposium on the Genetics of Industrial Microorganisms* (Kyoto) p. 66.

Lomovskaya, N. D., Chater, K. F., and Mkrtumian, N. M. (1980). Genetics and molecular biology of *Streptomyces* bacteriophages. *Microbiol. Rev. 44*:206–229.

Lomovskaya, N. D., Voeykova, T. A., and Mkrtumian, N. M. (1977). Construction and properties of hybrids obtained in interspecific crosses between *Streptomyces coelicolor* A3(2) and *Streptomyces griseus* Kr. 15. *J. Gen. Microbiol. 98*:187–198.

Makins, J. F., and Holt, G. (1981). Liposome-mediated transformation of streptomycetes by chromosomal DNA. *Nature 293*:671–673.

Mazieres, N., Peyre, M., and Penasse, L. (1981). Interspecific recombination among aminoglycoside producing streptomycetes. *J. Antibiot. 34*:544–550.

Okamoto, R., Nomura, H., Tsuchiya, M., Tsunekawa, H., Fukumoto, T., Inui, T., Sawa, T., Takeuchi, T., and Umezawa, H. (1979). The activity of 4"-acylated tylosin derivatives against macrolide-resistant gram-positive bacteria. *J. Antibiot. 32*:542–544.

Okanishi, M., Suzuki, K., and Umezawa, H. (1974). Formation and reversion of streptomycete protoplasts: cultural condition and morphological studies. *J. Gen. Microbiol. 80*:389–400.

Okanishi, M., Katagiri, K., Furumai, T., Takeda, K., Kawaguchi, K., Saitoh, M., and Nabeshima, S. (1983). Basic techniques for DNA cloning and conditions required for streptomycetes as a host. *J. Antibiot. 36*:99–108.

Pigac, J., and Alacevic, M. (1979). Mapping of oxytetracycline genes in *Streptomyces rimosus*. *Periodic. Biol. 81*:575–582.

Rhodes, P. M., Winskill, N., Friend, E. J., and Warren, M. (1981). Biochemical and genetic characterization of *Streptomyces rimosus* mutants impaired in oxytetracycline biosynthesis. *J. Gen. Microbiol. 124*:329–338.

Richardson, M. A., Mabe, J. A., Beerman, N. E., Nakatsukasa, W. M., and Fayerman, J. T. (1982). Development of cloning vehicles from the *Streptomyces* plasmid pFJ103. *Gene 20*:451–457.

Robinson, M., Lewis, E., and Napier, E. (1981). Occurence of reiterated DNA sequences in strains of *Streptomyces* produced by an interspecific protoplast fusion. *Mol. Gen. Genet. 182*:336–340.

Rudd, B. A. M., and Hopwood, D. A. (1979). Genetics of actinorhodin biosynthesis in *Streptomyces coelicolor* A3(2). *J. Gen. Microbiol. 114*:35–43.

Rudd, B. A. M., and Hopwood, D. A. (1980). A pigmented mycelial antibiotic in *Streptomyces coelicolor*: control by a chromosomal gene cluster. *J. Gen. Microbiol. 119*:333–340.

Sakaguchi, K., Ochi, K., Gunge, N., and Uchida, K. (1980). Protoplast fusion. In *Molecular Breeding and Genetics of Applied Microorganisms.* K. Sakaguchi and M. Okanishi (Eds.). Academic Press, Tokyo, pp. 85–105.

Schrempf, H. (1983). Genetic instability in *Streptomyces.* In *Proceedings of the Fourth International Symposium on Genetics of Industrial Microorganisms.* Y. Ikeda and T. Beppu (Eds.). Kodansha, Tokyo, pp. 56–60.

Schupp, T. and Nüesch, J. (1979). Chromosomal mutations in the final step of rifamycin B biosynthesis. *FEMS Microbiol. Lett.* *6*:23–27.

Sermonti, G. and Lanfaloni, L. (1982). Antibiotic genes—Their assemblage and localization in *Streptomyces.* In *Overproduction of Microbial Products.* V. Krumphanzl, B. Sikyta, and Z. Vanek (Eds.). Academic Press, London, pp. 485–497.

Spagnoli, R., Cappelletti, L., and Toscano, L. (1983). Biological conversion of erythronolide B, an intermediate of erythromycin biogenesis, into new "hybrid" macrolide antibiotics. *J. Antibiot.* *36*:365–375.

Thompson, C. J., Skinner, R. H., Thompson, J. Ward, J. M., Hopwood, D. A., and Cundliffe, E. (1982). Biochemical characterization of resistance determinants cloned from antibiotic-producing streptomycetes. *J. Bacteriol.* *151*:678–685.

Traxler, P., Schupp, Th., and Wehrli, W. (1982). 16,17-dihydrorifamycin S and 16,17-dihydro-17-hydroxyrifamycin S, two novel rifamycins from a recombinant strain C5/42 of *Nocardia mediterranei.* *J. Antibiot.* *35*:594–601.

Wesseling, A. C. (1982). Protoplast fusion among the actinomycetes and its industrial applications. *Devel. Ind. Microb.* *23*:31–40.

Wesseling, A. C. and Lago, B. (1981). Strain improvement of genetic recombination of cephamycin producers, *Nocardia lactamdurans* and *Streptomyces griseus.* *Devel. Ind. Microb.* *22*:641–651.

Wright, W. E. (1978). The isolation of heterokaryons and hybrids by a selective system using irreversible biochemical inhibitors. *Exp. Cell Res.* *112*:395–407.

Yamashita, F., Hotta, K., Kurasawa, S., Okami, Y., and Umezawa, H. (1982). Antibiotic formation by interspecific protoplast fusion in streptomycetes and emergence of drug resistance by protoplast regeneration. *Abstracts of the Fourth International Symposium on the Genetics of Industrial Microorganisms.* (Kyoto), p. 108.

INDEX